COLLEGE PHYSICS

third edition

Franklin Miller, Jr.

Kenyon College

COLLEGE
PHYSICS
third edition

HARCOURT BRACE JOVANOVICH, INC.

New York/Chicago/San Francisco/Atlanta

TITLE PAGE AND COVER PHOTOGRAPH:
A high-resolution electron microscope was used to make this photograph of carbon black particles used in the manufacture of rubber. The basic lamellar (platelike) structure is shown, with crystal planes of carbon atoms 3 Å (3×10^{-10} m) apart. Magnification about 3 000 000. (Micrograph by L. L. Ban and W. M. Hess, courtesy Columbian Carbon Company.)

LIBRARY OF CONGRESS CATALOG CARD NUMBER: 74-190465

ISBN: 0-15-511732-7

PRINTED IN THE UNITED STATES OF AMERICA

PREFACE

The author of a text in elementary physics is, properly, under some obligation to define his audience and justify his outlook, aims, and methods. My primary aim has been simply to prepare a text having maximum clarity and teach-ability. With this objective in mind, I have paid attention to certain aspects of the teaching of physics which are of particular concern to teachers and students alike in this time of expanded interest in the subject.

This book is intended to be used in an elementary physics course to which the student brings no mathematics beyond simple algebra. It reflects my con-viction that all students can and should become familiar with the basic ideas of physics; for this reason I have emphasized principles rather than technological applications. This is not to say, however, that the text has no points of contact with everyday life and practical situations. On the contrary, a great many worked examples, questions, and problems are scattered throughout the book. Each of these has been chosen to illustrate and "bring down to earth" some physical concept. Three of the chapters deal specifically with applications but retain close correlation with basic principles. These chapters—on Applied Electricity, Applied Optics, and Applied Nuclear Physics—are so titled partly to keep the record straight.

The organization of the elementary physics course is under active study by many teachers. The changes that may well come about within the next decade will not necessarily be along the lines one might predict today. For this reason, I have presented the topics of physics in a semitraditional sequence which preserves a logical structure found useful by many teachers over a period of many years. But I have introduced so-called "modern physics" early and often within this framework; in this way physics can be presented as a growing subject, and the intimate relationship of classical physics to modern physics can be made clearer. Throughout the book I have given emphasis to the con-servation laws and the concepts of waves, fields, and particles. Although I have not adopted a straight historical approach, I have included many brief historical references and in a few areas have gone rather fully into the development of physical concepts.

Physics is a rewarding subject which requires a student's best effort, and I have refrained from spoon-feeding a watered-down or sugar-coated subject matter. Nevertheless, every possible aid to digestion has been included in the book. Among these study helps are the numerous worked examples in the body of the text. At the end of each chapter are a summary, a check list of key concepts, a set of questions to help in reviewing on the verbal level, and a series of carefully graded problems (see the explanation on pp. 22–23) to assist the student in the mastery of principles through practice. An important study aid is the review of simple high school mathematics in the Appendix, which also contains a short course on the use of the slide rule. For added flexibility, sections labeled "For Further Study," with accompanying problems, follow many of the chapters; these are set apart in such a way that the continuity of the primary material is not disturbed. The better students should be encouraged to dip into these sections on their own; and, of course, a teacher is free to assign any of these sections to an entire class. The text is complete and self-sufficient without this extra material.

It is becoming increasingly apparent that for maximum insight into physics at any level a student should not be deprived of the help offered by the use of limits. I have, therefore, defined, illustrated, and used the derivative as the limit of a ratio and the definite integral as the limit of a sum. This much calculus is developed and explained as the need arises, and it requires only a knowledge of high school algebra. Graphical interpretations are emphasized. Formulas for derivative and integral are not used in the main body of the text; some formal calculus is developed and used in some of the optional "For Further Study" sections.

The organization of this third edition follows that of the second edition, with several exceptions. The conservation laws for energy and linear momentum are now given a more unified treatment in Chapter 6. Topics in relativity, previously scattered through several chapters, have been brought together in Chapter 7 in order to make the concepts of classical and modern relativity more easily grasped. In the treatment of geometrical optics, image location by the method of curvatures has been retained, but the traditional lens equation has been added as an alternative method for those who prefer to use it. Material on atomic and nuclear physics has been reorganized into four chapters to give a somewhat less concentrated presentation. In general, additional material for this edition has been kept to a minimum, in accordance with my desire to present fundamentals without an excessive array of applications. However, several topics have been added at the suggestion of users of the book: Galilean relativity; the human ear; binding energy; biological effects of radiation. Other topics of contemporary interest have also been included: gravity waves; holography; colliding beams.

Symbols and abbreviations recommended by the Commission on Symbols, Units, and Nomenclature of the International Union of Pure and Applied Physics are used throughout the text (cf. *Phys. Today*, June, 1962, p. 20). While a few of these may be unfamiliar at the start, their use will cause no real difficulty. The need for a standard set of symbols has long been recognized; only by adopting and using the internationally agreed-upon symbols will their eventual acceptance become universal.

While both mks and cgs units are used in mechanics, only mks units are used in electricity and atomic physics. I believe there is real merit in the use of familiar force units in mechanics. In some places, therefore, the pound is used as a unit of force; no British unit of mass need be defined, since the ratio equation can be applied where needed. Illustrative examples in mechanics are stated in about equal numbers in metric and in British units, so that any teacher's preference can be satisfied. For this text, I have deliberately limited (but not omitted) the use of the cgs system of units. No matter which system of units is used, the student is encouraged to look for dimensional consistency in equations and in solutions of problems. A convention regarding significant figures has been adopted, as discussed on p. 16.

Electric current is described in terms of conventional positive current, for two reasons. If, as is customary, electric field and potential difference are defined in terms of a positive test charge, then there is a logical advantage in using conventional current which is directed "downhill" from a region of high potential. Also, a significant number of students using this book will encounter conventional current in their later study in fields outside physics.

I have not felt that an encyclopedic text is desirable for the type of course for which this book is intended. I have preferred to approach fundamentals in a leisurely way, allowing full opportunity for development of background and assimilation of principles. To achieve this treatment in depth, the discussion of certain time-honored subjects has been greatly reduced or even omitted. In most cases, references at the end of each chapter will serve the needs of students wishing to explore such areas. The references also give opportunity for further reading, both in the history of science and in contemporary developments.

In this edition, the references at the end of chapters include mention of certain films. This is not the place to attempt a critical evaluation of longer films. References are included, however, to selected short "single-concept" films which are distributed in cartridges and which are, in many colleges and universities, available for selection and viewing by individual students outside regular class hours. Cartridge films are listed which are judged to be helpful at the level of this text. These films are available separately and in sets from many distributors.

It is fitting to acknowledge my great indebtedness to my teachers, colleagues, and students who have influenced me throughout the years. Of particular value has been the constructive criticism of many teachers, users of the second edition, who responded to my request for comments and suggestions. Thanks for their assistance are also due to all those mentioned in the preface to the second edition, and to the following: D. E. Abrahamson; L. W. Alvarez; J. L. Glathart; J. A. Johnson; E. Nussbaum; J. R. Stallard; H. P. Stephenson; L. C. Teng; J. Upatnieks; J. G. Williamson; to the Lawrence Berkeley Laboratory; and to all those who furnished photographs which are acknowledged in the text. None of these should be in any way charged with the residual errors in the book, which are entirely my own. Finally, to my wife, I again offer appreciation for her continuing support and encouragement.

Franklin Miller, Jr.

CONTENTS

The Nature of Physics

1-1 Science: endless frontier of the mind

The representation of science as a frontier of the mind is a useful description of one of man's finest intellectual activities. The geographical frontier—the boundary between known and unknown territory—has always existed. The pioneer establishes himself at the very edge of the wilderness, consolidates his position, and strikes out anew. During the centuries, the forms of geographical exploration have been as many and varied as have been the explorers: the Phoenician sailor, whose frontier was the sea; the Roman legionary, who garrisoned the distant islands of Britain; the Indian scout, who opened up the Western territories of the United States. Today, the geographical frontier is above our heads and extends to the moon and beyond.

It is no mere play upon words to liken scientific investigation to exploration. We shall, in this course, repeatedly see how a scientific beachhead is won by an intrepid and imaginative thinker; how the frontier is made secure by developments in related fields and by the application of new basic knowledge to practical problems; and how the scientific frontier is extended a bit at a time by scientists building upon past achievements. We see how appropriate are two phrases which have often been used to describe science. James B. Conant, chemist and former president of Harvard University, considers science to be a "series of conceptual schemes resulting from observation and experiment and leading to further experiment and observation."[*] Percy W. Bridgman, a Nobel prizewinner in physics, puts it more forcefully: science means "doing one's damnedest with one's mind, no holds barred."[■] Each of these statements, modified in phraseology, would be a perfect description of the act of geographical exploration.

Not many decades ago, it seemed that the geographical world was limited and that frontiers must soon disappear. We are now well past the threshold of a space age, and once again we are presented with what may be an endless series of frontiers. Similarly,

[*] *On Understanding Science,* Yale University Press, 1946.
[■] "Prospect for Intelligence," *Yale Review,* **34,** 450, 1945; reprinted in *Reflections of a Physicist,* Philosophical Library, 1950, p. 342.

Physics class at Columbia University. (Photo by Jan Obus)

about 80 years ago, some scientists thought that the last frontier of physical science had been reached; they thought all that remained was to work out the details and extend the accuracy of measurement another decimal place or two. How wrong this turned out to be! Unforeseen new frontiers were reached and opened to investigation. By 1900 the free electron, x rays, and radioactivity had been discovered, and quantum theory was being developed. Major break-throughs continue to be made; in the language of exploration, scouting parties are returning with exciting glimpses beyond the present frontier. There is every evidence that scientific frontiers, like geographical ones, form an endless sequence. How are these frontiers being penetrated? What methods are used in scientific investigation?

It is amazing and somewhat reassuring that, while the content of physics has become enlarged and specialized, there has been no corresponding radical change in the underlying methodology of scientific investigation since the days of Galileo (1564–1642). However, if we look for a tried and true routine which will always yield results, we shall be disappointed. It is just the essence of scientific methods that the intellect is *freed* from routine. Any scientist, including the physicist, must combine inspiration, intuition, concentration, and hard work if he is to advance knowledge. Reliance on set formulas or routine procedure is not likely to give rise to either new discoveries or new understanding of old discoveries. Yet the scientist would be foolish indeed to forgo the many helps and advantages that are his inheritance from a continuous chain of fellow-scientists reaching back several thousand years. You who are beginning the study of physics will find this point illustrated in the text. We shall discuss the basic laws of physics, which are relatively few in number, and use them continually to illuminate everyday experience. We shall approach new ideas as much as possible from the point of

view of their relationship to our earlier work.

In this book our illustrations of scientific methods will be by implication rather than by detailed exposition. There are many ways of approaching scientific knowledge—perhaps as many ways as there are scientists. But all of them have several things in common: collecting facts by observation; constructing a *hypothesis;*[*] testing the hypothesis. It will be seen that any method which incorporates these three procedures is especially suitable for the development of *natural* science, where facts can often be collected by controlled experiment. In the *social* sciences it often proves necessary to construct hypotheses without benefit of controlled experiment. The data of social science are those of human relationships and cannot be gathered by a series of experiments where one variable at a time is changed. Thus the social scientist almost always must rely for his data upon close observation of happenings not under his direct control. The natural scientist often suffers from a similar limitation, especially when a new field of investigation is opened up.

Whatever the source of a hypothesis—an act of imaginative thinking, an educated guess, or a carefully constructed working hypothesis based on hundreds of precise experiments—the next step is to test it. It is in the devising of tests that the skill and ingenuity of the experimenter lie. A hypothesis is rejected if even a single previously known fact is at variance with it,[*] but the scientist goes further. He actively seeks new facts, perhaps by performing experiments suggested by the hypothesis or by mathematically analyzing experiments performed by

---

[*] Webster defines *hypothesis* as "a proposition tentatively assumed in order to draw out its logical or empirical consequences and so test its accord with facts that are known or may be determined."

[*] The picturesque expression, "the exception proves the rule," does not mean what it seems to. The word "prove" originally meant "test," as in "proving ground." So the statement really reads, "the exception *tests* the rule"— and the exception proves the rule false, not true.

others. If A (our hypothesis) is true, then B should be true, so let us make an experiment to see whether or not B is true. If B is not true, then our hypothesis is incorrect and must be modified or abandoned. If B *is* true, then we are one step closer to verifying our hypothesis A, but we have not proved it. Indeed, no matter how many different B's are proved true, we have not proved the truth of A beyond all possible doubt.

When a hypothesis leads to the design of experiments which would earlier have been meaningless or never thought of, and when these experiments in turn lead to a new hypothesis which can similarly be tested, then we may speak of a *theory*. All scientific "principles" are really theories, for the scientist is always ready to modify his most cherished concepts if new experimental evidence turns up, or even if another concept will explain the evidence equally well. From hypothesis to principle of nature may be a long or a short trail, but the terms are different only in degree. In the fifth century B.C. the Greek Demokritos made a *conjecture,* without any experimental evidence, of the existence of small indivisible particles which he called atoms. In 1808 Dalton's atomic *hypothesis* made its appearance, based upon certain experimental evidence. In the late 1800's the atomic *theory* was refined and expanded; in the late 1900's the *principle* of atoms is completely accepted, but subject to experimentation and reinterpretation so severe that the very use of the word "atom" may be misleadingly naïve.

We have deliberately kept our discussion of scientific methods brief, for two reasons. First, people mean different things when they speak of "scientific method"; you will discover this for yourself if you read some of the references listed at the end of this chapter. Second, and more important, we hope that you will derive from this book and others your own understanding of what science is. We believe that scientific attitudes should grow out of work in science rather than be laid down as arbitrary, almost meaningless rules. So we leave this discussion with a final note: scientific methods have limitations, the most severe being an unspoken presumption that "knowledge" means "scientific knowledge." It would be most *un*scientific to assume that religious knowledge, or faith, cannot exist simply because it cannot be shown to be true by scientific methods. One should not expect a method devised specifically to cope with experimental or observed facts to do anything other than that for which it was invented. Nevertheless, when properly applied, scientific methods have proved to be the most dependable and fruitful means of dealing with the physical universe and have made possible the degree of control over our physical environment which is so characteristic of our civilization.

1-2 Physics as a natural science

Man's deep interest in his surroundings dates back to prehistory, and his attempts to understand and control the forces of nature are a characteristic trait of the human intellect. Yet only comparatively recently has the special discipline which we call *physics* been a distinct field of study. Aristotle, Archimedes, and even Galileo and Newton called themselves natural philosophers, with all nature for their province. As the body of knowledge grew, it became necessary to specialize, for not even a genius could hope to keep up with developments in astronomy, biology, medicine, geology, chemistry, mathematics, and a host of other specialties. Those who are concerned mainly with the applications of science to the betterment of human environment have been the engineers. Physicists and chemists generally have been concerned with the more basic aspects of nonliving matter. Chemistry deals primarily with molecular changes and the rearrangements of the atoms which form molecules, but the dividing line between chemistry and physics can hardly be drawn in a dogmatic fashion. A list of some journals currently published in the United States illustrates the delicate shades of emphasis among physical scientists. Running the gamut from "pure physics" to "pure chemistry" we have the *Physical Review, Journal of Chemical Physics, Journal of Physical Chemistry,* and *Journal of the American Chemical Society.* The titles of some other research journals indicate a great deal of specialization even within the broad field of physics: *Journal of the Institute of Metals* (Great Britain); *Revue d'Optique* (France); *Kristallografiya* (U.S.S.R.); *Revista de Geofísica* (Spain); *Przegląd elektrotechniczny* (Poland). Many journals cover the whole field of physics, including *Zeitschrift für Physik* (Germany) and the *Canadian Journal of Physics.* These are just a few of hundreds of journals through which original research in physical science is presented to the scientific community. In the year 1970, no less than 79 830 technical papers were summarized in *Physics Abstracts;* this flood of reports came from 1684 journals published throughout the world. Statistics can, of course, be twisted, and yet other lines of evidence converge to the same conclusion: the all-round scholar of the seventeenth and eighteenth centuries, a man of the first rank in several disciplines, could exist only rarely now, if at all. No one scholar can hope to assimilate all the rapidly growing source material of even one science such as physics.

We are thus led to the idea of partial understanding: the acceptance of limited, but concentrated, knowledge as being better than no knowledge at all. The greatest research physicists of today have probed deeply and brilliantly into their own fields of interest, but have acquired less profound knowledge outside their specialty.

Like other sciences, physics has developed through the combined efforts of great numbers of dedicated workers in many countries. We must realize that the greatest of these scientists were, each in his own day, *at the* then existing frontier of science. We tend to accept Newton's laws of motion, which we use so often, as "handed down from on high," forgetting that Newton had to come to grips with nature, "no holds barred," in order to arrive at his formulation of the laws of mechanics. Newton's laws were, in fact, "modern physics" in the early 1700's.

We see that modern physics is as old as physics itself. Yet today we give Galileo and Newton a prominent place in our study of mechanics not for historical reasons alone; Galileo and Newton happened also to be correct, and their methods are still tremendously useful. The present-day physicist—the modern modern physicist if you will—relies on Newtonian mechanics in all his work with large-scale (visible) objects moving at ordinary speeds (small compared with the speed of light). The orbits of space vehicles are exactly calculated from Newton's laws, which were formulated 300 years ago; the law of conservation of energy is still, after more than a century, a fundamental statement about nature which is used daily by research physicists and engineers. We would limit our understanding of physics if we neglected the contributions of those who went before us, and we would be unable to understand the generalizations into which some of the earlier laws of physics have evolved.

1-3 The nature of the modern physics course

One objective of a physics course is to impart useful knowledge. Those students who are going on in medicine, dentistry, agriculture, architecture, home economics, science teaching, psychology, or geology may find that this is their only college course in physics. They will find professional use for the facts and methods they study with this book. However, perhaps half the students for whom this book is written do not intend to be scientists and are studying physics for its "cultural" value. It is all very well to discuss scientific methodology as an interesting example of man's intellectual development, but this sort of scrutiny might equally well be given in a philosophy class. In our study of physics, we frankly stress the value of factual, detailed knowledge of laws, principles, and applications. Our culture is one in which applied science plays an ever-increasing role, and it is truly of cultural value to be able to understand a little of what goes on. Naturally, you will not be able to design a nuclear reactor as a result of diligent study of this book, but you *will* be able to appreciate some of the design problems. You will not be able to repair your TV set, but you *will* understand the basic laws of photoelectricity. Most important, you will come to appreciate the openness of scientific knowledge. What one group of men and women has discovered, another group can as easily find out. No country can have a monopoly on hydrogen bombs or TV sets. Whether you intend to be a writer, clergyman, politician, businessman, housewife, or musician, the factual knowledge of physics which you gain from your college course will help you feel at home in our present-day, science-oriented culture.

You will be aided in your study of physics by your teacher's lectures, your laboratory work, and, of course, this book. A physics textbook can be organized in many ways. To a large extent, we shall find it profitable to stay with certain specific aspects of physics long enough to gain some firm knowledge and ability to reason out the solutions to problems. Although we shall follow a time-honored sequence of subjects, we shall introduce ideas of "modern" physics wherever possible in the earlier work, rather than save them all for a final few chapters. To be sure, certain aspects of radiation, radioactivity, and nuclear physics should be studied in a concentrated and systematic way. But many other aspects of recent physics will be brought in as illustrative material for the "classical"

or old-fashioned physics. The dividing line should not be drawn sharply. After all, in 1687 the law of gravitation was "modern" physics, and in 1867 the molecular theory of heat was "modern" physics. In the second half of the twentieth century, we have arrived at a point at which much that has been segregated as modern physics can be assimilated into the standard areas of mechanics, heat, and the like.

We begin with the study of mechanics, since the ideas of mechanics are used over and over again in the development of the other branches of physics. Indeed, the sequence of mechanics, sound, heat, electricity, light, and atomics has become almost standard for good reasons. The phenomena of sound are those of the mechanical vibrations of material bodies, and so we should study mechanics before sound. Heat energy is largely due to the mechanical motions of molecules, and so we should study mechanics before heat. We can calculate the pressure exerted by the molecules of a gas, using the laws of motion developed in our study of mechanics. Electricity can also best be studied with the aid of mechanical concepts: the forces acting on moving electrons or protons in giant atom-smashers cause acceleration in the same way that the force of gravitation causes acceleration of a satellite or a planet. We study light, which exhibits the properties of an electromagnetic wave, after electricity; indeed, light waves and radio waves are identical in nature and differ only in wavelength. We use mechanics, wave theory, electricity, and optics, as well as quantum ideas, in the interpretation of the light emitted by atoms and molecules; thus interpreted, the spectra give a clue to the structure of the outer atom. Our text concludes with a close study of the inner atom (the nucleus), and a look at some applications of nuclear physics.

As you study physics, you will notice that it is a cumulative process. Much of what you learn is carried over and used in later parts of the course. This relatedness means that you should keep up to date in your assignments. Several study aids have been included in this book to help you; use them. The summary, check list, and questions at the end of each chapter will help you review and consolidate your knowledge. Problems are graded in difficulty; all serve the purpose of aiding you to master *principles* through *practice*. Sections marked "for further study" appear at the close of many of the chapters; these should give you a chance to spread your wings if you so desire. Don't overlook the references at the end of each chapter; the original writings of the pioneers of physics have a flavor and a lasting value that cannot be duplicated. Their biographies are revealing and often inspiring. Through other references, you can make contact with the growing edge—the frontier—of contemporary physics.

And now, let us get on with the task at hand, which is to learn the facts and methods of physics, a typical science. During the next few months you will many times approach the frontiers of physics. You will stand alongside pioneers of the past as they penetrate the barriers of their time, and you will, yourself, see where your contemporaries are striking forward, "no holds barred." May physics come alive for you, as it has for them.

1-4 Precision, measurement, and notation in physics

It is the function of science to correlate precise measurements of physical quantities. By such correlations basic laws are discovered. The basic units of measurement must be well defined, and it is helpful if measurements are made in a consistent set of units. For instance, if we wished to find the distance traveled by a car, we would not multiply its speed in miles per hour by the elapsed time in minutes. Common sense dictates that we either convert the speed to miles per minute, or convert the time to hours. Such quantities

as speed (miles/hour), area (square feet), etc., are *dimensional quantities;* a measured dimensional quantity such as the area of a field has different numerical values in different systems of units. The same area can be expressed as 9000 square feet, 1000 square yards, etc. The conversion of such quantities from one unit to another unit of the same dimensions will be made clear in illustrative examples throughout the book.

It is found that all mechanical quantities, such as energy, force, momentum, volume, and velocity, can be expressed in terms of *three* basic dimensional quantities. The choice of the three quantities can be made in several ways. In scientific work it is customary to take as the basic dimensions (written in square brackets) mass [M], length [L], and time [T]. Velocity, for example, can be measured in miles per hour, feet per second, or meters per second—always a length divided by a time. We write [velocity] = [L]/[T], where the square brackets indicate that the equation relates *dimensions* only, and is not necessarily a complete numerical equation. A dimensionless numerical factor may or may not be present. Thus the complete equation for the volume of a cylinder of radius r and height h is $V = \pi r^2 h$. The factor π is dimensionless, and the dimensions of r and h are each [L]. Hence we have the dimensional equation [volume] = [L²][L] = [L³]. If the cylinder has diameter d and height h, the volume formula is $V = (\pi/4)d^2 h$, in which the numerical factor is different but the dimensional equation is still [volume] = [L³], as it must be for *any* volume.

To cover *all* fields of physics, three additional dimensional quantities are used. The *International System of Units* (abbreviated SI, for Système International) uses the meter, kilogram, and second for mechanics; other basic SI units are the ampere (electric current), the kelvin (temperature), and the candela (luminous intensity).

Three systems of units are in common use. We shall discuss briefly the fundamental quantities upon which these systems are based. Careful and precise definitions are necessary both for scientific reasons and in order to provide legal standards for commerce and trade.

1. Scientists use the *meter-kilogram-second* (mks) *system* as the preferred system of units. In the mks system the standard of length was formerly defined as the distance between two marks on a carefully protected bar of platinum-iridium alloy preserved at the International Bureau of Weights and Measures at Sèvres, France. This "international prototype meter" and all copies of it in the various countries of the world were subject to unknown aging effects, oxidation, or even destruction through natural or man-made catastrophe. A new standard, imperishable and always reproducible, was adopted in 1960. The meter is now defined as exactly 1 650 763.73 wavelengths of a certain orange line in the spectrum of the krypton isotope of atomic mass number 86. For convenience, multiples and submultiples of the meter are used, such as the kilometer (1000 m), centimeter ($\frac{1}{100}$ m), and the millimeter ($\frac{1}{1000}$ m).

The important metric prefixes, shown in Table 1-1, place emphasis on units in successive ratios of 1000*; note that multiples greater than 1 are usually abbreviated with capital letters. As indicated in the examples, metric prefixes can be used for multiples and submultiples of any unit, even those not in the metric system itself.

The mks unit of mass is the kilogram, which is the mass of a certain block of platinum-iridium alloy known as the "international prototype kilogram" preserved at Sèvres. It is well known that all material bodies tend to stay at rest or, if in motion, tend to resist any change in the magnitude or direction of the motion. This attribute of matter is called inertia, and the *mass* of a body is a quantitative measure of its inertia. The *weight* of a body, on the other hand, is the force of gravity upon it. Weight and mass have different dimensions and should not be confused with each other. We shall discuss mass in detail in Chap. 4, and in Chaps. 6 and 9 we shall see how it is possible to compare two masses by observing changes in motion. Such methods of measuring mass

----

*If you are not completely familiar with the use of exponents and with scientific notation (in powers of 10), this would be a fine time to review Secs. A-2 and A-3 of the Mathematical Appendix.

TABLE 1-1   Important metric prefixes

Prefix	Abbreviation	Meaning	Typical examples
tera	T	$\times 10^{12}$	
giga*	G	$\times 10^{9}$	1 gigayear = 10^9 years
mega	M	$\times 10^{6}$	1 megaton (equivalent TNT strength of nuclear weapons) = 10^6 tons
kilo	k	$\times 10^{3}$	1 kilogram = 1000 g
deci	d	$\times 10^{-1}$	1 decibel = 0.1 bel
centi	c	$\times 10^{-2}$	1 centimeter = 0.01 m
milli	m	$\times 10^{-3}$	1 milliampere = 0.001 A
micro	μ	$\times 10^{-6}$	1 microvolt = 10^{-6} V
nano*	n	$\times 10^{-9}$	1 nanosecond = 10^{-9} s
pico*	p	$\times 10^{-12}$	1 picofarad = 10^{-12} F
femto	f	$\times 10^{-15}$	1 femtometer (approximate size of a proton) = 10^{-15} m
atto	a	$\times 10^{-18}$	

* In the older literature, certain double prefixes are used: kilomega (kM) for 10^9; millimicro (mμ) for 10^{-9}; micromicro ($\mu\mu$) for 10^{-12}. The micron (μ), a unit of length, is 1 μm = 10^{-6} m = 0.001 mm.

(inertia) are fundamental but are not capable of the highest precision. Fortunately, the weights of two bodies, at any given point on the surface of the earth, are in strict proportion to their masses (Sec. 4-5), and therefore masses can be compared by weighing. Two bodies which weigh the same on an equal-arm balance or a spring balance also have equal masses. Other metric units for mass are the gram ($\frac{1}{1000}$ kg) and the metric ton (1000 kg).

It is interesting to note the historical basis of the metric units. When the French authorities established the metric system in 1791, they intended that the meter should be exactly one 10-millionth of the distance from the equator to either pole; and it was intended that the gram should be the mass of one cubic centimeter of pure water at 4°C (a temperature near which the density of water is relatively independent of temperature, hence allowing the most precise measurement.) These goals, however, were not quite realized, and so the prototype meter and kilogram based upon these procedures did not have exactly the intended values.

The adoption of the metric system by the French revolutionaries was a manifestation of the Age of Reason; they wished to establish a "rational" basis for every aspect of life, and the decimal system fitted into such a scheme. In 1795 the centime was adopted (it should properly have been called the centifranc). The division of the right angle into 100 equal parts, instead of 90 degrees, was proposed, but never gained wide acceptance. The revolutionary calendar had months of 30 days, divided into three 10-day "decades" instead of 7-day weeks; this calendar was abolished by Napoleon in 1805. It is significant that the names of the months in the revolutionary calendar were based on the seasons and the fruits of the earth; how logical it seemed to turn to the earth itself for the basis of new systems of measurement!

The mks unit of time is the second, and indeed this unit is common to all systems of units and is familiar in daily life. Formerly, the second was defined as 1/86 400 of a mean solar day, being based upon the supposed constancy of the earth's rotation on its axis. In recent years evidence has accumulated pointing to slight irregularities in the rotation of the earth on its axis (Sec. 8-8); it has been found that the orbital motion around the sun is more nearly constant and therefore more suitable for defining a time unit. In 1955 the second was redefined as exactly 1/31 556 925.9747 of a mean tropical year.[*] However, this standard has been replaced by one based upon an "atomic clock." In an early (1952) form of such a clock, the nitrogen atom in an ammonia molecule (NH_3) is allowed to vibrate back and forth within the molecule. Being almost wholly unaffected by external conditions, the frequency of vibration is even more nearly constant than the period of the earth's revolution around the sun. The present standard, accurate to a few parts in 10^{11}, is based on the cesium atom of atomic mass number 133. The cesium standard was adopted in 1964, but is designated as "temporary" because research in the next few years may indicate that a standard based on some other atom such as hydrogen or thallium would be even more reproducible.

Of the three fundamental quantities, two —length and time—can now be defined in terms of atomic behavior: the wavelength of a certain light, and the frequency of a certain atomic vibration. It is natural to ask if a similar atomic definition of mass can be constructed. The masses of individual atoms are presumably suitable for the purpose, but it is too difficult to make an accurate count of the large number of atoms contained in a weighable amount of matter. The best we could do is to define a kilogram as the mass of 5.0188×10^{25} atoms of the carbon isotope

---

[*] Since the length of the year is slowly changing at a known rate, the definition was made in terms of the earth's motion as of January 1, 1900.

of atomic mass number 12. The experimental uncertainty in the count is about 1 part in 20 000. Since two masses can be compared with far greater accuracy by weighing, it appears that we shall have to wait for improved experimental methods before an atomic definition of the kilogram would be any improvement over the present definition in terms of the international prototype kilogram.

The mks system is called an absolute system of units, since the meter, kilogram, and second are defined in a way that has nothing to do with local variations in the earth's gravitational field. It is mass (inertia), not weight, that is used.

2. The *centimeter-gram-second* (cgs) *system* differs from the mks system only in using the centimeter and the gram, which are submultiples of the meter and the kilogram. It, too, is an absolute system of units.

3. The *foot-pound-second* (fps) *system* is another system used in English-speaking countries. The pound is legally (since 1893) defined as exactly $1/2.2046$ of the weight of an object whose mass is 1 kilogram. The foot was legally defined as exactly $\frac{1200}{3937}$ m, which makes the inch come out to be $2.54000508 \ldots$ cm. In recent years the American Standards Association has defined the *inch*, rather than the foot, making it exactly 2.54 cm (eliminating the tag end of the decimal). The origins of the foot, yard, and pound are lost in antiquity. Prototype yards and pounds were formerly maintained in England and the United States, but now the primary standards for all countries are metric, and the British standards are secondary standards, defined in terms of the metric standards. We may call the mks and cgs systems *metric absolute systems,* and the fps system (which we use occasionally in this book) the *British gravitational system.* The metric system is used in all countries for scientific work and is the common system of measurement in almost all major countries. In the English-speaking countries the fps

system is used in ordinary daily life, and it is the basis for the British engineering system which is used extensively in engineering work. However, the growth of international trade is a strong force toward adoption and full use of the metric system, already legally permitted in the United States. Business and industry leaders are becoming aware of the monetary losses, chiefly in loss of trade, caused by the complex fps system and other irrational units such as quarts, miles, inches, and tons. In 1965 Great Britain announced her intention to switch to the metric system by 1975. The United States and Canada will no doubt follow suit in the foreseeable future.

For two reasons, therefore—the already prevalent use of the metric system in science, and the strong possibility of a changeover in business and industry—we will stress the use of metric units in this book.

1-5 The equations and formulas of physics

The study of physics has the reputation of being highly mathematical and is therefore approached with misgivings by some students. True enough, we assume that you have an elementary knowledge of ninth-grade algebra and a little of tenth-grade similar triangles, but any additional mathematics needed will be explained as we go along, and you will find a review of simple algebra and geometry in the Appendix. The most serious difficulty is probably not the mathematics itself, but its application to physical situations. Here is where dimensional checks will be of great help in testing equations for possible error. In algebra you learned not to add "dollars" to "doughnuts" if you wanted a meaningful answer. In physics, if two quantities are to be added or subtracted, they must be of the same dimensions, and likewise the quantities on both sides of an equals sign must have like dimensions.

EXAMPLE 1-1

A student derives an equation for the distance traveled by a falling body:

$$s = \tfrac{1}{2}v_0 t^2 + vt$$

where s is the distance fallen, v_0 is the original velocity of the body, t is the elapsed time, and v is the final velocity of the body. Is the equation a possibly correct one?

Making a dimensional check and ignoring the factor $\tfrac{1}{2}$, which has no dimensions, we have:

Using fractions:

$$[L] \overset{?}{=} \frac{[L]}{[T]}[T^2] + \frac{[L]}{[T]}[T]$$

$$[L] \overset{?}{=} [LT] + [L]$$

Using negative exponents:

$$[L] \overset{?}{=} [LT^{-1}][T^2] + [LT^{-1}][T]$$

$$[L] \overset{?}{=} [LT] + [L]$$

Note that the dimensions, in square brackets, are combined just like ordinary algebraic quantities. We see that the dimensions are incorrect and so the equation is *wrong.* How to make it *correct* is another matter, something into which we are not yet prepared to go.

EXAMPLE 1-2

A falling raindrop reaches a steady speed which depends on its size and on the coefficient of viscosity of the air. Check the following equation for possible correctness:

$$v_1 = \frac{2}{9\pi}\frac{r^2 g d}{\eta}$$

where v_1 is the steady speed of the drop, in cm/s; r is the radius of the drop, in cm; g is the acceleration due to gravity, in cm/s²; d is the density, in g/cm³; and η (Greek letter eta) is the coefficient of viscosity of the air, in g/cm·s.•

Viscosity is fluid friction, to be discussed in Chap. 12. However, at this time we are merely checking the dimensions of a proposed equation and need not be concerned at all with its derivation or physical significance. Substituting the

---

• Often we present units in this form to save space; thus g/cm·s is understood to mean $\dfrac{g}{cm \cdot s}$.

dimensions of the various quantities into the proposed equation, we find that the check is successful:

Using fractions:

$$\frac{[L]}{[T]} \overset{?}{=} \frac{[L^2]\left[\dfrac{L}{T^2}\right]\left[\dfrac{M}{L^3}\right]}{\left[\dfrac{M}{LT}\right]}$$

$$\frac{[L]}{[T]} \overset{?}{=} \frac{[L]}{[T]}$$

Using negative exponents:

$$[LT^{-1}] \overset{?}{=} [L^2][LT^{-2}][ML^{-3}][LTM^{-1}]$$

$$[LT^{-1}] \overset{?}{=} [LT^{-1}]$$

The equation is *not wrong,* from a dimensional point of view. As a matter of fact, it is *not right,* either, since the numerical factor $2/9\pi$ is incorrect. Dimensional analysis never gives information about numerical constants, which must be found by detailed analysis from physical laws. In spite of this, dimensional analysis is a powerful tool when properly used.

Let us use this example as the occasion for some remarks about the way we write equations, and how we use the given information.

(*a*) The letters are *symbols,* which stand for observable quantities. They are a shorthand way of saying what factors influence the final speed, and how. Usually the symbols are chosen to remind one of the name of the quantity, such as v for velocity, r for radius, etc. Unfortunately, the demand for symbols is so great that we have to use upper- and lower-case letters, subscripts, and even Greek letters, as in this example.

(*b*) Careful study of the list just after the equation shows two sorts of quantities: *variables,* such as r, d, and v_1, and *constants* such as 2, 9, π, g, and η. There is no automatic way to tell constants from variables except to know the exact meaning of the symbols— that is why a list is a good idea. Often a subscript (such as the small 1 after the v) is used to single out some particular value of a variable, in this case the "final" or "steady" value of v which is eventually reached. An unadorned symbol v might represent the velocity at any time, before reaching the steady value v_1.

(*c*) We call g and η *constants* in this problem, since they would be the same regardless of the radius or density of the drop. They are surely dependent on other factors (for instance, g would be less on top of a high mountain), but they are constant for this problem. We imagine asking ourselves, "How does the final velocity depend on r and d, the factors which are likely to vary in a series of experiments?"

(*d*) Although you have not studied falling bodies or viscosity, you can determine the *dimensions* of g and η simply by looking at the units in which they are given. For instance, since η is given in g/cm·s (read as grams per centimeter per second), we know that the dimensions of η are [M]/[LT] or [ML^{-1}T^{-1}]. This is an example of a constant which has dimensions.

(*e*) The equation also contains *numerical constants*, such as 2, 9, and π, which are of definite value and are dimensionless, as are *all* numerical constants.

(*f*) We call η the *coefficient* of viscosity for air, meaning that it is a property of matter which can be tabulated. Elsewhere we will study the coefficient of thermal expansion, the coefficient of surface tension, temperature coefficient of electric resistance, and other coefficients. These depend on external circumstances such as temperature (the coefficient of viscosity for air becomes larger as temperature increases), but they are considered to be constants during any one experiment.

(*g*) With some justice, the relationship may be called an *equation* (since it has an equals sign), a *formula* for v_1 (since the desired quantity v_1 is all by itself and may be easily calculated by substituting values of the other variables such as r and d), or a *law* of physics. Perhaps the designation "formula" fits best in this case, as the term "law" is generally reserved for something of more general importance. In this case, we would expect that our formula (and many others) can be *derived* from basic laws of physics such as Newton's laws of motion, the atomic hypothesis, etc.

In our study of physics, we shall be concerned with the basic laws of physics, as revealed by experiments and described by numerous formulas which express the relationships between physical quantities.

■ SUMMARY

Physics deals mainly with the more fundamental aspects of energy and non-living matter. Physical knowledge comes to us through application of scientific methods: the gathering of data, formulation of hypotheses, and testing of hypotheses by means of controlled experiment. The division of physics into compartments such as mechanics, electricity, etc., is merely an aid to study, and we shall cross the boundaries many times in our further work.

Mechanical quantities are expressible in terms of three fundamental dimensional quantities such as mass, length, and time; or force, length, and time. Scientists are now able to define length and time in terms of atomic standards which are imperishable and universally reproducible, but such a procedure for mass is not yet practicable although theoretically possible. Dimensional analysis is often useful in checking equations for possible error.

■ CHECK LIST

hypothesis   kilogram   fps system
theory   second   pound
law   SI system   absolute system
dimensional quantities   mks system   equation
meter   cgs system   formula

■ QUESTIONS

1-1   Make a list of the titles of the physics journals in English which are carried by your college library. Check also for *Science, Nature, Scientific American, American Scientist, Sky and Telescope, Environment, Bulletin of the Atomic Scientists* (*Science and Public Affairs*), and *Isis*.

1-2   Suppose we could communicate with little green men on Mars only by means of radio signals. Compose a text for their message that would give enough information to enable us to tell how long their unit of length, the retem, is in comparison with our meter. (Although they spell their words backward, they are supposed to be not at all backward intellectually.) Could we find out the size of their unit of mass?

1-3   What is the formula for the area of a triangle? What is the dimensional formula for the area of a triangle?

1-4   The radius of a circle inscribed in any triangle whose sides are a, b, c, is given by

$$\frac{\sqrt{s(s-a)(s-b)(s-c)}}{s}$$

where s is an abbreviation for $\frac{1}{2}(a+b+c)$. Check this formula for dimensional consistency.

1-5   A student derives the following formula for the distance traveled by an accelerated body:

$$s = vt - \tfrac{1}{2}at^2$$

where s = distance, in ft; v = final velocity of body, in feet per second; t = elapsed time, in seconds; and a = acceleration, in feet per second per second. Check this formula for dimensional consistency. What statement can you make about the correctness of this formula?

1-6   The French centime was defined to have a mass of 2 g of copper. Explain how the mass of this coin is related to the circumference of the earth and the physical properties of water.

■ REFERENCES

At the end of each chapter of this book will be found references to sources where various aspects of the material covered in the text are explored more deeply. In addition, we list here several reference works which are applicable throughout the course.

GENERAL REFERENCES

1   Magie, W. F., *A Source Book in Physics* (New York: McGraw-Hill, 1935; Cambridge: Harvard University Press, 1963). Contains extracts from the original writings of the great experimental physicists from 1600 through the year 1900. A valuable reference with which every student should have at least a nodding acquaintance.

2 Cajori, F., *A History of Physics* (New York: Macmillan, 1906; Dover, 1929).

3 Butterfield, H., *The Origins of Modern Science: 1300–1800* (London: Bell, 1957; New York: Macmillan, 1957; Collier, 1962 [revised ed.]; Free Press, 1965).

4 Sarton, G., *A History of Science* (Cambridge: Harvard University Press, 1952).

5 Taylor, L. W., *Physics: the Pioneer Science* (Boston: Houghton Mifflin, 1941; New York: Dover, 1959).

6 Holton, G., and D. H. D. Roller, *Foundations of Modern Physical Science* (Reading, Mass.: Addison-Wesley, 1958). Excellent historical and philosophical background material.

7 Rogers, E. M., *Physics for the Inquiring Mind* (Princeton: Princeton University Press, 1960).

8 Stacy, R. W., D. T. Williams, R. E. Worden, and R. O. McMorris, *Essentials of Biological and Medical Physics* (New York: McGraw-Hill, 1955).

9 Burns, D. M., and S. G. G. Macdonald, *Physics for Biology and Pre-Medical Students* (Reading, Mass.: Addison-Wesley, 1970).

STANDARD TEXTS

The student should have available also some standard text which is designed for those specializing in physics, chemistry, and engineering. It will be of value to refer from time to time to a book of a more technical nature than this one, for more detail, more rigor, more mathematics, or more diversified applications. Some of the standard texts in the field are the following:

1 Feynman, R. P., R. B. Leighton, and M. Sands. *The Feynman Lectures on Physics* (Reading, Mass.: Addison-Wesley, 1963).

2 Sears, F. W., and M. W. Zemansky, *University Physics* (4th ed.; Reading, Mass.: Addison-Wesley, 1970).

3 Weidner, R. T., and R. L. Sells, *Elementary Classical Physics* (1965), *Elementary Modern Physics* (2nd ed., 1968) (Boston: Allyn & Bacon).

4 Halliday, D., and R. Resnick, *Physics* (New York: Wiley, 1966).

5 *Berkeley Physics Course,* Vol. I, *Mechanics,* Kittel and Knight; Vol. II, *Electricity and Magnetism,* Purcell (New York: McGraw-Hill, 1965).

This is not meant to be an all-inclusive list of good books in this classification; your instructor no doubt has his favorites which he will be glad to recommend.

REFERENCES, CHAPTER 1

1 Thomson, G. P., "Nature of Physics and Its Relation to Other Disciplines," *Am. J. Phys.,* **28,** 187 (1960).

2 Holton, G., "The Relevance of Physics," *Physics Today,* Nov., 1970, p. 40.

3 Butterfield, H., "The Scientific Revolution," *Sci. American,* Sept., 1960, p. 173.

4 Nicholson, M., "Resource Letter SL-1 on Science and Literature," *Am. J. Phys.,* **33,** 175 (1965). Gives 96 references, many of them available in college libraries.

5 Macurdy, L. B., "Standards of Mass," *Physics Today,* April, 1951, p. 7. A review of practices at the National Bureau of Standards.

6 Brouwer, D., "The Accurate Measurement of Time," *Physics Today,* Aug., 1951, p. 6. Discusses changes in the length of the year, wandering of the poles.

7 Astin, A. V., "Standards of Measurement," *Sci. American,* June, 1968, p. 50.

8 Ritchie-Calder, Lord, "Conversion to the Metric System," *Sci. American,* July, 1970, p. 17.

2

Structure and Properties of Matter

A significant development in science took place when man began to consider the properties of matter, rather than the behavior of particular objects. It is one thing to know the weight of an iron axhead or an iron suit of armor, but axheads are large and small, heavy and light, and it is surely of greater importance to study the nature of the substance iron than to limit oneself to specific iron objects, be they ball bearings or bridge girders. Substances have many properties; in addition to a host of mechanical characteristics, we can list heat conductivity, electrical conductivity, optical index of refraction, shielding effect for x rays, and neutron absorption, to name only a few. Tables of properties of matter are contained in handbooks which may run to several thousand pages. We presume that *all* properties of matter are, in principle, ultimately to be correlated with a relatively few basic *attributes* of matter, such as inertia, gravitational forces, electric forces, and nuclear forces. Let us take a preview of these fundamental attributes of matter. As in all preliminary surveys, certain terms will have to be used loosely. You have undoubtedly studied general science, and so you will have a feeling for words such as "force" and

"charge" which we shall use now, but cannot define precisely until later chapters.

2-1 What is matter?

A first and basic attribute of matter is that any piece of it, which we call a "body," tends to resist any change in its motion. This attribute of matter is called *inertia*. A forward force must be applied to a car to cause it to pick up speed. If the car is already moving, a backward force must be applied to cause it to slow down. Unless a force is applied, the car tends to remain as it is, either at rest or in uniform (steady) motion in a straight line. We call such an object a *material body,* by which we mean that all matter has inertia. In Chap. 4 we shall see how inertia can be measured quantitatively, and we shall use the word *mass* to describe the amount of inertia that a material body has.

A second attribute of matter is its weight. We all know that the *weight* of an apple is the *force of gravitation* between the earth and the apple. Weight is an *interaction* between two bodies, and gravitational force is known to fall off rapidly as the distance between the bodies increases. There is no unique weight

Silicon crystal showing images formed by atomic planes about 3×10^{-10} m apart. The regular arrangement of atoms in the crystal lattice is not maintained in the region of a dislocation defect. The electron micrographic technique gave a magnification of about 5×10^6. (Dr. Victor A. Phillips, General Electric Research and Development Center)

of a given apple—its weight depends on its position relative to a second body (the earth). It is a property of every particle of matter to exert gravitational attraction on every other particle of matter anywhere else in the universe. We shall discuss gravitation more fully in Chap. 4.

Most matter exerts *electric force* upon other matter.[*] Electric forces are varied and complex, but to a first approximation several distinct aspects of such forces can be studied separately. Coulomb's law, to be discussed in Sec. 17-5, describes the force between stationary charged bodies: like charges repel each other, unlike charges attract each other, and the magnitude of the force varies inversely as the square of the distance between the charges. Coulomb forces are also called *electrostatic forces.* Actually, as we shall see in Chap. 17, electrostatic forces are intrinsically far greater than gravitational forces. It is only because most ordinary matter is neutral (since atoms contain equal numbers of positive and negative charges) that the electric forces between ordinary objects cancel out sufficiently to unmask such relatively small gravitational forces as the weight of a 10 ton truck. Another aspect of electric force is presented by the force exerted between charges in motion. Such forces are called *magnetic forces;* they are evident in the behavior of magnets, electric motors, and many other engineering devices. *Elastic forces* involve a third aspect of electric force which has to do with the overlapping of the charged parts of atoms. When an archer stretches his bow, he is calling into play electric forces between neutral atoms. Such forces are, of course, effective only when atoms are nearly in contact with each other, say within about 5×10^{-10} m or less. We often call these forces *short-range electric forces,* because at greater distances they are negligible compared with electrostatic forces or magnetic forces. If you have studied chemistry you may know that many *chemical forces* (the covalent bonds) are also short-range electric forces. We see that electric forces of various kinds are characteristic attributes of matter. It is important to realize that electric forces and gravitational forces are exerted entirely independently of each other, and there is no known connection between them.[■]

Finally, two kinds of *nuclear forces* are exerted between particles of matter. These are extremely short-range forces and are negligibly small unless the particles are about 10^{-14} m or less from each other. Consequently, these forces are unfamiliar in the large-scale, everyday world. They are of vital importance, however, for without some sort of cohesive nuclear force the particles which make up the nuclei of atoms would not stick together. With modern high-energy apparatus it is possible to shoot nuclei at each other with sufficient speed to overcome the electric forces of repulsion and allow positively charged particles to approach closely the positively charged nuclei. In this way the nuclear forces are being studied in laboratories all over the world. As the picture unfolds, it is found that nuclear forces are at least as complex and varied as are the electric forces. In Chap. 30 we will discuss the two kinds of nuclear force, known as the *strong interaction* and the *weak interaction.*

These, then, are the attributes of what we call matter. Matter has inertia, and matter exerts gravitational, electric, and nuclear forces upon other matter. These are the underlying physical facts that, properly interpreted, must give us an account of the many tremendously varied properties of matter encountered in nature.

[*] Certain neutral subatomic particles such as the neutron and the neutral mesons exert no electric forces, although they do exert nuclear forces and (presumably) gravitational forces. All such particles which have a finite rest mass are unstable when far removed from other particles and cannot long exist by themselves (p. 675).

[■] The theoretical physicist Albert Einstein (1879–1955) devoted the last decades of his life to an unsuccessful search for such a connection.

2-2 Density and specific gravity

Perhaps the most obvious property of matter is its density. Aluminum is a "light" metal; gold is a "heavy" metal. We define *density d* as the mass per unit volume of a substance. Thus, since each cubic centimeter of water has a mass of 1 g, its density is 1 g/cm³. The density of this most familiar substance is expressed in the mks system as 1000 kg/m³. In Table 2-1 are given the densities of some representative solids, liquids, and gases. Just as for water, the density of any substance in kg/m³ is found by multiplying the cgs value by 1000.

EXAMPLE **2-1**

A block of metal has dimensions 5 cm × 7 cm × 20 cm, and its mass is 5 kg. Out of which of the various metals listed in Table 2-1 is the block probably made?

The density of the block must be expressed in grams per cubic centimeter in order to compare it with the table. We must first express the mass in grams. Since 1 kg = 1000 g, the mass is 5 × 1000 = 5000 g.

Volume = (5 cm)(7 cm)(20 cm) = 700 cm³

We now calculate the density:

$$\text{Density} = \frac{\text{mass}}{\text{volume}} = \frac{5000 \text{ g}}{700 \text{ cm}^3} = \boxed{7.14 \text{ g/cm}^3}$$

The block is probably made of zinc.

A note on significant figures. Strictly speaking, writing the mass of the block in Example 2-1 as 5 kg implies that the mass lies somewhere between 4.50 kg and 5.50 kg. (See the discussion of significant figures in Sec. A-4 of the Appendix.) However, in this text we arbitrarily assume that all data given in the examples and problems are known to three significant figures, unless otherwise stated. Thus, in this text, $m = 5$ kg may be thought of as $m = 5.00$ kg for purposes of computation. This is not good engineering practice, and it would not be allowed in a write-up of a laboratory experiment. How-

ever, our purpose in solving problems is to illustrate basic laws and methods of physics, not to illustrate laboratory practice. When we write $m = 5$ kg in this book, we mean that the mass may be considered to lie between 4.995 kg and 5.005 kg. This would be written as 5.00 kg in a laboratory report.

TABLE **2-1**   **Density**

Substance	Density g/cm³
Solids	
aluminum	2.70
copper	8.93
tin	7.29
brass	8.44
zinc	7.14
magnesium	1.75
iron	7.86
lead	11.3
gold	19.3
uranium	18.7
wood	0.35–0.9
concrete	2.7
ice	0.917
limestone	2.7
bone	1.7–2.0
diamond	3.51
cork	0.22–0.26
steel	7.8
Liquids	
water	1.00
gasoline	0.66–0.69
glycerin	1.26
mercury	13.6
alcohol (ethyl)	0.791
sea water	1.025
carbon tetrachloride	1.595
olive oil	0.918
Gases*	
air	1.293×10^{-3}
hydrogen	0.0899×10^{-3}
oxygen	1.429×10^{-3}
helium	0.1785×10^{-3}
carbon dioxide	1.977×10^{-3}
tungsten hexafluoride	12.9×10^{-3}
methane	0.717×10^{-3}

* All at 0°C and 1 atm pressure.

EXAMPLE 2-2

If the mass of an iron engine block is 157 kg, how much mass would an airplane designer save if the same block were cast in magnesium?

From Table 2-1, we determine that the density of iron is 7860 kg/m³. Since density = mass/volume, we can solve for the volume of the block:

$$\text{Volume} = \frac{\text{mass}}{\text{density}} = \frac{157 \text{ kg}}{7860 \text{ kg/m}^3} = 0.020 \text{ m}^3$$

When the block is cast in magnesium, its volume is still 0.020 m³.

$$\text{Mass} = (\text{density})(\text{volume})$$
$$= (1750 \text{ kg/m}^3)(0.020 \text{ m}^3) = 35 \text{ kg}$$

The designer will save 157 − 35 = $\boxed{122 \text{ kg}}$

For many purposes it is convenient to compare substances with one another. Pure water at 0°C (or at some other standard fixed temperature) is often used as a standard substance, and we define *specific gravity* as a ratio:

$$\text{Specific gravity} = \frac{\text{density of substance}}{\text{density of water}}$$

Any given volume of aluminum weighs 2.70 times as much as the same volume of water (both at 0°C), and its specific gravity is 2.70. Note carefully that specific gravity is a pure number, having no units, since it is a ratio of two similar quantities.* Because of its definition in terms of water, and because water has a density of 1.00 g/cm³, the specific gravity of any substance is *numerically equal to* the density in cgs units.

It is possible to define a *weight density* which is weight per unit volume. But weight is not an absolute property of a body; it depends on the strength of the gravitational field in which the body finds itself. For instance, what is the weight density of aluminum if measured on the moon? Surely the aluminum

weighs less—about $\frac{1}{6}$ as much as on earth—and its weight density is correspondingly less. Because of this ambiguity about weight density, we will find it preferable to define density as mass per unit volume. We will, however, continually keep in mind that at any given place weight (the pull of gravity) is proportional to mass (inertia), as will be discussed later. The weight density (on earth) of water at 0°C is 62.4 lb/ft³; the weight density of any other substance can be found by multiplying its specific gravity by 62.4 lb/ft³.

2-3 The structure of matter

We have seen the advantages of considering the *substance* iron, rather than a multitude of different iron objects of various sizes and shapes. Can we simplify still further and consider what, if anything, is common both to iron and aluminum, for instance? What is the difference between a block of iron and an equally large block of aluminum that makes the iron block weigh so much more? Let us compare iron and aluminum in detail.

As a result of much experimentation by many scientists, a picture of the structure of a substance such as iron has gradually unfolded. According to this picture, iron is a collection of atoms in a symmetrical arrangement called a *crystal*. The crystal structures of various elements, alloys, and compounds differ in both the arrangement and spacing of their atoms (Fig. 2-1). Thus the density of a solid depends in part on its crystal structure. By way of illustration, the atomic weight of iron is 56, and that of aluminum is 27, meaning that each iron atom is just over twice as heavy as each aluminum atom. Since the density of iron is almost 3 times that of aluminum (Table 2-1), we see that the fact that each iron atom is twice as heavy as each aluminum atom is not enough to account for the difference in density. Therefore, the iron atoms must be packed more closely.

---

* To avoid small numbers, specific gravity for gases is sometimes tabulated relative to a standard gas such as air at 0°C and 1 atmosphere pressure.

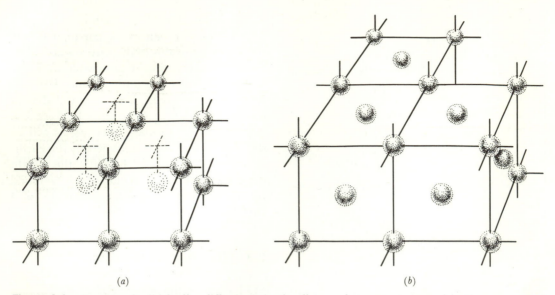

Figure 2-1 Crystals of solid bodies differ both in the distance between atoms and in the geometrical form of the crystal structure. (*a*) Iron; (*b*) aluminum. Approximate scale: 1 cm = 1.6×10^{-10} m.

Detailed x-ray analysis of the crystal structures of iron and aluminum shows this to be the case, and exact numerical check is obtained for the densities.

Going one step further, we ask why the atomic weights of iron and aluminum differ. Here the planetary model* of an atom is a good first approximation. The heavy *nucleus* contains more than 99.9% of the total mass (and weight) of the atom. The rest of the mass of an atom is due to a number of *electrons* which travel around the nucleus in orbits, like planets around the sun. The nucleus contains "heavy" particles known as *nucleons*, which may either be electrically neutral or have a positive charge. A neutral nucleon is called a *neutron*, and a positively charged nucleon is called a *proton*. These heavy particles, the neutron and the proton,

are considered to be merely different states of the same particle. They have almost identical masses, and it is entirely possible for a proton to change into a neutron, and vice versa. The probability of such a changeover within the nucleus depends on many factors, but when the change takes place spontaneously, the nucleus is said to be *radioactive*, or *unstable*.

Each electron has a charge that is negative and numerically equal to that of the proton. A neutral atom, therefore, has the same number of electrons in the electron cloud as it has protons in the nucleus. This number is called the *atomic number* of the element. Chemical properties are determined by the number of electrons in the cloud. The "iron-ness" of iron, for instance, is summarized by the many chemical reactions in which iron takes part, and these reactions are governed by the number of electrons in the electron cloud, that is, by the atomic number. Figure 2-2 shows schematically several common atoms. Note that two kinds of iron atom are shown. One has a total of 54 nucleons in its nucleus, and the other has a

---

* To the scientist, the term *model* does not usually mean a physical rendering, such as a model airplane; a model is a *representation* which allows us to think more concretely about some aspect of the real world. The model is not the reality; it simply offers us a useful way of dealing with reality. Models, like hypotheses, are always subject to correction as we gather more evidence from experiment or observation.

total of 56 nucleons. Both have 26 protons—the difference is in the number of neutrons. These atoms are called *isotopes* of iron; the atomic number of each iron isotope is seen to be 26. As a matter of notation, a symbol such as $^{54}_{26}Fe$ is used, where the *mass number,* equal to the total number of nucleons, is written as a superscript at the upper left of the chemical symbol, and the atomic number is written as a subscript at the lower left.[*] Normally, an iron nucleus draws to itself a cloud of 26 electrons, equal to the positive charge of the 26 protons within the nucleus. If, somehow, only 25 electrons surround a $_{26}Fe$ nucleus, we have a net positive charge of 1 unit, and the atom is called an *ion,* written as Fe^+ in this case. A doubly charged iron ion, written as Fe^{++}, would have only 24 electrons in its cloud, and so on. Naturally, since the electron cloud determines chemical properties, iron ions behave quite differently from neutral iron atoms. It is also possible to have negative ions; thus a $_{17}Cl^-$ ion has 18 electrons in its cloud, one more than the 17 that would be equal to the atomic number.

The isotopes of the simplest element, hy-

drogen, are of special interest. The ^1_1H nucleus is simply a proton; most (99.985%) of the hydrogen atoms in nature have nuclei of this sort. The nucleus of "heavy hydrogen," also called *deuterium,* is ^2_1H, made up of one proton and one neutron. (This combination is called a *deuteron.*) Finally, ^3_1H, or *tritium,* has one proton and two neutrons. However, tritium is radioactive, with a 50-50 chance that one neutron will change to a proton within about 12 years; when this happens, an electron is formed and ejected from the nucleus.

Many elements have more than one stable isotope. For instance, there are ten stable isotopes of tin, ranging from $^{112}_{50}Sn$ to $^{126}_{50}Sn$. The chemical atomic weight is the average value of the weights of all stable isotopes, with due regard for relative abundance (for tin, the average works out to 118.69, the value given in chemical tables). Chemically, the isotopes of an element—tin for instance—are identical, since the atomic numbers are the same. Separation of isotopes is thus difficult and depends upon physical rather than chemical methods.

It is impossible to draw atoms to scale. Most of the atom is empty space, with the distances between nucleus and electrons vastly larger than the sizes of the particles.

[*] Since the atomic number of any atom is known whenever the chemical symbol is known, the writing of the atomic number is superfluous and is often omitted.

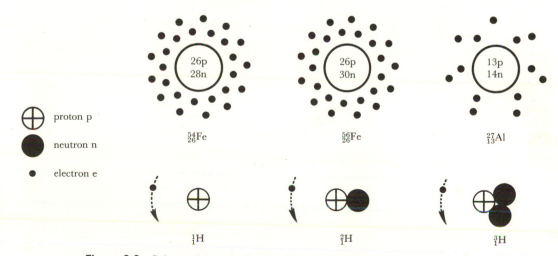

proton p

neutron n

electron e

$^{54}_{26}Fe$ $^{56}_{26}Fe$ $^{27}_{13}Al$

^1_1H ^2_1H ^3_1H

Figure 2-2 Schematic representation of several atoms (not drawn to scale).

It also follows that the density of the nuclear matter is enormous, for only in that way can the over-all density of matter be as great as it is. If the electron cloud of the iron atom ^{56}Fe were expanded to the size of a football field, the nucleus would be represented by a pea-sized ball 4 mm in diameter, weighing 6 million tons, and the electrons would be represented by 26 mosquitoes weighing 120 tons each, flying around the pea at distances ranging up to 50 meters. This is an extreme case of "modelitis"[*]—the uncritical acceptance of models as true descriptions of nature. In some respects the model is valid, especially as regards the rather definite boundary for the nuclear matter. On the other hand, the behavior of the electrons is often more closely approximated by a model of a continuous cloud of charge, rather than by a small group of mosquito-like or planet-like particles.

Experiment shows that the sizes of the nuclei increase as the number of particles increases. In fact, the *volume* of the ^{238}U nucleus is just about $\frac{238}{56}$ times as much as the volume of the ^{56}Fe nucleus. This means that at long last we have come upon a property of matter which is remarkably uniform from element to element: all nuclei have about the same density. Our model of a nucleus as a collection of incompressible neutrons and protons therefore represents this phase of nuclear behavior fairly well. The variations in density of various solids, which are shown in Table 2-1, are to be attributed to several causes: (1) the atomic weights of the elements vary from 1 to over 200; (2) the sizes of the individual atoms differ because of variations in the sizes of the electron clouds; and (3) atoms themselves are packed together in different ways to form crystals. The atoms in a crystal are held apart by electric forces, due to the repulsion between the like positive charges of neighboring nuclei. If somehow these repul-

sive forces could be annulled, the nuclei could be packed closely and matter of tremendous density would result. The nuclei of all the atoms in the earth would make a ball only 200 m in radius! Although such densities are unknown on earth, certain stars (known as "white dwarfs") have partially collapsed, and have densities ranging up to 50 000 g/cm^3 (almost a ton per cubic inch) —still far less than the universal density of nuclear matter, which is about 2×10^{14} g/cm^3.

2-4 The masses of individual atoms— Avogadro's number

So far, we have described atoms in relation to each other. When we say that ^{238}U has an atomic weight of 238 units and ^{56}Fe has an atomic weight of 56 units, we are merely saying that

$$\frac{\text{Weight of one } ^{238}\text{U atom}}{\text{Weight of one } ^{56}\text{Fe atom}} = \frac{238}{56}$$

This knowledge of the *ratio* of atomic weights comes from the data of chemistry—careful weighing experiments involving compounds containing uranium and iron. Now, bodies that have more inertia also weigh more, in strict proportion. Therefore, the chemical data allow us to say that

$$\frac{\text{Mass of one } ^{238}\text{U atom}}{\text{Mass of one } ^{56}\text{Fe atom}} = \frac{238}{56}$$

In this way the *ratios* of atomic masses were known many years ago. It is quite a different thing, however, to find out the mass of an individual atom.

Suppose we have 56 g of iron (^{56}Fe) in one lump, and 238 g of uranium (^{238}U) in another. These two chunks of matter are each said to contain 1 *mole* of the metal; this term simply means that in each case the mass in grams numerically equals the atomic weight. We also know that *each chunk contains the same*

---

[*] See p. 88 for further description of this dread disease.

number of atoms. To see that this statement is true, let us imagine two large sealed boxes *A* and *B* with box *A* containing golf balls and box *B* containing marbles. The total mass of box *A* is measured to be 238 kg, and the total mass of box *B* is 56 kg. Suppose we reach in and remove one golf ball from box *A* and find its mass to be 23.8 g. Then we pull out one marble from box *B* and find that its mass is 5.6 g. The ratio of the individual masses is 23.8 g/5.6 g which may also be written as 238:56. We conclude that there are as many golf balls in *A* as there are marbles in *B*, the greater total mass of box *A* being exactly accounted for by the greater mass of the individual balls. Similarly, there must be some definite number of atoms in a mole of any element. This number is called Avogadro's number, N_A.

For a chemical compound, the word "mole" is used to refer to an amount whose mass in grams equals the *molecular weight*. A mole of molecules contains N_A molecules, just as a mole of atoms contains N_A atoms. For instance, the atomic weight of carbon is 12, and that of oxygen is 16. The molecular weight of CO_2 is $12 + 16 + 16 = 44$. Therefore 1 mole, or 44 g, of CO_2 contains N_A molecules; there are N_A atoms of C and $2N_A$ atoms of O, a total of $3N_A$ atoms in the N_A molecules of CO_2.

Although the *existence* of Avogadro's number was known as soon as the existence of identical atoms was postulated, the *measurement* of the number is difficult, and Avogadro himself had no clear idea of its value. One early method was to let a drop of oil spread out on the surface of water, forming a thin film. As the film spreads, it gets thinner, until finally it becomes a monolayer, one molecule thick. We know the original volume of the drop (found from its weight and density) and the area of the film, so the thickness of the monolayer can be found. This gives the diameter of a single molecule, and hence the volume of a single molecule can be estimated.

In this way the mass of a single molecule can be estimated, and therefore the number of molecules in a known mass of the oil can be found. Modern measurements, more precise but less direct, give for Avogadro's number the value[*] $N_A = 6.02 \times 10^{23}$ atoms per mole of element, or molecules per mole of compound.

The mass m of a single iron atom can be found by a proportion:

$$\frac{56 \text{ g}}{6.02 \times 10^{23} \text{ atoms}} = \frac{m}{1 \text{ atom}}$$

$$m = 9.3 \times 10^{-23} \text{ g}$$

The cgs system is commonly used for such calculations, since chemical atomic and molar weights are usually expressed in grams.

EXAMPLE **2-3**

How many atoms are contained in a speck of colloidal silver that can just be seen under a microscope? Assume the speck to be a cube 10^{-6} cm on an edge. The atomic weight of silver is 107, and its specific gravity is 10.5.

First we find the mass of the silver from its volume and density.

$$\text{Volume} = (10^{-6} \text{ cm})^3 = 10^{-18} \text{ cm}^3$$
$$\text{Mass} = (\text{density})(\text{volume})$$
$$= (10.5 \text{ g/cm}^3)(10^{-18} \text{ cm}^3)$$
$$= 10.5 \times 10^{-18} \text{ g}$$

We make a proportion, knowing that one mole (107 g) has 6.02×10^{23} atoms:

$$\frac{6.02 \times 10^{23} \text{ atoms}}{107 \text{ g}} = \frac{N \text{ atoms}}{10.5 \times 10^{-18} \text{ g}}$$

$$N = \left(\frac{10.5 \times 10^{-18} \text{ g}}{107 \text{ g}}\right)(6.02 \times 10^{23} \text{ atoms})$$

$$= 0.59 \times 10^5 \text{ atoms} = \boxed{5.9 \times 10^4 \text{ atoms}}$$

This small piece of silver contains 59 000 atoms.

---

[*]A more precise value of Avogadro's number is $N_A = 6.0225 \times 10^{23}$ atoms per mole, with some uncertainty in the last digit. In this book we usually round off the values of physical constants to three significant figures.

■ SUMMARY

Matter is characterized by inertia and by the occurrence of gravitational, electric, and nuclear forces between material particles. Gravitational force and some types of electric force act at long range, but other types of electric force are effective only at short range and are responsible for elastic forces and certain cohesive forces between atoms in molecules. Nuclear forces, which act at still shorter range, are not yet fully understood.

All the many properties of matter would be, in theory at least, predictable if these basic attributes were fully understood. Mathematical difficulties are so formidable that approximate models are often used as an aid to (but not a substitute for) the description of nature. One such model is discussed in this chapter: the atomic model for correlating the densities of substances.

Density is mass per unit volume. Specific gravity is a dimensionless quantity equal to the density of a substance divided by the density of a standard substance, usually water. Differences in density of various substances are not entirely due to differences in the masses of the individual atoms. The sizes of atoms, determined mainly by the sizes of the electron clouds, and the differences in crystal structure also affect the density.

The planetary model of an atom has proved valuable. The nucleus contains a close-packed cluster of nucleons, known as neutrons and protons, with total neutrality of charge for the atom as a whole maintained by an electron cloud which contains as many electrons as there are protons in the nucleus. This number is the atomic number of the element. Chemical properties of an atom are determined by the number of electrons in the cloud. Isotopes of a given element are chemically alike, have the same nuclear charge, and differ only in mass due to different numbers of neutrons in the nucleus. The nuclei of various elements differ in size in such a way that the density of nuclear matter is approximately constant for all elements.

Avogadro's number is the number of atoms in a mole of any element, or the number of molecules in a mole of any compound. The value of this important constant is 6.02×10^{23} atoms per mole.

■ CHECK LIST

inertia	nuclear forces	isotope
mass	density	ion
weight	neutron	deuterium
matter	proton	tritium
Coulomb electric forces	nucleon	modelitis
short-range electric forces	atomic number	Avogadro's number

■ QUESTIONS

Note: Each chapter in this book is concluded with a set of questions and problems. The *questions* are designed to help fix ideas, on a verbal level. The *problems* are arranged into three groups: Type A problems are the easiest and often merely test knowledge of the definitions and equations given in the text. Type B problems usually require some understanding of physical laws and methods. They are "standard" problems, well within the grasp of the student who has understood the chapter. Type C problems are of several kinds: they may be more difficult or time-consuming, or may depend upon more mathematics than do the others, or they may follow and deal with the additional text which is presented at the end of some chapters under the heading "For further study" (see p. 53).

2-1 What are the dimensions of density?

2-2 Consider the copper isotope $_{29}^{63}$Cu. How many electrons are in the electron cloud, if the atom is neutral? How many if the atom is a doubly charged positive ion? How many neutrons are in this nucleus? How many neutrons are in the nucleus of $_{29}^{65}$Cu? How many protons are in the nucleus of $_{29}^{65}$Cu? How many nucleons are in the nucleus of $_{29}^{65}$Cu? What is the atomic weight of $_{29}^{65}$Cu? What is the atomic number of $_{29}^{63}$Cu? Of $_{29}^{65}$Cu?

2-3 Since like charges repel each other, why don't the 26 positively charged protons in an iron nucleus fly apart instead of remaining together in a small region of space?

2-4 How many moles of water, H_2O, are in a glass containing 180 g of water? How many water molecules are contained in this same glass?

2-5 How many atoms of hydrogen are contained in 180 g of water, H_2O? How many atoms of oxygen are contained in this same amount of water?

2-6 Which has the greater number of atoms—a kilogram of lead or a kilogram of iron? (See Periodic Table on p. 721 for the atomic weights.)

2-7 The deuterium nucleus contains two particles. Name the particles; what is the nature of the attractive force between them?

2-8 What is the difference between the structure of the uranium isotopes $_{92}^{235}$U and $_{92}^{238}$U?

■ PROBLEMS

Note: Use the Periodic Table on p. 721, when necessary, to find atomic weights.

2-A1 Express the number of seconds in a year (p. 8) in the form 3.156×10^x.

2-A2 The circumference of the earth at the equator is 24 902 mi. Express this in inches, in the form $x.xx \times 10^x$.

2-A3 The density of air at standard conditions is 1.29×10^{-3} g/cm^3. Express this in kg/m^3.

2-A4 What is the specific gravity of carbon dioxide, relative to air? (Use Table 2-1.)

2-A5 What is the mass of the helium in a balloon, if the volume is 50 m^3 and the specific gravity (relative to air) is 0.14?

2-A6 What is the specific gravity relative to air of the "heaviest" gas in Table 2-1?

2-A7 What is the approximate volume of an elephant bone of mass 5.60 kg?

2-A8 The Great Pyramid of Cheops has a volume of 2.4×10^{12} cm^3 and is made of limestone. What is the mass of the pyramid   (*a*) in grams?   (*b*) in kilograms?   (*c*) in metric tons?

2-A9 If a cube of material 0.01 mm on an edge (barely visible to the naked eye) had the density of nuclear matter, what would be its mass? Could you lift it?

2-B1 How many atoms are in an iron nail of volume 0.200 cm^3?

2-B2 It is estimated that there is 10^{21} kg of water in the oceans of the earth. If a sailor near Australia throws a thimbleful (2 g) of dye (molecular weight 200) into the ocean, and the dye molecules eventually become distributed uniformly among all the water molecules of the oceans, how many of the dye molecules will be contained in a 360 g sample of sea water collected near Bermuda? (*Hint:* Use the molecular weight of water to find the total number of water molecules; then find the ratio of total number of dye molecules to total number of water molecules. This same ratio will be preserved in the water at Bermuda, if perfect mixing has occurred.)

2-B3 The atomic weight of ^1_1H (ordinary hydrogen) is 1.008 g/mole. Use Avogadro's number to compute the mass of a proton.

2-B4 Compute the mass of a million atoms of the isotope of mercury that is designated by the symbol $^{200}_{80}Hg$.

2-B5 If 10 liters of water and 5 liters of glycerin are mixed to form 15 liters of antifreeze solution, find the specific gravity of the mixture, assuming that no chemical reaction takes place.

2-B6 Of the two stable isotopes of boron, $^{10}_5B$ occurs 19% of the time and $^{11}_5B$ occurs 81% of the time. What is the average atomic weight of boron? Check your answer by referring to the Periodic Table on p. 721.

2-B7 A bottle contains 1000 cm^3 of liquid carbon tetrachloride (CCl_4). (*a*) What is the mass of the liquid? (*b*) How many molecules of CCl_4 are in the bottle?

2-B8 Calculate the mass, in kg, of a single uranium atom $^{238}_{92}U$.

2-C1 To illustrate how certain forces can be "effective" at short range but "not effective" at long range, assume that the force between two particles obeys the following law:

$$F = F_1 + F_2 = \frac{10^{-10}}{x^2} + \frac{10^{-80}}{x^9}$$

where x is the distance between the particles, in meters. Calculate F_1, F_2, and the total force F for the following values of x: 10^{-8} m, 10^{-9} m, 10^{-10} m, 10^{-11} m, and 10^{-12} m. (*a*) Which component of the force, F_1 or F_2, is the "short-range" one which dominates the expression for the shortest distances x? (*b*) For what separation of the two particles are the forces F_1 and F_2 equal? (*c*) For a separation that is 10^{-9} m, what error is made in the total force if the short-range part of the force is neglected?

2-C2 A teaspoonful of an organic oil (volume 5.00 cm^3) when dropped on the surface of a quiet lake spreads out to cover an area of 4000 m^2, almost an acre. What was the thickness of the film? Explain why this monolayer (one molecule thick) is thicker than the atomic plane separations of Fig. 2-1.

2-C3 Stars known as pulsars show variations in light during times much shorter than 1 s, apparently related to rapid rotation. They are therefore assumed to be very small and dense. Calculate the diameter of a completely collapsed neutron star of density equal to that of nuclear matter, if the star's mass is equal to that of the sun, 2×10^{30} kg.

■ REFERENCES

1   Nash, L. K., *The Atomic-Molecular Theory* (Cambridge, Mass.: Harvard University Press, 1950); also in Harvard Case Histories in Experimental Science, Vol. 1 (Cambridge, Mass.: Harvard University Press, 1957), pp. 215–321.

3

Kinematics
—the Description
of Motion

We continue our study of physics by considering *motion*. We first attempt to develop a way of describing the motions of physical bodies, ignoring for the time being the possible causes of motions. The study of motion as such, divorced from its origin or cause, is called *kinematics*.[*] Fortunately, our study of kinematics will prove useful in its own right, since some interesting problems can be solved without considering the causes of motion. Even more important, without the ability to think clearly about motion itself, we would be prevented from getting an adequate grasp of *dynamics*,[*] the study of forces and the changes in motion which they cause.

The Italian natural philosopher Galileo Galilei (1564–1642) became the most gifted experimental physicist of his time. Indeed, it was largely because of the special insight and example of Galileo that experimental methods became widely accepted as a part of physical science. He argued eloquently for the Copernican theory of the solar system —which asserted that the earth and the planets are revolving around the sun, as opposed to the Ptolemaic system, according to which the earth was the center of the universe—and was persecuted for his teachings. Typically, although Galileo did not invent the telescope, it was he who first used it to observe the satellites of Jupiter—experimental evidence for the Copernican theory. Galileo's experimental outlook, his reliance on observation as the final authority, and his skill in explaining his ideas to others make him one of the great men in the history of physics. His outlook pervades all of present-day science.

Galileo realized that an adequate description of motion requires a quantitative, or mathematical, outlook, but he was hampered by the primitive condition of mathematics in his day. Even with the simple algebra which is the common property of every high-school student, we are better off than Galileo. We will find that for maximum insight into the nature of motion we must also use the concepts of functions and limits. The branch of mathematics that deals with these concepts is called *calculus,* and we shall develop a useful amount of both differential and integral calculus as we go along.

After we have expressed the laws of kine-

[*] From the Greek *kinein,* to move, the same root from which come *cinema,* the motion picture; *Kinescope,* the TV picture tube; *kinetic,* pertaining to motion.

[*] From the Greek *dynamis,* power, the same root from which come *dynamo, dynamite, dynasty.*

Small lights attached to the runner's body give a picturesque description of motion. Can you interpret the details of the shapes of the various curves? (Popperfoto, Pictorial Parade)

matics in algebraic form, we will be able to proceed to the study of dynamics, with its numerous applications to basic science and to everyday life. The study of dynamics is a foundation for all that is to come, including jet aircraft and cyclotrons, as well as older subjects such as pendulums and collisions between molecules.

3-1 Types of motion

Consider these motions:

(*a*) A coffee cup near the edge of a table slips off, falls to the floor, and breaks.

(*b*) A parachutist falls freely for 800 ft, then opens his chute and slows down to a steady speed of 25 mi/h; he continues to fall at this speed and eventually strikes the ground and comes to rest.

(*c*) A golf ball dropped from a window rebounds from the sidewalk a number of times, always rising to successively lesser heights, and eventually comes to rest on the sidewalk.

(*d*) A hockey puck glides across the ice, traveling in a straight line toward the goal without any appreciable loss of speed.

(*e*) The earth moves around the sun at variable speed, in an elliptical orbit, moving slightly faster when it is slightly closer to the sun (in January).

(*f*) A swimmer dives off a diving board, and the end of the board continues to vibrate up and down for a while.

(*g*) A truck driver avoids a collision by jamming on his brakes and comes to a screeching halt without swerving to either side.

(*h*) An electron leaves the electron gun in the neck of a television picture tube, passes immediately through the deflection coils, where it is bent sideways, and then proceeds at constant speed in a straight line until it strikes the face of the tube and makes a spot of light.

(*i*) An artificial satellite travels at a steady speed of 15 700 mi/h around the earth, in a circular orbit of radius 5000 mi.

(*j*) A boy on a swing travels along a circular arc, picking up speed as he approaches the lowest point in his motion, and losing speed as he rises again.

We have only scratched the surface; countless additional types of motion easily suggest themselves for analysis.

Which of these motions is "simplest"? The first thing to notice is that most of the descriptions involve several distinct types of motion. For instance, in (*b*) there are four distinct periods: (1) while the man is freely falling; (2) the short interval of time while the parachute has just opened and he is slowing down; (3) the remainder of the trip in the air; and (4) the small fraction of a second during which his speed is reduced to zero while he is striking the earth. Considering the *whole* motion, we must say that (*d*) is perhaps the simplest, since neither direction of motion nor speed changes. Motion (*g*) is probably the next most simple (assuming from the meager description that the truck loses speed at a steady rate). Motion such as (*d*) is called *uniform motion in a straight line*, and motion such as (*g*) is called *uniformly accelerated motion.** Motions, such as (*a*) and (*c*) are combinations of the two basic motions described above, but the descriptions (*e*), (*f*), (*h*), (*i*), and (*j*) involve changing directions and are more complicated. Motion (*b*) is hard to describe exactly, since just after the parachute opens the man loses speed, but not the same amount each second.

It is tempting to contemplate the causes of some of the motions described. For instance, in (*b*), is the effect of air resistance constant? Such questions belong in the realm of dynamics, and will be studied in Chap. 4.

---

*In this case "deceleration" might seem a more descriptive word, but in physics the term "negative acceleration" is usually preferred.

3-2   Uniform motion in a straight line

The quantitative description of uniform motion in a straight line is simple: by definition, velocity is the time rate of change of position, that is, displacement divided by time. In symbols, for *uniform* velocity,

$$v_0 = \frac{s}{t}, \quad \text{or} \quad s = v_0 t \qquad (3\text{-}1)$$

The meaning of the symbols is important, since we shall use the same symbols later on. Here t is the *elapsed time,* s is the *displacement* (the distance measured from the starting point to the final point which the body reaches in the time t), and v_0 is the *constant velocity* of the body. Since the body's speed* does not change during uniform motion, v_0 is also the initial velocity; we shall use v_0 for initial velocity in our later study of nonuniform motion. Equation 3-1 is valid only for *uniform* motion at constant velocity.

The dimensional equation for velocity is [velocity] = [L]/[T] = [LT^{-1}]. Velocities are measured in m/s, ft/s, mi/h, and so on—always a length unit divided by a time unit.

A simple example of uniform motion will repay close study.

EXAMPLE **3-1**

In a television tube, an electron travels in a straight line from the electron gun to the fluorescent screen, a distance of 18 in., in 6 nanoseconds (6 ns) (see the table of prefixes on p. 7). What is its velocity in feet per second? In miles per hour?

$$\boxed{\begin{array}{l} s = 18 \text{ in.} \\ t = 6 \text{ ns} \\ v_0 = ? \end{array}} \quad (1 \text{ ns} = 10^{-9} \text{ s})$$

---

*In Sec. 3-8 we shall distinguish carefully between *speed,* which simply describes how fast a body is moving, and *velocity,* which describes both how fast and in what direction it is moving. In the next few sections we deal only with motion along a straight line. Such motion can be in only two directions, forward and backward (or up and down, etc.), so we can take care of change in direction by the use of algebraic signs, + and −. See Example 3-4.

$$s = v_0 t$$

$$v_0 = \frac{s}{t} = \frac{(18 \text{ in.})(1 \text{ ft}/12 \text{ in.})}{(6 \text{ ns})(10^{-9} \text{ s}/1 \text{ ns})}$$

$$= \boxed{2.50 \times 10^8 \text{ ft/s}}$$

To change to mi/h, we make use of the fact that 60 mi/h is the same as 88 ft/s:

$$v_0 = (2.50 \times 10^8 \text{ ft/s})\left(\frac{60 \text{ mi/h}}{88 \text{ ft/s}}\right)$$

$$= \boxed{1.70 \times 10^8 \text{ mi/h}}$$

We have used this simple problem to illustrate some of the "tricks of the trade" in problem solving:

(*a*) First we read the problem carefully and make a list of knowns and unknowns, setting them off in a box.

(*b*) We then write down an equation, preferably a basic one that actually defines a quantity that is of interest in the problem. Here such an equation is $s = v_0 t$; the linear relationship between s and t is the basis for defining uniform velocity.

(*c*) Since we are asked not for s but for v_0, which is buried in the equation, we next solve *algebraically* for the desired unknown, in this case obtaining $v_0 = s/t$. We usually do this before substituting any numbers.

(*d*) We substitute into the revised equation the information from the box containing the knowns and unknowns. The factor 1 ft/12 in. in the numerator doesn't change the value of the answer, since 1 ft *is the same as* 12 in., and hence 1 ft/12 in. is just another way of writing 1. The solver thinks somewhat like this: "I have 18 in. in the numerator, but I want ft in the numerator. Therefore I write

$$18 \text{ in.} \times \frac{\text{ft}}{\text{in.}}$$ so that I can later cancel out

the inches and have feet left. Finally, I fill in the numbers, knowing that 1 ft equals 12 in., getting 18 in. × (1 ft/12 in.)."* Note that, for the purposes of canceling units, it is

---

*In this particular example we *could* have changed the 18 in. to 1.5 ft at the very start; we followed the routine involving 1 ft/12 in. chiefly to encourage a habit that will prove useful later on in more complicated problems.

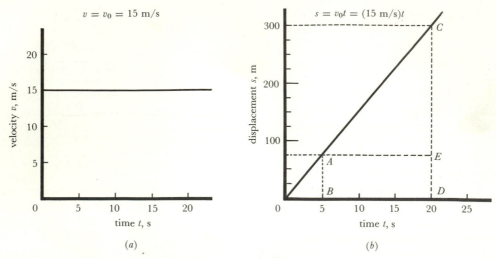

Figure 3-1 (*a*) Velocity graph for uniform motion; (*b*) displacement graph for uniform motion.

better to write $\dfrac{\text{ft}}{\text{s}}, \dfrac{\text{mi}}{\text{h}}$, etc., rather than use the slant line to indicate division, as ft/s (although for convenience of typesetting we shall often use the latter style in this text).

(*e*) The final answer is clearly indicated by some device such as a box or underlining.

It is simply taking advantage of past experience and study to use a basic formula which you fully understand, such as $s = v_0 t$. It would be wrong, however, to rely on memory and blind substitution into formulas. Problem solving is intended to help you master the meaning of physics; it should not become a battle of wits between the student and a collection of formulas pulled out of a magician's hat. *The most important single step is the first one—careful study of the problem to see what information is pertinent and just what is asked for.* The expert may carry these bits of information in his head for as long as it takes to come up with just the right equation or equations; the beginner would do well to emulate the example and place the given data in a list, perhaps enclosed in a box.

Uniform motion can be given a graphical interpretation, if we plot velocity $v\ (= v_0)$ or displacement s as a function of time t (see

the Mathematical Appendix, p. 706, for a discussion of graphs). The time is plotted horizontally (axis of abscissas) since time is considered the *independent variable*. That is, we choose a certain time at will and then determine the corresponding velocity or displacement, which we call *dependent variables*. Consider first the case of uniform motion in a straight line (Fig. 3-1); say, a train moving at 15 m/s. The velocity is constant, so the velocity graph is a horizontal straight line (Fig. 3-1*a*). The displacement increases steadily as time goes by, so the displacement graph is a straight line passing through the origin (Fig. 3-1*b*). There is a relationship between these two graphs: the *slope** of the displacement graph is a constant, and this slope equals the velocity v_0. In Fig. 3-1*a* the train's velocity is a constant 15 m/s. The displacement 5 s after the time $t = 0$ is 75 m, as read from the graph of Fig. 3-1*b*, and the displacement 20 s after the time $t = 0$ is

---

*The slope of a straight line is the increase in the dependent variable, divided by the corresponding increase in the independent variable. Thus in Fig. 3-1*b*, the slope is 300 m/20 s = 15 m/s. Slopes usually have units; in this case, m/s. The angle that the line makes with the horizontal axis depends on the scale to which the graph is plotted. This is arbitrary and has nothing to do with slope as here defined.

300 m. The slope of segment AC of the displacement graph is

$$v_0 = \frac{300 \text{ m} - 75 \text{ m}}{20 \text{ s} - 5 \text{ s}} = \frac{225 \text{ m}}{15 \text{ s}} = 15 \text{ m/s}$$

The slope of a straight-line graph is everywhere the same; the slope of segment OA is also 15 m/s, as can easily be calculated. (Note that triangle OAB is similar to triangle ACE.)

3-3 Average velocity

As an example of a motion which is more complicated than that in Example 3-1 but which is not unusual, consider a car on a straight turnpike where "mile markers" indicate distances along the road, measured from the state border. If the car moved from the 144 mi marker to the 184 mi marker, the displacement was 184 mi − 144 mi = 40 mi. If this portion of the journey began at 11:10 A.M. and ended at 11:58 A.M., the elapsed time was 48 min, or 0.8 h. If the car moved at a steady rate its velocity was constant with magnitude 40 mi/0.8 h = 50 mi/h. If the car did not move at a steady rate, the velocity we have just calculated is called the *average velocity*. In general, average velocity $\bar{v}$ or v_{av}, is given by

$$\bar{v} = v_{av} = \frac{s - s_0}{t - t_0} \qquad (3\text{-}2)$$

where s_0 is the initial displacement (at time t_0) and s is the final displacement (at time t). Usually we choose to measure displacement in such a way that s_0 and t_0 are both zero; this amounts to calling $s = 0$ when $t = 0$. For instance, at 11:10 A.M. we could set our watch to read 0 and could renumber the mile markers so that the 144 mi marker, which we are passing at this instant, is "0." Then we would pass the "40" marker at "0:48" by our watch.

With $s_0 = 0$ and $t_0 = 0$, Eq. 3-2 becomes

$$\bar{v} = \frac{s}{t} \quad \text{or} \quad s = \bar{v}t \qquad (3\text{-}3)$$

Note that we now have two equations for displacement. The equation $s = \bar{v}t$ is always true, for *any* motion; the similar equation $s = v_0 t$ is true for *uniform* motion, where v_0 is constant.

Let us study the motion of the car along the turnpike more closely. So far, we have found the average velocity to be 50 mi/h, but we know nothing about the details of the motion. Such details are available in the graph of Fig. 3-2a which gives the displace-

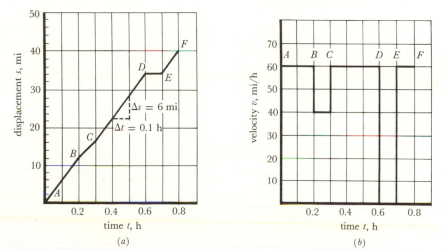

Figure 3-2

ment of the car as a function of time. The journey is idealized in such a way that the displacement graph is a series of straight-line segments.* During any interval, the car's displacement increased steadily, and its velocity was constant. This constant velocity was also the average velocity during the interval. According to the definition expressed in Eq. 3-2,

$$\bar{v} = \frac{\Delta s}{\Delta t}$$

where $\Delta s/\Delta t$ ▪ is the slope of a straight segment of the displacement graph in Fig. 3-2a. Thus, during any part of the interval CD, the velocity was 60 mi/h, as given by the illustrated values of Δs and Δt; here $\bar{v} = \Delta s/\Delta t = 6 \text{ mi}/0.1 \text{ h} = 60 \text{ mi/h}$.

The constant value $\bar{v} = 60$ mi/h is plotted as the horizontal segment CD of the velocity graph in Fig. 3-2b. As may easily be verified from the graphs, the car's velocity was 40 mi/h in a tunnel (BC), 0 mi/h during a 0.1 h stop for a coffee break (DE), and 60 mi/h during segments AB and EF. The average velocity for the whole journey was 40 mi/0.8 h = 50 mi/h, although (as it happens) for no segment of the journey was the velocity equal to this value.

The discussion above makes it evident that a velocity graph, such as Fig. 3-2b, can be found by taking the slope of the displacement graph at several points. This statement has, of course, been proved so far only for motion with constant velocity; what to do if the displacement graph is not made up of straight-line segments is still an open question.

3-4   Instantaneous velocity

If the displacement graph is curved, the ratio $\Delta s/\Delta t$ is not constant, and hence the average velocity depends on the magnitude of the time interval used. The *instantaneous velocity* is defined by a limiting process in which smaller and smaller time intervals are used, each interval containing the instant at which the velocity is desired. As smaller time intervals are used, both Δt and Δs approach zero, but their ratio $\Delta s/\Delta t$ approaches a limit. This means that Δt can be made small enough so that a further decrease of Δt will change the value of $\Delta s/\Delta t$ by an amount which is less than any arbitrary pre-assigned value, however small. We define instantaneous velocity v by such a limit:

$$v = \lim_{\Delta t \to 0} \left(\frac{\Delta s}{\Delta t} \right) \qquad (3\text{-}4)$$

which is read "v equals the limit of $\Delta s/\Delta t$ as Δt approaches zero." This limit, if it exists,▲ is so useful that it is given a special name, the *derivative* of s with respect to t, written as ds/dt. Thus we have

$$\left(\begin{array}{c} \text{definition} \\ \text{of derivative} \end{array} \right) \qquad \lim_{\Delta t \to 0} \left(\frac{\Delta s}{\Delta t} \right) = \frac{ds}{dt} \qquad (3\text{-}5)$$

In another notation often used, if s is some function of t written as $s(t)$, then the derivative of s is another (related) function of t called $s'(t)$; $s'(t)$ and ds/dt mean exactly the same thing. Using our new terminology, we say that *instantaneous velocity is the time derivative of displacement.* Graphically, the derivative is the slope of the tangent to the displacement curve, drawn at the time in question.

EXAMPLE **3-2**

The displacement of a body varies according to the equation $s = 10t^3$, where s is in cm and t is in s. (Such a displacement function is possible,

---

* Physically, this amounts to assuming that changes in velocity are abrupt, taking place during intervals of, say, a few seconds, too short to show clearly on the graph.
▪ The symbol Δ (Greek delta) is used to indicate a change in any quantity. Thus Δs (read "delta s") means the change in s. We shall use this notation many times in our future work.

▲ All functions used in elementary physics are sufficiently "well behaved" so that ratios such as we are considering *do* approach a limit.

although not very common in physics.) What is the instantaneous velocity at $t = 2$ s?

First approximation: If we choose $\Delta t = 0.1$ s, we find

t	$s = 10t^3$
2.0	$10(2.0)^3 = 80.00$
2.1	$10(2.1)^3 = 92.61$

$$\bar{v} = \frac{\Delta s}{\Delta t} = \frac{92.61 - 80.00}{0.1}$$

$$= \frac{12.61}{0.1} = 126.1 \text{ cm/s}$$

This is surely somewhat too large, since Fig. 3-3 shows that the chord PQ whose slope we have obtained is steeper than the tangent PR.

Second approximation: As a next try, we choose $\Delta t = 0.01$ s:

t	$s = 10t^3$
2.00	$10(2.00)^3 = 80.00000$
2.01	$10(2.01)^3 = 81.20601$

$$\bar{v} = \frac{\Delta s}{\Delta t} = \frac{81.20601 - 80.00000}{0.01}$$

$$= \frac{1.20601}{0.01} = 120.601 \text{ cm/s}$$

Third approximation: For a still smaller Δt, we choose $\Delta t = 0.001$ s:

t	$s = 10t^3$
2.000	$10(2.000)^3 = 80.00000000$
2.001	$10(2.001)^3 = 80.12006001$

$$\bar{v} = \frac{\Delta s}{\Delta t} = \frac{80.12006001 - 80.00000000}{0.001}$$

$$= \frac{0.12006001}{0.001} = 120.06001 \text{ cm/s}$$

The calculations seem to indicate that $\Delta s/\Delta t$ does indeed approach a limit, and we conjecture that this limit is, for $t = 2$ s, exactly 120 cm/s. Whatever the exact value of the limit, we know that it *does* exist, because we know that the tangent PR has a definite slope. We can calculate this limit as precisely as we wish by choosing a smaller and smaller Δt. Thus we have (laboriously) found *one* point on the velocity graph: when $t = 2$ s, $v = 120$ cm/s. We could now repeat

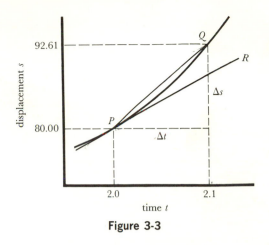

Figure 3-3

the whole process to find $\lim\limits_{\Delta t \to 0} (\Delta s/\Delta t)$ at any other time, and thus build up the entire velocity graph. The result, v as a function of t, is shown in Fig. 3-4.

3-5 Calculating the displacement

We have found that there is a close relationship between displacement and instantaneous velocity; in Sec. 3-3 we saw that whenever the displacement graph is known, we can construct the velocity graph. Each ordi-

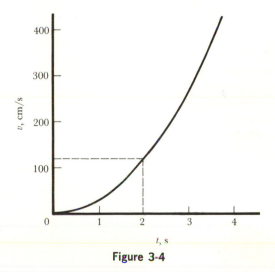

Figure 3-4

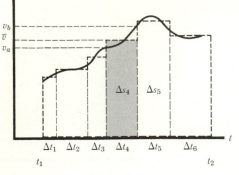

Figure 3-5

nate of the velocity graph equals the slope of the displacement graph at the corresponding time. This construction amounts to finding the derivative of the displacement with respect to time, which can be done graphically, if necessary.[*]

It is natural to ask if the process can be reversed: given the velocity graph, can the displacement graph be found? The answer is yes, and we illustrate the method in Fig. 3-5 and the following discussion. Once again we use a limiting process, but this time we find the limit of a sum.

To find the total displacement during the time interval extending from t_1 to t_2, we divide the total interval into many sub-intervals having durations $\Delta t_1, \Delta t_2, \ldots, \Delta t_n$. From the definition of average velocity (Eq. 3-3), it follows that the displacement during any sub-interval is given by $\Delta s = \bar{v} \Delta t$, where $\bar{v}$ is "in the neighborhood of" v_a (start of sub-interval) and v_b (end of sub-interval). (Note that $\bar{v}$ for a sub-interval does not necessarily lie between v_a and v_b; for instance, during the fifth sub-interval $\bar{v}$ is greater than both v_a and v_b.) The displace-

ment during any sub-interval is thus represented by the area of a rectangle, as shown by the shaded area for the fourth sub-interval. The total displacement from t_1 to t_2 is the sum of the areas of the rectangles. The crux of the problem is in choosing the proper values for the average velocities $\bar{v}$. The difficulty is lessened if we use very many segments, letting each Δt approach zero, for then v_a, v_b, and $\bar{v}$ tend to coincide. The total displacement equals the sum[*] of the displacements during the sub-intervals:

$$s = \sum_{i=1}^{n} \Delta s_i = \Delta s_1 + \Delta s_2 + \cdots + \Delta s_n$$

$$= \bar{v}_1 \Delta t_1 + \bar{v}_2 \Delta t_2 + \cdots + \bar{v}_n \Delta t_n$$

$$= \sum_{i=1}^{n} \bar{v}_i \Delta t_i \qquad (3\text{-}6)$$

For $n = 6$, this sum of rectangles is the area under the broken line in Fig. 3-5. In the limit, as $n \to \infty$ and each $\Delta t \to 0$, the broken line coincides with the curve, and the sum in Eq. 3-6 equals the area under the curve. Thus we have shown that the displacement equals the area under the velocity graph. We indicate that the displacement equals the limit of a sum by writing

$$s = \lim_{\substack{n \to \infty \\ \Delta t_i \to 0}} \sum_{i=1}^{n} v_i \Delta t_i \qquad (3\text{-}7)$$

where v_i is *any* value of v in the ith interval. As a matter of terminology, the limit of the sum, as written in Eq. 3-7, is called the "definite integral of v with respect to t, between the limits t_1 and t_2," and is denoted by

$$\left(\begin{matrix} \text{definition} \\ \text{of integral} \end{matrix}\right) \quad \lim_{\Delta t \to 0} \sum v \, \Delta t = \int_{t_1}^{t_2} v \, dt \qquad (3\text{-}8)$$

The integral sign $\int$ is a script "s," the initial letter of the word "sum."

By way of illustration, let us compute the displacement graph of Fig. 3-2a, starting with the velocity graph of Fig. 3-2b. First

---

[*] If the displacement is a known algebraic function of the time, the velocity function (derivative) can be found by using a formula. Such methods, known as formal calculus, are illustrated in Sec. 3-12, but we will not need to use any formal calculus in the main sections of this book. It is important to understand the meaning of the basic concepts of calculus, but our emphasis will be on graphical interpretations rather than on formulas.

---

[*] The symbol Σ (Greek sigma) indicates a summation process.

let us find the value of s at the time $t = 0.6$ h. The area under the velocity graph (from $t = 0$ to $t = 0.6$ h) consists of three rectangles, and thus

$$s = v_1 \Delta t_1 + v_2 \Delta t_2 + v_3 \Delta t_3$$
$$= (60 \text{ mi/h})(0.2 \text{ h}) + (40 \text{ mi/h})(0.1 \text{ h})$$
$$+ (60 \text{ mi/h})(0.3 \text{ h})$$
$$= 12 \text{ mi} + 4 \text{ mi} + 18 \text{ mi} = 34 \text{ mi}$$

Thus we have found point D on the displacement graph: when $t = 0.6$ h, $s = 34$ mi. Every other point on the displacement graph can be found by a similar summing of the area under the velocity graph, and the entire displacement function becomes known, starting with the known velocity function.

Our brief study of differential and integral calculus centers around two related functions: the displacement function $s(t)$, and the velocity function $v(t)$. The relationship between these two functions can be expressed in two equivalent ways:

$$v(t) = \frac{ds}{dt} \qquad (3\text{-}9a)$$

and

$$s(t) = s(0) + \int_0^t v \, dt \qquad (3\text{-}9b)$$

Here $s(0)$ is the value of s when $t = 0$; this initial displacement could also be written as s_0. As pointed out at the start of Sec. 3-3, we usually choose our reference point so that $s = 0$ when $t = 0$; for this choice $s(0) = 0$. Graphical methods can always be used to calculate the approximate numerical value of a derivative or an integral; but if one function (s or v) is given in algebraic form, tables such as those in the Mathematical Appendix can often be used to find the related function (v or s). In summary,

The derivative of a function at any point is the slope of its graph at that point.

The definite integral of a function between any two points is the area under that part of its graph which extends between the two points.

Graphical interpretations such as these are probably the most useful way of expressing the basic concepts of calculus.

3-6 Uniformly accelerated motion

In common language, "acceleration" is the picking up of speed. The technical definition involves the *rate* at which velocity changes. If the velocity changes rapidly, the acceleration is greater than if the same velocity change takes place during a longer time interval. Acceleration is the time rate of change of velocity and is therefore equal to the derivative of velocity with respect to time. In symbols,

$$a(t) = \lim_{\Delta t \to 0} \left(\frac{\Delta v}{\Delta t} \right)$$

or

$$a(t) = \frac{dv}{dt} = v'(t) \qquad (3\text{-}10)$$

Note that acceleration is, in general, a function of the time. However, in *uniformly accelerated motion* the acceleration is constant and for this case $dv/dt = \Delta v/\Delta t$ regardless of the size of the interval Δt. The velocity at a time t seconds after the start of an interval is found from

$$a = \frac{dv}{dt} = \frac{\Delta v}{\Delta t} = \frac{v - v_0}{t - 0}$$

which can be rearranged to give

$$v = v_0 + at \qquad (3\text{-}11)$$

where a is the acceleration, v_0 is the initial velocity, and v is the final* velocity after a time t has elapsed. The dimensional equation for acceleration is

$$[\text{acceleration}] = \frac{[\text{velocity}]}{[\text{time}]}$$
$$= \frac{[LT^{-1}]}{[T]} = [LT^{-2}]$$

---

*The adjective "final" refers to the end of the time interval. The motion may actually continue for an indefinite time.

Acceleration is measured in m/s², ft/s², mi/h·s, etc., always a distance unit divided by two time units.*

EXAMPLE 3-3

A sports car traveling on a straight road has an initial velocity of 10 m/s; it gains velocity steadily, and 5 s later its velocity is 25 m/s. What is the acceleration?

First we read the problem carefully and make a list of knowns and unknowns, setting them off in a box. We decide that the statement of the problem indicates that the car moves with uniform acceleration (it gains velocity "steadily").

$v_0 = 10$ m/s
$v = 25$ m/s
$t = 5$ s
$a = ?$

$v = v_0 + at$

Solving for the unknown a, we get

$$a = \frac{v - v_0}{t}$$

$$= \frac{25 \text{ m/s} - 10 \text{ m/s}}{5 \text{ s}}$$

$$= \frac{15 \text{ m/s}}{5 \text{ s}} = \boxed{3 \text{ m/s}^2}$$

After a time interval t, a body is found to be displaced from its original position. If the acceleration is positive (velocity increasing), the displacement is greater than it would have been if the velocity had remained constant at its initial value. We can calculate the displacement by integrating the velocity function. For uniformly accelerated motion, the velocity graph is a straight line of slope $\Delta v / \Delta t = a$ (Fig. 3-6). To find the displacement during a time interval from $t = 0$ to $t = t$, we need to calculate

$$s = \lim_{\Delta t \to 0} \sum v \, \Delta t$$

or

$$s = \int_0^t v \, dt$$

---

* The unit m/s² is read as "meters per second per second," or sometimes as "meters per second squared." The unit mi/h·s is read as "miles per hour per second." An area unit such as m² is usually read as "square meters."

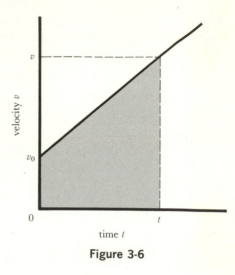

Figure 3-6

Evaluation of the definite integral is easy; it is the area of the shaded trapezoid in Fig. 3-6. Using plane geometry, we obtain the formula

$$s = \frac{v + v_0}{2} t = \bar{v}t \qquad (3\text{-}12)$$

where $\bar{v}$ is the average velocity.

Actually, the two equations for v and s (Eqs. 3-11 and 3-12) are sufficient to solve all problems concerning uniformly accelerated motion. However, for convenience in solving problems two additional equations are often used. They can be obtained by algebraic manipulation of the basic equations and give us two new equations relating v, v_0, s, t, and a:

Solve Eq. 3-11 for t and substitute into Eq. 3-12, getting

$$s = \left(\frac{v + v_0}{2} \right) \left(\frac{v - v_0}{a} \right)$$

which simplifies to

$$v^2 = v_0{}^2 + 2as \qquad (3\text{-}13)$$

Again, substituting Eq. 3-11 for v into Eq. 3-12, we find

$$s = \frac{(v_0 + at) + v_0}{2} t$$

which simplifies to

$$s = v_0 t + \tfrac{1}{2}at^2 \qquad (3\text{-}14)$$

Let us collect these four equations together:

$$v = v_0 + at \qquad (3\text{-}11)$$

$$s = \frac{v_0 + v}{2}t \qquad (3\text{-}12)$$

$$v^2 = v_0{}^2 + 2as \qquad (3\text{-}13)$$

$$s = v_0 t + \tfrac{1}{2}at^2 \qquad (3\text{-}14)$$

The student will note a certain systematic property of these equations. We have four different variables which are related to each other (v_0 is not a variable; the initial velocity is a constant for any given problem). The variable s is missing from Eq. 3-11, a from Eq. 3-12, t from Eq. 3-13, and v from Eq. 3-14. Try to select a suitable equation from the set 3-11 through 3-14, to avoid as much waste motion as possible.

Uniformly accelerated motion includes uniform motion as a special case. In uniform motion there is no acceleration, and by

setting a equal to zero in Eq. 3-14 we get $s = v_0 t$, which is the same as Eq. 3-1.

In uniformly accelerated motion, the velocity graph (plotted from Eq. 3-11) is a straight line, and the displacement graph (plotted from Eq. 3-14) is a parabola. These graphs are shown in Fig. 3-7 for the motion of a stock car being given an acceleration test on a drag strip; the car has initial velocity 10 ft/s and accelerates at 20 ft/s².

3-7 Acceleration due to gravity

By far the most familiar example of uniformly accelerated motion is that of a freely falling body. If air resistance can be ignored,[*] a

---

[*] Air resistance *can* be ignored if the falling body is heavy and has a relatively small surface area. Air resistance can certainly not be ignored for a falling raindrop, or for a man-parachute combination. But in these cases, the motion is not uniformly accelerated. Our study of dynamics in later chapters will reveal the effects of friction, including air resistance, more clearly.

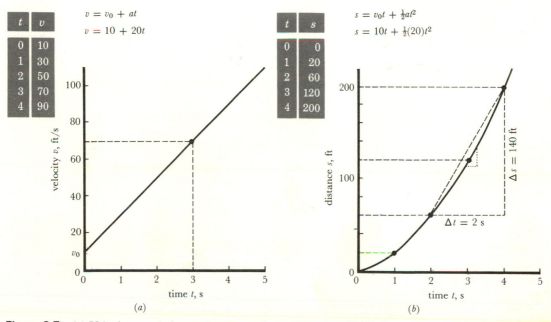

Figure 3-7 (*a*) Velocity graph for uniformly accelerated motion; (*b*) displacement graph for uniformly accelerated motion. The instantaneous velocity at $t = 3$ s is 70 ft/s. Between $t = 2$ s and $t = 4$ s, the car travels 140 ft.

falling body picks up speed at the rate of about 9.8 m/s each second (32 ft/s each second). This constant is called the *acceleration due to gravity* and is denoted by the symbol g. The value of this constant varies slightly from place to place on the earth; at 45° latitude and sea level, g is 9.80600 m/s² (32.169 ft/s²). In problem work g is taken to be exactly 9.80 m/s² or 32.0 ft/s² unless otherwise stated.

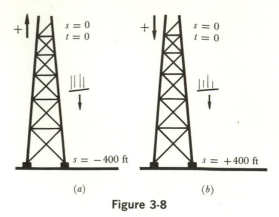

(a)   (b)

Figure 3-8

EXAMPLE **3-4**

On Jan. 16, 1972, a steel TV tower was under construction. At 4:15 P.M. a girder broke loose and fell 400 ft to the ground. How long did it take to reach the ground?

First we make a table of knowns and unknowns, assuming the positive direction to be *upward;* we also assume that $s = 0$ at the top of the tower, 400 ft above the ground, since that is where the girder is when $t = 0$. For this choice of positive direction, the acceleration, which is downward, is *negative*. The displacement, which is downward from the starting point, is also negative. We therefore use -32 ft/s² for a, and -400 ft for s.

$$s = -400 \text{ ft}$$
$$t = ?$$
$$a = -32 \text{ ft/s}^2$$
$$v_0 = 0$$

We indicate these facts in a diagram (Fig. 3-8a). We use Eq. 3-14:

$$s = v_0 t + \tfrac{1}{2}at^2 = 0 + \tfrac{1}{2}at^2$$
$$t = \sqrt{\frac{2s}{a}} = \sqrt{\frac{2(-400 \text{ ft})}{-32 \text{ ft/s}^2}}$$
$$= \sqrt{25 \text{ s}^2} = \boxed{5 \text{ s}}$$

Do not worry about making the proper choice of signs; merely take care that $s = 0$ at the location of the object at the start of the problem, when $t = 0$. Either direction may be taken as positive providing you stick with it throughout the problem. For instance, the

same problem can be solved in this way (Fig. 3-8b):

$$s = +400 \text{ ft}$$
$$t = ?$$
$$a = +32 \text{ ft/s}^2$$
$$v_0 = 0$$

$$t = \sqrt{\frac{2s}{a}} = \sqrt{\frac{2(+400 \text{ ft})}{+32 \text{ ft/s}^2}}$$
$$= \sqrt{25 \text{ s}^2} = \boxed{5 \text{ s}}$$

If you are confident that the correct answer will be obtained regardless of the mechanical work of algebra and signs, you are free to concentrate on the really important part of problem solving—"setting up the problem." Cultivate the ability to grasp the *physical* essentials and ignore the irrelevant details (such as the time of day, the fact that the tower is made of steel, etc.). Our next example shows the advantage of choosing the proper equation in order to make the algebraic work easiest.

EXAMPLE **3-5**

A ball is thrown straight up with initial velocity 20 m/s. After reaching its maximum height, on the way down the ball strikes a bird which is 10 m above the ground. How fast is the ball moving when it strikes the bird?

The problem could be solved in three steps: first find the maximum height (it turns out to be

20.4 m), then find the time required for the ball to fall 10.4 m down to the 10 m level starting from rest at the highest point (it turns out to be 1.46 s). Finally, the velocity acquired in this time interval can be computed. This method of solution hides the essential simplicity of the action which takes place. The motion is in one continuous sequence, and the acceleration is constant both in magnitude and direction, being 9.8 m/s² downward at all times. Even when the ball is (momentarily) motionless at the highest point, the acceleration is 9.8 m/s² downward, since the velocity is changing from upward to downward at this instant. We solve the problem in one step, without finding the maximum height:

We choose the positive direction to be *upward*, with $s = 0$ at the ground level. We seek the velocity of the ball when the displacement is $+10$ m.

$$v_0 = 20 \text{ m/s}$$
$$v = ?$$
$$s = +10 \text{ m}$$
$$a = -9.8 \text{ m/s}^2$$

$$v^2 = v_0{}^2 + 2as$$
$$= \left(20 \, \frac{\text{m}}{\text{s}}\right)^2 + 2\left(-9.8 \, \frac{\text{m}}{\text{s}^2}\right)(+10 \text{ m})$$
$$= 400 \text{ m}^2/\text{s}^2 - 196 \text{ m}^2/\text{s}^2$$
$$v = \sqrt{204 \text{ m}^2/\text{s}^2} = \boxed{\pm 14.3 \text{ m/s}}$$

We interpret the $\pm$ sign as follows: $v = +14.3$ m/s when the ball is at the $+10$ m level, on the way up, and $v = -14.3$ m/s when the ball is at the $+10$ m level on the way down. In view of the wording of the problem, we select the $-$ sign for our answer; $v = -14.3$ m/s when the ball strikes the bird.

3-8 Vectors

Our study of simple kinematical problems would hardly be complete if we discussed only motion in a straight line. The motion of a baseball after it leaves the bat on its way into home-run territory; the motion of a particle of water after it leaves a nozzle on

its way to a blazing building—these are examples of projectile motion, in which the body has two *simultaneous* velocities. The up-and-down part of the motion is due to gravity, and the forward motion is practically constant (if air resistance can be neglected). Other types of motion also involve two simultaneous velocities; for example, a man walking across a moving train or swimming across a river. To study the resulting motion, each part going on independently of the other, we need to use the concept of vectors and components of vectors.

A *vector* quantity has both magnitude and direction, whereas a *scalar* quantity has only magnitude. We speak of an *upward* force, an *eastward* velocity, a displacement *toward the northwest;* these are vector quantities. On the contrary, it is meaningless to think of "4 hours toward the south" or "100 cubic feet downward"; these quantities are scalars. A partial list of vector and scalar quantities is given in Table 3-1.

A speedometer on a car is correctly named, because it indicates only the *magnitude* of the velocity. You cannot tell by looking at the speedometer what the *direction* of the velocity is. Likewise, when a salesman extols the pick-up of a car, he doesn't care whether the acceleration is acting northward or eastward; he is interested only in its magnitude. On the

TABLE 3-1
Vector and scalar quantities[*]

Vector quantities		Scalar quantities
displacement	↔	distance
velocity	↔	speed
acceleration	↔	pick-up
force		time
momentum		volume
torque (moment of force)		work
magnetic field strength		mass (inertia)

[*] The first three of these vector quantities are related to the corresponding scalar quantities in the second column; for instance, speed is the magnitude of the velocity. Such a correlation cannot be made for the remaining vectors.

other hand, weight (the downward gravitational force on a body) is a vector, like all forces. Vectors are usually represented by arrows. The length of the arrow represents (to some scale) the magnitude of the vector, and the direction of the arrow indicates the direction of the vector. In writing equations involving vectors, **bold face** type is used to indicate vectors,* and ordinary *light face* type is used to indicate the magnitudes of vectors, as well as scalars. Thus the vector equation $\mathbf{s} = \mathbf{v}_0 t$ conveys more information about uniform motion than does the scalar equation $s = v_0 t$; since $\mathbf{s}$ and $\mathbf{v}_0$ are related by the scalar factor t, they have the same direction and are parallel vectors.

The essential feature of vectors is that the combined effect (known as the *resultant*) of two or more vectors may be found by a geometrical rule known as *vector addition*. We place the vectors head to tail, maintaining their magnitudes and directions, and the resultant is the vector drawn from the tail of the first vector to the head of the final vector. Applied to displacements, it works out like this:

EXAMPLE 3-6

A boy delivering papers covers his route by traveling 3 blocks west, 4 blocks north, then 6 blocks east. What is his final displacement? What is the total distance he travels?

The displacements are laid off on a scale diagram (Fig. 3-9) and, by measurement of the length of the resultant (or by simple geometry), the displacement is found to be of magnitude 5 blocks. A protractor is used to find that the direction of the resultant is 53° north of east. (Can you see how to apply the Pythagorean theorem to this example?) The displacement is a vector and must be fully described: the boy's displacement is

    5 blocks, 53° N of E

The distance traveled by the boy is a total of

---

* In handwritten work, a vector quantity is denoted by an arrow above the symbol: $\vec{s} = \vec{v}_0 t$.

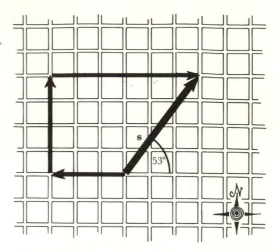

Figure 3-9 Vector addition of displacements.

13 blocks, found by adding up the various legs of his journey. This distance is a scalar quantity, and no particular direction can be associated with the figure of 13 blocks.

This example illustrates the distinction between displacement, a vector, and distance, a scalar. Another example: After a summer of vacation driving, the displacement is zero as the family car eases back into its garage, although the distance covered (and the gas consumption!) may be considerable.

Vector addition of velocities is necessary if a body has two or more velocities simultaneously, due perhaps to separate causes.

EXAMPLE 3-7

A helicopter heads due northeast and has an air speed of 70 mi/h. Simultaneously, a wind of 30 mi/h blows from the north. What is the displacement of the helicopter 2 h after leaving the heliport?

In 2 h the helicopter would have gone 140 mi due to its own air speed, and 60 mi due to the wind. The two displacements take place simultaneously. Laying off the two displacements to scale (Fig. 3-10), we find (graphically) the resultant displacement to be

    106 mi, 23.5° N of E

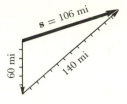

Figure 3-10

EXAMPLE **3-8**

Let us look at the same problem from a slightly different viewpoint. Instead of adding displacements, we can add (vectorially) the two velocities to get the magnitude and direction of the resultant velocity (Fig. 3-11). In this case the

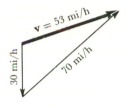

Figure 3-11

"sum" of 30 mi/h and 70 mi/h is only 53 mi/h. The magnitude of the resultant displacement is found from

$$s = v_0 t$$
$$= (53 \text{ mi/h})(2 \text{ h}) = \boxed{106 \text{ mi}}$$

as before.

Any number of vectors can be added by the head-to-tail method. Thus in Fig. 3-12, **R** is the sum of vectors **A**, **B**, **C**, **D**, and **E**. The order of addition of these vectors is immaterial. If only two vectors are involved, it is often convenient to originate two arrows at the same point. Then the diagram looks

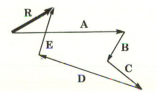

Figure 3-12   Head-to-tail method for addition of vectors.

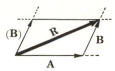

Figure 3-13   Parallelogram method for addition of vectors.

like Fig. 3-13. The resultant is the diagonal of a parallelogram, and this method is called the *parallelogram method*. It is not so general as the head-to-tail method, since it can be used for only two vectors at a time.

3-9   Relative velocity

We are all familiar with the idea of relative velocity. A motorcycle officer traveling at 80 mi/h overtakes a car traveling at 65 mi/h; we say that the velocity of the motorcycle relative to the car is 80 − 65, or 15 mi/h. In this case, speeds are subtracted. Sometimes speeds are added, as in the case of a head-on collision. For a complete treatment of relative velocity, useful for situations in which the motions are not parallel to each other, vector methods are required. A ship is going southward at 30 mi/h, and a deck-walker moves northward at 2 mi/h relative to the deck. It is not hard to see that the velocity of the walker relative to the ocean is 28 mi/h, southward. To avoid confusion in more complicated cases, let us put this simple problem into a standard form. We call northward the positive direction:

velocity of man relative to ocean

= (velocity of man relative to ship)
      + (velocity of ship relative to ocean)

= (+2 mi/h) + (−30 mi/h)

= −28 mi/h (i.e., southward)

In general such additions are to be carried out in a vector sense. If we use the symbol $\mathbf{v}_{MO}$ for the velocity of the man (M) relative to the ocean (O), etc., we can write the vector sum as

$$\mathbf{v}_{MO} = \mathbf{v}_{MS} + \mathbf{v}_{SO} \qquad (3\text{-}15)$$

Note carefully the order of the subscripts. This scheme can be extended to include any number of bodies in relative motion, as in our next example.

EXAMPLE **3-9**

A train T is moving eastward at 8 m/s, a waiter W is walking toward the rear of the train at 1 m/s; and a fly F is charging toward the north across the waiter's tray at 1.5 m/s. What is the velocity of the fly relative to the earth E?

$$\mathbf{v}_{FE} = \mathbf{v}_{FW} + \mathbf{v}_{WT} + \mathbf{v}_{TE}$$

The vector diagram (Fig. 3-14) expresses the vector equation written above; if a carefully constructed large (full-page) diagram is used, the velocity of the fly relative to the earth measures out to be

7.2 m/s, directed 12° N of E

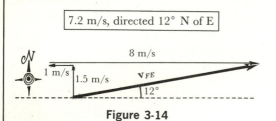

Figure 3-14

Another example involving relative velocities is thought-provoking, mainly because one of the given vectors must be reversed in direction before it is used in a vector sum.

EXAMPLE **3-10**

A bicyclist moves northward at 15 m/s, and a wind of 20 m/s is blowing directly from the east. What is the apparent magnitude and direction of the wind experienced by the bicyclist?

The actual wind may be considered as the velocity $\mathbf{v}_{AE}$ of the air A relative to the earth E. The apparent wind is the velocity $\mathbf{v}_{AB}$ of the air A relative to the bicyclist B. We want $\mathbf{v}_{AB}$, so we build a vector sum with $\mathbf{v}_{AB}$ on the left side; there is only one order of subscripts that is suitable for the right side of the equation.

$$\mathbf{v}_{AB} = \mathbf{v}_{AE} + \mathbf{v}_{EB}$$

We are given $\mathbf{v}_{BE}$, the velocity of the bicyclist relative to the earth (15 m/s northward), but the equation calls for $\mathbf{v}_{EB}$, the velocity of the earth relative to the bicyclist. However, $\mathbf{v}_{EB} = -\mathbf{v}_{BE}$, so we rewrite the vector equation

$$\mathbf{v}_{AB} = \mathbf{v}_{AE} + (-\mathbf{v}_{BE})$$

We add $\mathbf{v}_{AE}$ and $(-\mathbf{v}_{BE})$, laying them off to scale as in Fig. 3-15, with $(-\mathbf{v}_{BE})$ equal to

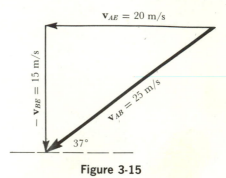

Figure 3-15

15 m/s *southward*. From measurement of the drawing we obtain the magnitude and direction of $\mathbf{v}_{AB}$:

25 m/s, coming from 37° N of E

3-10 Components of a vector

When one vector such as $\mathbf{C}$ is the resultant of two other vectors such as $\mathbf{A}$ and $\mathbf{B}$, then we call $\mathbf{A}$ and $\mathbf{B}$ the *components* of $\mathbf{C}$. Any given vector can be resolved into components in infinitely many ways (Fig. 3-16). Usually we find it useful to look for components that are perpendicular to each other. For instance, if a ball is traveling through the air in a direction making an angle θ with the horizontal, we can show in a diagram (Fig. 3-17) the horizontal component of $\mathbf{v}$ (denoted by $\mathbf{v}_x$) and the vertical component of $\mathbf{v}$ (denoted by $\mathbf{v}_y$). Since the components are at right angles to each other, the Pythagorean theorem can be used, and the magnitude of $\mathbf{v}$ is given by $v = \sqrt{v_x^2 + v_y^2}$.

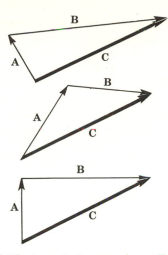

Figure 3-18

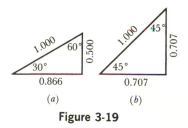

(a)   (b)

Figure 3-19

Figure 3-16   In each diagram, **A** and **B** are components of **C**.

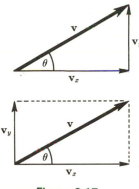

Figure 3-17

At this point we introduce the sum total of the trigonometry to be required of the student in this text, namely, the *definitions* of the sine, cosine, and tangent (abbreviated sin, cos, and tan). In the right triangle[*] having sides a, b, c (Fig. 3-18), we simply use sin θ as an abbreviation for the ratio a/c (opposite side over hypotenuse). Similarly, cos θ is an abbreviation for b/c (adjacent side over hypotenuse), and tan θ is an abbreviation for a/b (opposite side over adjacent side). In *any* 30°-60°-90° triangle (Fig. 3-19a), the short side is half the hypotenuse, thus sin 30° = 0.500. Other sines, cosines, and tan-

[*] These definitions are true *only for right triangles.*

gents of common angles can be read directly from the figures; for other angles, "trig tables" or slide rules are used.

Let us emphasize that when we use sin θ etc., we are merely using a ratio of two sides. Everything that can be done by use of trigonometry can also be done by using a scale diagram, but with limited accuracy. The sine and cosine are particularly useful in finding components of vectors, as in the following examples.

EXAMPLE **3-11**

A home-run ball moving at 140 ft/s is caught by a fan in the bleachers. Its path as it approaches the fan makes an angle of 30° with the horizontal (Fig. 3-20). What are the hori-

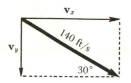

Figure 3-20   Components of velocity.

zontal and vertical components of the ball's velocity?

These components can, of course, be found by laying off the velocity to scale and drawing a rectangle with horizontal and vertical sides. However, if we use trigonometry, we need only use the definitions of sine and cosine, and no scale measurements are needed.

$$\frac{v_x}{140 \text{ ft/s}} = \cos 30°$$

$$v_x = (140 \text{ ft/s})(\cos 30°)$$

$$= (140 \text{ ft/s})(0.866)$$

$$= \boxed{121 \text{ ft/s}}$$

$$\frac{v_y}{140 \text{ ft/s}} = \sin 30°$$

$$v_y = (140 \text{ ft/s})(\sin 30°)$$

$$= (140 \text{ ft/s})(0.500)$$

$$= \boxed{70 \text{ ft/s}}$$

In this example, v_y would be called -70 ft/s or $+70$ ft/s, depending on which direction had been chosen as the positive direction.

EXAMPLE 3-12

A sled coasts down a 40° hill with an acceleration of 5 m/s² (Fig. 3-21). What is the vertical component of its acceleration?

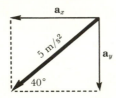

Figure 3-21 Components of acceleration.

$$\frac{a_y}{5 \text{ m/s}^2} = \sin 40°$$

$$a_y = (5 \text{ m/s}^2)(\sin 40°)$$

$$= (5 \text{ m/s}^2)(0.643)$$

$$= \boxed{3.22 \text{ m/s}^2}$$

With a little practice, you will be able to get the component directly as a product, skipping the step involving the proportion. Think of sin θ and cos θ as fractions which are used to calculate the sides of a right triangle when you know the hypotenuse. The side is always less than the hypotenuse, and the sine or cosine is always less than 1. To get the side *opposite* the angle, simply multiply the hypotenuse by the *sine* of the angle. To get the side *adjacent* to the angle, multiply the hypotenuse by the *cosine* of the angle. Thus, in Fig. 3-22, the marked component is found in each case by multiplying the magnitude of the vector by a sine or a cosine.

The tangent is useful in finding the direction of a vector when its components are known, as in the following example.

EXAMPLE 3-13

Find the direction of the velocity of the fly relative to the earth in Example 3-9.

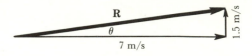

Figure 3-23

Here we know the forward component of the fly's velocity, which is 7 m/s eastward, and the sideways component of the velocity, which is 1.5 m/s toward the north. Putting these together in a vector triangle (Fig. 3-23), we find

$$\tan \theta = \frac{1.5 \text{ m/s}}{7 \text{ m/s}} = 0.214$$

Looking this up in a table of trig functions, we find that the angle θ is about 12°. (We use the angle in the table whose tangent is nearest to

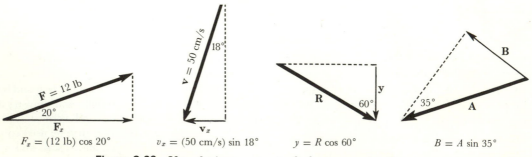

$F_x = (12 \text{ lb}) \cos 20°$ $v_x = (50 \text{ cm/s}) \sin 18°$ $y = R \cos 60°$ $B = A \sin 35°$

Figure 3-22 Use of trigonometry to find components of vectors.

0.214.) Thus we get the same answer that was obtained in Example 3-9 by a scale diagram—the fly moves at an angle of

$$\boxed{12° \text{ N of E}}$$

3-11 Projectiles

In projectile motion we really have two motions occurring at the same time. The only thing the two motions have in common is that they take place *during the same time interval*. The horizontal part of the motion is uniform motion in a straight line, since there is no horizontal resisting force. The vertical part of the motion is uniformly accelerated motion, since it is governed by the laws of falling bodies. Thus it is merely necessary to resolve the velocity into its horizontal and vertical components at any instant, and treat each motion separately.

EXAMPLE 3-14

A stone is thrown horizontally from a cliff 100 ft high. The initial velocity is 20 ft/s. How far from the base of the cliff does the stone strike the ground (Fig. 3-24)?

Consider first the *vertical* motion in order to find how long the stone is in the air. The vertical component of the initial velocity is zero, since the stone is thrown horizontally; therefore $v_0 = 0$, as far as the vertical motion is concerned. We choose *upward* as the positive direction for this part of the problem, and show the choice in the diagram.

$$\boxed{\begin{array}{l} v_0 = 0 \\ a = -32 \text{ ft/s}^2 \\ s = -100 \text{ ft} \\ t = ? \end{array}}$$

$$s = v_0 t + \tfrac{1}{2}at^2$$
$$-100 \text{ ft} = 0(t) + \tfrac{1}{2}(-32 \text{ ft/s}^2)(t^2)$$
$$t = 2.50 \text{ s}$$

Next consider the *horizontal* motion. Since $a = 0$ (there is no horizontal acceleration since air resistance is neglected), $s = v_0 t$ for this motion,

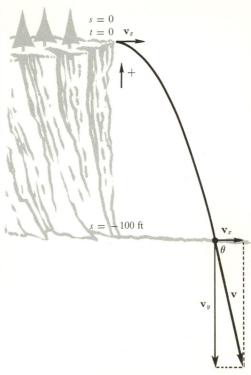

Figure 3-24

and v_0 is the full 20 ft/s since it is directed horizontally.

$$s = v_0 t = (20 \text{ ft/s})(2.50 \text{ s}) = \boxed{50 \text{ ft}}$$

EXAMPLE 3-15

In the above example, calculate the velocity of the stone as it hits the ground.

Since velocity is a vector, we must find both its magnitude and its direction in order to have a complete answer.

Vertical motion:

$$\boxed{\begin{array}{l} v_0 = 0 \\ a = -32 \text{ ft/s}^2 \\ v = ? \\ s = -100 \text{ ft} \end{array}}$$

$$v^2 = v_0^2 + 2as$$
$$v^2 = 0^2 + 2(-32 \text{ ft/s}^2)(-100 \text{ ft})$$
$$v = \pm 80 \text{ ft/s}; \quad v_y = -80 \text{ ft/s}$$

We interpret this answer as v_y, the vertical component of the final velocity.

Horizontal motion: The horizontal component of the final velocity is the original horizontal component, which has remained unchanged since air resistance is neglected. The magnitude of the resultant final velocity is found by combining v_x and v_y:

$$v = \sqrt{v_x{}^2 + v_y{}^2} = \sqrt{20^2 + (-80)^2} = 82.5 \text{ ft/s}$$

The angle θ is found from a scale diagram or by trigonometry:

$$\cos\theta = \frac{20 \text{ ft/s}}{82.5 \text{ ft/s}} = 0.242$$

$$\theta = 76°$$

Thus our answer is

$$\mathbf{v} = \boxed{\begin{array}{c} 82.5 \text{ ft/s at an angle} \\ 76° \text{ below horizontal} \end{array}}$$

In Example 3-14, we used the same symbols for different quantities. Thus in the box $v_0 = 0$, whereas in the last line of the same problem, $v_0 = 20$ ft/s. Similarly, in the box $s = -100$ ft, while on the last line $s = 50$ ft. These apparent contradictions will be cleared up if you remember that we are actually solving two separate problems when we solve a projectile problem. The only symbol that is necessarily the same for the two motions is t, the time interval, which is 2.50 s for both parts of Example 3-14.

EXAMPLE 3-16

A bomb is released from a jet plane flying horizontally with a speed of 300 m/s at an altitude of 2000 m. Where will the bomb strike the ground?

In this case the bomb carries with it the horizontal velocity of the plane. Just after it is released, the bomb is traveling forward at 300 m/s.

We consider the vertical and horizontal motions separately, as in all projectile problems:

Vertical motion:

$$\boxed{\begin{array}{l} v_0 = 0 \\ s = -2000 \text{ m} \\ a = -9.8 \text{ m/s}^2 \\ t = ? \end{array}}$$

$$s = v_0 t + \tfrac{1}{2}at^2$$

Since the original velocity has no vertical component, $v_0 = 0$, and so

$$t = \sqrt{\frac{2s}{a}} = \sqrt{\frac{2(-2000 \text{ m})}{-9.8 \text{ m/s}^2}} = 20.2 \text{ s}$$

Horizontal motion: Since the horizontal acceleration is zero,

$$s = v_0 t = (300 \text{ m/s})(20.2 \text{ s}) = \boxed{6060 \text{ m}}$$

The bomb strikes 6060 m ahead of the point directly below the point of release.

Since the plane's velocity is 300 m/s and the bomb's horizontal component of velocity is also 300 m/s, the bomb always remains directly beneath the plane as it travels. We can say that the bomb always has the same *horizontal* position as the plane does, because its *horizontal* velocity relative to the plane is always zero.

So far, our treatment of projectiles has been confined to cases in which the initial velocity is horizontal. If this is not the case, we resolve the initial velocity into its horizontal and vertical components. The vertical component is used to find the time in the air, and the horizontal component of the initial velocity is used to find the horizontal distance that the projectile travels.

EXAMPLE 3-17

A baseball is thrown at a velocity of 80 ft/s at an angle of 30° above the horizontal. Calculate (*a*) the maximum height of the ball; (*b*) the time in the air; (*c*) the horizontal distance (range) to the point where the ball strikes the ground.

Resolving the initial velocity into its components (Fig. 3-25), we find that v_0 in the vertical direction is 80 sin 30° = 40 ft/s.

(*a*) We choose the upward direction to be positive and make a table of knowns and un-

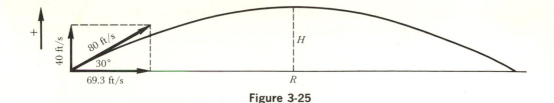

Figure 3-25

knowns for the vertical motion of the ball. At the highest point, the vertical velocity of the ball is 0.

$$
\boxed{
\begin{aligned}
s &= \,? \\
v &= 0 \\
v_0 &= +40 \text{ ft/s} \\
a &= -32 \text{ ft/s}^2
\end{aligned}
}
$$

$$v^2 = v_0{}^2 + 2as$$

$$s = \frac{v^2 - v_0{}^2}{2a}$$

$$= \frac{0 - (+40 \text{ ft/s})^2}{2(-32 \text{ ft/s}^2)} = \boxed{25 \text{ ft}}$$

This is the maximum height, H.

(b) The ball is in the air for a time sufficient for its initial upward velocity component $+40$ ft/s to become 0 (at the highest point) and then become -40 ft/s as the ball strikes the ground.

$$
\boxed{
\begin{aligned}
v_0 &= +40 \text{ ft/s} \\
v &= -40 \text{ ft/s} \\
a &= -32 \text{ ft/s}^2 \\
t &= \,?
\end{aligned}
}
$$

$$v = v_0 + at$$

$$t = \frac{v - v_0}{a}$$

$$= \frac{(-40 \text{ ft/s}) - (+40 \text{ ft/s})}{-32 \text{ ft/s}^2} = \boxed{2.5 \text{ s}}$$

(c) The horizontal component of the initial velocity is $80 \cos 30° = 69.3$ ft/s. Since we are neglecting air resistance, the horizontal motion is uniform, and the distance to the point where the ball strikes the ground is

$$s = v_0 t = (69.3 \text{ ft/s})(2.5 \text{ s})$$

$$= \boxed{173 \text{ ft}}$$

This is the range, R.

Suppose a gunner wishes to make a table showing how the horizontal range depends on the angle of elevation θ of his gun. Let us derive an equation for this purpose and check the dimensions on both sides of our final equation.

Although the mathematical solution of this problem is not difficult, let us first look at the physics of the situation. There are two extreme cases: (a) If the elevation is 90° (gun pointed straight up), the range R is obviously zero. (b) If the gun is pointed nearly horizontally ($\theta = 0$), the bullet is in the air for only a short time interval t before striking the ground, since the vertical component of the initial velocity is so small. Therefore the forward velocity component acts for only a very short time, and the range will be very small in this case also.

In some respects it is easier to work with symbols than with numbers. We follow the same scheme as in Example 3-17, using symbols instead of numbers.

Vertical motion:

$$v_y = v_0 \sin \theta \qquad\qquad (3\text{-}16)$$

$$t = \frac{v_0 \sin \theta - (-v_0 \sin \theta)}{g}$$

$$= \frac{2v_0 \sin \theta}{g}$$

Horizontal motion:

$$v_x = v_0 \cos \theta \qquad\qquad (3\text{-}17)$$

$$R = v_x t = (v_0 \cos \theta)\left(\frac{2v_0 \sin \theta}{g}\right)$$

$$= \frac{2v_0{}^2}{g} \cos \theta \sin \theta \qquad\qquad (3\text{-}18)$$

This is the equation we wished to derive. We now proceed to check our result in two ways. Any physical equation must have the same dimensions on both sides of the equals sign. In this case, we make the check as follows (cos θ and sin θ are ratios of lengths, and hence have no dimensions):

$$[L] \stackrel{?}{=} \frac{[LT^{-1}]^2}{[LT^{-2}]}$$

$$\stackrel{?}{=} \frac{[L^2T^{-2}]}{[LT^{-2}]}$$

$$[L] = [L]$$

By this check, we have shown that our result is at least not obviously wrong, but we have of course only verified its dimensions, not the exact form. At any rate, we have passed the first hurdle. Our answer is "not wrong" dimensionally, and is the more likely to be "right."

A second check on our derivation is to see whether the answer is "reasonable." Does it fulfill the expected behavior listed under (a) and (b) at the beginning of our solution? If the angle is 90°, condition (a) is fulfilled, since cos 90° = 0, and the factor cos θ therefore makes R become zero. Likewise, condition (b) is fulfilled since the factor sin θ becomes zero when θ is 0°. Thus both of the preliminary expectations (a) and (b) are fulfilled, another indication that our derivation is probably correct.

Similar checks should be made for any physical derivation.

In actual practice, the effects of air resistance are by no means negligible in projectile problems. Air resistance lengthens the time in the air and also causes the horizontal velocity to decrease somewhat from the original value. The exact study of projectile motion is called *ballistics,* and its mathematical complexities are such that electronic computers are often used.

We have by no means exhausted the study of kinematics, for we have concerned ourselves chiefly with constant motion and with uniformly accelerated motions, or combinations of such motions. We can immediately extend our analysis of straight-line motion to include some aspects of curvilinear motion. Consider a car moving on a winding turnpike. Even if its speed is constant, its velocity (a vector) changes since its direction changes. Acceleration is defined as the rate of change of velocity, and so we see that curvilinear motion always is accelerated motion. However, if we consider only components of displacement, velocity, and acceleration measured along the curved turnpike, the one-dimensional equations 3-11 through 3-14 are valid, just as if the path were stretched out into a straight line. We shall return to kinematics in Chapter 8, when we investigate circular motion. For the present, we have a sufficiently precise description of motion to be able to proceed to dynamics, the study of the causes of motion.

■ SUMMARY

Kinematics is the description of motion, without regard to causes. Dynamics (to be studied in Chap. 4) is the study of the causes of motion, in terms of force and inertia. For straight-line motion, average velocity is displacement divided by elapsed time; it equals the slope of a chord drawn between two points on the graph of displacement versus time. Instantaneous velocity is the limit of the average velocity as the time interval approaches zero; it equals the slope of the tangent to the displacement graph drawn at the point representing the time at which the velocity is desired. Instantaneous velocity is the time derivative of displacement, and displacement is the integral of the velocity function.

In uniformly accelerated motion, the variables are related to each other by the following equations:

$$v = v_0 + at \qquad s = \frac{v + v_0}{2}t = \bar{v}t \qquad v^2 = v_0{}^2 + 2as \qquad s = v_0t + \tfrac{1}{2}at^2$$

Motion with constant velocity is a special case of accelerated motion, in which $a = 0$. The acceleration of a freely falling body is approximately 9.80 m/s² or 32 ft/s², in the absence of air resistance. For uniformly accelerated motion the displacement graph is a portion of a parabola, and the velocity graph is a straight line whose slope is the acceleration.

Vector quantities have both magnitude and direction; scalars have magnitude only. The resultant of any number of vectors is found by the head-to-tail rule; for two vectors at a time the parallelogram rule is also convenient. Components of vectors are found by scale drawings or by the use of the sine and cosine.

If $\mathbf{v}_{AB}$ represents the velocity of A relative to B, etc., then the velocity of A relative to another body Z can be found by vector addition:

$$\mathbf{v}_{AZ} = \mathbf{v}_{AB} + \mathbf{v}_{BC} + \mathbf{v}_{CD} + \cdots + \mathbf{v}_{YZ}.$$

The motion of a projectile may be regarded as two motions that are independent and yet simultaneous. In the absence of air resistance, the vertical motion is uniformly accelerated, and the horizontal motion is one of constant velocity.

■ CHECK LIST

kinematics	acceleration	$\mathbf{v}_{AC} = \mathbf{v}_{AB} + \mathbf{v}_{BC}$
dynamics	pick-up	resultant
vector	average velocity	component
scalar	instantaneous velocity	sine
displacement	slope of a graph	cosine
distance	derivative	tangent
velocity	definite integral	
speed	relative velocity	

■ QUESTIONS

3-1 Which of the following equations cannot be correct, because of dimensional inconsistency? Are the others necessarily correct?

(a) $s = v_0t + 3at^2$ (b) $v = v_0{}^2 - \tfrac{1}{2}as^2$ (c) $s + \tfrac{1}{2}at^2 = vt$ (d) $at^2 + v/t = 2s^2/t^3$

3-2 How would you measure "the velocity" of a jet airplane at the instant it passes by a control tower in which you are stationed?

3-3 Every automobile is equipped with an odometer on the dashboard. Does this measure a scalar or a vector quantity?

3-4 What is the description, in words, of the motion of the falling bomb in Example 3-16 as seen from the plane? What is the description, as seen from a speeding automobile moving at 40 m/s, parallel to the plane's motion and in the same direction? The automobile was directly beneath the plane when the bomb was released.

3-5 A base runner sometimes takes a jump at the bag as he approaches first base. Is this more likely to result in an out than if he makes a level approach?

■ PROBLEMS

Note: Use $g = 9.80$ m/s^2 or 32.0 ft/s^2. It is also useful to remember that 60 mi/h $= 88$ ft/s.

3-A1 A driver travels the length of a turnpike which is 243 miles long in 4 h 30 min. What is his average speed (*a*) in mi/h? (*b*) in ft/s?

3-A2 If, while you are driving along at 45 mi/h, your attention wanders for 0.25 s, how far (in ft) do you travel "blind" during that quarter-second?

3-A3 A snail traveling at a snail's pace (12 ft/day) decides to slow down to only 5 ft/day and allows himself 2 min in which to make the change. (*a*) Express his initial velocity in mi/h; in in./year; in ft/s. (*b*) Compute the acceleration in ft/day·min; in ft/s^2.

3-A4 The displacement graph for a subway train traveling from one station to another along a straight track is shown in Fig. 3-26. (*a*) Estimate the instantaneous velocity 40 s after leaving the station. (*b*) What is the instantaneous velocity 100 s after leaving the station? (*c*) What is the average velocity during the first 100 s?

3-A5 Using Fig. 3-26, (*a*) what is the average velocity of the train during the first 50 s? (*b*) What is the instantaneous velocity 75 s after leaving the station?

3-A6 The velocity graph for the motion of a playful dolphin is shown in Fig. 3-27. (*a*) At what time was the dolphin's velocity greatest? (*b*) At what time was the acceleration greatest? (*c*) Was the acceleration constant at any time during the motion? (*d*) What was the acceleration at $t = 1.5$ s? (*e*) Use graphical integration to estimate the total distance traveled by the dolphin during the first 4 s.

3-A7 At 3:15 P.M. a motorboat was moving at 40 ft/s eastward; at 3:18 P.M. its velocity was 28 ft/s eastward. What were the magnitude and direction of the average acceleration of the motorboat?

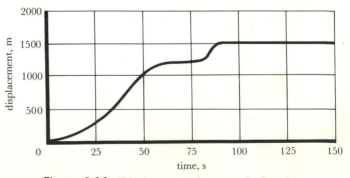

Figure 3-26 Displacement-time graph for the motion of a subway train.

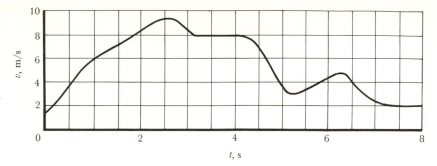

Figure 3-27   Velocity graph for the motion of a dolphin.

3-A8   (*a*) What is the instantaneous velocity of a falling body at the end of 2 s of its fall? (*b*) What is the velocity at the end of 3 s of fall? (*c*) What is the average velocity, averaged over the third second of the fall? (*d*) What is the distance traveled during the third second?

3-A9   A sprinter was timed at 10.2 s for the 100 meter dash; another runner in the same race was timed at 10.7 s. (*a*) Calculate the average speed of each runner. (*b*) How far apart were the two runners when the winner crossed the finish line?

3-A10   A ball is thrown upward at 50 ft/s. Calculate the magnitude and direction of the ball's velocity (*a*) 1 s after being thrown; (*b*) 2 s after being thrown.

3-A11   In Example 3-7, add the same two vectors, in the other sequence, starting with the 140 mi displacement.

3-A12   Using numerical values of sine or cosine, calculate the magnitudes of the components F_x and v_x in Fig. 3-22.

3-A13   On copies of each triangle of Fig. 3-22, draw an arrowhead on the remaining leg of the triangle so that the hypotenuse is the vector sum of the two components, arranged head-to-tail.

3-A14   Using a careful scale diagram, add the vectors of Fig. 3-12 in the sequence

$$\mathbf{A} + \mathbf{C} + \mathbf{B} + \mathbf{E} + \mathbf{D}$$

3-A15   A child in a city where the blocks are square starts from the corner of West Third Avenue and North Second Street, and walks south 5 blocks along West Third Avenue, crossing Main Street and reaching South Third Street. She then walks 4 blocks eastward, crossing Center Avenue to reach East First Avenue. She then runs to the corner of North First Street and East First Avenue, and finally she goes along North First Street to West Fourth Avenue. (*a*) What is the distance traveled? (*b*) What is the displacement for the entire trip?

3-A16   An airplane in a power dive is moving at 200 m/s downward at an angle of 60° below the horizontal. If the sun is directly overhead, how fast is the plane's shadow moving along the ground?

3-A17   A plane is traveling toward the southwest at 200 mi/h. (*a*) What is the westward component of its velocity? (*b*) What is the eastward component of its velocity?

3-A18   Two children 30 ft apart are playing catch on board a train which is moving at 90 ft/s in a straight line. Boy *A* throws the ball horizontally at 40 ft/s relative to himself toward the rear of the train. (*a*) What is the velocity of the ball relative to boy *B*? (*b*) What is the velocity of the ball relative to a stationary observer on the ground? (*c*) How long a time is required for the ball to go from *A* to *B*?

3-A19 All points on the equator are moving eastward at about 470 m/s due to the earth's rotation. (*a*) If a rocket is launched at the equator, aimed horizontally toward the east with a velocity of 3600 m/s relative to the earth, what will its velocity be relative to the fixed stars? (*b*) Repeat, assuming the rocket is aimed toward the west. (*c*) Why are most U.S. space launchings made from the Florida coast?

3-B1 A bullet is fired horizontally at 700 m/s and strikes a target which is 175 m from the gun. If the marksman hears the sound of the impact on the target 0.75 s after he fires the gun, what is the speed of sound?

3-B2 A child drops a stone into a well that is 100 ft deep. If the speed of sound is 1100 ft/s, how long after dropping the stone does he hear the splash?

3-B3 Bob Feller pitched a baseball which traveled from the pitcher's mound to home plate (a distance of 18.5 m) in 0.419 s. (*a*) What was the speed of the ball, in m/s? In mi/h? (*b*) If the catcher allowed his mitt to recoil backward 8.0 cm while catching the ball, what was the negative acceleration of the ball while it was being slowed down by the catcher?

3-B4 A truck traveling a mile a minute is slowed down at a uniform rate of 4.4 ft/s². How far will it travel before its speed is half the original value?

3-B5 A speedboat increases its speed at the rate of 2 m/s². How much time is required for the speed to increase from 9 m/s to 15 m/s? How far does it travel during this time?

3-B6 A stone is dropped from a bridge, and 1 s later another stone is dropped. How far apart are the two stones by the time the first one has reached a speed of 14.7 m/s?

3-B7 A ball is thrown vertically upward at 48 ft/s. (*a*) How high does it rise? (*b*) How long a time is required for it to reach its maximum height? (*c*) How long is it in the air? (*d*) With what speed does it strike the ground?

3-B8 A small mailbag is released from a helicopter which is descending steadily at 6 ft/s. After 2 s, (*a*) what is the velocity of the bag? (*b*) How far is it below the helicopter?

3-B9 A small mailbag is released from a helicopter which is rising steadily at 6 ft/s. After 2 s, (*a*) what is the velocity of the bag? (*b*) How far is it below the helicopter?

3-B10 With what upward velocity should a package be thrown in order to be caught easily by a person on a balcony 25 ft above the ground?

3-B11 Two boys are on a balcony 80 ft above the street. One boy throws a ball vertically downward at 64 ft/s; at the same instant the other boy throws a ball vertically upward at 64 ft/s; the second ball just misses the balcony on the way down. (*a*) How long after one ball strikes the street does the other ball strike? (*b*) With what velocity does each ball strike the street? (*c*) How far apart are the balls 1 s after they are thrown?

3-B12 A book is held 3 in. above a table top and released. How long is the book in the air?

3-B13 In a test of automobile bumpers, a car moving at 20 mi/h makes a head-on collision with a stone wall; from what height would the car have to fall straight down in order to make an equally hard collision?

3-B14 A ball is thrown vertically to a height of 81 ft and allowed to strike the ground. If it loses one-fourth its speed while in contact with the ground, how high does it rise on the rebound?

3-B15 A certain car has a maximum positive acceleration (when starting up) of +6 ft/s² and a maximum negative acceleration (when stopping) of −8 ft/s². The speed limit is 44 ft/s. What is the shortest legal time in which the driver of the car can go from one stop sign to another, a distance of 800 ft?

3-B16 A football is given a forward velocity of 40 ft/s by a passer whose arm moves through a horizontal distance of 3 ft. Compute the average forward acceleration of the ball.

3-B17 A speeding car passes a highway patrol check-point, and then decelerates at a constant rate; 5 s later, the car is 225 m from the check-point, and its speed is then 30 m/s. (a) What was the car's velocity when it passed the check-point? (b) What was the acceleration of the car?

3-B18 A sailor drops his pocket knife from the top of a mast on a ship sailing eastward at 10 m/s. The mast is 19.6 m high. Where does the knife hit the deck?

3-B19 A plane has an air speed of 100 mi/h, and the pilot notices that, although he is headed due east, he is actually traveling northeast, and covers 70.7 miles in 30 min. What is the magnitude and direction of the velocity of the wind that is blowing him off course?

3-B20 A helicopter headed due north has an air speed of 40 mi/h, and the wind is 30 mi/h from the west. (a) What are the magnitude and direction of the plane's velocity relative to the ground? (b) How far does the helicopter travel in 12 min?

3-B21 A swimmer can swim 4 mi/h in still water. (a) Starting from the west bank of a river which flows southward at 2 mi/h, in what direction should he head (somewhat upstream) so that he will travel directly across? (*Hint:* Velocity of swimmer relative to earth must be a vector directed toward the east.) (b) What will be the magnitude of the swimmer's resultant velocity? (c) How much time will be required for the trip, if the river is 0.2 mi wide? (d) If the swimmer swims 0.2 mi directly downstream and then swims back again to his starting point, how does his time for the round trip compare with what it would have been if he had swum to a point directly across from his starting point and back again, which is also a round trip of 0.4 mi?

3-B22 If the swimmer of Prob. 3-B21 heads straight across the river, so that he lands somewhat downstream, how long will it take him to cross?

3-B23 A ping-pong ball rolls with a speed of 0.6 m/s toward the edge of a table which is 0.8 m above the floor. The ball rolls off the table; how long is it in the air? How far out from the edge of the table does the ball hit the floor?

3-B24 Why does a hunter raise the barrel of his rifle when aiming at a distant target? If he aims directly at a target 300 m away, by how much will he miss the target if the muzzle velocity of the bullet is 600 m/s?

3-B25 A bullet strikes a target that is 300 ft from a gun, and level with it. During its flight, the maximum height of the bullet above the horizontal line between gun and target was 1.00 in. Calculate the muzzle velocity of the bullet.

3-B26 A tail-mounted gunner fires a bullet horizontally with a muzzle velocity of 620 m/s directly backward from a plane flying at an altitude of 160 m with a ground speed of 200 m/s. How far behind the plane should a target on the ground be when the gun is fired?

3-B27 A diver leaves a springboard horizontally with a velocity of 2.5 m/s. The board is 3 m above the surface of the pool. (a) How far out from a point below the end of the springboard does he strike the water? (b) What are the horizontal and vertical components of the diver's velocity when he strikes the water?

3-B28 A stone is thrown horizontally from a bridge 100 ft above the water. (a) How long is the stone in the air? (b) What must be the initial velocity of the stone if the line joining the bridge and the splash is inclined 45° below the horizontal? (c) At what angle does the stone strike the water?

3-B29 A boy can throw a ball a maximum horizontal distance R on a level field. How high can he throw the same ball vertically upward? Assume that his muscles give the ball the same speed in each case (is this a valid assumption?).

3-B30 A boy throws a ball horizontally at 20 ft/s from a window in a building; the ball is caught by a friend on the street who is 30 ft from the base of the building. How high above the street is the window?

3-B31 A shell leaves a gun with a horizontal velocity component of 200 m/s and a vertical velocity component of 98 m/s. Calculate   (*a*) the time in the air;   (*b*) the horizontal range;   (*c*) the maximum height of the shell;   (*d*) the initial speed of the shell.

3-B32 On level ground, a ball is thrown forward and upward. The ball is in the air 2 s, and strikes ground 120 ft from the thrower. With what speed, and at what angle, was the ball thrown?

3-B33 A broad jumper takes off at an angle of 20° with the horizontal and reaches a maximum height of 60 cm at mid-flight.   (*a*) What is his forward velocity?   (*b*) How far does he jump in the forward direction?

3-B34 On the moon the acceleration due to gravity is $\frac{1}{6}$ that on the earth. Prove that a broad jumper can jump 6 times as far on the moon as on the earth, if he takes off with the same speed at the same angle.

3-B35 A toy rocket is fired upward at an initial speed of 160 ft/s. At the very top of its rise, a second motor fires briefly, giving the rocket a horizontal velocity of 160 ft/s.   (*a*) How long is the rocket in the air?   (*b*) Where does it strike the ground?   (*c*) Where is the rocket 7 s after it left the ground?

3-C1 A late passenger, sprinting at 8 m/s, is 30 m away from the rear end of a train when it starts out of the station with an acceleration of 1 m/s². Can the passenger catch the train if the platform is long enough? (*Note:* This problem requires solution of a quadratic equation. Can you explain the significance of the two answers you get for the time?)

3-C2 A mile runner traveling at constant speed finds himself with 1480 ft still to go when 3 min 10 s have already elapsed. If for his home-stretch "kick" he accelerates at 1.00 ft/s² for 10.0 s and then holds the new speed until the finish line, can he still run a 4-minute mile?

3-C3 A car and a truck are each traveling at 70 ft/s, and the car is 75 ft behind the truck. The car driver decides to pass the truck, and he steps on the gas, producing an acceleration of 6 ft/s².   (*a*) How long will it be before the car is alongside the truck?   (*b*) How fast will the car be moving relative to the truck when they are side by side?   (*c*) How far will the car travel while reaching the truck?   (*d*) How far will the truck travel during the same time?

3-C4 The equation $s = vt - \frac{1}{2}at^2$ was studied *dimensionally* in Ques. 1-5 and can be shown to be *not incorrect* since the dimensions on each side of the equals sign are those of length. By algebraic manipulation of the equations of Sec. 3-6, show that the given equation is correct not only dimensionally, but also algebraically. (The symbols are as defined in Ques. 1-5.)

3-C5 A ball is thrown vertically upward at 64 ft/s.   (*a*) How long a time is required for it to reach a level 48 ft above its starting point?   (*b*) How fast is it then moving?   (*c*) Explain your two answers to part (*a*).

3-C6 A flower pot topples from a penthouse window sill. A tenant on a lower floor observes that the pot passes a picture window 6 ft tall in 0.1 s. How far above the top of the window is the penthouse window sill?

3-C7 A diver takes off with a speed of 8.00 m/s from a diving board 3 m high, at an angle of 30° above the horizontal. How much later does he strike the water?

3-C8 A pilot cuts loose his fuel tanks in an effort to gain altitude. At the time of release, he was 420 ft above ground and traveling upward at an angle of 30° above the horizontal, with a speed of 256 ft/s. For how long were the tanks in the air?

3-C9 On a ski jump, a skier takes off in a direction 30° above the horizontal with a speed of 24 m/s. He lands at a point 20 m below the level of the take-off point. (*a*) Through what distance, measured horizontally, does the skier jump? (*b*) With what speed does he strike the ground?

3-C10 A basketball player shoots toward a basket which is 20 ft horizontally from him and 12 ft above the floor. The ball leaves his hand 7 ft above the floor at an angle 60° above horizontal. What speed should the player give the ball? (*Hint:* Set up a pair of equations, one for horizontal motion and one for vertical motion, with v_0 and t as unknowns.)

3-C11 A ball is projected horizontally from the edge of a table which is 1 m high, and it strikes the floor at a point 1.2 m from the base of the table. (*a*) What is the initial velocity of the ball? (*b*) How high is the ball above the floor when its velocity makes an angle of 45° with the horizontal?

3-C12 A cat jumps off a piano that is 4 ft high. The initial velocity of the cat is 10 ft/s, at an angle 37° above the horizontal. How far out from the edge of the piano does the cat strike the floor?

3-C13 Prove that the path of the stone in Fig. 3-24 is a parabola. [*Hint:* Eliminate t from the equations for $x(t)$ and $y(t)$, to obtain y as a function of x.]

3-C14 Derive a formula for the maximum height of a projectile shot at an elevation θ with a muzzle velocity $\mathbf{v}_0$. Check your answer dimensionally.

3-C15 A projectile is shot at an elevation θ, and the line joining the gun to the point where the projectile has maximum height makes an angle ϕ with the horizontal. Derive a trigonometric relationship between θ and ϕ.

3-C16 For what elevation angle θ above the horizontal is the range R of a projectile a maximum, for a given muzzle velocity $\mathbf{v}_0$?

For Further Study

Note: In sections titled "For Further Study" at the ends of many chapters, we shall go beyond the material required of most students. These additional sections are of different types: some are fairly rigorous proofs of statements assumed without proof earlier in the chapter; some are interesting topics that are sidelines rather than integral parts of the course; and some are extensions of material in the text, more difficult or more mathematical in nature.

3-12   Formal calculus

Differentiation.   Whenever a well-behaved function is given, its derivative is, in principle, also known. At worst, we may plot the function and graphically measure the slope of the tangent at any desired point, or we could go through a limiting process as in Example 3-2. However, there are useful formulas for finding derivatives of most functions encountered in physics and engineering. To illustrate such methods, let us derive a formula for the derivative of $s(t)$, when $s(t)$ is given as the function $s(t) = At^n$, with A a constant.

For $t = t_0$, the value of s is $s(t_0) = At_0{}^n$

For $t = t_0 + \Delta t$, the value of s is $s(t_0 + \Delta t) = A(t_0 + \Delta t)^n$. Hence,

$$\frac{\Delta s}{\Delta t} = \frac{s(t_0 + \Delta t) - s(t_0)}{(t_0 + \Delta t) - t_0}$$

$$= \frac{A(t_0 + \Delta t)^n - At_0{}^n}{\Delta t}$$

We now use the binomial theorem to expand $(t_0 + \Delta t)^n$.

$$\frac{\Delta s}{\Delta t} = \frac{A\left[t_0^n + nt_0^{n-1}\Delta t + \dfrac{n(n-1)}{2!}t_0^{n-2}\Delta t^2 + \dfrac{n(n-1)(n-2)}{3!}t_0^{n-3}\Delta t^3 + \cdots + \Delta t^n\right] - At_0^n}{\Delta t}$$

$$= A\left[nt_0^{n-1} + \frac{n(n-1)}{2!}t_0^{n-2}\Delta t + \frac{n(n-1)(n-2)}{3!}t_0^{n-3}\Delta t^2 + \cdots + \Delta t^{n-1}\right] \qquad (3\text{-}19)$$

Note that the sum in brackets has $n+1$ terms; we assume that n is a positive integer or zero. This is the slope of the chord PQ in Fig. 3-28. To find the value of the derivative at $t = t_0$, we must find $\lim\limits_{\Delta t \to 0}\left(\dfrac{\Delta s}{\Delta t}\right)$; as $\Delta t \to 0$, the chord PQ approaches the tangent at P. In Eq. 3-19, all terms except the first one approach zero as $\Delta t \to 0$. Hence, in the limit, we have

$$\lim_{\Delta t \to 0}\left(\frac{\Delta s}{\Delta t}\right) = \frac{ds}{dt} = Ant_0^{n-1}$$

This is the magnitude of the derivative, evaluated at the point $t = t_0$. Similarly, at another value of t, say t_1, the derivative is Ant_1^{n-1}. The derivative is thus a function of t, since its value is known whenever t is known. In general, for an arbitrary value of t, we make the statement

$$\text{if } s(t) = At^n, \quad \text{then} \quad \frac{ds}{dt} = s'(t) = Ant^{n-1}$$

$$(3\text{-}20)$$

This derivative formula is valid for all real values of n, although our proof has been

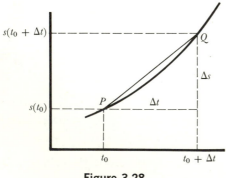

limited to the case where n is a positive integer or 0. The letters in Eq. 3-20 are of no particular significance from a strictly mathematical point of view; Eq. 3-20 could equally be written thus: if $f(x) = Bx^m$, then $f'(x) = Bmx^{m-1}$, and so on.

We can now verify the conjecture made at the end of Sec. 3-4. Our function is $s(t) = 10t^3$, and hence Eq. 3-20 tells us that $s'(t) = 30t^2$. Substituting $t = 2$ gives the instantaneous velocity

$$v(2) = s'(2) = 30(2)^2 = 120 \text{ cm/s}.$$

The derivatives of other functions can be worked out by a limiting process such as we have used for At^n. Further discussion and a brief table of derivatives are contained in the Mathematical Appendix, p. 708.

Integration. In Sec. 3-6 we collected several equations relating velocity v, displacement s, and acceleration a in uniformly accelerated motion.

A more elegant solution for $v(t)$ and $s(t)$ can be given, if we first derive a formula for $\int_A^B t^n\, dt$. Such *formal integration* uses the "fundamental theorem" of calculus. Study of Eqs. 3-9 reveals a broad generality which we approach by noting that Eq. 3-9b can be written as

$$\int_0^t v(t)\, dt = s(t) - s(0) \qquad (3\text{-}21)$$

It is not hard to see that if the limits of integration are A and B instead of 0 and t, then

$$\int_A^B v(t)\, dt = s(B) - s(A) \qquad (3\text{-}22)$$

Since $v(t) = s'(t)$, this means that to inte-

Figure 3-28

grate a function we look for another function whose derivative is the function which is to be integrated. We call such a function an *antiderivative;* $s(t)$ is the antiderivative of $v(t)$ because $s'(t)$ equals $v(t)$. The value of the integral between the limits A and B is found by evaluating the antiderivative function at $t = B$ and at $t = A$, then forming the difference according to Eq. 3-22. This is the fundamental theorem of calculus. In terms of a single function, the theorem may be written

$$\int_A^B s'(t)\,dt = s(B) - s(A) \qquad (3\text{-}23)$$

For example, let us compute $\int_A^B t^n\,dt$. By trial, we find the antiderivative function to be $\dfrac{t^{n+1}}{n+1}$, since, according to Eq. 3-20, the derivative of $\dfrac{t^{n+1}}{n+1}$ is $\dfrac{(n+1)t^{(n+1)-1}}{n+1}$, or simply t^n. Therefore

$$\int_A^B t^n\,dt = \left[\frac{t^{n+1}}{n+1}\right]_A^B = \frac{B^{n+1}}{n+1} - \frac{A^{n+1}}{n+1}$$
$$(3\text{-}24)$$

Integral tables usually give the antiderivative function thus: $\int t^n\,dt = \dfrac{t^{n+1}}{(n+1)}$ without indicating limits; when so written the antiderivative function is called the *indefinite integral.* Although our proof has been limited to $n = 0$ or a positive integer (see the derivation of Eq. 3-20), the formula for $\int t^n\,dt$ is valid for all real values of n except $n = -1$.

Two rather obvious properties of integrals follow directly from the idea of the limit of a sum. Any constant multiplier can be factored out of each term of the sum; therefore a constant C can be taken out from under the integral sign:

$$\int Cf(t)\,dt = C\int f(t)\,dt$$

Similarly, regrouping the elements making up the sum allows us to see that the integral of the sum of two functions is the sum of the integrals:

$$\int [f_1(t) + f_2(t)]\,dt = \int f_1(t)\,dt + \int f_2(t)\,dt$$

A brief table of antiderivatives (integrals) is given in the Mathematical Appendix, p. 709.

We have now derived *all* of the differential and integral calculus to be used in the sections titled "For Further Study." This simple calculus, an extension of algebra, is used because it gives us greater insight into some of the basic ideas of motion. Let us use our calculus to derive some of the equations of uniformly accelerated motion in a systematic fashion. We start with a definition: acceleration $= v'(t) = a =$ constant. Now we apply the fundamental theorem of calculus, Eq. 3-23, with the function s replaced by the function v, $A = 0$, and $B = t$. We obtain[*]

$$v(t) = \int_0^t v'(t)\,dt + v(0) = \int_0^t a\,dt + v(0)$$
$$= a\int_0^t t^0\,dt + v(0) = a\left(\frac{t^1}{1} - \frac{0^1}{1}\right) + v(0)$$

or

$$v = v_0 + at$$

where $v(0) = v_0 =$ initial velocity. This is Eq. 3-11. Proceeding, we find $s(t)$ by a similar integration:

$$s(t) = \int_0^t s'(t)\,dt + s(0) = \int_0^t v(t)\,dt + s(0)$$
$$= \int_0^t (v_0 + at)\,dt + s(0)$$
$$= \int_0^t v_0 t^0\,dt + \int_0^t at^1\,dt + s(0)$$
$$= \frac{v_0 t^1}{1} + \frac{at^2}{2} + s(0)$$

If we choose $s(0) = 0$, as is usual, we have

$$s(t) = v_0 t + \tfrac{1}{2}at^2$$

This is Eq. 3-14. Thus, starting with the acceleration, we have, in sequence, found the

---

[*] In this derivation we need $\int_0^t dt$; since $t^0 = 1$, we can write $\int_0^t dt = \int_0^t t^0\,dt$ and then use Eq. 3-24 to evaluate the integral.

velocity and the displacement by integration. In reverse, we can by differentiation verify that $s'(t) = v(t)$, and $v'(t) = a$. The acceleration is the derivative of the derivative of the displacement, and is called the second derivative of s with respect to t, and written as $s''(t)$, or as d^2s/dt^2, or as $\dfrac{d^2}{dt^2}(s)$.

■ PROBLEMS

3-C17 For the displacement function $s(t) = 4t^2 - 7t$, evaluate $\Delta s/\Delta t$ at $t_0 = 3$ s by the method of Example 3-2, using (a) $\Delta t = 0.1$ s; (b) $\Delta t = 0.01$ s; and (c) $\Delta t = 0.002$ s. Does $\Delta s/\Delta t$ approach a limit? (*Hint:* Organize your calculations in neat tabular form as you work. Check your result by differentiation using Eq. 3-20.)

3-C18 Use a limiting process to estimate the numerical value of $y'(2)$ if $y(x) = \log_{10} x$. Check your result using the derivative formula in the Appendix. (*Hint:* Use a five-place table of logarithms; start with $\log_{10} 2.1 - \log_{10} 2.0$.)

3-C19 Use the table of derivatives in the Appendix to calculate the following: (a) $f'(x)$ if $f(x) = 8x^3 + 8/x^3$; (b) $y'(2)$ if $y(u) = 5e^u + 5u^5$; (c) dz/dy if $z = (4 - y^2)^{-1/2}$; (d) dv/dt at $t = 1$, if $v(t) = 41t^2 + 41t - 41$; (e) dy/dx if $y = x^2 e^{-3x}$.

3-C20 Compute the second derivative $p''(2)$ if $p(u) = 5u^3 - u + 1$.

3-C21 A particle moves according to the equation $v(t) = t^2 - 3$ (in m/s). Estimate graphically the displacement of the particle from $t = 2.0$ s to $t = 3.0$ s. (*Hint:* Use ten sub-intervals Δt, each of size 0.1 s.) Check your result by formal integration, using Eq. 3-24.

3-C22 Use the table of integrals in the Appendix to calculate the following:

(a) $\displaystyle\int_1^3 (x^2 - x + 1)\, dx$; (b) $\displaystyle\int_0^B \tfrac{1}{2}\sqrt{x}\, dx$; (c) $\displaystyle\int_0^5 (x^4 + x)\, dx$; (d) $\displaystyle\int_0^{\pi/2} \sin x\, dx$;

(e) $\displaystyle\int_0^1 e^{-3x}\, dx$; (f) $\displaystyle\int_a^b \ln x\, dx$. [*Hint:* In (d) the limit is expressed in radians; $\pi/2$ radians is 90°. In (f) $\ln x$ stands for $\log_e x$, the natural logarithm of x.]

3-C23 Is it possible for two different functions to have the same derivative? Is it possible for two different functions to have the same value of the definite integral evaluated (for each function) between the same limits a and b?

■ REFERENCES

1 Cohen, I. Bernard, "Galileo," *Sci. American,* Aug., 1949, p. 40. A biography.

2 Lindsay, R. B., "Galileo Galilei, 1564–1642, and the Motion of Falling Bodies," *Am. J. Phys.* **10**, 285 (1942).

3 Magie, W. F., *A Source Book in Physics* (McGraw-Hill, 1935) (also, Harvard University Press, 1963). Selections from Galileo's *Two New Sciences:* pp. 2–17, acceleration and laws of falling bodies; and pp. 19–22, projectile motion.

4 Holton, G., and D. H. D. Roller, *Foundations of Modern Physical Science* (Addison-Wesley, 1958), pp. 24–28, 34–38. Selections from Galileo's writings, with commentary.

5 Chapman, S., "Catching a Baseball," *Am. J. Phys.,* **36,** 868 (1968).

6 Friedman, F. L., *The Velocity Vector; The Acceleration Vector; Velocity and Acceleration in Free Fall* (films).

7 Stull, J., *Constant Velocity and Uniform Acceleration* (film).

4

Dynamics

In our study of certain simple motions in Chap. 3, we made a beginning in the study of kinematics, a science for which Galileo (1564–1642) was largely responsible. Galileo studied projectile motion in detail and had a clear idea of the concept of inertia. His methods of investigation placed emphasis upon experiment and upon the mathematical description of physical laws such as those concerned with the motions of falling bodies and projectiles. These methods of Galileo are, of course, characteristic of science as we know it today. Building upon the work of Galileo and others and making full use of the "new" outlook in his methods, Isaac Newton (1642–1727) created a system of mechanics that ranks as one of the great intellectual achievements of all time. In this chapter we shall study the laws of motion as set forth by Newton, according to which the foundations of the science of dynamics were laid.

Born in the year of Galileo's death, Isaac Newton was the most influential scientist of modern times. According to his own opinion, Newton was "in the prime of my age for invention" during the plague years 1665 and 1666, when he left Cambridge University and studied mathematics, mechanics, and optics at his farm home in Lincolnshire. When he returned to the university, his own teacher, Isaac Barrow, resigned his professorship so that the 26-year-old Newton could have it. By this time the young genius had made profound discoveries in many fields, but became so involved in controversy concerning his first published work (*Theory About Light and Colors*) that he came to care little about public recognition for his work. In 1684 he was persuaded by his friend, the astronomer Halley, to organize and publish his system of mechanics, including the theory of gravitation. Thus in 1687 Newton's *Principia* made its appearance—probably the most important book in physics ever written. Newton lived for 40 years after the *Principia* and was a legend in his own time, but he devoted himself mainly to theology and to the conscientious performance of his duties as Warden (later Master) of the Mint. The story of Newton's life is best told at length; consult the references at the end of this chapter.

Astronaut David R. Scott dropped a feather and a hammer to the surface of the moon in this modern version of an experiment proposed by Galileo; both objects were seen to have the same downward acceleration in the absence of any retarding force of air resistance. Drawing by Robert McCall made at Manned Spacecraft Center at Mission Control during live telecast from the moon. (Courtesy NASA)

4-1 Force, the cause of acceleration

The role of force in everyday life is a familiar one. Indeed, it seems almost superfluous to try to define such a self-evident concept as "force." Pushes and pulls of all sorts—especially those exerted by muscles—are easily tagged with the word. A horse pulls a wagon; a baby throws a rattle; a motor raises an elevator cage. These are common examples of force involving motion. However, the existence or nonexistence of motion is really no indication of the existence of force. The horse can exert a force and yet not move a heavy block of stone; the baby can (and often does) push on an immovable wall and produce no motion; and there must be a force (tension) exerted by the elevator cable to support the cage even when it is stopped at the sixth floor.

If we are to get anywhere in our study of dynamics, we must first have a precise definition of force. It is obvious that forces *can* cause motion, but they do not seem to be *always* correlated with motion. Where is the motion, if any, "caused" by the pull of gravity on a parked tractor-trailer? And where is the force, if any, causing the motion of the star Regulus as it drifts majestically through space? Questions such as these puzzled the natural philosophers of many centuries. Newton clearly stated that *change* of motion is caused by forces and only by forces; unchanging, "uniform" motion requires no force. However, all bodies have inertia, by which we mean they "resist" being accelerated. Let us then define *force* as *that which causes the acceleration of a material body.*

No object is ever acted on by only a single force, although this situation is often closely approximated when one of the acting forces is much larger than all the others. The essential factor causing acceleration is the *net force,* which is the unbalanced force, or the resultant force. For instance, if two equal and opposite forces act on a body, there is no net force to cause acceleration, and the body's velocity remains constant. The condition of constant velocity is called *equilibrium;* this definition includes uniform motion in a straight line ($\mathbf{v}$ = constant $\neq 0$) as well as a state of rest ($\mathbf{v}$ = constant = 0).

4-2 Newton's first law

The motion of a body on which no net force acts is described by Newton's first law, often called the law of inertia:

Newton's first law of motion: A body remains at rest, or if in motion it remains in uniform motion with constant speed in a straight line, unless it is acted on by an unbalanced external force.

The first law is, of course, merely a statement of the qualitative definition of force which we have already given and is a recognition of inertia as an essential property of matter. Nevertheless, Newton's first law was of great value, for it refuted a common misconception which had been prevalent since the time of Aristotle. It had been supposed that the "natural" state of matter was one of no motion; indeed this seemed borne out by experiments in which an object sliding across a floor came to rest with no apparent force acting on it. Newton interpreted this same observation by saying that the force of friction caused the change of motion, or acceleration. When a horse pulls a wagon along the road with constant velocity, the older view was that the force exerted by the horse is needed in order to *cause* the (uniform) motion. Newton's interpretation was that the forward force of the horse is balanced by a backward force of friction, the net force is zero, and hence the wagon continues to move forward because of its inertia.

4-3 Newton's second law

We have seen that Newton's first law tells us what doesn't happen when no net force is applied to a body. To extend the laws of motion, we naturally seek a law describing what *does* happen when a net force *is* applied. Our first step is to quantify the concept of force. To do this, we assume that the magnitude of a force is measured by how much acceleration it can impart to a given body. Thus, if a net force F gives a certain body 6 times as much acceleration as does a net force F', then we *define* F to be 6 times as large as F'. In general, the ratio of two forces is defined to be equal to the ratio of the accelerations they produce. In symbols, for a given test body,

$$\frac{(\text{net } F)}{(\text{net } F)'} = \frac{a}{a'} \qquad (4\text{-}1)$$

This equation gives us a way of comparing the magnitudes of two forces, by letting them act, one after the other, on the same test body. We are, in effect, defining the magnitude of force by describing an experimental procedure for measuring it. This "operational" definition is a satisfying and useful one, since experiment shows that it gives the same ratio for any given pair of forces F and F', no matter what "test body" is used for the acceleration experiments. Now we multiply both sides of Eq. 4-1 by $(\text{net } F)'/a$, obtaining

$$\frac{(\text{net } F)}{a} = \frac{(\text{net } F)'}{a'}$$

$$= \text{constant for any test body}$$

To complete the definition of force, we define the direction of a force as being parallel to the acceleration it gives an object. Thus net $\mathbf{F}$ and $\mathbf{a}$ are parallel vectors, and we can write, for any test body,

$$\frac{\text{net } \mathbf{F}}{\mathbf{a}} = \text{constant} \qquad (4\text{-}2)$$

Our next step is to quantify the concept of inertia. The constant in Eq. 4-2 is surely somehow related to the inertia of the test body, for, if the body has a large inertia, a large net force is needed to produce a given acceleration, and so $(\text{net } \mathbf{F})/\mathbf{a}$ is large. Let us then *define* the constant in Eq. 4-2 as being proportional to the *mass m* of the body. That is,

$$\frac{\text{net } \mathbf{F}}{\mathbf{a}} \propto m \qquad (4\text{-}3)$$

where we use a proportion (symbolized by $\propto$) rather than an equality, since we have not yet established units for $\mathbf{F}$ and m. Mass is a quantitative measure of inertia, and by our definition the masses of two bodies can be compared by acceleration experiments. For example, suppose we wish to find the value of the mass of an unknown object in relation to the mass of some standard object, such as that of the international prototype kilogram. Our experiment consists in applying a given force (any force) to each mass in turn, and observing the accelerations produced. If the force gives the standard mass an acceleration of 3 m/s^2, and later it is found that the same force gives the unknown mass an acceleration of 10 m/s^2, we use Eq. 4-3 to conclude that the unknown body has less inertia than the standard body; its mass is $\frac{3}{10}$ of that of the standard body.

We have now seen how to compare two forces with each other and how to compare two masses with each other, although we have said nothing about the units in which force and mass might be measured. Placing the emphasis upon acceleration, we can rewrite Eq. 4-3 as

$$\mathbf{a} \propto \frac{\text{net } \mathbf{F}}{m} \qquad (4\text{-}4)$$

In words, we have

Newton's second law of motion: The acceleration produced by an unbalanced force acting on a body is proportional to the magnitude of the net force, in the same direction as the force, and inversely proportional to the mass of the body.

In Eq. 4-4 $\mathbf{a}$ and net $\mathbf{F}$ represent vectors which are parallel to each other, and m is a scalar quantity.

The student is earnestly advised to learn why we take the trouble to write the word "net" in front of the force in Newton's second law. Failure to write (or worse yet, to think) *net* force is the most prevalent single source of error in problem solving. For example, if a railroad engine pulls with a force of 20 tons in a forward direction and there is a retarding frictional force of 3 tons acting toward the rear of the train, the net force is a vector sum, or resultant, of four forces. The downward pull of gravity is balanced by the upward push of the tracks on the train, and the remaining two forces act in the same line. Therefore, ordinary algebraic addition can be used: $20 + (-3) = 17$ tons. The use of the word "net" or the summation symbol Σ will help remind you that you must look carefully for all the forces on a body, not just the obvious one. Many simple problems apparently involving only one force are really highly simplified by neglecting friction. This is permissible, but we should know what we are neglecting. Then, too, in many problems the weight of a body is balanced out by an upward push of floor, table, or earth, as in the case just discussed. We tend to forget that these forces are acting on the body, since they cancel out in so many cases.

Let us look for a moment at the relationship between Newton's first and second laws. In the proportion $\mathbf{a} \propto (\text{net } \mathbf{F})/m$, if we substitute 0 for the net force, we get an acceleration equal to 0. Thus Newton's second law predicts no change in motion if no net force acts. But this is Newton's first law. Hence we say that the first law is just a *special case* of the second law; it is obtained from the second law by using a special value of the net force, namely zero.

4-4 Newton's third law— action and reaction

Forces occur in pairs. Everyone who has played tug of war knows from experience that he can exert force only if there are people on the other end of the rope. Replace the opposing team by a post fixed in the ground, and we have a one-man tug of war. The man exerts force on molecules at his end of the rope, and the rope molecules pull back on him. Facts such as these are summarized in

Newton's third law of motion: Whenever one body exerts a force upon a second body, the second body exerts a force upon the first body; these forces are equal in magnitude, and oppositely directed.

Newton called these paired forces "action" and "reaction." It is essential to realize that the two forces mentioned in Newton's third law always act on different bodies, and hence cannot add up to zero. Only if forces act on the same body can they be combined into a single net force.

Pairs of forces (action-reaction pairs) are all around us, but sometimes the reaction force is not obvious. Any action-reaction pair can be described in the form: "A acts on B and B acts on A." If a book rests on a table, the book presses down *on the table,* and the table pushes up *on the book.* When a strong wind blows against a wall, molecules of air push against the wall, and the wall pushes back against the molecules of air. A more complicated (and subtle) example is that of a 10 lb ball hanging by a rope from the ceiling. Very many molecular action-reaction pairs are involved here, but if the rope is not accelerated, or if it has negligible mass, the net effect is that the rope "transmits" the force of the ball to the ceiling, and vice versa. We say that there are two complete action-reaction pairs involved here, a "tension" pair and a "weight" pair (four forces in all, each equal to 10 lb). In Fig. 4-1, $\mathbf{T} = -\mathbf{T}'$ and $\mathbf{W} = -\mathbf{W}'$, in accordance with Newton's third law. (Test them by reading the description of $\mathbf{T}$ backward, and see if it properly describes $\mathbf{T}'$.) However, in spite of the fact that $\mathbf{T}$ and $\mathbf{W}$ are equal and opposite, they do not consti-

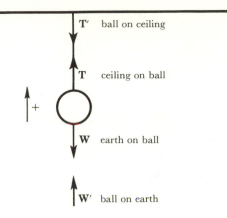

T' ball on ceiling

T ceiling on ball

+

W earth on ball

W' ball on earth

Figure 4-1 Action-reaction pairs.

tute an action-reaction pair. Bear in mind that action and reaction always act on different bodies, whereas **T** and **W** both act on the ball. Hence they must be equal and opposite for some other reason, not the third law. This reason is Newton's second law: the ball is in equilibrium, its acceleration is zero, and hence the vector sum of the forces acting *on the ball* must be zero. Only two forces, **T** and **W**, act on the ball, and hence **T** + **W** = 0, or **T** = −**W**. This has nothing whatever to do with the third law. If now we pull down slightly on the ball, the rope stretches slightly, and **T** (hence also **T'**) increases, say to 12 lb. Releasing the ball, we find that it is no longer in equilibrium, since net **F** = −10 + 12 = 2 lb (upward). But even though the ball is not in equilibrium, Newton's third law still holds. **T** = 12 lb, **T'** = −12 lb; **W** = −10 lb, **W'** = 10 lb. If we cut the rope, so that the ball is freely falling, it still weighs 10 lb, and it still exerts an upward force of 10 lb on the earth. It will help you identify pairs of action-reaction forces if you cast the description of each force in the form "_____ acting on _____." Thus, in the example of the ball hanging from the ceiling, the force **W** is that of *earth acting on ball,* and its reaction force **W'** is the force of *ball acting on earth.* This is in the standard form, "*A* acts on *B*, and *B* acts on *A*."

Newton's third law is an important statement about the nature of force. In general terms, a force is an *interaction* between two bodies. Whatever the nature of the interaction, be it gravitational, electrostatic, magnetic, nuclear, or other, we must recognize that it takes two bodies to make a force. For example, during a tug of war, short-range electric forces act between neighboring atoms of the rope. Such forces, called elastic forces in Chap. 9, occur as action-reaction pairs. Action and reaction are simply two aspects of the same force—two sides of the same coin, so to speak. It is impossible to have one without the other. The word "*interaction*" describes this situation nicely. Either force could equally well be called the "action," and the other force would then be the "reaction."

4-5   Units for force and mass

To be useful, Newton's second law must be made into an equation. We can do this by first solving Eq. 4-4 for net force: net **F** ∝ m**a**, then writing net **F** = Km**a**, where K is a constant of proportionality. You will be relieved to know that it is possible to make this constant K equal to 1, by suitable choices of force units described below. Therefore we shall write Newton's second law as

$$\text{net } \mathbf{F} = m\mathbf{a} \qquad (4\text{-}5)$$

In many applications of Newton's second law, motion is in a straight line, and the vector directions can be taken care of by use of + and − signs. In straight-line motion, therefore, we can write Newton's second law as an equation between the magnitudes of force and acceleration:

$$\text{net } F = ma \qquad (4\text{-}6)$$

We will use this equation, in which $K = 1$, to define the units of force in the two systems of metric units that we use.

The mks system is based upon the meter,

kilogram, and second, and the newly defined force unit is the *newton* (N). *A newton is that force which will give a mass of one kilogram an acceleration of one meter per second per second.*

To gain a feeling for this unfamiliar force unit, the newton, let us imagine a 1 kg ball falling freely, acted on only by the force of gravity (its weight). We assume negligible air friction. The ball has a mass of 1 kg. Experiment shows that the acceleration of this freely falling body is 9.8 m/s². Hence

$$\text{net } F = ma$$
$$W - 0 = (1 \text{ kg})(9.8 \text{ m/s}^2)$$
$$W = 9.8 \text{ kg·m/s}^2 = \boxed{9.8 \text{ N}}$$

We see that the weight of the ball is 9.8 N.[*]

The dimensional equation for force is found by substituting dimensions for mass and acceleration into Newton's second law:

$$\text{net } F = ma$$
$$[F] = [M][LT^{-2}]$$

Thus $[F] = [MLT^{-2}]$ for any system of units. In particular, for the mks system, we have just seen that the collection of units kg·m/s² is equivalent to the single unit "newton." Since we know that a 1 kg ball weighs about 2.2 lb, our argument shows that 9.8 N is about 2.2 lb, or 1 N is about 0.23 lb. Typical

---

[*] We do not say that "1 kg equals 9.8 N." This is incorrect, since the kilogram is a unit of mass, and the newton is a unit of force. However, it *is* correct to say that a body whose mass is 1 kg weighs 9.8 N on the earth.

example: a quarter-pound stick of butter weighs about one newton.

Our first example illustrates the use of mks units when solving a simple problem in dynamics with the help of Newton's second law.

EXAMPLE **4-1**

A car having a mass of 1000 kg comes to a stop in 40 m. If the initial speed was 20 m/s, what average stopping force (assumed to be constant) was supplied by the road acting on the car?

First we assume the positive direction to be toward the right (Fig. 4-2). Then, using the equations of kinematics (Chap. 3), we find the acceleration:

$v_0 = +20$ m/s	$v^2 = v_0^2 + 2as$
$v = 0$	$a = \dfrac{v^2 - v_0^2}{2s}$
$s = +40$ m	
$a = ?$	$= \dfrac{0 - (20 \text{ m/s})^2}{2(+40 \text{ m})}$
	$= -5$ m/s²

Now, using Newton's second law, we find the backward force of friction f exerted by the road on the car (Fig. 4-2).

$$\text{net } F = ma$$
$$f = (1000 \text{ kg})(-5 \text{ m/s}^2)$$
$$= \boxed{-5000 \text{ N}}$$

The answer appears as a negative force, which indicates that it acts in the negative direction, or backward. If we had chosen the positive direction to be toward the left, the force would have come out positive, but the acceleration

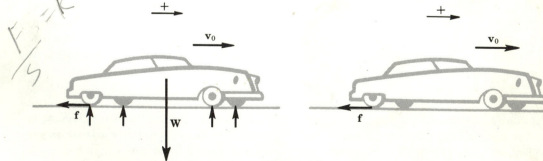

Figure 4-2

would also have been positive. The *physical meaning* of the answer is the same regardless of which direction is chosen as positive.

The first diagram in Fig. 4-2 shows the downward force of gravity (the weight **W** of the car) and also the upward forces exerted on the four wheels of the car by the pavement. The sum of these four upward forces exerted by the pavement is equal to the downward force **W**, and so all these vertical forces have a vector sum equal to zero. The second diagram, which is simplified by omitting these vertical forces which cancel out, shows only the force of friction, **f**. In either diagram, the net force is **f**, and it is **f**, whose magnitude is *f*, which is used in applying Newton's second law.

During a rescue at sea, a man of mass 70 kg dangles on the end of a rope attached to a helicopter (Fig. 4-3). If the helicopter accelerates upward at 400 cm/s², what tension must the rope be able to withstand?

Before setting up an equation, we decide to use mks units. We find the weight of the man in newtons, by multiplying his mass by 9.8, and change the acceleration to meters per second per second by dividing by 100. Then, from the second law,

$$\text{net } F = ma$$
$$N = (kg)(m/s^2)$$
$$T - 70(9.8) = (70)(+4.00)$$
$$T = \boxed{966 \text{ N}}$$

This force is greater than the weight of the man —an expected result, since there would have to be a net upward force on the man in order to cause his upward acceleration.

In this example, note that by Newton's third law the downward force of the man on the rope is equal to the upward force of the rope on the man. Hence the man must be able to grip the rope with a force of 966 N,

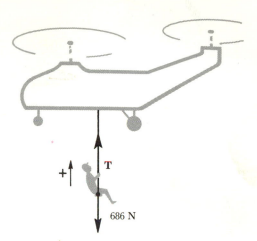

Figure 4-3 The net force on the man is $T - 686$ N.

which is considerably greater than his own weight.

The cgs system is based upon the centimeter, gram, and second, and the *dyne* (dyn) is the newly defined force unit. *A dyne is that force which will give a mass of one gram an acceleration of one centimeter per second per second.*

The dyne is a very small unit of force. A medium-sized pea has a mass of about a gram. If the pea is allowed to fall freely, acted on only by gravity, the only appreciable force acting will be the weight W of the pea.

$$\text{net } F = ma$$
$$W - 0 = (1 \text{ g})(980 \text{ cm/s}^2)$$
$$W = 980 \text{ g·cm/s}^2 = \boxed{980 \text{ dyn}}$$

Thus a pea weighs almost a thousand dynes! In fact, a good-sized mosquito weighs about a dyne. Our discussion also shows that the collection of units g·cm/s² is equivalent to the single unit "dyne."

The cgs system of units was formerly used for almost all physical work in the metric system, but it is gradually being replaced by the mks system. However, cgs units are often still used where very small forces are involved, as in our next example.

EXAMPLE 4-3

In an approximate model of mammalian heart action, the speed of about 20 cm³ of blood increases from 30 cm/s to 40 cm/s during a portion of the heartbeat 0.08 s long. What force is exerted by the heart muscle?

We assume the density of blood to be 1.1 g/cm³. Then the mass of blood is

$$m = (20 \text{ cm}^3)(1.1 \text{ g/cm}^3) = 22 \text{ g}$$

The average acceleration is

$$a = (40 \text{ cm/s} - 30 \text{ cm/s})/0.08 \text{ s} = 125 \text{ cm/s}^2$$

$$\text{net } F = ma$$
$$= (22 \text{ g})(125 \text{ cm/s}^2) = \boxed{2750 \text{ dyn}}$$

Force and mass have different dimensions and can never have the same units. Consistent sets of units for use in Newton's second law may be summarized as follows:

$$\text{net } F = ma$$
$$\text{N} = (\text{kg})(\text{m/s}^2)$$
$$\text{dyn} = (\text{g})(\text{cm/s}^2)$$

In the metric system, these units, and these only, should be used in Newton's second law or in any equation derived from the second law. However, in common language people *do* say "the mouse weighed 250 g at the start of the experiment" or "a weight of 50 kg is applied at one end of a lever," etc. These statements are wrong, and you should guard against such looseness of expression. The mouse had a *mass* (inertia) of 250 g, and his weight, a force, was

$$250 \times 980 = 245\,000 \text{ dyn}$$

A formula for weight. Imagine a body of mass m which is acted upon only by the force of gravity (its weight, a downward force **W**). Such a body is a freely falling body, and its acceleration is **g**, a downward vector. We now apply Newton's second law to this hypothetical situation in order to derive a formula for the weight of the body: Substituting into

$$\text{net } \mathbf{F} = m\mathbf{a}$$

gives

$$\mathbf{W} = m\mathbf{g} \qquad (4\text{-}7)$$

Although derived by a "thought experiment," the result is general. In any system of units, the magnitude of the weight of a body is given by $W = mg$. For instance, the weight of a 10 kg turkey is $10 \times 9.8 = 98$ N; a 5 g ping-pong ball weighs $5 \times 980 = 4900$ dyn.

4-6 The ratio equation

It would be possible, of course, to define an absolute fps system of units based on the foot, pound, and second in exactly the same way that we have defined the mks and cgs systems. However, we can get around the necessity of using a new and unfamiliar force unit by expressing Newton's second law in the form of a ratio. This alternative way of writing Newton's second law is of some value in problem solving, for it (almost) allows one to forget about questions of consistency of units. In any system of units, we have shown that the weight of a body is given by $W = mg$. Thus an object of mass 1 kg weighs 9.8 N, etc. If this equation is solved for the mass,

$$m = \frac{W}{g} \qquad (4\text{-}8)$$

Newton's second law then becomes

$$\text{net } F = \frac{W}{g}a \qquad (4\text{-}9)$$

$$\frac{\text{net } F}{W} = \frac{a}{g} \qquad (4\text{-}10)$$

This proportion has a simple interpretation. The weight W, if acting alone, would produce an acceleration g, for then the body would be freely falling. The net force, however, produces an acceleration a. It is the essence of Newton's second law that *for any given object*, the accelerations produced are in proportion to the forces that act (Eq. 4-1). Thus the ratio equation expresses Newton's second law without actually mentioning mass.*

---

* In the British engineering system of units, the mass of a body is measured in slugs. According to Eq. 4-8, the mass is given by W/g; the weight of a body (in pounds)

In this book the pound (lb) is used only as a unit of force. Thus the statement "a man weighs 200 lb" means "the gravitational force of the earth on the man is 200 lb." Usually in problem-solving the numerical value of the mass is not required, and questions about the acceleration of a body can be answered by use of the ratio equation, as in the following examples.

EXAMPLE **4-4**

A man whose weight is 200 lb is on ice skates and is pushed horizontally by a net force of 20 lb by the wind. What is his horizontal acceleration?

From the ratio equation,

$$\frac{\text{net } F}{W} = \frac{a}{g}$$

$$\frac{20 \text{ lb}}{200 \text{ lb}} = \frac{a}{32 \text{ ft/s}^2}$$

$$a = \boxed{3.2 \text{ ft/s}^2}$$

To put it as simply as possible, the acceleration is $\frac{1}{10}g$, since the net force is $\frac{1}{10}$ the man's weight.

EXAMPLE **4-5**

What is the horizontal force on a bowling ball which weighs 16 lb, if the bowler gives it a velocity of 20 ft/s in 0.40 s?

The acceleration is found first:

$$
\begin{array}{l}
v = 20 \text{ ft/s} \\
v_0 = 0 \\
a = ? \\
t = 0.40 \text{ s}
\end{array}
\qquad
\begin{aligned}
v &= v_0 + at \\
a &= \frac{v - v_0}{t} \\
&= \frac{20 \text{ ft/s} - 0}{0.40 \text{ s}} \\
&= 50 \text{ ft/s}^2
\end{aligned}
$$

We now use Eq. 4-9:

$$\text{net } F = \frac{W}{g} a$$

$$= \left(\frac{16 \text{ lb}}{32 \text{ ft/s}^2}\right)(50 \text{ ft/s}^2)$$

$$= \boxed{25 \text{ lb}}$$

divided by the acceleration of gravity (in ft/s²) is the mass of the body (in slugs). We will not use this terminology in this book, and will rely on the ratio equation when British units are to be used in Newton's second law.

In this case, the net force is *numerically equal* to $\frac{25}{16}$ the weight of the ball, because the acceleration is *numerically equal* to $\frac{25}{16}g$. The net force on the ball is, of course, horizontal, since the acceleration is horizontal.

These problems have nothing to do with falling bodies, even though g appears in the ratio equation. The ratio equation has all the advantages of proportions and ratios, including a freedom from worries about units. It is only necessary that a and g be in the *same* units and that net F and W be in the *same* units. Then the ratios are dimensionless numbers. The ratio equation also has the disadvantages of proportions, one of which is that you don't have to think very much and are thereby encouraged to use other (possibly wrong) proportions in other types of problems.

Our next example shows how the ratio equation can be used even for a body whose weight is zero.

EXAMPLE **4-6**

In a region far removed from the earth's gravitational attraction, workmen are assembling a space station. A section which on earth weighed 8 tons is floating through space at a speed of 1 ft/s toward a worker who is 3 ft in front of the very massive section of the station which has already been assembled (Fig. 4-4). If the worker

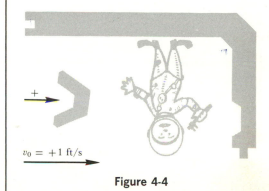

$v_0 = +1 \text{ ft/s}$

Figure 4-4

can exert a force of 150 lb, should he try to stop the oncoming section, or should he quickly move out of the way? First, we find the acceleration of the moving section, using the ratio equation:

$$\frac{a}{32 \text{ ft/s}^2} = \frac{-150 \text{ lb}}{16\,000 \text{ lb}}$$

whence

$$a = -0.3 \text{ ft/s}^2.$$

While the section is being brought to a stop, it travels a distance

$$s = \frac{v^2 - v_0{}^2}{2a} = \frac{0 - (1 \text{ ft/s})^2}{2(-0.3 \text{ ft/s}^2)} = \boxed{+1.67 \text{ ft}}$$

The worker has no need to worry.

4-7 Isolation of bodies in problem solving

Newton's second law applies to any complete system, or to any part of a system of bodies. When solving any specific problem, it is essential that you decide clearly and definitely just *what body* or system of bodies is being considered. Obviously, if all forces on a system are considered, the total mass of the system must be used; if forces on only part of a system are considered, then the mass of only that part must be used. Let us illustrate this point by considering a typical pulley system known as Atwood's machine.

EXAMPLE **4-7**

A cord connecting objects of mass 10 kg and 6 kg passes over a light, frictionless pulley (Fig. 4-5). (*a*) What is the acceleration of the system? (*b*) What is the tension in the cord?

First, we consider the system as a whole. The total mass that must be set into motion is 16 kg. Two external forces act on the system: $+(10 \text{ kg})$ $(9.8 \text{ m/s}^2) = +98.0 \text{ N}$ and $-(6 \text{ kg})(9.8 \text{ m/s}^2) =$ -58.8 N, where the signs are in accordance with the assumed positive direction shown in the diagram. The net force on the system is $98.0 \text{ N} - 58.8 \text{ N} = 39.2 \text{ N}$. The tension is an internal force, acting between parts of the system, and does not enter into the calculation of the net force *on* the system.

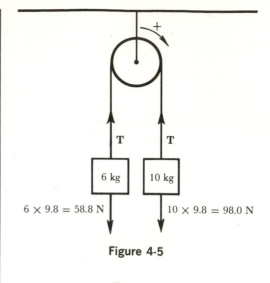

$$6 \times 9.8 = 58.8 \text{ N} \qquad 10 \times 9.8 = 98.0 \text{ N}$$

Figure 4-5

$$\text{net } F = ma$$
$$98.0 \text{ N} - 58.8 \text{ N} = (16 \text{ kg})(a)$$
$$a = \frac{39.2 \text{ N}}{16 \text{ kg}} = \boxed{2.45 \text{ m/s}^2}$$

Next, to find the tension, we must consider one or the other object by itself. To indicate this procedure, we draw a dotted line around the 10 kg object (Fig. 4-6*a*) and apply Newton's second law. We disregard everything outside the dotted line. From this point of view, we do not know whether the tension is caused by a 6 kg

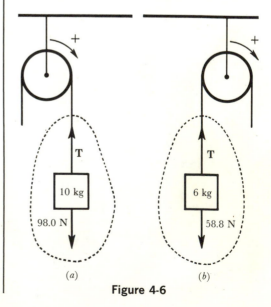

(*a*)   (*b*)

Figure 4-6

object on the other side of the pulley, or by a spring attached to the cord, or perhaps, for all we know, a little green man from Venus situated just outside the dotted line and pulling upward on the 10 kg object. The essential point is that we don't care *what* causes the tension; we merely wish to compute its value, whatever the cause. Applying the second law to the 10 kg object, we find

$$\text{net } F = ma$$
$$98.0 \text{ N} - T = (10 \text{ kg})(2.45 \text{ m/s}^2)$$
$$T = \boxed{73.5 \text{ N}}$$

As a check, we can compute the tension by isolating the 6 kg object (Fig. 4-6*b*).

$$\text{net } F = ma$$
$$T - 58.8 \text{ N} = (6 \text{ kg})(2.45 \text{ m/s}^2)$$
$$T = \boxed{73.5 \text{ N}}$$

Let us look at some of the features of the solution to Example 4-7. (1) It should scarcely be surprising that the tension in the cord on the left side of the pulley is numerically the same as that on the right side of the pulley. According to Newton's third law, the force of the 6 kg object on the 10 kg object is equal and opposite to the force of the 10 kg object on the 6 kg object. The cord merely serves to transmit this force.[*] (2) The ratio equation is illustrated in Example 4-7, where the

---

[*]If the cord had mass, then the tension would not be constant, since each particle of the cord would require a net force to accelerate it. If the pulley had mass, the tensions on the two sides of the pulley would have to be different, in order to accelerate the particles of the pulley. In this text, unless stated otherwise, all connecting cords and pulleys have negligible mass, and under these circumstances the tension is uniform.

acceleration is $\frac{1}{4}g$, and each object receives a *net* force of $\frac{1}{4}$ its weight. Thus the 10 kg object experiences a net downward force of $98.0 \text{ N} - 73.5 \text{ N} = 24.5 \text{ N}$, which is $\frac{1}{4}$ its weight. Likewise, the 6 kg object experiences a net upward force of $73.5 \text{ N} - 58.8 \text{ N} = 14.7 \text{ N}$, which is numerically equal to $\frac{1}{4}$ *its* weight, although directed upward, as it must be to cause upward acceleration. (3) An alternative solution to the problem is to use two unknowns, the acceleration a and the tension T, and set up two equations, one for each object:

$$98.0 \text{ N} - T = (10 \text{ kg})(a)$$
$$T - 58.8 \text{ N} = (6 \text{ kg})(a)$$

Solving these equations simultaneously[*] gives $a = 2.45 \text{ m/s}^2$ and $T = 73.5 \text{ N}$.

EXAMPLE 4-8

A tractor of mass 1000 kg is attached by means of a horizontal, massless chain to a log whose mass is 400 kg. The tension in the chain is 2000 N, and the backward force of friction exerted by the ground on the log is 800 N. How far will the log move in 2 s, starting from rest?

Our problem is really two problems: first we compute the acceleration, using Newton's second law; then we have a straightforward problem in kinematics to find the displacement.

First we make a diagram (Fig. 4-7), showing on it all the known forces and masses, as well as other forces whose values we do not know. Then we isolate one of the bodies for consideration; we choose the log for this purpose, since all the horizontal forces on it are known. We also choose a positive direction and indicate it on the

---

[*] See Sec. B-2 of the Mathematical Review in the Appendix.

Figure 4-7

diagram. Now we apply Newton's second law to the log:

$$\text{net } F = ma$$
$$2000 \text{ N} - 800 \text{ N} = (400 \text{ kg})(a)$$
$$a = 1200 \text{ N}/400 \text{ kg} = +3 \text{ m/s}^2$$

To find the displacement, we use the equations of kinematics:

$s = ?$
$v_0 = 0$
$t = 2 \text{ s}$
$a = +3 \text{ m/s}^2$

$$s = v_0 t + \tfrac{1}{2}at^2$$
$$= 0 + \tfrac{1}{2}(3 \text{ m/s}^2)(2 \text{ s})^2$$
$$= \boxed{6 \text{ m}}$$

Note that the mass of the tractor is not used in this problem, since we are considering the *log's* motion. Similarly, the forward force of the ground on the tractor is not used, since this force does not act *on the log*.

4-8 Gravitation

In 1798 Henry Cavendish completed an experiment which demonstrated directly the gravitational force of attraction between small bodies a few centimeters apart. Two lead balls of diameter about 5 cm were attached to the ends of a light rod suspended in a horizontal position by a long, fine wire about 100 cm in length. (Figure 4-8 shows a modern, compact version of Cavendish's apparatus.) Two larger lead balls, each about 20 cm in diameter, were placed almost touching the small balls. The gravitational attraction between the balls caused the moving system to rotate to a new equilibrium position determined by the stiffness of the wire. By carefully shielding the apparatus from air currents, Cavendish was able to measure the tiny force of gravitation (amounting to less than $\frac{1}{20}$ dyne). Without specially designed apparatus of this sort, the gravitational forces between ordinary objects are masked by frictional forces of one kind or another, or by the tremendously larger force of gravitation due to the earth.

Cavendish's experiment, the first terrestrial measurement of gravitational force, came more than a century after Newton had announced his *law of universal gravitation* in 1687:

Every body in the universe attracts every other body with a force which is directly proportional to the product of the masses of the two bodies and inversely proportional to the square of the distance between them.

It was a triumph of inductive reasoning that Newton was able to arrive at this law from considering the application of his laws of motion to the paths of the moon and the planets, as described by the German astronomer Johannes Kepler (1571–1630).

Kepler's laws:

1. The orbit of any planet around the sun is an ellipse, with the sun at one focus of the ellipse.

2. The line joining any planet to the sun sweeps out equal areas in equal times.

3. For any two planets in the solar system, the squares of the periods of revolution are in the same proportion as the cubes of their average distances from the sun.

We shall discuss the derivation and significance of Kepler's laws more fully in Chapter 8, after studying the kinematics of motion in a curved path. For the present, we note that two of Kepler's kinematical laws, combined with Newton's own laws of motion, lead to facets of Newton's law of gravitation; knowledge of falling bodies is also needed. The evolution of the law of gravitation may be summarized schematically as follows:

Kepler's second law + Newton's laws of motion:
  Force between planet and sun is either an attraction or a repulsion, directed along the line joining planet and sun.•

• This is not so obvious as it seems. The force might well have been tangent to the orbit, or perpendicular to the orbit, or in some other direction.

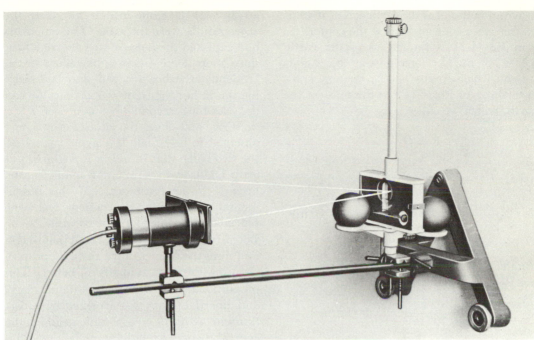

Figure 4-8 Modern form of Cavendish apparatus. Two small balls (one visible in photo) are on a bar suspended by a fine vertical metal wire. When the massive stationary lead balls are shifted, the gravitational force on the small balls changes. The resulting motion is made evident by the shift of a beam of light reflected from a small mirror attached to the moving system.

Kepler's first law + Newton's laws of motion:

Force between planet and sun is an attraction, varying inversely as the square of the distance.

Acceleration g of falling body independent of mass of body + Newton's laws of motion:

Force between planet and sun is proportional to the product of the masses of planet and sun.

Putting all these statements together, we write Newton's law of universal gravitation as

$$F = G\frac{mm'}{r^2} \qquad (4\text{-}11)$$

where G is a constant of proportionality, and m and m' are the masses of any two bodies separated by a distance r. It is assumed that the bodies are "particles," i.e., their sizes are small compared with their separation r.

The Cavendish experiment dealt with a law of nature already known to be true from astronomical observations. The significance of the experiment was in the determination of the constant G ("Newton's constant"). In Eq. 4-11, Cavendish used his measured values of F, m, m', and r to compute G, obtaining the value 6.7×10^{-8} dyn·cm^2/g^2. (The peculiar unit for G is necessary to make Eq. 4-11 dimensionally correct. This is an example of a constant that has units. The velocity of light, the acceleration due to gravity, and other well-known constants also have units.) In the mks system G has the value 6.7×10^{-11} N·m^2/kg^2. Cavendish then proceeded to use his delicate measurement of a force less than the weight of the ink required to print the word "gravitation" to determine the mass of the earth itself. Consider a 1 kg object at the surface of the earth. Even though the earth is not a "particle," the gravitational pull of the earth acts as if all its matter were concentrated at the center,[*] so r in Eq. 4-11 is the radius of the earth, 6.4×10^6 m. The force on the object is its weight, which equals 9.80 N. We now know everything in Eq. 4-11 but the mass of the earth, which we call m'.

$$F = G \frac{mm'}{r^2}$$

9.8 N

$$= (6.7 \times 10^{-11} \text{ N} \cdot \text{m}^2/\text{kg}^2) \frac{(1 \text{ kg})(m')}{(6.4 \times 10^6 \text{ m})^2}$$

$$m' = 6.0 \times 10^{24} \text{ kg}$$

Finally, Cavendish concluded that the average density of the earth is

$$\frac{6.0 \times 10^{24} \text{ kg}}{\frac{4}{3}\pi(6.4 \times 10^6 \text{ m})^3} = 5500 \text{ kg/m}^3$$

or 5.5 times the density of water. Since the density of the upper crust is known to be only about 2.5 times the density of water, it it evident that the earth must have a dense core, to bring the average density up to the calculated value.

There is a philosophical question about mass which needs answering. Are we talking about the same thing when we speak of "mass" as inertia (Newton's first and second laws) and when we speak of "mass" as capable of exerting force at a distance upon another such mass (Newton's law of gravitation)? Not necessarily; it is a matter for experiment to show whether a body which has 3 times as much weight as another also has exactly 3 times as much inertia. Galileo's famous (but probably apocryphal) experiment at the Leaning Tower of Pisa is such an experiment. We are all familiar with the bare facts: a heavy ball and a light ball dropped at the same time reached the ground at (almost) the same instant; therefore the accelerations were the same. The onlookers had expected the heavier ball to strike first, since more force (its weight) acted on it. Evidently, the heavier ball also had more inertia, in just the right proportion. We can use Newton's second law: $a = (\text{net } F)/m = W/m$, and since the accelerations were observed to be equal, $W_1/m_1 = W_2/m_2$ for the two balls; that is, *weight is proportional to mass*. Galileo's experiment was inaccurate, because of air resistance, but this fundamental and crucial experiment has been repeated in modified form by Newton, Bessel, Eötvös, and others, with the result that weight and mass are known to be in strict proportion, to within 3 parts in 10^{11} or better.[*] This proportionality between "gravitational mass" and "inertial mass" was accepted by Albert Einstein (1879–1955) as a basic axiom of the general theory of relativity.

4-9   Gravitational field

Let us for a moment imagine that we are in a spacecraft exploring the solar system. At any given point (for instance, halfway between Mars and Jupiter) we can measure the net gravitational force $\mathbf{F}_{\text{grav}}$ acting on a "test body" of mass m in the spacecraft. This force, or weight, is the resultant of all gravitational forces due to the sun, Mars, Jupiter, and all other astronomical bodies everywhere in the universe. If we measure the weights of various test bodies in the spacecraft, we find, of course, that the force of gravitation $\mathbf{F}_{\text{grav}}$ is proportional to the mass of the test body that we use. We conclude that a vector can be drawn at this point in space whose magnitude is F_{grav}/m and whose direction is the direction of the resultant gravitational force on the test body.

---

[*] Newton delayed publication of his law of gravitation for more than a decade while he worked out a satisfactory proof of this statement, which is true for a *spherical* earth whose density at any point depends only on the distance of that point from the center of the earth.

[*] This means that, if the mass of one body is, say, 1.00000000000 times that of another body, the ratio of their weights is also 1.00000000000, not 0.99999999997 or 1.00000000003. See Reference 7 at the end of this chapter.

This ratio, force per unit mass, is called the *field strength* at the point in question.

Now we proceed to a different point in space and measure the field strength there; it will no doubt have a different magnitude and direction and be represented by a different vector. The *gravitational field* is a collection of vectors, one located at every point in space, which give the magnitude and direction of the gravitational force per unit mass. Gravitational field strength is measured in newtons per kilogram, and is indicated by the symbol γ (Greek letter gamma). Thus

$$\gamma = \frac{\mathbf{F}_{\text{grav}}}{m}$$

If we know the location of all external bodies which exert a significantly large gravitational force on the test body, we can calculate the field using Newton's law of gravitation. The mass of the test body cancels out. Thus, in the simple case of a test body of mass m in the gravitational field of a single other body of mass m_1 which is at a distance r,

$$F_{\text{grav}} = \frac{Gmm_1}{r^2}$$

The magnitude of γ at the location of the test body is

$$\gamma = \frac{F_{\text{grav}}}{m} = \frac{Gm_1}{r^2}$$

and the direction of γ is toward the body whose mass is m_1.

EXAMPLE **4-9**

Find the resultant gravitational field at the location of a spacecraft which is at a point P (Fig. 4-9) between the earth and the moon. It is new moon, with the moon on a line between earth and sun; the sun is 1.5×10^{11} m from point P. Masses: $m_{\text{earth}} = 6.0 \times 10^{24}$ kg; $m_{\text{moon}} = 7.4 \times 10^{22}$ kg; $m_{\text{sun}} = 2.0 \times 10^{30}$ kg.

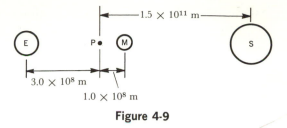

Figure 4-9

The field is the resultant of three fields:

$$\gamma_{\text{sun}} = \frac{Gm_{\text{sun}}}{r_1{}^2}$$

$$= \frac{(6.7 \times 10^{-11}\,\text{N}\cdot\text{m}^2/\text{kg}^2)(2.0 \times 10^{30}\,\text{kg})}{(1.5 \times 10^{11}\,\text{m})^2}$$

$$= 6.0 \times 10^{-3}\,\text{N/kg (toward sun)}$$

$$\gamma_{\text{earth}} = \frac{Gm_{\text{earth}}}{r_2{}^2}$$

$$= \frac{(6.7 \times 10^{-11}\,\text{N}\cdot\text{m}^2/\text{kg}^2)(6.0 \times 10^{24}\,\text{kg})}{(3.0 \times 10^{8}\,\text{m})^2}$$

$$= 4.5 \times 10^{-3}\,\text{N/kg (toward earth)}$$

$$\gamma_{\text{moon}} = \frac{Gm_{\text{moon}}}{r_3{}^2}$$

$$= \frac{(6.7 \times 10^{-11}\,\text{N}\cdot\text{m}^2/\text{kg}^2)(7.4 \times 10^{22}\,\text{kg})}{(1.0 \times 10^{8}\,\text{m})^2}$$

$$= 0.5 \times 10^{-3}\,\text{N/kg (toward moon)}$$

The resultant field is a vector sum, with the direction of the field toward the sun taken as positive:

$$\gamma = \gamma_{\text{sun}} + \gamma_{\text{earth}} + \gamma_{\text{moon}}$$

$$= (6.0 - 4.5 + 0.5) \times 10^{-3}\,\text{N/kg}$$

$$= 2.0 \times 10^{-3}\,\text{N/kg (toward sun)}$$

The weak gravitational fields found in Example 4-9 are to be compared with the much stronger gravitational field at the surface of the earth. At the earth's surface, $\mathbf{F}_{\text{grav}} = m\mathbf{g}$, and hence $\gamma = (m\mathbf{g})/m = \mathbf{g} = 9.8$ N/kg (toward center of earth).

We may, if we wish, interpret a field as a property of space itself. Thus we can say that the region of space where P is located has somehow been changed by the presence of distant gravitating bodies like the earth and the moon; this region of space now has the property that a test body placed there experiences a force. This viewpoint about fields may or may not be helpful; but from any viewpoint, gravitational field is an observable quantity, operationally defined in terms of the procedure used for measuring it.

In later chapters we will discuss other kinds of fields. All field strengths are defined as force per unit "something," and in the case of gravitational field that "something" is mass. Other test bodies are used to explore and measure electric and magnetic fields.

4-10   The nature of scientific theories

We have illustrated how Kepler's laws led Newton to the law of gravitation. In the detailed working out of the theory, Newton did not obtain Kepler's third law in exactly the form in which it was stated by Kepler. Newton obtained the following proportion (written for the case in which the two planets compared are Mars and the earth):

$$\frac{T^2_{\text{Mars}}\left(1 + \frac{m_{\text{Mars}}}{m_{\text{sun}}}\right)}{T^2_{\text{earth}}\left(1 + \frac{m_{\text{earth}}}{m_{\text{sun}}}\right)} = \frac{d^3_{\text{Mars}}}{d^3_{\text{earth}}} \qquad (4\text{-}12)$$

Here T is the period, the time for one revolution of the planet around the sun, and d is the average distance from the sun. Kepler's form of the proportion (stated in words in Sec. 4-8) omits the factors in parentheses involving the masses of the planets and the sun. Does this mean that Kepler was "wrong" and Newton "right"? Not at all. Kepler used existing data (collected during a lifetime of observations, chiefly of Mars, by his teacher Tycho Brahe) to construct a theory which agreed with the experimental facts within the limit of observational error. Other theories might agree with observation equally well, and such theories would be equally "correct." It is to be expected that, at any time, our most cherished laws of physics may suffer the same fate that Kepler's third law suffered. But note that the new law which replaces the old one always must contain the old law as a special case. For example, Newton's form of Kepler's third law (Eq. 4-12) reduces to the Kepler form, if the factors in brackets are set equal to 1; and this is very nearly true in view of the special circumstance that the sun is so much more massive than any of the planets. We say that Kepler's form is a first approximation to the law of planetary periods, and Newton's form is a second approximation. Are there third, fourth, . . . nth approximations? We face the prospect with equanimity, knowing that this is the way science progresses, each new step including all that went before.

As a matter of fact, Newton's law of gravitation has already received similar treatment at the hands of Einstein, and the Einstein law of gravitation includes Newton's form as a special case. Almost every prediction of the two laws of gravitation is the same, within experimental error. As discussed in Sec. 8-9, Mercury's orbit around the sun, however, is just noticeably different on the basis of the two laws, and the Einstein law agrees with observation while the Newton law does not. This does not mean that we have "repealed" Newton's law of gravitation. It is a very useful and almost exact first approximation, valid for the situations in , which we apply it.

Another example of the changing validity of a "law" of physics is supplied by Newton's second law itself. The "breakdown" of Newton's second law became evident in the early twentieth century, as atomic phenomena became more clearly understood. The finishing touch was given in 1926 by Schrödinger, Born, Heisenberg, and others, who developed a new mechanics (quantum mechanics) to apply to a single atom.* This

---

* For a single particle, the quantum-mechanical version of Newton's second law is quite different from net $F = ma$, but, when applied to the motion of a large collection of particles, this quantum-mechanical law reduces to the familiar net $F = ma$, just as Newton's form of Kepler's third law reduces to Kepler's form.

does not detract in the least from the validity of Newton's laws as we use them in this text and as engineers use them for the most complicated calculations. We always deal with large-scale objects (marbles, cars, even dust particles) which consist of many billions of atoms grouped closely together. We do not expect a political poll to shed any light on a specific individual's voting preference; yet the poll does reflect the average behavior of a large group. Similarly, we should not expect a description of the average behavior of a group of atoms (Newton's laws) to predict the behavior of a single electron or atom (quantum mechanics). A single atomic particle's behavior is, of course, basic; but it is nonclassical in the sense that all of the familiar (classical) large-scale laws of Newtonian mechanics are not applicable. We will return to this subject briefly in Secs. 9-6 and 16-8, and in Chaps. 27 and 28 where both the classical and nonclassical properties of electrons are studied in some detail.

■ SUMMARY

Newton's first law serves as a qualitative definition of the concepts of force and inertia. Mass is a quantitative measure of inertia—that property of matter which causes it to resist linear acceleration. Newton's second law, net $F = ma$, tells us how much unbalanced or "net" force is required to give acceleration to an object having mass. This kind of mass is called inertial mass. Newton's third law states that forces occur in pairs. Action and reaction always act on different bodies.

Problem solving requires a clear understanding of exactly which body or system of bodies is considered in Newton's second law. This body or system must be carefully isolated, and *all* forces acting *on* the body must be considered. Forces exerted *by* the body or system are to be ignored. Consistent sets of units must be used when applying Newton's second law. In the mks system, such units are the newton, kilogram, and meter per second per second. In the cgs system, the acceptable units are the dyne, gram, and centimeter per second per second. Force is not mass, and mass is not force; these quantities cannot have the same units in an equation. The ratio equation can be used when solving problems in which British units are used.

Newton's law of universal gravitation was derived by the application of the laws of motion to an explanation of Kepler's laws of planetary motion. The gravitational force of one body on another depends upon the product of the two masses and is also inversely proportional to the square of the distance between the bodies. This kind of mass is called gravitational mass. Experiment shows that gravitational mass and inertial mass for any body are proportional; since this is so, the same units (kg, g) are used for these two kinds of mass. The constant in the law of gravitation was first measured by Cavendish.

Gravitational field strength is defined as force per unit mass, and in the mks system is measured in newtons per kilogram.

Laws of physics are based upon experiment, and may be superseded by other, more general laws, as new experimental facts become known. The new laws must include the old laws as special cases.

Newton's first law
Newton's second law
Newton's third law
newton
dyne
pound

Kepler's laws
Newton's law of universal gravitation
inertial mass
gravitational mass
gravitational field strength

■ QUESTIONS

4-1 Is the following statement true or false? "When a book is at rest lying on a table, the downward force of gravity on the book is equal and opposite to the upward force of the table on the book. This is an example of Newton's third law." Explain your answer.

4-2 Two sleds, A and B, are connected by a light but stiff spring and then separated far enough apart, on smooth ice, so that the spring is stretched somewhat. The sleds and their passengers are then released, and the spring contracts, pulling the unequally loaded sleds together. During the motion of A toward B and B toward A, (a) are the forces on the sleds equal and opposite? If so, is this an example of Newton's third law? (b) Do the sleds have equal and opposite accelerations? (c) Do the sleds acquire equal speeds? (d) How can the mass ratio of the two sleds be found?

4-3 A horse is pulling a heavy cart, and the cart (and horse) are both being accelerated. Is the force of the cart on the horse equal and opposite to that of the horse on the cart? If your answer is "yes," explain how it is that these equal and opposite forces give rise to the "net force" which is necessary to cause acceleration. If your answer is "no," explain whether Newton's third law is true while the cart is accelerating.

4-4 In Ques. 4-3, explain how the horse becomes accelerated. What is the origin of the "net force" upon him? In which direction is the net force on the horse?

4-5 Which of the following units are suitable for expressing the mass of a body? Which are suitable for weight? Kilogram, pound, dyne, gram, ton, newton, ounce.

4-6 If Galileo dropped two objects of unequal weight from the Leaning Tower, why is it that the heavier one did not reach ground first, since its greater weight would be expected to cause it to be accelerated more rapidly?

4-7 How would you go about determining whether two bodies had the same weight? How would you go about determining whether two bodies had the same inertial mass? the same gravitational mass?

4-8 Two small ball bearings are placed 1 cm apart on a smooth, level metal surface. Will they accelerate toward each other because of gravitational attraction, and, if so, will they eventually touch each other? Answer this question for two conditions: (a) on earth; (b) in interstellar space, where there is no normal force to cause friction.

4-9 If a body of mass 36 kg were taken to the moon, where the acceleration due to gravity is $\frac{1}{6}$ as much as on the earth, what would be its weight? What would be its mass?

4-10 A freight train of 100 cars each weighing 10 tons is held together by couplings between the cars. Ignore friction as far as the cars are concerned. Is the tension in the coupling between the third and the fourth cars the same as the tension in the coupling between the thirty-third and the thirty-fourth cars? Answer this question for two conditions: (a) if the train has constant velocity; (b) if the train is accelerating.

4-11 In Example 4-8, where is the reaction force which, by Newton's third law, is equal and opposite to the 800 N force of friction? Express your answer in the form "_____ acting on _____."

4-12 An iron ball is hanging by a thread, and a similar thread is attached to the lower side of the ball and hangs loosely. If the loose thread is given a sharp jerk, it may break, while if it is given a slow and steady pull, the top thread breaks. Explain.

4-13 A light rope hangs over a frictionless pulley, and a 30 lb monkey is counterbalanced by a 30 lb mirror that is just opposite the monkey (Fig. 4-10). As the monkey climbs up and down his side of the rope, he cannot escape his reflection, since the mirror is always just opposite him. Explain. [After thorough discussion of this problem, you may wish to verify your reasoning by reference to the *Am. J. Phys.* **16**, 248 (1948) and **16**, 320 (1948).]

4-14 A spaceship is far away from the sun, moving with practically constant velocity in an orbit that is practically a straight line. In Fig. 4-11, *AB* and *CD* represent motions each taking place in one day's time. Use plane geometry to show that Kepler's second law is true in this case.

4-15 What is the unit for gravitational field strength in the cgs system?

Figure 4-10

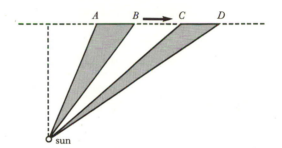

Figure 4-11   Kepler's law of areas for the special case of uniform motion in a straight line.

■ PROBLEMS

4-A1 What is the net force acting on a spider whose weight is 10 000 dynes, if he is partially supported by a strand of silk which exerts a force of 8000 dynes?

4-A2 What is your own mass, in kilograms? What is your own weight, in newtons?

4-A3 During a satellite re-entry, the satellite, which weighs 1200 lb, is acted on by air resistance which supplies an upward force of 5000 lb. What is the net force on the satellite?

4-A4 A body of mass 4 kg is acted on by a net force of 25 N. What acceleration is produced?

4-A5 A force of 400 dyn acting on a body produces an acceleration of 50 cm/s². What is the mass of the body?

4-A6 What is the force, in newtons, required to give a ball of mass 2 kg an acceleration of 3 m/s²?

4-A7 What net force is needed to give a mass of 50 g an acceleration of 40 cm/s²?

4-A8 A child can exert a forward push of 10 lb. What acceleration can he give to a loaded go-cart which weighs 64 lb?

4-A9 A 640 lb midget car is powered by an engine which can give rise to a maximum forward force of 360 lb. What is the greatest possible acceleration of the car?

4-A10 A sled of mass 30 kg coasts over the ice, with an acceleration of -0.5 m/s^2. What is the retarding force of friction?

4-A11 A space explorer is 1 billion miles away from a certain star, and he observes that the gravitational force of attraction exerted on his spaceship by the star is 100 lb. What will the force become when the ship has approached a position half a billion miles from the star?

4-A12 The radius of the moon is 1080 mi. What is the downward force (due to the moon's gravitational attraction) on an astronaut hovering in an orbit 1080 mi above the moon's surface, if his weight on the surface of the moon is 30 lb?

4-B1 A falling coconut has a mass of 2 kg, and the upward force of air resistance is 8.6 N. What is the acceleration of the coconut?

4-B2 A parachutist who weighs 160 lb is partially supported by a 120 lb upward force of air resistance. What is the acceleration of the parachutist?

4-B3 An elevator weighs 8000 lb, and the upward tension in the supporting cable is 10 000 lb. (a) What is the acceleration?   (b) Starting from rest, how far will the elevator rise in 2 s?

4-B4 A truck of mass 2000 kg is moving at 15 m/s and is acted on by two forces (assumed constant): the forward force of 800 N due to the engine, and a 300 N retarding force of friction.   (a) At what rate is the truck gaining speed?   (b) How far will it travel in 4 s?

4-B5 A car of mass 1000 kg traveling at 10 m/s is 15 m from the start of an intersection 55 m wide when the traffic light changes from green to yellow; 6 s later the light changes to red. Can the driver avoid being in the intersection on a red light? Available forces are 3000 N (in a panic stop) and 1000 N (for forward acceleration).

4-B6 A golf ball of mass 60 g is struck by a golf club, and acquires a speed of 80 m/s during the impact, which lasts for 2×10^{-4} s. What force (assumed constant) is exerted on the ball?

4-B7 A rifle bullet of mass 10 g has a muzzle velocity of 900 m/s and the length of the rifle barrel is 80 cm. What is the net force accelerating the bullet, assuming it to be constant?

4-B8 During a test of seat belts, a car of weight 3200 lb traveling at 40 ft/s collides with a heavy truck. The car moves 20 ft forward while being brought to rest.   (a) What force (assumed constant) is exerted on the car by the truck?   (b) What force (assumed constant) is exerted on a 160 lb driver by the seat belt?

4-B9 An 80 kg aviator in "free fall" acquires a velocity of 60 m/s and then opens his parachute. After falling an additional 30 m, his velocity has been reduced to 6 m/s.   (a) What is the average acceleration of the aviator while his fall is being checked?   (b) What is the average force exerted by the parachute?

4-B10 A girl weighing 100 lb wishes to elope by sliding down an improvised rope made of nylon stockings tied together. The maximum upward force that the stockings can exert without tearing is 60 lb.   (a) Can the girl slide down at constant speed?   (b) What is the least acceleration with which the girl can slide down the rope?

4-B11 A vertical rope is attached to a trunk of mass 40 kg. What tension in the rope is needed to cause the trunk to acquire an upward velocity of 3 m/s in 0.6 s?

4-B12 A 30 kg sled is coasting with constant velocity at 5 m/s over perfectly smooth, level ice. It enters a rough stretch of ice 20 m long in which the force of friction is 12 N. With what speed does the sled emerge from the rough stretch?

4-B13 A 16 lb iron ball is jerked upward by a rope; it rises 0.5 ft in 0.1 s. Assuming uniform acceleration, calculate the tension in the rope.

4-B14 A rocket which weighs 96 000 lb is acted on by an upward thrust of 108 000 lb. If the rocket is 80 ft tall, how long a time is required for it to rise off the launching pad a distance equal to its own height?

4-B15 An elevator of mass 1000 kg is supported by a cable that can sustain a force of 12 000 N. What is the greatest upward acceleration that can be given the elevator without breaking the cable?

4-B16 A tractor of mass 1000 kg is attached by a horizontal massless chain to a log whose mass is 300 kg. The backward force of friction exerted by the ground on the log is 800 N. The system has a forward acceleration of 2 m/s². Calculate   (*a*) the tension in the chain; (*b*) the forward force of the ground on the tractor.

4-B17 A horse weighing 1000 lb is setting into motion a loaded stoneboat weighing 600 lb (Fig. 4-12). The force of friction is 100 lb.   (*a*) What forward force *F* is exerted on the horse by the ground, if the system has a forward acceleration of 8 ft/s²? (*Hint:* Consider the whole system in solving this part of the problem.)   (*b*) What is the tension in the connecting rope? (*Hint:* Consider either the horse *or* the stoneboat for this part; as a check, do it both ways.)

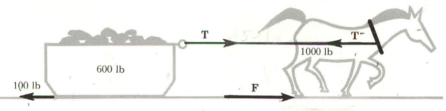

Figure 4-12

4-B18 A bag of feed weighing 100 lb is being hoisted at a steady speed of 5 ft/s by a farmer who uses a 60 lb counterweight and also applies a downward 40 lb force to a rope which passes over a pulley. If the farmer releases the rope, what will be the velocity of the feed bag after 3 s? (Ignore friction in the pulley.)

4-B19 Objects of mass 3 kg and 2 kg are connected by a light cord which passes over a horizontal frictionless rod.   (*a*) What is the acceleration of the system?   (*b*) What is the tension in the cord on the 3 kg side?   (*c*) What is the tension in the cord on the 2 kg side?

4-B20 In Fig. 4-13 the frictional force between the 30 kg block and the table is negligible. (*a*) If the peg is removed, what will be the acceleration of the system?   (*b*) How long will it take the block to hit the pulley?   (*c*) What is the tension in the cord while the block is moving?   (*d*) What is the tension in the cord after the block ceases to move?

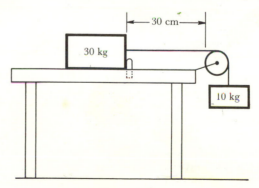

Figure 4-13

4-B21 Objects weighing 3 lb and 5 lb are connected by a light cord which passes over a friction-less pulley. (*a*) How long a time will be required for the 5 lb object to fall 9 ft, starting from rest? (*b*) What is the tension in the cord on the 5 lb side? (*c*) What is the tension in the cord on the 3 lb side?

4-B22 A 160 lb track star, at the start of a sprint, pushed on the ground with a measured force of 400 lb at an angle of 60°, as shown in Fig. 4-14. What forward acceleration was produced?

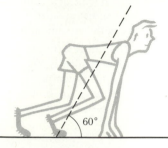

Figure 4-14

4-B23 Calculate the gravitational force of attraction between a mole of iron and a mole of aluminum, separated by 1 m.

4-B24 Compute the gravitational force between a proton and an electron in the $_1^1$H atom, using the following data: mass of proton, 1.67×10^{-27} kg; mass of electron, 9.11×10^{-31} kg; radius of electron orbit, 5.29×10^{-11} m.

4-B25 What is the force of gravitation (in dynes) between two touching spherical iron balls each 10 cm in radius?

4-B26 A small research rocket of mass 100 kg is released from rest at a point one earth radius above the earth's surface. (*a*) What is the rocket's weight at this point? (*b*) With what initial acceleration does it start to fall?

4-B27 Prove that the mass m_P of a planet is given by $m_P = r_P^2 g_P / G$, where r_P is the planet's radius and g_P is the acceleration due to gravity on the planet's surface.

4-B28 What is the dimensional formula for the gravitational constant G, in terms of [M], [L], and [T]?

4-B29 The sun's mass is 2×10^{30} kg. What is the ratio $\dfrac{m_{\text{earth}}}{m_{\text{sun}}}$? What is the value of the factor $\left(1 + \dfrac{m_{\text{earth}}}{m_{\text{sun}}}\right)$ which occurs in Eq. 4-12, the Newtonian form of Kepler's third law? (Express in decimal form.) Is this factor sufficiently close to unity to be neglected, if observations are accurate to 1 part in a million?

4-B30 Using data from Example 4-9, calculate the magnitude and direction of the resultant gravitational field when the explorer's spacecraft is halfway around the moon, 1.0×10^8 m from the moon, observing the hidden side of the moon's surface from a point on the extension of the line joining earth and moon. As in Example 4-9, it is new moon.

4-C1 The driver of a 1200 lb sports car, heading directly for a railroad crossing 1000 ft away, applies the brakes in a panic stop. The car is moving at 90 mi/h, and the brakes can supply a force of 300 lb. (*a*) How fast will the car be moving when it reaches the crossing? (*b*) Will the driver escape collision with a freight train which, at the instant the brakes are applied, is blocking the road and still requires 11.0 s to clear the crossing?

4-C2 Two boxes weighing 40 lb and 24 lb respectively are in contact, on a horizontal, frictionless surface. If a horizontal force of 30 lb is applied to the 40 lb box, pushing it against the other box, (*a*) compute the acceleration of the system; (*b*) compute the force with which the two boxes push against each other.

4-C3 Objects of weight 5 lb and 15 lb are connected by a light cord which passes over a frictionless pulley that is 7 ft above the floor. Initially the weights are each at rest, 2 ft above the floor, with the cord taut. After the system is released, what maximum height above the floor does the 5 lb object reach?

4-C4 A paint bucket of mass 12 kg is attached to another paint bucket of mass 9 kg by a light rope 21 m long. The light bucket is at rest on the ground, and the rope goes up and over a small frictionless pulley and is attached to the heavy bucket, which is on a window sill 11 m above the bucket on the ground. There is no slack in the rope. If the heavy bucket slips off the window sill, after what time will the first big splash occur when the buckets collide?

4-C5 Solve Problem 4-B20, under the assumption that the table is rough and an 18 N force of friction acts toward the left.

4-C6 Two small balls of equal mass are connected by a string 4 ft long and are laid out on a smooth table top with one ball (*A*) just at the edge of the table and the other ball (*B*) 4 ft from the edge. The table is 2 ft high. Ball *A* is nudged gently over the edge of the table, and things begin to happen. (*a*) How much time is required for ball *B* to strike the floor? (*b*) How far from the base of the table does ball *B* strike?

4-C7 Three cars of a model train, each of mass 0.8 kg, are connected together and are pulled by an engine of mass 1.6 kg. The track supplies a forward force of 12 N on the engine, and there is a negligible frictional drag. (*a*) Calculate the acceleration of the system. (*b*) Calculate the tension in the coupling which connects the front car to the middle car.

4-C8 The system shown in Fig. 4-5 is held with the 6 kg block just opposite the 10 kg block, both blocks being 2 m above the floor and 3 m below the pulley. The 6 kg block is then given a downward initial velocity sufficient to cause it to just reach the floor. How long after the initial push will the blocks again be opposite each other?

4-C9 The two blocks in Fig. 4-15 are connected by a heavy chain which weighs 10 lb. An upward force of 32 lb is applied to the top block. (*a*) How far and in which direction will the system move in 0.50 s, starting from rest? (*b*) What is the tension at the top link of the chain?

4-C10 If an Olympic athlete could throw the shot 18.31 m in Helsinki (latitude 60.2° N), where g is 981.91 cm/s^2, how much farther could he throw it in Los Angeles (latitude 34.1° N), where g is 979.69 cm/s^2? (*Hint:* Consider the fractional change in g.)

4-C11 A 64 lb sack of feed is hung from a rope which passes over a frictionless pulley to a man on the ground. If the breaking strength of the rope is 100 lb, what is the shortest time in which the man can raise the bag 9 ft, starting from rest?

4-C12 Bob and Joe, two construction workers on the roof of a building, are about to raise a keg of nails from the ground

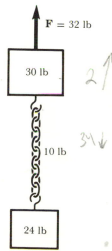

Figure 4-15

by means of a rope passing over a pulley 36 ft above the ground. Bob weighs 210 lb, Joe 140 lb, the keg 70 lb, and the nails 140 lb. They slip off the roof, and the following unfortunate sequence of events takes place: Bob and Joe, hanging on the rope together, strike the ground just as the keg hits the pulley. Unnerved by his fall, Bob lets go of the rope, and the keg pulls Joe up to the roof where he cracks his head against the pulley but gamely hangs on. However, the bottom falls out of the keg of nails when it strikes ground, and the empty keg rises as Joe returns to the ground. Finally, Joe has had enough and lets go of the rope and remains on the ground, only to be hit by the empty keg again. Ignoring the possible mid-air collisions which merely add insult to injury, how long did it take this little drama to unfold?

■ REFERENCES

1 Cohen, I. Bernard, "Newton," *Sci. American,* Dec., 1955, p. 73. A biography.

2 Sullivan, J. W. N., *Isaac Newton, 1642–1727* (New York: Macmillan, 1938).

3 Metzdorf, Robert F., "Sir Isaac Newton, 1642–1727—A Study of a Universal Mind," *Am. J. Phys.,* **10,** 293 (1942).

4 Hanson, Norwood Russell, "History and Philosophy of Science in an Undergraduate Physics Course," *Physics Today,* Aug., 1955, p. 4. See especially remarks on pp. 7 and 8 on the law of inertia.

5 Sciama, Dennis, "Inertia," *Sci. American,* Feb., 1957, p. 99. A defense of Mach's position that inertia depends on the interaction between any body and the rest of the universe, rather than being a property of a body.

6 Cavendish, Henry, "The Density of the Earth," *Philos. Trans.,* Vol. 17 (1798). The original account of the measurement of *G.* Reprinted in W. F. Magie, *A Source Book in Physics* (New York: McGraw-Hill, 1935), pp. 105–11 (also, Cambridge: Harvard University Press, 1963).

7 Dicke, R. H., "The Eötvös Experiment," *Sci. American,* Dec., 1961, p. 84. See also Roll, P. G., R. Krotkov, and R. H. Dicke, "The Equivalence of Inertial and Passive Gravitational Mass," *Ann. Phys.,* **26,** 442 (1964).

8 Gamow, G., "Gravity," *Sci. American,* Mar., 1961, p. 94.

9 Beams, J. W., "Finding a Better Value for *G,*" *Phys. Today,* May, 1971, p. 34.

10 Miller, F., *Measurement of "G"—The Cavendish Experiment* (film).

11 Stull, J., *Newton's First and Second Laws* (film).

12 Stull, J., *Newton's Third Law* (film).

5

Statics

5-1 Equilibrium

We have seen that Newton's laws of motion are adequate to determine the behavior of a body or a system of bodies, since the acceleration can be calculated from a knowledge of the net force and the mass on which it acts. In particular, we now ask what are the conditions necessary to cause a body to have zero acceleration. Such a body is said to be in *equilibrium,* and the study of forces in equilibrium is called *statics.* We have seen that Newton's first law is a special case of Newton's second law; thus statics is a special case of dynamics, the case in which acceleration = 0. For the engineer who is concerned with the stability of structures and the safe loads that can be carried by girders, it is an important special case, and the problems of statics are of vital concern to him. For us, however, the important thing is to recognize that the study of statics presents nothing really new; it is merely a systematic application of one aspect of Newton's second law.

Equilibrium is the condition of a body whose velocity is constant in magnitude and direction. This includes, of course, the case where the velocity is zero and remains so. If a body remains at rest, it is said to be in *static*

equilibrium. If a body remains in steady motion in a straight line, it is said to be in *dynamic equilibrium.* These are *definitions* of equilibrium; it is one thing to define a condition, and another to state how to obtain that condition. In order to obtain equilibrium, we must see to it that no unbalanced force is applied to the body in question. It is rare indeed that *no* force acts on a body. Such a situation could exist only in interstellar space, where the gravitational attraction of the stars would be negligibly small. Far more commonly, a number of forces are acting, but they are balanced to give a resultant force of zero.

We shall first consider the situation in which all the acting forces pass through a single point (the dimensionless "particle" of the philosophers and mathematicians). In general, Newton's second law says that net $\mathbf{F} = m\mathbf{a}$; if the acceleration is zero, then

$$\text{net } \mathbf{F} = m(0)$$

which means that

$$\text{net } \mathbf{F} = 0$$

or

$$\Sigma \mathbf{F} = 0$$

This vector equation gives the condition for

Engineers must solve many problems in statics to carry through the design and construction of a project such as the Gateway Arch in St. Louis, Missouri. (Arteaga Photos)

equilibrium of a particle. In words, a particle is in equilibrium if the vector sum of all forces acting on it is zero.

5-2   Center of gravity

Actual physical objects are never mathematical points; for instance, a solid plastic phonograph record is composed of 10^{23} or so atoms, on each of which there is a downward force of gravity. In Fig. 5-1a are shown a few of these parallel downward forces. The *center of gravity* (c.g.) is defined as the point in the body through which a single downward force equal to the weight of the body may be considered to act. For the *uniform* disk in Fig. 5-1b, the center of gravity is seen to be at the geometrical center of the disk. In other words, if $\mathbf{w}_1$, $\mathbf{w}_2$, $\mathbf{w}_3$, ... are the weights of the individual molecules and $\mathbf{W}$ is the total weight (equal to $\mathbf{w}_1 + \mathbf{w}_2 + \mathbf{w}_3 + \cdots$), then the many tiny forces $\mathbf{w}_1$, etc., are equivalent to the single force $\mathbf{W}$ acting through the center of gravity. For a uniform body of sufficient symmetry, such as a disk, sphere, cube, square or rectangle, the c.g. is at the geometric center. (We shall see in Sec. 5-6 how to compute the position of the c.g. of more complicated bodies.) This trick of replacing many parallel forces by a single force often makes it possible to apply the condition for equilibrium of a particle (net $\mathbf{F} = 0$) to an actual body whose c.g. is known.

A simple example will show how we tend

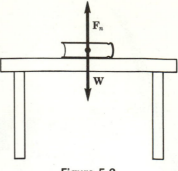

Figure 5-2

to take these ideas for granted. Consider a book lying at rest on a table. We say that the downward force of gravity is balanced by the upward force of the table. Actually, the downward force of gravity is the sum of all the downward forces of gravity on the individual molecules; this "effective" or "resultant" force passes through the c.g. Likewise the upward force $\mathbf{F}_n$ of the table is the resultant of the upward forces exerted by many molecules of the table top pressing upward on the many places where the book is in contact with the table; the line of action of the resultant of all these upward forces also passes through the c.g. of the book. We can draw a picture (Fig. 5-2) showing only two forces acting on one particle, even though we really have many forces acting on many particles.

5-3   Equilibrium of a particle

In Chap. 3 we learned how to add vectors by the head-to-tail method, as well as by the component method. The resultant force $\Sigma\mathbf{F}$ is a vector, and for equilibrium this vector is zero. If $\Sigma\mathbf{F} = 0$, the forces must form a closed polygon (a triangle if there are only 3 forces) (Fig. 5-3). A vector can be zero only if each of its perpendicular components is zero;* thus the single vector equa-

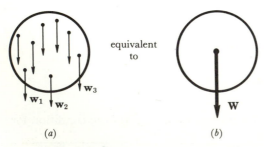

(a)   *(b)*

Figure 5-1   The center of gravity of a uniform disk is at its center.

---
* By the Pythagorean theorem, the magnitude of a vector is given by the square root of the sum of the squares of its perpendicular components. Since the squares of the

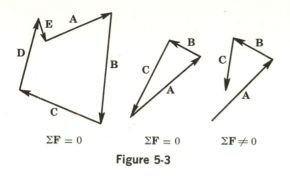

$$\Sigma\mathbf{F} = 0 \qquad \Sigma\mathbf{F} = 0 \qquad \Sigma\mathbf{F} \neq 0$$

Figure 5-3

tion $\Sigma\mathbf{F} = 0$ is equivalent to three component equations:

$$\Sigma F_x = 0 \qquad \Sigma F_y = 0 \qquad \Sigma F_z = 0$$

When a body is in equilibrium and all but one of the forces acting on it are known, the unknown force can be found by a graphical construction, trigonometry, or the component method.[*] Let us solve a typical equilibrium problem by each method. It is a matter of choice (and convenience) which method to use.

EXAMPLE **5-1**

An 80 lb boy sits in a swing. What force $\mathbf{F}$, applied horizontally, will hold the swing and boy out at an angle of 30° with the vertical? What is then the total tension in the ropes?

To avoid confusion, we use two diagrams. The *space diagram* shows the tree limb, ropes, etc. Engineers call such a diagram a *freebody* diagram. The forces involved are all shown acting *on* the body whose equilibrium we are consider-

---

components will always be positive, regardless of their directions, $\sqrt{F_x{}^2 + F_y{}^2 + F_z{}^2}$ can be zero only if F_x, F_y, and F_z are each zero.

Our preferred unit of force in applying Newton's second law ($\Sigma\mathbf{F} = m\mathbf{a}$) is the newton in the mks system of units; we use the dyne ($= 10^{-5}$ N) in the cgs system. We have somewhat more freedom in statics, where $\mathbf{a} = 0$, for then $\Sigma\mathbf{F} = 0$ is true for *any* force units, even the pound, ton, ounce, etc. A conversion factor between the pound and the newton, for instance, would cancel out because of the 0 on the right side of the equation. Therefore, in this chapter, we will frequently use the pound as a force unit in problems involving equilibrium. In this way we can more easily make connection with everyday experience.

ing—in this case, the boy. There are three such forces: (1) the weight of the boy (80 lb acting downward, through the c.g. of the boy); (2) the force $\mathbf{F}$ (of unknown magnitude and directed horizontally along a line passing through the c.g.); and (3) the total tension $\mathbf{T}$ exerted by the ropes (of unknown magnitude, directed along the ropes and passing through the c.g. of the boy). All these forces act at the same point, the c.g., and so we can apply the condition for equilibrium of a particle. To construct the *force diagram*, we place the three force vectors head to tail to make a closed triangle. *The fact that the forces form a closed triangle expresses the condition* $\Sigma\mathbf{F} = 0$. It is now easy to find the unknown forces, from the force diagram.

First method: Graphical solution (Fig. 5-4): We lay off the force diagram to scale—letting 80 lb be represented by 8 cm, for instance—and find by measurement that F is about 4.6 cm, which

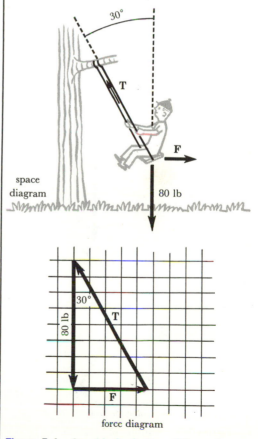

Figure 5-4 Graphical solution of Example 5-1.

corresponds to

$$F = \boxed{46 \text{ lb}}$$

and T is about 9.2 cm, which corresponds to

$$T = \boxed{92 \text{ lb}}$$

Second method: Using trigonometry (Fig. 5-5): From the definitions of tan θ and cos θ (p. 704) we have

$$\frac{F}{80 \text{ lb}} = \tan 30°$$

$$F = (80 \text{ lb})(\tan 30°)$$

$$= (80 \text{ lb})(0.577) = \boxed{46.2 \text{ lb}}$$

$$\frac{80 \text{ lb}}{T} = \cos 30°$$

$$T = \frac{80 \text{ lb}}{\cos 30°} = \frac{80 \text{ lb}}{0.866} = \boxed{92.3 \text{ lb}}$$

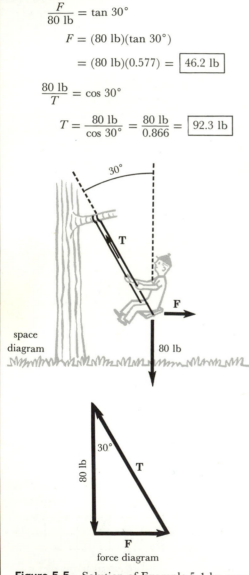

Figure 5-5 Solution of Example 5-1 by use of simple trigonometry.

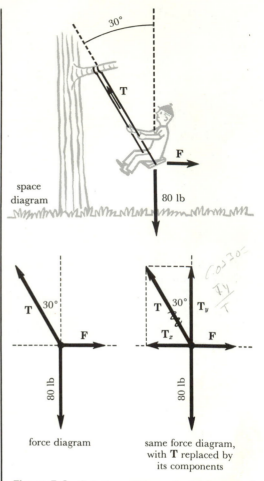

Figure 5-6 Solution of Example 5-1 by use of components of force.

Third method: Using components (Fig. 5-6): Let us look at this same problem from the point of view of components of force. The space diagram is the same as before. For the force diagram, we choose a set of axes with the x axis horizontal and the y axis vertical, and we lay out all the forces acting *on* the boy. We then replace the tension $\mathbf{T}$ by its components, T_x and T_y. To indicate that $\mathbf{T}$ itself is no longer to be considered, we draw a wavy line through the vector $\mathbf{T}$ and ignore it from now on. To solve the problem, we use the conditions for equilibrium in the component form (there is no z-component of force, so $\Sigma F_z = 0$ is automatically satisfied and is not even written down).

From $\Sigma F_y = 0$,

$$T_y - 80 \text{ lb} = 0$$

$$T_y = 80 \text{ lb}$$

but, since $T_y = T \cos 30°$,

$$T = \frac{T_y}{\cos 30°} = \frac{80 \text{ lb}}{\cos 30°} = \frac{80 \text{ lb}}{0.866} = \boxed{92.3 \text{ lb}}$$

Using this value for T, we obtain

$$T_x = T \cos 60° = (92.3 \text{ lb})(0.500) = 46.2 \text{ lb}$$

Now from $\Sigma F_x = 0$,

$$F - T_x = 0$$

$$F = T_x = \boxed{46.2 \text{ lb}}$$

A body need not be at rest to be in equilibrium. All that is necessary is that the acceleration be zero; but this does not require the velocity to be zero. In our next example, the conditions for equilibrium are applied to a moving body which is in dynamic equilibrium.

EXAMPLE 5-2

A trunk weighing 500 N is being pushed up a smooth inclined plane making an angle of 30° with the horizontal. What force, applied parallel to the plane, is necessary to allow the trunk to move up the plane with a steady speed of 1 m/s?

Since the motion is uniform, the trunk is in dynamic equilibrium, and the same diagrams are to be used as if the trunk were held stationary. The magnitude of the constant speed, 1 m/s, is of no importance for the solution of the problem; nor is it of any consequence for us to know how and by what forces the trunk was accelerated to this speed.

The first space diagram (Fig. 5-7a) is incomplete, because one force was overlooked. A little reflection will show that the trunk cannot be in equilibrium if the weight **W** and the force **F** are the only forces acting. These two forces are not in the same straight line, and hence, regardless of their magnitudes, they could never cancel each other. The third force is not hard to find. The plane helps hold up the trunk. Since the plane is smooth, the force **F**$_n$ of the plane on the trunk is normal (perpendicular) to the surface of the plane (i.e., there is no frictional drag,

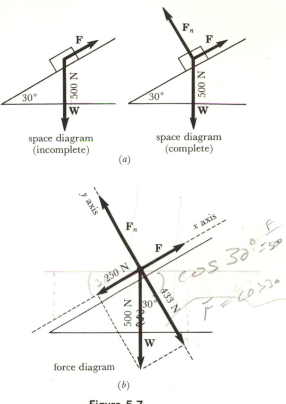

Figure 5-7

which would be parallel to the surface). We may simplify the vector diagram by choosing the x axis along the plane and the y axis perpendicular to the plane (Fig. 5-7b). Then the components of the weight are $(500 \text{ N})(\sin 30°) = 250 \text{ N}$ parallel to the plane, and $(500 \text{ N})(\cos 30°) = 433 \text{ N}$ perpendicular to the plane. Hence,

from $F_x = 0$ we get $F = \boxed{250 \text{ N}}$

from $F_y = 0$ we get $F_n = \boxed{433 \text{ N}}$

In solving problems, Newton's third law is of great help. It often happens that we are asked for a certain force, but the equal and opposite reaction force is the one which acts on the body in question and which can be computed from the conditions for equilibrium. In this case we compute the force *on* the body, and then use the third law to find the force exerted *by* the body.

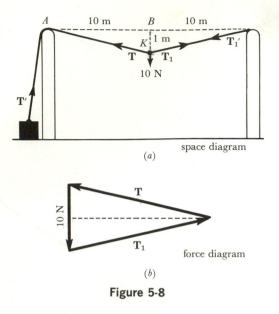

space diagram

(a)

force diagram

(b)

Figure 5-8

EXAMPLE **5-3**

A rope attached to one wall is passed over an equally high smooth wall 20 m away, and is then attached to a heavy box on the ground (Fig. 5-8). If a downward force of 10 N is applied at the center of the rope as shown and the rope sags 1 m, what force is exerted by the rope on the box?

Although we are asked for the force T' *on* the box, we can compute the force T exerted *by* the box on the knot K, which is the particle whose equilibrium we are considering. Newton's third law tells us then that T and T' are numerically equal, since they are an action-reaction pair. Putting the forces head to tail, we make the force diagram a triangle, half of which is similar to the space triangle ABK. By the Pythagorean theorem, $AK = \sqrt{10^2 + 1^2} = 10.05$ m. Hence, using similar triangles, we have

$$\frac{5 \text{ N}}{T} = \frac{1 \text{ m}}{10.05 \text{ m}}$$

$$T = \boxed{50.3 \text{ N}}$$

This is the force of the box on the knot. By Newton's third law, the force of the knot on the box (transmitted by the rope) is also 50.3 N. Note also that, since the force triangle is isosceles, T_1 is also 50.3 N.

5-4 Friction

When one body slides or tends to slide over another body, the force which acts to oppose the tendency to move is called the force of friction. This frictional force is always *parallel* to the surfaces which are in contact. The experimental facts about friction may be summarized by the "laws" of friction:[*]

1. The maximum frictional force on a body resting upon another body is proportional to the normal force pushing the two surfaces together. In symbols,

$$f_s = \mu_s F_n \quad \text{(bodies at rest)} \qquad (5\text{-}1)$$

where f_s is the *maximum* force of static friction, F_n is the *normal force* (resultant of all forces perpendicular to the surfaces), and μ_s (the Greek letter mu) is the *coefficient of static friction*. In words, coefficient of static friction:

$$\mu_s = \frac{\text{maximum force of friction}}{\text{normal force}}$$

2. If one body is sliding upon another body, the force of friction is constant, independent of the relative velocity of the two surfaces.

$$f_k = \mu_k F_n \quad \text{(bodies in motion)} \qquad (5\text{-}2)$$

where f_k is the *constant* force of kinetic friction (also called sliding friction), F_n is the normal force, and μ_k is the *coefficient of kinetic friction*. Coefficient of kinetic friction:

$$\mu_k = \frac{\text{force of friction}}{\text{normal force}}$$

To a good approximation, static and kinetic frictional forces are independent of the area of the surfaces which are in contact. Kinetic friction is also almost independent of the

[*] These are not laws in the same sense as Newton's first law, which is exactly true under all circumstances. The "laws" of friction, like many so-called laws of physics, merely represent conveniently certain experimental facts well enough for making a start in design of machines, etc. For an interesting discussion of the failures of the laws of friction, especially the second law, see Reference 7 at the end of the chapter.

relative velocity of the surfaces, if the velocity is not too great. A coefficient of friction, which is the ratio of two forces, is a dimensionless quantity and has no units. Because Eqs. 5-1 and 5-2 are very similar, you will have to be careful to distinguish between static and sliding friction in any given situation and use the proper coefficient of friction. The coefficient of kinetic friction is less than the coefficient of static friction (Table 5-1). This means that the moment a body starts to slip, the force of friction decreases, usually by a considerable amount.

The force of *rolling friction* is much less than that of kinetic friction. When a ball rolls on a table, the table and ball are slightly deformed at the points of contact, with the result that the ball is perpetually climbing a small hill as it rolls along. The hill exerts a slight backward force on the ball, which accounts for the friction. Imperfect recovery of the ball or table after the point of contact moves on to another part of the ball is also responsible for some losses due to internal molecular friction. Reduction of friction by means of ball bearings often makes the difference between success and failure of modern machines.

To make the laws of friction more reasonable, let us look at the phenomenon of static friction from a microscopic point of view. Consider a 100 lb block resting on a table, and let the coefficient of static friction be 0.30. Then the maximum force of friction is, according to our definition,

TABLE 5-1
Coefficients of static and kinetic friction*

Surfaces	Coefficient of static friction	Coefficient of kinetic friction
steel on steel, dry	0.6	0.3
steel on wood, dry	0.4	0.2
steel on ice	0.1	0.06
wood on wood, dry	0.35	0.15
metal on metal, greased	0.15	0.08

*Typical values; actual coefficients of friction vary greatly and depend on condition of surfaces, moisture, grain of wood, lubrication, duration of contact, etc.

$$f_s = \mu_s F_n = (0.30)(100 \text{ lb}) = 30 \text{ lb}$$

How can such a force arise? On account of the roughness of the surface, contact is actually made at only a relatively few points. At these points, we can imagine little hook-like protuberances which come into action as soon as any force, however small, is applied (Fig. 5-9). The hooks are deformed slightly, and they exert elastic forces against each other. The hooks are probably actually "welded" together where they are in intimate contact. If, say, a force of 5 lb is applied to the right, the block remains in static equilibrium, which means that all the little elastic frictional forces f_1, f_2, etc., add up to exactly 5 lb. If now the applied force is increased to 10 lb, the block does not start to move, but the small hooks "give" a little; being more deformed than before, they exert a correspondingly larger horizontal force,

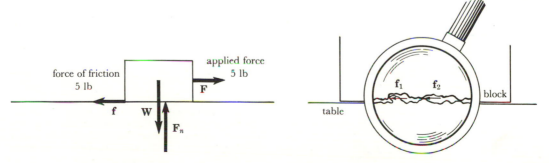

Figure 5-9   The force of friction acts at the surface between two bodies.

amounting now to exactly 10 lb. (We know that the force of friction must be *exactly* 10 lb; if it were 10.01 or 9.99 lb, for instance, there would be a resultant unbalanced force of 0.01 lb one way or the other on the block, and it would start to move.) As the applied force is increased, there finally comes a time when the little hooks break down. This happens when the total force reaches 30 lb, and so we say that 30 lb is the "maximum" force of static friction.

If the block weighs 200 lb instead of 100 lb, it sinks into the table a little further, and (presumably) twice as many little hooks come into play. Hence the maximum force of friction is twice as much as before; in general, we expect the maximum force of static friction to be in proportion to the normal force pressing the surfaces together. We have arrived at an interpretation of the relationship $f_s = \mu_s F_n$. Our procedure has been to make a "model" of a physical phenomenon (static friction) in terms of a more easily visualized or better-understood phenomenon (in this case, elasticity). The making of models is a favorite pastime for physicists. For instance, in Chap. 2 we described the atom in terms of a familiar model, a miniature solar system. We must bear in mind that the models are not in themselves reality, but are merely suggestive. They are to be used only as long as their use helps predict new phenomena and correlate old observations. Sooner or later, we expect that even a good model may break down, for by definition a model is large-scale, drawn from everyday experience, and it is not necessarily true that small-scale phenomena must follow the large-scale laws.[*] Often, of course, a model is only an approximate analogy (with limited uses) and fails to describe the phenomenon fully even at the outset.

In considering problems involving frictional force, it is necessary to remember that the normal force is the resultant of all forces acting normal to the surfaces, not merely the weight or a component of the weight.

EXAMPLE 5-4

A box that weighs 100 N is being steadily dragged along the floor by a rope that makes an angle of 30° with the horizontal (Fig. 5-10). If the tension in the rope is 40 N, what is the force of friction? What is the coefficient of friction?

Since the box is actually in motion, the force of kinetic friction is operative here. The box is in equilibrium (constant speed), and so $\Sigma F_x =$

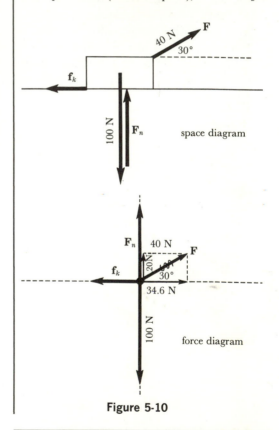

Figure 5-10

---

[*] The model of an atom as a miniature solar system broke down in 1926. This does not mean that the atom really "is" or "is not" like a solar system. We still use the model and speak of the nucleus around which the electrons move, but now we realize that the model has its limitations and some predictions of the model are simply not experimentally true, nor need they be. The planetary model is an approximation, but a useful approximation. Uncritical acceptance of models as reality (a disease known in some circles as "modelitis") has led to numerous blind alleys in physics.

0 and $\Sigma F_y = 0$. The condition $\Sigma F_x = 0$ tells us that the horizontal component of the force of the rope $(40\text{ N})(0.866)$ equals the force of friction f_k. Hence $f_k = 34.6$ N. Considering next the vertical forces, we find several such forces: the weight, the upward component of the tension in the rope, and the upward force of the floor on the box. It is the latter force F_n which is the normal force causing the frictional drag. The condition $\Sigma F_y = 0$ gives us $F_n + 20\text{ N} = 100$ N, whence $F_n = 80$ N. We may say that the upward component of the tension in the rope somewhat "eases up" the force between box and floor. If we know both f_k and F_n, it is a simple matter to find the coefficient of kinetic friction:

$$\mu_k = \frac{f_k}{F_n} = \frac{34.6\text{ N}}{80\text{ N}} = 0.43$$

EXAMPLE 5-5

An otter of weight **W** is at rest on a 30° plane (Fig. 5-11). If the otter is about to slip, what is the coefficient of static friction?

From the space diagram, it is evident that the otter is acted on by three forces: the weight **W**, the normal force $\mathbf{F}_n$ of the plane on the otter, and the force of static friction $\mathbf{f}_s$. We resolve **W** into two components as shown, and then apply the conditions for equilibrium:

$$F_n = W \cos 30°$$
$$f_s = W \sin 30°$$

Hence

$$\mu_s = \frac{f_s}{F_n} = \frac{W \sin 30°}{W \cos 30°} = \frac{0.500\ W}{0.866\ W} = 0.577$$

Note that, in the solution of this problem, the weight of the otter canceled out. Only

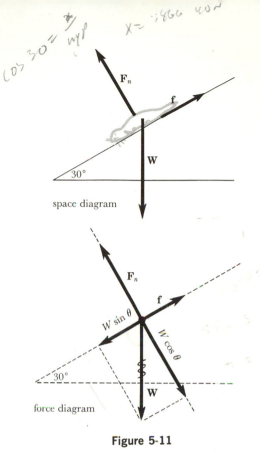

Figure 5-11

the angle of the plane was important. This illustrates a common way of measuring the coefficient of static friction: place the object on a plane and increase the angle of the plane until the object just begins to slip. The angle so measured is called the *angle of repose,* and the coefficient of static friction equals the tangent of the angle of repose (Fig. 5-11). If the otter in Example 5-5 moves along a

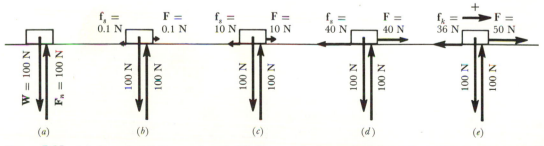

Figure 5-12 (a) through (d) Force of static friction equals the applied force; $f_s \leqslant 0.40$ N. (e) Force of kinetic friction is constant; $f_k = 0.36\ F_n = 36$ N.

plane without acceleration, it is in dynamic equilibrium, and the tangent of the angle of the plane equals the coefficient of kinetic friction.

Bear in mind that the equation $f_s = \mu_s F_n$ gives the *maximum* possible force of static friction for any given pair of surfaces and for a given normal force. Unless motion is taking place, or about to take place, the force of friction is less than $\mu_s F_n$. Until the maximum value is reached, the force of friction is self-adjusting, only as much as is required to give equilibrium. For the block and table in Fig. 5-12, the coefficient of static friction is 0.40, and the coefficient of sliding friction is 0.36. The normal force is 100 N in each case, equal to the weight of the block. The static force of friction varies from 0 to 40 N, as the applied force **F** increases. Of course, in Fig. 5-12e the block is no longer in equilibrium, since net **F** $\neq$ 0. Since it is sliding, we know that we are obtaining the force of kinetic friction, which is 0.36 × 100 N = 36 N. In this case, the unbalanced force of 14 N causes an acceleration which can be computed using Newton's second law.

EXAMPLE 5-6

Compute the acceleration of the block in each case of Fig. 5-12. The block weighs 100 N, and its mass is 100/9.8 kg.

For (a), (b), (c), and (d), the net force is zero, and Newton's second law gives zero for the acceleration. The block is in equilibrium.

For case (e),

$$\text{net } F = ma$$
$$50\text{ N} - 36\text{ N} = (100/9.8\text{ kg})(a)$$
$$a = \frac{14\text{ N}}{100/9.8\text{ kg}} = \boxed{1.37\text{ m/s}^2}$$

The following example illustrates how the force of friction which acts *on* a body tends to oppose the motion or tendency to move, and yet may still cause an acceleration relative to an outside observer.

EXAMPLE 5-7

An iron casting of mass 2000 kg is being transported by a truck which is moving at a steady velocity (Fig. 5-13a). The driver sees a red light and applies his brakes, causing the truck to decelerate (i.e., have negative acceleration) at a rate of 3 m/s². Will the casting remain in place, or will it slide forward and crush the driver? The coefficient of static friction is 0.4, and the coefficient of sliding friction is 0.3.

The question really is, *can* the casting be decelerated at 3 m/s²? (All velocities and accelerations are relative to the ground, as seen by an outside observer standing on the sidewalk.) The casting tends to slide forward on the bed of the truck, but the frictional force acts to oppose this tendency. Consider the forces *on the casting* (Fig. 5-13b): only one horizontal force acts on the casting, namely, the force of friction.

$$\Sigma F = ma$$
$$f = (2000\text{ kg})(-3\text{ m/s}^2)$$
$$= -6000\text{ N} \quad \text{(backward)}$$

A frictional force of 6000 N would be needed to cause the casting to decelerate along with the truck. The weight of the casting is (2000 kg)(9.8 m/s²) = 19 600 N. The maximum force

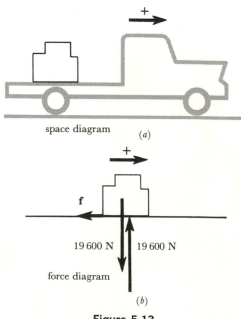

space diagram

(a)

f

19 600 N   19 600 N

force diagram

(b)

Figure 5-13

of static friction is (0.4)(19 600 N) = 7840 N, so there is no reason for alarm. The force of friction in this example is self-adjusting, and equal to 6000 N, which is all that is required; the maximum value of 7840 N is not needed.

5-5 Equilibrium of a rigid body: torque

We have considered the equilibrium of an idealized "particle" and have found that a single vector equation, $\Sigma \mathbf{F} = 0$, expresses the conditions that must be fulfilled in order to obtain equilibrium. We also have seen that, in many cases in which all applied forces pass through the c.g., a body may be considered to be a particle even though it has extension (i.e., has a size and shape). Let us consider the types of motion possible for a body which has measurable extension. Such a body may have rotation as well as translation. In *translational motion* all particles of a rigid body travel in parallel paths (not necessarily straight lines), while in *rotational motion* the particles describe circles of various sizes about some axis. Often the two types of motion are combined, as in a football which is end-over-ending (Fig. 5-14). The only type of motion having meaning for an ideal particle is translation. Our previously discussed condition that the resultant force be zero is a condition for translational equilibrium.

In considering rotational equilibrium (no change in rotational velocity), we find from experience that a new quantity called *torque* is the deciding factor. The effectiveness of a force in changing the rotational state of a body is governed not only by the magnitude of the force but also by its direction and its point of application. The *lever arm* is the perpendicular distance from the axis of rotation to the line of action of the force. The product, force × lever arm, is defined as the torque caused by the force; in order to have rotational equilibrium it is necessary that the torques of all forces add up to zero. The dimensional equation for torque is [torque] = [F][L], or [torque] = [MLT^{-2}][L] = [ML^{2}T^{-2}]. Torques are expressed in units such as N·m, dyn·cm, lb·ft, and oz·in. —the product of any force unit and a distance unit. No entirely satisfactory symbol for torque has been universally agreed upon; we shall use the symbol τ (the Greek letter "tau") for torque to avoid confusion with symbols for tension or time.*

The torque caused by a given force depends greatly on the point of application and the direction of the force. In Fig. 5-15 is shown a piece of plywood 4 ft on an edge with an axis of rotation through one corner. This axis is perpendicular to the plane of the plywood, and the force in each case is applied parallel to the plane of the plywood. In each case the magnitude of the force is 8 lb, but the lever arms are quite different, and so are the torques. In (*a*), the distance from the axis to the point of application of the force is 5 ft, but it is the distance to the *line of action* (3 ft) that counts. We shall consider torques as plus or minus, according to whether they tend to cause counterclockwise

---

* Torque is often called "moment of force," but this term sometimes leads to confusion. If you learned about "moments" in a high school course, rest assured that torque and moment of force are identical in meaning and use.

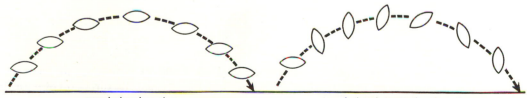

translational motion   translational + rotational motion

Figure 5-14

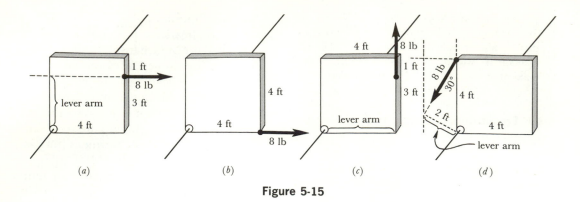

Figure 5-15

or clockwise rotation, respectively. If you have trouble in determining whether a given force produces a clockwise or counterclockwise torque, draw an imaginary circle tangent to the line of action (Fig. 5-16) and with its center coincident with the axis of rotation. The radius of this circle is the lever arm, and the direction in which a point on the circumference of the circle would be moved by the force is the direction associated with the torque. Thus in Fig. 5-16, the direction of the torque is clockwise, or negative.

Torque may be given a vector interpretation, with the direction of a torque vector being parallel to the axis of rotation. All the torques of Fig. 5-15 can be represented by vectors directed along the axis of rotation, either toward or away from the reader. To make this more definite we use the *right-hand screw rule* to define the direction of a torque

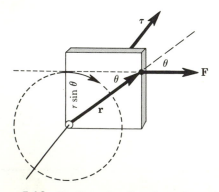

Figure 5-16 Finding the direction of a torque.

vector. We imagine turning an ordinary (right-handed) screw either clockwise or counterclockwise in the direction of rotation produced by the torque. The direction of advance of the screw is defined as the direction of the torque vector. Thus in Fig. 5-16 the torque vector is in the direction shown, away from the reader, which is the direction a screw would advance if turned clockwise.

With this background, we now give a formula for torque. In Fig. 5-16 the vector **r** is drawn from the axis of rotation to the point of application of the force **F**; the magnitude of the lever arm is less than r and is given by $r \sin \theta$, where θ is the angle between the direction of **r** and that of **F**. The magnitude of the torque, force × lever arm, is

$$\tau = (r \sin \theta)(F)$$

and the direction of the torque is perpendicular both to **r** and **F**, as specified by the right-hand rule. To give a single compact formula for both magnitude and direction of the torque we use the *cross product* of two vectors:

$$\tau = \mathbf{r} \times \mathbf{F}$$

where the "cross" multiplication symbol is an abbreviation for the factor $\sin \theta$, and it is understood that the direction of τ is given by the right-hand rule for the advance of a screw rotated from the direction of **r** toward the direction of **F**. (The order of writing the factors is important: $\mathbf{F} \times \mathbf{r}$ is the negative of

$\mathbf{r} \times \mathbf{F}$ because the screw would advance in the opposite direction if turned from $\mathbf{F}$ toward $\mathbf{r}$.) This vector interpretation of torque will be useful in Chap. 8 where we consider rotational dynamics; other cross products will be useful in our study of electromagnetism. In our present study of statics, we do not need the full power of vector methods, because in simple problems the torques are all along one axis and can be represented algebraically as $+$ or $-$. (We did much the same thing in Chap. 3 where velocities and accelerations, all along the same line, were given $+$ and $-$ values.)

For a rigid body to be in equilibrium, two conditions must be satisfied. Net force must be zero, and net torque must be zero.* We can write these conditions in the form of two vector equations:

Translational equilibrium: $\quad \Sigma \mathbf{F} = 0$

Rotational equilibrium: $\quad \Sigma \boldsymbol{\tau} = 0$

These are equivalent to *six* simultaneous equations. $\Sigma \mathbf{F} = 0$ means $\Sigma F_x = 0$, $\Sigma F_y = 0$, and $\Sigma F_z = 0$ (forces along each of three mutually perpendicular directions must add up to zero); and similarly $\Sigma \boldsymbol{\tau} = 0$ means $\Sigma \tau_x = 0$, $\Sigma \tau_y = 0$, and $\Sigma \tau_z = 0$ (torques around each of three mutually perpendicular axes of rotation must add up to zero). However, in most practical problems, we need consider torques about only one axis, since usually all the forces either act in one plane or can be replaced by forces acting in a single plane. If we call this plane the xy-plane, all axes of rotation are parallel to the z axis, and so we have just three equations for equilibrium of a rigid body:

Horizontal forces to the right = horizontal forces to the left:

$$\Sigma F_x = 0$$

---

*Although we have stated the torque condition as based on "experience" (p. 91), it is not an independent law of nature. The torque condition can be derived from Newton's first and third laws, with the additional (very reasonable) assumption that the force between any pair of particles of a rigid body acts along the line joining those two particles.

Upward forces = downward forces:

$$\Sigma F_y = 0$$

Clockwise torques = counterclockwise torques:

$$\Sigma \tau = 0$$

Since we have three equations to be solved simultaneously, each problem could have as many as three unknowns (either forces or angles). The solution of problems in statics involving forces and torques calls for ingenuity and common sense. There are no simple rules of procedure, and the following examples will give an idea of some of the "tricks of the trade." The most common source of error is failure to identify the object whose equilibrium is being considered. You must learn to consider *all* the forces acting *on* the body and to ignore forces *exerted by* the body. Of course, Newton's third law will be of great help when you are given (or asked for) forces exerted by the body. Use the third law to obtain the corresponding forces on the body, and use this information in the solution of the problem.

EXAMPLE 5-8

A load weighing 60 N is to be supported by a force $\mathbf{F}$ applied at the end of a weightless lever, as shown in Fig. 5-17a. What force is necessary? What is the force on the fulcrum when the lever is in equilibrium?

The body whose equilibrium is being considered is the lever. Draw in all the forces acting *on the lever* (including the upward force exerted on the lever by the fulcrum) (Fig. 5-17b).

From $\Sigma \tau = 0$,

Clockwise torques = counterclockwise torques

$$(60 \text{ N})(40 \text{ cm}) = (F)(120 \text{ cm})$$

$$F = \frac{(60 \text{ N})(40 \text{ cm})}{120 \text{ cm}}$$

$$= \boxed{20 \text{ N}}$$

To find the force on the fulcrum, we must first find the force exerted *by* the fulcrum on

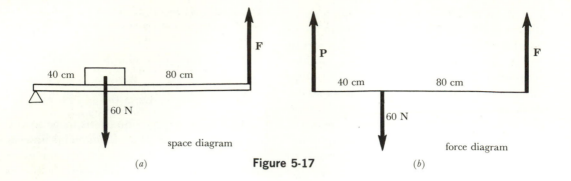

space diagram

(a)

Figure 5-17

force diagram

(b)

the lever. We have just found that $F = 20$ N. From $\Sigma F_y = 0$,

Upward forces = downward forces

$$P + 20 \text{ N} = 60 \text{ N}$$

$$P = 40 \text{ N (upward)}$$

Now by Newton's third law, the force exerted by the lever on the fulcrum is equal and opposite to P.

$$P' = \boxed{40 \text{ N (downward)}}$$

We do not show P' in the diagram because it does not act *on* the lever whose equilibrium we are considering.

EXAMPLE **5-9**

An upraised forearm (Fig. 5-18) is pulling a window shade downward with a force of 10 lb. The biceps muscle is attached to the radius as shown. For simplicity neglect the weight of the radius (about $\frac{1}{4}$ lb) and ignore the effect of the ulna. Calculate the force exerted by the biceps muscle.

It is the radius that we are concerned with, and in the vector diagram we are careful to show only the forces on that bone. We ignore forces due to the radius. Thus at the elbow, **H** and **V** are the unknown horizontal and vertical components of the force exerted jointly (!) by the humerus and the ulna on the radius. The

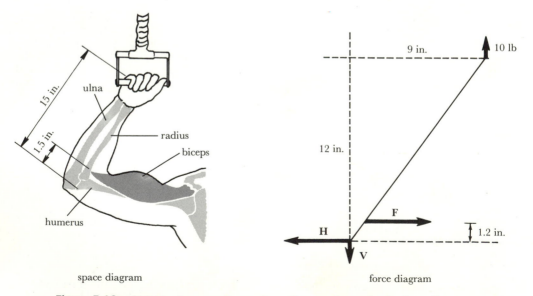

space diagram

force diagram

Figure 5-18 Only the forces acting *on* the radius are drawn in the force diagram.

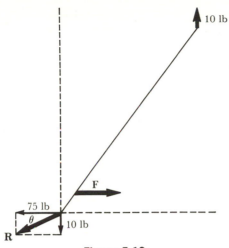

Figure 5-19

resultant force of the elbow on the radius will be downward and to the left. The window shade exerts an upward force of 10 lb.

Although it is actually immaterial where the axis of rotation is assumed to be, in practice a judicious choice of the axis will simplify the work considerably. By putting the axis at the joint, two of the three unknowns (**H** and **V**) have no lever arm and hence exert no torque. The torque equation then involves only one unknown force:

Counterclockwise torque = clockwise torque

$$(10 \text{ lb})(9 \text{ in.}) = F(1.2 \text{ in.})$$

$$F = \boxed{75 \text{ lb}}$$

Now using $\Sigma F_x = 0$ gives

$$H = F = \boxed{75 \text{ lb}}$$

and $\Sigma F_y = 0$ gives

$$V = \boxed{10 \text{ lb}}$$

If we wish, we can combine the two forces due to the joint into a single force (Fig. 5-19):

$$R = \sqrt{H^2 + V^2} = \sqrt{75^2 + 10^2} = 75.7 \text{ lb}$$

$$\tan \theta = \frac{10 \text{ lb}}{75 \text{ lb}} = 0.133$$

$\theta = 8°$ below the horizontal

Either way we look at it there are three unknowns in this problem. Either we are asked for

F, H, and V, or we are asked for F, R, and θ. We are assured of being able to solve the problem, since we have three general equations out of which to solve for any three of the unknowns.

Let us note one interesting feature of the solution. The force of the joint on the radius is not directed along the bone. The bone makes an angle of 53° with the horizontal (the angle whose tangent is $\frac{12}{9}$), while the force of the joint makes an angle of 8° with the horizontal (Fig. 5-19).

5-6 Computation of the center of gravity

In Sec. 5-2 we stated that the downward forces due to the weights of the individual molecules of a body may be "replaced" by a single downward force acting through the center of gravity; the magnitude of this single force is the total weight of the body. The c.g. of a geometrically regular body is at its center (for instance, the c.g. of a uniform meter stick is at the 50 cm mark). The c.g. of an irregular body composed of a few "heavy spots" at definite places can be found by the law of torques.

EXAMPLE **5-10**

A uniform meter stick 100 cm long weighing 3 N is loaded with two small heavy objects as follows: A load of 2 N is at the 30 cm mark, and a load of 4 N is at the 80 cm mark (Fig. 5-20). Where is the center of gravity of the system?

Imagine the meter stick supported horizontally, with an axis through the 0 cm mark.

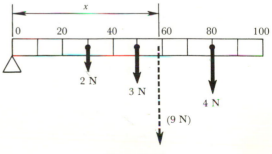

Figure 5-20

We equate the torque due to the weight of the stick and its two loads to the torque that would act if all the weight were at the center of gravity, at an unknown lever arm x.

$$(2 \text{ N})(30 \text{ cm}) + (3 \text{ N})(50 \text{ cm})$$
$$+ (4 \text{ N})(80 \text{ cm}) = (9 \text{ N})(x)$$
$$x = \frac{530 \text{ N} \cdot \text{cm}}{9 \text{ N}} = \boxed{58.9 \text{ cm}}$$

The c.g. of the loaded meter stick is at the 58.9 cm mark.

The initial choice of axis at the 0 cm mark was entirely arbitrary. If a body is in equilibrium about one axis, it must also be in equilibrium if the torques are computed about any other axis. To illustrate this, let us solve Example 5-10 using an axis through the 20 cm mark instead of the 0 cm mark (Fig. 5-21).

$$(2 \text{ N})(10 \text{ cm}) + (3 \text{ N})(30 \text{ cm})$$
$$+ (4 \text{ N})(60 \text{ cm}) = (9 \text{ N})(x)$$
$$x = \frac{350 \text{ N} \cdot \text{cm}}{9 \text{ N}} = \boxed{38.9 \text{ cm}}$$

This is the location of the c.g. *measured from the 20 cm mark*. Physically, it is the same point that was obtained previously, 58.9 cm from the 0 cm mark.

The c.g. of a body having a continuous distribution of weight (such as a semicircular sheet of plywood) can be computed in a similar way, except that, since there are infinitely many weight particles instead of just a few as in Example 5-10, the methods of

integral calculus must be used in the summing up of the torques due to gravity.

5-7 Types of equilibrium

If a body is supported at a single point, it is acted on by two forces: the force of gravity (its weight), acting at the c.g., and the upward force of the pivot. The body will be in equilibrium if its c.g. is somewhere on the vertical line passing through the point of support, for then it is possible for the two forces to balance. There are three possible cases:

1. *Stable equilibrium.* If the c.g. is directly below the point of support, the body tends to return to its original position after being given a slight displacement. When displaced, the weight, acting at the c.g., gives rise to a restoring torque. A typical example of a body in stable equilibrium is a pendulum bob at its lowest point (Fig. 5-22).

2. *Unstable equilibrium.* If the c.g. is directly above the point of support, the body tends to move further away from its original position when given a slight displacement. When displaced, the weight, acting at the c.g., gives rise to a torque which tends to increase the displacement. A typical example of a body

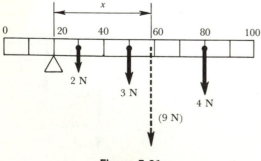

Figure 5-21

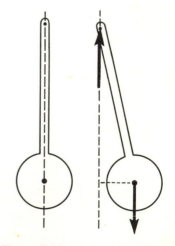

Figure 5-22 Stable equilibrium.

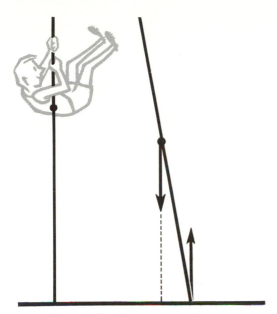

Figure 5-23 Unstable equilibrium.

in unstable equilibrium is a pole vaulter at the very top of his motion (Fig. 5-23).

3. *Neutral equilibrium.* If the c.g. coincides with the point of support, the body remains in whatever position it is placed. The weight, acting at the c.g., produces no torque since the lever arm is zero. A well-balanced wheel on a horizontal axis is an example of a body in neutral equilibrium (Fig. 5-24).

An example of a different type of equilibrium is afforded by a chair in its normal

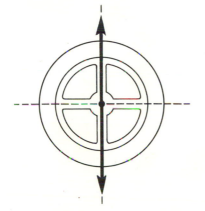

Figure 5-24 Neutral equilibrium.

position. The vertical line through the c.g. intersects the floor at a point which lies within the area defined by the points of support (the legs). The chair is in stable equilibrium with respect to tilting but in neutral equilibrium with respect to sliding along the floor. Unstable equilibrium can easily be obtained (but not maintained) by tilting back the chair.

If a body has uniform velocity in a straight line, it is in dynamic equilibrium. Here too, stable, unstable, and neutral equilibrium can be discussed. As an example of stable dynamic equilibrium, consider a raindrop falling steadily. If for some reason its speed increases, air resistance also increases (the drop strikes more air molecules per second), and so the speed returns to the original value.

5-8 Solution of problems involving center of gravity

In solving problems the weight of a body, really consisting of almost infinitely many tiny forces of gravitation, is replaced by a single force acting at the c.g. From this point of view, discussed more fully in Sec. 5-2, the weight then becomes a single force, to be treated like any other force. A little reflection will show that this is a tremendously important simplification, although one that is usually taken for granted. Our next examples make use of this simplifying concept.

EXAMPLE **5-11**

A 200 lb man climbs to the top of a 20 ft ladder which leans against a smooth wall at an angle of 60° with the horizontal. The ladder weighs 100 lb, and its c.g. is 6 ft from the bottom end (Fig. 5-25). What must be the coefficient of static friction at the ground, if the ladder is not to slip?

In our mind's eye, let us isolate the ladder and consider all forces *on the ladder*. These are: the weight of the man; the weight of the ladder; the push of the wall **P**; and the two compo-

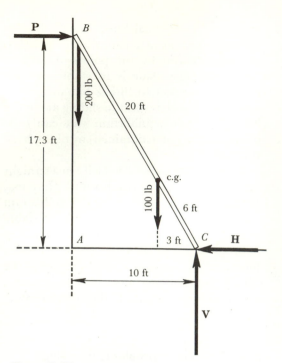

Figure 5-25 Only the forces acting *on* the ladder are drawn in the diagram.

nents of the force of the ground on the ladder, **H** and **V**. Since the wall is smooth, there can be no frictional force parallel to the surface of the wall, so the push **P** must be horizontal, as shown. We next fill in the dimensions in the space diagram, using the length of the ladder and the known 60° angle. For instance, the vertical height AB is found from $AB/20$ ft = sin 60°; $AB = 17.3$ ft.

By taking torques about an axis passing through C, we eliminate the torques due to **H** and **V**.

$$(200 \text{ lb})(10 \text{ ft}) + (100 \text{ lb})(3 \text{ ft}) = P(17.3 \text{ ft})$$

$$P = \frac{2300 \text{ lb} \cdot \text{ft}}{17.3 \text{ ft}} = \boxed{133 \text{ lb}}$$

Now, from $\Sigma F_x = 0$,

$$H = P = \boxed{133 \text{ lb}}$$

and, from $\Sigma F_y = 0$,

$$V = 200 \text{ lb} + 100 \text{ lb} = \boxed{300 \text{ lb}}$$

We interpret H, the force exerted by the ground parallel to the surface of the ground, as the

frictional force. Our solution shows that we need 133 lb of frictional force. Since, by the wording of the problem, the ladder is about to slip, we know that the 133 lb of friction represents the maximum force of static friction. Hence

$$\mu_s = \frac{\text{maximum force of static friction}}{\text{normal force holding surfaces together}}$$

$$= \frac{133 \text{ lb}}{300 \text{ lb}} = \boxed{0.44}$$

There are two points to note in the solution to Example 5-11. When we isolated the ladder, we carefully excluded forces exerted by the ladder. There *is* a horizontal force to the left, at the top of the ladder, but this force acts on the wall, not on the ladder; hence we ignore it. Newton's third law tells us that the ladder pushes against the wall with a force of 133 lb toward the left, since P is 133 lb toward the right. Also, we were able to calculate the coefficient of static friction only because the ladder was just about to slip. If the coefficient of static friction were 0.50, the ladder would still have been in equilibrium,

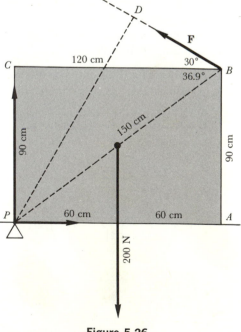

Figure 5-26

and the force of static friction would still have been 133 lb. The maximum possible force of static friction would have been $\mu_s F_n = (0.50)(300 \text{ lb}) = 150 \text{ lb}$, but we would have been getting only as much friction as was actually needed for equilibrium.

Sometimes, as in the following example, it is convenient to resolve all forces into horizontal and vertical components before computing torques.

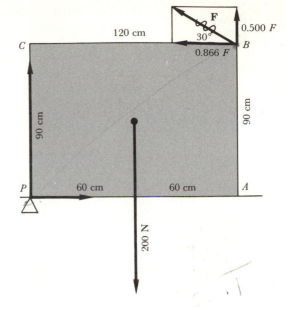

Figure 5-27

EXAMPLE **5-12**

A uniform piece of plywood weighing 200 N is in a vertical plane, with a 120 cm side resting on the floor and a 90 cm side vertical (Fig. 5-26). What force **F** applied as shown will cause the plywood to just start to rotate about the pivot at the corner P?

By use of a little trigonometry we find the lever arm for the force **F** about an axis through P. The diagonal PB is 150 cm, and the angle PBC is found from $\sin PBC = 90 \text{ cm}/150 \text{ cm} = 0.600$; $\angle PBC = 36.9°$. Hence the angle PBD is 66.9°, and $PD = (150 \text{ cm})(\sin 66.9°) = 138 \text{ cm}$. Now the torque equation is used:

$$(200 \text{ N})(60 \text{ cm}) = F(138 \text{ cm})$$

$$F = \frac{12\,000 \text{ N} \cdot \text{cm}}{138 \text{ cm}} = \boxed{87 \text{ N}}$$

Perhaps you will find it easier to solve this problem by first resolving **F** into its compo-

nents before applying the torque equation. The advantage is that now the lever arms are more easily found (Fig. 5-27).

$$(200 \text{ N})(60 \text{ cm}) = (0.866F)(90 \text{ cm}) + (0.500F)(120 \text{ cm})$$

$$12\,000 \text{ N} \cdot \text{cm} = (78F)\text{cm} + (60F)\text{cm}$$

$$= (138F)\text{cm}$$

$$F = \frac{12\,000 \text{ N} \cdot \text{cm}}{138 \text{ cm}} = \boxed{87 \text{ N}}$$

■ SUMMARY

Equilibrium is the condition in which the magnitude and direction of the velocity of a body remain constant. In static equilibrium, the velocity remains zero, and the body is at rest. In dynamic equilibrium, the body moves with constant velocity in a straight line.

For a particle to be in equilibrium, it is necessary for the vector sum of the forces on the particle to be zero. In many cases, it is possible to treat a body which is composed of many particles as a single particle, provided all externally applied forces act or can be considered to act through the center of gravity of the body. This is true because the c.g. is defined as that point of a body through which the resultant of all the tiny parallel forces of gravity passes.

Frictional forces act parallel to surfaces which are in contact and are dependent upon the normal force pushing the surfaces together. For any given normal force, the force of static friction is variable, of whatever amount is needed, up to a maximum force which is equal to the coefficient of static friction $\times$ the normal force. After motion starts, the force of kinetic friction is constant, regardless of the velocity, and is equal to the coefficient of kinetic friction $\times$ the normal force. The force of kinetic friction is usually considerably less than the maximum force of static friction.

Torque, defined as force $\times$ lever arm, is a vector quantity whose direction is given by the right-hand screw rule. The cross product is used to define the magnitude and direction of a torque vector.

For a rigid body to be in equilibrium, two conditions must be fulfilled: the vector sum of the forces on the body must be zero (as for a particle), and also the vector sum of the torques on the body must be zero. In applying the second condition, any convenient axis may be chosen when computing the torques due to the applied forces.

In solving problems in statics, it is vital to be sure just which particle or body it is whose equilibrium is being considered. In applying the conditions for equilibrium, only the forces acting *on* the body are to be considered.

■ CHECK LIST

equilibrium (definition)
static equilibrium (definition)
dynamic equilibrium (definition)
equilibrium of a particle (conditions)
translational motion
rotational motion
torque
lever arm
right-hand screw rule
cross product

center of gravity
equilibrium of a rigid body
   (conditions)
stable, unstable, and neutral equilibrium (definitions and conditions)
coefficient of static friction
coefficient of kinetic friction
normal force
angle of repose

■ QUESTIONS

5-1 A freight train moves along a straight track with a constant speed of 50 ft/s. Is it in equilibrium? Explain why the condition of equilibrium of a particle can be applied to a mile-long freight train.

5-2 A baseball is thrown straight upward, and momentarily comes to rest at the highest point in its path. Is it in equilibrium at this instant?

5-3 Give several examples of stable equilibrium; of unstable equilibrium; of neutral equilibrium.

5-4 Discuss the nature of the equilibrium of the leaning tower of Pisa.

5-5 In Fig. 5-12e what force **F** would cause the block to be in dynamic equilibrium while sliding toward the right? In which direction should the force be applied to cause equilibrium?

5-6 List some beneficial uses of the force of static friction.

5-7 Upon what does the force of friction between two surfaces depend?

5-8 Upon what does the coefficient of friction between two surfaces depend?

5-9 In Sec. 5-2 the c.g. of a uniform, *solid* phonograph record was stated to be at the center. Where is the c.g. of a 45 rpm record which is 17.3 cm in diameter and has a central hole 3.8 cm in diameter?

5-10 A sports car goes over the top of a hill at 70 mi/h. Is the motion of the car one of translation, rotation, or both?

5-11 Give an example of a body which is in motion, yet is in equilibrium. Give an example of a body which is at rest, yet is not in equilibrium.

5-12 What is meant by the statement that statics is a special case of dynamics?

■ PROBLEMS

Note: Force and torque are vector quantities. Hence an answer to a problem in which force or torque is calculated is, in general, incomplete unless both magnitude and direction are specified.

5-A1 A horizontal force of 60 N acts toward the northeast. (*a*) What is the component of this force along an east-west line? (*b*) What is its component along a line running toward a point 15° north of east?

5-A2 A plane in level flight has a velocity of 200 mi/h at an angle of 30° S of W. (*a*) What is the component of this velocity along an east-west line? (*b*) What is the velocity component along a line which runs northeast-southwest?

5-A3 In an experiment in atomic physics, an electron is acted on by two forces: an electric force of 30 units, directed horizontally, and a magnetic force of 40 units, directed upward. What are the magnitude and direction of the net force acting on the electron?

5-A4 A boat is acted on by a force due to the motor of 80 lb directed toward the north, and by an eastward force of 60 lb caused by the wind. What are the magnitude and direction of the net force on the boat?

5-A5 A vertical force of 30 N is required to raise the upper sash of a school window. How much force will be needed if the force is applied by a pole which makes an angle of 25° with the window?

5-A6 Whenever an electric power line follows a bend in a road, a guy wire must be used to prevent the pole from leaning over. Try to observe this in your community. Show by a vector diagram the three forces that act at the top of the pole, and explain how the horizontal component of the force of the guy wire balances the two forces exerted on the pole by the power line.

5-A7 A guy wire 6 m long extends from the top of a telephone pole to a point on the ground 3 m from the base of the pole. The tension in the wire is 800 N. What is the horizontal component of the force on the pole?

5-A8 A 120 lb boy sits on a sled which weighs 15 lb. If the coefficient of kinetic friction between the steel runners and the ice is 0.08, with how much force must a playmate pull horizontally in order to keep the boy and sled moving over the level ice at constant velocity?

5-A9 In the preceding problem, it is found that a horizontal force of 13.5 lb is needed to start the boy and sled moving. What is the coefficient of static friction?

5-A10 With what force must an ice skater push a fellow skater who weighs 250 N in order to start her gliding over the ice? (Use data from Table 5-1.)

5-A11 Suppose that the plywood in Fig. 5-15 weighs something. Is the torque due to the weight of the plywood about the indicated axis a clockwise or counterclockwise torque?

5-A12 Again referring to Fig. 5-15, compute the torque due to the 8 lb force in each case, if the axis passes through the upper right-hand corner of the plywood square.

5-A13 Calculate the torque of the 100 lb force about an axis through B in Fig. 5-25. Is this torque clockwise or counterclockwise? What is the direction of the vector representing this torque?

5-A14 What is the torque exerted by a jeweler using a micro-wrench, if his fingers supply a force of 4 oz at a distance of 3 in. from the axis of rotation? The force is applied in a direction making an angle of 70° with the handle of the wrench.

5-A15 When a rigid body is in equilibrium, $\Sigma\tau = 0$ when the torques are computed about *any* axis. In Example 5-11 the axis was chosen to pass through the bottom of the ladder. Using the values of H, V, and P found in the solution for Example 5-11, verify that $\Sigma\tau = 0$ when torques are computed about an axis through point A.

5-B1 Three horizontal forces act on a particle: 4.0 dyn toward the north, 3.0 dyn toward the east, and 2.0 dyn toward the southwest. Using a scale of 1 dyn = 2 cm or 1 in., lay off the forces head to tail and find the magnitude and direction of the resultant force on the particle. (Use graph paper and a protractor.)

5-B2 Repeat Prob. 5-B1, laying off the vectors in a different order.

5-B3 Repeat Prob. 5-B1, using the component method. Resolve each of the given forces into x- and y-components; find the resultant x-component by addition; find the resultant y-component by addition; and combine these two vectors by the Pythagorean theorem to find the resultant of all the forces. (Express the direction of the resultant with respect to east.)

5-B4 Add graphically the following vectors: 100 lb toward the north, 60 lb toward the west, and 70 lb toward the southwest.

5-B5 Repeat Prob. 5-B4, using the component method. Resolve each of the given forces into x- and y-components; find the resultant x-component by addition; find the resultant y-component by addition; and combine these two vectors by the Pythagorean theorem to find the magnitude and direction of the resultant of all the forces.

5-B6 Solve Example 5-2 by the head-to-tail (triangle) method.

5-B7 A 50 lb traffic light is suspended by two wires as in Fig. 5-28. What is the tension in each wire?

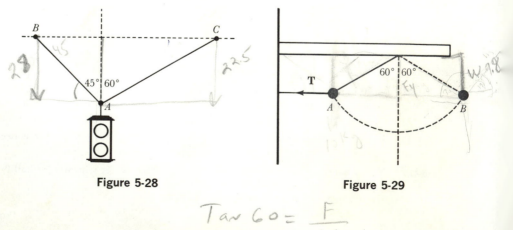

Figure 5-28   **Figure 5-29**

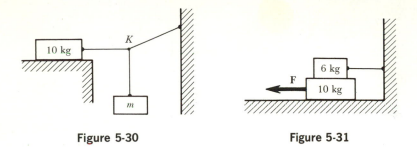

| Figure 5-30 | Figure 5-31 |

5-B8   A light horizontal wire is stretched between two posts 100 ft apart. A bird weighing 1 lb alights at the center of the wire, and the wire sags 0.2 ft. What is the tension in the wire when the bird is sitting on it?

5-B9   In Fig. 5-29, the ball's mass is 10 kg. Calculate the tension T in a horizontal string which holds the ball in position A.

5-B10 Suppose that the trunk in Example 5-2 is moving *down* the plane at a constant speed of 1 m/s. What force F, directed upward parallel to the plane, will allow this to happen?

5-B11 Solve Example 5-2, if the applied force is horizontal instead of parallel to the plane.

5-B12 A 10 kg block on a table for which $\mu_s = 0.4$ is connected to a knot K by a horizontal string (Fig. 5-30). Another block hangs from point K, and a string at 30° goes upward to a fixed wall. What maximum value can m be before block A starts to move to the right?

5-B13 In Fig. 5-31, the coefficients of friction for both surfaces are $\mu_s = 0.6$ and $\mu_k = 0.4$. What force F is needed to keep the bottom block moving to the left with a constant acceleration of 2 m/s²?

5-B14 What force, applied upward and at an angle of 30° with the vertical, is needed to move a 5 lb scrub brush upward with uniform velocity along a vertical wall for which $\mu_k = 0.2$ and $\mu_s = 0.3$?

5-B15 A coin weighing 0.5 N rests on a wooden table top. The table is 100 cm on an edge, and the coefficient of static friction is 0.4. The table is then tilted so that one edge is 60 cm higher than the opposite edge, and the coin begins to slide. With how much force should one press against the coin, perpendicular to the surface of the table, in order to keep it from sliding? (*Hint:* First find out how much frictional force is needed.)

5-B16 The center of gravity of an empty automobile is 3 ft ahead of the rear axle, and the empty car weighs 1500 lb. A 100 lb driver sits 5 ft ahead of the rear axle, and three 200 lb passengers sit 2 ft ahead of the rear axle. How far from the rear axle is the c.g. of the loaded car?

5-B17 A collapsible fishing pole consists of three sections, each 50 cm long and uniform. The sections weigh 6 N, 4 N, and 2 N, respectively. When assembled with the 4 N section in the middle, how far from the heavy end is the c.g. of the system?

5-B18 Tom and Dick are carrying their young friend Harry on a uniform horizontal plank 8 m long and weighing 125 N. Harry, who weighs 300 N, is sitting 5 m from Tom and 3 m from Dick. How much weight does each man support?

5-B19 A 200 lb movie stunt man walks out to the end of a uniform horizontal plank which projects perpendicularly over the edge of a roof. The plank is 20 ft long and weighs 100 lb. How far from the roof can the plank overhang?

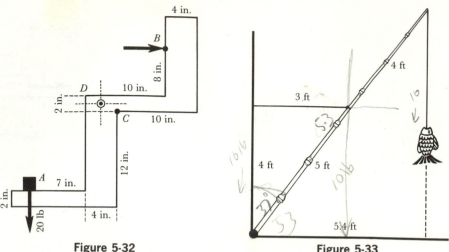

Figure 5-32   **Figure 5-33**

5-B20 The lever of Fig. 5-32 is a flat piece of metal in a vertical plane. The lever weighs 15 lb with its c.g. at C. What horizontal force applied at B will hold the 20 lb load at A in equilibrium?

5-B21 What vertical force applied at B in Fig. 5-32 (replacing the horizontal one) will hold the 20 lb load at A in equilibrium?

5-B22 A trunk weighing 600 N is on a frictionless inclined plane which makes a 30° angle with the horizontal. A force of 400 N is applied, directed upward along the plane.   (*a*) Calculate the net force along the plane.   (*b*) What is the acceleration of the trunk?   (*c*) What is the normal push of the plane upon the trunk?

5-B23 A sledge weighing 400 lb is being dragged along a rough road, for which the coefficient of kinetic friction is 0.7. The forward tension in the rope is 380 lb. Compute the resultant force on the sledge. Is the sledge in equilibrium? If not, what is its acceleration?

5-B24 A mountain climber weighing 800 N is held on a 60° slope by a rope attached to a tree at the top of the peak. The tension in the rope is 500 N. Compute   (*a*) the normal force on the climber;   (*b*) the magnitude and direction of the force of friction on the climber.

5-B25 A box weighing 200 lb is on a rough horizontal pavement for which the coefficient of kinetic friction is 0.3. The box is acted on by gravity and by a horizontal rope in which the tension is 110 lb. How far does the box move in 3 s, starting from rest?

5-B26 A box weighing 200 lb is on a rough horizontal pavement for which the coefficient of kinetic friction is 0.3. The box is acted on by gravity and by a rope in which the tension is 110 lb. The rope makes an angle of 30° with the horizontal. How far does the box move in 3 s, starting from rest?

5-B27 A light bamboo fishing pole 9 ft long is supported by a horizontal string as shown in Fig. 5-33. A 10 lb fish hangs from the end of the pole, and the pole is pivoted at the bottom. What is the tension in the supporting string, and what are the components of the force of the pivot on the pole?

5-B28 A 130 lb cylindrical barrel of diameter 26 in. is lying on a pavement with its curved surface snugly against a curb 8 in. high.   (*a*) Make a careful diagram, and use the Pythagorean theorem to find the distance from the foot of the curb to the bottom of the barrel.   (*b*) What force, applied horizontally at the top of the barrel, is needed to just ease it up off the pavement, so that the barrel is pivoted on the edge of the curb?

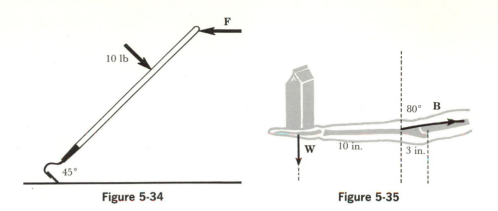

Figure 5-34   **Figure 5-35**

5-B29 A gardener using a hoe 6 ft long applies with one hand a horizontal force of **F** lb at the top of the handle, and with the other hand he applies a force of 10 lb perpendicular to the handle and 2 ft from the top (Fig. 5-34). The handle makes an angle of 45° with the ground. (*a*) What is the value of *F*? (*b*) What is the horizontal (useful) component of the force exerted on the ground by the hoe? (Ignore the weight of the hoe.)

5-B30 A housewife holds a 2 kg carton of milk at arm's length (Fig. 5-35). What force **B** must be exerted by the brachialis muscle? (Ignore the weight of the forearm.)

5-B31 A uniform pole 8 m long weighs 500 N and is attached at one end to a vertical wall. A load of 400 N hangs from the other end of the pole, and a horizontal guy wire attached to the outer end of the pole holds it at an angle of 30° above the horizontal. (*a*) Find the tension in the guy wire. (*b*) What is the resultant force of the wall on the pole?

5-B32 A uniform ladder 5 m long weighing 200 N is leaning against a smooth vertical wall with its base 3 m from the wall. The coefficient of static friction between the bottom of the ladder and the ground is 0.4. How far, measured along the ladder, can a 600 N man climb before the ladder starts to slip?

5-B33 A ladder which is 20 ft long weighs 60 lb, with its c.g. 5 ft from the lower end. The ladder leans against a smooth wall and makes a 30° angle with the horizontal. What coefficient of static friction is needed to keep the ladder from slipping? (*Hint:* First compute the components of the force exerted by the ground on the ladder.)

5-B34 A 100 lb block rests on a rough table for which the coefficient of static friction is 0.40 and the coefficient of kinetic friction is 0.30. The block is counterbalanced as shown, by a 50 lb weight and a weight *W* (Fig. 5-36). For what range of values for *W* can the system remain at rest?

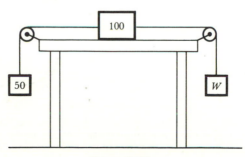

Figure 5-36

5-B35 In Fig. 5-36, assume the weight W to be 10 lb and the coefficient of kinetic friction between the 100 lb block and the table to be 0.25. What speed will the system acquire after moving 4 ft, starting from rest?

5-B36 A uniform rectangular sign 4 ft tall and 8 ft wide weighing 125 lb is held in a vertical plane, perpendicular to a wall, by a horizontal pin through the top inside corner, and by a guy wire which runs from the outer top corner of the sign to a point on the wall 6 ft above the pin. (*a*) Calculate the tension in the wire. (*b*) Calculate the magnitude and direction of the force on the pin.

5-B37 Show that the casting of Example 5-7 will not slip, whatever its mass (other data for the problem remaining the same).

5-C1 Using the definitions of sine and cosine given on p. 704, prove that for any angle $\tan \theta = \sin \theta / \cos \theta$. Check this equation by referring to the table of sines, cosines, and tangents on p. 720, using any angle of your choice.

5-C2 Prove that the tangent of the angle of repose for a block resting on an inclined plane is equal to the coefficient of static friction (see Example 5-5).

5-C3 Tell how the coefficient of kinetic friction is measured by determining the angle of a plane down which a block will slide at constant velocity.

5-C4 Calculate the torque of the 8 lb force in Fig. 5-15*d* if the axis passes through the lower right-hand corner of the plywood square.

5-C5 Given a table with four light legs equally spaced around the circumference of a uniform circular top weighing 200 N, find the smallest weight which when placed on the table will be able to upset it.

5-C6 A four-legged desk weighing 100 lb is being dragged along a rough floor at uniform velocity (Fig. 5-37). The top surface of the desk is 30 in. above the floor, and the distance

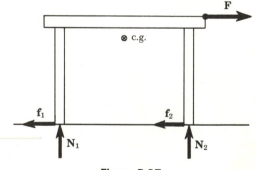

Figure 5-37

between the front legs and back legs is 50 in. The coefficient of sliding friction is 0.20. The c.g. of the desk is several inches below the center of the top, and the force is applied horizontally at the center of one edge. (*a*) What force is needed to drag the desk? (*b*) What is the normal force exerted by the floor on each leg of the desk? (*c*) What are the magnitude and direction of the force of one of the front legs upon the floor?

5-C7 A ball on the end of a light string (Fig. 5-29) is held at a 60° angle by means of the horizontal thread T. The thread is burned, and the ball swings over to position B, also 60° from the vertical. What is the ratio of the tension in the string when the ball is at position A, before burning the thread, to the tension when the ball has swung to position B?

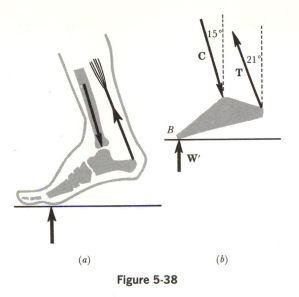

Figure 5-38

5-C8 A 200 lb man stands on tiptoe on one foot so that all his weight is borne by the ground beneath the ball of the foot (Fig. 5-38a). If the foot and ankle are considered as an isolated body (Fig. 5-38b), the three forces which are in equilibrium are the reaction $\mathbf{W}'$ of the ground, the pull $\mathbf{T}$ of the Achilles tendon (exerted by the gastrocnemius muscle), and the compression $\mathbf{C}$ of the tibia. The force $\mathbf{C}$ is downward, at 15° from the vertical, and the force $\mathbf{T}$ is upward, at 21° from the vertical. $\mathbf{W}'$ is an upward force of 200 lb. (a) Make a carefully drawn scale diagram, and estimate the values of C and T. (b) Calculate the values of C and T. (*Hint:* Use two simultaneous equations, one expressing the equilibrium of horizontal components of force, and the other expressing the equilibrium of vertical components of force. An alternative method of solution is to use the law of sines; see the Mathematical Appendix, p. 704.)

5-C9 A cake of ice starting from rest slides down a 20 ft long chute that is inclined 30° to the horizontal; the ice reaches the bottom of the chute with a speed of 23 ft/s. Calculate the value of the appropriate coefficient of friction.

5-C10 A wood block is on a horizontal wood plank (see Table 5-1). The plank is tilted until the block starts to move. The plank is then held at this angle; how far down the plank does the block move in 2 s?

5-C11 A small boy is playing inside a horizontal empty cylindrical storage tank of radius 10 m. If the coefficient of static friction between his sneakers and the tank surface is 0.51, how high above the bottom of the tank (measured vertically) can the boy climb without slipping?

5-C12 A dog of weight W is on icy ground (coefficient of friction 0.05) and sees a squirrel in a tree 20 ft away. How long will it take him to reach the tree, if he accelerates as rapidly as possible?

5-C13 Two ropes support a bag of feed which weighs 100 lb. The ropes are perpendicular to each other, and one rope has twice the tension of the other. Find the tension in each rope, and the angle that each rope makes with the vertical.

5-C14 Three small masses are glued to a sheet of light cardboard. m_1 is at (x_1, y_1), m_2 is at (x_2, y_2), and m_3 is at (x_3, y_3). Prove that the x coordinate of the c.g. is given by $x_{c.g.} = (m_1 x_1 + m_2 x_2 + m_3 x_3)/(m_1 + m_2 + m_3)$. What is $y_{c.g.}$?

For Further Study

5-9 Center of mass

Let us look a little more closely at our statement of Newton's second law (Sec. 4-3). We used the purposely vague term "body" when we said that "the acceleration produced by an unbalanced force acting on a body is proportional to the magnitude of the force ...." In actual practice we do not often deal with particles, but rather with large collections of particles on which various forces are exerted, and which may exert various forces upon each other. For instance, a book falls through the air after being dropped from some height. In Fig. 5-39 we show just 7 of the 10^{23} or so atoms in the book, and we show the externally acting forces on these atoms: due to gravity, and due to air resistance. We can combine all these small forces into one *net force* ($\Sigma \mathbf{F}$) which is vertical, having magnitude $W - f$. We can also combine the masses of all the molecules into a total mass (Σm). It can be shown that there exists a point called the *center of mass* (c.m.), such that, if the net external force $\Sigma \mathbf{F}$ is applied at the c.m., then the c.m. moves just as would a *particle* of mass Σm if acted on by a force $\Sigma \mathbf{F}$. Thus the center of mass is *defined* as that point at which all the *mass* of a body could be concentrated, when using Newton's second law. This definition is not to be confused with that for the center of gravity, which is that point at which all the *weight* of a body could be concentrated, when applying the torque condition or when calculating the resultant force of gravity.

It is for such reasons that we have been free and easy about treating extended bodies (trains, cars, balls, etc.) as if they were particles. Such treatment is valid; although we omit the proof, we use the concept of center of mass implicitly throughout the book. For instance, a bomb bursts in mid-air (Fig. 5-40). At B, fragments of the bomb start to move in various directions. The fragments are of unequal mass. However, the c.m. of the fragments continues to move along the original parabolic path. Since no new *external* force has been applied to the bomb, the fragments behave just as if the force of gravity were still acting on the total mass concentrated at the c.m. of the system. Newton's first law (a special case of the second law) also should be stated in terms of center of mass: "The center of mass of a body remains at rest, or moves with uniform velocity in a straight line, unless acted on by an unbalanced external force."

For all practical purposes, the c.m. is located at the same point as the c.g. However, this is strictly true only if the body is in a *uniform* gravitational field. Consider a uniform cube of stone (a mountain) one mile on an edge. The c.m. is at the exact geometrical

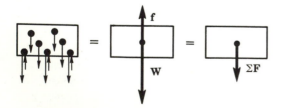

Figure 5-39 Forces acting on the molecules of a falling body, equivalent to a single force $\Sigma \mathbf{F}$, acting at the center of mass of the body.

Figure 5-40 After the explosion, the center of mass of the fragments continues to move in a parabolic path.

center of the cube, but the c.g. is slightly below the center, since the bottom half of the cube is closer to the center of the earth and consequently weighs more than the upper half. Fine points such as these help sharpen our thinking, and help point out the real difference in concept between center of mass and center of gravity.

■ PROBLEMS

5-C15 A body (A) of mass 3 kg is on a smooth horizontal floor at the origin of coordinates, and a body (B) of mass 5 kg is on the x axis, 8 m east of A. (*a*) Where is the c.g. of the system? (*b*) If a force of -6 N acts on A for 4 s, and a force of $+5$ N acts on B for 4 s, where is the new c.g.? (*c*) Calculate the acceleration of the c.g., using $s = \frac{1}{2}at^2$. (*d*) Verify for this case that the c.m. is at the same point as the c.g. by showing that the acceleration found in (*c*) is also given by $(\Sigma F)/(\Sigma m)$.

5-C16 State Newton's second law in a general form which applies to the motion of the center of mass of a system of particles.

■ REFERENCES

1 Steinman, David B., "Bridges," *Sci. American,* Nov., 1954, p. 60. Excellent photographs, including one of the Tacoma Narrows bridge in the act of collapsing.

2 Strait, L. A., V. T. Inman, and H. J. Ralston, "Sample Illustrations of Physical Principles Selected from Physiology and Medicine," *Am. J. Phys.,* **15,** 375 (1947). Also *Am. J. Phys.,* **19,** 173 (1951). An analysis of the action of the human body, considered as a system of levers and forces applied through muscles and tendons.

3 Sutton, R. M., "Two Notes on the Physics of Walking," *Am. J. Phys.,* **23,** 490 (1955).

4 Palmer, Frederic, "What About Friction?" *Am. J. Phys.,* **17,** 181 (1949); **17,** 327 (1949); **17,** 336 (1949).

5 Maney, Charles A., "Experimental Study of Sliding Friction," *Am. J. Phys.,* **20,** 203 (1952).

6 Palmer, Frederic, "Friction," *Sci. American,* Feb., 1951, p. 54.

7 Rabinowicz, Ernest, "Stick and Slip," *Sci. American,* May, 1956, p. 109.

8 Magie, W. F., *A Source Book in Physics* (New York: McGraw-Hill, 1935), p. 22 (also, Cambridge: Harvard University Press, 1963). An early treatment of the inclined plane, by Stevin.

In this chapter we shall study two of the three[*] great conservation laws of mechanics which deal with isolated systems (systems not acted on by external forces). Wherever possible we use conservation laws as statements of basic and general truths about the physical world. They are indeed of great philosophical significance for an understanding of the symmetries of space and time. A more practical reason for using conservation laws is that they apply to whole *systems* of particles, making it possible to get useful results without being overwhelmed by detailed consideration of every particle or part of the system individually.

The law of conservation of momentum was known to Newton, and it is still considered to be valid, even in areas where Newton's original formulation of the laws of mechanics has been replaced by Einstein's more inclusive theory. As a general principle, the law of conservation of momentum continues to be of utmost value in many fields of physics and engineering, including space navigation, quantum mechanics, and relativity.

[*]Conservation of angular momentum is the subject of Sec. 8-8.

Another unifying principle, the law of conservation of energy, was harder to come by. Not until about 1845 was it realized that the energy content of an isolated system remains constant. For example, when a falling ball approaches the earth, and the earth moves slightly to meet the ball, the system (earth + ball) has a constant total amount of energy; the energy of the system is simply transformed from one kind into another during the process. Similarly, if a sufficiently broad definition of energy is used, the total energy of a system is conserved during a chemical reaction between two atoms, or during the collision of two particles in a nuclear reactor.

6-1 Definition of momentum

The momentum of a moving body is so fundamental that early writers such as Newton called it "quantity of motion" or simply "motion." We define the linear momentum of a body as the product of its mass and its velocity:

$$\text{Momentum} = (\text{mass})(\text{velocity})$$

Since velocity is a vector quantity, momen-

Conservation of energy at the bowling alley: some of the kinetic energy of the ball has been transformed into potential energy of the pins which have been raised up. (American Machine and Foundry Company)

tum is also a vector quantity, having the same direction as the velocity of the body. We use the symbol **p** for linear momentum and write

$$\mathbf{p} = m\mathbf{v}$$

In many problems concerned with motion in a straight line a separate symbol for momentum is not needed, and the vector nature of momentum can be taken care of by writing $+mv$ and $-mv$. The dimensions of momentum are

$$[\text{momentum}] = [\text{M}][\text{LT}^{-1}] = [\text{MLT}^{-1}]$$

There are no special names for units of momentum; kg·m/s or g·cm/s may be used—any mass unit multiplied by a velocity unit.

6-2 Newton's second law in terms of momentum

We may restate Newton's second law in terms of change of momentum. In fact, Newton's own statement of the law involved rate of change of "motion," which was his term for momentum.

Let us consider the special case of a constant force, which gives rise to a constant acceleration. As we saw in Chap. 3, for constant acceleration the equation

$$\mathbf{a} = \frac{\mathbf{v} - \mathbf{v}_0}{t - t_0} = \frac{\Delta\mathbf{v}}{\Delta t}$$

is valid for a time interval Δt of any size, however large.

$$\text{net } \mathbf{F} = m\mathbf{a}$$

$$\text{net } \mathbf{F} = m\frac{\mathbf{v} - \mathbf{v}_0}{t - t_0} = m\frac{\Delta\mathbf{v}}{\Delta t}$$

$$\text{net } \mathbf{F} = \frac{m\mathbf{v} - m\mathbf{v}_0}{t - t_0} = \frac{\Delta\mathbf{p}}{\Delta t} \qquad (6\text{-}1)$$

Thus *net force equals the rate of change of momentum.** Multiplying both sides by the

*If the force is not constant, but is instead some function of time, the conclusion is still true that net force equals

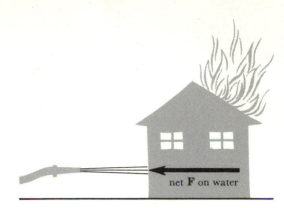

Figure 6-1

time interval Δt, we obtain

$$(\text{net } \mathbf{F})\,\Delta t = m\mathbf{v} - m\mathbf{v}_0$$
$$= \mathbf{p} - \mathbf{p}_0 = \Delta\mathbf{p} \qquad (6\text{-}2)$$

This product of net force and the time during which it acts is defined as *impulse;* Eq. 6-2 tells us that the impulse equals the change of momentum of a body. Impulse is measured in units of force multiplied by time, such as N·s, dyn·s, or lb·s. We see from Eq. 6-2 that these units are also suitable for momentum. As far as total change of momentum is concerned, a large force acting for a short time may produce the same effect as a small force acting for a long time. In combination with Newton's third law, we have here an interpretation of several common phenomena, such as the impact of a stream of water, the force on the piston of an automobile cylinder, the propulsion of a jet rocket, and even the pressure due to the impact of gas molecules on the wall of a container.

EXAMPLE 6-1

A fire hose sends 20 kg of water each second on a burning building (Fig. 6-1). If the water leaves the nozzle at 60 m/s and does not bounce back

the rate of change of momentum. To prove this we take the limit of $\Delta\mathbf{p}/\Delta t$ as $\Delta t \to 0$ and obtain a derivative: net $\mathbf{F} = d\mathbf{p}/dt$. Similarly, Eq. 6-2, in the limit, leads to a definite integral for impulse:

$$\int_{t_0}^{t}(\text{net } \mathbf{F})\,dt = m\mathbf{v} - m\mathbf{v}_0.$$

from the wall, what is the force (assumed constant) on the wall of the building?

In 1.00 s a certain amount of momentum of the water is canceled; to do this, the wall exerts a force to the left.

$$\text{net } F = \frac{mv - mv_0}{\Delta t}$$

$$= \frac{(20 \text{ kg})(0) - (20 \text{ kg})(60 \text{ m/s})}{1.00 \text{ s}}$$

$$= -1200 \text{ kg} \cdot \text{m/s}^2 = -1200 \text{ N (to the left)}$$

This is the force of the wall on the water. By the third law, the force of the water on the wall is equal and opposite:

$$\boxed{1200 \text{ N to the right}}$$

EXAMPLE 6-2

The force of an explosion on the piston of an engine is due to the change of momentum of the gas particles. What must be the speed of the molecules of gas, if 0.6 g of gas exerts a force of 1400 N on the piston, in an explosion that lasts 0.001 s?

To have a consistent set of mks units, we express the mass of 0.6 g as 0.6×10^{-3} kg. Each molecule has its velocity changed from $+v$ to $-v$ by the piston, and the change of velocity is therefore $\Delta v = v - (-v) = 2v$.

$$\text{net } F = \frac{\Delta(mv)}{\Delta t} = \frac{m \, \Delta v}{\Delta t}$$

$$1400 \text{ N} = \frac{(0.6 \times 10^{-3} \text{ kg})(2v)}{10^{-3} \text{ s}}$$

$$v = \boxed{1170 \text{ m/s}}$$

EXAMPLE 6-3

A rocket of mass 10^4 kg, starting from rest, is acted on by a net force of 2×10^5 N for 20 s. What is the final velocity of the rocket?

$$F \, \Delta t = mv - mv_0$$

$$(2 \times 10^5 \text{ N})(20 \text{ s}) = (10^4 \text{ kg})(v) - 0$$

$$v = \boxed{400 \text{ m/s}}$$

In practice, the mass of the rocket would decrease as fuel is used up; our solution is approximate, based on an *average* mass of 10^4 kg.

Do not imagine that a rocket depends upon the presence of air for its operation. It does not need air to "push against" and would actually work better in a vacuum. It is the molecules of ejected gas that push forward on the rocket. The gas is accelerated backward by a force exerted by the rocket, and Newton's third law tells us that the gas molecules must push forward on the rocket. It is this reaction force that propels the rocket. The idea of jet propulsion is not new. The first steam engine of recorded history was that of the Greek philosopher Hero (130 B.C.). Steam escaping through the nozzles of his device was the ancestor of the exhaust escaping from a jet bomber. In modern times certain lawn sprinklers operate on the same principle. In some models a gear system is even provided to convert some of the rotational energy into linear KE, and the sprinkler "walks" across the lawn. The locomotion of the squid, a cephalopod, depends largely upon the reaction force of ejected liquid, and the locomotion of an astronaut during a "space walk" depends upon reaction force of ejected gas.

6-3 Conservation of momentum

To derive the law of conservation of momentum, we shall consider first an isolated system consisting of just two bodies. Suppose A and B, of mass m_1 and m_2, respectively, are traveling toward the right at velocities $\mathbf{v}_1$ and $\mathbf{v}_2$, with A overtaking and colliding with B (Fig. 6-2). When they collide, A exerts a force $(+\mathbf{F})$ on B, and B exerts an equal and opposite force $(-\mathbf{F})$ on A. The bodies receive different accelerations, since their masses are different. Suppose that the duration of the collision is Δt. Then, as we have shown in the previous section, $-\mathbf{F} \, \Delta t$ equals the change in momentum of body A, and $+\mathbf{F} \, \Delta t$ equals the change in momentum of body B. These changes are numerically equal, since the forces are equal and opposite and the time

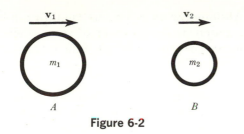

Figure 6-2

initial velocities (before collision) and $\mathbf{V_1}$ and $\mathbf{V_2}$ for the final velocities (after collision):

$$m_1\mathbf{v_1} + m_2\mathbf{v_2} = m_1\mathbf{V_1} + m_2\mathbf{V_2}$$

If all motions take place along the same line, we can use $+$ and $-$ signs to designate directions and vector notation is not needed for straight-line collision problems.

of contact is the same. Thus the gain in momentum of one body exactly equals the loss of momentum of the other body; the total momentum of the system has remained constant during the interaction between the two bodies.

It is not hard to extend the proof to include a system containing any number of interacting bodies, for instance, a collection of 10^{23} molecules in a tank of compressed air. No part of the system can undergo a change of momentum unless another part undergoes an equal and opposite change; the justification for this statement is Newton's third law. In general, we state the law of conservation of momentum as follows:

The total linear momentum of an isolated system of bodies remains constant.

The importance of this law arises from its generality. It is true whether or not some mechanical energy has been transformed into heat energy, sound energy, etc., as in an inelastic collision (see Sec. 6-11). It is true for any number of bodies making up a system. For instance, when a grenade explodes in mid-air, the fragments move in various directions with various speeds, but the vector sum of the momentum of all the fragments equals the momentum of the original body. The reason for the quite general validity of the law of conservation of momentum is that it follows from Newton's third law, which is always true.

To express mathematically the fact that the total momentum of the two bodies remains constant, we use $\mathbf{v_1}$ and $\mathbf{v_2}$ for the

EXAMPLE 6-4

A car of mass 600 kg traveling at 20 m/s collides with a stationary truck of mass 1400 kg. The two vehicles are locked together after collision; what is their combined velocity?

After the collision the car and truck form a single object of combined mass 2000 kg.

Total momentum before impact
= total momentum after impact

$$m_C v_C + m_T v_T = m_{C+T} v_{C+T}$$

$$(600 \text{ kg})(20 \text{ m/s}) + (1400 \text{ kg})(0)$$
$$= (2000 \text{ kg})(v_{C+T})$$

$$v_{C+T} = \boxed{6 \text{ m/s}}$$

EXAMPLE 6-5

What is the recoil velocity of a 3 kg gun which fires a bullet of mass 0.060 kg at a velocity of 200 m/s? Let v stand for velocities before firing; V for velocities after firing.

$$m_G v_G + m_B v_B = m_G V_G + m_B V_B$$

$$(3 \text{ kg})(0) + (0.06 \text{ kg})(0)$$
$$= (3 \text{ kg})(V_G) + (0.06 \text{ kg})(200 \text{ m/s})$$

$$V_G = \boxed{-4 \text{ m/s}}$$

The law of conservation of momentum can be used as the basis of an experiment to measure mass. In a collision between two objects, let $\mathbf{v_1}$ and $\mathbf{v_2}$ be their respective velocities before collision, and let $\mathbf{V_1}$ and $\mathbf{V_2}$ be their velocities after collision.

$$m_1\mathbf{v_1} + m_2\mathbf{v_2} = m_1\mathbf{V_1} + m_2\mathbf{V_2}$$

$$m_1(\mathbf{v_1} - \mathbf{V_1}) = m_2(\mathbf{V_2} - \mathbf{v_2})$$

$$\frac{m_1}{m_2} = \frac{\mathbf{V_2} - \mathbf{v_2}}{\mathbf{v_1} - \mathbf{V_1}} \qquad (6\text{-}3)$$

Conservation of momentum **113**

The velocities $\mathbf{v}_1, \mathbf{v}_2, \mathbf{V}_1, \mathbf{V}_2$ can be measured (for instance, with a movie camera); hence we have found one way of measuring the mass m_2 in terms of some "standard" mass m_1 such as the standard kilogram. As we said in Sec. 4-3, *to measure a property is to define it*; the kind of mass we are talking about here is inertial mass (Sec. 4-8), and we may say that the law of conservation of momentum provides a way of *defining* inertial mass. Applying these ideas to the straight-line collision described in Example 6-4, we have $v_1 = 20$ m/s, $v_2 = 0$, $V_1 = 6$ m/s, $V_2 = 6$ m/s. Therefore,

$$\frac{m_1}{m_2} = \frac{6-0}{20-6} = \frac{6}{14}$$

If an observer had measured only the velocities, knowing nothing at the start about the masses, he would have been able to deduce the *ratio* of the two masses, and he would have found $6:14$, as is actually the case (600 kg : 1400 kg).

6-4   Work

Whatever the literary or biological usages of the term may be, in a physical sense work is done only if a force $\mathbf{F}$ causes a displacement $\mathbf{s}$ of a particle or other object. It is only when an object is moved by a force which has a component acting in the direction of motion that work is done; if θ is the angle between the directions of $\mathbf{F}$ and $\mathbf{s}$, the magnitude of the component of $\mathbf{F}$ parallel to $\mathbf{s}$ is $F \cos \theta$. The work done during a displacement is defined as a product:

Work = (force component)(displacement)

$$W = (F \cos \theta)(s)$$

If, as is often the case, the force is in the same direction as the displacement, then $\theta = 0$, $\cos \theta = 1$; in this special case we can use the simpler formula

$$W = Fs$$

Although work is a scalar quantity, it is calculated as the product of two vectors. This type of product is called the *dot product*:

$$W = \mathbf{F} \cdot \mathbf{s}$$

The "dot" multiplication symbol includes the factor $\cos \theta$. The equation $W = \mathbf{F} \cdot \mathbf{s}$ reduces to $W = Fs$ if $\mathbf{F}$ and $\mathbf{s}$ are parallel.

In any system of units, work units are products of the force unit and the length unit; in the metric system certain work units have received special names. In the mks system, work and energy are measured in newton·meters (joules); and one *joule* (J) is the work done when a body is moved one meter against an opposing force of one newton. Similarly, in the cgs system, one *erg* (one dyne·centimeter) is the work done when a body is moved one centimeter against an opposing force of one dyne. No special name is given to the *foot·pound*, which is the work done when a body is moved one foot against an opposing force of one pound. As with all mks and cgs units, there is a simple relationship (involving only powers of 10) between the joule and the erg:

$$\begin{aligned} 1\,\mathrm{J} &= (1\,\mathrm{N})(1\,\mathrm{m}) \\ &= (10^5\,\mathrm{dyn})(10^2\,\mathrm{cm}) \\ &= 10^7\,\mathrm{dyn \cdot cm} = 10^7\,\mathrm{ergs} \end{aligned}$$

A joule is thus 10 million ergs.

Whatever the system of units, the *dimensions* of work are the same: [work] = [F][L], or [work] = $[\mathrm{MLT}^{-2}][\mathrm{L}] = [\mathrm{ML}^2\mathrm{T}^{-2}]$. Note that the mass or weight of the object moved is of no consequence for the definition of work, unless it so happens that the force of gravity is the force against which the work is being done. In general, the only requirement for work to be done is that a force be exerted and something or some point be moved in a direction parallel to the force or some component of the force. Newton's third law is important here; whenever work is done by a force, something exerts an equal and opposite force on the agent that does the work. Thus we can equally well think of work as being done *by* an applied force or *against* an opposing force. The two viewpoints are equivalent.

EXAMPLE 6-6

How much work is done by gravity on a sled weighing 200 N which slides 40 m down a hill (measured along the road) whose angle with the horizontal is 30°?

The angle between **F** and the direction of the displacement is 60° (Fig. 6-3); hence the component of force in the direction of the displacement is

$$F \cos \theta = (200 \text{ N})(\cos 60°)$$
$$= (200 \text{ N})(0.500) = 100 \text{ N}$$

The work done by gravity while the sled slides 40 m is

$$W = (100 \text{ N})(40 \text{ m}) = \boxed{4000 \text{ J}}$$

6-5 Work done against a variable force

In defining work by the equation $W = (F \cos \theta)s$, we have tacitly assumed that the force **F** is constant in magnitude and direction throughout the displacement **s**. This is far from being true in many actual situations, even when $\cos \theta = 1$. For instance, when an astronaut is lifted to an altitude of 200 km, the force of gravity against which work is done varies according to the inverse-square law, and so the work to lift him cannot be

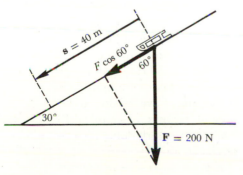

Figure 6-3

(figure labels: $s = 40$ m, $F \cos 60°$, 60°, 30°, $F = 200$ N)

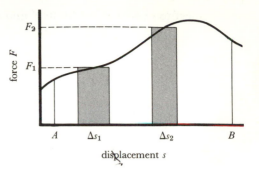

Figure 6-4 Work done during displacement against a variable force.

(figure axis labels: force F, F_2, F_1, A, Δs_1, Δs_2, B, displacement s)

computed from $(F \cos \theta)s$. The value of F changes during the displacement. More prosaically, we can ask how much work is done in pushing a thumbtack into a bulletin board, if the opposing force due to the board is not constant.

We can use a graphical interpretation to approach this problem. Suppose the force varies in some fashion (Fig. 6-4) as the displacement increases (for simplicity we assume that the force and displacement are parallel to each other). During any short displacement interval Δs_1, the force is practically constant in magnitude, and the work done is approximately $F_1 \Delta s_1$. This work is numerically equal to the area of the shaded rectangle shown on the graph. For another short interval Δs_2, the work equals the area $F_2 \Delta s_2$. For the whole motion from $s = A$ to $s = B$, the total work is approximately equal to the sum of the areas of all the rectangles. You will recognize that this involves much the same technique as was used in Sec. 3-5 for finding the displacement of a car which has a variable velocity (see Fig. 3-5). In the limit, as each Δs approaches zero and the number of rectangles approaches infinity, the sum of the rectangular areas approaches the area under the curve. Thus *the work done is equal to the area under the force-displacement curve.* Using the language of calculus, we can say that the work to move a body from $s = A$ to $s = B$ is the definite integral of F

with respect to s between the limits A and B:

$$W = \lim_{\substack{n \to \infty \\ \Delta s \to 0}} \sum_{i=1}^{n} F_i \, \Delta s_i$$

or

$$W = \int_A^B F \, ds$$

EXAMPLE **6-7**

A car is pushed by a forward force which varies according to the graph in Fig. 6-5, where the curved portion when plotted to the scale shown is one-quarter of a circle with center at D. What work is done in moving the car 7 m, starting from $s = 0$?

The area of a circle of radius R is πR^2, which is $\frac{1}{4}\pi$ times the area of a circumscribing square. Hence the area under the quadrant AC is $\frac{1}{4}\pi$ times the area $ABCD$.

$$W = W_1 + W_2$$

$$= \int_{s=0}^{s=4} F \, ds + \int_{s=4}^{s=7} F \, ds$$

$$= \tfrac{1}{4}\pi(30\ \text{N})(4\ \text{m}) + (30\ \text{N})(3\ \text{m})$$

$$= 94\ \text{N·m} + 90\ \text{N·m} = \boxed{184\ \text{J}}$$

6-6 Energy

Whenever work is done on a body, it gains *energy*. Thus if 7 J of work is done on a body, the body has gained 7 J of energy. We may speak of the energy of a body as "stored work." It is possible to distinguish between two kinds of mechanical energy, depending

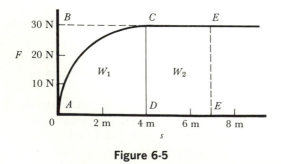

Figure 6-5

on how the work was done on the body. *Potential energy* is the energy of a system of bodies due to the relative position of the parts of the system. We often say that potential energy (PE) is "due to position," but some examples will make it clear that it takes at least two bodies (i.e., a system) in order to have PE. When a bricklayer gives a brick PE by lifting it, he does work against gravity. The bricklayer performs work on a system, and two bodies are involved—brick and earth. When you wind your watch, you change the relative position of the various atoms of iron in the watch spring. The spring (a system of atoms) is distorted, and you have done work against elastic forces. Chemical and electrical phenomena also illustrate the idea that PE involves a system of bodies. A lump of coal is said to have PE; we can burn coal in a steam engine and obtain some mechanical work from it. Note, however, that the chemical PE is that of a system: carbon atoms and oxygen atoms. The chemical equation

$$C + O_2 = CO_2 + \text{heat energy}$$

is symmetrical as regards carbon and oxygen, and shows that carbon by itself has no more chemical PE than does oxygen by itself. The *system* has PE because the atoms are separated from each other, and energy is released when the atoms, under the influence of electric forces of attraction, rearrange their positions to form CO_2. When a parallel-plate capacitor is charged and stores energy, the PE of the charges is due to their relative positions. Some free electrons have been moved from one plate to the other, so that there is an excess of negative charge on one plate and a deficiency of negative charge on the other. There are many ways in which a system may have PE due to the relative positions of its parts; mechanical PE is, however, confined to work done against gravitational or elastic forces.

Kinetic energy is the energy a body has because of its motion. To give velocity to a body, it must be accelerated. By Newton's second law this requires a force, and by Newton's third law an equal and opposite force (called the reaction force, or the force of inertia) acts back on whatever applies the forward force. Therefore, to give a body a velocity, work must be done to move it against the reaction force. All bodies in motion possess kinetic energy (KE). The energy of the winds and the seas is kinetic; work done to speed up the molecules is stored as KE. Heat energy, to the extent to which it is due to random motions of the molecules within a body, thus is really KE.

It is important to understand that no work is done if there is no motion, no matter how much force is applied. If a 10 ton boulder rests for 100 000 years on top of another rock, a force is exerted, but no work is done. From a physical viewpoint no work is required to hold a 3 lb book at arm's length, since there is no displacement.* Also, no work is done if the force is applied perpendicularly to the direction of motion. Thus if a stationary sled weighing 200 N is on frictionless level ice ($\mu_k = 0$) and a child pulls with a horizontal force of 3 N, the sled starts to move. By Newton's third law, the sled pulls back with a force of 3 N against the child, and, if the child maintains a steady 3 N force until the sled moves 4 m, the work done by the child on the sled is (3 N)(4 m) = 12 J. The sled therefore acquires 12 J of KE. No work is done against gravity, however, since the 200 N force of gravity acts downward, perpendicularly to the motion; in fact, the weight of the sled has no bearing on the situation so far. If now the sled comes to a level stretch of ice having $\mu_k = 0.015$, the frictional force is $\mu_k F_n = (0.015)(200 \text{ N}) =$

---

* No work is done on the book; however, work is done within the muscles. It is impossible for a normal muscle to "lock in" and become rigid. Muscles contract slightly, then release, then contract again, and so on indefinitely as long as force is applied by the muscle. Work, in a physical sense, is done since these displacements are in the direction of the force. This energy is dissipated within the muscle.

3 N, and the child's forward force is balanced by the force of friction. The sled is in equilibrium, and its already acquired velocity remains constant. As the sled moves another 4 m, the child, still exerting 3 N of force, does another 12 J of work. This time the work is done against the force of friction and becomes heat energy as the ground and sled become slightly warmer. Finally, work *will* be done against gravity if the sled enters upon a gently sloping frictionless hill which rises 1.5 m for every 100 m measured along the ground (Fig. 6-6). The downhill component of the force of gravity on the sled is 3 N, and the sled is in equilibrium (hence the velocity remains constant). As the child pulls the sled 4 m, he again does 12 J of work, this time against gravity. The sled gains 12 J of PE. From start to finish the child has done 36 J of work; at the end of the trip the sled has 12 J of KE and 12 J of PE, and there is 12 J of heat energy distributed between the sled and the ground.

6-7 Formulas for mechanical energy

A professional engineer's handbook contains many formulas, to cover applications too numerous and specialized to be worth remembering. This is as it should be, but our

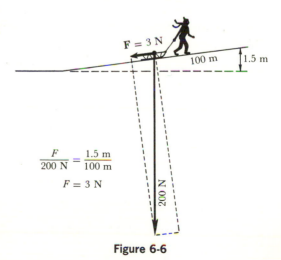

Figure 6-6

use of formulas is quite different. We derive (and use) only a few formulas, chosen to throw into prominence the essential physical factors upon which energy, for instance, depends. Formulas such as we are about to derive are basic and are used over and over again.

If a body of mass m is raised vertically a height h, the force of gravity is mg; hence the work done is mgh. (We assume that the body remains near the surface of the earth, so that **g** is constant in magnitude and direction.) The formula is true in any system of units, but in discussing gravitational PE we prefer to emphasize weight rather than mass. Therefore we write the force of gravity as W, the weight of the body, and the work to raise it a vertical distance h simply as Wh. Thus we have the formula

$$\text{Gravitational PE} = mgh = Wh \qquad (6\text{-}4)$$

$$J = (N)(m)$$

$$\text{erg} = (\text{dyn})(\text{cm})$$

$$\text{ft}\cdot\text{lb} = (\text{lb})(\text{ft})$$

Whether or not the object is lifted straight up, the factor h in Eq. 6-4 refers to the *vertical* rise measured from some reference level where the gravitational PE is assumed to be zero. In any practical problem it is always *change* in PE that is required, so the reference level for zero PE can be quite arbitrary— sea level, the floor of a room, or the top of a table.

EXAMPLE **6-8**

A roller coaster car of mass 3000 kg proceeds from point A (Fig. 6-7) to point B and then to C. What is its PE at C? How much PE did the car lose in going from A to C?

First we note that the weight of the car is 3000×9.8 N. At C the PE (relative to the ground level) is

$$\text{PE} = (3000 \text{ kg})(9.8 \text{ m/s}^2)(15 \text{ m})$$

$$= 4.3 \times 10^5 \text{ N}\cdot\text{m} = \boxed{4.3 \times 10^5 \text{ J}}$$

In going from A to C, the car falls 18 m.

$$\Delta\text{PE} = W(h_2 - h_1)$$

$$= (3000 \text{ kg})(9.8 \text{ m/s}^2)(18 \text{ m})$$

$$= 5.3 \times 10^5 \text{ N}\cdot\text{m} = \boxed{5.3 \times 10^5 \text{ J}}$$

Note that the exact nature of the car's path (hills, loop, etc.) as it goes from A to C has no bearing at all on the net change in PE during the over-all motion.

EXAMPLE **6-9**

A butterfly raises its four wings from horizontal to vertical positions; each wing is approximately rectangular, 3 cm tall by 4 cm long, of mass 0.020 g. What is the increase in PE?

The c.g. of each wing is raised by 1.5 cm.

$$\text{PE} = 4mgh = 4(0.020 \text{ g})(980 \text{ cm/s}^2)(1.5 \text{ cm})$$

$$= 118 \text{ dyn}\cdot\text{cm} = \boxed{118 \text{ erg}}$$

Kinetic energy is given to a body when it speeds up. When a body is at rest it has no KE. Let us apply a force **F** to a body of mass m, which is initially at rest ($v_0 = 0$), and let us allow this force to operate for a sufficient time for the body to acquire a velocity v. Let s be the displacement, the distance through which the body moves. Here $\Sigma\mathbf{F}$ is simply **F**, since all other forces on the body are balanced. We calculate the displacement by first finding the acceleration.

$$a = \frac{F}{m}$$

$$v^2 = v_0^2 + 2as$$

$$s = \frac{v^2}{2a} = \frac{v^2}{2(F/m)} = \frac{v^2 m}{2F}$$

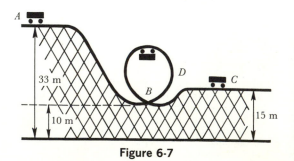

Figure 6-7

By Newton's third law, while we are applying the force $\mathbf{F}$, the body reacts on us with an equal and opposite force $-\mathbf{F}$. We do work in moving the body against this reaction force, and the work done by us is the product of the force and the distance moved. By definition, work done against this reaction force (we may call it "work done against inertia") is the KE stored in the body.

Work done against inertia

$$= \text{(reaction force)(displacement)}$$

$$\text{KE} = (F) \left(\frac{v^2 m}{2F} \right)$$

or

$$\text{KE} = \tfrac{1}{2}mv^2 \qquad (6\text{-}5)$$

$$\text{J} = \text{(kg)(m/s)}^2$$

$$\text{ergs} = \text{(g)(cm/s)}^2$$

It may be objected that our derivation of $\text{KE} = \tfrac{1}{2}mv^2$ is limited to a uniformly accelerated body acted on by a constant force. Fortunately, it can be shown by the use of calculus that the work done against inertia is given by $\tfrac{1}{2}mv^2$ regardless of the way in which the body reached its final velocity (see footnote following Eq. 7-9 on p. 160).

In the fps system, KE is usually written as $\tfrac{1}{2}(W/g)v^2$, since Eq. 4-8 allows us to substitute (W/g) for m.

Note that the displacement and the force applied do not appear in the formula for the kinetic energy. If a smaller force had been applied, the acceleration would have been less, and the body would have had to move farther in order to attain the same speed v. The product Fs would have remained the same. Our two formulas for gravitational PE and KE, Eqs. 6-4 and 6-5, are easily remembered, and they serve to emphasize the significant physical quantities: weight and vertical rise for gravitational PE, mass and velocity for KE.

EXAMPLE **6-10**

A boy on a motorbike is scooting along at

20 m/s. If the boy's mass is 50 kg, what is his KE?

$$\text{KE} = \tfrac{1}{2}mv^2$$

$$= \tfrac{1}{2}(50 \text{ kg})(20 \text{ m/s})^2$$

$$= 10\,000 \text{ kg} \cdot \text{m}^2/\text{s}^2 = \boxed{10\,000 \text{ J}}$$

EXAMPLE **6-11**

What is the KE of a 4 ton truck moving down a turnpike at 60 mi/h? How does this compare with the PE that would be stored in the truck if it were lifted 100 ft straight up?

In this example we use W/g for the mass of the truck. We also recall that 60 mi/h = 88 ft/s. Then

$$\text{KE} = \tfrac{1}{2}mv^2 = \tfrac{1}{2}(W/g)v^2$$

$$= \frac{1}{2} \left(\frac{8000 \text{ lb}}{32 \text{ ft/s}^2} \right)(88 \text{ ft/s})^2$$

$$= \boxed{968\,000 \text{ ft} \cdot \text{lb}}$$

If the truck were lifted 100 ft straight up, its PE would be

$$(100 \text{ ft})(8000 \text{ lb}) = \boxed{800\,000 \text{ ft} \cdot \text{lb}}$$

6-8   Conservation of energy

The concept that energy can neither be created nor destroyed represents one of the great generalizations of the nineteenth century. Before the work of Mayer, Rumford, Helmholtz, and especially the English physicist James Joule (1818–1889), the possibility of conversion of energy to heat and vice versa was known, but the *quantitative* equivalence of heat energy and other forms of energy was obscure. Joule showed experimentally that whenever a certain amount of heat energy (1 Btu*) disappears or appears, the

---

*Joule used British thermal units (Btu's); today the preferred heat units in scientific work are calories or kilocalories. Many workers use joules for thermal as well as other forms of energy.

same amount of mechanical work (778 ft·lb) appears or disappears. This law of conservation of energy was extended by Joule and others to include "all" of the then-known forms of energy (including electrical, chemical, etc.), and was first stated somewhat as follows: *The total amount of mechanical, thermal, chemical, electrical, or other energy in any isolated system remains constant.*

In 1905 Einstein broadened the law still further to include the equivalence of mass and energy. This will be discussed in Chap. 7. In everyday life, however, we usually need to consider only chemical, thermal, electrical, or mechanical energies. The law of conservation of energy, even in this limited form, is a profound generalization, and we shall see, in Sec. 6-9, how to use it in solving problems.

The logical significance of the law of conservation of energy has been variously interpreted by authors. Some feel that we are using the law when we say that kinetic energy equals reaction force times displacement, for otherwise, without such a postulate, how can we "know" that the work is stored in the body? Our viewpoint is that it is fairly obvious that work is done against a reaction force, or "inertial force," and it is merely a matter of definition to say that this work is stored as kinetic energy. We consider the law of conservation of energy to refer to something far broader than mechanical processes. The generalization that *all* forms of energy (including thermal energy, electrical energy, mass energy, and so forth) are quantitatively exchangeable with mechanical energy seems to us to be the essence of the energy principle that we call the law of conservation of energy.

6-9 The energy principle in solution of problems

It is often possible to solve mechanical problems by use of the law of conservation of energy. This is scarcely surprising, since our formula for kinetic energy (Eq. 6-5) came directly from Newton's second law. When we use this formula in a problem or a derivation, we are really only using Newton's second law in a "predigested" way which has already taken care of the kinematical part of the problem.

EXAMPLE **6-12**

In the problem of the anxious spaceman (see Example 4-6) we can use the energy principle to find how far the moving section moves before stopping. The man is pushed back by the section, and work is done *on* him by the section.

$$\begin{pmatrix} \text{Work done on} \\ \text{man by section} \end{pmatrix} = \begin{pmatrix} \text{loss of KE of} \\ \text{moving section} \end{pmatrix}$$

$$(150 \text{ lb})(s) = \tfrac{1}{2}\left(\frac{16\ 000 \text{ lb}}{32 \text{ ft/s}^2}\right)(1 \text{ ft/s})^2$$

$$s = \boxed{1.67 \text{ ft}}$$

This is the same answer that was obtained in Chap. 4, but here we got the result without computing the acceleration and without using the equations of uniformly accelerated motion.

EXAMPLE **6-13**

A bowling ball is returned to the bowler by a mechanism which places it on a return ramp 100 cm above the alley level (Fig. 6-8). If the ball is given an initial speed of 0.6 m/s at point A, what is its speed when it reaches the bowler on the short level stretch DE? Ignore friction; ignore the rotational KE of the ball.

To have a consistent set of mks units, we first convert the heights from centimeters to meters. Then we apply the energy principle:

$$\text{KE}_1 + \text{PE}_1 = \text{KE}_2 + \text{PE}_2$$

Figure 6-8

$$\tfrac{1}{2}(m \text{ kg})(0.6 \text{ m/s})^2 + (m \text{ kg})(9.8 \text{ m/s}^2)(1.0 \text{ m})$$
$$= \tfrac{1}{2}(m \text{ kg})v^2 + (m \text{ kg})(9.8 \text{ m/s}^2)(0.4 \text{ m})$$

The mass cancels out, and the solution of the equation gives

$$v = \boxed{3.48 \text{ m/s}}$$

A direct solution of this problem would be more complicated. You would have to use a force triangle to find the acceleration of the ball in the slope AB, and the final velocity (at B) would become the initial velocity for a second problem to cover the interval BC. The energy method skips some of the steps and gives an answer for the final velocity. Note, however, that the energy method cannot be used to find times or accelerations.

EXAMPLE **6-14**

A stone of mass 2 kg is thrown at 5 m/s upward at a 30° angle from a cliff 20 m high (Fig. 6-9). How fast will the stone be moving when it strikes the ground? (Ignore air resistance.)

The direct method would make this into a projectile problem (Sec. 3-11), for which you would have to find the horizontal and vertical components of the velocity at C. Combining these components by the parallelogram rule would give the magnitude of the velocity at C. The energy method skips the intermediate steps. The weight of the stone is 19.6 N.

$$KE_1 + PE_1 = KE_2 + PE_2$$
$$\tfrac{1}{2}(2 \text{ kg})(5 \text{ m/s})^2 + (19.6 \text{ N})(20 \text{ m})$$
$$= \tfrac{1}{2}(2 \text{ kg})v^2 + 0$$
$$25 \text{ J} + 392 \text{ J} = (1 \text{ kg})v^2 + 0$$
$$v = \sqrt{417 \text{ J/1 kg}} = \boxed{20.4 \text{ m/s}}$$

The angle of throw has nothing to do with this particular problem, although of course the angle *does* affect the maximum height of rise, time in the air, and distance out from the cliff. If the angle had been 20° instead of 30°, these latter quantities would have been different, but the final speed would still have been 20.4 m/s, as is indicated by the fact that all the quantities in the energy equation would have been the same.

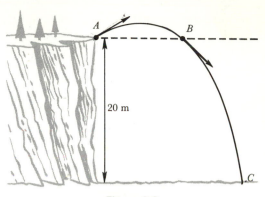

Figure 6-9

6-10 Power; efficiency

James Watt (1736–1819) in his pioneering investigations assumed that an exceptionally powerful dray horse could lift 550 lb a distance of 1 ft in 1 s and could perform work at this rate more or less steadily during a working day. Horses of lesser quality required a longer time to do the same amount of work. The quantity that interested Watt was *power*, defined as the rate of doing work, and Watt gave the name *horsepower* (hp) to the "standard" rate of 550 ft·lb/s. In the mks system the unit of power is the *watt* (W), a rate of doing work equal to one joule per second. Most people think of the watt as a unit of electric power, and it is true that electric or any other power can be measured in watts. However, James Watt was a mechanical engineer, and the watt is fundamentally a mechanical unit. It appears in electricity only because the electrical unit "volt" is an mks unit defined in the first place in terms of the mechanical work unit, the joule. (One reason for the popularity of mks units is their tie-in with basic electrical units.) It is sometimes useful to know that 746 W = 1 hp.

Power is the rate of doing work and is expressed as the time derivative of work. Thus

$$P = \lim_{\Delta t \to 0} \frac{\Delta W}{\Delta t}$$

or

$$P = \frac{dW}{dt}$$

A simple special case arises when work is done by a constant force which moves a body against the force of friction. The total work during a displacement **s** is **f·s**; frictional force is always parallel to the displacement, so the dot product becomes simply fs.

$$W = fs$$

$$\frac{dW}{dt} = f\frac{ds}{dt}$$

or

$$P = fv$$

This equation gives the power required to move a body at speed v against a constant opposing force f.

EXAMPLE **6-15**

What is the power P of an engine which pulls a 500 000 kg train at a steady speed of 40 m/s along a horizontal track for which the coefficient of friction is 0.02?

The force of friction is given by

$f = $ (500 000 kg)(9.8 m/s²)(0.02) = 98 000 N

$P = fv = $ (98 000 N)(40 m/s)

$\quad = 3.92 \times 10^6$ N·m/s $ = 3.92 \times 10^6$ J/s

$\quad = 3.92 \times 10^6$ W $= \boxed{3920 \text{ kW}}$

It is often useful to consider the *efficiency* of a device, defined as the work output divided by the work input.

$$\text{Efficiency} = \frac{\text{work output}}{\text{work input}}$$

If, as is usual, the input and output take place during the same time interval, the efficiency is also the *power* output divided by the *power* input. Thus an 80 watt fluorescent lamp may have an output of visible energy at the rate of 8 watts; its efficiency is 8 W/80 W = 10%. The efficiency of some types of simple machines is discussed in Sec. 6-14.

6-11 Energy change in collisions

Let us look closely at Example 6-4, in which we studied the collision of a car and a truck. Before collision, all the KE of the system was in the car:

$$\tfrac{1}{2}mv^2 = \tfrac{1}{2}(600 \text{ kg})(20 \text{ m/s})^2 = 120\ 000 \text{ J}$$

After collision, the system has less kinetic energy:

$$\tfrac{1}{2}mv^2 = \tfrac{1}{2}(2000 \text{ kg})(6 \text{ m/s})^2 = 36\ 000 \text{ J}$$

At first glance it seems that the law of conservation of energy has been violated. There has been no obvious increase of potential energy, since all the motions are horizontal. Where, then, did the other 84 000 J of energy go?

We can account for the "missing" energy if we realize that *some* force must have been applied to the car in order to slow it down. At first, work was done against the force of the bumpers, etc. Then the metal crumpled and work was done against internal friction. The point is, regardless of the exact process, *some* force had to be exerted to slow the car down, and this force must have been exerted through *some* distance, however short; the car's velocity could not decrease instantaneously unless an infinite force acted on it. If this force had not been present, the car would have sailed right through the truck and there would have been no collision! This hidden energy may properly be called PE if it is recoverable, but generally if KE is lost in a collision it is transformed mostly into heat energy as the colliding bodies are permanently deformed. A minute fraction of the KE may go into sound energy, light energy (if sparks fly), etc.

We define an *inelastic impact* as one in which some KE is "lost" by transformation into other forms of energy. There are various degrees of inelastic impact; a collision such as discussed in Example 6-4 in which the bodies stick together is said to be completely inelastic. The essential feature of a completely inelastic or a partially elastic impact is that some of the KE of the system is transformed during the collision.

Turning now to the other extreme, we define an *elastic impact* as one in which no KE is lost. A bouncing ball may be tremendously deformed during impact, but if the elastic limit (Sec. 9-1) has not been exceeded, the ball returns exactly to its original shape. Momentarily the ball has much PE, but this is returned to the form of KE as the ball separates from the ground. No energy remains in the ball as either elastic PE or heat.

A familiar example of elastic impact is the head-on collision of two billiard balls of equal mass (Fig. 6-10). To derive a general result, we work with symbols rather than with specific numbers. Let v_0 be the initial velocity of a moving ball of mass m heading straight for a stationary second ball with the same mass. We have two unknowns: the velocities V_1 and V_2 of the two balls after collision. We set up two equations, one for the conservation of momentum (always true) and one for the conservation of KE (true for elastic impact).

Conservation of momentum:

$$mv_0 + m(0) = mV_1 + mV_2$$

Elastic impact:

$$\tfrac{1}{2}mv_0^2 + \tfrac{1}{2}m(0)^2 = \tfrac{1}{2}mV_1^2 + \tfrac{1}{2}mV_2^2$$

The first equation gives $V_2 = v_0 - V_1$; substituting this value of V_2 into the second equation, we obtain, after simplifying,

$$V_1^2 - V_1v_0 = 0$$

or

$$V_1(V_1 - v_0) = 0$$

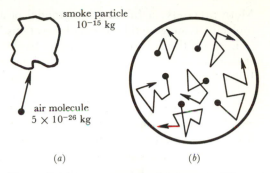

Figure 6-11 (*a*) Brownian motion. (*b*) View of smoke particles in microscope.

One root[*] of this quadratic equation is $V_1 = 0$; whence $V_2 = v_0$. This means that the first ball stops dead, and the second ball takes up the entire velocity v_0 of the first ball. This result is well known to pool players, although in practice rotational and frictional effects must also be considered.

On the microscopic level, conservation of momentum is well illustrated by *Brownian motion* (Fig. 6-11). This is often demonstrated by viewing particles of cigarette smoke through a microscope; the rapid zigzag motion of the smoke particles is the result of incessant random bombardment by fast-moving molecules of nitrogen, oxygen, and other components of air. The molecules themselves are, of course, far too small to be seen even in the most powerful microscope, but as a result of statistical fluctuations in the average effect of many collisions per second, the momentum of a smoke particle changes perceptibly in an erratic fashion. It is somewhat as if a battleship were bombarded by BB shot—one could deduce the existence of the BB shot by observing the motion of the ship, in principle, even though the recoil velocity of the ship is small because of its great mass. The Brownian motion (discovered by botanist Robert Brown in 1827) gives visual evidence for the existence of air molecules.

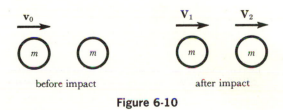

before impact   after impact

Figure 6-10

---

[*] The other root $V_1 = v_0$ merely says that it is possible for the first ball to retain its original velocity (it could miss the second ball completely).

More precisely, a billiard-ball model of a molecule is suggested by these observations.[*]

Even on the sub-sub-microscopic level of smallness involved in nuclear collisions, a billiard-ball model of a nucleus is good enough to predict behavior with some degree of accuracy. (See also Sec. 2-3, where the relatively sharp boundary for nuclear matter was discussed.) The basic reaction in a nuclear reactor is the capture of a neutron by a ^{235}U nucleus, which results in the highly unstable ^{236}U nucleus. The ^{236}U divides (by fission) into two nearly equal fragments, plus several neutrons. One of these neutrons can keep the reaction going if it is captured by another ^{235}U nucleus. The technological problem lies in the fact that the emitted neutrons are fast (say 10^6 m/s), whereas only slow neutrons (say 10^3 m/s) are effectively captured. Hence the use of a "moderator," a substance added to the pile to slow down the neutrons by successive collisions. Commonly used moderators are deuterons (heavy hydrogen nuclei contained in heavy water) and carbon (contained in graphite). Which of these is more effective? The relative masses are neutron 1, deuteron 2, carbon nucleus 12. If a neutron traveling at a known velocity v_0 collides elastically with a stationary deuteron, it is a routine problem in conservation of momentum to find the velocities of neutron V_N and deuteron V_D after collision.

Conservation of momentum:

$$1v_0 + 2(0) = 1V_N + 2V_D$$

Elastic impact:

$$\tfrac{1}{2}(1)v_0{}^2 + \tfrac{1}{2}(2)(0)^2 = \tfrac{1}{2}(1)V_N{}^2 + \tfrac{1}{2}(2)V_D{}^2$$

from which we obtain

$$V_N = -\tfrac{1}{3}v_0 \quad \text{and} \quad V_D = +\tfrac{2}{3}v_0$$

That is, the neutron rebounds with $\tfrac{1}{3}$ its original velocity. After two collisions it will have $\tfrac{1}{3} \times \tfrac{1}{3}$ or $\tfrac{1}{9}$ its original velocity, and, if $v_0 = 10^6$ m/s, after only 7 collisions the velocity will be less than 10^3 m/s, as desired. Repeating the problem with carbon as the target, we obtain $-\tfrac{11}{13}v_0$ for the neutron's velocity after one collision, not very much less than its original velocity. Obviously, heavy water[■] is a more effective moderator than graphite, and heavy-water reactors are smaller and more compact than those using graphite. In the economics of reactor design, other factors must also be considered. Heavy water is expensive; but even the carbon is relatively costly, since it must be painstakingly purified to avoid capture of neutrons by "impurity atoms" before they have a chance to hit another ^{235}U nucleus. One might think that the protons in ordinary water would be still more effective; for, as in the case of the two equal billiard balls the neutron would be completely stopped[▲] at the first collision. Unfortunately, the proton has a strong affinity for capturing a neutron; in fact, a proton plus a neutron makes a deuteron, according to the equation $^1_1\text{H} + ^1_0\text{n} = ^2_1\text{H}$. Thus with ordinary water many of the neutrons would be absorbed, and would not be available for continuing the chain reaction with ^{235}U.

Our next example shows how W/g can be used for m in a problem in British units.

EXAMPLE 6-16

A ball weighing 6 lb slides down a frictionless chute (Fig. 6-12) and then strikes a stationary 10 lb block on a horizontal table. The coefficient

---

[*] The mathematical theory of the Brownian motion is due to Einstein (in a series of five papers published in the first decade of the twentieth century).

[■] It was Germany's all-out collection of heavy water from occupied Norway in the early 1940's that apprised the scientists of other countries that the Germans were attempting to release nuclear energy from uranium. Norwegian scientists had been studying heavy water, which is produced as a by-product of the great hydroelectric installations of that country.

[▲] This assumes a head-on collision with a stationary proton. In practice, not all collisions would be head-on; the target protons would also be in motion at some 10^3 m/s due to thermal agitation. For both these reasons, after many collisions the average KE of the scattered neutrons will become equal to that of the protons among which they move.

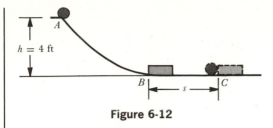

Figure 6-12

of kinetic friction between the block and the table is 0.1. If the ball and the block stick together, how far do they slide before coming to rest?

We solve the problem in several steps. Throughout, we use W/g for mass.

(a) Conservation of energy for path AB: The PE at A is Wh; the KE at B is $\frac{1}{2}mv^2 = \frac{1}{2}(W/g)v^2$.

$$KE + PE = KE + PE$$

$$0 + (6\ lb)(4\ ft) = \frac{1}{2}\left(\frac{6\ lb}{32\ ft/s^2}\right)v^2 + 0$$

$$v = \boxed{16\ ft/s}$$

This is the speed of the ball just before impact.

(b) Conservation of momentum during impact:

$$m_{ball}v_{ball} = m_{ball+block}V_{ball+block}$$

$$\left(\frac{6\ lb}{32\ ft/s^2}\right)\left(16\frac{ft}{s}\right) = \left(\frac{16\ lb}{32\ ft/s^2}\right)(V)$$

$$V = \boxed{6\ ft/s}$$

(c) Finally we apply the law of conservation of energy to the path BC: The KE of the system just after impact is dissipated as work is done against friction. The force of friction is

$$f = \mu_k F_n = 0.1(16\ lb) = 1.6\ lb$$

$$KE\ at\ B = work\ against\ friction$$

$$\frac{1}{2}\left(\frac{16}{32\ ft/s^2}\right)(6\ ft/s)^2 = (1.6\ lb)s$$

$$s = \boxed{5.63\ ft}$$

6-12 The significance of the conservation laws

We have by now come across two important laws of physics, both of them expressed in the form of conservation laws. Taken together, the laws of conservation of mechanical energy and conservation of momentum are adequate to give answers to most of the questions that might be asked about the mechanics of physical objects. These conservation laws are equivalent to Newton's laws; as a matter of fact, we derived them from Newton's laws of motion in the first place (Secs. 6-3 and 6-7). It would be entirely possible to do it the other way around: assume the conservation laws, and derive Newton's laws from them. Let us therefore ask the question, which is more fundamental—Newton's second law in the form net force = (mass)(acceleration) or Newton's second law in the form net force = (change of momentum)/(time)? So far, we have seen no reason to prefer one form of the law to the other. Situations arise, however, in which the second way (involving change of momentum) is distinctly preferable. What if the mass of a body changes while it is being acted on by a force? There could then be a change of momentum even if velocity were constant. Acceleration would be zero; the equation net $F = ma$ would not apply.

EXAMPLE 6-17

A train of empty coal cars is being pulled along a straight, level piece of track at 20 m/s. If rain starts to fall, and the cars receive 300 kg of water per minute, how much additional force must the engine supply to keep the train moving at a constant speed of 20 m/s?

Here the momentum changes because the mass changes. The amount of momentum that is added in each minute is (300 kg)(20 m/s) = 6000 kg·m/s. We now apply Newton's second law in the momentum form:

$$F = \frac{\Delta(mv)}{\Delta t}$$

$$= \frac{6000\ kg\cdot m/s}{60\ s} = \boxed{100\ N}$$

A rather small force, to be sure, but note that the equation net $F = ma$ would have been inadequate to cope with the problem without a number of simplifying assumptions.

The momentum of a body is defined as the product of its mass and its velocity. Newton's second law may be restated in terms of change of momentum: The net force acting on a body equals the rate of change of its momentum. The product of net force and the time during which the force acts, defined as impulse, is equal to the total change of momentum.

Newton's third law may be used to derive the law of conservation of momentum: The total momentum of an isolated system remains constant.

Rockets are propelled by the reaction force of the expelled exhaust gases against the body of the missile, and do not require a medium such as air for their operation. In a vector sense, the total momentum of the system (rocket + exhaust gases) remains zero.

Work is done on a body whenever it is moved against an opposing force. Work equals displacement times the component of force in the line of motion ($W = Fs$). Energy is stored work. Units for work and energy are the joule, the erg, and the foot·pound. Potential energy (PE) is due to the relative position of the parts of a system of bodies, and this includes gravitational, elastic, chemical, and some electrical energy, as well as nuclear energy. Kinetic energy (KE) is due to motion, and this includes much of what is known as heat energy. Some useful formulas for mechanical energy are:

$$\text{Gravitational PE} = Wh = mgh \qquad \text{KE} = \tfrac{1}{2}mv^2$$

The energy of an isolated system remains constant. The laws of conservation of momentum and of energy include the content of Newton's laws, and are general enough to be useful in many non-Newtonian aspects of contemporary physics. The statement that net force equals the rate of change of momentum serves as a general definition of force.

In collisions, momentum is always conserved, but kinetic energy is conserved only in elastic collisions. In an inelastic collision some KE is transformed to heat as the colliding bodies are permanently deformed, or else some of the KE is stored as PE in one or more of the colliding bodies.

Power is the rate of doing work. One watt is 1 joule per second; one horsepower is 550 ft·lb/s.

■ CHECK LIST

momentum
impulse

$$\Sigma \mathbf{F} = \frac{\Delta(m\mathbf{v})}{\Delta t} = \frac{\Delta \mathbf{p}}{\Delta t}$$

law of conservation of
 momentum
work
energy

PE
KE
joule
erg
foot·pound
law of conservation of
 energy
elastic impact
inelastic impact

power
watt
horsepower
efficiency
$W = Fs$
gravitational PE =
 $Wh = mgh$
KE $= \tfrac{1}{2}mv^2$
$P = fv$

6-1 When an apple falls to the ground and strikes the earth without rebound, what "becomes of" the momentum of the apple?

6-2 Can a man in a rowboat "create" momentum by his unaided efforts?

6-3 A prospector carrying a bag of valuable uranium ore is trapped on absolutely smooth ice on a frozen lake in northern Canada. Is there any way in which he can move across the ice to safety, or must he freeze to death?

6-4 When a jet plane lands on the deck of a carrier, is this an elastic or an inelastic "collision"? What becomes of the momentum of the jet? What becomes of its KE?

Figure 6-13

6-5 Is it possible for a rocket to attain a speed greater than the velocity with which the exhaust gases leave it?

6-6 A man stands on a wooden plank which in turn rests on a concrete floor. If he hammers one end of the plank with a heavy mallet, can he make the "bumpmobile" move? If so, explain how. (See Fig. 6-13.)

6-7 A car is coasting on a level stretch. If the driver turns on his headlights, where does the energy come from   (a) if they are connected to a battery;   (b) if they are connected directly to a generator geared to the wheels? Does the car slow down in either case?

6-8 Is there any source of energy available to us on earth that does not eventually trace back to the sun's energy? What about muscular energy? What is the source of the sun's energy?

6-9 When an artificial satellite is moving in an orbit around the earth, it has more energy than it had when it was on earth before launching. Where did this energy come from? Into what forms has this energy been transformed when the satellite is circling the earth? What is the eventual fate of this energy when the satellite spirals toward the earth and burns up?

6-10 Is some of the sun's energy used up in causing the earth to move in an orbit (assumed circular for convenience)?

6-11 The tides in some parts of Norway rise some 40 ft or more, and when the tide goes out the water dashes against the rocks and beach and flows slowly out to sea, ready to be raised again by the moon in $12\frac{1}{2}$ hours as the earth turns. What becomes of the PE that has been stored in the water? If the water were trapped in a dam, the PE could be used by man, and tidal energy stations have actually been constructed. What is the ultimate source of this energy?

6-12 What is wrong with this sentence? "The motor did 12 000 watts of work while pumping 90 gallons of water from the well."

6-13 Show that the kilowatt·hour (kW·h) is a unit of energy. How many joules equals 1 kW·h?

6-14 In Example 6-14, is the speed at A the same as that at B? Is the velocity at A the same as that at B?

6-15 A flash bulb is carefully weighed on an analytical balance; then it is "fired," and after cooling off it is weighed again. Will the two observed weights be the same?

6-16 The dimensions of both torque and work are $[ML^2T^{-2}]$. Does this mean that torque and work are the same?

6-17 An open freight car is coasting along on a frictionless track, and rain starts to fall, thereby increasing the mass of the car. Does the velocity of the car remain constant?

■ PROBLEMS

6-A1 Compute the momentum, in kg·m/s, of a fullback who has a mass of 100 kg and is moving at 3 m/s.

6-A2 Compute the momentum of a golf ball which has a mass of 60 g and is moving with a velocity of 70 m/s.

6-A3 (a) Compare the momentum of a 10 ton truck moving at 5 ft/s with that of a 1 oz bullet moving at 4000 ft/s. (b) Compare the KE's of the same objects.

6-A4 If in Prob 6-A2 the impact between the golf club and the ball lasted for 2×10^{-4} s, what was the rate of change of momentum? What force acted on the ball? What force acted on the club?

6-A5 A boy holds a 5 lb air rifle loosely and fires a bullet which weighs 0.01 lb. The muzzle velocity of the bullet is 1000 ft/s. What is the recoil velocity of the gun?

6-A6 If the boy holds the rifle of Prob. 6-A5 tightly against his body, the recoil is less. Explain. Calculate the new recoil velocity, if the boy weighs 95 lb.

6-A7 In a freight yard a train is being made up. An empty freight car, coasting at 10 m/s, strikes a loaded car which is stationary, and the cars couple together. Each of the cars has a mass of 3000 kg when empty, and the loaded car contains 12 000 kg of bottled soft drinks. With what speed does the combined mass start to move?

6-A8 An astronaut of mass 80 kg carries an empty oxygen tank of mass 10 kg. He throws the tank away from himself with a speed of 2 m/s (measured relative to the fixed stars). With what velocity does the astronaut start to move through space?

6-A9 Use the energy principle, in conjunction with the definition of elastic impact, to prove that if a falling body makes *elastic* collision with the ground, it bounces back to the same height from which it was originally dropped.

6-A10 A child who weighs 200 N is sitting on a horizontal floor. (a) How much work is required to move him 0.6 m against an opposing force of friction equal to 30 N? (b) How much work is required to lift him 0.6 m straight up?

6-A11 How much work must be done to push a thumbtack 0.8 cm into a bulletin board against an opposing force of 20 000 dyn?

6-A12 A ladder 20 ft long weighs 50 lb, and its c.g. is at the center. What is the increase of PE of the ladder when it is raised from a horizontal positon to a vertical position?

6-A13 How much work is done by a train's engine which pulls a train of mass 5×10^6 kg 1 km over a level track, if the coefficient of rolling friction is 0.01?

6-A14 When a 100 lb girl walks up a hill which is 400 ft long, her PE increases by 20 000 ft·lb. What is the vertical rise of the hill?

6-A15 A student who weighs 600 N sleeps in a bed 80 cm above the floor. The student's c.g. is

100 cm from the soles of his feet. What is the increase in his PE when he gets up in the morning? Where does this energy come from?

6-A16 While pulling down a window shade, a housewife exerted a downward force increasing steadily from 5.2 N to 17.2 N and the length of the shade was extended by 1.5 m. (*a*) What average force opposed the motion? (*b*) What was the shade's increase of PE?

6-A17 A wild goose of mass 5 kg is flying at a speed of 4 m/s. What is the bird's KE?

6-A18 What is the KE of an 8 ton truck moving at 24 ft/s up a hill which makes an angle of 20° with the horizontal?

6-A19 What is the KE of a neutron which has a mass of 1.67×10^{-27} kg and is moving at 300 m/s?

6-A20 Calculate the KE of a 1000 kg car moving at 3 m/s.

6-A21 How fast is a 128 lb sprinter running if his KE is 1800 ft·lb?

6-A22 A boy pushes with a steady force of 80 N on another boy who is on frictionless roller skates. How far must he push in order to give his friend 400 J of KE?

6-A23 A fly takes off horizontally with KE equal to 125 ergs. What force was exerted by the fly's legs, if the force was exerted through a distance of 2 mm?

6-A24 The fly of Prob. 6-A23 weighs 24.5 dyn. What is its mass? What is its velocity just after take-off?

6-B1 At 4:40 P.M. the momentum of a speedboat was measured as 30 000 kg·m/s, and at 4:43 P.M. it was 120 000 kg·m/s. Assuming that the boat was moving along a straight course and that its speed increased steadily (constant force), what forward force was exerted on the boat? By what was this force exerted? (Do not find the acceleration.)

6-B2 What force, acting for 0.001 s, will change the velocity of a 100 g baseball from 30 m/s eastward to 40 m/s westward? (Do not find the acceleration.)

6-B3 What forward force, applied for 1.5 s, is needed to give a 16 lb bowling ball a speed of 18 ft/s? (Do not find the acceleration.) (*Hint:* Use W/g for the mass of the ball.)

6-B4 A fire hose sends 900 kg of water every minute against a burning building. The water strikes the building with velocity +20 m/s and does not bounce back. (*a*) What is the rate of change of momentum of the water? (*b*) What force does the building exert on the water? (*c*) What force does the water exert on the building? (*Note:* Be careful to affix proper signs, + or −, to your answers to this problem.)

6-B5 A ventilator fan moves 100 m³ of air per minute, and the blast of air strikes a wall at a speed of 20 m/s. What force is exerted on the wall by the moving air? (See Table 2-1 for the density of air.) Assume that the air molecules make inelastic impact at the wall.

6-B6 A 180 lb fullback is moving at 20 ft/s; he is tackled head-on by a 220 lb linebacker who is approaching him at 10 ft/s. With what speed, and in which direction, do the pair of players move after the tackle?

6-B7 A 4 kg duck sitting on a pole 8 m tall is struck by a 20 g bullet which is moving horizontally at 200 m/s. The bullet remains embedded in the bird. (*a*) What horizontal velocity does the bird acquire? (*b*) How far from the base of the pole does the duck strike the ground?

6-B8 One way of measuring the muzzle velocity of a bullet is to fire it horizontally into a massive block of wood placed on a cart. Assuming no friction, we then measure the velocity with which the wood (containing the bullet) starts to move. In one experiment the bullet had a mass of 50 g, and the wood and its cart had a mass of 20 kg. After the shot, a stop watch was used to determine that the cart, wood, and bullet moved at constant velocity, traveling 8 m in 0.40 s. What was the original speed of the bullet?

6-B9 A life raft of mass 200 kg carries two swimmers of mass 40 kg and 70 kg, respectively. The raft is initially floating at rest; then the swimmers simultaneously dive off opposite ends of the raft, each with a horizontal velocity of 4 m/s. With what speed and in what direction does the raft start to move?

6-B10 A basketball player jumps into the air with both arms extended vertically overhead. Just as he reaches maximum height he pulls one arm down so that his c.g. is moved to a point 1.5 in. closer to his toes. Do the finger tips of the other hand go higher or lower as a result of this maneuver? How much?

6-B11 Ten books, each 2 in. thick and weighing 4 lb, are lying horizontally on a table. What work is required to construct from them a stack of books 20 in. high?

6-B12 Repeat the derivation leading up to Eq. 6-5, but do not assume that $v_0 = 0$. Show that the work done against the reaction force of inertia is $\frac{1}{2}mv^2 - \frac{1}{2}mv_0^2$. Interpret your answer physically.

6-B13 A .22 caliber bullet of mass 0.0026 kg is shot into a massive wooden target at 300 m/s. It penetrates 0.07 m into the target; what is the average force exerted on the bullet by the wood? (Use the energy principle; do not find the acceleration.)

6-B14 A 160 lb boy, starting from rest, slides down a 30° hill which is 150 ft long. He arrives at the bottom with a speed of 30 ft/s. How much heat energy has been shared between the surface of the hill and the seat of his pants? (Use the energy principle; do not find the acceleration.)

6-B15 A ping-pong ball of mass m rolls off the edge of a table which is 3 ft high. When the ball strikes the floor, its speed is 16 ft/s. How fast was it rolling when it left the table? (Use the energy principle.)

6-B16 A soccer player gives a ball of mass 1 kg an initial horizontal speed of 10 m/s. The ball slides 15 m along the turf against an average force of 2.5 N. What is the final speed of the ball? (Use the energy principle.)

6-B17 Solve Prob. 4-B12, using the energy principle.

6-B18 (a) Use the energy principle to find with what speed the roller coaster car of Example 6-8 reaches point B starting from rest (ignore friction).   (b) Find the speed at point C.

6-B19 A stone of mass m is thrown horizontally at 30 ft/s from a cliff 25 ft high. With what speed does the stone strike the ground below the cliff? (Use the energy principle.)

6-B20 A sack of mail of mass 40 kg slides down a post-office chute 6 m long which has a vertical drop of 2.5 m. The force of friction (parallel to the chute) is 100 N.   (a) What is the loss of PE of the bag?   (b) How much work is done against friction?   (c) With what speed does the bag reach the bottom of the chute if it is given an initial speed of 9 m/s? (Use the energy principle.)

6-B21 An egg weighing 49 kilodynes is dropped from a building 20 m high.   (a) Find the egg's KE just as it strikes the pavement.   (b) The target is a thick sheet of foam rubber, and the egg penetrates 20 cm into the rubber while stopping. What force is exerted on the egg?

6-B22 A boy weighing 500 N hangs on the end of a rope 6 m long. How much PE does he gain when a friend pulls him aside so that the rope makes an angle of 30° with the vertical?

6-B23 A boy of mass m swings back and forth on the end of a rope 20 ft long which is attached to the ceiling of a gymnasium. If he approaches within 8 ft of the ceiling during each cycle, what is his speed as he passes through the lowest point of his swing? (Use the energy principle.)

6-B24 A heavy chain 80 cm long is stretched out on a frictionless table top, with one end of the chain just at the edge and the length of the chain perpendicular to the edge. The chain is displaced so that a tiny bit projects over the edge; the chain starts to pick up speed and slides off the table. What is its speed when the last link of the chain is just leaving the table? (*Hint:* Assume the mass of the chain to be *m.*)

6-B25 A car of mass *m* is moving with speed *v* along a level road for which the coefficient of static friction between road and tire is μ_s. Use the energy principle to derive a formula for the minimum stopping distance without skidding in terms of *v*, μ_s, and *g*. Why is the coefficient of *static* friction appropriate here?

6-B26 A certain machine performs 60 J of work when a force of 4 N is applied through a distance of 20 m. What is the efficiency of the machine?

6-B27 Out of a total of 300 J of work put into a machine, an amount equal to 60 J produced heat energy in overcoming friction. What is the efficiency of the machine?

6-B28 How much energy must be supplied to a machine whose efficiency is 80% if 400 J of useful work is to be obtained?

6-B29 A storage battery lights up a 100 W lamp and also runs a 1 hp motor. What is the battery's loss of chemical PE in 10 min?

6-B30 What is the power of an engine which keeps a boat moving at a steady speed of 30 m/s if water resistance exerts a constant drag of 8000 N on the boat?

6-B31 What must be the input of electrical power, in watts, to a motor which is 80% efficient and delivers useful work at the rate of 0.5 horsepower?

6-B32 Three men, each of whom develops 0.3 horsepower, lift an 800 lb piano to a penthouse 110 ft above the street. If the pulley system is 70% efficient, how much time is required for the job?

6-B33 A 50 ton rocket reaches a speed of 6000 ft/s in 50 s after launching. (*a*) What is the rocket's KE? (*b*) Neglecting work done against gravity and against air resistance, what average power (in horsepower) is supplied by the rocket engine?

6-B34 Lindbergh flew the Atlantic in 1927 with a 230 hp motor in his single-engine plane. The trip took 33 h, 30 min, 30 s. (*a*) What was the power of his engine, in kilowatts? (*b*) Compute the total work performed by the engine. What assumption did you make in solving this problem? (*c*) What force of air resistance opposed Lindbergh's plane at a time when his 230 hp engine was causing the plane to move at 50 m/s?

6-B35 According to James Watt, a horse can keep a weight of 150 lb moving upward at a constant speed of $2\frac{1}{2}$ mi/h. Show that this statement is in agreement with the definition of 550 ft·lb/s for the horsepower.

6-B36 A motor whose useful output is 10 hp is used to lift an elevator and its load (weighing altogether 3000 lb) to a height of 33 ft. The motor is 60% efficient. (*a*) How much work is done? (*b*) How much energy must be supplied to the motor? (*c*) How much time will be required for the trip?

6-B37 Make a detailed energy check for the elastic collision of a neutron and a deuteron considered on p. 124. The total KE before impact is that of the proton, $\frac{1}{2}(1)v_0^2$ units. Using the final velocities calculated on p. 124, show that the total KE of the system after impact is also $\frac{1}{2}(1)v_0^2$ units.

6-B38 How much KE is lost during the inelastic collision at point *B* in Fig. 6-12?

6-B39 A ball of mass 1 kg moving at 8 m/s westward overtakes and collides with a ball of mass 3 kg moving at 4 m/s westward. The balls stick together after impact. (*a*) Calculate the magnitude and direction of the velocity of the combined mass after impact.   (b) Calculate the total KE of the system before and after impact, and find how many joules of KE were lost during the collision.

6-B40 A neutron of mass 1 unit, having velocity v_0, makes a head-on elastic collision with a stationary oxygen nucleus of mass 16 units.   (a) What forward velocity does the oxygen nucleus receive?   (b) What fraction of the KE of the neutron is transferred to the oxygen nucleus?

6-B41 A ball of mass 3 kg, moving at 2 m/s eastward, strikes head-on a ball of mass 1 kg which is moving at 2 m/s westward. The balls stick together after the impact.   (a) What is the magnitude and direction of the velocity of the combined mass after collision?   (b) Compute the total KE of the system before and after impact, and find how many joules of KE were lost during collision.

6-B42 Solve Prob. 6-B41, if the collision is perfectly elastic. Find the magnitude and direction of each ball's velocity after the collision.

6-C1 Two objects of mass m_A and m_B are held together by a strong, light thread and are also acted on by a light spring which is compressed as shown in Fig. 6-14. When the restraining thread is burned, the two objects fly apart with velocities $-v_A$ and $+v_B$. Use the law of conservation of momentum to solve for the ratio of the velocities, v_A/v_B. Then compute the ratio of the kinetic energies, KE_A/KE_B, and prove that the ratio of the KE's of the two objects is the reciprocal of the ratio of their masses.

6-C2 The arrangement for measuring the speed of a bullet described in Prob. 6-B8 is not very practical, since friction would prevent the cart from rolling at constant velocity, and the measurement of the short interval of time might prove difficult. A more precise arrangement is the ballistic pendulum (Fig. 6-15). After the bullet lodges in the wooden block, the block swings over and the height of rise is measured.   (a) In one experiment, the string was 5 m long and the rise was 7 cm. The mass of the bullet was 6 g, and the mass of the block of wood was 4 kg. What was the forward velocity of the block at its lowest position,

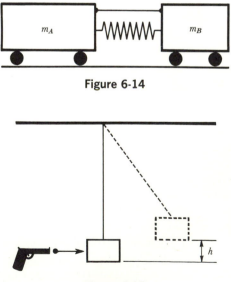

Figure 6-14

Figure 6-15

just after impact? (*Hint:* Use the energy principle.)   (*b*) Using the velocity found in (*a*), apply the law of conservation of momentum to find the original velocity of the bullet.

6-C3 A roll of heavy but perfectly flexible paper is initially at rest at the top of a rough inclined plane. The roll is released, and it unwinds to form a single long strip of paper lying at rest stretched out along the plane. Show that there is a loss of PE for the paper. What becomes of this PE? After thorough discussion of this problem, you may wish to refer to Reference 7 at the end of the chapter.

6-C4 An astronaut of mass 80 kg (which includes his back pack) is on a trajectory far above the moon's surface, moving 20 m/s upward and at an angle of 60° with the horizontal. A short burst of compressed nitrogen is ejected horizontally at 100 m/s (all velocities are relative to the moon's surface). After this maneuver the astronaut is moving vertically upward.   (*a*) What mass of gas was ejected?   (*b*) What is the astronaut's new speed? (*Hint:* Horizontal and vertical components of momentum are each conserved.)

6-C5 What is the magnitude and direction of the force exerted by the stream of water of Example 6-1 if the water strikes the roof horizontally and is deflected without loss of speed to be 30° from horizontal, parallel to the roof surface? (*Hint:* Consider the horizontal and vertical components of momentum separately.)

6-C6 The horizontal component of initial velocity of a projectile is v_{x0}, the projectile reaches a maximum height H, and the range is R. If, when the projectile is at its maximum altitude, it explodes into two equal fragments, with one fragment falling vertically downward, prove that the other fragment strikes the ground at a distance $\frac{3}{2}R$ from the gun.

6-C7 A 6 kg ball makes a perfectly elastic head-on collision with a stationary ball and rebounds with $\frac{1}{5}$ of its original speed. What is the mass of the struck ball?

For Further Study

6-13 Partially elastic collisions

Our study of collisions in Sec. 6-11 was limited to two rather special cases—completely *inelastic* impact and perfectly *elastic* impact. To deal with intermediate cases, we define a constant known as the *coefficient of restitution* e, which depends on the nature of the colliding bodies:

$$e = \frac{\text{relative velocity of separation}}{\text{relative velocity of approach}}$$

For a completely inelastic impact, in which the bodies stick together, the relative velocity of separation is 0 and $e = 0$. For a perfectly elastic collision $e = 1$, and the bodies separate as fast as they come together. The proof, for a head-on collision, is based on the definition of elastic collision (no loss of KE). Let

the masses of the bodies be m_1 and m_2, their initial velocities be v_1 and v_2, and their velocities after collision be V_1 and V_2. Then the relative velocity of approach before collision is $v_1 - v_2$. The relative velocity of approach after collision is $V_1 - V_2$, hence the relative velocity of *separation* after collision is $V_2 - V_1$.

Elastic impact:

$$\tfrac{1}{2}m_1v_1{}^2 + \tfrac{1}{2}m_2v_2{}^2 = \tfrac{1}{2}m_1V_1{}^2 + \tfrac{1}{2}m_2V_2{}^2$$

Conservation of momentum:

$$m_1v_1 + m_2v_2 = m_1V_1 + m_2V_2$$

Rearrangement of these equations gives

$$m_1(v_1{}^2 - V_1{}^2) = m_2(V_2{}^2 - v_2{}^2)$$
$$m_1(v_1 - V_1) = m_2(V_2 - v_2)$$

Dividing one equation by the other, we obtain

$$v_1 + V_1 = V_2 + v_2$$

or

$$v_1 - v_2 = V_2 - V_1$$

Thus we have proved that the relative velocity of separation equals the relative velocity of approach, and hence $e = 1$ for a perfectly elastic collision. For a good grade of steel ball bearing bouncing on a steel plate, e may be as high as 0.9; for an orange bouncing on the floor, e may be as low as 0.1 or less. Most collisions are partially elastic (or partially inelastic, if you prefer); this means that some mechanical energy is "lost" as the bodies are permanently deformed or otherwise changed (Fig. 6-16).

Figure 6-16 Partially elastic impact of a bouncing tennis ball, photographed at intervals of $\frac{1}{60}$ s. Relative velocity of separation is less than relative velocity of approach. Note momentary storage of potential energy as ball is deformed during impact.

EXAMPLE 6-18

What is the coefficient of restitution in the collision of Prob. 3-B14?

The relative velocity of approach is the velocity with which the ball struck the ground; call it v_0 (the detailed solution gives 72 ft/s for this velocity, but the value is not needed). By the terms of the problem, the ball lost one-fourth of its speed in the collision, so the relative velocity of separation was $\frac{3}{4}v_0$:

$$e = \frac{\frac{3}{4}v_0}{v_0} = \boxed{0.75}$$

In a partially inelastic collision, as in a completely inelastic collision, some mechanical energy is lost, as seen in the following example.

EXAMPLE 6-19

A car of mass 600 kg overtakes and makes a rear-end collision with a truck of mass 3000 kg. Before impact the car is moving at 25 m/s, and the truck at 10 m/s. (*a*) If the coefficient of restitution is 0.6, what are the velocities of the car and truck after collision? (*b*) How much KE has been lost during the collision?

(*a*) We let V_C be the final velocity of the car, and V_T be the final velocity of the truck. Thus the velocity of separation of car and truck is

$V_T - V_C$. There are two equations and two unknowns. One equation is the expression of the law of conservation of momentum, and the other equation is the expression of the definition of the coefficient of restitution.

Conservation of momentum:

$$(600 \text{ kg})(25 \text{ m/s}) + (3000 \text{ kg})(10 \text{ m/s})$$
$$= (600 \text{ kg})V_C + (3000 \text{ kg})V_T$$

Coefficient of restitution:

$$0.6 = \frac{V_T - V_C}{25 \text{ m/s} - 10 \text{ m/s}}$$

Simplifying these two equations, we obtain

$$\left.\begin{array}{r} 45\,000 = 3000V_T + 600V_C \\ 9 = V_T - V_C \end{array}\right\}$$

whence

$$V_C = \boxed{+5 \text{ m/s}}$$

$$V_T = \boxed{+14 \text{ m/s}}$$

Both velocities are in the forward direction after the collision.

(*b*) To find the KE's before and after collision, we use the original velocities and the new velocities just computed.

Before collision:

$$KE_C = \tfrac{1}{2}(600 \text{ kg})(25 \text{ m/s})^2 = 187\,500 \text{ J}$$
$$KE_T = \tfrac{1}{2}(3000 \text{ kg})(10 \text{ m/s})^2 = 150\,000 \text{ J}$$
$$KE_{C+T} = 337\,500 \text{ J}$$

After collision:

$$KE_C = \tfrac{1}{2}(600 \text{ kg})(5 \text{ m/s})^2 = 7500 \text{ J}$$
$$KE_T = \tfrac{1}{2}(3000 \text{ kg})(14 \text{ m/s})^2 = 294\,000 \text{ J}$$
$$KE_{C+T} = 301\,500 \text{ J}$$

We see that the car lost 180 000 J of KE and the truck gained 144 000 J. The net loss of KE was

$$\boxed{36\,000 \text{ J}}$$

According to the law of conservation of energy, the KE in Example 6-19 was not "lost," but was transformed into other forms of energy, mostly heat energy. Considering the process in detail, the solid parts of the truck and car were distorted during the collision and thus elastic PE was momentarily stored in the bumpers and elsewhere. However, after the vehicles separated, the metallic parts probably vibrated, perhaps invisibly, and, as these vibrations died away due to internal molecular friction, the elastic PE was transformed into heat energy. A minute amount of the energy of the collision was transformed into sound energy, and possibly also into radiant energy (light).

6-14 Machines

Typical problems of modern industrial life involve the use of machines of greater or less complexity which allow us to exert large

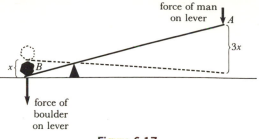

force of man on lever

A

3x

x

B

force of boulder on lever

Figure 6-17

forces by means of the application of much smaller forces. As in the use of a lever to pry loose a boulder, the applied force often is much less than the force exerted on the load. Figure 6-17 illustrates such a simple machine. The lever arms are assumed to be in a 3 : 1 ratio; therefore the point A moves down three times as far as point B moves up. The force at A is $\tfrac{1}{3}$ that at B (from the torque condition). The *work* done by the lever on the stone equals the work done on the lever by the man, if it is assumed that no energy is transformed into heat due to friction.

In general, however, because of friction the mechanical work done by the machine is less than the work done on the machine. We must now define some terms: The *actual mechanical advantage* (AMA) is the ratio of the output force to the input force. The *ideal mechanical advantage* (IMA) is the ratio of the input distance (i.e., the distance through which the input force moves) to the output distance (i.e., the distance moved by the load). The *efficiency* of the machine is the ratio of the work output to the work input.

A typical machine is illustrated schematically by the box in Fig. 6-18 which contains a collection of gears, pulleys, and levers the exact construction of which we do not need to know. Since $\mathbf{F}_1$ and $\mathbf{s}_1$ are parallel, the work input is $F_1 s_1$; similarly, the work output is $F_2 s_2$. Suppose we apply a force of 2 N and push the input button inward a distance of 0.02 m, and suppose that as a result the output platform moves 0.002 m, exerting a force of 15 N on the load.

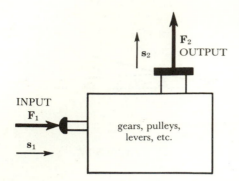

Figure 6-18 A simple machine (schematic).

Work input $= F_1 s_1$

$$= (2 \text{ N})(0.02 \text{ m}) = 0.04 \text{ J}$$

Work output $= F_2 s_2$

$$= (15 \text{ N})(0.002 \text{ m}) = 0.03 \text{ J}$$

Efficiency

$$= \frac{\text{work output}}{\text{work input}} = \frac{0.03 \text{ J}}{0.04 \text{ J}} = 0.75 = 75\%$$

$$\text{AMA} = \frac{\text{force output}}{\text{force input}} = \frac{15 \text{ N}}{2 \text{ N}} = 7.5$$

$$\text{IMA} = \frac{\text{input distance}}{\text{output distance}} = \frac{0.02 \text{ m}}{0.002 \text{ m}} = 10$$

Note that

$$\text{Efficiency} = \frac{\text{work output}}{\text{work input}}$$

$$= \frac{(\text{force output})(\text{output distance})}{(\text{force input})(\text{input distance})}$$

$$= \frac{\text{force output}}{\text{force input}} \Big/ \frac{\text{input distance}}{\text{output distance}}$$

$$= \text{AMA/IMA}$$

This is illustrated in the above example:

$$\frac{\text{AMA}}{\text{IMA}} = \frac{7.5}{10} = 75\%$$

At the most (100% efficiency) the distance ratio equals the force ratio and the AMA equals the IMA.

EXAMPLE 6-20

A tennis net is tightened by means of a crank (Fig. 6-19). The rope is wound around a spindle of radius 0.5 in., and the crank handle moves in a circle of radius 6 in. If the efficiency is 60%, what is the force in the rope when a player exerts a force of 25 lb on the handle in the most effective direction?

The distances moved by the crank and by the spindle are found by considering one complete revolution of the crank:

$$\text{IMA} = \text{distance ratio} = \frac{2\pi(6 \text{ in.})}{2\pi(0.5 \text{ in.})} = 12$$

Since efficiency = AMA/IMA,

$$0.60 = \text{AMA}/12$$

$$\text{AMA} = 7.2$$

But AMA = force ratio, so

$$7.2 = F/(25 \text{ lb})$$

$$F = \boxed{180 \text{ lb}}$$

EXAMPLE 6-21

In the pulley system shown in Fig. 6-20*a*, the weight of the load is borne equally by four ropes. The downward rope on the right in (*a*) doesn't count, because it merely changes the direction of the force **F**; see the equivalent dia-

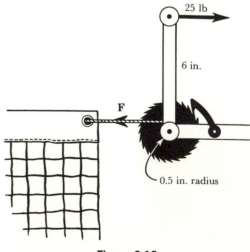

Figure 6-19

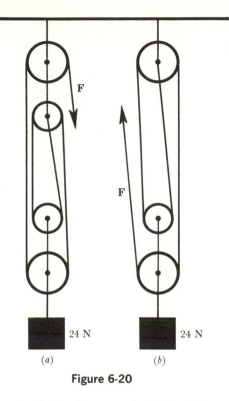

(a)   (b)

Figure 6-20

gram in (b). The distance ratio is 4:1, which can be verified experimentally or by a little geometrical reasoning. (We know that the distance ratio must be 4:1, since the load is shared among four ropes and hence IMA = 4.) If a load of 24 N is lifted by application of 10 N,

$$\text{Efficiency} = \frac{\text{AMA}}{\text{IMA}} = \frac{24 \text{ N}/10 \text{ N}}{4} = \frac{24}{40}$$

$$= \boxed{60\%}$$

The IMA and AMA may be less than 1, for a machine which is designed to convert small motions into large ones. A bicycle pedal is an example; the foot of the rider moves about 2.3 in. for every foot moved by the rim of the wheel. Here the applied force must be greater than the output force, even for 100% efficiency. Convenience is the goal, rather than magnification of force.

An inclined plane, while it is not usually thought of as a machine, does have a mechanical advantage; by means of a plane a heavy weight may be raised by application of a smaller force. Indeed, the erection of the Pyramids, a prodigious feat accomplished through use of human labor, was made possible through ingenious use of ramps and inclined planes.

EXAMPLE **6-22**

A team of movers use an inclined plane to get a piano which weighs 3600 N over a doorstep (Fig. 6-21). The plane is 3.0 m long and the step is 0.27 m high. If the movers apply a force of 500 N, what are the IMA, AMA, and efficiency?

$$\text{IMA} = \text{distance ratio} = \frac{3 \text{ m}}{0.27 \text{ m}} = \boxed{11.1}$$

$$\text{AMA} = \frac{3600 \text{ N}}{500 \text{ N}} = \boxed{7.2}$$

$$\text{Efficiency} = \frac{\text{work output}}{\text{work input}} = \frac{(3600 \text{ N})(0.27 \text{ m})}{(500 \text{ N})(3 \text{ m})}$$

$$= \frac{972 \text{ J}}{1500 \text{ J}} = \boxed{65\%}$$

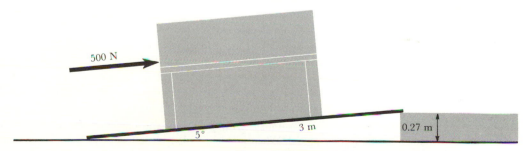

Figure 6-21

The efficiency could also be found by dividing the AMA by the IMA: $7.2/11.1 = 65\%$.

The work done against friction = (input of mechanical work) − (useful output of mechanical work):

$$W = 1500\,\text{J} - 972\,\text{J} = 528\,\text{J}$$

This work against friction becomes heat energy, and the plane and piano get slightly warmer.

Within the system (men + piano + plane + earth), energy has been conserved. We start with chemical PE (assuming the men ate a good nourishing breakfast). After the operation, this chemical PE has become partly gravitational PE, as the piano was lifted (i.e., as the piano and the earth were separated from each other), and partly thermal KE, as the molecules of plane and piano were speeded up, and as heat energy was released in the muscles of the men during their effort.

■ PROBLEMS

6-C8 Make some measurements on Fig. 6-16, and determine the coefficient of restitution for the bouncing tennis ball.

6-C9 A billiard ball of mass m moving at $+2$ m/s collides head-on with an equal, stationary ball. The coefficient of restitution is 0.8. Compute the velocity of each ball after the collision.

6-C10 What fraction of the original KE of the system is lost in the collision of Prob. 6-C9?

6-C11 What is the coefficient of restitution of a golf ball which, when dropped from a height of 100 cm, rebounds to a point which is 19 cm below the starting point?

6-C12 A 4 lb body slides along a horizontal surface for which the coefficient of kinetic friction is 0.4. When its speed is 12 ft/s, it strikes a stationary 5 lb body. The second body slides 2.5 ft before coming to rest. Calculate the coefficient of restitution for the impact.

6-C13 The inflation of the official basketball is standardized so that when dropped from a distance of 6 ft (measured from the bottom of the ball) it will rebound to a height of 49 to 54 in., measured from the top of the ball. The ball is 10 in. in diameter. What is the coefficient of restitution of a ball which is of the maximum permissible "liveness"?

6-C14 Solve Prob. 6-B41, if the coefficient of restitution is 0.8.

6-C15 A ball is dropped from a tower of height h at the same instant that another ball of equal mass is projected upward from the bottom of the tower, with a velocity just enough to raise it to the top of the tower. Show that if the balls collide head-on, the falling ball will rise, on the rebound, to a height which is $(3 + e^2)(h/4)$ above the ground, where e is the coefficient of restitution. Discuss the answer when $e = 1$.

6-C16 A windlass whose efficiency is 80% is used to raise a bucket of water from a well. The diameter of the axle is 4 in., and the radius of the crank handle is 10 in. The bucket weighs 5 lb empty, and it holds 10 gallons of water ($1.34\,\text{ft}^3$). (*a*) What is the IMA? (*b*) What is the AMA? (*c*) What force, applied in the most advantageous way, must be applied to the crank handle in order to raise the bucketful of water?

6-C17 A chain hoist in a garage has pulleys arranged so that the load rises 1 cm for every 60 cm moved by the worker's hand. When an engine block weighing 1000 N is hoisted 80 cm, the mechanic expends 1600 J of energy. Calculate (*a*) the IMA; (*b*) the efficiency; (*c*) the AMA; (*d*) the force exerted by the worker.

6-C18 A trunk weighing 800 N is pulled up an inclined plane at a steady speed by a force of 100 N parallel to the plane. The vertical rise is 0.20 m, and the plane is 4.00 m long. Calculate   (*a*) the IMA;   (*b*) the AMA;   (*c*) the efficiency.

6-C19 What is the IMA of the lever system consisting of the radius bone pictured in Fig. 5-18 (p. 94) if the load is vertical as in the figure.

■ REFERENCES

1 Ordway, F. I., "Principles of Rocket Engines," *Sky and Telescope,* **14,** 48 (1954).

2 Stacy, R. W., D. T. Williams, R. E. Worden, and R. O. McMorris, *Essentials of Biological and Medical Physics* (New York: McGraw-Hill, 1955), pp. 382–383; Lewis, H. W., "Ballisto-cardiography," *Sci. American,* Feb., 1958, p. 89. Conservation of momentum in the study of heart action. The subject lies on a movable platform, and, as the heart ejects blood, the platform recoils.

3 Mayer, Julius R., "The Conservation of Energy," *Phil. Mag.,* **24,** 371 (1862). Translation in Magie, *A Source Book in Physics,* pp. 196–203.

4 Joule, James P., "On the Mechanical Value of Heat," *Phil. Mag.,* various memoirs in the period 1843–50. Excerpts in Magie, *A Source Book in Physics,* pp. 203–11.

5 Mackay, R. S., "On the Strength of Insects," *Am. J. Phys.,* **26,** 499 (1958).

6 Ferguson, E. S., "The Measurement of the 'Man-Day'," *Sci. American,* Oct., 1971, p. 96.

7 Freeman, I. M., "The Dynamics of a Roll of Tape," *Am. J. Phys.,* **14,** 124 (1946).

8 Stull, J., *Conservation of Momentum—Inelastic Collisions; Conservation of Momentum—Elastic Collisions; Conservation of Energy* (films).

7-1 Frames of reference

To study the motion of a billiard ball, a player uses the table top as his frame of reference. In physics, a *frame of reference* is a set of three mutually perpendicular axes, such as the $x, y,$ and z axes, relative to which positions in space can be measured. The space coordinates of a body depend on the location of the origin of the frame of reference, and on the orientation of the axes. They also depend on the time t if the body is in motion; until the start of the twentieth century it was assumed that the rate of flow of time is the same for observers in different frames of reference even if these frames are in relative motion.

The relative motion of two frames of reference is illustrated by a numerical example (Fig. 7-1). At an airport, a suitcase is on a conveyor belt 1 ft above the floor, moving horizontally with a uniform velocity of 2 ft/s relative to the earth. The luggage is 5 ft from a seam O' on the belt which is considered to be the origin of the moving frame of reference FR'. The origin of the stationary frame of reference FR is a paint mark O on the floor.[*] At time $t = 0$ the seam is 7 ft to the right of the paint mark. A child was at the suitcase at $t = -2$ s, and ran at 0.5 ft/s forward along the belt. You can easily answer questions about the suitcase and the child, using "common sense" plus the relative velocity equation 3-15. We summarize the results for $t = 0$ and for $t = 3$ s. You should check your understanding by verifying the entries in the table. Note that t (in frame FR) and t' (in frame FR') are assumed to be always equal to each other. This example illustrates the way Galileo and Newton dealt with relative motion. It would not be difficult to use these methods to derive equations, called the *Galilean transformation,* which relate the coordinate values in the two frames.

We have stated earlier that Newton's laws apply to any system or any part of a system of bodies. An *inertial frame* is a frame relative to which Newton's laws are valid. It turns out that any other frame in uniform motion relative to an inertial frame is also an inertial

---

[*] Of course, it is only relative motion that is important; we could equally well call the belt stationary and the floor moving.

Joseph Hafele and Richard Keating pose with two atomic clocks prior to taking off on a round-the-world flight to test the time dilation predicted by the Einstein theory of relativity. They found that the rates of the moving clocks depended on their velocities relative to a clock on the ground. (Wide World Photos)

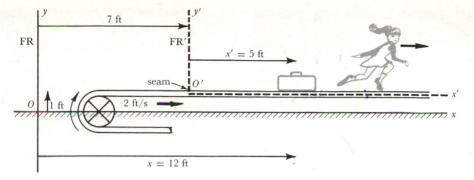

Figure 7-1 Two frames of reference which have a relative velocity of 2 ft/s, shown at $t' = t = 0$.

	(In frame FR')				(In frame FR)			
	t'	x'	y'	v'	t	x	y	v
Coordinates of suitcase	0	5	0	0	0	12	1	2
	3	5	0	0	3	18	1	2
Coordinates of running child	0	6	0	0.5	0	13	1	2.5
	3	7.5	0	0.5	3	20.5	1	2.5

frame. This powerful and simplifying result arises from the fact that Newton's second law is concerned not with velocity as such, but with acceleration, which is the rate of change of velocity. No accelerations of any space coordinates are introduced by constant relative motion of two frames of reference. In our example, if the earth is an inertial frame, then so is the conveyor belt.

Not all frames are inertial; for instance, a frame attached to the rotating earth is, strictly speaking, not an inertial frame of reference. This is illustrated by the circular motion of winds in a hurricane. A large air mass veers toward the east as it moves northward from the equatorial region (Fig. 7-2). This violates Newton's first law, as viewed by someone whose frame of reference is attached to the earth, for the air molecules are subject to no net force, and yet they fail to move with uniform velocity in a straight line. However, an observer from outer space, whose frame of reference might be the "fixed" stars, would see no conflict, because he would see that the earth is turning, and the eastward velocity of the earth at the equator (about 1000 mi/h) is greater than the eastward velocity (about 700 mi/h) at the latitude of the target, Newfoundland. Any air mass leaving the West Indies has a greater eastward component of velocity than does the target, so the air "gets ahead of" the target and moves toward the right. We see

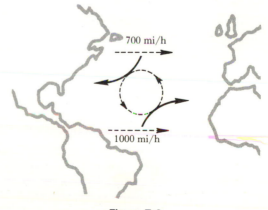

Figure 7-2

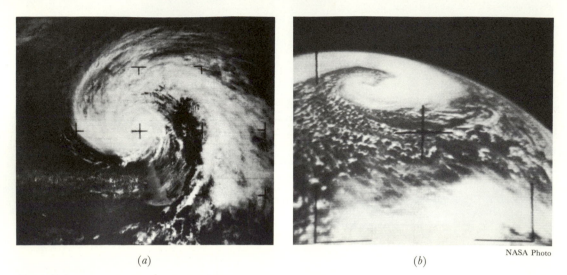

(a)   (b)

Figure 7-3 Satellite photographs showing hurricane wind patterns (a) counterclockwise in the Northern Hemisphere and (b) clockwise in the Southern Hemisphere. In (a) the coastline of Florida is seen south of the "eye" of Hurricane Alma, June 9, 1966. The storms "wind up" as viewed from above.

that there is an "unexplained" or "fictitious" force on the air, from the point of view of the observer whose frame of reference is attached to the rotating earth, and therefore his frame is not an inertial one. Similarly, air moving southward veers toward the west because the target is moving eastward faster than the air. The resulting pattern of winds is counterclockwise in the Northern Hemisphere; the same argument shows that hurricanes in the Southern Hemisphere are clockwise (Fig. 7-3).

Fictitious forces arising from the rotation of the earth are often called *Coriolis forces*, after the French physicist Gaspard Coriolis who, in 1835, first studied them experimentally. In daily life the effects of Coriolis forces are small, and we can consider the earth to be an inertial frame of reference for most purposes. Only when large distances and times are involved do the small fictitious forces due to the rotation of the earth produce significant effects. Fictitious forces are also called *inertial forces* or *pseudoforces*; in the next section we will see that inertial forces of considerable magnitude can arise

when a frame of reference has a large acceleration.

7-2 Weight, weightlessness, and artificial gravity

The weight of a body whose mass is m is the downward force of gravitation upon it, given by Newton's law of gravitation. We call this the *true weight* of the body because it does not depend on the acceleration of our frame of reference. Thus, if we consider only the earth's gravitational field and ignore the moon, the sun, and the other planets, a body's weight can never be zero except at the center of the earth. At the earth's center, the gravitational force due to an off-center element of mass Δm is balanced by an equal and opposite gravitational force due to an equal mass $\Delta m'$ located symmetrically opposite to Δm (see Fig. 7-4). Assuming the earth to be of uniform density,[*] as a body moves

---

[*] This is not in fact true, and the weight of a body in a mine can only be calculated accurately by taking account of the dense core of the earth.

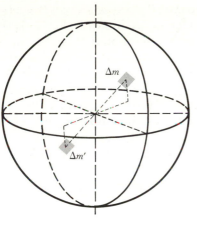

Figure 7-4

through a hypothetical tunnel up toward the surface of the earth, its weight increases and reaches a maximum value at the surface. This maximum weight is mg, and also equals GmM/R^2, where M is the mass of the earth and R is the radius of the earth. Above the earth's surface the weight decreases and is given by GmM/r^2 at points where $r > R$. It is sometimes stated that a space probe can get far enough away from the earth to "escape the earth's gravitational pull," but this statement is wrong. At any finite distance r there is *some* gravitational pull; only as $r \to \infty$ does the true weight of a body approach zero.

EXAMPLE **7-1**

The mass of the moon is 7.4×10^{22} kg, and its radius is 1740 km (about 1080 mi). Calculate the magnitude of the acceleration due to gravity (*a*) on the moon's surface and (*b*) 1000 km above the moon's surface. Ignore the gravitational pull of the earth.

(*a*) The weight of a test body of mass m is mg.

$$mg = G\frac{mM}{r^2}$$

$$g = \left(6.7 \times 10^{-11}\frac{\text{N}\cdot\text{m}^2}{\text{kg}^2}\right)\frac{(7.4 \times 10^{22} \text{ kg})}{(1.74 \times 10^6 \text{ m})^2}$$

$$= \boxed{1.6 \text{ m/s}^2}$$

This is about $\frac{1}{6}$ of the value of g at the earth's surface.

(*b*) At an altitude of 1000 km, the distance from the moon's center is 2740 km. Since the force on m is inversely proportional to the square of the distance, we use a proportion:

$$\frac{m(g')}{m(1.6 \text{ m/s}^2)} = \frac{(1740 \text{ km})^2}{(2740 \text{ km})^2}$$

$$g' = \boxed{0.65 \text{ m/s}^2}$$

Although the true weight of a body depends only on its position relative to other gravitating bodies such as the earth, the moon, and the sun, the *apparent weight*—the sensation of weight—can be radically affected by the acceleration of the observer's frame of reference. For example, an astronaut whose true weight at the earth's surface is 600 N is sitting on a seat cushion in his capsule prior to launch. He is in equilibrium, and the downward 600 N of gravitational force (his true weight) is balanced by an upward 600 N caused by the compression of the seat cushion. The apparent weight is the actual force between the astronaut and the cushion which supports him; at this moment his apparent weight equals his true weight. Now suppose a malfunction occurs during lift-off and the astronaut is ejected upward, still strapped in his seat. The astronaut and seat are now projectiles having acceleration **g**, where **g** is a downward vector. There is no force between the astronaut and the seat; he has no sensation of weight and he is "weightless." He regains his apparent weight as soon as his parachute opens; the lift of the chute (which we imagine to be attached to the seat) causes the seat cushion to be compressed against him, giving the astronaut the sensation of weight. Of course, the sensation of weightlessness includes effects within the body; while he is weightless the astronaut's internal organs no longer press against each other in the normal fashion. Physiological changes due to prolonged weightlessness may be severe.

For a more familiar example of the change in weight caused by the acceleration of a frame of reference, we consider a man in an elevator.

EXAMPLE **7-2**

A man of mass 100 kg stands on a scale in an elevator which has an upward acceleration of 2 m/s². What is his apparent weight?

The apparent weight of the man equals the push P of the platform of the scale against the soles of his feet (Fig. 7-5a). The man's true weight, mg, is $100 \times 9.8 = 980$ N.

From Newton's second law applied to the man,

$$\text{net } F = ma$$

$$P - 980 \text{ N} = (100 \text{ kg})(+2 \text{ m/s}^2)$$

$$P = 980 \text{ N} + 200 \text{ N} = \boxed{1180 \text{ N}}$$

The man apparently weighs 200 N more than when at rest—a gain of about $\frac{1}{5}$ of his true weight. The reading of the scale changes from 980 N to 1180 N during the acceleration.

There are two equivalent ways of analyzing the forces in Example 7-2. (a) To an outside observer the acceleration is known to exist (relative to an inertial frame), and the problem is solved by applying Newton's

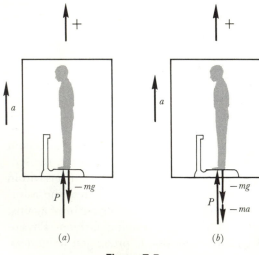

(a)   (b)

Figure 7-5

second law: $P + (-mg) = ma$; $P = m(g + a)$. These forces are shown in Fig. 7-5a. For an upward acceleration $a > 0$, $P > mg$, so the floor (scale platform) pushes upward with a force greater than the man's weight. According to Newton's third law, the man pushes downward on the scale with a force of magnitude P, and therefore the scale registers a force which is greater than mg. Similarly, when $a < 0$ (downward acceleration), the apparent weight is less than mg, giving rise to a "sinking feeling" in the pit of the stomach.

(b) If the man does not know that the elevator is accelerating, he considers himself to be in equilibrium under the action of two forces (Fig. 7-5b): the upward push of the scale platform P, and a downward "gravitational" force $-m(g + a)$. The fictitious inertial force $-ma$ which has arisen because of the (unknown to him) acceleration of his frame of reference is in every respect equivalent to a gravitational force. The man is at liberty to say either "someone accelerated the elevator upward" or "someone turned on an extra downward gravitational force."

We can generalize to say that an inertial force arises whenever a frame of reference is accelerated. In fact, if the frame's acceleration is **a**, an inertial ("fictitious") force arises given by $-m\mathbf{a}$, in a direction opposite to the acceleration of the frame of reference. The inertial force is an "artificial gravity" to a person in the accelerated frame, and is in no way distinguishable in its effects from "real gravity." Albert Einstein made this equivalence a cornerstone of his theory of gravitation in the general theory of relativity.

7-3 Galilean relativity

In his *Dialogs Concerning Two New Sciences* (1638) Galileo describes a "thought experiment" in which he imagines that a sailor drops an object from the top of a mast of a moving ship. The ship is moving horizontally with constant velocity; where does the object strike the deck? Let us study Galileo's example using our terminology about frames of reference. The ship and the earth are both inertial frames (ignoring the small Coriolis acceleration due to the earth's rotation). The two frames have a *constant* relative velocity. In each frame of reference the laws of motion are perfectly normal. Our Problem 3-B18 (p. 51) is just such a problem. A knife fell 19.6 m (which required 2 s) and during this time the ship (and knife) moved 20 m eastward (Fig. 7-6). Thus the knife hit the deck at the foot of the mast. In the ship's frame of reference (denoted by FR_{ship}) the knife was released from rest, and it fell straight down. In the earth's frame of reference FR_{earth} the knife was projected horizontally at 10 m/s, and it followed the usual parabolic path downward and to the east. Al-though the appearance of the trajectory differed in the two frames, the same laws of motion—Newton's laws—were applicable in each of the two inertial frames of reference.

Newton clearly stated that the *form* of any law of mechanics is not changed by the relative *uniform* velocity of an observer. Although a general proof is not difficult, we will be content with a simple illustration. In Sec. 6-12 we saw that during a *perfectly elastic* collision between a moving ball and an identical stationary ball, the balls simply exchange velocities; both the momentum and the kinetic energy of the system remain constant. These conclusions ("laws of mechanics") are true in the laboratory frame of reference FR_{lab} (Fig. 7-7):

Momentum:

$$mv_0 + 0 \stackrel{?}{=} 0 + mv_0$$

$$mv_0 = mv_0$$

Kinetic energy:

$$\tfrac{1}{2}mv_0^2 + 0 \stackrel{?}{=} 0 + \tfrac{1}{2}mv_0^2$$

$$\tfrac{1}{2}mv_0^2 = \tfrac{1}{2}mv_0^2$$

Now let's look at the same event from a different frame of reference. The center of mass of the system is always midway between the two equal balls; thus the c.m. frame is moving to the right with velocity $\tfrac{1}{2}v_0$ relative to the laboratory frame. Figure 7-8 shows what the collision looks like in the new frame of reference ($FR_{c.m.}$). Note that the relative velocity of approach is v_0 in either frame, and the relative velocity of separation is

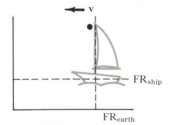

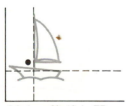

Motion of knife in FR_{ship}

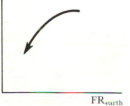

Motion of knife in FR_{earth}

Figure 7-6

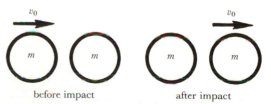

before impact   after impact

Figure 7-7 Collision of equal balls in laboratory frame FR_{lab}.

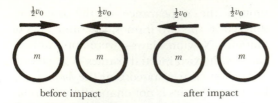

$\frac{1}{2}v_0$ $\frac{1}{2}v_0$ $\frac{1}{2}v_0$ $\frac{1}{2}v_0$

m m m m

before impact   after impact

Figure 7-8 Collision of equal balls in c.m. frame $FR_{c.m.}$.

also v_0 in either frame. Now we see if the momentum and the KE of the system have changed during the collision.

Momentum:

$$m(\tfrac{1}{2}v_0) + m(-\tfrac{1}{2}v_0) \overset{?}{=} m(-\tfrac{1}{2}v_0) + m(\tfrac{1}{2}v_0)$$

$$0 = 0$$

Kinetic energy:

$$\tfrac{1}{2}m(\tfrac{1}{2}v_0)^2 + \tfrac{1}{2}m(-\tfrac{1}{2}v_0)^2$$
$$\overset{?}{=} \tfrac{1}{2}m(-\tfrac{1}{2}v_0)^2 + \tfrac{1}{2}m(\tfrac{1}{2}v_0)^2$$
$$\tfrac{1}{4}mv_0^2 = \tfrac{1}{4}mv_0^2$$

In the two frames of reference the numerical values of the total momentum are different (mv_0 in FR_{lab} and 0 in $FR_{c.m.}$). Likewise the numerical values of the total KE are different ($\tfrac{1}{2}mv_0^2$ in FR_{lab}, $\tfrac{1}{4}mv_0^2$ in $FR_{c.m.}$). Nevertheless, the *forms* of the laws are unaltered: during a perfectly elastic collision, in any inertial frame, the total momentum is constant and the total KE is constant.

Galilean relativity, then, states that the forms of the laws of mechanics are the same in all inertial systems. This means that all inertial frames are equivalent. You would be unable to perform any experiment which would single out any one inertial frame as being "at rest." For example, it is perfectly possible to play ping-pong on a jet plane, an inertial frame moving at 800 ft/s relative to another inertial frame on the earth. You would not have to adjust your muscular reflexes to take account of the motion. The laws of mechanics would be the same as on the ground, and you would, in fact, be una-

ware of the plane's uniform motion. It would be impossible to tell which system was "actually" in motion: the plane moving forward at velocity v, with the earth stationary, or the ground moving backward at velocity v, with the plane stationary. This is what is meant by the statement that the relativity theory gives up the possibility of measuring absolute space and time. The plane could easily become a non-inertial frame of reference, if it becomes accelerated relative to the earth or some other inertial frame. You would hardly be able to avoid knowing if the plane banked to the left, hit an air pocket, or accelerated along a runway. Similarly, fictitious or inertial forces arise in non-inertial frames such as the elevator of Sec. 7-2.

Newton himself was greatly concerned about relativity, for he saw that the implications were tremendous. He felt it necessary to state what we have taken for granted—that space and time are "absolute." This means that even if two observers are in relative motion, they will always agree on the length of an object, or on the duration of a time interval. Nothing arose to challenge these very reasonable assumptions until late in the 19th century, when an attempt was made to fit the new science of electromagnetism—especially light waves—into this system.

7-4 Einsteinian relativity

Historically, the modern relativity theory originated in an effort to dispose of certain difficulties in the Maxwell equations. These equations describe the "field theory" which links light, radio waves, and x rays together as various forms of electromagnetic radiation. The measured speed of light and other electromagnetic waves in a vacuum is

$$c = 2.997925 \times 10^8 \text{ m/s}$$

This constant which we round off to 3 significant figures as

$$c = 3.00 \times 10^8 \text{ m/s} \quad \text{or} \quad 186\,000 \text{ mi/s}$$

was actually predicted by Maxwell[*] from his theory of electricity and magnetism. However, it was recognized in the period 1890–1905 that fundamental inconsistencies existed between the Maxwell equations and the concepts of space and time used by Galileo, Newton, and their successors. False starts were made by a number of physicists; Einstein succeeded in showing that Newton, not Maxwell, was at fault. It was not a question of some one particular equation being wrong; rather, underlying concepts about space and time were found to be unnecessarily naïve.

A century ago, as today, the wave nature of light was an accepted fact, but physicists then were thinking exclusively in *mechanical* terms. All *mechanical* waves need a medium in which to propagate. Thus, water waves exist on the surface of a medium, the water; sound waves are transmitted through gases, liquids, or solids, but not through a vacuum. So, too, it was felt necessary to postulate a medium—the "ether"—in which electromagnetic waves could propagate. This substance would have to be miraculous indeed, for it must be completely transparent to light, have negligible density, and yet be so very rigid that vibrations as rapid as 10^{15} per second could take place in it. Whatever the properties of the ether, it seemed reasonable to use Galilean relativity to measure the velocity (or "drift") of the earth through this medium. Perhaps by knowing the extent of our motion through the ether we could understand it a little better.

Such an experiment was devised in 1887 by Michelson and Morley,[*] using the Michelson interferometer (Sec. 25-7). Suppose that an observer on earth is moving toward the east through a stationary ether (Fig. 7-9).

[*] James Clerk Maxwell (1831–1879), one of the great British physicists of the mid-nineteenth century.
[*] Albert A. Michelson (1852–1931) and Edward W. Morley (1838–1923), American physicists who collaborated while at Case Institute of Technology and the neighboring Western Reserve University, respectively.

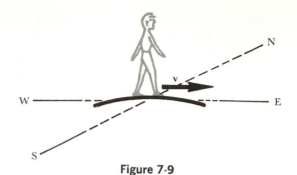

Figure 7-9

Relative to the observer, an "ether wind" is blowing past him, from east to west, and his measurements of the speed of light would be expected to depend on the ether wind's magnitude and direction. Michelson and Morley found that the velocity of a beam of light moving from east to west is the same as that of a beam of light moving from north to south or in any other direction. (Compare Prob. 3-B21, in which we found that a swimmer going upstream and back took longer for the trip than did the same swimmer going across the river and back.) It is as if the vector law of velocity addition (Sec. 3-10) is at fault, since one could surely find the speed of a river by measuring a motorboat's "upstream" and "downstream" velocities relative to the river bank. For light, no absolute space ("river bank," or ether) exists!

Albert Einstein (1879–1955) was led to a similar conclusion by quite different reasoning. Apparently unaware of the Michelson-Morley experiment, he reasoned that the velocity of light, relative to an observer, could never be zero—as would be predicted if the observer moved away from the source at a speed c. So Einstein was led to postulate that the velocity of light is unaffected by the motion of the observer (and everybody agreed that it would be unaffected by the motion of the source). This, too, gives up the idea of absolute space and absolute time, and makes the ether unnecessary.

We give this brief discussion about the

origin of the relativity theory in order to indicate why c, the speed of *light*, enters into what seem to be essentially *mechanical* problems. If we remember that c is the speed of *any* electromagnetic wave, and if we realize that the Maxwell theory of electromagnetism had to be accepted on the basis of its internal consistency, we can see how the constant c (essentially an electromagnetic constant) might be expected to creep into the new equations of mechanics which were designed specifically to avoid the conflict of mechanics with the existing electromagnetic theory of Maxwell. The constant c happened to be first discovered by workers in the field of electricity, long before electromagnetic waves were known to exist; Maxwell gave a *second* interpretation of this constant when he found it to be the speed of light and other electromagnetic waves; Einstein gave a *third* interpretation of this constant, as the maximum possible velocity of any body relative to any observer; further interpretations may well be made by future investigators.

For these reasons, a rigorous study of relativity theory must require thorough knowledge of electricity and magnetism; such study is beyond the scope of this book. However, it will be worthwhile to write down the two postulates upon which Einstein based his theory of special relativity:

1. **The forms of the laws of physics are the same in all inertial frames of reference.**

2. **Measurements of the speed of light always give the same numerical value, regardless of the relative velocity of the observer and the source.**

The first postulate is, of course, Galilean relativity—but with an important change. Instead of referring to the laws of *mechanics,* Einstein broadened Galilean relativity to include *all* laws of physics, including the laws of electromagnetism.

The second postulate specifically gives up the idea of absolute space and absolute time,

but it is in agreement with the results of the Michelson-Morley experiment. The radical nature of the second postulate is apparent if we consider a "thought experiment" in which we imagine two searchlights which are uncovered for a brief instant by two experimenters A and B who are both at the same point in space (Fig. 7-10). A is stationary, and B is in motion to the right with a speed of 18 600 mi/s, one-tenth the speed of light. After 1 s, a pulse of light has spread out from A's lamp a distance 186 000 mi and has reached point X. However, after 1 s the light from B has spread out 186 000 miles from B, and, since B has moved to B' during the second, the light has reached Y', which is 204 600 miles from A. (Note that the second postulate requires that the velocity of the light away from B must be 186 000 mi/s, regardless of B's velocity.) From A's point of view, the light from B's searchlight should only have reached to Y at this same instant, for otherwise its velocity relative to A would have been more than 186 000 mi/s. How can the light be at both Y and Y' at the same time? Einstein's postulate seems to lead to contradiction; in the resolving of such contradictions Einstein was led to reconsider space-time relations in general. He found that A and B are measuring their time intervals of 1 s by clocks that don't agree with each other, and the light was not "simultaneously" at Y and Y', because of this difference in the rates of ticking of their clocks. Eventually, a set of physical laws was found which was consistent with the two postulates. It is because of the second postulate that we are justified in speaking of "the" speed of light, equal to 3.00×10^8 m/s or 186 000 mi/s regardless of the motion of the source or the observer.

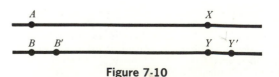

Figure 7-10

7-5 Time dilation

One major consequence of the Einstein relativity is that the duration of a time interval between two events depends on the frame of reference in which the events are observed. Newton explicitly denied this possibility when he wrote "Absolute, true, and mathematical time, of itself, and from its own nature, flows equably and without regard to anything external, and by another name is called duration." This view of the nature of time turned out to be wrong, but we cannot fault Newton, for the effect is observationally significant only for speeds comparable with that of light.

As a "thought experiment," let us imagine two boys playing a game. One boy, A, is on the sidewalk (frame of reference FR) while the other boy, B, is on a truck (frame of reference FR′) that is moving past A eastward at a uniform velocity v. The situation as seen by A is shown in Fig. 7-11a. The distance between the boys (at closest approach) is a, and they are equipped with identical clocks with which they can measure time intervals. The object of the game is for

a short pulse of light from A's flashlight to be reflected from B's mirror back to A. We have two events—when A sends the pulse, and when A receives the reflected pulse. What is the duration of the time interval between these two events?

Suppose that A turns on his flashlight at just the right time to illuminate B when B is opposite A. The velocity of light relative to either A or B is c, independent of the motion of the source or the observer (Einstein's second postulate). In order to hit his target, the boy A must turn on his light just before B is opposite him (B at the dotted position in Fig. 7-11a); the light travels a distance $2a$ at velocity c, so the time required for the round trip, t_0, is given by

$$t_0 = \frac{\text{distance}}{\text{speed}} = \frac{2a}{c} \tag{7-1}$$

This is A's measurement of the interval between the two events. We call t_0 the *proper time* between the two events, since it is the duration between two events which take place at the same place in some frame of reference (A's frame, in this case).

On the other hand, B describes the same

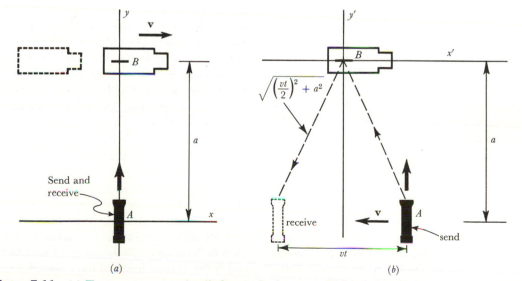

Figure 7-11 (a) Two events as seen in A's frame of reference. (b) Same two events as seen in B's frame of reference.

experiment somewhat differently. As B sees it (Fig. 7-11b), *he* is stationary and A is moving backward (to the west) with a velocity v. As before, A points his flashlight northward and, as before, he turns it on just before he is opposite B. The resultant velocity of the light relative to B is found by B to be equal to c, as it must according to Einstein's second postulate. According to B, his friend A has moved a distance vt during the interval between the two events, where t is the duration of the interval *measured in B's frame*. B now uses the Pythagorean theorem to find the distance the light has traveled during this time; he obtains $2\sqrt{(vt/2)^2 + a^2}$. Since the speed of light in B's frame is c, B concludes that the time interval is given by

$$t = \frac{\text{distance}}{\text{speed}} = \frac{2\sqrt{(vt/2)^2 + a^2}}{c} \qquad (7\text{-}2)$$

Now a little elementary algebra is needed to solve Eq. 7-2 for t. We obtain

$$t = \frac{2a}{\sqrt{c^2 - v^2}} = \frac{2a}{c\sqrt{1 - v^2/c^2}}$$

$$= \frac{t_0}{\sqrt{1 - v^2/c^2}} \qquad (7\text{-}3)$$

Even though their clocks are identical, A and B find different answers for the duration of the interval between the "sending" and the "receiving" events; t is greater that t_0. There *is* a difference between their frames, however, because the two events do not take place at the same place in B's frame (see Fig. 7-11b). Thus only t_0, measured at the same place in A's frame, is a "proper time." A considers that B's clock is "wrong," measuring an interval that is "too long." In an exactly similar experiment B could hold the flashlight and A the mirror. Now the interval measured in B's frame would be the proper time, and B would claim that A's clocks were giving too long an interval. Fortunately, we don't have to decide the argument, for it is impossible to tell which of the observers is "really" at rest (Einstein's first postulate).

We are sure of one thing—there is no "absolute" time, and A and B are both correct, each according to his own standards.

In general terms, we conclude that if an experiment has a duration t_0 as measured at some point in a frame of reference, then the same experiment has a longer duration t as measured by an observer in a frame of reference which is in uniform motion relative to the first frame.

In our example, the ratio of the two time intervals representing the same experiment is

$$\frac{t}{t_0} = \frac{1}{\sqrt{1 - v^2/c^2}} \qquad (7\text{-}4)$$

This same ratio, which is always greater than 1, holds true for any experiment viewed by two observers who are moving relative to each other. This is what is meant by *time dilation*, the stretching out of time.

The relativity of time has been experimentally observed. For example, an unstable particle called a muon can be created in the upper atmosphere by the action of cosmic rays (Sec. 30-10). A slow muon has a life expectancy t_0 of about 2×10^{-6} s, on the average, and then it decays into an electron and a neutrino. A very fast muon, traveling at nearly the speed of light, could, on the average, move only about 600 m in 2×10^{-6} s. However, many fast muons are observed to travel as far as 6000 m or more before decay. The theory of relativity helps explain this result. The muon's own "clock" would register the proper time, a short interval t_0 ($= 2 \times 10^{-6}$ s), while the duration t measured by an observer on the earth (a frame of reference which is moving relative to the muon) might be 20×10^{-6} s because of the time dilation. We see that the clock attached to the muon is running slow, and to the earth-bound scientists the muon reaching sea level is "younger" (hence more likely to be still "alive") than it would have been if it had been at rest. In a similar fashion, a space traveler moving at nearly the speed of light would not age as rapidly as his twin brother left behind on earth. All bodily processes would slow down (as judged by the stay-at-home twin), although the traveling twin would be unaware of the difference until he greeted his gray-bearded brother after a seemingly short trip! A traveling-clock experiment is shown on p. 140.

7-6 Relativistic mass increase

We shall use the law of conservation of momentum and the relativistic time dilation to derive the Einstein mass formula $m = m_0/\sqrt{1 - v^2/c^2}$. This equation concerns the *measurement of mass*, since m is the mass as measured by a moving observer, or by a stationary observer who looks at a moving body.[*]

For our thought experiment let us perform a collision experiment with two identical, perfectly elastic, spherical balls. The two boys A and B arm themselves with balls instead of flashlights and throw the balls with equal speeds so as to cause them to collide just when they are halfway between the truck and the sidewalk. Even though the balls are identical, their masses may not seem to be the same as judged by the two observers; each boy considers his own ball to have a mass m_0 (its *rest mass*) and his friend's ball to have a mass m.

As A describes the event (Fig. 7-12), the initial y-component of the velocity of ball b' is $-u'$, and the y-component of velocity of his own ball b is $+u$. After impact, the y-components are reversed and are $+u'$ for ball b' and $-u$ for ball b. The observer A now applies the law of conservation of momentum[*] to the elastic collision:

[*] It is the essence of relativity that only *relative* motion is important; the two measurements must give the same result, regardless of who is "stationary" (Einstein's first postulate).

[*] This is in accord with Einstein's first postulate, according to which the *forms* of the laws of physics do not depend on the relative motion of the two frames of

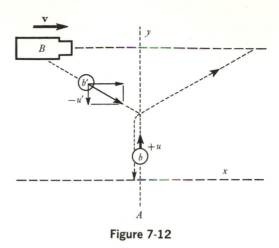

Figure 7-12

Total momentum in y-direction before impact = total momentum in y-direction after impact

$$m(-u') + m_0(u) = m(u') + m_0(-u)$$

$$2m_0 u = 2mu' \qquad (7\text{-}5)$$

Now the boys agreed to throw the balls at the same speed u, but A believes that B's clocks are running too slowly, and so he believes that B failed to give the proper velocity to his ball. Hence he deduces that u' is less than u; in fact, the boy A concludes that

$$u' = u\sqrt{1 - v^2/c^2}$$

where the square root (which is less than 1) is the reciprocal of the time factor given in Eq. 7-4. Putting this value for u' into Eq. 7-5, we get

$$2m_0 u = 2mu\sqrt{1 - v^2/c^2}$$

$$m = \frac{m_0}{\sqrt{1 - v^2/c^2}} \qquad (7\text{-}6)$$

This is the expression for the increase of mass with velocity. Here m_0 is the rest mass, as seen by the stationary observer A, and m is the mass of the ball which is moving relative to him.

It follows that "mass" is not a unique property of a body. Two observers, equally

reference. All momentums are judged by *one* observer, namely A.

competent, will measure the inertial mass of a body differently, if their velocities relative to the body are different. Note that Eq. 7-6 predicts that the mass becomes infinitely great as the relative velocity approaches c, the speed of light. Therefore, according to Newton's second law, an ever-increasing force is required to accelerate the body, and a physical body can never reach the speed of light, since an infinite force is not possible.

EXAMPLE **7-3**

A hypothetical spaceship travels away from the earth at a velocity of 2×10^8 m/s. A projectile is shot forward from the spaceship at 2×10^8 m/s relative to the spaceship. What is the velocity of the projectile relative to the earth?

We do not have to know the equation for relative velocity in order to look at certain features of the solution. Of one thing we can be sure—the velocity of the projectile relative to the earth is certain to be less than 3×10^8 m/s. Offhand, we would have expected 4×10^8 m/s, but the simple vector treatment of relative velocity (Sec. 3-9) is not valid at high velocities. The relativity result is 2.77×10^8 m/s (see Prob. 7-C6).*

The increase in mass of a moving electron had actually been observed by Bucherer and Kaufmann several years before the theory of relativity appeared. No satisfactory explanation was made until the mass equation, Eq. 7-6, appeared as a sort of unexpected by-product of Einstein's fundamental re-examination of space and time and relative motion.

7-7 Mass and energy

The most "practical" consequence of the theory of relativity is the possibility of inter-conversion between matter and energy. Einstein showed that whenever mass disappears and is converted into energy and vice versa, the same amount of energy (9×10^{16} J) always buys the same amount of mass (1 kg), and vice versa. We in the present generation are familiar with the conversion of matter into energy, for that is the source of the vast energy of the nuclear bombs. All the fragments, gases, ash, and so on, of a nuclear explosion would, if gathered together, have slightly less mass than the original uranium or hydrogen. The Einstein equation relating mass and energy is

$$\Delta E = (\Delta m)c^2 \tag{7-7}$$

$$J = (kg)(m/s)^2$$

where c is the fundamental constant of nature which, among other things, equals the speed of light in a vacuum (3.00×10^8 m/s).*

Einstein reasoned thus: By giving a body kinetic energy, we increase its mass according to Eq. 7-6. He concluded that, if *increase* of mass corresponds to *gain* in energy, the mass of a body even at rest must be equivalent to some energy. His conclusion (1905) was verified within less than 20 years, when Rutherford (1919) observed the conversion of mass into energy during the first experiments in transmutation of elements.

There is no room for doubt that the theory of relativity correctly describes the interconversion of mass and energy. Nuclear physicists have verified the equation $\Delta E = (\Delta m)c^2$ in countless precise experiments. The law of conservation of energy must, therefore, be worded in such a way as to include the energy equivalent of mass. The law has been generalized in another way. An isolated system is

_____

* Relative to the people on earth, the projectile is moving faster than it is moving relative to the people on the ship. As judged by the earthlings, the projectile therefore has more mass than as judged by the space travelers,

and so the same force (reaction force of the fuel) produces less acceleration. The projectile hence receives less final velocity, as viewed from earth, than would have been expected if the mass had not changed.

* If the history of physics had gone in a different way, or if man had evolved without sight, c might well have first been discovered and *defined* as the square root of the ratio of energy to mass equivalent.

one that does not have any interactions with objects outside the system, and the universe is, by definition, an isolated system. By extrapolation, and as a matter of scientific faith, the law of conservation of energy has been assumed to apply to the entire universe. Thus, finally, we state the law of conservation of energy in its most general form:

The total equivalent amount of energy in the universe remains constant.

This generalization, evolved by many scientists during a period of several centuries, has proved to be of tremendous importance for every branch of physical science. We shall use it frequently in our future work.

The conversion of mass into energy is of utmost importance to our understanding of nature. Aside from the spectacular controlled release of energy in the nuclear reactor and in uranium and hydrogen bombs, a less obvious but more important conversion of mass into energy takes place continually in the sun and other stars. As the end result of a chain reaction, 5×10^{19} kg of hydrogen is converted into nearly the same amount of helium in 1 s (see Sec. 30-5). The mass balance is not exact, however, and the sun's mass continually decreases, at the rate of 4×10^9 kg per second. This is the mass that is converted into energy, keeping the sun hot. The sun has been radiating energy (and losing mass) at this rate for several billion years. It speaks for the tremendous mass of the sun (1.97×10^{30} kg) that, had it been losing 4×10^9 kg (4 million metric tons) every second for a billion years, over 99.99% of the sun's mass would still remain!

In everyday life, however, the effects of relativity and the increase of mass with velocity are too small to be noticed.

EXAMPLE **7-4**

A rocket ship of rest mass exactly 4×10^3 kg is moving at 3000 m/s. What is its increase in mass? What is its KE?

Its new mass m is given by Eq. 7-6:

$$m = \frac{m_0}{\sqrt{1 - v^2/c^2}}$$

$$= \frac{4000.0000000 \text{ kg}}{\sqrt{1 - \frac{(3 \times 10^3 \text{ m/s})^2}{(3 \times 10^8 \text{ m/s})^2}}}$$

$$= \frac{4000.0000000 \text{ kg}}{\sqrt{1 - \frac{9 \times 10^6 \text{ m}^2/\text{s}^2}{9 \times 10^{16} \text{ m}^2/\text{s}^2}}}$$

$$= \frac{4000.0000000 \text{ kg}}{\sqrt{1 - 10^{-10}}} = \frac{4000.0000000 \text{ kg}}{0.99999999995}$$

$$= 4000.0000002 \text{ kg}$$

$$\Delta m = 4000.0000002 \text{ kg} - 4000.0000000 \text{ kg}$$

$$= \boxed{2 \times 10^{-7} \text{ kg}} = 0.2 \text{ mg}$$

This increase is about equal to the mass of the ink used to write an address on a rocket-mail letter carried by the ship.

The KE of the rocket ship is

$$\text{KE} = \tfrac{1}{2}mv^2$$

$$= \tfrac{1}{2}(4 \times 10^3 \text{ kg})(3 \times 10^3 \text{ m/s})^2$$

$$= \boxed{18 \times 10^9 \text{ J}}$$

We may say that 18×10^9 J of KE is equivalent to 2×10^{-7} kg of mass. This checks with $\Delta E = (\Delta m)c^2$, as you can easily verify; each kilogram of mass increase corresponded to 9×10^{16} J of energy increase, as it should.

This example clearly shows that the mass changes accompanying ordinary energy changes are tiny indeed. It is for this reason that we can apply the law of conservation of energy to the solution of problems with a clear conscience, ignoring these small relativity effects. In other words, we usually need to consider only chemical, thermal, electrical, or mechanical energies. The law of conservation of energy, even in this limited form, is a profound generalization, and we saw, in Sec. 6-9, how to use it in solving problems.

Let us look critically at the statement "mass can be converted into energy." We shall find that it is *rest mass* which can be so converted; the total (relativistic) mass of the universe remains constant, and in this sense mass is conserved, and also energy is conserved. Unless we use words carefully and consistently, we shall run into some difficulty in interpreting the law of conservation of mass and the law of conservation of energy.

Surely Example 7-4 seems to be no example of the conversion of mass into energy, or vice versa, since both the ship's mass *and also* its energy increased. This is like trying to keep books with all credits and no debits. In fact, the theory of relativity considers that *any* energy increase ΔE, not just a kinetic energy increase, is accompanied by an increase of mass Δm. The Einstein equation says that these changes are related by $(\Delta m)c^2 = \Delta E$, or $\Delta m = (\Delta E)/c^2$. Considering the whole system, ship plus fuel, we would find that the fuel had more mass before it was burned than afterward, because of its original PE.[*] In such problems mass and energy are each conserved, if by "mass" we mean total (relativistic) mass, including the mass equivalent of KE or PE.

Consider a 100 kg safe that is raised to the top of a building 100 m high. The *system*, earth + safe, gains PE, calculated from mgh as $(100 \text{ kg})(9.8 \text{ m/s}^2)(100 \text{ m}) = 98\,000 \text{ J}$. Thus the change in mass of the system is

$$\Delta m = \frac{\Delta E}{c^2} = \frac{9.8 \times 10^4 \text{ J}}{(3.00 \times 10^8 \text{ m/s})^2}$$

$$= 1.1 \times 10^{-12} \text{ kg}$$

The safe and earth are at rest in an inertial frame, but their rest mass has increased by 1.1×10^{-12} kg. Here a change in energy has been accompanied by a change in rest mass.

You may ask, "Where did the extra 1.1×10^{-12} kg of mass come from in the first place?" Suppose the safe had been hoisted to the top floor by a pulley system activated by the muscular efforts of several workmen. The mass of the workmen must have *decreased* during the process, for they used up chemical PE, perhaps from a ham-and-eggs breakfast. In a sense, mass "flowed" from men to safe as it was raised. No matter how far back you carry the questioning, you will find that the eventual source of the energy (and mass) was nuclear—more than likely from solar energy which caused the plants to grow which certain hogs and hens converted into ham-and-eggs-on-the-hoof, and so forth. But we know that the sun's energy comes to us as a result of nuclear changes. Thus it would seem that all or nearly all[*] of the energy changes on the earth ultimately come from mass changes in nuclear reactions, either in man-made nuclear reactors or in the gigantic natural reactor we know as the sun. What, then, is the difference, if any, between the changes taking place in the safe or in chemical reactions, and the changes taking place in a nuclear reactor? The difference is one of degree only.

Let us compare the common chemical heating reaction mentioned on p. 116 with some nuclear heating reactions to be discussed in Chap. 30:

Chemical fuel:

$$C + O_2 = CO_2 + \text{energy}$$

Nuclear fuel (fusion):

$$^1H + {}^3H = {}^4He + \text{energy}$$

Nuclear fuel (fission):

$$^{235}U = \text{fragment } A + \text{fragment } B + \text{energy}$$

Before combustion, the C and the O_2 are at rest and have no KE, and they have PE because of their separation. Because of this PE, they also have some extra rest mass.

---

[*] In an exact treatment of the rocket-ship problem, we would have to consider also the KE and the mass changes of the exhaust gases.

[*] The energy of the tides comes from the rotation of the earth; no one can yet say with certainty what was the original source of the earth's rotational KE, but this, too, may very likely have been a result of nuclear changes.

Therefore the rest masses of the C and O_2 add up to more than the rest mass of the CO_2, and we say that some of the rest mass of the universe has changed to energy. Similarly, on a much grander scale, in nuclear fusion the rest mass of a ^4He nucleus is less than the total rest mass of the ^1H and ^3H nuclei, which have PE because of their separation. In nuclear fission the ^{235}U nucleus is unstable, which means that PE is stored in the nucleus ready to be "triggered off." This stored energy means that ^{235}U has more rest mass than its fragments after fission. In all these cases, it is *rest mass* which has been converted into energy; total (relativistic) mass (which includes the mass equivalent of all forms of energy) has been conserved. We use the term *mass energy* to represent m_0c^2, where m_0 is the rest mass of a body. The mass energy of a 100 g cup of water, for instance, is tremendous, for it equals the energy that could be obtained if *all* the rest mass of the water were converted to energy. However, the changes that occur in mass energy are only a tiny fraction of the total mass energy of a body, except in certain special cases (to be discussed in Chaps. 29 and 30, where we study nuclear physics in more detail).

We have written the relationship between mass changes and energy changes as

$$\Delta E = (\Delta m)c^2$$

where the increased energy of a body is accompanied by an increase in mass. We can also write

$$E = mc^2$$

where E is the total energy (including the mass energy) and m is the total (relativistic) mass given by $m = m_0/\sqrt{1 - v^2/c^2}$. This equation places energy in a prominent place, without distinguishing too closely between the various forms that energy can take. The kinetic energy due to motion is then (total energy) $-$ (rest energy). Thus

$$KE = mc^2 - m_0c^2 \qquad (7\text{-}8)$$

This is the relativistic formula for KE; it is derived in Sec. 7-9. We expect that at velocities which are small relative to c, Eq. 7-8 will reduce to the familiar $\frac{1}{2}m_0v^2$ (see Prob. 7-C4). In fact, *all* relativity formulas must include the classical formulas of physics as special cases (Sec. 4-10). We see that the KE of a body is not a definite quantity, since it depends upon m, which is not the same for all observers. This dependence on m is new; but even in the classical relativity of Galileo and Newton the KE of a body depends on the observer. For example, consider an artificial satellite which is ejected in a forward direction from a launching rocket. The KE of the satellite is very much less as viewed from the rocket than as viewed from the earth, and, of course, is zero as viewed from the satellite itself.

7-8   Relativistic length contraction

Consider once more the muon described in Sec. 7-5. Let v be the magnitude of the velocity of the muon relative to the earth; v is also the magnitude of the velocity of the earth relative to the muon. How far did the muon travel (from point A to point B) during its decay period? There are two answers to this question. In 2×10^{-6} s (the proper time t_0, measured in its own frame of reference) the muon traveled $t_0v = 2 \times 10^{-6} v$ meters. Experimenters on the earth saw the muon move for a much longer time $t \ (= t_0/\sqrt{1 - v^2/c^2})$ equal to 20×10^{-6} s. Earthbound observers thus found the distance to be $tv = 20 \times 10^{-6} v$ meters. This is the proper length, since observers on earth were at rest relative to both A and B. In the muon's frame of reference, in which A and B were moving, the distance was only $t_0v = 2 \times 10^{-6} v$ meters. We see that there has been a *length contraction* along the direction of the muon's motion.

$$\frac{L}{L_0} = \frac{\text{measured length}}{\text{proper length}} = \frac{t_0 v}{tv} = \frac{t_0}{t}$$

$$= \sqrt{1 - v^2/c^2}$$

Thus

$$L = L_0 \sqrt{1 - v^2/c^2}$$

Note that v cancels out because it is the relative speed—that's the only thing the two observers can agree on! The length contraction ratio is numerically equal to the time dilation ratio.

Another way of looking at this is to consider it from the muon's viewpoint. The earth is rushing toward it at a speed v; because of the relative motion the 6000 m is contracted to only 600 m, and the muon can easily negotiate this foreshortened distance in the allotted time (2×10^{-6} s, by its own reckoning).

It is interesting to note that a length contraction, indeed of the identical amount derived here, was proposed about 1900 by Fitzgerald, to account for the negative result of the Michelson-Morley experiment, and independently by Lorentz, in an attempt to fix the theory of electromagnetism. However, their proposed contraction was of an *ad hoc* nature, not related to the fundamental concepts of space and time with which Einstein dealt about 5 years later.

In summary, we have seen that the time interval between two events is longer when measured by an observer who is moving relative to the frame in which the events take place. This time dilation is the basis for the relativistic mass increase, through the following argument: we must retain the form of the law of conservation of linear momentum, but because of time dilation the observed speeds of colliding bodies are altered. Therefore a corresponding change in the observed mass of a moving body is predicted, to conserve the quantity mv. The mass increase, in turn, leads to the equivalence of mass and energy, and thus to a profound generalization of the law of conservation of energy. We also found that the observed length of an object suffers a contraction along the direction of motion. All of these relativistic refinements of Newtonian mechanics which we have considered in this chapter are significant only for large relative velocities, where v^2/c^2 is appreciable. Thus Einsteinian relativity contains within itself the older Galilean-Newtonian relativity; it is entirely possible that Einsteinian relativity is a special case of some still more comprehensive theory yet to be discovered.

Before we leave the subject of relativity, a further point should be noted. Our study of Einsteinian relativity has been limited to the *special theory of relativity,* which applies to inertial frames of reference in *uniform* relative velocity (i.e., no acceleration of one observer relative to another). By experiments performed inside a steadily moving train, you cannot tell whether the trees and ground outside your window are moving backward or whether you are moving forward. The laws of physics are the same within the train as on the ground. But try to remain standing upright in a bus which suddenly starts up, or try to eat soup in a plane which swerves in a sharp unbanked curve (accelerated motion, as we shall see in Chap. 8). The laws of physics appear quite out of the ordinary in such situations, and the observer can explain the strange happenings in either of two ways: (*a*) his frame of reference is accelerated, or (*b*) a force acts upon him or on the soup. The *general theory of relativity* is an extension of the special theory to deal with accelerated motion, and in the general theory it is postulated that these two ways of looking at any experiment involving force are entirely equivalent. For instance, if a (weightless) astronaut in outer space is enclosed in a sealed box which is accelerated upward at 9.8 m/s², he experiences all the sensations of being on the earth subject to the downward force of gravity. No experiment he can perform inside the box can help him decide whether his frame of reference has been accelerated, or whether gravity has suddenly been "turned on." The general theory of relativity tells us much about the nature of gravitation and other forces, but this is far beyond the scope of this text.

■ SUMMARY

An inertial frame of reference is one in which Newton's laws of motion hold true; an accelerated frame of reference is not an inertial frame. Inertial forces are fictitious forces which arise because of the acceleration of the frame of reference. In an accelerated frame the apparent weight of a body differs from its true weight.

In the classical relativity of Galileo and Newton the forms of the laws of mechanics are the same in all inertial frames, unaffected by the uniform relative velocity of the observer and the frame in which the experiment is performed. Einstein broadened the theory of relativity to apply to all laws of physics, including electromagnetism. This implies that the speed of light in a vacuum is a constant c for any observer who is in uniform motion relative to the source of light.

Some consequences of relativity are: time dilation, length contraction, mass increase, and the equivalence of energy and mass. When v/c is small compared to 1, Einsteinian relativity reduces to the earlier Galilean relativity.

The mass energy of a body is $m_0 c^2$, where m_0 is the rest mass; the total (relativistic) energy is mc^2. When an isolated system undergoes change, the total (relativistic) mass of the system remains unaltered, and the total energy content of the system remains unaltered. These are the laws of conservation of mass and conservation of energy. No energy can be created or destroyed unless a corresponding amount of rest mass disappears or appears; this conversion of matter into energy is governed by the Einstein equation $\Delta E = (\Delta m)c^2$.

■ CHECK LIST

inertial frame of reference
inertial force
Coriolis force
true weight
apparent weight
Galilean relativity
Einstein's postulates
relativistic time dilation
proper time
relativistic mass increase

rest mass
relativistic length contraction
energy equivalence of mass
$t = t_0 / \sqrt{1 - v^2/c^2}$
$m = m_0 / \sqrt{1 - v^2/c^2}$
$L = L_0 \sqrt{1 - v^2/c^2}$
$\Delta E = (\Delta m)c^2$
$KE = mc^2 - m_0 c^2$

7-1 A passenger in a car is moving with constant velocity at 20 ft/s toward the south. As he passes a child on the sidewalk, the child throws a ball straight up at 32 ft/s. (a) How long is the ball in the air? (b) Relative to the car, where does the ball strike the ground? (c) Describe the path of the ball relative to the child. (d) Describe the path of the ball relative to the passenger in the car.

7-2 If a heavy ball is dropped from a great height, the rotation of the earth causes it to miss, by a small amount, the spot directly underneath the point of release. Does the ball alight to the north, south, east, or west of the expected point? In which direction is the fictitious force on the ball?

7-3 You are a passenger on an interstate bus; the bus is well lighted inside but it is a very dark night outside. A child at the rear of the bus rolls a steel ball down the center of the aisle. As you watch the motion of the ball, you observe it to move in a straight line until it is just past your seat, then it swerves to the left as it moves forward and strikes a mysterious black suitcase resting at the side of the seat of a traveling salesman. Make two equivalent statements as to the possible cause of the ball's behavior.

7-4 A body of mass m experiences a fictitious force due to the acceleration of its frame of reference. Does this fictitious force involve the inertial mass or the gravitational mass of the body?

7-5 The projectile of Example 7-3 emits a beam of light which can be seen from both the spaceship and the earth. What is the velocity of this light as measured by a scientist on the spaceship? What is the velocity of the light as measured by a scientist on earth?

7-6 Since a formula for the speed of light in a vacuum can be derived from the laws of electromagnetism, does this mean that Einstein's second postulate is in fact just a special case of the first postulate which concerns the invariant form of *all* laws of physics?

7-7 Do you think that the traveling twin (Sec. 7-5) upon his return to earth would "really" be younger than his brother?

7-8 An electron's speed cannot be greater than c; is there some upper limit to the momentum that an electron can have?

7-9 If mass is a form of energy, is it then a true statement that a wound-up watch has more mass than does the same watch when it has run down?

7-10 In principle, is there *any* change of weight (however slight) when the flash bulb of Ques. 6-15 (p. 128) is fired?

7-11 A nuclear particle in a large circular accelerator is traveling at $0.99c$; is it possible to double the kinetic energy of this particle by speeding it up?

7-12 Is the equivalence of mass and energy a consequence of the special theory of relativity or of the general theory of relativity?

■ PROBLEMS

7-A1 How long did it take the girl in Fig. 7-1 to run from the suitcase to the position shown in the figure?

7-A2 In Fig. 7-1, what are the values of x' and x at $t' = t = 5$ s?

7-A3 An electron of rest mass m_0 is moving at a speed $0.6c$. What is its relativistic mass?

7-A4 How fast is a cosmic ray particle moving when its mass is double its rest mass?

7-A5 How much energy would be obtained if the rest mass of 10 milligrams of coal were to be entirely converted into energy? How high could this much energy lift a battleship of mass 100 000 tons?

7-B1 In Fig. 7-1, the girl's weight is 64 lb. (a) What is her KE relative to the floor? (b) What is her KE relative to the suitcase?

7-B2 A 128 lb girl is in an elevator which is accelerating downward at 8 ft/s^2. (a) What is the net force on the girl? (b) What is the apparent weight of the girl (i.e., the upward force of the floor upon her)?

7-B3 What upward acceleration of an elevator will cause a passenger's apparent weight to increase from 800 N to 900 N?

7-B4 A crate of glassware of mass 200 kg is on the smooth, flat, horizontal rear deck of a pickup truck which is traveling westward at 14 m/s. The truck approaches a traffic light and slows down to 5 m/s in 3 s. What is the magnitude and direction of the fictitious inertial force on the crate?

7-B5 The Great Pyramid of Cheops has a mass of about 6.5×10^9 kg (see Prob. 2-A8), and its c.g. is about 37 m above the base. (a) What is the weight of the pyramid? (b) What is the PE of the stones which make up the pyramid? (c) How much mass flowed into the pyramid while the workmen erected it in about 2890 B.C.? Where did this mass come from?

7-B6 How fast would a rocket of mass 10^7 kg have to move in order for its mass to increase by 2 g?

7-B7 A space traveler takes off at 1.5×10^8 m/s, leaving his twin brother behind on the earth. One day later (measured on the ship), the traveling twin's heart has made a normal 100 000 beats. (a) What is the duration of the "experiment" as measured by an observer on the earth? (b) How many heartbeats has the stay-at-home twin's heart made?

7-C1 Electrons move through a high-energy linear accelerator 3 km long at a speed so close to c that their masses are 10^4 times their rest masses. (a) How long is required for the trip, as measured in the laboratory frame of reference (use $v \approx c$)? (b) In the electron's frame of reference, what is the length of the tube through which they move? (c) What is the proper time for the trip?

7-C2 A proton of rest mass 1.67×10^{-27} kg has a relativistic mass three times its rest mass. What is its KE?

7-C3 In 1969 James A. McDivitt and others spent 10 days in orbit around the earth, moving at about 7.8×10^3 m/s. (a) What was the time dilation factor? (*Hint:* Use $1/\sqrt{1-x} \approx 1 + \frac{1}{2}x$ for $x \ll 1$.) (b) Show that when he splashed down, Col. McDivitt was about 300 microseconds younger than he would have been if he had not been in orbit.

7-C4 Combining Eqs. 7-8 and 7-6, show that KE $= m_0 c^2 (1 - v^2/c^2)^{-1/2} - m_0 c^2$. Simplify this expression, using the binomial theorem, and show that KE $= \frac{1}{2} m_0 v^2 + (\ ) + \cdots$. To prove that the correct expression for KE reduces to the classical formula as a special case, you will have to prove that the second and other terms in the infinite series become negligibly small compared with the first term, as v/c approaches zero.

7-C5 An electron at rest has mass 9.11×10^{-31} kg. (a) Calculate its mass when it is moving at a constant speed of 2.4×10^8 m/s. (b) Calculate the increase of mass (Δm) of the electron due to its motion. (c) Calculate from Δm the KE of the electron, using the Einstein formula $\Delta E = (\Delta m)c^2$. (d) Verify numerically that the formulas $\frac{1}{2}m_0v^2$ and $\frac{1}{2}mv^2$ both give incorrect answers for the KE of this "relativistic" particle.

For Further Study

7-9 Derivation of equivalence of mass and energy

Einstein himself felt that the relationship between mass and energy was one of the most important results of the theory of relativity. Only a modest use of derivatives is needed to derive the relativistic formula for kinetic energy (Eq. 7-8) from which we can generalize the result that $E = mc^2$.

We begin by reaffirming some definitions which are already familiar from Newtonian mechanics. Momentum is *defined* as $p = mv$, the product of the relativistic mass and the velocity. Thus a body's momentum approaches infinity as its speed approaches c. Just as in Newtonian mechanics, force is *defined* as $F = dp/dt$, the rate of change of momentum. When a force acts on a body, its momentum changes. The change in kinetic energy is the work done by the force.

$$dE = F\,dx = \frac{dp}{dt}dx = \frac{dx}{dt}dp = v\,dp \qquad (7\text{-}9)$$

Since $p = mv$, we have $dp = m\,dv + v\,dm$. Thus Eq. 7-9 becomes[*]

$$dE = mv\,dv + v^2\,dm \qquad (7\text{-}10)$$

To find $mv\,dv$, we differentiate the relativistic mass formula, using the chain rule (Appendix E).

---

[*] In Newtonian mechanics m would be constant and $dm = 0$. Then Eq. 7-10 would be $dE = mv\,dv$, yielding, upon integration, kinetic energy $= E = m\int v\,dv = \frac{1}{2}mv^2 - \frac{1}{2}mv_0^2$. This derivation does not assume uniformly accelerated motion, as was the case in Prob. 6-B12.

$$m = m_0\left(1 - \frac{v^2}{c^2}\right)^{-1/2}$$

$$\frac{dm}{dv} = -\frac{1}{2}m_0\left(1 - \frac{v^2}{c^2}\right)^{-3/2}\left(-\frac{2v}{c^2}\right)$$

$$\frac{dm}{dv} = \left[m_0\left(1 - \frac{v^2}{c^2}\right)^{-1/2}\right]\left(1 - \frac{v^2}{c^2}\right)^{-1}\left(\frac{v}{c^2}\right)$$

$$\frac{dm}{dv} = m\left(1 - \frac{v^2}{c^2}\right)^{-1}\left(\frac{v}{c^2}\right)$$

$$\frac{dm}{dv} = \frac{mv}{c^2}\left(\frac{c^2 - v^2}{c^2}\right)^{-1} = \frac{mv}{c^2 - v^2}$$

$$mv\,dv = (c^2 - v^2)\,dm \qquad (7\text{-}11)$$

Now we substitute (7-11) into (7-10):

$$dE = (c^2 - v^2)\,dm + v^2\,dm$$

or

$$dE = c^2\,dm \qquad (7\text{-}12)$$

This relates the increase of kinetic energy of a body to the increase of its mass. To obtain the kinetic energy, we integrate (7-12), starting from rest.

$$\int_{E=0}^{E=\text{KE}} dE = c^2\int_{m=m_0}^{m=m} dm$$

or

$$\text{KE} = c^2(m - m_0)$$

This is Eq. 7-8.

7-10 A formula for relative velocity

What if two bodies approach each other, each moving with a speed $0.8c$ relative to their center of mass? "Common sense" would give 1.6c as the velocity of one body relative to the other, but we know this is impossible, for no relative velocity can exceed c, the speed of light. In the Einstein theory of relativity, the formula for relative velocity is

$$v_x = \frac{v_x' + v}{1 + \dfrac{v}{c^2} v_x'} \qquad (7\text{-}13)$$

In this equation* v_x is the velocity of a body as seen by an observer in one frame of reference FR; v_x' is its velocity as seen by an observer in another frame of reference FR'. The velocity of frame FR' relative to frame FR is v, and c is the speed of light. Let us see how this works out in a simple example where the velocities are small (see Sec. 3-9).

| EXAMPLE 7-5

A ship is moving southward at 15 m/s, and a deckwalker moves northward with a velocity of 1 m/s relative to the ship. What is the man's velocity relative to the water?

Here the water is one frame of reference

---

* We assume that the relative velocity of the two frames is along the x axis, and we also assume that v_y' and v_z' are both zero, so that motion is along the x axis in each frame.

(FR) and the ship is the other frame (FR'). Calling northward the positive direction, we have

$$\text{frame FR} = \text{water; frame FR'} = \text{ship}$$

v_x = velocity of man relative to FR = ?

v_x' = velocity of man relative to FR' = +1 m/s

v = velocity of FR' relative to FR = −15 m/s

c = velocity of light = 3×10^8 m/s

Plugging these values into the formula, we get

$$v_x = \frac{(+1 \text{ m/s}) + (-15 \text{ m/s})}{1 + \dfrac{(-15 \text{ m/s})(+1 \text{ m/s})}{(3 \times 10^8 \text{ m/s})^2}}$$

$$= -14.000000000000002 \text{ m/s}$$

In carrying through the calculation, we see that the denominator is so close to 1 that it can be ignored; indeed we are not justified in carrying out the division to so many decimal places. Thus, on account of the smallness of v and v_x' in comparison with the velocity of light, the equation is essentially $v_x = v_x' + v$, the same as the *classical* result which would have been expected without relativity theory. Had either v_x' or v been comparable in magnitude with c, the result would have been quite different (Prob. 7-C6).

■ PROBLEMS

7-C6 Prove that the projectile's velocity relative to the earth in Example 7-3 is 2.77×10^8 m/s.

7-C7 The answer to Prob. 7-C6 turned out to be less than c, as it should. Show that this will always be the case. That is, show that if v_x' and v are less than c, then the "sum" of v_x' and v, given by Eq. 7-13, is also less than c. [*Hint:* Let $v_x' = ac$ and $v = bc$, where a and b are positive and less than 1. The proof reduces to showing that $(a + b)/(1 + ab) < 1$.]

7-C8 Suppose that the projectile of Example 7-3 has a rest mass of 1000 kg. (*a*) What is its KE relative to the ship? (*Hint:* Use Eq. 7-8.) (*b*) What is its KE relative to the earth? (*c*) What is the value of the quantity $\frac{1}{2}m_0v^2$, where $v = 2.77 \times 10^8$ m/s? (*d*) Does this latter quantity have any physical significance?

■ REFERENCES

1 "The Amateur Scientist," *Sci. American,* Apr., 1960, p. 183. A discussion of Coriolis forces.

2 Bauman, R. P., "Visualization of the Coriolis Force," *Am. J. Phys.,* **38,** 390 (1970).

3 Lieber, Lillian R., *The Einstein Theory of Relativity* (New York: Holt, Rinehart & Winston, 1945). Our proof of the relativistic mass formula is based on material in this remarkable little book.

4 The "twin paradox" has been a subject of debate among relativists. (Identical twins are separated at a certain instant when one of them climbs aboard a spaceship which leaves Earth at a speed nearly equal to c. When the traveling twin returns, will he actually be physically and biologically younger than his twin who stayed at home?) See *Sci. American,* Dec., 1956, p. 58; Mar., 1957, p. 63; July, 1957, p. 68; Frye, R. M. and V. M. Brigham, "Paradox of the Twins," *Am. J. Phys.,* **25,** 553 (1957); Builder, G., "Resolution of the Clock Paradox," *Am. J. Phys.,* **27,** 656 (1959); Bronowski, J., "The Clock Paradox," *Sci. American,* Feb., 1963, p. 134.

5 "The Relativistic Clock," *The Physics Teacher,* **9,** 416 (1971). A short report on the experiment of J. C. Hafele.

6 Miller, F., *Inertial Forces—Translational Acceleration* (film).

7 Shapiro, A., *The Bathtub Vortex* (film). Coriolis force arising from the earth's rotation causes counterclockwise rotation when the drain valve of a tank of water is opened.

8

Rotation

It has been said that nature is in unending motion. In Chap. 3 we studied the kinematics of linear motion—a description of uniformly accelerated motion. Looking at the motions characteristic of our civilization, as distinct from those occurring naturally, one is struck by the increased emphasis on motions which are *periodic*, that is, in which a particle or other object repeatedly returns to its original position. The progress of civilization involves man's increasing control of energy, and control generally means keeping flywheels, pistons, piano strings, and pendulums in motion and yet at a stationary location. This is not to say that there are no periodic motions in uncivilized nature; on the contrary, the revolution of the earth about the sun and the vibration of a meadowlark's vocal cords are eloquent reminders of the periodicities which have existed since before man's emergence upon the planet. The vibrations and rotations of molecules are an important part of the atomic model of the structure and behavior of matter. In this chapter and the next we shall study the simplest form of rotation (uniform circular motion) and the simplest form of vibration (simple harmonic motion).

8-1 Angular quantities

In spite of a somewhat forbidding notation, there is nothing particularly difficult about the equations which describe circular motion, for they are strictly analogous to the familiar equations for linear motion (Sec. 3-6). In fact, the chief difficulty—and one that can be easily overcome—may well be the use of the Greek alphabet. It is useful to employ separate symbols to keep linear and angular concepts distinct and yet comparable.

A particle in circular motion (Fig. 8-1) moves through an angle θ when it moves a distance s measured around the circumference of the circle. We call the arc s the *linear displacement*[*] and θ the *angular displacement*.

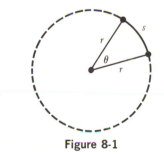

Figure 8-1

[*] Even though the path is curved, it is a "linear" displacement since the arc is measured in linear units such as feet, meters, or centimeters.

Friction supplies centripetal force horizontally to the right as these cyclists round a curve. (Wide World Photo)

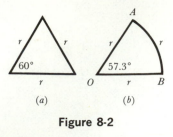

Figure 8-2

We shall find it convenient to measure angles (and angular displacement) in a new unit, the radian. A *radian* (rad) is the angle (Fig. 8-2*b*) whose arc equals the radius of the circle; it is equal to about 57.3°. [Bulge out one side of an equilateral triangle (Fig. 8-2*a*) to be an arc of radius r (Fig. 8-2*b*). It is reasonable that the angle AOB, which is a radian by definition, is a little less than 60°.] There are 2π rad in a complete circle, since the circumference is 2π times the radius, and 2π arcs each equal to the radius can be fitted around the circumference. We calculate that the size of a radian is $(1/2\pi)$ times a complete circle = $(1/2\pi) \times 360° = 57.3°$. If the arc subtended is $\frac{1}{2}r$, then the angle is $\frac{1}{2}$ rad, etc. In general,

$$\theta = \frac{s}{r}$$

$$s = r\theta \quad (\theta \text{ in radians})$$

Angular velocity (denoted by ω, the Greek letter omega) is the rate of change of angular displacement:

$$\omega = \lim_{\Delta t \to 0} \left(\frac{\Delta\theta}{\Delta t}\right) = \frac{d\theta}{dt}$$

(Compare Eq. 3-4, which gives an analogous definition of linear velocity as the time rate of change of linear displacement.) Angular velocity can be measured in revolutions per minute, degrees per second, etc.; however, if radian measure is used, linear velocity is related to angular velocity in the same way that linear displacement is related to angular displacement:

$$v = r\omega \quad (\omega \text{ in rad/s})$$

(The radian was defined in the first place so that this would be so.) Likewise, *angular acceleration* (denoted by α, the Greek letter alpha) is the rate of change of angular velocity, given by

$$\alpha = \lim_{\Delta t \to 0} \left(\frac{\Delta\omega}{\Delta t}\right) = \frac{d\omega}{dt}$$

(Compare Eq. 3-10, which gives an analogous definition of linear acceleration as the time rate of change of linear velocity.) Angular acceleration can be measured in revolutions per minute per second, degrees per minute per hour, etc.; if radian measure is used, linear and angular acceleration are related in a simple fashion:

$$a = r\alpha \quad (\alpha \text{ in rad/s}^2)$$

With only the change of Latin symbols to corresponding Greek symbols, all the equations of Sec. 3-6 are equally valid in angular measure. These similarities are summarized below:

$$s = r\theta \tag{8-1}$$

$$v = r\omega \tag{8-2}$$

$$a = r\alpha \tag{8-3}$$

LINEAR	ANGULAR	
$v = v_0 + at$	$\omega = \omega_0 + \alpha t$	(8-4)
$s = \dfrac{v_0 + v}{2}t$	$\theta = \dfrac{\omega_0 + \omega}{2}t$	(8-5)
$v^2 = v_0^2 + 2as$	$\omega^2 = \omega_0^2 + 2\alpha\theta$	(8-6)
$s = v_0 t + \frac{1}{2}at^2$	$\theta = \omega_0 t + \frac{1}{2}\alpha t^2$	(8-7)

Equations 8-1 to 8-3 *require* use of radian measure, since linear and angular quantities appear in the same equation. However, Eqs. 8-4 through 8-7 are valid in any consistent set of angular units.

EXAMPLE **8-1**

What is the angular size, in radians, of a 6 ft football player viewed by a spectator 100 yards away (Fig. 8-3)? We know that the angle is

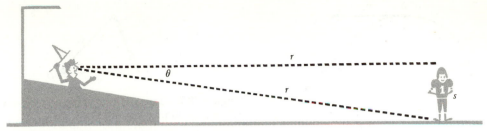

Figure 8-3

small, and so we make the simplifying assumption that the arc s is equal to the player's height:

$$\theta = \frac{s}{r} = \frac{6 \text{ ft}}{300 \text{ ft}} = \boxed{0.02 \text{ rad}}$$

Note that an angle in radians is a pure number (length/length) and has no dimensions.

EXAMPLE **8-2**

A ventilator fan is turning at 600 rev/min when the power is cut off, and it turns through 1000 rev while coasting to a stop. Calculate the angular acceleration and the time required to stop.

Just as we did earlier for problems in linear motion, we make a box with knowns and unknowns and select a suitable equation from our list.

$\omega_0 = 600$ rev/min
$\omega = 0$
$\alpha = ?$
$\theta = 1000$ rev

$$\omega^2 = \omega_0{}^2 + 2\alpha\theta$$

$$\alpha = \frac{\omega^2 - \omega_0{}^2}{2\theta}$$

$$= \frac{0 - (600 \text{ rev/min})^2}{2(1000 \text{ rev})}$$

$$= \boxed{-180 \text{ rev/min}^2}$$

To find the time,

$$\omega = \omega_0 + \alpha t$$

$$t = \frac{\omega - \omega_0}{\alpha} = \frac{0 - 600 \text{ rev/min}}{-180 \text{ rev/min}^2}$$

$$= \boxed{3.33 \text{ min}}$$

Note that in this problem we did not have to use radian measure, since no linear and angular

quantities appeared together in the same equation. For convenience, we used "revolutions" as the unit for angles.

EXAMPLE **8-3**

A boy on a bicycle is traveling at 10.0 m/s. What is the angular speed of a point on the tire of the bicycle, if its radius is 34.0 cm?

Since the tire is always in contact with the road, the linear velocity relative to the axle of any point on the tire equals the forward velocity of the bicycle. We must use radian measure, since both linear and angular quantities are involved.

$$v = r\omega$$

$$\omega = \frac{v}{r} = \frac{10.0 \text{ m/s}}{0.340 \text{ m}} = 29.4 \frac{(\ )}{s}$$

$$= \boxed{29.4 \text{ rad/s}}$$

The dimensionless "radian" has been filled in to make a complete angular velocity unit. Recall that use of $v = r\omega$ ensures that the angular velocity will be in rad/s.

8-2 Uniform circular motion

Consider the motion of a speck of dust on the rim of a phonograph record turning at 45 rev/min. We cannot call this "uniform motion," for the *direction* of the motion is continuously changing. Velocity is a vector, and the velocity of the dust speck is not constant since only the magnitude (speed) is constant. We see that in spite of the apparent simplicity and constancy of the motion, the particle

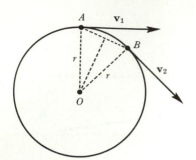

Figure 8-4 Change of velocity during a finite time interval.

is actually accelerated, since the velocity changes.

A useful formula for the acceleration of a particle in uniform circular motion can easily be derived. In Fig. 8-4 the linear velocity of the particle at a certain instant is $\mathbf{v}_1$; after a short time interval Δt the particle has moved from A to B, and the velocity is $\mathbf{v}_2$ (numerically the same as $\mathbf{v}_1$, but directed differently). As always, the acceleration is the rate of change of velocity, that is, change of velocity divided by the time required for the change. We use the vector equation for average acceleration during an interval Δt,

$$\overline{\mathbf{a}} = \frac{\mathbf{v}_2 - \mathbf{v}_1}{\Delta t}$$

Even though the magnitude of the velocity has not changed, the *vector* difference $\mathbf{v}_2 - \mathbf{v}_1$ is not zero. We have not studied vector subtraction, but we can write $\mathbf{v}_2 - \mathbf{v}_1 = \mathbf{v}_2 + (-\mathbf{v}_1)$, and use ordinary vector addition. Since $-\mathbf{v}_1$ is the same as $\mathbf{v}_1$, but oppositely directed, the vector diagram is constructed by laying off vectors parallel to $\mathbf{v}_2$ and $\mathbf{v}_1$ of Fig. 8-4. We find that $\mathbf{v}_2 - \mathbf{v}_1$ is a vector perpendicular to the dashed line of Fig. 8-5, which is parallel to the average direction of $\mathbf{v}$.[*] Hence the average acceleration (proportional to the change in velocity $\mathbf{v}_2 - \mathbf{v}_1$) is directed toward the center of the circle; it is called a *centripetal acceleration*

---

[*] The altitude of an isosceles triangle bisects the vertex angle.

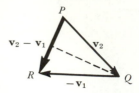

Figure 8-5 The vector difference $\mathbf{v}_2 - \mathbf{v}_1$ is found by adding $\mathbf{v}_2$ and $-\mathbf{v}_1$.

("seeking the center"). The acceleration is not constant any more than the velocity is constant. The direction of each is continuously changing, with the average acceleration always perpendicular to the average velocity.

To find the instantaneous acceleration, we must use a very short time interval (see Sec. 3-4). In the limit, as $\Delta t \to 0$, the *direction* of the instantaneous acceleration is perpendicular to the direction of the instantaneous velocity $\mathbf{v}$. The *magnitude* of the acceleration is found from $a = $ limit of $\Delta v / \Delta t$, as $\Delta t \to 0$. For small angles, the triangles AOB and PQR in Figs. 8-4 and 8-5 may be replaced by sectors (Fig. 8-6). The vertex angle $\angle AOB$ equals the angular speed times the time interval; i.e., $\omega \Delta t$. Then $\angle PQR$ also equals $\omega \Delta t$ (similar triangles, with the sides mutually perpendicular). The velocities $\mathbf{v}_1$ and $\mathbf{v}_2$ each have magnitude v, so we label PQ and RQ equal to v, and the magnitude of the change in v is PR, which we now call Δv. Finally, from the sector PQR we have the approximate equation

$$\Delta v / v \approx \omega \Delta t$$

or

$$\Delta v / \Delta t \approx \omega v$$

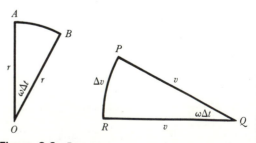

Figure 8-6 Instantaneous acceleration is found by letting Δt approach zero.

Passing to the limit, as $\Delta t \to 0$, we see that the magnitude of the centripetal acceleration is

$$a_c = \omega v \qquad (8\text{-}8)$$

This is a most interesting equation for the centripetal acceleration, involving as it does both angular velocity and linear velocity. We can interpret it as follows: The factor v tells us how much velocity *magnitude* there is to change, and the factor ω tells us how rapidly the *direction* of v is changing. For computational purposes, it is convenient to use either ω or v, not both. Thus, since $v = r\omega$,

$$a_c = \omega v = \omega(r\omega) = \omega^2 r \qquad (8\text{-}9)$$

$$a_c = \omega v = \left(\frac{v}{r}\right)v = \frac{v^2}{r} \qquad (8\text{-}10)$$

Other formulas for centripetal acceleration involve the *period* T (time for one revolution, s/rev) and the *frequency* ν (number of revolutions per second, rev/s, denoted by a Greek nu).

$$a_c = \omega v = \frac{2\pi}{T}\left(\frac{2\pi r}{T}\right) = \frac{4\pi^2 r}{T^2} \qquad (8\text{-}11)$$

or, since ν is the reciprocal of T,

$$a_c = 4\pi^2 r \nu^2 \qquad (8\text{-}12)$$

These five formulas for centripetal acceleration are entirely equivalent.

EXAMPLE 8-4

Calculate the centripetal acceleration of a speck of dust on the rim of a phonograph record 7 inches in diameter turning at 45 rev/min. The radius is 3.5 in. = 0.292 ft, and the circumference is $2\pi(0.292 \text{ ft}) = 1.83$ ft.

First method: The angular velocity is

$$\omega = \left(45 \,\frac{\text{rev}}{\text{min}}\right)\left(\frac{2\pi \text{ rad}}{1 \text{ rev}}\right)\left(\frac{1 \text{ min}}{60 \text{ s}}\right) = 4.71 \,\frac{\text{rad}}{\text{s}}$$

$$a_c = \omega^2 r = (4.71 \text{ rad/s})^2(0.292 \text{ ft})$$

$$= \boxed{6.48 \text{ ft/s}^2}$$

Second method: The linear velocity of the dust particle is

$$v = \left(\frac{1.83 \text{ ft}}{1 \text{ rev}}\right)\left(45 \,\frac{\text{rev}}{\text{min}}\right)\left(\frac{1 \text{ min}}{60 \text{ s}}\right) = 1.37 \,\frac{\text{ft}}{\text{s}}$$

$$a_c = \frac{v^2}{r} = \frac{(1.37 \text{ ft/s})^2}{0.292 \text{ ft}} = \boxed{6.48 \text{ ft/s}^2}$$

If the acceleration of the earth (for instance) is toward the sun, why then does not the earth eventually fall into the sun? The truth of the matter is that it *does* fall toward the sun—but it never gets any closer. In the absence of the centripetal acceleration, the earth would continue moving in a straight line, i.e., would fly off on a tangent (Fig. 8-7), and after a time t would have arrived at some point such as C. The earth actually arrives at B, and the distance CB represents a fall toward the sun. In fact, it may be verified that the laws of falling bodies apply, and $CB = \frac{1}{2}at^2$, where a is the centripetal acceleration previously derived, equal to v^2/r. (See Prob. 8-C6.)

8-3 Centripetal force

Having studied the kinematics (description) of uniform circular motion, we are now able to examine the causes of such motion (the problem of dynamics). Circular motion is no exception to the general rule that forces are needed to cause accelerations of bodies having mass. The *centripetal force*, which causes centripetal acceleration, acts on the body

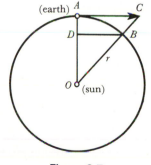

Figure 8-7

which is in motion and is directed toward the center of the circle. Any type of force can serve as a centripetal force. In the following examples we illustrate how elastic, frictional, gravitational, and magnetic forces can cause uniform circular motion of a body having mass.

EXAMPLE 8-5

A 10 kg block rests on a smooth surface and is attached to a vertical peg by a rope (Fig. 8-8). What is the tension T in the rope if the block moves in a horizontal circle of radius 2 m at an angular speed of 100 rev/min?

First we find the magnitude of the centripetal acceleration:

$$a_c = \omega^2 r$$
$$= \left[\left(100 \frac{\text{rev}}{\text{min}}\right)\left(\frac{1 \text{ min}}{60 \text{ s}}\right)\left(\frac{2\pi \text{ rad}}{1 \text{ rev}}\right)\right]^2 (2 \text{ m})$$
$$= 219 \text{ m/s}^2$$

The block is acted on by three forces: its weight, the upward force of the surface, and the tension in the rope (an elastic force). The vertical forces balance, and the net force equals the tension in the rope:

$$\text{net } F = ma$$
$$T = (10 \text{ kg})(219 \text{ m/s}^2) = \boxed{2190 \text{ N}}$$

The direction of T is the same as that of a_c, namely, toward the center of the circular path. We have found the force *on* the block; by Newton's third law the block exerts an equal and opposite (centrifugal) force on the pivot.

Frictional forces often supply the necessary centripetal force, as for an automobile going around a level curve.

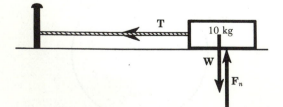

Figure 8-8 Centripetal force supplied by tension in a rope.

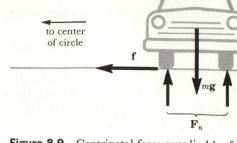

Figure 8-9 Centripetal force supplied by friction.

EXAMPLE 8-6

What is the maximum speed with which a car of mass m can go around an unbanked curve of radius 40 m, if the coefficient of static friction between tires and road is 0.7 (Fig. 8-9)?

As in the preceding example, the weight of the car mg is balanced by the upward push F_n of the road. The net force is that due to friction, and $f = 0.7F_n = 0.7mg$, since the car is presumably about to skid and the maximum force of friction is being obtained. Here it is convenient to use $a = v^2/r$ for the centripetal acceleration.

$$\text{net } F = ma = m\frac{v^2}{r}$$
$$(0.7)(m \text{ kg})(9.8 \text{ m/s}^2) = (m \text{ kg})\left(\frac{v^2}{40 \text{ m}}\right)$$
$$v = \boxed{16.6 \text{ m/s}}$$

If this speed is exceeded, there will be insufficient force to cause the car to maintain the acceleration, unless v^2/r is kept constant by an increase in r. The car skids, in an arc of greater radius of curvature. (Note that the mass of the car does not appear in the answer.)

To illustrate centripetal acceleration caused by gravitational force, consider a car going over the top of a hill.

EXAMPLE 8-7

What is the force of a 700 kg sports car on the road surface, as it goes at 10 m/s over the crest of a hill having a radius of curvature of 35 m measured in a vertical plane (Fig. 8-10)?

We are asked for the force exerted by the car,

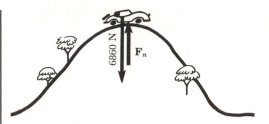

Figure 8-10 Centripetal force supplied by gravitation.

but Newton's second law must be applied to the motion of the car, and so we consider the forces *on* the car. The weight of the car is 700 kg × 9.8 m/s² = 6860 N.

$$\text{net } F = ma = m\frac{v^2}{r}$$

$$6860 \text{ N} - F_n = (700 \text{ kg})\frac{(10 \text{ m/s})^2}{35 \text{ m}}$$

$$6860 \text{ N} - F_n = 2000 \text{ N}$$

$$F_n = 4860 \text{ N} \quad (\text{upward})$$

This is the upward push of the road on the car. Newton's third law tells us that the force of the car on the road is equal and opposite, i.e.,

$$\boxed{4860 \text{ N downward}}$$

As the car goes over the crest of the hill, it tends to "take off" and presses less against the road than when on a level stretch.

Magnetic force on a moving charge is responsible for the circular motion of a charged particle in an atom-smasher.

EXAMPLE **8-8**

In a cyclotron a beam of protons (each of rest mass 1.67×10^{-27} kg) is moving in a circle of radius 80 cm. If an electromagnet supplies a force of 8.00×10^{-13} N directed toward the center of revolution, what is the velocity of the protons? What is the KE of each proton?

$$\text{net } F = ma = m\frac{v^2}{r}$$

$$v = \sqrt{\frac{Fr}{m}} = \sqrt{\frac{(8.00 \times 10^{-13} \text{ N})(0.80 \text{ m})}{1.67 \times 10^{-27} \text{ kg}}}$$

$$= 1.96 \times 10^7 \text{ m/s}$$

$$= \boxed{19\ 600 \text{ km/s}}$$

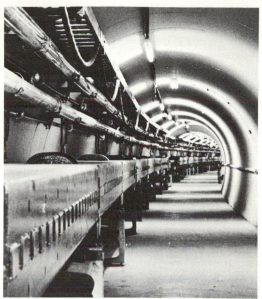

National Accelerator Laboratory

Figure 8-11 Bending magnets in interior of main accelerator ring at the National Accelerator Laboratory in Batavia, Illinois. The magnets supply centripetal force on protons moving in a circle of radius 1.00 km. By adding additional "piggyback" superconducting magnets in the space above the present magnets, a much larger centripetal force can later be obtained.

$$\text{KE} = \tfrac{1}{2}mv^2$$

$$= \tfrac{1}{2}(1.67 \times 10^{-27} \text{ kg})(1.96 \times 10^7 \text{ m/s})^2$$

$$= \boxed{3.20 \times 10^{-13} \text{ J}}$$

Although 3.20×10^{-13} J may seem to be too small to bother with, it is actually, as we shall see in Chap. 29, a great deal of energy for a single atomic particle to have. Of course, if spread out among the 10^{23} or so atoms that make up a marble or a bullet, 3.20×10^{-13} J would be insignificant indeed.

The circular path of particles in a large accelerator is shown in Fig. 8-11.

8-4 Centrifugal force

If there is a force directed *toward* the center *on* a rotating body, by Newton's third law there must be a force directed *away* from the center exerted *by* the rotating body. This force, exerted on the axis or whatever other body supplies the inward force, is called the *centrifugal force* ("fleeing the center"). Centripetal force and centrifugal force are an action-reaction pair. Like all such pairs, they never act on the same body, and although equal and opposite they never cancel each other. For instance, in Example 8-5, the block pulls outward on the peg with a centrifugal force of 2190 N. In Example 8-6 the car pushes outward against the road with a horizontal force of 0.7 times its weight. These are correctly termed centrifugal forces. However, some of the popular statements about centrifugal force are confusing, if not actually incorrect. Thus it is stated that in a clothes drier "centrifugal force throws the water outward" as the drum containing wet clothes rotates. From the point of view of the housewife, an outside observer, this statement is incorrect. A better way to describe what happens is as follows: "The molecular forces between water and fiber supply the necessary centripetal force to keep the water drops moving in a circle. When the speed is increased, the required force is increased, and sooner or later the molecular forces become insufficient. The water then flies off on a tangent, and moves away from the center of the drum because of lack of sufficient force." In other words, centrifugal force doesn't *cause* the water to fly outward. The water will do that anyway, due to its inertia, since its natural tendency is to move in a straight line (Newton's first law) *unless* acted on by a centripetal (or other) force.

There is, however, one way in which common usage and scientific accuracy may be combined. From the point of view of an "insider"—one whose frame of reference is the rotating system itself—the rotation is not obvious. To him there actually seems to be a centrifugal force causing certain objects to move away from center. This is why we on the earth (a rotating frame of reference) can talk about centrifugal force causing a decrease in weight of a body at the equator. The upward force is a "fictitious" or "inertial" force, arising from acceleration of the frame of reference;[*] in this respect it is analogous to a Coriolis force (Sec. 7-2). In general, however, you will avoid much confusion if you take an outsider's view of things, and place the emphasis on the centripetal force which acts on the rotating body to cause its acceleration toward the center.

Many practical applications of the centripetal-centrifugal action-reaction pair come to mind. In the centrifuge, the heavy liquid (milk) can be collected at the rim of the rotating dish of the cream separator; corpuscles can be separated from blood serum. The effect is the same as if the gravitational field of the earth were increased. In the ultracentrifuge, rotational speeds up to 600 000 rev/min are used to obtain separation of complex protein molecules which would require thousands of years to separate under the action of ordinary gravity.

8-5 Banking of curves

Let us consider the predicament of a bicyclist going around a level (unbanked) curve. If he remains vertical, the forces acting on him are (1) his weight $\mathbf{W}$ ($=m\mathbf{g}$); (2) the normal force $\mathbf{F}_n$ of the road; and (3) a frictional force $\mathbf{f}$ acting toward the center of the circle. These forces are not in equilibrium, and the resultant force (which is simply the frictional force, since the others balance) is the centripetal force which causes the centripetal acceleration. Without sufficient friction, the cyclist is unable to go around the curve and will skid in a curve of greater radius, as described in Example 8-6.

Even when the coefficient of friction is adequate, the cyclist is in danger of toppling

---

[*] See remarks on inertial frames of reference in Sec. 7-2.

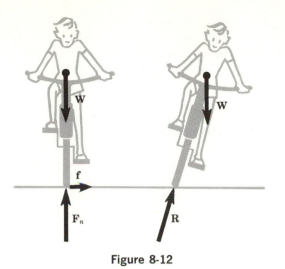

Figure 8-12

over, and the common practice of cyclists is to lean inward (Fig. 8-12) as they go around a curve, in order to avoid this. The vector sum of **f** and $\mathbf{F}_n$ represents the force **R** of the ground on the cycle; if the cyclist does not lean over, **R** does not pass through the center of gravity, and the result is a counterclockwise torque tending to push the wheels out from under the rider. Leaning at just the proper angle causes **R** to act along the frame of the bicycle, and no instability results.

EXAMPLE **8-9**

The combined mass of a motorcycle and its rider is 100 kg. What is the necessary force of friction if the cyclist is to go around a curve of 80 m radius at 20 m/s (Fig. 8-13)? If the coefficient of friction is 0.6, will the cyclist negotiate the curve successfully? At what angle should he lean over to avoid a spill?

$$\text{Centripetal force} = m\frac{v^2}{r}$$

$$= (100 \text{ kg})\frac{(20 \text{ m/s})^2}{80 \text{ m}}$$

$$= \boxed{500 \text{ N}}$$

The cyclist's weight is 100 kg × 9.8 m/s² = 980 N.

$$\text{Maximum force of friction} = \mu_s F_n$$

$$= (0.6)(980 \text{ N})$$

$$= \boxed{588 \text{ N}}$$

Since the required force is less than the maximum possible force of friction, the cyclist will be able to make it.

$$\tan \theta = \frac{f}{F_n} = \frac{500 \text{ N}}{980 \text{ N}} = 0.511$$

$$\theta = \boxed{27°}$$

Curves on highways are banked, computed to be safe at some particular speed. The normal force exerted by the roadbed has a horizontal component which serves as the centripetal force, and the vertical component of the normal force serves to support the weight. Since the road itself furnishes the centripetal force as one component of the compressional force (the normal force), no frictional force (parallel to the surface) is needed. A car can negotiate a properly banked curve even on glare ice for which $\mu_s \approx 0$.

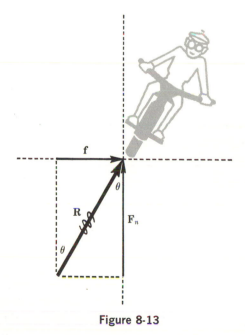

Figure 8-13

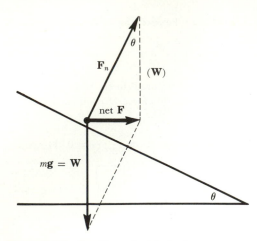

Figure 8-14 Banking of a frictionless curve; centripetal force supplied by a component of the road's normal force F_n.

The force diagram (Fig. 8-14) shows only two external forces acting on the car. They are not balanced, of course, and the resultant force is horizontal, directed toward the center of the circle. This net F serves as the centripetal force. From the diagram, (net F)/W = tan θ, and so net $F = W$ tan $\theta = mg$ tan θ. From Newton's second law,

$$\text{net } F = ma$$

$$mg \tan \theta = m \frac{v^2}{r}$$

$$\tan \theta = \frac{v^2}{rg}$$

This formula shows that the angle of banking depends on the speed and on the radius of the curve. The absence of m or W from the formula shows that all cars and trucks (and bicycles) require the same banking on the same curve at any given speed, regardless of their weights.

Of course, if the car goes around a banked curve too fast, some frictional force will be needed to supplement the horizontal component of the normal force. If it goes around the curve too slowly, frictional force is again needed to oppose some of the horizontal component of the normal force, to keep the car from sliding downhill into the ditch. On a very slippery road, it is as bad to go too slowly as too fast around a banked curve. At the speed for which the curve is designed, friction is not needed, and road conditions are immaterial.

8-6   Rotational inertia

Just as the tendency of a body to resist change in its linear velocity is called its linear inertia, so its tendency to resist change in its angular velocity is called its *rotational inertia*. To apply Newton's laws to rotational acceleration, we need quantities corresponding to force and mass; we look for an equation of the form net ( ) = ( )α, which will correspond to net $F = ma$ in linear motion.

Bodies having rotational inertia, such as a propeller or a door on its hinges, are accelerated only when a torque is applied. Force must be applied, but there must be an effective lever arm if there is to be rotational acceleration. We see intuitively that torque takes the place of force, when rotational motions are considered. Consider a body consisting of a single mass m at a distance r from an axis, acted on by a single force F as shown in Fig. 8-15. The torque on the body is Fr. We use Newton's second law, remembering that $a = r\alpha$:

$$\text{net } F = ma$$

$$F = mr\alpha$$

$$\tau = Fr = mr^2\alpha$$

Thus we infer that the quantity mr^2 is for rotation what the mass m is for linear motion.

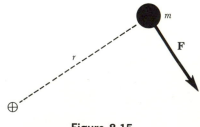

Figure 8-15

If the body contained many particles, with various masses located at various distances from the axis, the total torque would be the sum of the various torques, each computed as above. Then

$$\text{net } \tau = (\Sigma mr^2)\alpha$$

In this equation, Σmr^2 represents the sum of the masses making up the body, each multiplied by the square of its distance from the axis of rotation. For convenience, we call this sum the *moment of inertia* of the body and denote it by the symbol I. The dimensions of moment of inertia are $[ML^2]$; moments of inertia are expressed in $kg \cdot m^2$, $g \cdot cm^2$, etc. We may now write

$$\text{net } \tau = I\alpha \qquad (8\text{-}13)$$

which is Newton's second law for rotation. Note how similar this equation is to the familiar linear form net $F = ma$.

EXAMPLE 8-10

A fisherman starts his outboard motor by applying a steady tangential force of 100 N by means of a starter rope. The flywheel has a moment of inertia of 0.05 kg·m² and a diameter of 0.2 m. Ignoring the effects of friction and the opposing torque due to the compression of the motor, calculate the angular speed of the flywheel after it has turned through half a revolution starting from rest.

We first find the angular acceleration, then we use the equations of uniformly accelerated angular motion (Eqs. 8-4 to 8-7) to find the final speed.

$$\text{net } \tau = I\alpha$$
$$(100 \text{ N})(0.1 \text{ m}) = (0.05 \text{ kg} \cdot \text{m}^2)\alpha$$
$$\alpha = 200 \text{ rad/s}^2$$
$$\omega^2 = \omega_0^2 + 2\alpha\theta$$
$$= 0 + 2(200 \text{ rad/s}^2)(\pi \text{ rad})$$
$$= 1260 \text{ rad}^2/\text{s}^2$$
$$\omega = 35.4 \text{ rad/s}$$
$$= (35.4 \text{ rad/s})(1 \text{ rev}/2\pi \text{ rad})$$
$$= \boxed{5.64 \text{ rev/s}}$$

The moment of inertia of a body depends upon how the masses are distributed relative to the axis of rotation. A simple example will make this clear.

EXAMPLE 8-11

A thin, light wooden meter stick (Fig. 8-16) is loaded with two small masses as follows: 3.0 kg at the 20 cm mark, and 5.0 kg at the 70 cm mark. What is the moment of inertia about an axis through the zero end of the stick?

$$I = \Sigma mr^2$$
$$= (3 \text{ kg})(0.2 \text{ m})^2 + (5 \text{ kg})(0.7 \text{ m})^2$$
$$= 0.12 \text{ kg} \cdot \text{m}^2 + 2.45 \text{ kg} \cdot \text{m}^2$$
$$= \boxed{2.57 \text{ kg} \cdot \text{m}^2}$$

What is the moment of inertia of the same system about an axis through the 100 cm end of the meter stick?

$$I = \Sigma mr^2$$
$$= (3 \text{ kg})(0.8 \text{ m})^2 + (5 \text{ kg})(0.3 \text{ m})^2$$
$$= 1.92 \text{ kg} \cdot \text{m}^2 + 0.45 \text{ kg} \cdot \text{m}^2$$
$$= \boxed{2.37 \text{ kg} \cdot \text{m}^2}$$

The second moment of inertia is less than the first one. It can be proved that, for all possible axes parallel to each other, the least moment of inertia is obtained if the axis passes through the center of gravity.

If mass is distributed continuously, as in the case of most real bodies, it is a problem of integral calculus to perform the summation indicated in Σmr^2 (see Sec. 8-10). Results for some simple regular bodies are summarized in Table 8-1. In each case, the dimensions of the formula for moment of inertia are $[ML^2]$, as they should be to agree with the dimensions of Σmr^2.

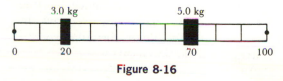

Figure 8-16

TABLE **8-1** **Moments of inertia for some simple regular bodies**

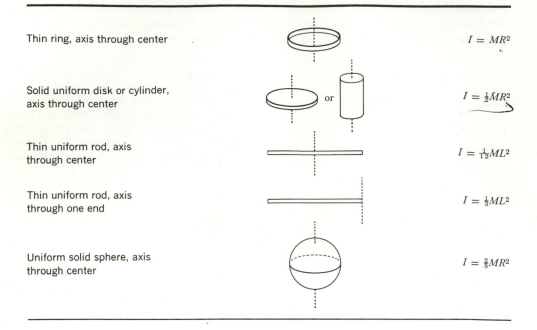

Thin ring, axis through center	$I = MR^2$
Solid uniform disk or cylinder, axis through center	$I = \frac{1}{2}MR^2$
Thin uniform rod, axis through center	$I = \frac{1}{12}ML^2$
Thin uniform rod, axis through one end	$I = \frac{1}{3}ML^2$
Uniform solid sphere, axis through center	$I = \frac{2}{5}MR^2$

8-7   Linear and rotational analogies

From what has been said about Newton's second law, it may be suspected that most, if not all, of the concepts and laws of linear motion may be carried over into rotational motion. This is indeed the case. We pass over the proofs of these theorems and state a few results. In all cases, to obtain an angular form of some linear law or formula, replace *force* by *torque, mass* by *moment of inertia,* and of course *linear* displacement, velocity, or acceleration by *angular* displacement, velocity, or acceleration. For example, *rotational KE* is given by $\frac{1}{2}I\omega^2$ by analogy with the linear form $\frac{1}{2}mv^2$; *angular momentum* is given by $I\omega$ by analogy with the linear form mv. *Work* is done whenever a body is rotated against an opposing torque (compare the linear statement on p. 114) and hence work = (torque)(angular displacement) = $\tau\theta$ by analogy with the linear formula work = (force)(linear displacement) = Fs.

Another analogy concerns the statement of Newton's second law. If mass is not constant, we saw in Chap. 6 that $F = ma$ must be replaced by the more general form $F = d(mv)/dt$; force equals the rate of change of linear momentum. Similarly, the rotational equation $\tau = I\alpha$ (which assumes constant I) must be, in case I varies, replaced by $\tau = d(I\omega)/dt$; torque equals rate of change of angular momentum. Change of moment of inertia is by no means as unusual as change of mass, as we will see in Secs. 8-8 and 8-9.

8-8   Conservation of angular momentum

The law of conservation of angular momentum may be stated as follows:

The total angular momentum of an isolated system remains constant.

For instance, a flywheel rotating at constant angular speed has constant angular momentum unless it is acted on by an external torque, such as might be caused by a motor or friction in the bearings. The product $I\omega$ remains constant.

Applications of the law of conservation of angular momentum are among the most interesting in the field of mechanics. Some of the conclusions are unexpected, because of the possibility of changing moment of inertia without changing mass, by a redistribution of mass. A student standing on a frictionless rotating table (Fig. 8-17) can change his own speed of rotation by changing his moment of inertia. If he draws in his arms, his moment of inertia decreases, and hence, since $I\omega$ remains constant, his angular velocity ω increases. The effect is more pronounced (sometimes too much so for safety) if he carries a mass in each hand to make a larger change in moment of inertia. If the student holds an electric drill vertically while stationary on the (nonrotating) platform, the system has zero angular momentum until the current is turned on. When he turns the drill on, giving it angular momentum $+I\omega$, the total $I\omega$ must still be zero, so the student's angular momentum must become $-I\omega$. He starts to rotate in a direction opposite to that of the drill. Since the student's moment of inertia is so much larger than that of the

moving part of the drill, his angular velocity is correspondingly less, in order for his $I\omega$ to be numerically equal to (but opposite in direction to) the drill's $I\omega$.

On a grand scale, much the same phenomenon may account for minute variations in the length of the day. Due to internal rearrangements of the earth's matter, hundreds or thousands of miles below the surface, its moment of inertia varies. The variations in I, although quite large, are only a small fraction of the total I for the earth. Nevertheless, present-day clocks have been developed to such a degree of perfection that reliable indications of erratic variations in the earth's rotation have been obtained. For instance, during the 20 year period from 1910 to 1930, the length of the day changed by 0.0047 s, a rather abrupt change on a geological time scale. It is because of such variations in the period of the earth's rotation that the unit of time, the second, is no longer defined as 1/86 400 of a mean solar day, as was the case for many centuries (Sec. 1-4).

A diver or acrobat makes good use of the conservation of angular momentum. If the diver wishes to execute a double somersault, she pulls in her arms and doubles up her legs, thus decreasing her moment of inertia. According to the law of conservation of angular momentum, her angular velocity increases correspondingly, enabling her to make the turns in the short time available. Even a cat seems to know about conservation of angular momentum; it is through intricate changes of moment of inertia (chiefly through maneuvering the tail) that a falling cat arranges to alight feet first.

Finally, the body of a helicopter with one set of rotors would tend to rotate in a direction opposite to that of the rotors, to keep the total angular momentum of the system zero. To counteract this, some helicopters have two sets of lift rotors, revolving in opposite directions, so that the total $I\omega = 0$ without requiring rotation of the body of the helicopter; more commonly, a small auxiliary

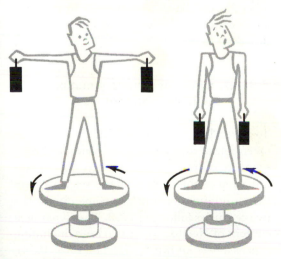

Figure 8-17 Conservation of angular momentum.

Figure 8-18 The small propeller at the tail of the helicopter keeps the ship from rotating about a vertical axis.

propeller at the tail supplies an auxiliary horizontal force, to give a torque which opposes the reaction torque of the main rotors (Fig. 8-18). We have here an illustration of Newton's third law for rotation: the forward torque on the rotors is supplied by the body of the helicopter through the motor, and therefore the rotors supply an equal and opposite reaction torque on the ship.

The "automatic pilot" of an airplane illustrates the vector nature of angular momentum. A wheel having considerable moment of inertia is mounted in bearings so that its axis is free to point in any direction, with the torque due to gravity balanced out. The wheel is kept rotating by a small motor, and its axis remains pointing in the same direction in space, since no external torque is applied. If the plane's course changes, the wheel's axis turns relative to the plane, initi-

ating an action to return the plane to its course. Thus a reliable control is obtained, based essentially on Newton's first law. Another illustration of the vector nature of angular momentum is afforded by the gyroscope, discussed in Sec. 8-12.

8-9 Dynamics of planetary motions —the Newtonian synthesis

To Isaac Newton (1642–1727) we owe the extension of mechanics to astronomical phenomena, a synthesis which was both a culmination of the work of many scientists and the foundation for much to come during the two centuries following 1700. The dynamical content of Newton's first two laws, including $F = ma$, was known to Galileo (1564–1642), and the kinematical description of planetary motions had been found by Kepler (1571–1630). There was, by 1665, a widespread belief among scientists that some unifying principle was awaiting discovery. In the introduction to his *Principia,* Newton wrote that his intention had been to induce the "forces of nature" [the laws] from the "phenomena of motion" [experimental observations], and from these laws to "deduce the motions of the planets, the comets, the moon, and the sea." It was his genius to have done just this, at about 22 or 23 years of age, using information that was known to many of his scientific contemporaries including Christopher Wren, Robert Hooke, and Edmund Halley.

At the outset, the fact that planets move in curved paths rather than in straight lines means that they are not in equilibrium. Therefore some net force acts on a planet at every point in its orbit to give it a centripetal acceleration. Newton's first law, the law of inertia, led him to look for the source (or sources) of such a force. Earlier workers, especially Descartes (1596–1650) proposed that a planet is caught up in vortices (whirlpools) in an invisible fluid; but no detailed agreement with experiment was possible. (Newton later worked out a mechanical theory of fluids, showing that a vortex theory is untenable.) Kepler, not realizing the nature of inertia, thought that a "magnetic" force from the sun might keep a planet moving forward in its orbit. Although his mechanical ideas were wrong, Kepler's precise description of planetary motions gave Newton the necessary clues from which he developed the law of gravitation. We can apply the ideas developed in this chapter to gain an insight into the dynamics underlying Kepler's laws.

Kepler's second law (the law of areas) states that the line joining any planet to the sun sweeps out equal areas in equal times. This law is a consequence of the fact that the force on the planet is always directed toward

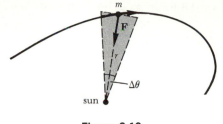

Figure 8-19

a center (the sun). Such a force is called a *central force.* The exact nature of the central force (an inverse-square gravitational force in this case) is not involved in the proof of the law of areas. In Fig. 8-19, a body of mass m is moving in an orbit, subject to a force directed toward the sun. The force **F** has no lever arm relative to the sun, hence the torque on m is always zero. Newton's second law in its general form says that the torque equals the rate of change of angular momentum.*

$$\tau = \frac{d}{dt}(I\omega)$$

$$0 = \frac{d}{dt}(mr^2\omega)$$

The rate of change of $mr^2\omega$ is zero, hence $mr^2\omega$ is a constant. But m is constant (at nonrelativistic speeds), therefore

$$r^2\omega = \text{constant} \qquad (8\text{-}14)$$

Both r and ω may change as the planet moves, but the product $r^2\omega$ remains constant. Next we calculate the rate of change of area. From Fig. 8-19 we see that $\Delta A \simeq \frac{1}{2}(\text{base}) \times (\text{altitude}) \simeq \frac{1}{2}(r\,\Delta\theta)(r)$. (This approximation becomes better as $\Delta\theta \to 0$.) Thus

$$\Delta A \simeq \tfrac{1}{2}r^2\,\Delta\theta$$

$$\frac{\Delta A}{\Delta t} \simeq \tfrac{1}{2}r^2\,\frac{\Delta\theta}{\Delta t}$$

---

* The moment of inertia of the planet, relative to the sun, is $I = mr^2$. Net $\tau = I\alpha$ is correct at any instant, but this form of Newton's second law is not very useful here, because I is not constant (r varies).

and in the limit, as $\Delta t \rightarrow 0$ and $\Delta\theta/\Delta t \rightarrow \omega$,

$$\frac{dA}{dt} = \tfrac{1}{2}r^2\omega$$

Using Eq. 8-14 we obtain

$$\frac{dA}{dt} = \text{constant} \qquad (8\text{-}15)$$

This is Kepler's second law. Note that the force could equally well have been a repulsive one. Such a motion occurs when a positively charged helium nucleus moves past a positively charged target nucleus, Fig. 8-20 (see Rutherford scattering experiment, Sec. 28-1). The law of areas holds for this motion, even though the moving nucleus is not in a closed orbit; what is important here is that the force on the moving body is central, i.e., it is directed toward or away from a center.

Our analytical proof of Kepler's law of areas is essentially based on the first two of Newton's laws of motion. Newton himself gave an equivalent geometrical proof. The *magnitude* of the central force is not specified by the argument so far; it might vary inversely as r ($F \propto 1/r$), be proportional to $1/r^2$, $1/r^3$, or vary in some other way. The law of areas shows a central force, directed toward the sun, of some as yet unspecified magnitude.

Kepler's first law (the law of orbits) states that the orbit of any planet around the sun is an ellipse, with the sun at one focus of the ellipse. The proof of this law requires that we assume a special form of central force—in fact, we must assume the inverse-square law of gravitation. Newton's proof is omitted, since it involves some formal calculus. There

are only a few force laws that give a bounded orbit in which the planet stays in the solar system for an infinitely long time. Newton investigated several such laws, including the inverse-square force law which was conjectured (without proof) by his close friend Halley and others. Curiously, Newton found that $F \propto 1/r^5$ also gives a closed orbit, in fact a circle, but the center of force (the sun) is *on* the orbit, rather than at the focus (center of the circle) as observed. (A planet in such an orbit would also be disastrously hot once a year as it passed through the sun!)

Our discussion of the Newtonian synthesis is now almost complete: we have seen that the law of areas and the law of elliptical orbits, coupled with Newton's second law of motion, give an inverse-square law of gravitation, in which $F \propto 1/r^2$, and F is directed toward the sun.

The role of mass in the law of gravitation which Newton developed is simple: the force is proportional to the product of the masses of the interacting bodies. Galileo had found that the acceleration of any falling body at a given point on the earth is the same, independent of the mass m of the body. Newton devised experiments to substantiate this to a high degree of precision. Since g is a constant (by experiment), $F = mg$ means that $F \propto m$. The force is proportional to the mass of the falling body. The *earth* is also "falling" toward the center of mass of the (earth + body) system, and the force F' on the earth is proportional to *its* mass m'; thus $F' \propto m'$. Now by Newton's third law $F = F'$, so F is proportional to m' as well as to m. The mutual force between two interacting bodies is, therefore, proportional to both masses. Thus, finally, we have arrived at Newton's law of gravitation:

$$F = G\frac{mm'}{r^2}$$

In Sec. 4-8 we discussed the measurement of the constant G.

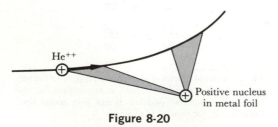

He^{++}

Positive nucleus in metal foil

Figure 8-20

Kepler's third law (the law of periods) states that for any two planets in the solar system, the squares of the periods of revolution are in the same proportion as the cubes of their average distances from the sun. It was a gratifying check on his work that Newton found the law of periods to be a consequence of the law of gravitation which he had derived using only Kepler's first two laws. We can easily derive Kepler's third law for the special case of *circular* orbits (a circle is a special case of an ellipse, with the two foci coinciding at the center of the circle). The planet's centripetal acceleration is caused by the force of gravitation of the sun's mass M acting on the mass m_1 of the planet which is at a distance r_1 from the sun (see Eq. 4-11). We apply Newton's second law to planet 1:

$$\text{net } F = ma$$

$$G\frac{Mm_1}{r_1{}^2} = m_1\frac{v_1{}^2}{r_1}$$

The orbital velocity for a circular orbit is $v_1 = (\text{circumference})/(\text{period}) = (2\pi r_1)/T_1$. Thus

$$G\frac{Mm_1}{r_1{}^2} = \frac{m_1}{r_1}\frac{4\pi^2 r_1{}^2}{T_1{}^2}$$

Rearranging, we get

$$\frac{T_1{}^2}{r_1{}^3} = \frac{4\pi^2}{GM} \tag{8-16}$$

For another planet of period T_2 at a radius r_2, we obtain by the same reasoning

$$\frac{T_2{}^2}{r_2{}^3} = \frac{4\pi^2}{GM}$$

The constant $4\pi^2/GM$ is the same for all planets revolving around the *same* sun, and so we conclude that

$$\frac{T_1{}^2}{r_1{}^3} = \frac{T_2{}^2}{r_2{}^3}$$

and hence

$$\frac{T_1{}^2}{T_2{}^2} = \frac{r_1{}^3}{r_2{}^3} \tag{8-17}$$

for any two planets revolving around the sun. This is Kepler's third law.[*] Note that the mass of the planet canceled out. This means that even a tiny marble-sized "planet" 93 000 000 miles from the sun would have a period of revolution equal to 365 days, the same as the earth or any other planet at the same distance from the sun.

Kepler's third law (Eq. 8-17) is valid for an elliptical orbit if r represents the *mean distance* from planet to sun, defined as the average of the closest approach (at perihelion) and farthest distance (at aphelion).

EXAMPLE 8-12

The asteroid Icarus is a small planet, several hundred meters in radius, whose orbit is oriented in such a way that a close approach to the earth is possible. The perihelion distance is 17.4×10^6 mi, and the aphelion distance is 183.1×10^6 mi. Calculate the period of revolution of Icarus.

First we find the mean distance r_1 by averaging the perihelion and aphelion distances:

$$r_1 = \tfrac{1}{2}(17.4 + 183.1) \times 10^6 \text{ mi}$$
$$= 100.2 \times 10^6 \text{ mi}$$

The earth's mean distance is $r_2 = 92.9 \times 10^6$ mi, and the earth's period is $T_2 = 365$ days. By Kepler's third law,

$$\frac{T_1{}^2}{(365 \text{ days})^2} = \frac{(100.2 \times 10^6 \text{ mi})^3}{(92.9 \times 10^6 \text{ mi})^3}$$

$$T_1 = \boxed{409 \text{ days}}$$

The Newtonian synthesis combined the terrestrial laws formulated by Galileo and Newton with the astronomical laws of Kepler into one general framework which included the radical and powerful law of universal gravitation. Newton worked out many consequences of his gravitational theory. He gave a quantitative explanation of the tides, which are

---

[*] For an elementary extension of this discussion to take account of the motion of the sun itself, as well as the elliptical nature of orbits, see Reference 4 at the end of this chapter. See also Sec. 4-10 on p. 72.

caused by the gravitational attraction of the moon and the sun. He showed that comets are members of the solar system, having elliptic, parabolic, or hyperbolic orbits with the sun at one focus. The moon's motion is very complicated because of the gravitational attraction of bodies such as the sun and Jupiter; Newton gave the first reliable predictions of the moon's motion, of great practical value for navigation. Since Newton's time, the law of gravitation has been extended far beyond the solar system, for example to the mutual revolution of a pair of stars about their center of mass. With all these successes, Newton never thought he had "explained" gravitation, because he had no experimental evidence for how the force can be transmitted through space between, say, the earth and the sun. Only in recent decades are testable theories for the transmission of gravitational force being proposed (see Secs. 18-3 and 23-10), but experiments are extremely difficult.

In some ways Newtonian mechanics is incomplete. It is precisely applicable to large-scale bodies moving at low speeds. The extension of Newtonian mechanics to the realm of the very small, known as quantum mechanics, is of overriding importance on the atomic scale of things. The special theory of relativity (Chapter 7) extends Newtonian mechanics to the realm of very high speeds.

At low speeds, for large bodies, both quantum mechanics and relativity theory reduce to "ordinary" Newtonian mechanics to an incomparably high degree of precision. In fact, as far as planetary motions are concerned, only one difference between the predictions of the Newtonian gravitation and Einsteinian general relativity is large enough to be observed. Close study of the motion of the innermost planet Mercury reveals that its orbit is not a simple ellipse. Instead, there is a slow *advance of perihelion* (Fig. 8-21). The perihelion point P (the closest approach to

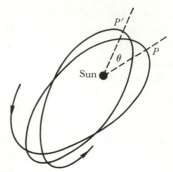

Figure 8-21 Advance of perihelion of Mercury's orbit (greatly exaggerated for clarity; actual angle θ is about 10^{-8} degree for perihelia of successive orbits).

the sun) moves at the slow rate of about 574 seconds of arc (574″, or 0.16°) per century. The orbit does not close up on itself. Such an advance of perihelion is predicted by Newtonian gravitational theory, because the other planets (Venus, Earth, and especially Jupiter) exert small forces called *perturbations* which are not in the same direction as the main force exerted by the sun. Classical (Newtonian) mechanics, taking account of all known sources of perturbation (including frictional drag of interplanetary matter) predicted an advance of 531″ per century. This is, however, 43″ less than the observed value. It is most encouraging that the theory of general relativity gives the required 43″ excess rate of advance of perihelion. In fact, possible inadequacies in general relativity can be tested, if the observed 43″ excess cannot precisely be explained within observational limits of error. This is an active field of current research, of tremendous theoretical importance, but of essentially no practical importance for observable every-day mechanics. The effect on Mercury's orbit is large enough to be noticed because the sun's gravitational field, proportional to $1/r^2$, is strongest there for any natural planet. A solar probe, circling the sun much closer than Mercury, would be most helpful in the further study of gravitational theories.

■ SUMMARY

The kinematics of uniformly accelerated rotational motion is quite analogous to the kinematics of uniformly accelerated translational motion, and the same equations are used, with appropriate changes in symbols. Radian measure is not needed except where linear and angular quantities appear in the same equation. A complete circle contains 2π radians.

A particle moving in a circular path has a linear acceleration toward the center, since the linear velocity is changing in a vector sense due to change in direction. This centripetal acceleration is given by $a_c = \omega v$, where v is the linear velocity and ω is the angular velocity; equivalent forms of this equation are $a_c = v^2/r$, $a_c = \omega^2 r$, and $a_c = 4\pi^2 r/T^2$. The necessary centripetal force to cause the change in direction of a body moving in a circle is given by $F_c = m\omega v = mv^2/r = m\omega^2 r = 4\pi^2 rm/T^2$.

Centrifugal force is the reaction force exerted *by* the moving body. From the point of view of an outside observer, the centrifugal force never acts *on* the revolving body.

Curves are banked so that the normal force of the road, due to compression, has an inward horizontal component which can serve as the centripetal force. In this way the need for frictional force is lessened or eliminated.

In rotation, moment of inertia takes the place of mass, and torque takes the place of force. Newton's second law for rotation reads net $\tau = I\alpha$. Angular kinetic energy and work are computed by formulas analogous to those used for similar linear quantities: $\text{KE} = \frac{1}{2}I\omega^2$ and $W = \tau\theta$. The angular momentum $I\omega$ of an isolated system remains constant in magnitude and direction.

Kepler's second law, the law of areas, is true for any orbit which is described under the action of an attractive or a repulsive central force; Kepler's law of orbits requires that the central force be an attractive inverse-square force such as that of gravitation. The Newtonian synthesis extended the range of physical laws to the realm of astronomical phenomena.

■ CHECK LIST

angular displacement θ	LINEAR FORMULAS	ANGULAR FORMULAS
radian	net $F = ma$	net $\tau = I\alpha$
angular velocity ω	$\text{KE} = \frac{1}{2}mv^2$	$\text{KE} = \frac{1}{2}I\omega^2$
angular acceleration α	$W = Fs$	$W = \tau\theta$
centripetal	linear momentum $= mv$	angular momentum $= I\omega$
acceleration a_c		
period T	$a_c = \dfrac{v^2}{r} = \dfrac{4\pi^2 r}{T^2} = 4\pi^2 rv^2 = \omega^2 r$	
frequency ν		
centripetal force		
centrifugal force		
moment of inertia		
angular momentum		

■ QUESTIONS

8-1 A fan rotates through 7π rad. After this rotation, what is the position of a blade which was originally pointing straight up?

8-2 A penny is laid on the rough surface of a phonograph turntable, near the outer edge, and the motor is turned on. The penny does not slip, and it turns with the turntable with a constant speed of $33\frac{1}{3}$ rev/min. Is the velocity constant? Is the penny in equilibrium?

8-3 Is it possible for a body to be accelerated if its speed is constant?

8-4 Consider an atom of aluminum near the rim of a phonograph turntable turning at 45 rev/min. Does any centripetal force act on the atom? If so, what is this force caused by?

8-5 What is the source of the centripetal force on the pilot of a plane which is executing a vertical loop-the-loop, when the plane is at the bottom of the loop, curving upward?

8-6 When an eagle flying westward starts to swerve to the left, what is the source of the centripetal force on the bird? What is the direction of the force?

8-7 "The earth doesn't fall into the sun because centrifugal force pushes it outward." Criticize this sentence.

8-8 A heavy fan with its blades attached to a horizontal axle is resting on a table, supported by four symmetrically placed short vertical legs, each leg bearing one-fourth the weight of the fan. If the fan is turned on, are the forces in the legs still equal? When the fan has reached constant speed, are the forces in the legs equal?

8-9 In Example 8-8, was it legitimate to use the Newtonian form of the KE formula ($\frac{1}{2}mv^2$) for the KE of the proton, or should we have used the relativistic formula, taking account of the increase in mass due to its velocity? (*Hint:* How does the proton's velocity compare with the speed of light?)

8-10 In a game of marbles played on a perfectly smooth level surface, is it possible for a marble to roll in an arc of a circle?

8-11 In the example of the student holding the drill (Sec. 8-8), what will happen if the student turns the drill (while it is running) so that it points down instead of up?

8-12 What happens to the length of the day when a sprinter starts running in an easterly direction? What happens to it when he stops running? What happens to it when an office worker takes an elevator to the top floor? Explain why these changes in the earth's angular speed are too small to observe.

■ PROBLEMS

8-A1 Express the following angles in radians: (*a*) 180°; (*b*) 30°; (*c*) 90°; (*d*) 40°; (*e*) 0.30 rev.

8-A2 Express the following angles in degrees: (*a*) 1.50 rad; (*b*) 0.21 rad; (*c*) $\pi/6$ rad; (*d*) 0.40 rev; (*e*) 2π rad.

8-A3 What is the linear velocity of a point on the rim of a flywheel 4 ft in diameter, if the wheel is turning at an angular velocity of 3 rad/s?

8-A4 What is the angular velocity in rad/s of a 45 rev/min phonograph turntable?

8-A5 A phonograph turntable rotates at $33\frac{1}{3}$ rev/min for 3 min. Through what distance, measured along the arc, has a point on the rim of a 12 in. record moved?

8-A6 The moon has a diameter of 2160 mi and is 240 000 mi from the earth. Calculate its angular size (a) in radians; (b) in degrees.

8-A7 What is the angular size, in radians, of a football player whose image, viewed 10 ft from a television screen, is 2 in. high?

8-A8 In aviation, a "standard turn" for level flight of a propeller-type plane is one in which the plane makes a complete circular turn in 2 min. If the speed of the plane is 200 m/s, (a) what is the radius of the circle? (b) What is the centripetal acceleration of the plane?

8-A9 (a) Using the length of the day, compute the angular velocity of the earth, in rad/s. (b) Compute the magnitude of the centripetal acceleration of a point on the earth's equator (radius of earth is 6.37×10^6 m).

8-A10 A car goes around a curve at 50 ft/s. The radius of the curve is 200 ft. Calculate the centripetal acceleration of the car.

8-A11 (a) Calculate the angular velocity (in rad/s) of the minute hand of a clock. (b) Calculate the magnitude of the centripetal acceleration of a paint speck on the minute hand, 20 cm from the axis of rotation.

8-A12 Calculate the moment of inertia about its axis of a bicycle wheel of mass 2 kg and radius 0.30 m. Assume the wheel's mass to be concentrated at the rim.

8-A13 In Sec. 6-10 we saw that the power required to keep a body moving at velocity v against an opposing force f is $P = fv$. Use the method of analogy (Sec. 8-7) to write down the corresponding rotational equation. State the meaning of each symbol appearing in your equation.

8-A14 A small ball of mass 25 kg is swinging on the end of a light wire 10 m long. What is the moment of inertia of the ball? (You may assume all the mass of the ball to be concentrated at one point.)

8-A15 Calculate the angular momentum of a body of moment of inertia 20 kg·m² which is turning at 2 rev/s.

8-A16 What are the dimensions of angular momentum? of torque?

8-A17 Suppose an astronaut goes into a circular orbit about the sun with an orbital radius of 93 000 000 miles. How long would be required for the astronaut to make one revolution around the sun?

Note: In solving problems where data are expressed in British units, use W/g for the mass.

8-B1 The specifications of a certain hi-fi turntable call for it to reach its final angular velocity of $33\frac{1}{3}$ rev/min while turning through $\frac{1}{4}$ rev, starting from rest. What is the angular acceleration of such a table? (Express your answer in rev/s² and in rad/s².)

8-B2 A flywheel turning at 900 rev/min slowed down to 300 rev/min in 5 min. (a) Calculate the angular acceleration of the flywheel, in rev/min². (b) How many revolutions did it make during this time interval?

8-B3 An amusement park "ride" consists of a flat turntable which rotates about a vertical axis while people try to sit on it. The turntable reaches its final speed of 0.5 rev/s in 40 s. Assuming the angular acceleration to be constant, calculate (a) the value of the angular acceleration; (b) the number of revolutions turned during the 40 s.

8-B4 As a yo-yo is let fall, its angular speed increases steadily with an angular acceleration of 40 rad/s². Through what angle has it turned when its final angular speed is 50 rad/s?

8-B5 A fly of mass 2 g is sunning itself on a phonograph turntable at a location which is 5 cm from the axis. The turntable is turned on and rotates at 45 rev/min. Calculate the inward (centripetal) force needed to keep the fly from slipping.

8-B6 A car of mass m goes around an unbanked curve at 20 m/s. If the radius of the curve is 50 m, what is the least coefficient of friction that will allow the car to negotiate the curve without skidding?

8-B7 A car goes around a level unbanked curve of radius R. The coefficient of static friction is μ_s. Derive a formula for the maximum safe speed, expressing v in terms of μ_s, g, and R. Does the safe speed depend on the mass of the car?

8-B8 What centripetal force must act on a 3200 lb truck which rounds a curve of radius 200 ft at 70 ft/s?

8-B9 In a certain ultracentrifuge a G-field of 300 000 times gravity is realized (this means that the acceleration is 300 000 g). If the rotor is 10 cm in radius, what must the rotational speed be?

8-B10 A 50 kg boy is swinging on a rope 5 m long. He passes through the lowest position with a speed of 3 m/s. What is then the tension in the rope? (*Hint:* Use net $F = ma$. The net force is the resultant of the weight and the tension; cf. Example 8-7.)

8-B11 A carnival ride has cars that move at constant speed in a vertical circle of radius 20 m. (*a*) What must the angular velocity be so that a 90 kg passenger will be "weightless" when at the highest point A (Fig. 8-22)? (*b*) What will the passenger's apparent weight be when he is at the bottom point B?

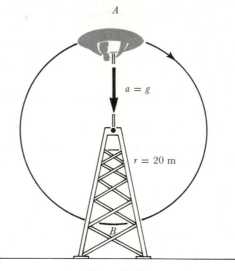

Figure 8-22

8-B12 A plane comes out of a power dive, turning upward in a curve whose center of curvature is 4000 ft above the plane. The plane's speed is 800 ft/s. (*a*) Calculate the upward force of the seat cushion on the 200 lb pilot of the plane. (*b*) Calculate the upward internal force on a 0.2 lb sample of blood in the pilot's brain. What happens if the heart muscles are unable to supply this force?

8-B13 A 100 lb skier starting from rest coasts down a frictionless 30° hill that is 120 ft long, enters a frictionless horizontal stretch, then turns sharply in a horizontal circle of radius 50 ft. What horizontal force is exerted (by what?) on the skier during the turn?

8-B14 In the "Rotor Ride" at an amusement park, riders are pressed against the inside vertical wall of a rotating drum 2.5 m in radius which has an angular speed of 27 rev/min about a vertical axis. (a) Viewed from a stationary frame of reference outside the drum, what is the magnitude and direction of the force of the wall on the rider whose mass is 60 kg? (b) While the drum is rotating, the floor is removed. If the coefficient of static friction between the rider and the wall is 0.6, will he slip or will he be held pressed against the vertical wall? (See Reference 6 at the end of this chapter for a film depicting this motion.)

8-B15 A level road curves with a radius of 1000 ft. At what angle should it be banked for cars traveling at 60 mi/h?

8-B16 What is the proper speed for a car to go around a slippery curve of radius 70 m if the road is banked at an angle of 20°?

8-B17 Compute the moment of inertia of a pole vaulter's pole which is vertical and is pivoted about an axis on the ground through its lowest point. The pole is 5 m long and has mass 4 kg.

8-B18 Compute the moment of inertia of the loaded meter stick of Example 8-11 about an axis through the 60 cm mark on the stick.

8-B19 A playground carousel is in the form of a solid uniform horizontal disk of radius 4 ft and weight 200 lb. If a child applies a force of 20 lb for 3 s, tangent to the rim, what will be the final angular velocity of the carousel, starting from rest?

8-B20 A farmer, fleeing from an enraged bull, seeks safety by going through a gate which is standing ajar at an angle of 90° with the fence, and closing it behind him. The gate has a moment of inertia of 2000 kg·m², and the farmer exerts a force of 180 N at a point 5 m from the hinge. As the gate swings shut, the farmer adjusts the direction in which he pulls, always applying force in the most advantageous way. How long will it take him to shut the gate?

8-B21 What is the rotational KE of a drum majorette's baton which is rotating about an axis through its center at an angular speed of 3 rev/s? Assume the baton to be a uniform rod 0.6 m long, of mass 0.5 kg. Would the light rubber ball at one end have much effect?

8-B22 A man on a platform (Fig. 8-17) has a small 2 kg object in each hand. Initially, his arms are extended, each object being 1 m from the axis of rotation, and he is rotating at 10 rad/s. The platform and man's body are assumed to have a constant moment of inertia of 1 kg·m². (a) Calculate the total angular momentum of the system (man plus platform plus objects). (b) The man now draws his arms in until each object is 0.3 m from the axis. What is the new angular velocity of the system? (c) Calculate the total KE of the system in part (a) and also in part (b). How do you account for the increase of KE? Has there been any change in PE?

8-B23 A uniform solid grindstone wheel weighs 32 lb and its radius is 0.5 ft. (a) Calculate the rotational KE of the grindstone when it is turning at 1800 rev/min. (b) After the grindstone's motor is turned off, a knife blade is pressed against the outer edge of the grindstone with a perpendicular force of 3 lb. The coefficient of kinetic friction is 0.8. What will be the angular speed of the grindstone 5 s later?

8-B24 A roller coaster car (Fig. 6-5) of weight W starts 23 m above the bottom of a loop. If the loop is 15 m in diameter, calculate the downward force of the rails on the car when it is upside down at the top of the loop. (*Hint:* First use the energy principle to find the value of v^2 for the car at this point.)

8-B25 When a satellite revolves in an orbit around the earth, the centripetal force equals the gravitational attraction. (a) Derive a formula for the period T (time for one revolution) of a satellite in a circular orbit. (b) Using data from p. 70, calculate the period of a satellite in a circular orbit of radius 7000 km.

8-B26 A certain small planet (asteroid) revolves about the sun with a period of 8 years. What is its average distance from the sun? (Express your answer in astronomical units; 1 a.u. = 93 000 000 mi, the average distance of the earth from the sun.)

8-B27 A planet's closest distance to the sun is a, and its farthest distance is b (Fig. 8-23). Use Kepler's second law to prove that $v_A/v_B = b/a$.

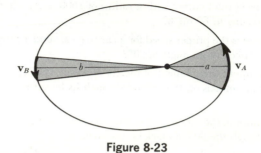

Figure 8-23

8-B28 (*a*) How fast must a satellite of mass m be launched horizontally at the surface of the earth if its orbit is to be a circle just grazing the highest mountains? (Assume no air resistance.) (*b*) What time would elapse between successive passes of this imaginary lowest satellite? (*Hint:* Use data from p. 70. Equate the centripetal force to the gravitational pull of the earth. The mass of the satellite cancels out.)

8-B29 (*a*) Calculate the mass of the sun, given the length of the year to be 365.3 days, mean radius of the earth's orbit 1.50×10^8 km, and Newton's gravitational constant $G = 6.67 \times 10^{-11}$ N·m²/kg². (*Hint:* Solve first in symbols; the mass of the earth cancels out.) (*b*) The sun's radius is 6.95×10^8 m; calculate its average specific gravity.

8-B30 The planet Venus is unusual in having practically no spin angular momentum about its own axis (rotation on axis once in 243 days; Earth turns on axis in 1 day). It has been suggested that Venus originally was spinning like the earth and its angular momentum was canceled when it captured a small backward-moving planet which later fell to the surface. Calculate the mass of the supposed planet, if it moved at 10 km/s tangent to the surface of Venus before capture. (Data: Assume Venus and Earth to be identical uniform spheres, each of mass 6×10^{24} kg and radius 6×10^6 m. The angular momentum of the incoming planet would be negative; consider that for it $I\omega$ is the same as $mr^2\omega$, or mrv.)

8-B31 Make a calculation to show that the gravitational force of the sun on the moon is always greater than the force of the earth on the moon. Thus at new moon the net force is away from the earth. This being so, how is it that the moon goes around the earth and not around the sun? Or does it? (Data: Mass of sun, 2.0×10^{30} kg; mass of earth, 6.0×10^{24} kg; mass of moon, 7.4×10^{22} kg; average distance from earth to sun, 1.5×10^{11} m; average distance from moon to earth, 3.8×10^8 m.)

8-B32 The acceleration of a falling body near earth's surface, at a distance R from the earth's center, is 9.80 m/s². (*a*) Use a suitable proportion to calculate the acceleration toward the earth of a falling body which is $60R$ from the earth's center. (*b*) The moon is in an orbit of radius $60R$, with a period of revolution 27.3 days. Show, as did Newton, that the centripetal acceleration of the moon toward the earth agrees with the answer to part (*a*). For this part you need to know that the earth's radius R is 6.37×10^6 m.

8-C1 The door of a bank vault has a mass of 800 kg and may be considered to be many thin rods, stacked one above the other, each pivoted at one end. The door is 1.5 m wide and

3 m high. (a) What is the moment of inertia of the door? (b) If a clerk pulls with a force of 100 N in the most effective way, through what angle will the door move in 2 s, starting from rest?

8-C2 Verify (partially) the statement at the end of Example 8-11. To do this, first find the c.g. of the weighted stick (Sec. 5-6), and then compute the moment of inertia about an axis through the c.g. Your answer should turn out to be less than either of the answers to Example 8-11, or that to Prob. 8-B18.

8-C3 Refer to Prob. 5-C7. Prove that the tension in the string when the ball passes through its lowest position is the same as the tension in the string in position A before the thread is burned. (*Hint:* Use the energy principle to find the ball's velocity at its lowest point.)

8-C4 A pole vaulter releases his pole without giving it a sideways shove, and the pole, starting from an upright position, topples over and falls to earth. The pole has length L and mass M. Show that the linear velocity with which the end of the pole strikes the ground is given by $\sqrt{3gL}$. (*Hint:* Use the energy principle; first find the angular velocity as the pole strikes the ground.)

8-C5 A boy on a swing "pumps up" by a process idealized with the help of the following model: The boy, of mass 50 kg, is swinging in an arc of radius 3.0 m and rises each time to a maximum height 2.0 m above his lowest point. (a) What is his linear velocity at the lowest point (use conservation of energy). (b) What is his angular velocity at the lowest point? (c) At an instant when he is passing through the lowest point, the boy by muscular effort raises his body's center of mass 10 cm so that he is now swinging in an arc of radius 2.9 m. What is his new KE? (*Hint:* Find his new velocity, using conservation of angular momentum.) (d) How high does the boy rise after this maneuver?

8-C6 In Fig. 8-7, prove that CB is approximately equal to $\frac{1}{2}at^2$ where $a = v^2/r$, thus making a connection between uniform circular motion and the kinematics of falling bodies. [*Hint:* First use plane geometry to show that, if $\angle AOB$ is small, $(DB)^2 \simeq (AD)2r$. Then make use of the fact that $CB \approx AD$ and $DB \approx AB$.]

For Further Study

8-10 Calculation of moment of inertia for solid bodies

By definition, $I = \Sigma mr^2$; that is, moment of inertia equals the sum of a number of products, each mass multiplied by the square of its distance from the axis of rotation. In Example 8-11 we saw how to calculate I when there were only a few masses to consider. In a real body of ordinary size, say a meter stick, there are perhaps 10^{23} or 10^{24} separate molecules, each at its own distance from the axis. To take account of so many particles calls for the limiting process of calculus known as *integration*, which we studied in Secs. 3-5 and 3-12. In what follows, we shall use algebra to derive one of the formulas of Table 8-1 as an illustration of the definite integral. Let us find the moment of inertia of a slender, uniform rod, of total mass M and length L, with the axis through one end of the rod.

1. As a first approximation, we could assume all the mass concentrated at the center (Fig. 8-24). Then, although the definition calls for a sum, Σmr^2, there is only one term

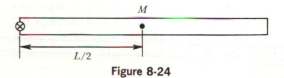

M

L/2

Figure 8-24

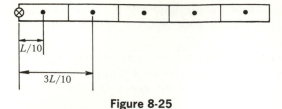

Figure 8-25

in the series, and $I = M(L/2)^2 = \frac{1}{4}ML^2$. This does not agree with the formula in Table 8-1, but we could hardly expect agreement because our assumption that the mass is concentrated at the center is certainly false.

2. For a second approximation, let us divide the rod into 5 equal segments, each $M/5$ in mass, and let us concentrate the mass of each segment at its center (Fig. 8-25). We form the sum as follows:

$$I = \Sigma mr^2 = \frac{M}{5}\left(\frac{L}{10}\right)^2 + \frac{M}{5}\left(\frac{3L}{10}\right)^2$$

$$+ \frac{M}{5}\left(\frac{5L}{10}\right)^2 + \frac{M}{5}\left(\frac{7L}{10}\right)^2 + \frac{M}{5}\left(\frac{9L}{10}\right)^2$$

$$= \frac{M}{5}\left(\frac{1}{100} + \frac{9}{100} + \frac{25}{100} + \frac{49}{100} + \frac{81}{100}\right)L^2$$

$$= \frac{165}{500}ML^2 = \frac{99}{300}ML^2$$

Already we see that our approximate answer is very close to the correct value. We have a factor 99/300, whereas the correct answer has a factor 100/300.

3. To make a precise calculation, we will have to approach a limit, using infinitely many segments, each of them infinitesimally small. Let us divide the rod into n segments, and suppose for simplicity that the mass of each segment is concentrated at its right-hand edge (Fig. 8-26). (This assumption is permissible since we are going to let n approach infinity, and hence the distance between the center and the edge of any given tiny segment will be negligible.) The mass of

each segment is M/n, and so the sum Σmr^2 becomes

$$I = \Sigma mr^2 = \frac{M}{n}\left(\frac{L}{n}\right)^2 + \frac{M}{n}\left(\frac{2L}{n}\right)^2$$

$$+ \frac{M}{n}\left(\frac{3L}{n}\right)^2 + \cdots + \frac{M}{n}\left(\frac{nL}{n}\right)^2$$

$$= \frac{ML^2}{n^3}(1^2 + 2^2 + 3^2 + \cdots + n^2)$$

It is shown in algebra books that the sum of the squares of the first n integers is

$$[n(n + 1)(2n + 1)]/6$$

hence our sum is

$$I = \frac{ML^2}{n^3}\left[\frac{n(n + 1)(2n + 1)}{6}\right]$$

$$= \frac{ML^2}{6}\left(\frac{2n^3 + 3n^2 + n}{n^3}\right)$$

$$= \frac{ML^2}{6}\left(2 + \frac{3}{n} + \frac{1}{n^2}\right)$$

This is surely slightly too large, but it will become a better and better approximation if we divide the rod into more and more segments. As we do this, n approaches infinity, and the terms $3/n$ and $1/n^2$ become negligible in comparison with the term 2. Therefore, *in the limit,*

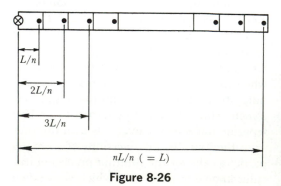

Figure 8-26

$$I = \frac{ML^2}{6}(2 + 0 + 0) = \tfrac{1}{3}ML^2$$

This is the formula given in Table 8-1 for the moment of inertia of a thin rod pivoted at one end.

In general, whenever the mass of a body is distributed in some fashion, as in a disk or sphere or rod, the moment of inertia can be found by taking the limit of a sum, i.e., as a definite integral. It is by such methods of integral calculus (refined and made less cumbersome) that the formulas of Table 8-1 have been found. As an example of the power of formal integration, let us calculate I for a uniform thin rod, using the methods of Sec. 3-12. The general formula for moment of inertia, replacing $I = \Sigma mr^2$, is

$$I = \lim_{\Delta m \to 0} \Sigma r^2 \, \Delta m = \int r^2 \, dm$$

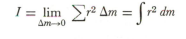
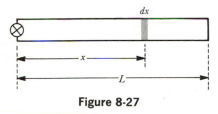

Figure 8-27

In Fig. 8-27, the mass of the short segment (whose length is dx) is $dm = (dx/L)M$, and here the distance r from the axis is simply x. Hence (see Sec. 3-12, especially Eq. 3-24)

$$I = \int r^2 \, dm = \int_{x=0}^{x=L} (x^2)\left(\frac{dx}{L}\right)M$$

$$= \frac{M}{L} \int_0^L x^2 \, dx = \frac{M}{L}\left[\frac{L^3}{3} - \frac{0^3}{3}\right]$$

$$= \tfrac{1}{3}ML^2$$

This is the same result obtained in the preceding paragraphs by evaluating the limit of the sum.

8-11 Pulleys with inertia

We are now in a position to study certain pulley problems in which the mass of the pulley is not negligible. Turning back to Example 4-7, we recall that at that time we were forced to assume that the pulley was "light." That is, its moment of inertia was assumed to be zero. Let us solve the same problem, giving the pulley a radius of 0.2 m and a moment of inertia $0.4 \text{ kg} \cdot \text{m}^2$. As before, we set up a system of simultaneous equations.

It is no longer true that the tensions on each side of the pulley are equal (Fig. 8-28). T_1 must be greater than T_2, for only in that way can there be the net clockwise torque which is required if the pulley is to receive a clockwise angular acceleration. We apply Newton's second law to each object in turn. Also, we use the fact that the angular acceleration of the pulley equals a/r, where a is the linear acceleration of the string. This assumes that the string is not slipping.

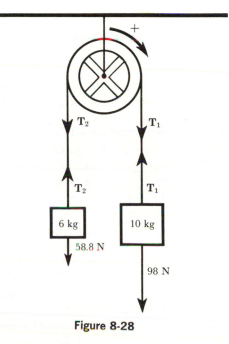

Figure 8-28

Pulley:

$$\text{net } \tau = I\alpha$$

$$T_1(0.2 \text{ m}) - T_2(0.2 \text{ m})$$
$$= (0.4 \text{ kg} \cdot \text{m}^2)(a/0.2 \text{ m})$$

6 kg object:

$$\text{net } F = ma$$

$$T_2 - 58.8 \text{ N} = (6 \text{ kg})(a)$$

10 kg object:

$$98 \text{ N} - T_1 = (10 \text{ kg})(a)$$

We now have three equations in three un-knowns: T_1, T_2, and a. Solving, we obtain $T_1 = 82.9$ N; $T_2 = 67.9$ N; $a = 1.51$ m/s². As expected, T_1 turns out to be greater than T_2, but less than 98 N in order to allow the right-hand object to accelerate downward.

The problem can be extended in several ways: suppose the string itself has mass, or there is a frictional torque at the axle, or the string is sliding slowly over the pulley. We shall not pursue these questions further; they are best studied in advanced work making full use of the calculus.

8-12 More about gyroscopes and tops

Small boys and girls have played with tops for centuries, and the fact of precession can hardly have escaped your notice (Fig. 8-29). A top is a symmetrical body which spins about an axis; it is observed that, if the point of support is not at the c.g., the direction in which the axis points slowly changes. This rotational motion of the axis is called *precession*. If the top spins on a sharp point, and if its angular speed is constant, its axis maintains a constant angle with the vertical, and the top end of the axis describes a circle. In the usual cases the precessional motion is relatively slow compared with the spin motion of the top itself, and we shall assume this to be so in what follows.

The questions we have to answer are these: (1) Why doesn't the top fall down? and

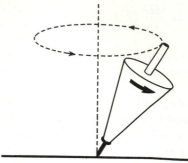

Figure 8-29 Precession of the axis of a top.

(2) What factors determine the direction and magnitude of the precessional motion of the axis? Let us fix our attention upon a particularly simple top, consisting of a bicycle wheel mounted on an axle. The wheel has a moment of inertia I about its axis, and its spin velocity is ω rad/s. We wish to apply Newton's second law to the motion of this system. First, we have to consider the vector interpretation of angular velocity and torque. In Fig. 8-30, by convention the spin velocity ω is represented by a vector pointing along the axis to the right. This is the direction that an ordinary (right-handed) screw would advance, if a screwdriver were turned in the direction of spin. (If the wheel were spinning in the other direction, the vector ω would be directed along the axis to the left.) Angular accelerations are also represented by vectors according to the right-hand screw rule. Similarly, a torque vector is directed in the

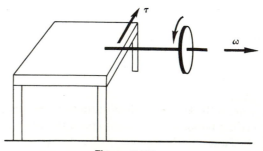

Figure 8-30

same direction as the acceleration which the torque tends to cause. Disregarding the spin for the moment, in Fig. 8-30 the force of gravity would pull the wheel down to the right, that is, rotate it around the edge of the table; the right-hand screw rule tells us that the vector τ due to gravity is along the edge of the table in the direction shown.

Suppose now the torque due to gravity acts for a very short time interval Δt, giving rise to an angular impulse $\tau \Delta t$. By Newton's second law, in the momentum form (by analogy with Eq. 6-2),

$$\tau \Delta t = \Delta(I\omega) \qquad (8\text{-}18)$$

In other words, the torque causes a change in angular momentum. After the time Δt has elapsed, the final angular momentum of the wheel is different from its initial value; the final value equals the initial value plus the change:

$$(I\omega)_2 = (I\omega)_1 + \Delta(I\omega)$$

Substituting from Eq. 8-18, we get

$$(I\omega)_2 = (I\omega)_1 + \tau \Delta t \qquad (8\text{-}19)$$

This is a vector equation. Let us apply it to the top, for two situations: (a) If the wheel is not spinning, and (b) if the wheel is spinning. Even though the *change* in $I\omega$ is the same in each case, the final $I\omega$'s are different because the initial $I\omega$'s are different.

(a) *Wheel not spinning:* In this case, $(I\omega)_1 = 0$; hence, applying Eq. 8-19, $(I\omega)_2 = \tau \Delta t$. This says that the wheel's final angular momentum is a vector parallel to the torque vector τ. In other words, the whole system rotates clockwise around the table edge as an axis, and falls down, just as would have been expected.

(b) *Wheel spinning:* In this case, the wheel has an initial angular momentum $(I\omega)_1$, and Eq. 8-19 tells us that the final angular momentum is to be found by vector addition (Fig. 8-31). The final angular momentum $(I\omega)_2$ turns out to be a horizontal vector, practically equal in magnitude to $(I\omega)_1$; the

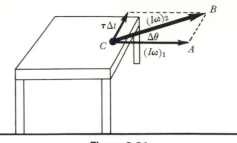

Figure 8-31

direction of the axis changes, but the axis remains horizontal since both $(I\omega)_1$ and $\tau \Delta t$ are vectors in a horizontal plane. The wheel does not fall over, but precesses, at a slow rate. The magnitude of the precessional angular velocity will be derived in Prob. 8-C13.

It is impossible to go into the numerous ramifications of the theory of tops and gyroscopes,* even with the simplifying assumption of slow precession. We content ourselves with two applications.

First, it is well known that the earth is not a perfect sphere, but bulges slightly at the equator owing to the centrifugal force acting while the earth was still molten. The gravitational attraction of the moon (and to a lesser extent, the sun) tends to "straighten out" the axis of the earth because of the unequal forces on the two parts of the bulge. At any instant, the force acting on the bulge on one side of the earth is greater than that on the bulge on the other side, since the two bulges are at different distances from the moon. Thus a net torque tends to "straighten up" the earth's axis (Fig. 8-32). However, since the earth is spinning, the torque causes a precession to take place instead of decreasing the angle θ. This precessional motion is very slow, about 26 000 years for one complete cycle, but the effect of this motion was actually observed by ancient Greek astronomers. The axis of the earth points now approximately to Polaris, the pole star; by 14 000 A.D. the axis will point more or less

---

*A gyroscope is, technically, a top which is mounted in such a way as to be subject to no external torque.

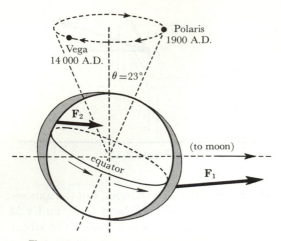

Figure 8-32 Precession of the earth's axis.

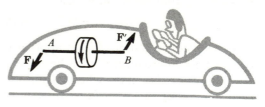

Figure 8-33

toward the star Vega, which is now 46° from the pole.

Second, if, while a wheel is spinning as shown in Fig. 8-33, one attempts to increase the precessional motion by applying a force $\mathbf{F}'$ to the axle, this force gives rise to a torque τ' directed upward (right-hand screw rule). Upward torque means that $\Delta(I\boldsymbol{\omega})$ is upward, and the axle rises at the B end. Therefore a *horizontal* force gives rise to an *upward* motion! The spinning wheel "reacts" in a most peculiar way. The heavy flywheel of a racing car has a distinct gyroscopic action of this sort (Fig. 8-34). When the driver turns the car toward the left, forces $\mathbf{F}$ and $\mathbf{F}'$ are applied

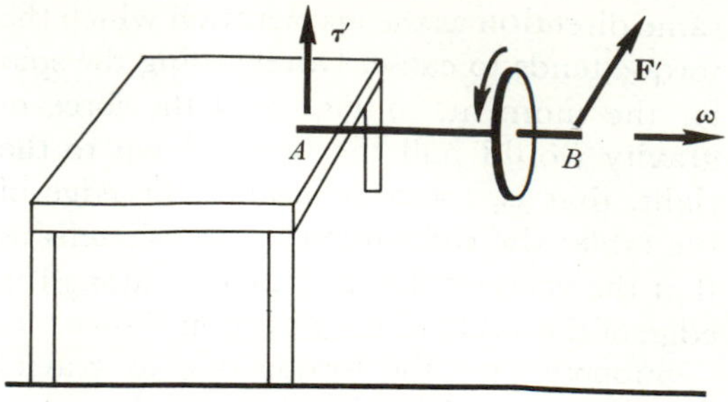

by the frame of the car. The rear end of the shaft, at B, moves upward and the front end, A, moves downward. The car "digs in" as it rounds the curve. During a turn toward the right, the front end of the car "takes off" and tends to rise. It is to prevent this that speedway races are run with the cars continually turning toward the left rather than toward the right. All this assumes that the flywheel turns counterclockwise as viewed from the rear of the car, as is the case for almost all cars in the United States.

Figure 8-34

■ PROBLEMS

8-C7 Carry through step 3 of the derivation in Sec. 8-10, assuming that the mass of each segment is concentrated at the left-hand edge instead of the right-hand edge. Take the limit of your sum, and show that you get the same result, namely, $I = \frac{1}{3}ML^2$. (*Hint:* You need the sum of the squares of the first $n-1$ integers.)

8-C8 Use a definite integral (from $r = 0$ to $r = R$) to find the moment of inertia of a solid disk of total mass M and radius R. (*Hint:* The mass of a ring of radius r and width dr is $dm = [(2\pi r\, dr)/\pi R^2]M$.) Check your answer by referring to Table 8-1.

8-C9 A basket of ripe tomatoes weighing 64 lb is being hoisted by a windlass. The rope is wrapped around an axle which is a solid cylinder of wood having radius 6 in. and weight 32 lb. The mass of the crank handle is negligible. The operator lets go the handle when the basket is 20 ft above the ground, and the rope unwinds. With what linear speed does the basket strike the ground? (*Hint:* Use the energy principle.)

8-C10 Objects of mass 6 kg and 10 kg are connected by a weightless string which passes over a frictionless pulley whose moment of inertia is 0.4 kg·m² and whose radius is 0.2 m. It is shown in Sec. 8-11 that the acceleration of the system is 1.51 m/s². (*a*) Calculate the velocities v and ω ($= v/r$) 3 s after starting from rest. (*b*) Calculate the net loss of PE of the system in 3 s, and show that it equals the gain of KE of the system.

8-C11 A solid sphere of mass M and radius R rolls without slipping down a plane of length L and height H. What is the linear velocity of its center of mass when the ball reaches the bottom of the plane? (*Hint:* The KE at the bottom is partially translational and partially rotational. At all times, $v = r\omega$. Use the energy principle, and solve for v.) (*b*) What would the speed be if the sphere slides down a smooth plane of length L and height H, without rolling? (*Hint:* In this case the energy is all translational.)

8-C12 The flywheel of an airplane spins counterclockwise as viewed from behind. When the plane makes a right turn, does the plane tend to nose up or nose down? Would a jet plane be subject to this effect?

8-C13 Let the slow angular velocity of precession be denoted by Ω (capital omega) and the spin velocity by ω (small omega). Prove that $\Omega = \tau/I\omega$. (*Hint:* In Fig. 8-31, Ω is the rate at which the axis is changing direction. Thus $\Omega = \lim \Delta\theta/\Delta t$. You can consider the triangle *BCA* to be a sector of a circle of radius $I\omega$.)

8-C14 Use the result of Prob. 8-C13 to find the rate of precession of the bicycle wheel in Fig. 8-30. Data: The mass of the wheel is 10 kg, with practically all the mass concentrated in the rim. The radius of the wheel is 35 cm, and it is spinning at 5 rev/s = 31.4 rad/s. The axle is supported on the table at a point which is 25 cm from the c.g. of the wheel.

8-C15 (*a*) From the data given in Sec. 8-12 show that the observed rate of precession of the earth's axis Ω (defined in Prob. 8-C13) is about 7.7×10^{-12} rad/s and the spin rate ω is 7.3×10^{-5} rad/s. For (*b*) through (*g*) make an order-of-magnitude check of the rate of precession, as outlined in each step. Such calculations, made without expectation of precise agreement, often give physical insight and are useful as a starting point for more detailed study. Use data from Table 1 of the Appendix as needed. (*b*) To apply $\Omega = \tau/I\omega$ we need the moment of inertia I of the spinning earth; assuming uniform density, show that $I \approx 1.0 \times 10^{38}$ kg·m². (*c*) The equatorial radius of the earth is 6378 km and the polar radius is 6357 km. Thus each bulge (shaded volume in Fig. 8-32) is roughly a slab 10 km thick, with a base area roughly 6000 km × 6000 km. Assuming the specific gravity of the crust to be about 3, calculate the mass of each bulge. (*d*) The moon's mass is given in Example 4-9; calculate the force F_1 on the "near" bulge (3.84×10^8 m from the moon) and the force F_2 on the "far" bulge (3.96×10^8 m from the moon). (*e*) Find the net torque τ, using a lever arm of about 3000 km for each force. (*f*) Show that $\Omega = \tau/I\omega$ gives a result which is about 0.1 the observed rate. (*g*) Two factors that were neglected would modify the computed Ω. Tell whether each of these refinements would change Ω in the proper direction: the earth is not of uniform density (Sec. 4-8), and the sun also exerts a torque on the bulges.

■ REFERENCES

1 Gray, George W., "The Ultracentrifuge," *Sci. American*, June, 1951, p. 42.

2 Beams, J. W., "Ultrahigh-Speed Rotation," *Sci. American*, Apr., 1961, p. 134.

3 Singer, S. F., "Satellites for Physicists," *Physics Today*, Apr., 1956, p. 21. Describes the precession of an artificial satellite due to the earth's bulge.

4 Miller, Franklin Jr., "Kepler's Third Law and The Mass of the Moon," *Am. J. Phys.,* **34,** 53 (1966).

5 McDonald, James E., "The Coriolis Effect," *Sci. American,* May, 1952, p. 72.

6 Miller, Franklin Jr., *Inertial Forces: Centripetal Acceleration* (film).

7 Friedman, Francis L., *Velocity and Acceleration in Circular Motion* (film).

The role of elasticity in everyday affairs is usually greatly underestimated. The short-range interatomic electric forces we call elastic forces are all around us, often taken for granted. A drugstore clerk wraps up a package and secures it with a rubber band. We are accustomed to calling the force exerted by the rubber band an elastic force because we can easily see the distortion produced in the rubber (a stretch in this instance) and can feel the reaction to the force which we apply to cause the distortion. If the clerk wraps the package with string, applying the same force, we tend to ignore the stretch of the string because it is not readily visible. However, the two situations differ only in degree. The string must stretch, be it ever so little, in order to give rise to the contractile force it exerts. A gnat alights on the flight deck of a 60 000-ton carrier, and the deck buckles (slightly) in order to supply the upward force necessary to support the gnat's weight. When you pry the cap off a bottle of soda, the opener must be slightly distorted to supply the necessary force on the bottle cap. In Sec. 5-4 we made a model for static and kinetic friction based on the ideas of elasticity.

Another example of "hidden" elastic force is afforded by the force of the rails beneath the wheels of a freight car. The car sinks into the rails just far enough so that the upward elastic force exerted by the rails is just enough to balance the weight of the car. If the upward force were insufficient, the car would experience a net downward force, and so it would move downward until the forces were balanced again. Except for the minuteness of the motion, this behavior is in no way different from the behavior of a letter placed on a spring-operated postage scale. The platform of the scale sinks down just far enough.

9-1 Elastic constants—Hooke's law

One of the most important reasons for studying elasticity is that it gives us an approach to the study of vibrational motion, such as the up and down motion of a massive object suspended by a rubber band. First, however, we will consider how the stretch of a rubber band, or a wire, depends on the material of which it is made, as well as on its length and cross-sectional area.

The vibrating system of this unusual 400-day clock is a disk suspended by a fine, flat wire; the period of vibration is determined by the moment of inertia of the disk and the stiffness of the wire. The two adjustable balls allow fine control of the moment of inertia. The letters A and R refer to the French words *avance* (fast) and *retard* (slow). The clock is from the Horolovar collection and is owned by Charles Terwilliger. (Photo by Edith Reichmann)

Just as specific gravity is a constant for a material, not depending on the particular size and shape of the object which has been fashioned from the material, so each substance has a number of elastic constants which do not depend on the particular size and shape involved. Robert Hooke (1635–1703) found that the stretch of an iron wire was proportional to the applied force, up to some maximum force. This proportionality between force and stretch is known as *Hooke's law*. All other things being equal, it is evident that a fine iron wire will stretch more than a thick iron wire, and yet the substance, iron, is the same in each case. It is possible to define an elastic property of the material itself, called a *modulus of elasticity*, and there are as many elastic moduli as there are types of distortion. We shall consider three simple types of distortion, and define a modulus for each.

Stretch modulus. Change in length ΔL of a wire or rod is caused by a change in stretching force ΔF (see Fig. 9-1). The ratio $\Delta F/\Delta L$ is constant, for any given wire, but of course it varies from wire to wire depending not only on the elastic modulus of the material, but also on the length L and the cross section A. When we consider wires of the same substance, but of different cross-sectional area and length, experiment shows that the amount of stretch is proportional to the original length of the wire. A wire 5 m long stretches twice as much as a wire 2.5 m long, other things being equal. Also, experiment shows that, to produce a given stretch in wires of equal length, the required force is proportional to the cross-sectional area. All this is summarized in one equation

$$\frac{\Delta F}{A} = E\frac{\Delta L}{L}$$

$$\text{Stress} = (\text{modulus})(\text{strain})$$

where *stress* is the force per unit cross-sectional area, and strain is the stretch per unit length. Stress has the dimensions of pressure, being force/area; strain is a pure number, being length/length. E is the *stretch modulus*—also called *Young's modulus*—for the material.

$$\text{Stretch modulus} = \frac{\text{stretching stress}}{\text{stretching strain}}$$

$$E = \frac{\Delta F/A}{\Delta L/L}$$

The stretch modulus, like all moduli, has the dimensions of force/area.

EXAMPLE **9-1**

A telephone wire 125.00 m long and 1.00 mm in radius is stretched to a length 125.25 m when a force of 800 N is applied. What is the stretch modulus for the material?

The cross-sectional area of the wire is

$$\pi(10^{-3} \text{ m})^2 = 3.14 \times 10^{-6} \text{ m}^2.$$

$$\text{Stress} = \frac{\Delta F}{A} = \frac{800 \text{ N}}{3.14 \times 10^{-6} \text{ m}^2}$$

$$= 2.54 \times 10^8 \text{ N/m}^2$$

$$\text{Strain} = \frac{\Delta L}{L} = \frac{0.25 \text{ m}}{125 \text{ m}}$$

$$= 0.002 = 2 \times 10^{-3}$$

(The stretch is 0.2% of the original length.)

$$E = \frac{\text{stress}}{\text{strain}} = \frac{2.54 \times 10^8 \text{ N/m}^2}{2 \times 10^{-3}}$$

$$= 1.27 \times 10^{11} \text{ N/m}^2$$

$$= \boxed{12.7 \times 10^{10} \text{ N/m}^2}$$

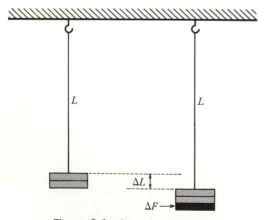

Figure 9-1 Stretch of a wire.

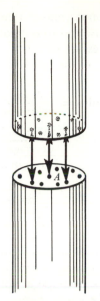

Figure 9-2 Separation of atomic planes during stretch of a wire.

To interpret this experimental behavior, let us make a model of the wire (Fig. 9-2). Consider a stretched wire; the force arises from the cohesion between the atoms. That is, the atoms in the crystals are so close together that short-range electric forces are considerable. We can imagine the wire to consist of many parallel planes of atoms stacked on top of each other. Due to a load, each pair of neighboring planes separates slightly. The total stretch is the sum of the individual stretches. The longer the wire, the more planes are separated, each by the same small amount, so we expect ΔL, the total stretch, to be proportional to L, as experiment shows. Likewise, the larger the cross section A, the more atoms act together at each layer. Hence we expect the required force to be proportional to A. We see that the model accounts satisfactorily for the main experimental facts.

Our simple model does not account for some of the details of the stretching process. As we know, if sufficient force is applied to a wire, it no longer remains in equilibrium but stretches a great deal and breaks. Before this happens, however, a more subtle discrepancy appears. If the applied stress is somewhat greater than some well-defined amount called the *elastic limit*, the wire does not break, but neither does it return to its original length when the stress is removed. The wire acquires a permanent distortion. In the design of structures, the elastic limit must never be exceeded. In fact, architects and engineers usually allow a "safety factor" and design for not more than $\frac{1}{5}$ to $\frac{1}{3}$ the elastic limit, especially in machinery where stress is repeatedly applied and removed.

Shear modulus—rigidity. We have considered the stretch modulus in some detail because the basic ideas are applicable to any elastic modulus, including a second form of distortion called *shear*. There is no change of any dimension during shear, but the *shape* of the body changes. For instance (Fig. 9-3), a book lying on a table is given a sideways shear by means of a force applied to the top cover (and an equal and opposite force of friction applied by the table to the bottom cover). Each page moves relative to its neighbor. In this case the stress is the applied force divided by the area of one of the pages; the strain is the horizontal distance moved divided by the height of the book.

$$\text{Shear modulus} = \frac{\text{shearing stress}}{\text{shearing strain}}$$

$$n = \frac{\Delta F/A}{\Delta x/h}$$

A model based on the mutual cohesion of adjacent layers of atoms (Fig. 9-4) will give a qualitative interpretation of the experimental facts: the amount of shear is inversely

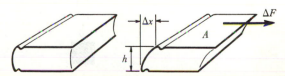

Figure 9-3 Shear of a solid body.

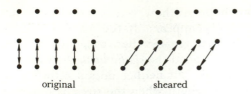

original   sheared

Figure 9-4 Separation of atomic planes during shear of a solid body.

proportional to the *tangential* area A, and it is also directly proportional to the height of the book (i.e., to the number of parallel planes each of which is sheared the same tiny amount).

The shear modulus is also called the *rigidity modulus,* for it is this property of a substance which determines how well a body will retain its shape when a shearing stress is applied. Indeed, if it were not for the property that the shear modulus describes, a solid block of steel would collapse under its own weight like a block of melting butter. The action of a common machine screw illustrates nicely several stresses (Fig. 9-5). When properly tightened, the shaft of the screw must withstand a sideways shearing stress; also, as the shaft is compressed its length decreases, so the essential factor determining the force with which the surfaces A and B are pressed together is the stretch modulus.

Bulk modulus. For our third example of a simple elastic distortion, we consider a solid block of iron on the end of a long wire. When the block is lowered to a great depth in the ocean, the pressure of the water causes the volume (bulk) of the block to diminish.

$$\text{Bulk modulus} = \frac{\text{volume stress}}{\text{volume strain}}$$

$$= \frac{\text{increase in pressure}}{\dfrac{\text{decrease in volume}}{\text{original volume}}}$$

$$B = \frac{\Delta P}{\dfrac{-\Delta V}{V}} = -\frac{V\,\Delta P}{\Delta V}$$

Values of the bulk modulus are positive; the $-$ sign in the equation compensates for the fact that ΔV (a decrease in volume) is negative. The well-known fact that most solids are not so easily compressed as liquids is borne out by the figures in Table 9-1. The pressure required to produce a given small relative change in volume is about 70 times as much for steel as for water. The only elastic modulus which applies to a liquid or gas is the bulk modulus, since liquids and gases have no definite length or shape. Table 9-1 gives values of the elastic constants for various substances in mks units of newtons per square meter. These moduli can be expressed in cgs or British units by making use of the fact that $1 \text{ N/m}^2 = 10 \text{ dyn/cm}^2 = 1.45 \times 10^{-4}$ lb/in.^{2}. There are *elastic limits* for distortions of the shear and volume types just as for the stretch type.

Our discussion of elasticity has been limited to highly idealized cases of "pure" stretch, "pure" shear, etc. Actually, many elastic distortions are complex. A wire that stretches must also change shape, since it is now a cylinder of greater length and smaller diameter. The volume usually increases slightly, too. Exact treatment of elasticity is highly mathematical, but our attempt has been rather to point out some main features of elasticity, and to indicate how widespread are the applications.

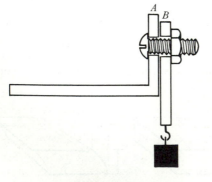

Figure 9-5

TABLE 9-1   Elastic constants*

Substance	Stretch modulus E, N/m^2	Shear modulus n, N/m^2	Bulk modulus B, N/m^2
Solids			
steel, annealed	20×10^{10}	8.0×10^{10}	14×10^{10}
iron, wrought	19	7.5	14
iron, cast	12	4.6	9.0
copper	10	3.9	11
aluminum	7.0	2.5	7.5
magnesium	4.1	1.6	3.1
lead	1.5	0.6	3.4
brass	9.0	3.4	8.3
glass	5.8	2.4	4.5
diamond	▪	▪	62
fused quartz	5.6	2.5	2.7
Liquids			
water			0.20×10^{10}
mercury			2.5
ethyl alcohol			0.10
glycerin			0.45
sea water			0.21
Gases▲			
air			1.01×10^5
hydrogen			1.01
helium			1.01
carbon dioxide			1.01

* Typical values; depend on heat treatment, cold work, etc.
▪ Since diamonds are anisotropic single crystals, the stretch and shear moduli depend on direction, hence no single value can be given. This effect averages out for the other solids listed, which are composed of many tiny crystals, randomly oriented.
▲ All gases at 1 atm pressure with no temperature change during compression or expansion.

It may seem that we have really explained nothing about elasticity, since we have put all the credit (or blame) on the mysterious interatomic forces which we call short-range electric forces. But actually "gravitational force" is an equally noncommittal term and equally unexplainable, and equally fundamental to the nature of things. Only our great practical familiarity with gravitational force (weight) leads us to think of it as better "explained" than the forces of elasticity. We are equally familiar with elastic forces, to be sure, and yet they seem more mysterious because no simple numerical formulas can be written comparable to Newton's elegant law of gravitation. Whether or not you are satisfied with this state of affairs is probably only a matter of degree of sophistication.

9-2   Simple harmonic motion

A *vibration* is a to-and-fro motion, generally along a straight line (linear vibration) or along an arc of a circle (angular vibration). Typical examples of linear vibration are those of a heavy load placed on the end of a spring or rubber band and the motion of an automobile engine's piston (Fig. 9-6a). A pendulum and the balance wheel of a watch illus-

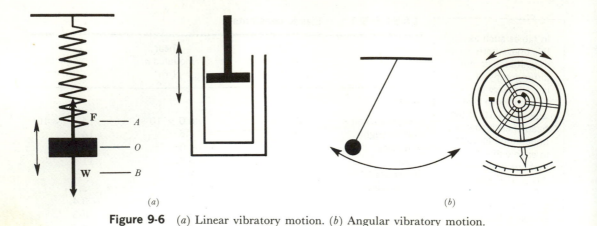

Figure 9-6 (*a*) Linear vibratory motion. (*b*) Angular vibratory motion.

trate angular vibratory motion (Fig. 9-6*b*). A brief look around us reveals many vibrations in nature. A raindrop strikes a leaf on a tree, and the leaf gyrates up and down; the tree itself sways in the breeze; the vocal cords of a bull moose vibrate many times per second, causing periodic compression of the air which in turn causes vibratory motion of his mate's eardrum. The ticking of a grandfather clock, the clang of a dropped piece of metal, the twang of a guitar string—all these are among the sounds which result from countless types of vibratory motion. Because of the importance of vibrations in music, the term "harmonic motion" has come to mean the same as "vibratory motion." One particularly simple form of harmonic motion occurs frequently both in the large-scale world and at the atomic level, and so we proceed to define and study *simple harmonic motion* (SHM), especially in its relation to Hooke's law for elasticity.

First of all, it is obvious that a material body must be acted on by a somewhat specialized force in order to vibrate. The body, after passing through its equilibrium position, must eventually stop and return; otherwise it would be lost forever, and the motion could not repeat itself. Hooke's law supplies just the right sort of force, as illustrated by a mass on the end of a spring (Fig. 9-6*a*). We assume that the spring has already been stretched

(by the force of gravity) until the mass hangs motionless, in equilibrium at some point O, and the force **F** in the spring is equal and opposite to the weight **W** of the load. If the mass is moved to a position A slightly above O, the force in the spring decreases, F becomes less than W, and so the net **F** is downward. If the mass is released while above O, the acceleration, given by (net **F**)$/m$, is downward, and the mass returns to O. Likewise, if the mass is pulled down below O to a position such as B, the force in the spring increases, F becomes greater than W, net **F** is upward, and if the mass is released while below O the acceleration is upward and once again the mass returns to O. Thus we see that there is a *restoring force,* always directed toward the position of equilibrium.

In practice, if the mass is pulled away from its equilibrium position and released, it accelerates toward O and reaches O with some velocity. It coasts through the equilibrium position and immediately starts to lose speed, since the acceleration changes direction at O. Eventually, the mass momentarily comes to rest, then starts to move back toward the equilibrium position O. The vibration continues indefinitely, unless there is external friction such as air resistance, or internal molecular friction in the material of the spring.

"Simple" harmonic motion is so called

because the force varies as smoothly as possible. The very simplest possible equation for a variable force is a direct proportion. Thus we define *simple harmonic motion* as motion for which the restoring force is directly proportional to the displacement away from the equilibrium position. We may write

$$\frac{F}{s} = k \qquad (9\text{-}1)$$

where s is the displacement, F is the restoring force (always directed opposite to the displacement), and k is the *force constant* of the spring. Any spring, wire, bar, etc., which obeys Hooke's law has a force constant.

EXAMPLE **9-2**

A certain spring is 30.0 cm long when a load of 40.2 N hangs from it and 37.0 cm long when the load is 45.2 N. What is the force constant of the spring?

The added force is 5.0 N, causing a stretch of 0.070 m:

$$k = \frac{F}{s} = \frac{5.0 \text{ N}}{0.070 \text{ m}} = \boxed{71 \text{ N/m}}$$

9-3 Period of simple harmonic motion: the reference circle

In our study of SHM, we consider a cart of mass m which is supported by frictionless wheels as it moves to and fro horizontally (Fig. 9-7). The net force on the cart is that caused by the springs, since its weight is balanced by the upward force of the table. The force constant of the spring system is k.

Several technical terms are used in the description of SHM. A *cycle* is one complete to-and-fro vibration (P to P' and back again), and the *period* T is the time required for this to take place. The *frequency* ν is the number of vibrations per unit time, and is the reciprocal of the period. Thus if the period is 0.01 s, the body makes 100 vibrations per second. The *displacement* **s** is the vector measured from the equilibrium point to the position at any time t, and the *amplitude* A is the maximum value of the displacement. The body moves a total distance of $4A$ in one cycle, but of course the velocity is by no means constant. The *phase* of the vibration is a measure of the position of the particle in its motion (Fig. 9-8). Two vibrations are in phase with each other if both particles pass through their equilibrium points at the same time, going in the same direction.

To derive a formula for the period of a SHM, we use a *reference circle* to make clear the relation of SHM to uniform circular motion. By definition, in SHM $F/s = k$. The restoring force F is the net force, hence $F = ma$, $ma/s = k$, or

$$\frac{\text{acceleration}}{\text{displacement}} = \frac{a}{s} = \frac{k}{m}$$

which is a constant for any given mass m on any given spring of force constant k. We can say that it is a characteristic of SHM that the acceleration is directed toward the center of the motion and is proportional to the displacement. This is equivalent to saying that *any* motion for which a/s is a constant is SHM.

Fortunately, it so happens that another simple motion has this same property. Consider a particle having uniform circular mo-

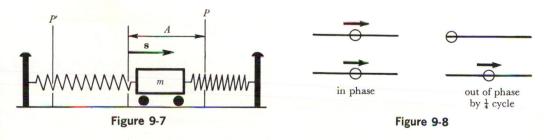

Figure 9-7

Figure 9-8

in phase

out of phase by $\frac{1}{4}$ cycle

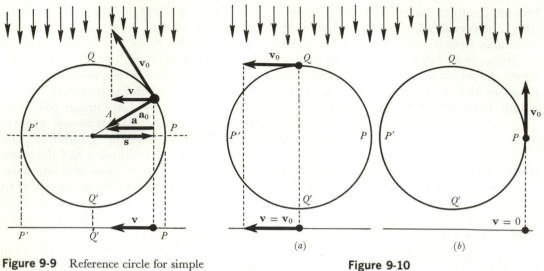

Figure 9-9 Reference circle for simple harmonic motion.

Figure 9-10

(a)

(b)

tion with velocity v_0 in a vertical circle of radius A. As in Fig. 9-9, let light shine vertically on the particle in this *reference circle*. The shadow (or projection) of the *reference particle* moves back and forth along a line parallel to the horizontal diameter. The shadow has maximum velocity at the midpoint of its motion, for then the reference particle is at Q (or Q'), and v, the horizontal component of v_0, equals v_0 itself (Fig. 9-10a). At the extreme points P and P', the horizontal component of v_0 is zero (Fig. 9-10b), which agrees with the fact that a vibrating particle is momentarily stationary as it changes direction of motion (its acceleration is not zero even though its velocity is zero; compare the motion of a vertically thrown baseball which has a downward acceleration of 32 ft/s² even when its instantaneous velocity is zero at the highest point).

We now prove that the motion of the shadow is SHM. Let v_0 be the constant speed of the reference particle. The acceleration a_0 of the reference particle is constant in magnitude (equal to v_0^2/A) and directed along the radius. The acceleration a of the shadow is the horizontal component of a_0. By similar triangles, $a/s = a_0/A =$ constant. Hence the

shadow moves in SHM, since the constancy of the ratio of acceleration to displacement is a general property of SHM. We also note that A, the radius of the reference circle, equals the amplitude of the SHM of the shadow. The time T for one complete vibration of the shadow is the same as the time for one complete revolution of the reference particle. Hence

$$T = \frac{\text{circumference of reference circle}}{\text{speed around the circle}} = \frac{2\pi A}{v_0}$$

$$v_0 = \frac{2\pi A}{T} \qquad (9\text{-}2)$$

Use of this formula for v_0 allows an easy calculation of the maximum velocity and maximum acceleration in any SHM.

EXAMPLE 9-3

A prong of a tuning fork vibrates in SHM, at a frequency of 500 vib/s. The prong moves through 2.00 mm, 1.00 mm on either side of center. What is its maximum velocity? What is its maximum acceleration?

The period is $\frac{1}{500}$ s = 0.002 s, and the amplitude is 1.00 mm = 1.00×10^{-3} m. At the midpoint of the motion,

$$v_0 = \frac{2\pi A}{T} = \frac{2\pi (1.00 \times 10^{-3} \text{ m})}{2 \times 10^{-3} \text{ s}}$$

$$= \boxed{3.14 \text{ m/s}}$$

At either end-point,

$$a_0 = \frac{v_0^2}{A} = \frac{(3.14 \text{ m/s})^2}{1.00 \times 10^{-3} \text{ m}} = \boxed{9870 \text{ m/s}^2}$$

The fact that the projection of uniform circular motion is SHM can be demonstrated as shown in Fig. 9-11. On a screen is cast the shadow of one ball attached to a rotating turntable, and alongside it is the shadow of another ball vibrating vertically on the end of a spring. When the period and phase of the turntable are properly adjusted, the two shadows move side by side at all times.

To derive a formula for the period of the SHM, we use the energy principle. During the SHM of a body, there is a continual transformation of energy from KE to PE and back again. Referring to Fig. 9-12, suppose the body has been moved by some outside agent from its equilibrium position 2

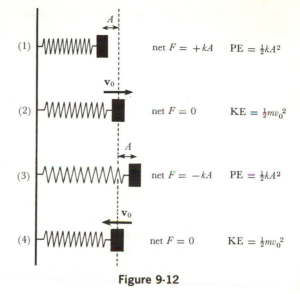

Figure 9-12

through a distance A to some new position 1, thus compressing the spring. The work done is stored in the spring as elastic PE. The work was done against a variable force, which was 0 at position 2 and was kA at position 1. As we saw in Sec. 6-5, work done against a variable force is an integral, represented by the area under the force-displacement curve. In this case, for a Hooke's law force, the area of the triangle of Fig. 9-13 is $\frac{1}{2}(kA)A$, or $\frac{1}{2}kA^2$. The work, and hence the PE stored, is

$$\text{PE} = \tfrac{1}{2}kA^2$$

If, now, the body is released, it moves back to position 2, and by the energy principle

Photo by Kenneth Jewell

Figure 9-11 The projection of uniform circular motion is simple harmonic motion.

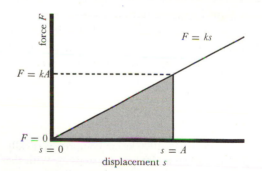

Figure 9-13 Work done during displacement against a Hooke's law force.

the total energy at 2 equals the total energy at 1:

$$KE_2 + PE_2 = KE_1 + PE_1$$

$$\tfrac{1}{2}mv_0^2 + 0 = 0 + \tfrac{1}{2}kA^2$$

Hence

$$\frac{A^2}{v_0^2} = \frac{m}{k}$$

$$\frac{A}{v_0} = \sqrt{\frac{m}{k}}$$

and, since $T = 2\pi A/v_0$,

$$T = 2\pi \sqrt{\frac{m}{k}} \qquad (9\text{-}3)$$

This is the basic formula for the period of any SHM, and it shows how the period is related to the essential physical attributes of the vibrating system. If friction is small, as it usually is, our application of the energy principle is correct, for the energy dissipated as heat is negligible. Note that the amplitude A cancels out, and the period of vibration depends only on the inertia of the object (represented by m) and the stiffness of the spring (represented by k). The period is independent of the amplitude of vibration. Equation 9-3 shows that, for any given load, a stiff spring (large k, i.e., large ratio F/s) has a shorter period T, which agrees with experience. The equation also shows that, for any given spring, a "heavy"• load vibrates more slowly, and T is larger. In Eq. 9-3, it is assumed that the force constant k is really constant. If the ratio F/s is not a constant, then the motion is not SHM in the first place, and Eq. 9-3 does not apply.

In solving problems in SHM, it will be sufficient to remember the definition of force constant, the idea of the reference circle (especially $v_0 = 2\pi A/T$), and the formula $T = 2\pi\sqrt{m/k}$.

EXAMPLE **9-4**

The springs of a 1000 kg car compress vertically 7.0 mm when a 100 kg man steps in. With the

•It is the inertia, or mass, and not the weight, that is important here.

man in the car, how many vibrations per second does the body of the car make after being jarred while going over a bump?

First we find the force constant:

$$k = \frac{F}{s} = \frac{mg}{s} = \frac{(100 \text{ kg})(9.8 \text{ m/s}^2)}{0.0070 \text{ m}}$$

$$= 1.40 \times 10^5 \text{ N/m}$$

$$T = 2\pi \sqrt{\frac{m}{k}}$$

$$= 2\pi \sqrt{\frac{1100 \text{ kg}}{1.40 \times 10^5 \text{ N/m}}} = 0.557 \text{ s}$$

The frequency of vertical vibration is the reciprocal of the period:

$$\nu = \frac{1}{T} = \frac{1}{0.557 \text{ s}} = \boxed{1.80 \text{ s}^{-1}}$$

The car body bounces up and down 1.80 times per second.

Actually cars are built with shock absorbers which provide frictional damping, and the vibration may last only one cycle or less.

EXAMPLE **9-5**

A spring having force constant 250 N/m is loaded with a mass of 10 kg. (a) Find the period of vibration. (b) If the mass is displaced 30 cm and then released, find the velocity with which it passes through the equilibrium position. (c) What is the total energy of the vibrating mass? (d) How much time is required for the mass to move the 30 cm from the end-point of the motion to the equilibrium position?

(a) $\quad T = 2\pi \sqrt{\dfrac{m}{k}}$

$$= 2\pi \sqrt{\frac{10 \text{ kg}}{250 \text{ N/m}}} = \boxed{1.256 \text{ s}}$$

(b) $\quad v_0 = \dfrac{2\pi A}{T}$

$$= \frac{2\pi(0.30 \text{ m})}{1.256 \text{ s}} = \boxed{1.50 \text{ m/s}}$$

(c) The total energy of the vibrating mass is the same at any point of the motion. There are two points at which calculation of this energy is simple. First, as the mass passes through the equilibrium position, the energy is all KE:

Maximum KE $= \frac{1}{2}mv_0{}^2$

$\qquad = \frac{1}{2}(10 \text{ kg})(1.50 \text{ m/s})^2$

$\qquad = \boxed{11.3 \text{ J}}$

Second, at the end-points, the energy is all PE:

Maximum PE $= \frac{1}{2}kA^2$

$\qquad = \frac{1}{2}(250 \text{ N/m})(0.30 \text{ m})^2$

$\qquad = 11.3 \text{ N·m} = \boxed{11.3 \text{ J}}$

(*d*) The time required to move the 30 cm is $\frac{1}{4}$ of the complete period, since the reference particle has moved through 90°, or $\frac{1}{4}$ of a complete circle. Hence

$$t = \tfrac{1}{4}T = \tfrac{1}{4}(1.256 \text{ s}) = \boxed{0.314 \text{ s}}$$

It would be incorrect to try to use the laws of uniformly accelerated motion (such as $s = \frac{1}{2}at^2$) to find the time, since acceleration is *not* constant. The very idea of SHM is one of *variable* acceleration, according to the proportion $a \propto s$. *The equations of uniformly accelerated motion cannot be used for SHM problems.*

An equation for the motion of a body in SHM can be derived, using the reference circle (Fig. 9-14). The angular speed ω of the reference particle, which is the angle swept out by the particle per unit time, is $\omega = (2\pi \text{ rad})/(T \text{ s}) = 2\pi/T$ rad/s. The angle θ, which is the angular displacement of the particle and which increases steadily, is given by $\theta = \omega t = 2\pi t/T$, where t is the time. Thus

the displacement of the projection, point P, is given by $y = A \cos \theta = A \cos \omega t$, or

$$y = A \cos \frac{2\pi t}{T} \qquad (9\text{-}4)$$

where A is the radius of the reference circle. In Eq. 9-4, it is assumed that time is measured from a point of maximum displacement (if $t = 0$, $\cos 0° = 1$, and $y = +A$). The significance of the period T in the denominator is as follows: imagine that t increases from 0 to T; then $2\pi t/T$ increases from 0 to 2π ($= 360°$), and the cosine function goes through one complete cycle of its variation.

Several equivalent equations for SHM can be written down, depending on the initial conditions. For instance, if we choose $t = 0$ to be an instant when the vibrating particle is at the center of its vibration, moving upward, a suitable equation would be

$$y = A \sin \frac{2\pi t}{T} \qquad (9\text{-}5)$$

since this makes $y = 0$ when $t = 0$. Also, the motion could be horizontal along the x axis, as in the following example.

EXAMPLE **9-6**

A particle is vibrating horizontally, to and fro along the x axis, making 30 vib/min. The total extent of its travel, from one extreme to the other, is 2.80 m. At $t = 0$, the particle is released from a point which is farthest to the left of

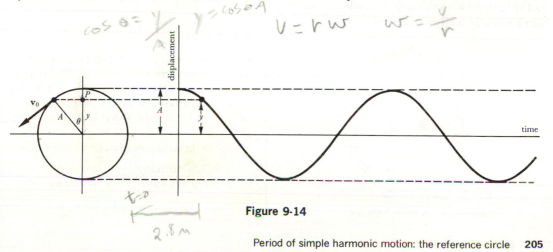

Figure 9-14

center. (*a*) Write an equation for the motion. (*b*) What is the displacement of the particle 1.00 s after it is released?

(*a*) The amplitude is 1.40 m. The period T is 2 s/vib, and therefore $2\pi/T = 2\pi$ rad/2 s = π rad/s. Since the particle has a maximum negative displacement at $t = 0$, we choose a negative cosine function. The motion is given by

$$x = -1.40 \cos (\pi t)$$

(*b*) At $t = 1.00$ s,

$$x = -1.40 \cos [(\pi \, \text{rad/s})(1.00 \, \text{s})]$$
$$= -1.40 \cos (\pi \, \text{rad}) = -1.40 \cos 180°$$
$$= -1.40(-1) = \boxed{+1.40 \text{ m}}$$

In general, we say that both the sine function and the cosine function are "sinusoidal" —their graphs are identical, except for a phase difference of $\frac{1}{4}$ cycle. Thus we conclude that a body which is acted on by a Hooke's law force moves in SHM, with its displacement a sinusoidal function of time.

In Sec. 6-3 we saw how the law of conservation of momentum could be used to measure the ratio of two masses by measuring the various velocities appearing in Eq. 6-3. The equation for the period of SHM gives another way of comparing masses without making use of gravitational forces. One form of "mass balance" is shown in Fig. 9-15. The unknown mass is placed on the platform which vibrates from side to side with a period determined by the stiffness of the flat springs and the inertia of the system. Weight has nothing to do with this balance, which would work equally well in interstellar space.

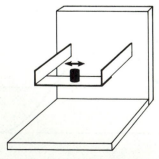

Figure 9-15 A mass balance.

EXAMPLE **9-7**

When a standard mass of 1.00 kg is placed on the platform, the vibration rate is 125 vib/min. What is the mass of an unknown object for which the vibration rate is 243 vib/min? Neglect the mass of the moving platform.

$$T = 2\pi \sqrt{\frac{m}{k}}$$

Squaring both sides gives

$$T^2 = \frac{4\pi^2 m}{k}$$

$$T^2 = (\text{constant})m$$

since k is constant for any particular spring system. Hence

$$\frac{T_1^2}{T_2^2} = \frac{m_1}{m_2}$$

Then, since the frequencies ν are the reciprocals of the periods T,

$$\frac{\nu_2^2}{\nu_1^2} = \frac{m_1}{m_2}$$

This equation can now be used to find the unknown mass:

$$\frac{(243 \text{ vib/min})^2}{(125 \text{ vib/min})^2} = \frac{1.00 \text{ kg}}{m}$$

$$m = \boxed{0.265 \text{ kg}}$$

In actual practice, the mass of the supporting platform would not be negligible. However, the mass of the platform can be found by timing the vibration with two known masses.

EXAMPLE **9-8**

When a standard mass of 1.00 kg is placed on the platform, the frequency of vibration is 120 vib/min. When two standard 1.00 kg masses are placed side by side on the platform, the rate is 90 vib/min. What is the mass of the platform?

We call the unknown mass of the platform m'.

$$\frac{(120 \text{ vib/min})^2}{(90 \text{ vib/min})^2} = \frac{m' + 2.00 \text{ kg}}{m' + 1.00 \text{ kg}}$$

$$m' = \boxed{0.286 \text{ kg}}$$

Knowing the mass of the platform, we could now proceed to measure the mass of any unknown object, as in the previous example.

We have now found two ways of measuring (or defining) inertial mass. Are we really defining the same thing? The momentum method is based on the law of conservation of momentum, and the SHM method just discussed is based on the energy principle. Ultimately, both the conservation of momentum and the energy principle are derived from Newton's laws of motion, as we have shown in earlier chapters, and so the two methods necessarily give identical results for the measurement of mass. Our conviction that this is so is a measure of our faith in the general validity of Newton's laws. The two ways of measuring inertial mass are entirely equivalent to each other.

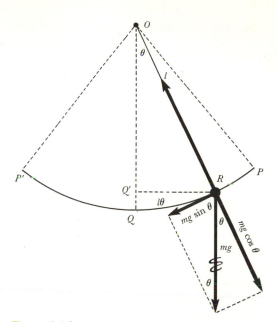

Figure 9-16 A simple pendulum moves in approximate SHM.

9-4 The simple pendulum

When a small, heavy pendulum bob is swinging at the end of a light rod or string, the pendulum is said to be "simple," since we assume that all moving parts that have inertia are concentrated at one place, all moving parts have the same speed, and the restoring force acts at a single point. These assumptions are never quite true, since even the lightest string has some mass, even the smallest bob has some distribution of mass, and all parts of the pendulum cannot be moving at the same speed. A real or "physical" pendulum (see Sec. 9-7) can be studied by methods similar to those we are about to use for the idealized *simple pendulum.*

The bob of the pendulum moves from P to Q to P' and back again (Fig. 9-16) with many of the characteristics of simple harmonic motion: the speed is maximum at Q; the acceleration is greatest at P and P' (when the velocity is momentarily zero, changing from $+$ to $-$ or from $-$ to $+$). To find out whether a pendulum does actually move in

SHM, we ask the usual question: Is the restoring force proportional to the displacement? In other words, *is* there a force constant?

The outward component of the weight, $mg \cos \theta$, is more than balanced by the tension in the string; there is a net force along RO to supply the necessary centripetal force on the mass. The component of the weight tangent to the arc is also unbalanced, and this unbalanced force (equal to $mg \sin \theta$) is the restoring force directed along the arc. The displacement is the length of arc, $l\theta$. If there is a force constant, then it is defined by

$$k = \frac{\text{restoring force}}{\text{displacement}} = \frac{mg \sin \theta}{l\theta} \qquad (9\text{-}6)$$

This is not quite constant, since $\sin \theta$ is not the same as θ. In the triangle ROQ', $\sin \theta =$ (chord RQ')$/l$ whereas the angle θ (in radians) is (arc RQ)$/l$. As the angle θ decreases, however, the arc and the chord become practically equal, and the ratio $(\sin \theta)/\theta$ approaches 1. Hence, if the angular displacement of a simple pendulum is small, the force constant is

$$k = \frac{F}{s} = \frac{mg}{l}\left(\frac{\sin\theta}{\theta}\right) \approx \frac{mg}{l} \qquad (9\text{-}7)$$

and, to the extent to which k is a constant, the motion is SHM. The period of the pendulum is

$$T = 2\pi\sqrt{\frac{m}{k}} \approx 2\pi\sqrt{\frac{m}{\dfrac{mg}{l}}} \qquad (9\text{-}8)$$

The mass m cancels out, and

$$T \approx 2\pi\sqrt{\frac{l}{g}} \qquad (9\text{-}9)$$

It is important to realize that this simple formula for the period of a simple pendulum is an approximation, but a very good one if the angle of swing is small.[*] It is also important to note that the symbol m canceled out, because weight (mg) is proportional to mass (m). This being so, the period of a pendulum is independent of the mass of the bob, depending only on the length of the string.[*] On the other hand, since g appears in the formula, we have a means of determining the value of the acceleration due to gravity at any place.

EXAMPLE **9-9**

What is the acceleration due to gravity on a planet where a space explorer's simple pendulum 40.0 cm long vibrates 100 times in 240 s?
The period is $(240 \text{ s})/100 = 2.40$ s.

$$T \approx 2\pi\sqrt{\frac{l}{g}}$$

$$T^2 = \frac{4\pi^2 l}{g}$$

$$g = \frac{4\pi^2 l}{T^2} = \frac{4\pi^2(40.0 \text{ cm})}{(2.40 \text{ s})^2} = \boxed{274 \text{ cm/s}^2}$$

Note that $T \approx 2\pi\sqrt{l/g}$ is a special formula, valid for a special case (simple pendu-

[*] The error is 0.1% if the angle θ_0 ($\angle\, POQ$) is 7°, and only 1% if θ_0 is as large as 23°.
[*] This property of the period of a pendulum was used by Bessel in his investigation of the proportionality of inertial mass and gravitational mass referred to on p. 70.

lum vibrating with not too large an amplitude), while the similar-appearing formula $T = 2\pi\sqrt{m/k}$ is a general formula for the period of *any* simple harmonic motion.

The ideas we have developed may be carried over into the study of angular SHM merely by substituting *moment of inertia* for *mass, torque* for *force*, etc. The "anniversary clock" (which is wound once a year) has two or four masses suspended from a stiff wire (see the photograph on p. 195). By analogy with the linear formula $T = 2\pi\sqrt{\dfrac{m}{F/s}}$, the period of oscillation is given by $T = 2\pi\sqrt{\dfrac{I}{\tau/\theta}}$, where τ is the torque required to cause a rotation of θ radians, and I is the moment of inertia. The period is controlled by adjusting the positions of the masses, thus varying I. The torsion constant τ/θ depends on the length and diameter of the wire and the shear modulus of the material of the wire.

9-5 Non-simple harmonic motion

The relative ease with which our study of SHM gave us useful formulas is impressive. We are also impressed by the ubiquitous occurrence of SHM, depending as it does only upon the ever-present elastic behavior of bodies as expressed by Hooke's law. However, not all vibrations are "simple," since the acceleration need not be proportional to the displacement as in SHM. Consider a ball bearing bouncing up and down on a highly elastic steel plate, always returning to the same height from which it fell. This vibration is less simple than that of a mass on the end of a spring, because the ball bearing's velocity changes abruptly (during impact) rather than smoothly. The two graphs in Fig. 9-17 compare SHM with the motion of a bouncing ball. The force acting in each case is represented by a dotted line. Note that for the bouncing ball, the force is a constant down-

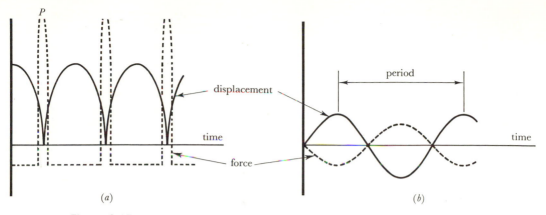

Figure 9-17 Two types of vibration. (*a*) Bouncing ball; (*b*) mass on a spring.

ward (negative) force equal to the weight of the ball, except for a very short fraction of each period, say 0.001 s, during which the ball is in contact with the base plate. During contact, the ball is acted on by a large upward (positive) force. Point *P* represents the instant of maximum distortion, corresponding to maximum upward force. Contrast this complex behavior of the bouncing ball with that of the mass on a spring. In SHM the displacement changes smoothly, and the force (and acceleration) are also free from sudden wild variations. In the SHM the force is always proportional to the displacement, but oppositely directed. Both force and displacement are zero at the same time, and the restoring force is negative when the displacement is positive, and vice versa.

Other vibrations which are not SHM occur because of the difference between static and kinetic coefficients of friction (Sec. 5-4). When a piece of chalk squeals as it is drawn across the blackboard, the vibration is due to the "stick-slip" phenomenon. The chalk sticks because of static friction until the applied force causes it to slip. Since kinetic friction, at the start, is less than static friction, the chalk's speed increases rapidly. Now the dependence of friction on speed comes into play. As the speed increases, eventually the chalk "sticks" and the process starts all over again. The frequency of this process depends

in a complicated way upon the properties of matter. The bouncing ball, the squeal of a car's brakes, the passage of a violin bow across the string, even the vibrations of human vocal cords—all illustrate so-called *relaxation vibrations* in which motions change abruptly.

9-6 Molecular rotation and vibration

It has proved possible to determine the frequencies of molecular vibrations and rotations by close study of the light emitted or absorbed by the molecules; such study is called *spectrum analysis* and will be discussed in Chap. 28. When the density is small, as for gases, the molecules are far apart and are not acted upon appreciably by their neighbors. They are then free to rotate or vibrate with certain characteristic frequencies. Consider the oxygen molecule O_2, consisting of two atoms, each of atomic mass 16 (Fig. 9-18). Because of their relatively small mass, we can for present purposes ignore the 16 electrons in the electron cloud surrounding the two nuclei and think of the molecule as a pair of identical balls connected by a stiff spring. This model is suggested by the known facts: molecules tend to resist both being compressed and being stretched, and there is some equilibrium distance of the one oxygen atom from the other. As a result of molecular

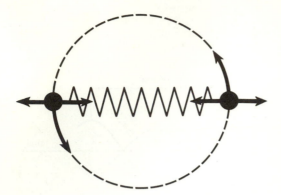

Figure 9-18 Simultaneous rotation and vibration of an oxygen molecule.

collisions in the gas, we expect that a typical molecule both rotates and vibrates. The spectrum analysis confirms this, and actually allows determination of the force constant for the hypothetical spring and the moment of inertia of the rotating system (thus allowing measurement of the separation of the atoms, since $I = \Sigma mr^2$). A very pretty interaction is evident, for experiment shows that those molecules which are rotating faster are also stretched a little bit, by a centrifugal force effect. The orders of magnitude involved are illustrated by the following data for an oxygen molecule: frequency of vibration 4.74×10^{13} vib/s; equilibrium distance between two oxygen atoms in a molecule 1.21×10^{-10} m; moment of inertia of molecule 1.95×10^{-46} kg·m²; KE of rotation 3.42×10^{-22} J; angular velocity of molecule 1.87×10^{12} rad/s.

Our model of the oxygen molecule in terms of tiny balls on the ends of a spring has proved successful in some details, but it fails miserably in other ways. Most spectacularly, experiment shows that not all rotational velocities are possible. The molecule either has no rotation at all, or else one of a series of definite angular velocities. For the oxygen molecule mentioned above, the angular velocity 1.87×10^{12} rad/s is one of the "allowed" velocities. As a rotating body speeds up, it gains one *quantum of energy* at a time, and, as

it slows down, it loses energy a quantum at a time. It might seem that a molecule could rotate at any angular velocity whatever, from zero up to infinity, with a continuous range of KE. Indeed, observation seems to show that an electric fan picks up speed uniformly when the current is turned on, with the velocity having every possible value up to some final velocity ω. Why then should an oxygen molecule behave differently? The fact is that the energy of *any* rotation is quantized, and the fan actually does speed up by fits and starts, gaining one tiny quantum of rotational energy at a time. The fan only *appears* to gain speed continuously because of the smallness of the quantum unit of energy relative to the total (observable) energy of rotation. When in the allowed state described by the above data, the rotating oxygen molecule has 3.42×10^{-22} J of rotational energy (corresponding to 3 quanta); the size of a quantum of rotational KE for the fan is even smaller.[*] Relatively speaking, the energy changes of the fan are as continuous as the water in the sea appears to be (even though we know that the water is made up of discrete molecules, it behaves as if continuous, for all ordinary purposes). We cannot picture what happens when the angular speed of a fan changes abruptly from ω_1 to ω_2 without having any of the in-between speeds, but this is simply modelitis[*] in reverse. It is as hard to make a model of a fan based on molecular behavior as it is to make a model of a molecule based on a fan's behavior. In general, any *periodic* motion such as rotation or vibration is quantized, but nonperiodic

---

[*] Do not suppose that all quanta of energy are of the same size; the size of the quantum varies from system to system. It so happens that the formula for the size of the first quantum jump in energy for a rotating body contains the moment of inertia in the denominator. Hence the first quantum of a rotating fan is many powers of ten smaller than that for a rotating molecule, because of the much larger moment of inertia for the fan. Even within a given system (molecule, fan, etc.) the quanta are often of different sizes. The important thing is the *discreteness* of energy changes.

[*] See footnote on p. 86.

motions are not quantized. A baseball pitcher can give the ball any desired linear KE within his power, but his choice of rotational KE's is restricted to a number of very closely spaced values. These "facts of nature" first became evident when experiments could be made on small particles for which one quantum of energy means a great deal. The study of the light emitted and absorbed by single atoms and molecules opened up the field of quantum mechanics, culminating in Bohr's theory of the atom (1913) and wave mechanics (1925), to be discussed in Chap 28.

In our study of rotation and vibration, we have seen how the "classical" model has proved inadequate. This is not a new experience; we have already seen how classical ideas of mechanics were modified by the introduction of Einsteinian relativity. Let us note that both relativity and quantum theory should be regarded as *extensions* of classical physics to realms of observation not accessible to earlier experimenters. Relativity is important in the realm of *high speeds,* but at low speeds the relativistic laws reduce to the classical laws of Newton and Galileo. Similarly, quantum effects are important in the realm of *small energy changes,* but for large-scale changes, so many quanta are involved that the averaged-out effect is just that to be expected on the classical model with which we are already familiar.

■ SUMMARY

Elastic behavior in all kinds of distortion is described by Hooke's law, which states that stress is proportional to strain. Hooke's law is not valid above the elastic limit. Stretch modulus, shear modulus, and bulk modulus are names for the ratio of stress to strain in special kinds of distortion. For liquids and gases the only modulus which is not zero is the bulk modulus.

Simple harmonic motion occurs whenever the restoring force on a body is proportional to the displacement of the body from its equilibrium position. Such a force is described by Hooke's law, and SHM occurs widely in nature. The ratio of restoring force to the displacement is the force constant: $k = F/s$. The period T of a SHM is the time for one complete vibration; the frequency ν is given by $\nu = 1/T$.

The projection of uniform circular motion upon any straight line is SHM. The radius of the reference circle equals the amplitude A of the SHM. The speed v_0 of the reference particle equals the maximum velocity of the SHM particle as it passes through center: $v_0 = 2\pi A/T$. In SHM, the displacement of a vibrating particle is a sinusoidal function of time.

During a vibration KE is continually being transformed to PE (usually elastic) and vice versa. The motion eventually dies out as the (KE + PE) total is transformed into heat energy through frictional losses.

The period of any SHM is given by $T = 2\pi\sqrt{m/k}$. As a special case, the motion of a simple pendulum turns out to be approximately SHM, if the amplitude of swing is not too great. The simple pendulum's period is given by $T \approx 2\pi\sqrt{l/g}$, an equation which can be used to measure the acceleration due to gravity.

In relaxation vibrations the restoring force varies abruptly. No general formula exists to give the period of all types of relaxation vibrations. One illustration is furnished by a bouncing ball; other relaxation vibrations are a result of the "stick-slip" phenomenon of friction.

The forces on atoms within molecules obey Hooke's law closely, and so the vibrations of molecules are nearly SHM. Rotational and vibrational motions are quantized, and only certain allowed energies are possible. Translational motion is not quantized. Quantum effects are too small to observe in experiments with objects large enough to be viewed in a microscope, but are of predominating importance in describing the behavior of small-scale objects such as single atoms and molecules.

■ CHECK LIST

Hooke's law	simple harmonic motion	relaxation vibration
stress	force constant	quantum of energy
strain	period	$F/s = k$
modulus of elasticity	frequency	
stretch modulus	displacement	$v_0 = \dfrac{2\pi A}{T}$
Young's modulus	amplitude	
shear modulus	phase	$T = 2\pi \sqrt{m/k}$
bulk modulus	reference circle	$T \approx 2\pi \sqrt{l/g}$
elastic limit	damping	
harmonic motion	simple pendulum	

■ QUESTIONS

9-1 Does the springiness of a telephone dial obey Hooke's law? Make a rough trial yourself, and try to tell whether it takes more force to hold the dial in the "0" position (all the way around) than it does to hold it with only a small displacement, when dialing "1" or "2."

9-2 Does a roller window shade obey Hooke's law?

9-3 Which of the three elastic moduli (stretch, shear, or bulk) is of chief importance in each of the following situations? (a) A boy stands on a long board which sags, with the ends remaining "square." Which modulus of wood is involved? (b) A helical spring made of steel, such as those shown in Fig. 9-7, is deformed by pulling on one end; each loop spreads slightly. Which modulus of the steel is involved? (c) A bicyclist pumps up his tire. He builds up a pressure by pressing down the handle of the pump, which forces the piston of the pump down. Which modulus of the air is involved? (d) A movie star gets a permanent wave. Which modulus of the hair is involved? (e) A worker uses a screwdriver to tighten a screw, and he twists the handle of the screwdriver while the blade of the tool is firmly seated in the slot of the screw. Which modulus of the shaft material of the screwdriver is involved?

9-4 When a body is vibrating in SHM, is its acceleration zero at any point in the motion?

9-5 A pendulum clock keeps perfect time when in Los Angeles. Will it gain or lose when it is taken to a summer resort in the mountains near the city?

9-6 The presence of oil beneath the earth is often indicated by a "salt dome" over the oil-bearing rock. The salt dome has a smaller specific gravity than the surroundings. Explain how a geophysicist on the surface can use a delicate pendulum apparatus to determine the probable location of oil.

9-7 An airplane loops the loop in a vertical circle, picking up speed as it approaches the bottom of the loop and losing speed as it approaches the top of the loop. Does the shadow of the plane move across the ground in SHM?

9-8 In what respect does the motion of a pendulum fail to be exactly SHM? That is, does it pick up speed too quickly or too slowly just after being released from an elevated position? (*Hint:* Is sin θ greater or less than θ?)

9-9 A mass is vibrating on the end of a spring. Is it in equilibrium at any point of the motion? Answer the same question for the bob of a pendulum.

9-10 Give an example of SHM other than those mentioned in the text. Give an example of relaxation vibration other than those mentioned in the text.

9-11 Which, if any, of the motions mentioned at the start of Sec. 3-1 are quantized?

9-12 "The amplitude of vibration of an oxygen molecule can change by any amount, no matter how small." Discuss this statement.

■ **PROBLEMS**

9-A1 A rubber band originally 20 cm long is stretched to a length of 22 cm by a certain load. What is the strain?

9-A2 What is the stress in a wire which is 20 m long and 0.01 cm² in cross section, if the wire bears a load of 100 kg?

9-A3 What is the strain in a wire cable of original length 200 ft whose length increases by 1 in. when a load is lifted?

9-A4 A piece cut from a bicycle inner tube is 2.0 ft long when it carries a load of 5 lb. The force constant is 2 lb/in. What will be the length of the piece of rubber when the load is 15 lb?

9-A5 The pan of a postal scale goes down $\frac{1}{16}$ in. when a 1 oz letter is weighed. What is the force constant of the spring, in lb/ft?

9-A6 A vibrating reed in a harmonica makes 200 vib/s. What is the period?

9-A7 What is the time required for 2400 vibrations of a loaded spring, if the frequency is 4 vib/s?

9-A8 The piston of an engine moves a total distance of 6 in. from one extreme point to the other. (*a*) What is the amplitude? (*b*) What is the displacement, when the piston is 3 in. from one end of its stroke?

9-A9 A pendulum is pulled aside +5 cm from its equilibrium position and allowed to vibrate. (*a*) What is the amplitude? (*b*) How far is the pendulum from its extreme position when the displacement is 0 cm?

9-A10 A pendulum has a period of 2.5 s. (*a*) What is the frequency? (*b*) How many vibrations does it make in 50 s?

9-A11 What are the dimensions of (*a*) period? (*b*) frequency? (*c*) force constant?

9-A12 A point on a violin string is vibrating at 400 vib/s with an amplitude of 1 mm. What is the speed of the string when it is passing through the central point of its motion?

9-A13 The piston of Prob. 9-A8 makes 2 vib/s. How fast is it moving when passing through the mid-point of its path?

9-A14 Compute the time for one vibration of a spring whose force constant is 16 N/m, if the load has a mass of 1 kg.

9-A15 Compute the length of a clock pendulum which ticks once each second. (*Hint:* It ticks twice during each complete vibration.)

9-A16 What would be the period of a pendulum suspended from the Empire State Building on a light string 1087 ft long?

9-B1 A crate weighing 4000 N is hoisted by a ship's derrick which has a wire cable 10 m long. The cable consists of 50 strands of steel each 4×10^{-6} m² in cross section. By how much does the cable stretch?

9-B2 An experimenter studying the stretch modulus of human bone reported in a journal that a load of 400 lb produced a stretch of 0.02 in. in a sample of femur 6 in. long. Young's modulus for bone is known to be 3×10^6 lb/in.². Can you deduce from these data the cross section of the bone specimen?

9-B3 A piece of beef tendon 25.00 cm long and of cross section 6.5 mm × 13 mm stretched to a length 25.19 cm when it supported vertically a load of 0.80 kg. Calculate the stretch modulus for the tendon material.

9-B4 A steel elevator cable 20 m long must supply an upward tension of 12 000 N. If the permissible stretch of the cable is 5 mm, what must be the total cross-sectional area of all the strands of the cable?

9-B5 A 20 000 lb truck on a lift in a service station is supported by a vertical steel column which is a solid cylinder 10 ft long and 4 in. in radius. By how much is the length of the column changed by the load? (*Hint:* Express the modulus in lb/in.² using the conversion factor on p. 198.)

9-B6 A 10 kg object is whirled in a horizontal circle on the end of a wire. The wire is 0.3 m long, has cross section 10^{-6} m², and is made of a material whose breaking stress is 4.8×10^7 N/m². Calculate the greatest angular speed the object can have.

9-B7 A camper is being pulled by a car (Fig. 9-19), and the coupling is required to transmit forces up to 5000 N to overcome friction, etc. The pin is 3 cm in diameter, 10 cm long, and is made of steel for which the maximum shearing stress is 8×10^7 N/m². Is this arrangement safe? (*Note:* A safety factor of 5 is generally allowed by design engineers.)

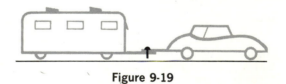

Figure 9-19

9-B8 A rubber eraser 1.5 in. × 0.4 in. × 3.0 in. is clamped at one end, with the 3-in. edge vertical. A horizontal force of 0.3 lb is applied at the free end, and the top of the eraser is displaced 0.1 in. horizontally. Calculate the shearing stress, the shearing strain, and the shear modulus.

9-B9 A jug contains 4000 cm³ of kerosene at atmospheric pressure (1.0×10^5 N/m²). When the cork is pushed in slightly, the pressure on the liquid increases to 4.9×10^5 N/m² and the volume of the liquid is 3999 cm³. Calculate the bulk modulus of kerosene.

9-B10 What increase in pressure is needed to cause the volume of a block of aluminum to decrease by 1%?

9-B11 A cast-iron anchor of mass 1580 kg is heaved overboard and settles to the bottom of the ocean, where the pressure is 1.8×10^6 N/m² greater than at the surface. Calculate the original volume of the anchor, and find the change in volume, in cm³, caused by the increased pressure.

9-B12 Artificial diamond crystals have been made by subjecting carbon in the form of graphite to a pressure of 1.5×10^{10} N/m² at a high temperature. Assuming that natural diamonds were formed at similar high pressures within the earth and that the bulk modulus and density at high temperature are roughly those given in Tables 9-1 and 2-1, what must have been the original volume of the fabulous Kohinoor diamond, which before cutting weighed about 160 g at the earth's surface?

9-B13 An object of mass 2 kg is moving with SHM of frequency 120 vib/min and amplitude 1.5 m. What is the restoring force when the object is at one of the end-points of its motion?

9-B14 A point on a horizontal vibrating piano string is moving up and down in SHM with a frequency of 200 vib/s. The maximum excursion of the point below its normal position is 0.8 mm. (*a*) What is the maximum velocity of the particle? (*b*) What is the maximum acceleration of the particle?

9-B15 A load of mass 100 g hangs from a long, light spring. When pulled down 20 cm below its equilibrium position and released, it vibrates with a period of 2 s. (*a*) What is its velocity as it passes through the equilibrium position? (*b*) By how much will the spring shorten if the load is removed?

9-B16 A load of mass 400 g is hanging from a light spring whose force constant is 10 N/m. The load is pulled down 10 cm from its equilibrium position and released. (*a*) How long is required for the load to reach the equilibrium position again? (*b*) What is then its velocity?

9-B17 A raft of weight 800 lb is floating in a pond. When a 200 lb man climbs on board, the raft sinks 1.5 inches deeper into the water to a new equilibrium position. If the man jumps off, how many vertical vibrations does the empty raft make in 5 s? (*Hint:* Use W/g for the mass of the raft.)

9-B18 Write an equation for the vertical motion of a body of mass 4 kg on a spring of force constant 100 N/m, if the motion has amplitude 2 m and the body passes downward through its equilibrium position at $t = 0$.

9-B19 In Example 9-5, what is the PE of the vibrating mass at an instant when its KE is 8.1 J?

9-B20 What energy must be given to a 20 kg mass suspended from a spring so that it will oscillate in SHM with a period of 0.5 s and amplitude 2 cm?

9-B21 A 32 lb weight is vibrating horizontally in a SHM of amplitude 5 ft with a period of 3.14 s. (*a*) Calculate the force constant. (*b*) Calculate the PE of the weight when it is at an extreme position. (*c*) Check your answer to part (*b*) by calculating the KE as the weight passes through its equilibrium position.

9-B22 A slingshot consists of a light leather cup containing a stone which is pulled back against the elastic force of two rubber bands. It takes a force of 20 N to stretch the bands 1 cm. (*a*) What is the PE of a 0.050 kg stone which is pulled back 20 cm from the equilibrium position? (*b*) With what speed does the stone leave the slingshot?

9-B23 In using the mass balance of Fig. 9-15, an experimenter finds the following data for the time required for 10 vibrations: with 50 g load, 5 s; with 350 g load, 10 s; with unknown load, 15 s. What is the mass of the unknown load? (*Hint:* First find the mass of the platform.)

9-B24 The platform of a mass balance (Fig. 9-15) has negligible mass. On earth, the period of vibration is 0.1 s when a 1 kg load is on the platform. What would be the period if the apparatus were in a space ship 4000 miles above the earth's surface, with a load of 9 kg on the platform?

9-B25 A small electronic computer placed on board a satellite must withstand a maximum acceleration of $10g$. To test this, the computer is attached to a horizontal table which is driven from side to side in SHM at 20 vib/s. What must be the amplitude of vibration of this shake-test apparatus in order to give a maximum acceleration of $10g$?

9-B26 A package is on a platform which vibrates in vertical SHM with period 0.1 s. (a) At what point in the motion is the package most likely to lose contact with the platform? (b) What is the maximum amplitude of SHM for which the package always remains in contact with the platform?

9-B27 Derive a formula for the maximum acceleration of a particle in SHM, in terms of the amplitude A and the period T. Check your answer for dimensional consistency.

9-B28 The value of gravitational acceleration on the moon is about $\frac{1}{6}$ that on the earth. What is the period on the moon of a simple pendulum whose period on the earth is 2.0 s?

9-B29 A perfectly elastic steel ball bearing is dropped from a height of 3 in., and bounces up and down on a steel plate. What is the frequency of this relaxation oscillation?

9-C1 A steel elevator cable must carry a total load of 1000 kg and accelerate it upward at 2.20 m/s^2. Taking a safe working stress for steel to be 1.20×10^8 N/m^2, calculate the necessary diameter of the cable. Express your answer in cm^2.

9-C2 A steel wire 30 m long has a cross section of 0.5 mm^2. Young's modulus for steel is 20×10^{10} N/m^2. (a) Compute the force constant of the wire. (b) If a 10 kg load is hung on the wire and is then pulled down a short distance below its equilibrium position, what will be the frequency of vertical vibration when the load is released?

9-C3 An object of mass m is supported by a wire of length L and radius r. When set into small vertical oscillation, the period is T. (a) Derive a formula for the stretch modulus of the wire, in terms of m, L, r, and T. (b) Make a dimensional check for your formula.

9-C4 A 10 kg load is suspended by a copper wire 10 m long, and is observed to vibrate vertically in SHM at a frequency of 10 vib/s. What is the cross section of the wire?

9-C5 A particle vibrates in SHM with period 6 s and amplitude 10 cm. (a) How long is required for the particle to move from one end of its path to a point 5 cm from the equilibrium position? (b) What is the particle's velocity at this point?

9-C6 An ancient military harbor was protected by an underwater reef which was 12 ft out of water at low tide and 4 ft under water at high tide. Friendly boats whose keels were 3 ft under water crossed the reef during the "clear" intervals of time when the water was deep enough. What was the duration of each clear interval? Assume the tides to be SHM, of period 12.5 h.

9-C7 A flat disk 15 cm in radius, of mass 8 kg, is suspended horizontally by a vertical wire attached to its c.g. It is found that a force of 3 N, applied tangentially at the circumference of the disk, rotates it through 45°. Calculate (a) the torsion constant in N·m/rad, and (b) the time for the disk to make 40 torsional vibrations.

9-C8 A light rubber band is hanging loosely from the ceiling. A load is attached to the free end of the rubber band, and it is observed that the load, when released, moves downward a distance of 40 cm before starting to rise again. What is the period of this SHM? (*Hint:* The mass is not given. Assume it to be m, and hope that it cancels out of your final equation for the period.)

9-C9 The moving system of an anniversary clock (pp. 195 and 208) may be approximated by four small balls, each of mass 250 g, with the center of each ball 10 cm from the vertical axis of rotation. The stiffness of the wire is such that a torque of 2000 cm·dyn causes the

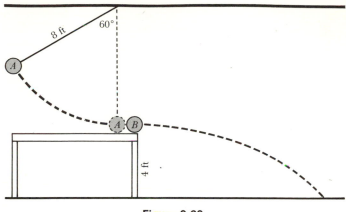

Figure 9-20

moving system to twist through an angle of 0.5 rad. What is the time required for one complete angular vibration of the system of balls?

9-C10 A steel ball A of mass $3m$ supported on a light cord 8 ft long is held out at a 60° angle from the vertical, as shown in Fig. 9-20. A second steel ball B of mass m rests just at the edge of the table, which is 4 ft high. Ball A is released and strikes ball B head on in a perfectly elastic collision. (a) Approximately how long after A is released does B strike the floor? (b) How fast is A moving when it strikes B? (c) How far out from the base of the table does B strike?

9-C11 (a) Using data from Sec. 9-6, calculate the force constant for the vibration of an oxygen atom in the O_2 molecule (*Hint:* Use Avogadro's number to find the mass of an oxygen atom.) (b) How much must the molecule stretch so that the Hooke's law force given by the force constant will supply the necessary centripetal force on an atom moving in the circular path described by the data?

For Further Study

9-7 The physical pendulum

We have already considered the simple pendulum in Sec. 9-4. No real pendulum is "simple," of course, for the entire bob cannot be at one point but must be spread out over a finite volume. Such a real pendulum is called a *physical pendulum*. To develop the theory of the physical pendulum we follow almost exactly the derivation made in Sec. 9-4 for the simple pendulum. It is only necessary to deal with angular SHM instead of linear SHM.[*] We are concerned with *angle*

instead of *arc, torque* instead of *force, moment of inertia* instead of *mass.*

Consider a body of any shape supported on an axis which is distant h from the center of gravity (Fig. 9-21). If the body is given an angular displacement θ away from its equilibrium position, there is an unbalanced torque equal to the force times the lever arm. The weight acts at a lever arm $h \sin \theta$, so that the restoring torque is $\tau = mg(h \sin \theta)$. The torsion constant τ/θ is

$$\frac{\tau}{\theta} = \frac{mgh \sin \theta}{\theta} \qquad (9\text{-}6')$$

which is not quite constant since $\sin \theta$ is

---

[*] It is instructive to compare the equations we are about to derive with the similarly numbered equations in Sec. 9-4 for the *linear* SHM of the simple pendulum.

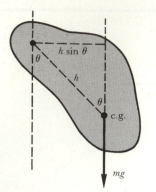

Figure 9-21 A physical pendulum.

not the same as θ. As the angle θ decreases, the ratio $(\sin\theta)/\theta$ approaches 1. Hence, if the angular displacement of the physical pendulum is small, the torsion constant is

$$\frac{\tau}{\theta} = mgh\left(\frac{\sin\theta}{\theta}\right) \approx mgh \qquad (9\text{-}7')$$

and, to the extent to which τ/θ is a constant, the motion is SHM. The period of the pendulum is

$$T = 2\pi\sqrt{\frac{I}{\tau/\theta}} \approx 2\pi\sqrt{\frac{I}{mgh}} \qquad (9\text{-}9')$$

This approximate formula has the same limitations as does the corresponding formula $T \approx 2\pi\sqrt{l/g}$ for a simple pendulum. It is a very good approximation if the angle of swing is small.

Although the period of a pendulum depends on the amplitude of swing, this does not affect the accuracy of a clock. It is the function of the mainspring, or of falling weights, to supply energy to the vibrating system to compensate for frictional losses. The amplitude of swing is thus held constant, and the period is constant, even though its value is slightly greater than that given by Eq. 9-9'.

■ PROBLEMS

9-C12 Show that the formula for the period of a physical pendulum contains, as a special case, the formula for the period of a simple pendulum. To show this, assume a *small* mass at the end of a *light* string of length l, and remember the definition of moment of inertia.

9-C13 What is the period of oscillation for a uniform meter stick suspended on a nail passing through a hole at one end? (Use Table 8-1 to find the moment of inertia of the stick.)

9-C14 Compute the period of oscillation of the loaded meter stick of Example 8-11 when suspended vertically about a horizontal axis passing through the 50 cm mark. (*Hint:* You must find the c.g. and also the moment of inertia about the given axis.)

■ REFERENCES

1 Magie, W. F., *A Source Book in Physics* (New York: McGraw-Hill, 1935), pp. 17–19, Galileo's discussion of the pendulum (also, Cambridge: Harvard University Press, 1963).

2 Friedman, Francis L., *Velocity in Circular and Simple Harmonic Motion; Velocity and Acceleration in Simple Harmonic Motion* (films).

3 Miller, Franklin, Jr., *The Wilberforce Pendulum* (film).

4 Stull, John, *Simple Harmonic Motion: The Stringless Pendulum* (film).

10

Wave Motion

We make contact with our surroundings through the five or more senses which we, and other sentient creatures, possess. One biology textbook[*] lists six major groups of specialized sense receptors—those sensitive to light, sound, touch and pressure, gravity and motion, taste and smell, and temperature. The function of a sense organ is to receive information about the outside world and transmit the information to the nervous system in order that the organism may respond in a coordinated fashion to the stimuli of its environment. From the point of view of physics, this process always involves transfer of energy. Now it is noteworthy that the two senses which can be considered most useful to higher organisms such as man—sight and sound—require the transmission of energy from source to receptor through a region of space which may range from a few inches up to many miles. Another point of similarity between sight and sound is that neither is instantaneous; we never see or hear anything at the very moment that it occurs. Thus we seek to describe and understand the flow of

*G. G. Simpson and W. S. Beck, *Life: An Introduction to Biology,* 2nd ed., Harcourt Brace Jovanovich, 1965, p. 362.

energy from one place to another at a finite speed.

A stream of machine-gun bullets or a blast of air molecules from a fan can carry energy; this *corpuscular* transfer of energy, called *convection,* will be discussed in Chap. 14. We turn our attention now to *wave motion,* the other important method by which energy moves from place to place.

10-1 Waves and disturbances

In the future we shall frequently use the concept of *waves.* Electromagnetic waves include radio waves, infrared radiation, visible light, ultraviolet radiation, x rays, and gamma rays. All of these travel at a speed of 3.00×10^8 m/s (186 000 mi/s) in a vacuum. We all have had some first-hand experience with other types of waves—sound waves, water waves, and waves upon stretched strings, as in a guitar or piano. Less familiar are earth waves (earthquakes), temperature waves (so-called heat waves or cold waves of popular meteorology), probability waves, and even matter waves. Perhaps enough has been said to indicate the value of studying

Photograph of shock wave pattern around a model of a space capsule (left) and its emergency escape tower (right). (NASA)

wave motion as such. What are the abstract properties of waves in general? What is common to the description of all these seemingly different types of waves? How is energy transmitted by a wave?

To most students the word "wave" conjures up a picture somewhat like water waves moving across a lake, or a flag undulating in the breeze. These are *periodic waves,* in which motions are repeated at regular intervals. But a wave can consist of only a single disturbance—never repeated at any given place. Such *shock waves,* or *pulses,* are in evidence when a high-speed bullet sets air molecules into rapid one-way motion, followed by a relatively slow diffusion of the molecules back to their original positions.

The key word in discussing waves is "disturbance." The reason there are so many types of waves is that there are so many ways of disturbing the physical state of a body or a substance. In water waves, it is the change of position of molecules which is the disturbance; the surface rises and falls. In sound waves, we can consider changes of pressure to be the disturbance; pressure does not remain constant at any given point. In general, a wave may be defined as the *propagation of a disturbance.* It is our purpose to find out what sorts of disturbances commonly exist, and how these disturbances are "handed on" from point to point as the wave travels through a medium.

We illustrate "disturbance" first by describing a *wave of enthusiasm* which passes through a crowd of devotees waiting outside a theater for a glimpse of their favorite. The word reaches the first few rows of the crowd, and then the disturbance spreads at a more or less definite rate toward the rear echelons. Here is a nonphysical wave; the disturbance is an emotional state rather than pressure, height, temperature, or some other physical state. The crowd is the "medium" through which the wave moves, and the mechanism by which it moves is rumor. As with most waves, there are (1) a disturbance, (2) a

medium• which can be disturbed, and (3) a connection between neighboring points of the medium—some means by which neighboring points can influence one another. All these features are present in the rather esoteric wave of enthusiasm (a shock wave) which we have described.

10-2   Shock waves of compression

On a molecular scale, a compression wave is started whenever a solid object is struck. For instance, when a hammer strikes the head of a vertical iron nail, the molecules of iron at the top surface are subject to a net force caused by the hammer. According to Newton's second law, this force causes an acceleration, and the molecules start to move downward. They, in turn, push on neighboring molecules, and so eventually a pulse is transmitted to the tip of the nail. It is important to note that, as the shock wave travels along the nail, a pressure is momentarily built up wherever the molecules are closer together than is normal. Indeed, it is this squeezing together of the molecules—a distortion—that gives rise to the elastic force that pushes the next molecules along.

On a larger scale, a model of such compression shock waves can be easily observed when a stationary freight train backs up (Fig. 10-1). Before the engine applies the initial force, the couplings are loose, with perhaps an inch or so of slack between each pair of cars. When the engine starts moving back-

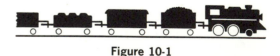

Figure 10-1

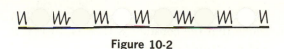

Figure 10-2

ward with a steady speed, one car after another is set into motion, and you can hear a "clank, clank, clank, . . ." as the shock wave travels the length of the train. If in a 100-car train the pulse has reached as far as the 30th car, this means that the couplings between the first 30 cars are tight, and these cars, moving as a rigid unit, are in the process of taking up the slack of the 31st coupling. The final 70 cars are still stationary at this instant. The velocity with which the pulse travels down the train is called the *wave velocity*. This is much faster than the speed of the engine; it will usually take only 5 or 10 s for such a shock wave to travel the length of a 100-car train. We have described a shock wave of compression; in similar fashion, if the engine starts forward, a shock wave of tension will be initiated.

If we replaced the cars by small heavy balls, and the loose couplings by stretched springs (Fig. 10-2), then our model would more closely resemble the structure of a solid body in which compression waves can travel.

10-3 Velocity of compression waves

We can make a shrewd guess as to the factors upon which the velocity of a compression wave should depend, using the freight-train model. Each car must, in turn, be accelerated to a velocity equal to that of the engine. This requires time. The more heavily loaded the cars are, the greater their mass (inertia) and, by Newton's second law, the less the acceleration produced by a given force and so the longer the time required. Thus we expect the wave velocity to be *less* for a loaded train. A similar argument holds for the system of balls in Fig. 10-2. Considering a set of cars (or balls) held together by stiffer springs, we would expect a *greater* wave velocity, since

there would be less "give" to the springs, and thus less lost motion.

Passing to the molecular scale of things, we then expect the wave velocity to depend somehow on the density of the material and its modulus of elasticity. A detailed analysis is possible, using the energy principle which in turn is based upon Newton's second law (Sec. 10-9). However, we can obtain a valuable insight by the method of *dimensional analysis*. We *assume* that the wave velocity v_w depends only on the density d and an elastic modulus E according to some simple formula such as $v_w = Ed$; $v_w = E/d$; $v_w = \sqrt{E/d}$; $v_w = E/d^2$; etc. To decide among the various possibilities, we write

$$v_w = KE^x d^y \qquad (10\text{-}1)$$

where x and y are two as yet unknown exponents and K is an unknown dimensionless constant. If Eq. 10-1 is true, it must be true dimensionally. Now the dimensions of velocity are

$$[v] = [LT^{-1}]$$

The dimensions of E are those of force/area (Sec. 9-1); hence

$$[E] = [FL^{-2}]$$
$$= [MLT^{-2}][L^{-2}] = [ML^{-1}T^{-2}]$$

Also, density is mass/volume, and the dimensions of d are

$$[d] = [ML^{-3}]$$

The constant K has been assumed to have no dimensions. Putting all these dimensions into Eq. 10-1, we have

$$[v] = [E]^x[d]^y$$
$$[LT^{-1}] = [ML^{-1}T^{-2}]^x[ML^{-3}]^y$$
$$[LT^{-1}] = [M^x L^{-x} T^{-2x}][M^y L^{-3y}]$$
$$[LT^{-1}] = [M^{x+y} L^{-x-3y} T^{-2x}] \qquad (10\text{-}2)$$

For this equation to be true, the exponents of L must be the same on each side, and also the exponents of T must be equal. This gives two equations:

Exponents of [L]: $\quad 1 = -x - 3y$

Exponents of [T]: $\quad -1 = -2x$

Solving these equations simultaneously for x and y gives the values $x = \frac{1}{2}$; $y = -\frac{1}{2}$.*
Therefore

$$v_w = KE^{1/2}d^{-1/2}$$

or

$$v_w = K\sqrt{\frac{E}{d}} \qquad (10\text{-}3)$$

This sort of dimensional analysis is far from being hocus-pocus; in the hands of experts it is a valuable tool. Without prior knowledge, and without detailed use of any physical laws, we have been able to find a plausible form of a physical law. What we had to work with was a "reasonable guess" as to which variables we expected to be significant. To be sure, we have not determined the constant K, but this could be found by analysis based upon Newton's laws. In any event, the constant K could be found by experiment, now that we know the form of the law to be $K\sqrt{E/d}$ and not $KE/\sqrt{d}$ or KEd^2 or some other collection of exponents. For many uses, the value of K is immaterial.

EXAMPLE **10-1**

If, by heat treatment, the stretch modulus of an iron rod is reduced to 81% of its original value, how will the speed of a compression wave in iron be altered?

We use our formula for wave velocity, and the unknown K cancels out.

$$\frac{v_w{}'}{v_w} = \frac{K\sqrt{E_2/d}}{K\sqrt{E_1/d}} = \sqrt{\frac{E_2}{E_1}} = \sqrt{\frac{0.81E_1}{E_1}}$$

$$= \sqrt{0.81}$$

$$= \boxed{0.90}$$

The new speed is 90% of the original value.

---

*It is gratifying to see that these values of x and y also make the exponents of [M] balance. The left side of Eq. 10-2 can be written $[\text{M}^0\text{LT}^{-1}]$, and for $x = \frac{1}{2}$,

As a matter of fact, analysis (Sec. 10-9) shows that the constant K is equal to 1, and so the formula can be written

$$v_w = \sqrt{\frac{\text{elastic modulus}}{\text{density}}} \quad \left(\begin{array}{c}\text{velocity of}\\\text{compression wave}\end{array}\right)$$

For a long solid rod in which the pulse travels along the length of the rod, the proper elastic modulus to use is the stretch modulus E, and we can write

$$v_w = \sqrt{\frac{E}{d}} \quad \left(\begin{array}{c}\text{velocity of compression}\\\text{wave in solid rod}\end{array}\right)$$

For a large body of liquid or gas, where the pulse spreads out in all directions, the proper elastic modulus to use is the bulk modulus B, and we can write

$$v_w = \sqrt{\frac{B}{d}} \quad \left(\begin{array}{c}\text{velocity of compression}\\\text{wave in liquid or gas}\end{array}\right)$$

In a liquid or gas, compression waves are called *sound waves* because, as compression reaches the eardrum, a complicated mechanical and neurological process is initiated which results in the sensation we know as sound. *Sound waves are compression waves.* Strictly speaking, only compression waves occurring in air should be called sound waves, unless the eardrum is actually in contact with another medium such as water (as for a skin diver). However, the term *sound* is often used for any compression wave, in any medium. We speak of the passage of sound waves through a plaster wall, or the propagation of sound waves along a steel railroad track.

EXAMPLE **10-2**

A railroad worker strikes a steel rail with a sledge hammer, and another worker hears the shock wave, which comes to him through the rail, 0.20 s after he sees the blow. How far apart are the workers?

$y = -\frac{1}{2}$, we have $x + y = 0$, and so the right side of Eq. 10-2 also contains $[\text{M}^0]$, a pure number. If this had *not* checked out, it would mean that our initial assumption as to the form of the law was wrong, and the method would have failed in this particular application.

To solve the problem we must first find the speed of sound in steel. From Table 2-1, we find that the specific gravity of steel is 7.8, so its density is 7.8 g/cm³ or 7800 kg/m³. From Table 9-1, the stretch modulus of steel is found to be 20×10^{10} N/m².

$$v_w = \sqrt{\frac{E}{d}} = \sqrt{\frac{20 \times 10^{10} \text{ N/m}^2}{7800 \text{ kg/m}^3}}$$

$$= 5.06 \times 10^3 \text{ m/s}$$

The distance s between the workers, measured along the rail, is

$$s = v_w t$$

$$= (5.06 \times 10^3 \text{ m/s})(0.20 \text{ s})$$

$$= 1.01 \times 10^3 \text{ m}$$

$$= \boxed{1.01 \text{ km}}$$

(We have assumed that the speed of light waves is so great that the time for the second worker to receive the message via light is negligibly small. A brief calculation, using 3×10^8 m/s for the speed of light, will justify this conclusion.)

Our next example involves compression waves in a liquid, and so we use the bulk modulus in computing the wave velocity.

EXAMPLE 10-3

A sailor strikes the hull of his ship with a hammer, and 0.54 s later he hears the echo, reflected from the ocean bottom directly below the ship. How far below him is the ocean bottom?

We must first find the speed of sound in sea water. The specific gravity of sea water is 1.025 (Table 2-1), and its density is

$$d = (1.025 \text{ g/cm}^3)\left(\frac{1 \text{ kg}}{1000 \text{ g}}\right)\left(\frac{100 \text{ cm}}{1 \text{ m}}\right)^3$$

$$= 1025 \text{ kg/m}^3$$

Using the value for the bulk modulus of sea water from Table 9-1

$$v_w = \sqrt{\frac{B}{d}}$$

$$= \sqrt{\frac{2.1 \times 10^9 \text{ N/m}^2}{1.025 \times 10^3 \text{ kg/m}^3}} = 1430 \text{ m/s}$$

The round trip is a distance of $2h$; hence

$$2h = (1430 \text{ m/s})(0.54 \text{ s})$$

$$h = \boxed{390 \text{ m}}$$

10-4 Longitudinal and transverse waves

Many (but not all) waves involve disturbances which have a direction. In the freight-train model (Sec. 10-2) each coupling is stretched parallel to the train's motion and each car is displaced from the neighboring cars, the displacement parallel to the direction of motion of the pulse; we call such a wave *longitudinal*. In a *transverse* wave the disturbances are directed at right angles to the direction of propagation of the wave. The most familiar longitudinal waves are sound waves, and the most familiar transverse waves are waves on strings. Water waves are approximately transverse, but water molecules move somewhat back and forth as well as up and down. Small ripples on the surface of a liquid *are* very nearly transverse, however, since any given molecule moves approximately in a vertical path, perpendicular to the direction of propagation of the wave. In the following paragraphs we shall use water waves to illustrate transverse waves, partly because they move slowly and are easily visualized.

If the disturbances of all particles in the medium are in the same plane, then the wave is said to be *plane-polarized*. The polarization of waves on a rope is shown in Fig. 10-3. It is evident that the plane of vibration (a vertical plane in the illustration) can be determined with the aid of a slot through which the rope passes. Only transverse waves can be polarized. A longitudinal wave, in head-on view, would show no preferred plane, and the orientation of a slot or other analyzing device would have no effect on the transmission or reflection of a longitudinal wave. In other words, polarization has no meaning for longitudinal waves. Historically, the ob-

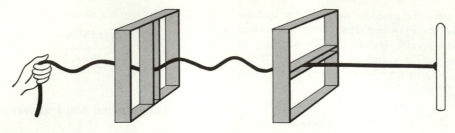

Figure 10-3

servation of polarization effects for x rays showed them to be transverse, and not longitudinal.

10-5 Graphical representation of waves

Let us now consider water waves on the surface of a lake. The medium consists of water molecules, and the disturbance is y, which is the displacement of the surface above or below the normal level of the water. To show the difference between "wave" and "vibration," we construct several graphs of the disturbance y. Let us suppose that a swimmer dives into the lake, and a short *wave train* of rather definite shape spreads out toward a fisherman whose line enters the water at A (Fig. 10-4). We start our stop watch at this

instant; $t = 0$ when the swimmer enters the water. As the wave train moves toward the left, it eventually moves past point A, and the cork A on the fishline starts to bob up and down. Two (related) graphs can be constructed: y as a function of x, at some given time, and y as a function of t, at some given point.

Figure 10-5a is a graph of y as a function of x, at the particular time $t = 11.5$ s. In other words, Fig. 10-5a might be a snapshot of the wave train, "freezing" the motion of the wave at $t = 11.5$ s. Such a graph is called a *wave-form* graph. As shown, for $x = 275$ cm (i.e., the cork A), the value of the displacement at this instant is $y = -3$ cm; but for other places (other corks at other values of x) the displacement may be quite different at this same time.

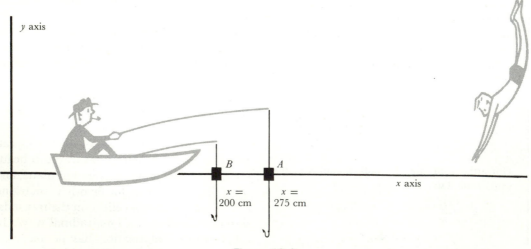

Figure 10-4

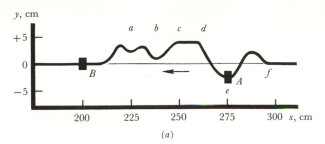

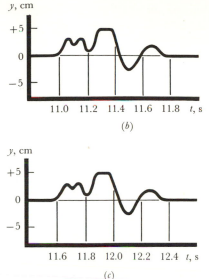

Figure 10-5 (*a*) Wave-form graph at $t = 11.5$ s; y as a function of x, at a given time. (*b*) Vibration graph for cork A at $x = 275$ cm; y as a function of t, at a given point. (*c*) Vibration graph for cork B at $x = 200$ cm.

If the fisherman pays attention only to his cork, not knowing that a swimmer has dived, he observes his cork bobbing up and down. Figure 10-5*b* is a graph of y as a function of t, at the point A ($x = 275$ cm); such a graph is called a *vibration graph* of the motion at a point. The cork remains stationary for 11.0 sec, until the leading edge (*a*) of the wave front hits it. The cork rises up a bit, wiggles up and down twice, then remains at the high level for 0.1 sec until the plateau (*cd*) has passed it. The cork then bobs down to a negative value of y as the trough (*e*) passes by, and finally, after a momentary $+$ value of y, the cork reaches zero level and remains at $y = 0$ as time increases indefinitely.

For a cork B at some neighboring point (at $x = 200$ cm), the same crazy motion of the cork is observed, but starting at a later time. The graph of B's motion is shown in Fig. 10-5*c*; this graph is the same as Fig. 10-5*b*, except for a different time axis.

Now it is immediately seen that the *shape* of Fig. 10-5*a* is exactly the same as the *shape* of Fig. 10-5*b*, even though these graphs have different meanings. A little reflection will show that it must always be so—if the snapshot reveals some characteristic kink in the wave form, that same kink will show up in the time graph of the motion at any point.

We see here the difference between a vibration and a wave. A *vibration* is a motion or disturbance of a single particle of the me-

dium, whereas a *wave* is an interrelated set of vibrations of many neighboring particles. We studied vibrations in Chap. 9 and found that some periodic vibrations, such as SHM, are simpler to describe than others, such as relaxation vibrations. In a general sense, even a single nonperiodic pulse may be called a vibration. But the term "wave" must be applied only to the totality of many vibrations. No single graph suffices to describe a wave. The wave-form graph moves in its entirety, so to speak, and a complete description of the wave would involve a series of instantaneous snapshots—a motion picture, actually—or else a number of recorders making time records at a sequence of points spaced along the path of the wave.

The *wave velocity* is the rate at which any given characteristic of the wave progresses from place to place. Thus we can fix our attention upon the trough (*e*) of the wave form, and we can deduce the wave velocity from the data given in the graphs. From Fig. 10-5*b*, the trough reached cork A at $t = 11.5$ s. (The same information is contained in Fig. 10-5*a*.) As shown in Fig. 10-5*c*, the trough reached cork B at $t = 12.1$ s. Since A and B are 75 cm apart, and the trough required 0.6 s to go from A to B, the wave velocity is 75 cm/0.6 s = 125 cm/s.

We have given a kinematical discussion of the water waves—a description of the motion, rather than a dynamical study of the causes in terms of force and mass. For large-scale water waves, such as those on the surface of the ocean, the connection between neighboring water molecules arises from gravity and from the elastic forces which are evidenced by the relative incompressibility of water. The force of gravity tends to "level out" any wave crest, but the inertia of the molecules keeps the water moving past the equilibrium point and so a crest becomes a trough. We thus expect the density of the water and the acceleration due to gravity to enter into an expression for the velocity of such water waves. For small ripples the surface of the water is highly curved, and surface tension forces enter into the picture. We shall not pursue further the dynamical study of water waves, since the mathematical complexities are considerable.

Nothing prevents us from making a graphical study of compression waves (sound waves) in the same way we have studied transverse waves. The disturbance is the pressure P rather than the displacement along the y axis. We need only plot P versus x, and P versus t, using pressure as the ordinates for the pair of graphs. Figure 10-6 represents a shock wave, or pulse, in air, resulting from a sudden compression in a limited region of

space where a test bomb was located; the air pressure before and after the pulse was 14.7 lb/in.². (Note the analogy with the swimmer who created a localized disturbance where he dived into the water.) As always, the vibration graph has the same shape as the waveform graph. The fact we wish to emphasize is that Fig. 10-6a is a *graph* of the pressure wave form and in no way represents any wave form in two dimensions which could be photographed, as was the water wave.

10-6 Periodic waves

We have deliberately avoided mention of periodic, sustained waves, since the general ideas apply equally well to pulses (shock waves). However, if the source of disturbance varies regularly, the wave form has additional symmetry, and the wave is said to possess a wavelength and a frequency.

Let us redraw Fig. 10-5a and b, this time assuming that instead of a diver and a single (complex) splash, some mechanism is periodically creating a + and − disturbance of the y-coordinate of the water surface at the point where previously we imagined the dive taking place. Assuming, for simplicity, that the disturbance is one of SHM, the two graphs will be sinusoidal in appearance. The wave-form graph (Fig. 10-7a) and the vibration graph (Fig. 10-7b) are identical in appearance, as we have seen must always be

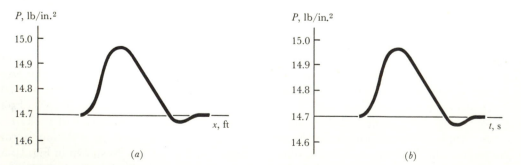

Figure 10-6 (*a*) Wave-form graph for a shock wave; P as a function of x, at a given time. (*b*) Vibration graph for same shock wave; P as a function of t, at a given point.

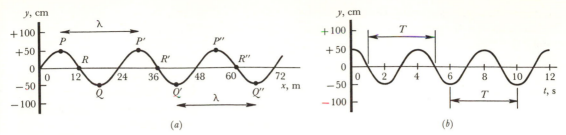

Figure 10-7 (a) Wave-form graph of a periodic water wave; snapshot at $t = 3.0$ s. (b) Vibration graph at one point of same periodic water wave; record of the motion at the point R' at $x = 36$ m.

the case. Successive particles P, P', P'', etc., are all at their highest points at the same instant (at $t = 3.0$ s, according to the legend accompanying the graph) and are momentarily at rest. Particles R, R', R'', etc., are all at their normal positions at this instant and are all moving in the same direction. (See Prob. 10-B3.) Particles which have the same displacement and are moving in the same direction with the same velocity are said to have the same *phase* of vibration. For instance, in Fig. 10-7a, particles P and P' are "in phase" with each other, as are particles P and P'', R and R', R and R'', Q and Q', Q and Q''. Particles P' and Q are exactly "out of phase"—sometimes referred to as "180° out of phase."•

To describe a periodic wave, we make use of five quantities, many of which we have already encountered in our study of vibration: (1) The *period* T is the time for one complete vibration of any given particle of the medium. (2) The *frequency* ν is the number of vibrations of any particle per unit time. As in any SHM, $\nu = 1/T$. (3) The *amplitude* A is the maximum displacement of any particle, measured from its equilibrium position. (4) The *wavelength* λ (lambda) is the distance between any two successive particles which have the same phase. (5) The *wave velocity* v_w is the velocity with which any specific phase of motion (crest, trough, compression,

etc.) is propagated through the medium.

An important relationship is true for periodic waves of all kinds. Study of Fig. 10-7 shows that, during the time T (one period), point P' moves down and up and makes one complete vibration; meanwhile, the wave crest has traveled from P' to P'', and the crest that was at P has moved over to P'. That is, the crest moves one wavelength in a time equal to the period of vibration of any individual particle. Thus

$$v_w = \frac{s}{t} = \frac{\lambda}{T}$$

For any periodic wave, $1/T = \nu$, hence

$$v_w = \nu\lambda \qquad (10\text{-}4)$$

Wave velocity = (frequency)(wavelength)

EXAMPLE **10-4**

The wavelength of sound waves emitted by a whistle is 5.5 ft. If the velocity of sound in air is 1100 ft/s, how many times per second does the pressure reach its maximum value above normal (atmospheric) pressure?

$$v_w = \nu\lambda$$

$$\nu = \frac{v_w}{\lambda} = \frac{1100 \text{ ft/s}}{5.5 \text{ ft}} = 200 \text{ s}^{-1}$$

$$= \boxed{200 \text{ vib/s}}$$

We interpret 200 s^{-1} as 200 "times" per second, or 200 vib/s. "Vibration" is simply a count, having no dimensions. One vibration is a "cycle," and frequency can also be

---

• If the SHM's of these two particles were represented with the aid of a single reference circle, the reference particles would be at opposite ends of a diameter of the circle, 180° apart.

expressed in *cycles per second*, or in *hertz* (Hz). These units for frequency are equivalent; we shall use them interchangeably:

$$1 \text{ vib/s} = 1 \text{ c/s} = 1 \text{ s}^{-1} = 1 \text{ Hz}$$

The hertz* is coming into increasing use, especially for frequencies of electromagnetic and acoustic waves.

EXAMPLE 10-5

What is the distance between adjacent regions of minimum electric field in an electromagnetic wave whose velocity in glass is 2.0×10^8 m/s and whose frequency is 10^{15} Hz?

The disturbance is evidently an entity known as "electric field," about which we need know nothing further at this time. Regions of minimum electric field are in the same phase of disturbance, and so the distance between adjacent regions of minimum electric field is equal to one wavelength.

---

*Named for Heinrich Rudolf Hertz (1857–1894), a pioneer experimental investigator of electromagnetic waves.

$$v_w = \nu\lambda$$

$$\lambda = \frac{v_w}{\nu} = \frac{2.0 \times 10^8 \text{ m/s}}{10^{15} \text{ s}^{-1}} = \boxed{2.0 \times 10^{-7} \text{ m}}$$

The pictorial representation of a periodic compression wave is not easy. We might imagine a large number of quick-acting pressure gauges spaced at close intervals along a pipe in which normal atmospheric pressure is 15.00 lb/in.² (Fig. 10-8). A loudspeaker creates periodic disturbances at one end of the pipe. As the wave progresses down the pipe, the reading of each gauge goes through a sequence of values, fluctuating between 15.01 lb/in.² in the *condensation* and 14.99 lb/in.² in the *rarefaction*. We have drawn a wave whose period is 0.004 s, so the gauges would have to be quick-acting indeed to follow such rapid changes. Gauges a and e are in phase, always reading the same at any given instant. The wavelength is the distance between any two adjacent in-phase pressure gauges, such as a and e, b and f, or c and g.

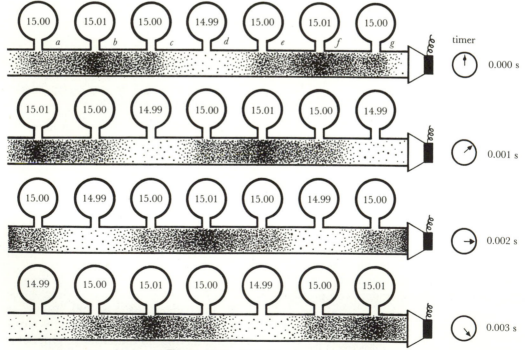

Figure 10-8 Periodic pressure wave.

As indicated in the figure, the pressure amplitude of the wave is 0.01 lb/in.², since this is the maximum excursion of the pressure from the normal value.

10-7 The superposition theorem—beats

If a wave A acting alone would create a certain disturbance at a point, and if wave B would create a (perhaps different) disturbance at the same point, the *superposition theorem* says that the resultant disturbance is the algebraic sum of the disturbances due to A and B. Stated in this way, the theorem seems trivial, but we should realize that we are making a definite assumption about a medium when we invoke the superposition theorem for waves of some sort in the medium. Let us illustrate the superposition theorem by reference to compression waves in an aluminum rod. Wave A might be capable of imparting a certain pressure P_A at some reference point. Wave B, acting alone, would be able to impart a pressure P_B. If both waves act simultaneously, will the pressure be $P_A + P_B$? The answer is "yes," provided we don't go beyond the elastic limit of aluminum. For compression waves having a very large pressure amplitude, each wave being almost enough to exceed the elastic limit, the superposition of the two waves might well cause the rod to deform permanently or even frac-

ture. In such a case the total disturbance certainly is not equal to the sum of the component disturbances.

Fortunately, in ordinary sound waves in air, the pressure amplitudes are small (± 0.001 atm for an intense sound). Under these circumstances we are justified in using the superposition theorem. Waves on strings (say, for a violin string being bowed) also obey the superposition theorem, unless the amplitude of vibration is excessively large.

As an illustration of the usefulness of the superposition theorem, consider the effect of two waves of different frequencies. We assume that both waves act on the same point, and we plot, in Fig. 10-9, the vibration graphs of the two waves and their superposition. The vibration graph of the resultant disturbance is the sum of the vibration graphs of the two waves. Figure 10-9 shows a 0.2 s portion of the vibration graphs of A (50 vib/s) and B (60 vib/s). If the waves are in phase at $t = 0$ s, the less rapidly vibrating wave (A) lags behind until at $t = 0.05$ s A and B are exactly out of phase; they are in phase again at $t = 0.10$ s, out of phase again at $t = 0.15$ s, etc. The amplitude of the resultant wave (C) rises and falls once during each 0.1 sec interval. These periodic changes in amplitude are called *beats*. The number of beats per second equals the difference in the frequencies of the two combining waves. In our example, waves of frequencies 50 and 60 vib/s com-

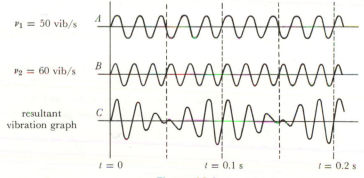

$\nu_1 = 50$ vib/s A

$\nu_2 = 60$ vib/s B

resultant vibration graph C

$t = 0$ $t = 0.1$ s $t = 0.2$ s

Figure 10-9

bine; they must be out of phase and cancel each other 10 times per second. It is equally true to say that they reinforce each other 10 times per second. The resultant amplitude fluctuates from twice that of either wave, to zero, and back again, 10 times per second.

The phenomenon of beats extends to all types of waves. For instance, the "squeals" heard on some radios are due to beats (also called heterodyning) between the frequencies of radio-frequency signals from different stations.

■ SUMMARY

A wave is the propagation of a disturbance. The nature of the disturbance may be mechanical, electrical, thermal, or even nonphysical. The wave velocity is the speed with which the disturbance is propagated. If a disturbance can exist only in a medium, the neighboring particles of the medium must somehow react upon one another for the disturbance to be propagated. In compression waves the elastic forces between molecules supply the connection between neighboring particles. The wave velocity of a compression wave depends on the elastic modulus and on the density of the medium. For waves in solids, the stretch modulus is used; for waves in liquids and gases, the bulk modulus is used.

In a longitudinal wave the disturbance is parallel to the direction of propagation. Sound waves and other compression waves are longitudinal waves. In a transverse wave the disturbance is in a direction at right angles to the direction of propagation. Waves on strings are examples of transverse waves, and waves on the surface of a liquid are approximately transverse. A transverse wave is said to be plane-polarized if all the disturbances are parallel to each other. Longitudinal waves cannot be polarized.

A wave is distinguished from a vibration by the fact that a wave consists of many vibrations of many particles, with definite phase relationships between the vibrations of neighboring particles. Two graphs are required to represent a wave: the vibration graph of any given particle (disturbance versus time), and the wave-form graph of the whole wave, at any given time (disturbance versus position). These two graphs necessarily have the same shape, although they are plotted with different variables.

Periodic waves are caused by a regularly repeated series of disturbances. The wavelength is the shortest distance between two particles having the same phase of vibration. For all periodic waves, of whatever kind, wave velocity = (frequency)(wavelength). The amplitude of a wave is the maximum value of the disturbance.

The superposition theorem states that the combined effect of two waves is the algebraic sum of the effects of the separate waves. The superposition of two periodic waves gives rise to beats, as the resultant amplitude rises and falls. The beat frequency is equal to the difference between the frequencies of the two combining waves.

wave
shock wave
wave velocity
$v_w = \sqrt{E/d}$
$v_w = \sqrt{B/d}$
longitudinal wave
transverse wave
plane-polarized wave

vibration graph
wave-form graph
vibrations in phase
vibrations out of phase
period
frequency
hertz

amplitude
wavelength
$v_w = \nu\lambda$
condensation
rarefaction
superposition theorem
beats

■ QUESTIONS

10-1 Give three examples of disturbances which can be propagated as a wave motion.

10-2 Does every wave have a wavelength? Does every wave have a wave velocity?

10-3 While a compression wave is moving past a point in a gas, is there any change in the density of the gas at this point?

10-4 The bulk modulus of air does not depend on temperature, but the density does (the latter decreases when the temperature rises). How would you expect the speed of sound in air to depend on the temperature?

10-5 Which elastic modulus enters into the formula for the speed of a compression pulse traveling through a liquid?

10-6 When we say that compression waves are longitudinal waves, do we mean all compression waves, or nearly all compression waves? Explain.

10-7 A fisherman is fishing from the end of a pier, in a dense fog, and all that he can see is the motion of the cork on his own fishline. The cork starts to bob up and down; can the fisherman be sure that a wave is moving past the cork's position?

10-8 In what sort of unit might you measure the amplitude of the wave of enthusiasm mentioned in Sec. 10-1? In what unit would you measure the wave velocity of this wave? How would you produce a periodic wave of enthusiasm? Could such a wave have a wavelength?

10-9 When a flag waves in the breeze, is this a wave in the strict physical sense of Sec. 10-5? If so, what is the disturbance? What is the medium? Approximately what is the period?

10-10 Two waves of equal frequency, wavelength, velocity, and amplitude are traveling in the same direction through a medium. Is the amplitude of the resultant wave, found by the superposition theorem, necessarily twice that of the individual waves?

10-11 What is the difference between a vibration and a wave?

■ PROBLEMS

Note: Use data from tables in Chaps. 2 and 9 where needed.

10-A1 Two metal rods have the same density, but the stretch modulus of one metal is four times that of the other. What is the ratio of the speed of compression waves in the two metals?

10-A2 Two liquids have the same density, but the volume modulus of one liquid is $\frac{1}{4}$ that of the other. What is the ratio of the speeds of compression waves in the two liquids?

10-A3 Calculate the speed of sound in a steel rod.

10-A4 Calculate the speed of sound in a brass rod.

10-A5 What is the speed of sound in water, in m/s? in ft/s?

10-A6 Calculate the speed of sound in mercury.

10-A7 An unidentified metal fragment is analyzed by measuring the speed of sound in a rod made out of the metal. The rod has mass 6.10 kg and volume 2.00×10^{-3} m³. The wave velocity in the rod is 4000 m/s. Calculate the stretch modulus of the metal.

10-A8 What is the wavelength of the compression waves shown in Fig. 10-8, if the pressure gauges are spaced at 5 ft intervals?

10-A9 How often does a compression pass an observer who is listening to a sound of wavelength 2 m and speed 340 m/s?

10-A10 What is the frequency, in vib/min, of the wave of Fig. 10-7? In which direction (to the right or the left) is the wave moving? What is the wave velocity?

10-A11 What is the shortest wavelength of sound which can be heard by a human if the frequencies of audible sound lie between 20 Hz and 20 000 Hz?

10-A12 What is the frequency of the compression wave shown in Fig. 10-8?

10-A13 What is the distance between rarefactions in a sound wave whose velocity is 1080 ft/s and whose frequency is 120 Hz?

10-A14 What is the wavelength (in meters) of waves broadcast by a TV station whose frequency is 60 MHz (megahertz)?

10-A15 At what time does cork A (Fig. 10-5a) start moving? At what time does cork B start moving? In what direction (up or down) is cork A moving at $t = 11.4$ s?

10-A16 Two whistles having frequencies 408 vib/s and 413 vib/s are sounded simultaneously. How many beats are heard per minute?

10-A17 Find the wavelength of an ultrasonic wave used by a bat for navigation purposes, if the frequency is 50 000 Hz and the speed of sound in air is 340 m/s.

10-B1 A 120-car train standing on a siding is started in motion by an engine. If there is 5 cm of slack between cars and the engine moves at a constant speed of 30 cm/s, how much time is required for the pulse to travel the length of the train?

10-B2 In Example 10-3 the combination of units $\sqrt{\dfrac{\text{N/m}^2}{\text{kg/m}^3}}$ is stated to yield m/s. Prove this statement.

10-B3 The points R, R', R'' are all in phase (Fig. 10-7a). Are they moving upward or downward at $t = 3.0$ s?

10-B4 What is the velocity of the compression wave of Fig. 10-8, if the pressure gauges are spaced at 5 ft intervals?

10-B5 A sound wave of velocity 600 m/s has a wavelength of 300 cm. At a certain instant one of the molecules of the medium is at its normal position. How long will it be before this same molecule is again at its normal position?

10-B6 Find the wavelength of an ultrasonic compression wave in a magnesium rod, if the frequency is 50 000 Hz.

10-B7 Calculate the frequency of compression waves in an aluminum rod if the wavelength is 20 cm.

10-B8 On a cold day when the speed of sound in air is 330 m/s, a blind man taps his cane on the sidewalk, and 0.050 s later he hears an echo from a building. How far is the man from the building?

10-B9 A boy standing on a steel bridge sees a construction worker strike a rivet and 0.1 s later feels the compression pulse at his feet. How far away is the worker?

10-B10 Periodic SHM waves spread out over the surface of a lake where two men are fishing, 72 ft apart. Each man's cork bobs up and down 20 times per minute. At a time when F's cork is at its crest, G's cork is at its lowest point, and there is one additional crest somewhere between the two men. What is the speed of the water waves?

10-B11 On a day when the speed of sound in air is 1080 ft/s, two whistles are blown simultaneously. The wavelengths of the sounds emitted are 8.00 ft and 9.00 ft, respectively. How many beats per second are heard?

10-B12 Two sources of sound are vibrating simultaneously with frequencies 2000 Hz and 2040 Hz. On a certain day the speed of sound is 1100 ft/s. (a) How many beats are heard? (b) At any instant, how far apart in space are the regions of maximum amplitude of vibration?

10-B13 A campus radio station is broadcasting on a frequency of 0.5800 MHz (= 0.5800 megacycles/second). (a) What is the wavelength of the electromagnetic waves? (b) If another station simultaneously broadcasts at 0.5802 MHz, how many beats per second are heard?

For Further Study

10-8 The Doppler effect

A phenomenon exhibited by all types of wave motion is the Doppler effect,[*] which has to do with the apparent change in frequency of a wave due to the relative motion of the source and the observer. Whether it is the source or the observer that is moving, the effect of relative motion is to make the apparent frequency greater, if the source and observer are approaching each other. On the other hand, the apparent frequency is lowered if the source and observer are moving away from each other. In our study of the Doppler effect for sound waves, we shall see that it makes a difference whether it is the source or the observer that moves. The interested student can derive general formulas, or (preferably) proceed directly from first principles in each problem, as illustrated in the following examples.

[*] The effect was first calculated in 1842 by Christian Doppler, who made experiments involving musicians on moving railroad cars.

EXAMPLE 10-6

A through train is receding from a station platform at 110 ft/s, blowing a whistle whose true frequency is 300 vib/s. The speed of sound in air is 1100 ft/s. What is the apparent frequency, as judged by an observer on the platform?

Since it is the train that is moving, a pulse, once emitted, travels through the air with the usual velocity of 1100 ft/s. However, the wavelength is increased (Fig. 10-10). In 1 s, the whistle emits 300 pulses, but the train has moved 110 ft during this 1 s. The last pulse is emitted at P' at the same instant that the first of the 300 pulses has reached the point Q. The wavelength is "stretched out"; we have 300 pulses in $(1100 + 110)$ ft = 1210 ft, and so

$$\lambda' = \frac{1210 \text{ ft}}{300} = 4.033 \text{ ft}$$

The apparent frequency is

$$\nu' = \frac{v_w}{\lambda'} = \frac{1100 \text{ ft/s}}{4.033 \text{ ft}} = \boxed{272.7 \text{ s}^{-1}}$$

Example 10-6 shows that, if the *source* is moving, the wavelength is altered but the

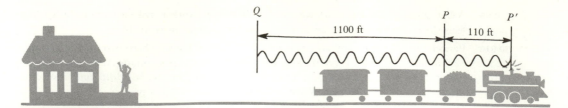

Q   P   P'
|←——————— 1100 ft ———————→|←— 110 ft —→|

Figure 10-10 Doppler effect, source in motion. The sinusoidal graph represents pressure variations.

wave velocity remains the same. On the other hand, if the *observer* moves, the wave velocity relative to him is altered, but the wavelength remains the same since the source is stationary.

EXAMPLE **10-7**

Solve Example 10-6, assuming that the whistle is stationary and the observer is moving away from it at 110 ft/s.

Here the wavelength is unaltered, but, since the observer is moving away from the source (Fig. 10-11), the condensations are not catching up to him as rapidly as if he were stationary. The wavelength is

$$\lambda = \frac{1100 \text{ ft/s}}{300 \text{ vib/s}} = 3.667 \text{ ft}$$

However, the apparent wave velocity is

$$v_w' = (1100 - 110) \text{ ft/s} = 990 \text{ ft/s}$$

Hence

$$v' = \frac{v_w'}{\lambda} = \frac{990 \text{ ft/s}}{3.667 \text{ ft}} = \boxed{270.0 \text{ s}^{-1}}$$

We can summarize the Doppler effect for sound waves in a simple way. The observed changes depend on v/v_w, the ratio of the relative velocity to the wave velocity. If the source moves, the *wavelength* changes by a relative amount equal to v/v_w; if the observer

moves, the *frequency* changes by a relative amount equal to v/v_w. In many cases, if v/v_w is small compared with 1, it makes no difference which method is used. If we follow the methods of Examples 10-6 and 10-7, we can derive equations for the apparent frequency. The results are:

$$v' = \frac{v}{1 - v/v_w} \quad \left(\begin{array}{c}\text{sound waves,} \\ \text{source in motion}\end{array}\right) \quad (10\text{-}5)$$

where v is the velocity of the source, and

$$v' = v(1 + v/v_w) \quad \left(\begin{array}{c}\text{sound waves,} \\ \text{observer in motion}\end{array}\right) \quad (10\text{-}6)$$

where v is the velocity of the observer.

According to Einstein's first postulate, discussed in Sec. 7-4, only relative motion is observable. The theory we have just developed seemingly violates the theory of relativity, for we could determine which of two bodies (source or observer) is "actually" in motion. For instance, in comparing Example 10-6 with 10-7, suppose we know the relative velocity to be 110 ft/s. If we measured the apparent frequency to be 272.7 vib/s, we could conclude that the source is moving relative to the air; if we measured the apparent frequency to be 270.0 vib/s, we could conclude that the observer is moving relative to the air. Thus a decision could be made as to which body is in motion. To resolve this

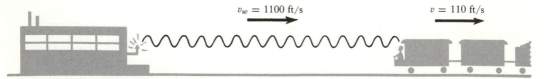

$v_w = 1100 \text{ ft/s}$ $v = 110 \text{ ft/s}$

Figure 10-11 Doppler effect, observer in motion. The sinusoidal graph represents pressure variations.

paradox, we should note that for sound waves there is a medium (air) which can serve for a reference set of axes; we can find out which body is moving relative to the air. For light waves, there is no such medium. The Doppler effect for light waves is calculated in the relativity theory and, as expected, only the relative velocity of source and observer enters into the equation. The result (for head-on motion) is

$$\nu' = \frac{\sqrt{1 + v/c}}{\sqrt{1 - v/c}} \, \nu \quad \text{(light waves)} \quad (10\text{-}7)$$

where v is the relative velocity of source and observer, and the symbol c (speed of light) is used instead of v_w. This interesting relativistic formula can be compared with the two formulas for sound waves; ν' is the geometric mean of the values given by the two acoustic formulas.

Equation 10-6 can be rearranged to give

$$\frac{\nu' - \nu}{\nu} = \frac{v}{v_w}$$

or

$$\frac{\Delta \nu}{\nu} = \frac{v}{v_w} \quad (10\text{-}8)$$

It can be shown (Prob. 10-C8) that the other exact equations for the Doppler effect (including the relativistic formula for light) also reduce to the form of Eq. 10-8 when v/v_w is small compared to 1.[*] Thus Eq. 10-8 serves as an approximate formula for any Doppler effect calculation, if the relative velocity of source and observer is small compared with the wave velocity.

EXAMPLE 10-8

A certain spectrum line of hydrogen has a frequency 4.56571×10^{14} Hz as measured in a laboratory on the earth. What is the relative velocity of the earth and a star, if the spectroscope shows that the same hydrogen line in the light reaching the earth from the star has a frequency 4.56711×10^{14} Hz?

[*] The answers to Examples 10-6 and 10-7 differ by about 1%. We expected a difference, since the ratio v/v_w is $110/1100 = \frac{1}{10}$, certainly not "small" compared with 1.

Here the relative velocity is very small compared with the wave velocity, since the relative change in frequency is so small.

$$\frac{\Delta \nu}{\nu} = \frac{0.00140 \times 10^{14} \text{ Hz}}{4.56571 \times 10^{14} \text{ Hz}} = 3.07 \times 10^{-4}$$

We can now use the approximate equation:

$$\frac{v}{c} = \frac{\Delta \nu}{\nu}; \qquad v = \frac{\Delta \nu}{\nu} c$$

$$v = (3.07 \times 10^{-4})(3.00 \times 10^8 \text{ m/s})$$

$$= 9.2 \times 10^4 \text{ m/s} = \boxed{92 \text{ km/s}}$$

Since the frequency is *raised*, we know that the star and the earth are *approaching* each other at 92 km/s. It is impossible to say which of the two is "really" in motion; the theory of relativity tells us that only the relative motion can be observed by this or any other means.

An interesting sidelight on the relativistic Doppler effect is the prediction of a shift in frequency when a sideways motion takes place. In the acoustic effect there is no change in frequency at the moment when the moving source is just opposite the observer, since at this instant it is neither approaching nor receding. However, there *is* a small effect in the case of light, given by $\nu' = \nu \sqrt{1 - v^2/c^2}$. This "transverse Doppler effect" is a consequence of the relativistic time dilation (Eq. 7-4). It has actually been observed in an experiment with moving ions. Since the factor v^2/c^2 is positive for either direction of motion (v^2 is always positive and less than c^2), the shift is always toward lower frequencies. There are still other experiments which show the validity of the relativistic concept of the Doppler effect.

10-9 Derivation of the equation for velocity of compression waves

By the method of dimensional analysis we have shown in Sec. 10-3 that the velocity of a compression wave along a solid rod is $K\sqrt{E/d}$, where E is the stretch modulus $\dfrac{\Delta F/A}{\Delta L/L}$. The dimensionless constant K could

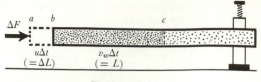

Figure 10-12

not be found by dimensional analysis; we stated without proof that its value is 1. There follows a derivation of the equation for the velocity of a compression wave based upon Newton's second law.

We shall use the expression of Newton's second law which is summarized in the energy principle. You will recall that we successfully applied Newton's laws in this form when we derived the period of the SHM of a body on a spring (Sec. 9-3). To derive the speed of a compression wave, we shall use the railroad-train model described in Sec. 10-2.

Consider a long rod of metal (Fig. 10-12) which has a density d. The cross section of the rod is A. With the rod initially at rest and clamped at one end, we suddenly apply a force ΔF at the other end of the rod, moving it to the right with a steady velocity u. After a time Δt has elapsed, the compression shock wave has reached a point c. During this time the metal to the left of c has been compressed by an amount $ab = u\,\Delta t$; this much metal is moving at the velocity u and is about to set into motion some additional metal beyond c. We let v_w be the wave velocity, and hence $ac = v_w\,\Delta t$. To apply the energy principle, we need to know the PE of the compressed metal. This PE was stored in the rod because of the compression by an amount $u\,\Delta t$. The work done to compress the rod, against a variable force which increases steadily from

0 to ΔF, is the average force times the displacement. Therefore the PE of the compressed portion of the rod is

$$\frac{(0 + \Delta F)}{2}(u\,\Delta t), \quad \text{or} \quad \tfrac{1}{2}\,\Delta F(u\,\Delta t)$$

The compressed part of the rod, bc, is in motion, and so it has some KE given by $\tfrac{1}{2}mu^2$. Its mass is $Vd = ALd\,(= Av_w\,d\,\Delta t)$, and the velocity is u. According to the energy principle, the work done by the applied force ΔF equals the gain in PE plus the gain in KE:

$$\Delta F(u\,\Delta t) = \tfrac{1}{2}\,\Delta F(u\,\Delta t) + \tfrac{1}{2}(Av_w\,d\,\Delta t)(u^2)$$

$$(10\text{-}8)$$

To simplify this we need to express u in terms of v_w. We have

$$u\,\Delta t = \Delta L$$

and

$$v_w\,\Delta t = L$$

Therefore $u/v_w = \Delta L/L$ and

$$u = v_w \frac{\Delta L}{L}$$

Putting this into Eq. 10-8 gives

$$\Delta F = Av_w{}^2 d\,\frac{\Delta L}{L}$$

Solving for the wave velocity v_w and rearranging terms gives

$$v_w = \sqrt{\frac{(\Delta F/A)/(\Delta L/L)}{d}}$$

The numerator is the stretch modulus E, as defined in Sec. 9-1. Hence

$$v_w = \sqrt{\frac{E}{d}}$$

■ PROBLEMS

Note: Use speed of sound in air $= 1100$ ft/s $= 340$ m/s, unless otherwise specified.

10-C1 A train approaches a station at a constant speed of 80 ft/s sounding a 1000 Hz whistle. (*a*) What frequency is perceived by an observer on the platform of the station? (*b*) After the train passes the station platform, what frequency is perceived by the observer as the train moves away?

10-C2 A gong at a railroad crossing emits a sound whose most intense component is at 400 Hz. (*a*) What is the apparent frequency heard by a passenger on a train which is approaching the crossing at 30 m/s?   (*b*) What is the apparent frequency after the train has passed the crossing and is moving away from the gong?

10-C3 A boy walks at 50 cm/s toward a stationary man. If the boy is whistling a note of frequency 550 Hz, what is the apparent frequency heard by the man?

10-C4 (*a*) With what speed should an observer in a plane fly toward a stationary source of sound of frequency 1000 Hz so that the apparent frequency becomes 1500 Hz? (Assume that at this altitude and temperature the speed of sound is 1000 ft/s.)   (*b*) If the source of 1000 Hz is on the plane, with what speed should the plane fly toward a stationary observer so that the apparent frequency becomes 1500 Hz?

10-C5 An observer and a source of sound waves are fixed relative to each other. If a strong wind blows toward the observer,   (*a*) is the wavelength changed?   (*b*) Is the wave velocity changed?   (*c*) Is the frequency changed?

10-C6 An observer in a mountain town hears a train whistle and 3 s later hears the start of an echo from a cliff. The echo's frequency is 0.9 that of the sound heard directly.   (*a*) How far is the train from the cliff?   (*b*) How fast, and in what direction, is the train moving?

10-C7 During a lunar landing, it is necessary to measure the speed of descent v. A beam of 2.50 GHz ($= 2.50 \times 10^9$ Hz) electromagnetic waves is emitted vertically downward from the spacecraft module. The waves are reflected from the lunar surface; the beam returns as if it were emitted from a source moving toward the module at $2v$. The reflected waves are superposed on the emitted wave and an on-board computer uses the beat frequency to calculate the speed.   (*a*) What is the wavelength of the radar waves emitted by the module?   (*b*) If the beat frequency is 1260 Hz, calculate the speed of descent.

10-C8 (*a*) Show that if v/v_w is small compared with 1, Eq. 10-5 reduces to $\nu' = \nu(1 + v/v_w + \cdots + \cdots)$, which is similar to Eq. 10-6 except for negligible terms.   (*b*) Show that if c is replaced by v_w and v/v_w is small compared with 1, the relativistic equation Eq. 10-7 also reduces to $\nu' = \nu(1 + v/v_w + \cdots + \cdots)$. Are the neglected terms the same in each case? [*Hint:* For convenience, let $v/v_w = x$, and use the binomial theorem, which can be written as $(1 + x)^n = 1 + nx + \dfrac{n(n-1)}{2 \cdot 1}x^2 + \dfrac{n(n-1)(n-2)}{3 \cdot 2 \cdot 1}x^3 + \cdots .$]

10-C9 Use values in Tables 9-1 and 2-1 to calculate the speed of a compression wave in air. Can you explain why your answer is about 20% lower than the observed value? (*Hint:* See footnote to Table 9-1. This will be studied in Sec. 16-10.)

■ REFERENCES

1   Laporte, Otto, "Shock Waves," *Sci. American,* Nov., 1949, p. 14.

2   Griffin, D. R., "How Bats Guide Their Flight by Supersonic Echoes," *Am. J. Phys.,* **12,** 343 (1944).

3   Strickland, James, *Doppler Effect; Formation of Shock Waves* (films).

4   Miller, Franklin, Jr., *Nonrecurrent Wavefronts* (film).

11

Interference and Stationary Waves

We have studied wave motion in general and have used sound and water waves to illustrate the propagation of a disturbance through a medium. So far, we have tacitly assumed that once a disturbance is created, it spreads out indefinitely, with no hindrance or absorption. In practice, we are often faced with the likelihood that two or more waves will be simultaneously passing through a region of space. Sound waves are reflected at a wall; water waves from several boats may reach a floating log at about the same time; waves on a stretched string sooner or later reach the end of the string and must be reflected or absorbed. We use the term *interference* to describe the superposition of several waves traveling through the same region of space. Interfering waves do not always tend to cancel or oppose one another, as might be expected from the nontechnical use of the word; there can be constructive as well as destructive interference, and intermediate cases of many sorts. The interference of two identical waves (having the same wavelength, amplitude, and speed), traveling in opposite directions gives rise to the phenomenon of *stationary waves,* which is of utmost importance in the production of musical and other sounds.

11-1 Interference

Graphical methods are adequate for a description of what happens when two identical waves interfere. Consider first transverse waves on a long string (Fig. 11-1). The wave represented by the dotted line is moving toward the right, and the dashed-line wave is moving toward the left. At the instant shown in the top graph (for which we have taken $t = 0$), the two waves happen to be exactly out of phase. At the point P, for instance, the dotted-line wave tends to move the string down, and the dashed-line wave tends to move the same particle of string upward. The net effect is zero displacement. A similar argument holds for every particle of the string, and therefore, at $t = 0$, the entire string is straight—in its natural, undistorted state. After a brief interval of time equal to $\frac{1}{4}$ period, each wave has moved $\frac{1}{4}$ wavelength. The dotted-line wave has taken the trough X to the right, and the dashed-line wave has taken the crest Y to the left. The two waves are now exactly in phase; they cooperate to make a king-sized wave form. (We are here using the superposition theorem of Sec. 10-7.) We can continue this graphical

Indian musicians of the Bolivian highlands. Principles of resonance and tuning of pipes are known in many cultures. (Linares from Monkmeyer)

procedure, $\frac{1}{4}$ period at a time, and obtain resultant wave forms as shown in Fig. 11-1. Thus at certain times ($t = 0$, $t = \frac{1}{2}T$, $t = T$, etc.) the string is undistorted, with no net disturbance anywhere along the string. At other times ($t = \frac{1}{4}T$, $t = \frac{3}{4}T$, $t = \frac{5}{4}T$, etc.) the string is violently disturbed. Also, we note that some points, marked N, never move from their normal positions. Such points of no disturbance are called *nodes*. The points marked A, which have maximum disturbance, are called *antinodes*. From the graphs, it is evident that *nodes are half a wavelength apart*, and also *antinodes are half a wavelength apart*. We call the whole phenomenon by the name *stationary wave*—but the two waves which interfere are not at all stationary. It is the location of nodes and antinodes that is stationary. This is shown in the time exposure

of Fig. 11-1, in which the nodes are clearly visible at fixed points along the string.

Although we have used waves on strings for our example, periodic waves of any sort can combine to give stationary waves. The discussion is general; it is not limited to waves on strings but applies to waves of compression, electromagnetic waves, and all other periodic waves. For instance, the acoustic properties of concert halls are strongly influenced by stationary wave patterns formed by the interference of waves reflected from the walls and ceiling.

11-2 Vibrating strings

We can now discuss the vibration of a string which is fixed at both ends. Such a string (or wire) might be found in a piano, on a guitar or violin, or between two telegraph poles along the railroad right of way. If any particle of the string is moved aside and released, it vibrates back and forth, perpendicular to the length of the string, and this disturbance spreads out along the string in both directions. The wave velocity depends on the tension in the string and on the string's mass per unit length. It can be shown (Sec. 11-8) that the velocity of a transverse wave along a string or wire is

$$v_w = \sqrt{\frac{F}{m/L}} \qquad (11\text{-}1)$$

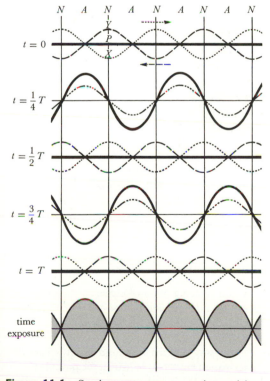

Figure 11-1 Stationary wave on a string, arising from interference of two identical waves traveling in opposite directions.

where F is the tension and m/L is the mass per unit length. This equation is a reasonable one. If the string is a "heavy" rope, the mass (inertia) of 1 ft of the rope is greater than the mass of 1 ft of a "light" string or thread. The greater inertia of the molecules causes neighboring particles to resist being set into motion, and the disturbance is handed on at a slower rate. This checks with the presence of m/L in the denominator of our formula. On the other hand, if the tension is greater, there is more net force acting to

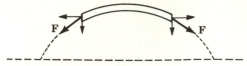

Figure 11-2 The net force on a short segment of string is the resultant of the two force vectors; the resultant is directed toward the equilibrium position.

accelerate any segment of the string (Fig. 11-2), and we expect larger accelerations. The neighboring particles are disturbed sooner, and the wave travels along the string faster. This checks with the presence of F in the numerator of our formula.

EXAMPLE **11-1**

A clothesline 20 m long has a mass of 0.80 kg and is stretched between two fixed posts with a tension of 400 N. If the rope is struck a sharp blow at one end, how much time will be required for the pulse to travel to the opposite post and back again?

First we find the wave velocity. The mass per unit length is

$$\frac{m}{L} = \frac{0.80 \text{ kg}}{20 \text{ m}} = 0.04 \text{ kg/m}$$

The wave velocity is

$$v_w = \sqrt{\frac{F}{m/L}} = \sqrt{\frac{400 \text{ N}}{0.04 \text{ kg/m}}}$$
$$= \sqrt{10\,000 \text{ N·m/kg}} = 100 \text{ m/s}$$

Now the time for the round trip can be found:

$$t = \frac{s}{v_w} = \frac{40 \text{ m}}{100 \text{ m/s}} = \boxed{0.4 \text{ s}}$$

The pulse of Example 11-1 will be reflected at each post and will travel back and forth, making one complete trip every 0.4 s, until finally it dies out because of internal molecular friction in the rope. Some energy will be transferred to the posts if they are not absolutely rigidly set into the ground.

An important concept has been introduced in Example 11-1. When the clothesline is struck near one end, the pulse returns again and again, with a definite frequency, 2.5

vib/s in this case. This frequency of vibration does not depend on the exact nature of the blow and is characteristic of the string. A *natural frequency* of a vibrating body is one which is characteristic of the body; we have just seen in Example 11-1 how one of the natural frequencies of a stretched rope depends on the length, mass, and tension of the rope.

Passing now to *periodic* waves traveling back and forth along a string, let us imagine a piano string on which sinusoidal displacement waves travel back and forth. Because of the reflections of the waves at the fixed ends, we will have interference between two identical waves traveling in opposite directions. That is, the conditions for obtaining stationary waves will be fulfilled. As a consequence, nodes and antinodes will be set up, spaced $\frac{1}{2}\lambda$ apart. One condition must be fulfilled: nodes *must* exist at each end of the string, for the string is surely never disturbed at these two points. There may or may not be nodes between the ends, and the string may vibrate in 1 segment, 2 segments, 3 segments, ..., etc. These are called *modes* of vibration. We can summarize by saying that for each different mode of vibration there is a different distribution of nodes, but, in all modes, nodes must exist at each end of the string. The possible frequencies of vibration are related to the number of segments, as shown in our next example. The lowest natural frequency is called the *fundamental* frequency, and the other frequencies are called *overtones*.

EXAMPLE **11-2**

A guitar string 50 cm long has a mass of 50 g and is under tension of 4×10^8 dyn. What are the wavelengths and frequencies of the first four modes of vibration?

The wavelengths are found directly from the diagrams (Fig. 11-3) on which the possible nodes are located. We make use of the fact that nodes are spaced $\frac{1}{2}\lambda$ apart. For instance, when the string is vibrating in 3 segments, with each seg-

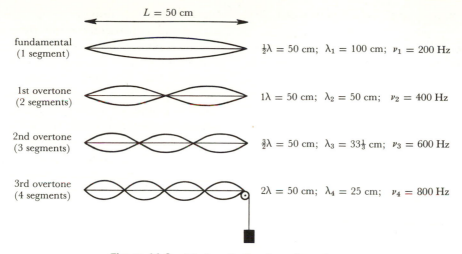

$\tfrac{1}{2}\lambda = 50$ cm; $\lambda_1 = 100$ cm; $\nu_1 = 200$ Hz

$1\lambda = 50$ cm; $\lambda_2 = 50$ cm; $\nu_2 = 400$ Hz

$\tfrac{3}{2}\lambda = 50$ cm; $\lambda_3 = 33\tfrac{1}{3}$ cm; $\nu_3 = 600$ Hz

$2\lambda = 50$ cm; $\lambda_4 = 25$ cm; $\nu_4 = 800$ Hz

Figure 11-3 Modes of vibration of a string.

ment $\tfrac{1}{2}\lambda$ in length, we have $3(\tfrac{1}{2}\lambda) = 50$ cm, whence $\lambda = \tfrac{100}{3}$ cm $= 33\tfrac{1}{3}$ cm. The final column is filled in from knowledge of the wave velocity:

$$v_w = \sqrt{\frac{F}{m/L}} = \sqrt{\frac{4 \times 10^8 \text{ dyn}}{50 \text{ g}/50 \text{ cm}}}$$

$$= 2 \times 10^4 \text{ cm/s} = 20\,000 \text{ cm/s}$$

Knowing the wave velocity, we can find the corresponding natural frequencies from $v_w = \nu\lambda$. The fundamental frequency is

$$\nu_1 = \frac{v_w}{\lambda_1} = \frac{20\,000 \text{ cm/s}}{100 \text{ cm}} = 200 \text{ Hz}$$

Similarly, the first overtone's frequency is

$$\nu_2 = \frac{v_w}{\lambda_2} = \frac{20\,000 \text{ cm/s}}{50 \text{ cm}} = 400 \text{ Hz}$$

and so forth for the frequency of any overtone.

It is evident from Example 11-2 that the natural frequencies of the allowed vibrations of a string are all integral multiples of the fundamental frequency. Such remarkable simplicity is a result of the fact that nodes on a string are equally spaced. If, as in this case, the overtones have frequencies that are integral multiples of the fundamental frequency, they are called *harmonics.*[*] For any string, the

[*] So called because musical scales and harmony are based upon integral ratios of frequencies.

natural frequencies are in the ratios $1:2:3:4:$ ... and hence are harmonics, as well as overtones.

EXAMPLE **11-3**

A horizontal string 120 cm long has a mass of 562 mg. It is attached at one end to a tuning fork producing 100 vib/s, and the other end of the string runs over a pulley and supports a light pan. What load should be placed in the pan to cause the string to vibrate in 4 segments?

The distance between nodes is $\tfrac{1}{4}$ the length of the string, or 30 cm. This means that $\tfrac{1}{2}\lambda = 30$ cm, or $\lambda = 60$ cm. We next find the wave velocity, from the frequency and wavelength:

$$v_w = \nu\lambda = (100 \text{ s}^{-1})(60 \text{ cm}) = 6 \times 10^3 \text{ cm/s}$$

Finally, we find the tension from the velocity formula which applies to strings:

$$v_w = \sqrt{\frac{F}{m/L}} = \sqrt{\frac{FL}{m}}; \qquad v_w{}^2 = \frac{FL}{m}$$

$$F = \frac{v_w{}^2 m}{L} = \frac{(6 \times 10^3 \text{ cm/s})^2(562 \times 10^{-3} \text{ g})}{120 \text{ cm}}$$

$$= 169 \times 10^3 \text{ dyn}$$

Using $m = W/g$, we conclude that the mass of the load in the pan is given by

$$\frac{169 \times 10^3 \text{ dyn}}{980 \text{ cm/s}^2} = \boxed{172 \text{ g}}$$

11-3 Vibrating air columns in pipes

The production of certain natural frequencies in a vibrating column of air was known to the ancients, who fashioned shepherd's pipes and flutes on this principle. It was found that the frequency of vibration depended on the effective length of the air column—hence the strategically placed air holes along the length of a flute or recorder. We can analyze the modes of vibration of pipes in much the same way as we discussed modes of vibration of strings. As before, the location of nodes is the key to understanding what happens. When a condensation approaches a closed end of the pipe (Fig. 11-4) the molecules at

Figure 11-4

P are free to move, and are swishing back and forth as pressure builds up or as a partial vacuum is created. However, molecules at Q are in contact with the solid end of the pipe. Hence they cannot move, and so Q must be a node. *The closed end of a pipe is a node.* On the other hand, if a condensation approaches an open end of the pipe, as at

Q', the molecules moving to the right have inertia and are carried out into the open air, leaving a partial vacuum behind. There is a maximum amount of to-and-fro motion at the open end of a pipe. In other words, *the open end of a pipe is an antinode.* Armed with these facts about the location of nodes and antinodes, we can study the natural frequencies of open and closed pipes.

For a pipe which is open at both ends (Fig. 11-5) there must be antinodes at each end. Remembering that antinodes are spaced $\frac{1}{2}\lambda$ apart, we can make diagrams of the various modes of vibration and determine the wavelength for any given mode by inspecting the diagrams. The calculations summarized in Fig. 11-5 show that, for an open pipe, the natural frequencies are integral multiples of the fundamental frequency. This is exactly the same result we found for vibrating strings—the natural frequencies are in the ratio $1:2:3:4:\ldots$. The reason for this similarity is not hard to find. Comparing Fig. 11-5 with Fig. 11-3, we see that the vibrating body (string or air column, as the case may be) is bounded at each end by similar conditions—either both ends nodes, or both ends antinodes. Replacing every N in Fig. 11-5 by A, and every A by N, would give a distribution of nodes and antinodes just like Fig. 11-3. In other words, the symmetry properties of

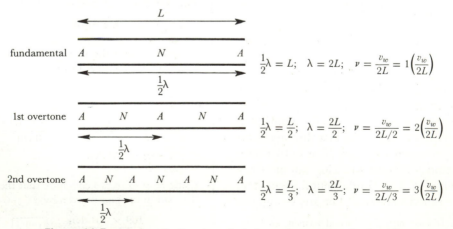

Figure 11-5 Modes of vibration of air in a pipe open at both ends.

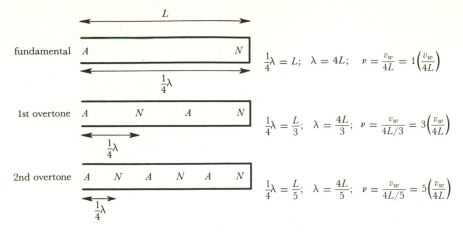

$$\tfrac{1}{4}\lambda = L; \quad \lambda = 4L; \quad \nu = \frac{v_w}{4L} = 1\left(\frac{v_w}{4L}\right)$$

$$\tfrac{1}{4}\lambda = \frac{L}{3}; \quad \lambda = \frac{4L}{3}; \quad \nu = \frac{v_w}{4L/3} = 3\left(\frac{v_w}{4L}\right)$$

$$\tfrac{1}{4}\lambda = \frac{L}{5}; \quad \lambda = \frac{4L}{5}; \quad \nu = \frac{v_w}{4L/5} = 5\left(\frac{v_w}{4L}\right)$$

Figure 11-6 Modes of vibration of air in a pipe closed at one end.

strings and open pipes are essentially the same, and so the modes of vibration, and the natural frequencies, would be expected to be similar.

If a pipe is closed at one end, there must be an antinode at the open end and a node at the closed end. We can prepare in the usual way a table of values for the natural frequencies of a closed pipe (see Fig. 11-6), remembering that from node to antinode is $\tfrac{1}{4}\lambda$. We see that for a closed pipe the possible frequencies are *odd* multiples of the fundamental frequency; even multiples of the fundamental frequency are not possible. The symmetry properties of a closed pipe differ from those of an open pipe or a string because the situation at one end is different from the situation at the other end. The natural frequencies for a closed pipe are in the ratio $1:3:5:7:\ldots$. Only odd harmonics occur; the 1st overtone is the 3rd harmonic, the 2nd overtone is the 5th harmonic, etc.

EXAMPLE 11-4

What is the fundamental frequency of a whistle which is 4 in. long from blowhole to open end? Assume the speed of sound in air to be 1100 ft/s.

We first draw a diagram of the pipe, Fig. 11-7, putting antinodes at each end, since the pipe is open at both ends. According to the statement

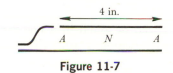

Figure 11-7

of the problem, we want the fundamental frequency, that is, the mode which has the fewest possible nodes and antinodes. This means that there is a single node, at the center of the pipe. From the diagram, $\tfrac{1}{2}\lambda = 4$ in.; $\lambda = 8$ in. $= \tfrac{2}{3}$ ft. Now, since the speed of sound is 1100 ft/s,

$$\nu = \frac{v_w}{\lambda} = \frac{1100 \text{ ft/s}}{\tfrac{2}{3} \text{ ft}} = \boxed{1650 \text{ Hz}}$$

EXAMPLE 11-5

By blowing harder, the whistler can cause the pipe to vibrate in its first overtone. What is the frequency of this overtone?

For the desired mode of vibration, the location of nodes and antinodes is as shown in Fig. 11-8.

$$\lambda = 4 \text{ in.} = \tfrac{1}{3} \text{ ft}$$

$$\nu = \frac{v_w}{\lambda} = \frac{1100 \text{ ft/s}}{\tfrac{1}{3} \text{ ft}}$$

$$= \boxed{3300 \text{ Hz}}$$

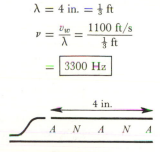

Figure 11-8

EXAMPLE **11-6**

What is the lowest frequency that can be produced in the pipe of Example 11-4, if the end of the pipe is sealed with a close-fitting plug?

The pipe is now a closed pipe, and the diagram of nodes and antinodes is shown in Fig. 11-9.

$$\tfrac{1}{4}\lambda = 4 \text{ in.} \quad \lambda = 16 \text{ in.} = \tfrac{4}{3} \text{ ft}$$

$$\nu = \frac{v_w}{\lambda} = \frac{1100 \text{ ft/s}}{\tfrac{4}{3} \text{ ft}} = \boxed{825 \text{ Hz}}$$

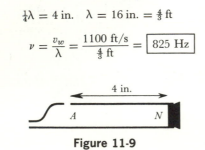

4 in.

A   N

Figure 11-9

The speed of sound in a gas depends on the density, which in turn depends on the temperature. As we will see in Sec. 16-10, v_w is proportional to the square root of the absolute temperature $(v_w \propto \sqrt{T})$, which implies that for air near room temperature v_w changes by about 0.60 m/s for every 1 C° change in temperature. For this reason the natural frequencies of any pipe depend on the temperature. Every player of a wind instrument knows how important it is to "warm up" his instrument. Temperature differences among the pipes of an organ may cause the organ to be badly out of tune. A change of frequency of as little as 1% would be distressingly noticeable, if some pipes were affected more than others due to uneven temperatures in the organ loft.

EXAMPLE **11-7**

A closed organ pipe is 0.750 m long and vibrating in its fundamental mode. (*a*) What is the frequency, on a day when the temperature is 15°C and the speed of sound in air is 340 m/s? (*b*) If the temperature rises to 25°C, what does the frequency become?

(*a*) The fundamental wavelength of the closed pipe is 4 times its length, or

$$\lambda = 4 \times 0.750 \text{ m} = 3.00 \text{ m}$$

Hence

$$\nu_1 = \frac{v_w}{\lambda} = \frac{340 \text{ m/s}}{3.00 \text{ m}} = \boxed{113.3 \text{ Hz}}$$

(*b*) At 25°C the speed of sound in air is $[340 + (0.60 \times 10)]$ m/s = 346 m/s. The wavelength is still 3.00 m (ignoring the negligible expansion of the pipe itself), since the locations of nodes and antinodes have not changed. The fundamental frequency is now

$$\nu_1' = \frac{v_w}{\lambda} = \frac{346 \text{ m/s}}{3.00 \text{ m}} = \boxed{115.3 \text{ Hz}}$$

11-4   Resonance

Our study of stationary waves on strings and in air columns has shown that each vibrating body has a certain set of natural frequencies, characteristic of the size and shape of the body and the nature of the medium. We have met with natural frequencies before: a simple pendulum has a single natural frequency, the reciprocal of its period. A mass on the end of a spring has a natural frequency of vibration. These examples from Chap. 9 are simpler than those discussed in this chapter, for the pendulum and the vibrating object on the spring each have only a single natural frequency instead of an infinite sequence of natural frequencies as do the string and the pipe.

A boy sitting in a swing can be set into vibration by a friend who applies impulses at the proper times. Experience tells us that even a small child can give a large amplitude of vibration to a heavy playmate if he applies repeated small impulses at just the right times. *Resonance* is the building up of a large vibration by repeated application of small impulses whose frequency equals one of the natural frequencies of the resonating body. In the example of the boy in the swing, a large amplitude is built up only if the impulses are applied in synchronism with the natural period of vibration of the swing. A slight mismatch of frequency would result in little or no vibration.

Acoustic resonance occurs in many musical instruments (Sec. 11-7) and in auditoriums. Resonance is not always helpful. For instance, if a loudspeaker enclosure has a natural frequency of 200 Hz, the sounds of this frequency will be "amplified" and the reproduced music will have an unpleasant boomy sound. Bridges have been known to collapse because of mechanical resonance as large vibrations were built up from gusts of wind (see Reference 14 at the end of this chapter). On the other hand, electrical resonance (Sec. 23-3) is helpful in radio receivers, which must respond with large amplitude to weak signals at the frequency of the desired station, but must reject (i.e., respond very little to) the signals of other stations which broadcast at other frequencies. A certain "rattle" in the family car may be noticed only at 33 mi/h, and not at 23 or 43 mi/h. The loose part evidently has a natural frequency which is "excited" by the vibration of the motor; the excitation is effective only when the motor supplies impulses at the resonant frequency of the vibrating part.

Resonance is related to physical limitation of a vibrating body or medium. A pipe of infinite length has no natural frequency of vibration. However, if a loudspeaker is placed in front of a pipe which is terminated (either by a closed end or an open end), it will cause a large disturbance of the air column if its frequency (the exciting frequency) closely matches one of the natural frequencies of the pipe. Thus every natural frequency of a body is also a resonant frequency.

A resonance is said to be "sharp" if the response is large only for exciting frequencies which are very close to the natural frequency (Fig. 11-10). The sharpness of a resonance depends upon the rate at which the vibration would die away if left to itself. In general terms, a resonance is sharp if the energy stored in the vibrating body is large compared with the energy which is dissipated per cycle.

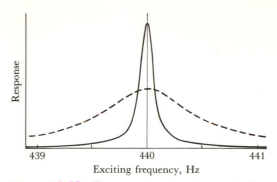

Figure 11-10 Resonance curves for two bodies, each with natural frequency 440.0 Hz. Solid curve: sharp resonance; dashed curve: broad resonance.

Let us illustrate this by discussing a simple experiment with the A string of a piano, which has a natural frequency of 440.0 Hz. If a singer emits a 440.0 Hz tone, the string can be heard to vibrate after the sung tone ceases. The response to 440.1 Hz is less than to 440.0 Hz, and the response to 441.0 Hz is so small as to be inaudible. This rather sharp resonance can be predicted, for we know that the vibration of a struck piano string endures for many seconds. In, say, 25 s, the string vibrates 11 000 times, and the energy lost per cycle is evidently a small fraction of 1%. The rattle of a part of a car, mentioned earlier, is a broader resonance; if the rattle is most pronounced at a frequency corresponding to 33 mi/h, it can also be heard if the car's speed is 32 mi/h or 34 mi/h. This broad resonance is correlated with the fact that the vibration, once started, would die out in less than a second if it were not continuously excited by the outside source of energy.

11-5 Musical sounds and their sources

Some of the most interesting and familiar applications of stationary waves and resonance involve musical instruments such as the piano, violin, trombone, saxophone, and even the musical saw. The primary distinction between music and noise is, of course, a cultural one. We are accustomed to calling certain sounds musical because they are "pleasing." Sharp differences of opinion exist as to whether certain sounds or combinations of sound are music or noise, and the opinions of one generation or cultural group are not necessarily the same as those of another group. However, in broad outline, the general attributes of the component sounds that make up music are not controversial. Musical sounds are characterized by pitch, loudness, and quality. These are psychological terms denoting sensations; we shall see that each of these three attributes is largely (but not entirely) dependent upon one of three physical attributes: frequency, intensity, and overtone structure, respectively.

PSYCHOLOGICAL SENSATIONS	PHYSICAL ATTRIBUTES
pitch	frequency
loudness	intensity
quality	overtone structure

Pitch and frequency. Musical sounds are periodic waves in air, and are characterized by wavelength and frequency. The sensation of *pitch* is closely related to the *frequency* of a sound wave; high-pitched sounds are of high frequency, and low-pitched sounds are of low frequency. Hence our first correlation is pitch (a sensation) and frequency (a physically measurable attribute of a wave). In these days of high fidelity and hearing aids, we are familiar with the fact that the human ear has frequency limits. The sensation of sound is not produced in the brain unless the frequency lies between limits which are roughly 20 Hz and 20 000 Hz. Even this range is not available to all persons throughout their lives; for most, the process of aging involves a loss of frequencies above, say, 10 000 Hz or even less. For some, a hearing defect may involve relative loss of low-frequency sounds.

Compression waves of frequencies above about 20 000 Hz are called *ultrasonic waves*. Although they do not affect the human ear, and hence have no pitch, such waves are in every other way similar to the sound waves we have discussed. For example, stationary ultrasonic waves are easily produced. Because of their high frequency (and short wavelength), ultrasonic waves produce some effects in matter which are little, if at all, evident at audible frequencies. (See Reference 1 at the end of the chapter.) At the other extreme of inaudible sounds, infrasonic waves of, say, 1 to 10 Hz are of great importance in engineering, for they can lead to destructively large vibrations if one of a structure's natural frequencies can be excited by a resonance type of interaction (Sec. 11-4).

As a matter of terminology, two sounds are said to be one *octave* apart if their frequencies are in the ratio $2:1$. Thus an interval of 7 octaves, representing almost the complete span of a piano, from lowest C to highest C, is a ratio $2 \times 2 \times 2 \times 2 \times 2 \times 2 \times 2$ $(=2^7)$, or $128:1$. The lowest C on the piano has frequency 32.70 Hz, and the highest C, 7 octaves higher, is 128×32.70, or 4186 Hz. It is seen that the human ear's sensitivity extends well beyond the piano frequencies, in the direction of both higher and lower frequencies. Another interval that is worth knowing is called a *fifth*. This interval corresponds to a frequency ratio $3:2$. Another simple frequency ratio, $5:4$, is called a *major third*, and $6:5$ is called a *minor third*. Note that the pleasing chord "do-mi-sol," exemplified by the tones C-E-G, contains a fifth (C-G), a major third (C-E), and a minor third (E-G). The frequencies in the chord are in the simple ratios $4:5:6$. Further information on musical scales can be found in the references at the end of the chapter.

Loudness and intensity. The *loudness* of a sound depends on the rate at which acoustic energy enters the ear of the listener. When a sound wave travels through the air, the individual gas molecules oscillate back and forth, but do not move far from their equilibrium positions. Nevertheless, energy is handed on from one molecule to the next,

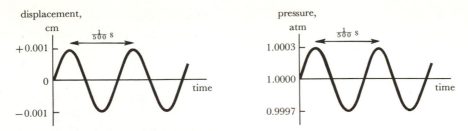

Figure 11-11 Two representations of a sound wave of frequency 500 Hz, having the maximum intensity which can be tolerated by the ear.

and this flow of energy can be measured by physical means. One direct method of measuring the energy flow is to place a sound-absorbing body in the path of the sound wave. As energy is absorbed, the temperature of the absorber rises at a steady rate. If we know the thermal characteristics of the absorber, we can compute the rate of energy flow. Of course, such measurements are extremely delicate, and the increases in temperature are very small. In actual practice, it would be more convenient to use a microphone (connected to an amplifier and a meter of some sort) for routine measurements, and calibrate the microphone once and for all by the more difficult, but more direct, thermal method. The rate of flow of energy in a wave, power per unit cross-sectional area, is called the *intensity I* of the wave. The mks unit for intensity is $J/s \cdot m^2$, or W/m^2.

Motions and pressures in a sound wave are surprisingly small. Even for the noise of a military jet plane, which is close to the threshold of pain, gas molecules are displaced only about 0.001 cm at the most (assuming a wave whose frequency is about 500 Hz). In other words, the displacement amplitude in such a wave is 0.001 cm. In the same wave, in the condensations in which the molecules are crowded together, the pressure is only 0.0003 atm above the normal atmospheric pressure of 1 atm. That is, the pressure amplitude is only 0.0003 atm (Fig. 11-11). These relatively small changes in pressure are enough to drive the eardrum beyond a "safe" limit. The sound wave de-

scribed in Fig. 11-11 has an intensity of about 1 W/m^2; at the other extreme, the faintest detectable sound[*] has an intensity of about 10^{-12} W/m^2.

The psychological sensation, loudness, is almost entirely determined by the physical attribute, intensity, other factors being equal.

The *intensity level* β of a sound wave is measured in *bels*. The definition of the bel is in terms of the ratio of two intensities: two sounds differ by one bel (named for Alexander Graham Bell) if their intensities are in the ratio 10:1. Usually a smaller unit, the decibel (dB) is used (1 dB = $\frac{1}{10}$ bel). A logarithmic scale is used to measure the intensity level in decibels:

$$\beta = 10 \log \frac{I}{I_0} \qquad (11\text{-}2)$$

where I_0 is some arbitrary reference level.

To find the difference between the intensity levels of two sounds whose intensities are I_1 and I_2, we can use the equation (see Prob. 11-A15)

$$\Delta\beta = \beta_2 - \beta_1 = 10 \log \frac{I_2}{I_1} \qquad (11\text{-}3)$$

The decibel is approximately the threshold of discrimination of the human ear under the most favorable circumstances. For instance, sounds differing by only 0.2 dB are indistinguishable in loudness.

---

[*] The faintest sound that can be heard has a pressure amplitude of only 0.0002 dyn/cm^2. Since atmospheric pressure is about 1×10^6 dyn/cm^2 (Sec. 12-2), this means that the ear responds to changes of $0.0002/10^6$, or 1 part in 5 billion!

Usually, when comparing two sounds the effective area A of the listener's ear is constant and then, since $I = P/A$, the ratio of the intensities is the same as the ratio of powers. We can thus use a power ratio instead of an intensity ratio when comparing two sounds:

$$\Delta\beta = 10 \log \frac{P_2}{P_1} \qquad (11\text{-}4)$$

EXAMPLE **11-8**

What is the difference in intensity level between the sound from 6 cheerleaders and the sound from 6000 cheering fans?

The power ratio is 6000:6, i.e., $P_2/P_1 = 1000$.

$$\Delta\beta = 10 \log 1000 = 10(3.0) = \boxed{30 \text{ dB}}$$

EXAMPLE **11-9**

A hi-fi set's power output is 40 W at some audible mid-frequency such as 100 Hz. The output is 3 dB less if the frequency is decreased to 20 Hz. What is the power output at 20 Hz?

We place the larger power in the numerator of the expression P_2/P_1, so that the ratio is greater than 1 and its logarithm is positive.

$$\Delta\beta = 3 = 10 \log \frac{40 \text{ W}}{P_1}$$

$$0.3 = \log \frac{40 \text{ W}}{P_1}$$

From a log table we find that $0.301 = \log 2.00$. Hence

$$2.0 = \frac{40 \text{ W}}{P_1}; \qquad P_1 = \boxed{20 \text{ W}}$$

Strictly speaking, the decibel is not an absolute unit, because it gives the *ratio* of two intensities, or two loudnesses. In many applications "0 dB" is defined as 10^{-12} W/m^2, the faintest detectable sound for the human ear. A sound level of 53 dB is, therefore, a sound which is 53 dB louder than the threshold of hearing. A calculation similar to that of Example 11-9 shows that such a sound has an intensity of 2×10^{-7} W/m^2.

Quality and overtone structure. Two sounds of equal loudness and pitch are often distinguishable from each other. A violin tone is said to have a *quality* which is different from that of a saxophone. What physical attribute is correlated with such obvious differences between sounds? Figure 11-12 shows an experimental setup for measuring the wave form of the pressure variations in a sound wave. As the wave spreads out, the pressure at the microphone varies in some fashion. The microphone, amplifier, and oscilloscope together perform the function of displaying on the screen a graph of pressure vs. time. Strictly speaking, the scope displays the vibration graph at a fixed point (the microphone). However, we know from Sec. 10-5 that the wave-form graph has the same shape as the vibration graph. We immediately notice that the wave forms of violin, trumpet, and clarinet (Fig. 11-13) are different, and so we correlate *quality* (a sensation) with *wave form* (a physically measurable characteristic).

The measurement of wave form in quantitative terms involves the superposition principle. It was shown by the French mathematician and physicist Joseph Fourier (1768–1830) that a complex wave form can be represented as a sum, or superposition, of SHM wave forms. Thus the complex wave form of Fig. 11-14 can be represented as the superposition of two simple waves: a fundamental vibration (light line) of amplitude 3 units and frequency 1000 Hz, and an over-

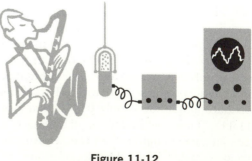

Figure 11-12

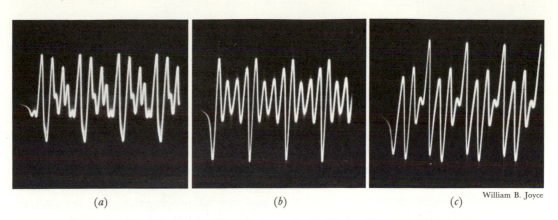

(a)   *(b)*   *(c)*   William B. Joyce

Figure 11-13   Wave forms of musical sounds. (*a*) Violin; (*b*) trumpet; (*c*) clarinet.

tone (dashed line) of amplitude 2 units and frequency 4000 Hz. The superposition of these two simple wave forms (Fig. 11-14) gives the observed wave form (heavy line). In actual practice, more than two components are often needed to represent a complex wave form. To represent a wave form, therefore, we construct a table showing the frequencies and relative intensities of the various overtones which together make up a

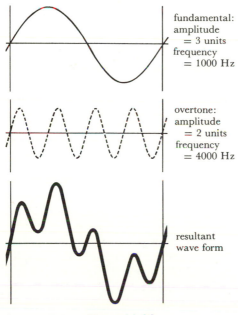

fundamental:
amplitude
= 3 units
frequency
= 1000 Hz

overtone:
amplitude
= 2 units
frequency
= 4000 Hz

resultant
wave form

Figure 11-14

complex wave.[*] Typical results for two sources of musical sounds are shown in Table 11-1. We call this table a table of *overtone structure* for each musical sound, and we summarize by saying that the sensation *quality* is closely related to the physical attribute *overtone structure*.

---

[*] If the same two simple waves shown in Fig. 11-14 were added, but with the wave of frequency 4000 Hz shifted by $\frac{1}{2}$ cycle so that it starts down at the instant the 1000 Hz wave starts up, an entirely different wave form would be obtained. In spite of this, the ear would hear little if any difference in the quality. This physiological fact (the relative insensitivity of the ear to phase differences among components of a complex tone) is the reason we prefer overtone structure to wave form as a means of describing the quality of a musical sound.

TABLE 11-1
Typical overtone structures for two musical sounds

CLARINET		FRENCH HORN	
Frequency, Hz	Relative intensity, %	Frequency, Hz	Relative intensity, %
400	36	100	3
800	0	200	22
1200	34	300	24
1600	9	400	44
2000	17	500	3
.	.	600	2
.	.	700	1
.	.		
4000	3		

Any psychological sensation of sound is not determined entirely by the corresponding physical attribute. A typical interrelation is that between pitch and loudness. A loud foghorn sounds "lower" in pitch than an exactly similar sound, of the same frequency, which is not so intense. In this case frequency is inadequate, by itself, to describe and predict the sensation of pitch which will be observed. As another example, consider two sounds of equal intensity (measured in W/m²). If sound A is of frequency 5000 Hz and sound B is of frequency 50 000 Hz, then B has much less loudness—zero loudness, in fact—since its frequency is beyond the range of audible sound. Quality and pitch are also interrelated, to a slight extent.

11-6 The human ear

A convenient way to summarize data about the ear's sensitivity to sounds of different frequency is through an audibility graph (Fig. 11-15). The *average* ear can just hear a sound whose frequency is 2000 Hz if the intensity level is about 10 dB (relative to the standard 0 dB level of 10^{-12} W/m²); a few persons (about 10% of the population) can hear a 0 dB sound at this frequency. Much more intensity is required to hear sounds of

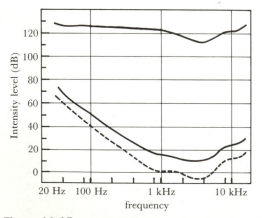

Figure 11-15 Audibility of sound as function of frequency. Solid curve: average human ear. Dashed curve: limit of audibility for person with exceptionally acute hearing.

frequency closer to the limits of audibility; for example, the graph shows that an intensity level of at least 50 dB is needed to detect a 100 Hz sound. By Eq. 11-2, this is 10^5 times the intensity of the faintest audible sound. The graph also shows a threshold of discomfort (or pain) at somewhat above 120 dB.

The fact that the sensation of loudness is only approximately proportional to the intensity of a sound indicates that the physical and neural processes in the ear and the brain are rather complicated. The dependence of pitch on frequency is similarly complex. A full description of the ear's remarkable anatomy and functioning would require a treatise; we will give an application of elementary physics to each of the three major divisions of the ear (Fig. 11-16).

The outer ear, open to the atmosphere, contains the ear canal, about 2.7 cm long, terminated by the tympanum (ear drum). Consider the canal to be a closed pipe vibrating in its fundamental mode; $\frac{1}{4}\lambda = 0.027$ m, from which $\lambda = 0.108$ m, and for $v_w = 350$ m/s the resonant frequency is about 3200 Hz. The ear canal thus enhances the pressure at the closed end (tympanum) for sounds in the mid-range of audible frequencies, 1 to 5 kHz. The resonance is fairly broad.

The middle ear transmits sound from the tympanum to the oval window which closes off one end of the inner ear. The coupling is through a delicate and intricate lever system made up of three tiny bones (ossicles). If they give a force multiplication (ideal mechanical advantage), it is not very great. The magnification of *pressure* (force/area), however, is large—up to 22—because the area of the tympanum is about that much greater than the area of the oval window. Thus a pressure wave is set up in the lymph (a liquid similar to spinal fluid) in the inner ear. A general physical principle is involved here. At an interface between two media, the energy of a wave is partially reflected,

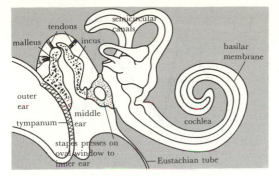

Figure 11-16 Middle and inner ear, approximately twice life size. The ossicles include the malleus, incus, and stapes. Adapted from *Life: An Introduction to Biology*, George Gaylord Simpson, et al., © 1957 by Harcourt Brace Jovanovich.

partially transmitted, and here the useful energy is that transmitted from the air in the outer ear to the liquid in the inner ear. High efficiency of energy transfer depends on the near equality for the two media of the product $v_w d$ of the wave velocity and the density (this product is called the acoustic impedance of the medium). For lymph versus air this ratio is as large as 3100; the ossicles, loosely bound by small tendons, serve the important purpose of matching the acoustic impedances of air and lymph, thus increasing the energy transfer into the cochlea.

The inner ear contains the semicircular canals (used for detection of orientation and acceleration) and the cochlea, a snail-shaped cavity divided along its length by the basilar membrane which supports the organ of Corti. Here is where sound energy is converted into electric energy, to go to the brain as nerve impulses; here, too, is where frequency discrimination takes place. In the organ of Corti, nerve fibers are somehow connected to about 30 000 sensory cells called hair cells, each of which has about 50 stereocilia (hairs) about 1 nm long. When the basilar membrane is distorted, nerve endings are stimulated. It would be an attractive theory (but not altogether correct) to suppose that the basilar membrane contains transverse

stretched entities like piano strings, one tuned to each observable frequency. For one thing, in the membrane the tension and mass per unit length do not vary enough to account for the wide range (almost 1000:1) of audible frequencies. It is generally accepted that for a pure tone of, say, 1000 Hz, stationary waves in the lymph and in the basilar membrane interact to give a broad antinode extending over a region of perhaps 50 hair cells. The combined effect of nerve impulses from these adjacent sensory cells is decoded in the brain to give the pitch sensation. This is primarily a "place" theory of pitch discrimination rather than wholly a "resonance" or a "telephone" theory. The basilar membrane is thinnest and under greatest tension at the fixed end nearest the oval window; the free end, at the apex of the cochlear spiral, is heavier and under less tension. The wave velocity along the membrane, given by $v_w = \sqrt{T/(m/L)}$, is largest near the oval window. Analysis is more difficult than our earlier study of a stretched wire's nodes and antinodes, since v_w is not constant. Nevertheless, scale models and theory both show that regions of greatest distortion occur along the basilar membrane closer to the oval window as the frequency is increased. In Fig. 11-17 is shown the

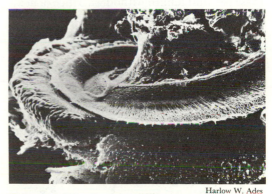

Figure 11-17 Complete degeneration (lower center of photograph) of part of guinea pig organ of Corti due to exposure to pure tone of 125 Hz at 148 dB for 4 hours (×80). [G. Bredberg and co-workers, *Science*, **170**, 863 (1970).]

destruction of a rather sharply defined region of the organ of Corti in an ear which had been subjected to prolonged intense sound of frequency 125 Hz.

The ear is an extremely sensitive detector of acoustic radiation. A motion of the tympanum of only 10^{-11} m (less than the size of a single atom) can give rise to a noticeable sensation.

11-7 Musical instruments

We now discuss briefly one of many musical instruments—the violin—not so much for its intrinsic importance as for the review and illustration of the ideas of this chapter which it affords.

In general, a musical instrument consists of a *generator* of sound and a *resonator,* which selects and accentuates certain of the overtones. In the violin the generator is a string of steel, nylon, or gut, "fixed" at the bridge and at the nut (Fig. 11-18). The string is under considerable tension, supplied by the pegs, which are held firmly in their holes by frictional forces. The mass per unit length is characteristic of the type of string used; the strings of lower pitch are wound with fine aluminum or silver wire to give them added mass per unit length without sacrifice of flexibility. Wave velocity along the string is given by the formula $v_w = \sqrt{F/(m/L)}$, and stationary waves can exist on the string with a node at the bridge and a node at the nut. Such waves are set up by the bow, which pulls the string aside a few millimeters until the force of static friction becomes inadequate

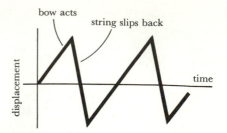

Figure 11-19 Vibration graph for a point on a violin string, idealized as a triangular relaxation oscillation.

and the string slips back. The bow repeatedly picks up and releases the string, and so the primary motion of the string is a relaxation oscillation with a more or less triangular wave form (Fig. 11-19). Such a wave form is not "simple"; it contains many overtones, all of which are among the natural frequencies of the string. The string vibrates simultaneously in many modes, each of which has a node at each end. The vibrations of the string are transmitted through the bridge to the belly of the violin, which in turn sets into vibration the enclosed air column. The sound wave spreads out into the room from the belly and the back of the violin, as well as from the vibrating air column enclosed in the body of the violin.

The making of music is an art. The violin-maker contributes to the art by constructing the resonator in such a way that certain overtones are accentuated, and others are weakened or suppressed. The performer contributes to the art by drawing the bow across the string at the right place, with the right pressure and speed, to further modify the overtone structure of the resulting sound. Although perhaps neither the maker nor the artist can describe his procedure in analytical terms, the goal is the efficient transfer of sound energy from the string into the surrounding air, and the modification of the relaxation oscillation of the generator, by means of the resonator, into a wave form having pleasing overtone structure. To play a melody, the artist changes the length of the

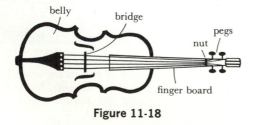

Figure 11-18

string by pressing a finger down somewhere between the nut and the end of the finger board. The distance between the finger and the bridge is one-half the fundamental wavelength. As the string is shortened, the fundamental frequency and all the overtones are raised in frequency, but the overtone structure remains about the same.

The four strings are tuned in fifths (Sec. 11-5), with each string's fundamental frequency three-halves that of the next lower string. The musician makes good use of the phenomenon of beats in tuning his instrument. First he tunes his A string to 440 Hz by comparison with a tone of standard pitch, played perhaps on a piano, or on the oboe of a symphony orchestra. The violinist adjusts the tension of his string, by means of the peg, until he hears no beats between his A and that of the standard, when both are sounded together. To tune his E string, he also uses beats, this time between overtones. The desired frequency is $\frac{3}{2} \times 440 = 660$ Hz. As seen from Table 11-2, the out-of-tune E produces 2 beats/s between the 1320 Hz overtone of the A and the 1322 Hz overtone of the E; other overtones produce 4 beats/s, 6 beats/s, etc. This makes a sensitive way of adjusting the two strings to an exact $3:2$

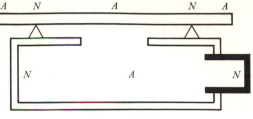

Figure 11-20

ratio of frequencies. The table also illustrates the basis of harmony: if the two tones are in a simple ratio ($3:2$ in this case), the lack of audible beats* between their overtones is very pleasing.

Among musical instruments are those with generators consisting of strings, air columns, vibrating bars, reeds, membranes, and plates. Resonators are usually air columns, although in the piano the resonator is a large thin wooden surface. The resonators may be untuned, as in the violin or piano, where no one frequency should be emphasized; or tuned, as for the dinner gong pictured in Fig. 11-20. In this gong the generator is a flat bar, supported at two nodal points near, but not at, the ends. For a solid bar, the distance from node to antinode is not constant, which indicates that the frequencies of the overtones are not integral multiples of the fundamental. That is, the overtones are not harmonics (Sec. 11-2); this lack of simple relationship between the overtones of a bar of metal accounts for the "clanging" sound when such a bar is dropped on a concrete floor. In order to remove the inharmonious overtones, the air chamber (a closed pipe) is adjusted in length (tuned) so that the fundamental frequency of the pipe resonates with the fundamental frequency of the bar. As seen from the table at the top of the following page, only the fundamental frequency of the bar is in resonance with any of the natural frequencies of the pipe; hence the

TABLE 11-2
Frequencies of fundamentals and overtones in violin strings

A string, Hz	E string in tune, Hz	E string slightly out of tune, Hz
440	660	661
880	**1320**	**1322**
1320	1980	1983
1760	**2640**	**2644**
2200	3300	3305
2640	3960	3966
.	.	.
.	.	.
.	.	.

*Beats between widely separated overtones in Table 11-2, such as 880 Hz and 660 Hz, are too rapid to be heard as distinct fluctuations in loudness.

NATURAL FREQUENCIES OF BAR, HZ	NATURAL FREQUENCIES OF CLOSED PIPE, HZ
400	400
1104	1200
2172	2000
.	.
.	.

total effect is a pure tone, of the single frequency 400 Hz. We may think of the resonator as being a selective amplifier, reinforc-

ing and building up just one out of the many frequencies present in the complex vibration of the bar.

We have discussed only a relatively few aspects of wave motion and sound, preferring to go deeply into fundamentals rather than superficially into applications. We hope that with this discussion of basic ideas you will be able to read further, according to your interests, in the fields of acoustics, musical sounds, speech and hearing, and ultrasonics.

■ SUMMARY

When two identical waves, having the same frequency, wavelength, and speed, travel in opposite directions through a medium, the resultant disturbance can be found by the superposition theorem. The combined effect of the two waves is to produce a stationary wave. Certain points in the stationary wave, called nodes, remain at rest, and other points, called antinodes, undergo vibration with maximum amplitude. The distance from node to node is $\frac{1}{2}\lambda$; from antinode to antinode it is $\frac{1}{2}\lambda$; from node to antinode it is $\frac{1}{4}\lambda$.

For a vibrating string, the wave velocity depends on the tension and the mass per unit length. The string can vibrate in 1,2,3,4, . . . segments, always with a node at each end. A vibrating body has many possible modes of vibration. The lowest frequency of vibration is called the fundamental frequency, and the higher frequencies are called overtones. An overtone frequency which is an integral multiple of the fundamental frequency is called a harmonic. The natural frequencies of vibration for a string are 1,2,3,4, . . . times the fundamental frequency. The modes of vibration of a column of air in a pipe are such that a closed end is a node and an open end is an antinode. If the pipe is open at both ends, the natural frequencies are 1,2,3,4, . . . times the fundamental frequency. If the pipe is closed at one end, the natural frequencies are 1,3,5,7, . . . times the fundamental frequency.

Resonance is the building up of a large vibration by repeated application of small impulses whose frequency equals one of the natural frequencies of the resonating body.

Musical sounds are characterized by the psychological sensations of pitch, loudness, and quality. Although there is some interaction between these sensations, to a large extent they depend upon frequency, intensity, and overtone structure in that order. Intensity is defined as power per unit area; differences of intensity level are measured in decibels. A musical instrument consists of a generator and a resonator. Most instruments vibrate simultaneously in many modes, leading to a complex overtone structure which is characteristic of the instrument. Two tones differ in pitch by an octave if their frequencies are in the ratio $2:1$; they differ by a fifth if their frequency ratio is $3:2$.

■ CHECK LIST

interference	fundamental frequency	intensity
stationary wave	overtone	decibel
node	harmonic	overtone structure
antinode	resonance	octave
$v_w = \sqrt{F/(m/L)}$	pitch	fifth
natural frequency	loudness	major third
mode of vibration	quality	minor third

■ QUESTIONS

11-1 What are the conditions necessary for the production of stationary waves?

11-2 A glass of water is on a kitchen table, and the floor of the room vibrates because of a laundry dryer in action. For certain speeds of rotation of the dryer, stationary concentric ripples are observed on the surface of the water. Explain.

11-3 Give an example of constructive interference.

11-4 A high-speed motion picture is taken of the string of a double bass which is being played loudly. Would you expect any frame of the sequence of photographs to show the string absolutely straight?

11-5 Observe the construction of piano strings, and try to decide whether the mass per unit length for a "low" string is the same as for a "high" string. Why is the "low" string not of solid construction?

11-6 State three factors on which the fundamental frequency of a string depends.

11-7 Are overtones of all musical sounds harmonics?

11-8 For a certain pipe, two out of the many overtones produced are separated by an octave. (*a*) Could this be an open pipe? (*b*) Could it be a closed pipe?

11-9 In what physical respects does a loud, high flute tone differ from a soft, low violin tone?

11-10 Does a clarinet behave like an open pipe or a closed pipe? (See Table 11-1.) What about a French horn?

11-11 Describe a musical instrument in which resonance is desirable. Describe a musical instrument in which resonance is undesirable.

11-12 For the human voice, considered as a musical instrument, what is the generator? What is the resonator? How can one distinguish between the voices of a soprano and a tenor?

11-13 Why does a column of marching soldiers "break step" when crossing a bridge?

11-14 Make the following experiments on a piano: press down the key of middle A without sounding the tone, thus releasing the damper for this string only. While holding the key down, (*a*) strike vigorously an A which is one or two octaves lower, and explain why the higher string starts to vibrate; (*b*) repeat the experiment, still holding the A key down and striking a lower A♭ or A♯; explain your result. (*c*) What if the lower A key is held down and the upper A key is struck?

11-A1 What is the velocity of transverse waves on a string of length 4 m which has a mass of 0.01 kg and is stretched with a tension of 16 N?

11-A2 Calculate the speed of transverse waves on a rope 40 ft long, stretched with a tension of 9 lb. The rope weighs 5 lb. (*Hint:* Use W/g for mass.)

11-A3 Stationary waves are produced on a string for which the velocity of transverse waves is 700 ft/s. The frequency of vibration is 600 Hz. How far apart are the nodes?

11-A4 A source vibrating with frequency 60 Hz sets up stationary waves on a string. The nodes are 20 cm apart. What is the wave velocity?

11-A5 What are the first three natural frequencies of a string whose fundamental frequency is 200 Hz?

11-A6 A cello A string 75 cm long is stretched with sufficient tension so that the fundamental frequency is 220 Hz. What is the velocity of transverse waves on this string?

11-A7 What is the fundamental frequency of an 85 cm pipe, open at both ends, in which the speed of sound is 340 m/s?

11-A8 The fundamental frequency of a closed pipe is 200 Hz, and the speed of sound is 1120 ft/s. How long is the pipe?

11-A9 What is the frequency of the next-to-the-lowest natural frequency of a closed pipe 2.5 ft long on a day when the speed of sound is 1100 ft/s?

11-A10 (*a*) A pipe open at both ends is 1.7 m long, and the speed of sound is 340 m/s. What is the fundamental frequency of the pipe? (*b*) If the pipe is now closed at one end, what does the fundamental frequency become?

11-A11 What are the first three natural frequencies of the string of Prob. 11-A6?

11-A12 What are the first three natural frequencies of the pipe of Prob. 11-A7?

11-A13 What are the first three natural frequencies of the pipe of Prob. 11-A8?

11-A14 The lowest tone of a violin has a frequency of 192 Hz. What is the frequency of a tone which is three octaves higher?

11-A15 Derive Eq. 11-3 from Eq. 11-2.

11-A16 Use Fig. 11-15 to determine the frequency below which the average ear cannot hear a sound whose intensity level is 30 dB above that of the most easily heard sound.

11-B1 Make a dimensional check for Eq. 11-1.

11-B2 A wire of length L has circular cross section of radius r, and the density of the material is d. The wire is stretched with a force F. (*a*) Derive a formula for the velocity of transverse waves on the wire. (*b*) Derive a formula for the frequency of the nth natural frequency of this wire. (*c*) Make a dimensional check for your answers to parts (*a*) and (*b*).

11-B3 A clothesline of total mass 1 kg is stretched between posts 10 m apart. It is observed that when one pole is struck by a lawnmower, the transverse pulse reaches the other pole in 0.2 s. What is the tension in the rope?

11-B4 A construction engineer wishes to determine the tension in an aluminum power cable, of cross section 5 cm², which is stretched between two towers. The length of cable is 200 m. A workman strikes the wire a sharp blow at one end and feels the return of the transverse wave pulse 4.00 s later. (*a*) What is the tension in the cable? (*b*) How much did the wire stretch while it was being hung? (Assume constant tension throughout; use data from tables in Chaps. 2 and 9.)

11-B5 In Example 11-2, what tension in the string would make it vibrate in 5 segments with frequency 200 Hz? Express your answer in megadynes.

11-B6 A string 100 cm long has mass 200 g and is stretched with a force of 450 megadynes. (a) Draw a diagram of the locations of nodes and antinodes for the string when it is vibrating in its next-to-the-lowest natural frequency.   (b) Calculate the frequency of this mode of vibration.

11-B7 To what tension should a violinist adjust his A string in order to tune its fundamental frequency to 440 Hz? The distance from bridge to nut is 30 cm, and the string's mass is 2 g.

11-B8 How far, and in which direction, should a cellist move her finger to adjust an out-of-tune A from 449 Hz to an in-tune 440 Hz? The string is 68 cm long, and the finger is 20 cm from the nut.

11-B9 When the tension is 9 N, a string 150 cm long has a fundamental frequency of 100 Hz. (a) What is the mass of the string?   (b) With what tension must the string be stretched so that it vibrates in three segments?

11-B10 A string under tension 800 N, of mass per unit length 0.00500 kg/m, has many resonant frequencies; one such frequency is 400 Hz and the next higher frequency is 500 Hz. How long is the string?

11-B11 Two identical mandolin strings under tension 200 N are sounding tones of frequency 500 Hz. The peg of one string slips slightly, and the tension becomes 190 N. How many beats are heard?

11-B12 On a day when the velocity of sound is 340 m/s, a saxophonist adjusts the keys of his instrument so that the distance from reed to first open hole is 85 cm. In operation, sound waves are reflected from the reed as from a fixed surface so that the saxophone behaves like a closed pipe. What are the frequencies of the fundamental and first two overtones?

11-B13 A bugler, by adjusting his lips correctly and blowing with the proper pressure, can cause his instrument to produce a sequence of tones, among which are the following: . . . 440, 660, 880, 1100, . . . Hz, all without changing the length of the air column.   (a) Does the bugle behave like an open pipe or a closed pipe?   (b) What is the effective length of this bugle? (Use 1100 ft/s for the velocity of sound.)

11-B14 Two identical organ pipes, both at 27°C, are sounding tones of frequency 600 Hz. The temperature of one pipe goes up to 33°C; how many beats per second are heard? (See p. 244.)

11-B15 Open organ pipes, 40.00 cm and 40.30 cm long, are sounded together on a day when the speed of sound is 340 m/s. How many beats per second are heard?

11-B16 A space explorer who does a little anthropology on the side observes Ronald Qrxxt, a native of a distant planet, who emits a mating call of frequency 200 Hz. This tone resonates with the fundamental frequency of the future Mrs. Qrxxt's intercranial cavity, which is a tube 60 cm long closed at one end, located midway between two of her heads. (a) Calculate the speed of sound in the planet's atmosphere.   (b) If Qrxxt is successful on a day when the speed of sound on the planet is as calculated in (a), what must be the speed of sound on a warmer day if his rival, Donald Prxxg, is to succeed in capturing the attentions of the lady in question? Prxxg's mating call is at 210 Hz.

11-B17 A metal rod 200 cm long is clamped at its center and set into longitudinal vibration. (a) Make a diagram of the locations of nodes and antinodes for the first three modes of vibration.   (b) Are the natural frequencies in the ratio $1:2:3:4: . . .$ or $1:3:5:7: . . .$? (c) When the rod is vibrating in its fundamental mode, the tone produced is in tune with a whistle of frequency 1250 Hz. What is the wave velocity of the compression waves in the rod?

11-B18 (*a*) Make a diagram of the locations of nodes and antinodes for the fundamental mode of vibration of the resonant air column in the gong pictured in Fig. 11-20, and explain with the aid of the diagram why the frequency 800 Hz does not occur in the table on p. 254. (*b*) Using 340 m/s for the velocity of sound in air, calculate the total length of the resonant air column, assuming it to be vibrating in its fundamental mode.

11-B19 A copper rod 0.6 m long is clamped at its center and stroked longitudinally so that its fundamental mode is in tune with a loudspeaker vibrating at 2875 Hz. (*a*) Sketch the locations of nodes and antinodes for the vibrating rod. (*b*) Calculate the speed of sound in copper. (*c*) Calculate the bulk modulus of copper. (Use data from Table 2-1.)

11-B20 A wire whose density per unit length is 10^{-2} kg/m is stretched with a tension of 100 N. The wire's fundamental frequency is in tune with the first overtone of a pipe 1.5 m long which is closed at one end. The speed of sound in air is 340 m/s. How long is the wire?

11-B21 A jet airplane inside a hangar is making a considerable noise (see p. 247), and sound energy is flowing out through a door of dimensions 10×2 m. If the energy flowing through the door were converted efficiently into electrical energy, could a 40 W lamp be operated at full brilliance?

11-B22 One sound is 400 times as intense as another sound. What is the difference in intensity level of these two sounds, in dB?

11-B23 By soundproofing a living room near an airport, a noise reduction of 43 dB is obtained. What ratio of intensities does this correspond to?

11-B24 The total effective area of the auricles of a listener is about 50 cm². How much power (in watts) enters the ears of a person enjoying a concert where the intensity level is 60 dB above the threshold of hearing? How long would be required for a microjoule of energy to enter his ears?

11-B25 Under favorable conditions of loudness and frequency, the ear can just determine two sounds to be of different intensity if one is 0.6 dB louder than the other. Prove that to be perceptibly louder, a sound's intensity must be increased by about 15%. (*Hint:* Calculate the ratio I_2/I_1.)

11-C1 A tuning fork of frequency 512 Hz is held above a vertical tube partly filled with water. For what lengths of air column will resonance occur? (Take velocity of sound as 340 m/s.)

11-C2 The fundamental frequency of vibration of a string depends on its mass, length, and tension according to the equation $\nu = Km^x L^y F^z$. (*a*) Determine the form of the equation, finding x, y, z, by the method of dimensional analysis (Sec. 10-3). (*b*) Derive the equation, using the known formula for the velocity of a transverse wave of a string, and thus find the value of the dimensionless constant K.

11-C3 A wire 3 ft long weighing 0.01 lb is stretched between two points, and an open pipe 4 ft long is placed nearby. The velocity of sound in air is 1100 ft/s. What should be the tension in the string in order for the 4th overtone of the string to be in resonance with the 3rd overtone of the pipe?

11-C4 A guitar string under tension 200 N, vibrating in its fundamental mode, gives 5 beats/s with a tuning fork. The player increases the string tension to 242 N and again gets 5 beats/s. What is the frequency of the tuning fork?

11-C5 A sudden handclap takes place in front of the opening of a tube of length L, closed at the far end. The pulse travels back and forth in the pipe, sometimes changing from condensation to rarefaction and vice versa. As implied by the discussion in the first paragraph of Sec. 11-3, such changes of phase take place at the open end of the pipe but not at the closed end. Show that, after the pulse has made two round trips, it is again starting down the pipe as a condensation. Thus prove that $\lambda = 4L$ for a closed pipe in its fundamental mode.

11-C6 What must be the stress in a stretched brass wire in order for the speed of longitudinal waves to be equal to 100 times the speed of transverse waves? (*Hint:* Solve the problem first in symbols.)

For Further Study

11-8 Derivation of equation for velocity of transverse wave on a string

For a transverse wave on a string, we have stated in Sec. 11-2 that the wave velocity is $v_w = \sqrt{F/(m/L)}$, where F is the tension and m/L is the mass per unit length. To prove this we consider a wave crest which is an arc of a circle. Of course, no part of the string itself is moving in a circular path. Individual particles of the string move up and down, transversely to the direction of propagation of the wave. To derive the wave velocity formula, we find it convenient to change our frame of reference. Instead of letting the crest move past us along a stationary string, let us climb aboard an observation car which has the same velocity as the wave (Fig. 11-21). Now, as we ride along, the crest is always just abreast of us, seemingly stationary, and the string is moving backward with a velocity

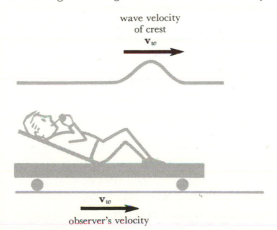

wave velocity
of crest
$\mathbf{v}_w$

$\mathbf{v}_w$
observer's velocity

Figure 11-21 Wave on a string, viewed from a moving frame of reference.

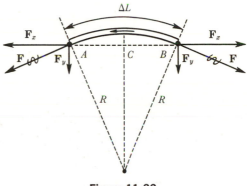

Figure 11-22

v_w relative to us, like a snake going around a tree. The advantage of this new viewpoint is that now we have a physical body (string particles) moving in an arc of a circle whose center is fixed. We can apply the equations of uniform circular motion to the particles of the string.

Consider a short segment of the string having length ΔL (Fig. 11-22). The mass of the segment is (mass per unit length)(length of segment) $= (m/L)\,\Delta L$. The net force on the segment is caused by the curvature of the string. At point A, the force $\mathbf{F}$ has a downward component which can be calculated by use of similar triangles:

$$\frac{F_y}{F} = \frac{\frac{1}{2}\Delta L}{R}$$

An important approximation has been made here: we assume that the string is not sharply curved, so that the chord $\overline{AB}$ is equal to the arc $\overparen{AB}$, and hence AC is approximately equal to $\frac{1}{2}\Delta L$. Solving for F_y gives $F_y = F(\Delta L/2R)$

for the downward component of force acting at each end of the segment. The total net downward force is, therefore, twice this amount, or $(F\,\Delta L)/R$. The horizontal components of the forces at the ends of the segment cancel each other. We now apply Newton's second law and solve for the wave velocity v_w. The centripetal acceleration is v_w^2/R, so

$$\text{net } F = ma$$

$$\frac{F\,\Delta L}{R} = \left(\frac{m}{L}\,\Delta L\right)\left(\frac{v_w^2}{R}\right)$$

$$v_w = \sqrt{\frac{F}{m/L}}$$

Our proof has several interesting features. First, we are limited to disturbances which are not too radically curved. This is not a serious limitation. In a typical case, a guitar string 20 in. long might be pulled aside $\frac{1}{4}$ in. at the center and then released. The string is never very much disturbed from its normal (straight) condition. Second, we note that in a complex wave form different parts of the string have different radii of curvature. However, since the radius of curvature cancels out and does not enter into the final result, we see that all parts of a disturbance travel at the same velocity. This means that a complex wave form is preserved in shape as it moves along the string. Another way of stating this same conclusion is to say that the wave velocity is independent of the wavelength of the disturbance.

■ REFERENCES

1 Henry, George E., "Ultrasonics," *Sci. American,* May, 1954, p. 54.

2 Bernstein, Joseph, "Tsunamis," *Sci. American,* Aug., 1954, p. 60. A description of water waves in the Pacific Ocean having wavelengths of from 100 to 600 mi and periods of vibration from $\frac{1}{4}$ to 1 h. See also the letter by William G. Van Dorn on p. 2 of the Jan., 1955, issue of *Sci. American.* A related phenomenon, the exceptionally high tides in the Bay of Fundy, is discussed briefly in the *Natl. Geograph. Mag.,* Aug., 1957, p. 156.

3 Békésy, Georg von, "The Ear," *Sci. American,* Aug., 1957, p. 66. An article of special interest to biologists. Physical mechanisms are discussed.

4 Saunders, Frederick A., "Physics and Music," *Sci. American,* July, 1948, p. 33.

5 Miller, Franklin, Jr., "Musical Sounds and the Science of Music," Chap. 33 in *Analytical Experimental Physics,* 2nd ed., Michael Ference, Jr., Harvey B. Lemon, and Reginald J. Stephenson, Univ. of Chicago Press, 1956.

6 Hagenow, C. F., "The Equal Tempered Musical Scale," *Am. J. Phys.,* **2,** 81 (1934).

7 Williamson, Charles, "Intonation in Musical Performance," *Am. J. Phys.,* **10,** 171 (1942).

8 Fletcher, Harvey, "The Pitch, Loudness, and Quality of Musical Tones," *Am. J. Phys.,* **41,** 215 (1946).

9 Herman, Robert, "Observations on the Acoustical Characteristics of the English Flute," *Am. J. Phys.,* **27,** 22 (1959). The English flute is a recorder.

10 Benade, A. H., "The Physics of Wood Winds," *Sci. American,* Oct., 1960, p. 145.

11 Hutchins, C. M., "The Physics of Violins," *Sci. American,* Nov., 1962, p. 79.

12 Blackman, E. D., "The Physics of the Piano," *Sci. American,* Dec., 1965, p. 88.

13 Knudsen, V. O., "Architectural Acoustics," *Sci. American,* Nov., 1963, p. 78.

14 Miller, Franklin, Jr., *Tacoma Narrows Bridge Collapse* (film).

Some of the most spectacular achievements of recent years have been concerned with the exploration of the upper atmosphere of the earth. Even at low altitudes, airplanes and missiles travel through a resisting air at speeds undreamed of a few decades ago, propelled by jets and rockets which make use of high-speed gases. Under the ocean, advanced types of submarines use the best possible streamlining as they cruise or remain at rest, securely suspended by the buoyant force of the water. The study of the ocean itself is proceeding at an increased pace, as scientists chart the flow of undersea "rivers" and study effects of waves and tides. All these and many other areas of applied science are concerned with fluids; in this chapter we study some basic facts about fluids at rest and in motion.

12-1 Definition of a fluid

The general term *fluid* includes any substance that has no rigidity; thus liquids and gases are fluids. The bulk modulus (Sec. 9-1) of a fluid correlates change in volume with change in pressure. A fluid cannot sustain a static (steady-state) shearing stress, and the shear modulus is zero for any fluid. Thus a fluid does not have a definite length or shape. We can make a distinction between liquids and gases, based upon the common observation that a liquid has a surface. In other words, the cohesive forces between the molecules of a liquid are sufficient to give a definite volume to a quantity of material. Gases have no such "natural" volume, but expand to fill completely the container in which they are placed. Both liquids and gases are more easily compressed than are solids, as shown by the smaller bulk modulus values in Table 9-1, and, as a result of gravitational forces, the density of a given sample of fluid is not strictly constant. It is well known that the air is more dense at sea level than at 10 000 feet altitude. The bulk modulus of air is so small that the weight of the air above any given level compresses the air beneath it appreciably and so increases its density. To a much smaller degree, the density of sea water at 24 000 feet depth is greater than at the surface.

One of the earliest discoveries about fluids is Archimedes' principle, which describes the phenomenon of buoyancy, or "lift"; it is applicable to both liquids (ships at sea) and

---

Ultrahigh speed photograph of a milk splash just after impact of a falling drop. Because of surface tension, droplets that have broken loose from the splash are spherical. (Harold E. Edgerton, Massachusetts Institute of Technology)

gases (balloons in the air). To study it in detail, however, we need to use the concept of pressure and its relationship to depth.

12-2 Pressure

Pressure is defined as force per unit area, the force understood to be perpendicular to the area:

$$P = \frac{F}{A}$$

Thus an object weighing very little can exert a tremendous pressure if the force acts only on a small surface area. On the other hand, the weight of 5.7×10^{16} tons of air creates only a relatively small pressure on the earth, since the force is spread out over the entire surface of the globe. Pressure units are force units divided by area units, such as N/m^2, dyn/cm^2, lb/ft^2, $lb/in.^2$, etc.[*] For practical purposes, pressures are often expressed as so many atmospheres, where one *atmosphere* (atm) is normal atmospheric pressure,[*] approximately 14.7 $lb/in.^2$, or 1.013×10^5 N/m^2, or 1.013×10^6 dyn/cm^2. Another common unit of pressure, used especially in meteorology, is the *bar,* which is 10^6 dyn/cm^2. A millibar (mbar) is 10^3 dyn/cm^2. The millimeter of mercury (mm Hg) is used in many gas-law problems and is defined as $\frac{1}{760}$ atm. It is the pressure due to a column of mercury 1 mm high. The mm Hg is also known as the *torr,* named for Evangelista Torricelli, who invented the barometer in 1643. The *inch of water* is used in engineering for measuring relatively small deviations from normal atmospheric pressure, as in a household vacuum cleaner. These units can be readily manipulated with the aid of a few conversion factors:

---

[*] The dimensional equation for pressure is [pressure] = $[FL^{-2}]$ or $[ML^{-1}T^{-2}]$.

[*] The normal atmosphere is defined as the pressure at the bottom of a column of mercury exactly 76 cm high, if the mercury is at such a temperature that its density is 13.5951 g/cm^3 and it is located at a point where $g = 980.665$ cm/s^2.

$$
\begin{aligned}
1 \text{ atm} &= 14.7 \text{ lb/in.}^2 \\
&= 1.013 \times 10^5 \text{ N/m}^2 \\
&= 1.013 \times 10^6 \text{ dyn/cm}^2 \\
&= 1.013 \text{ bar} \\
&= 76 \text{ cm Hg} \\
&= 760 \text{ mm Hg} \\
&= 760 \text{ torr} \\
&= 34 \text{ ft of water}
\end{aligned}
$$

EXAMPLE 12-1

What is the pressure on the pavement if a 10 ton truck's weight is supported by 6 wheels, each having 18 $in.^2$ of surface in contact with the concrete?

$$P = \frac{F}{A} = \frac{20\,000 \text{ lb}}{6 \times 18 \text{ in.}^2} = \boxed{185 \text{ lb/in.}^2}$$

EXAMPLE 12-2

Compute the pressure exerted by the stylus on a phonograph record, if the groove has a width of 0.001 in. (one mil) and the recommended stylus force is 0.0157 lb.

We assume that contact is made over a circular area of diameter 0.001 in.

$$P = \frac{F}{A} = \left[\frac{0.0157 \text{ lb}}{\pi (0.0005 \text{ in.})^2} \right]$$

$$= \boxed{2.00 \times 10^4 \text{ lb/in.}^2}$$

Comparing these examples, we see that the pressure due to the stylus is 20 000 $lb/in.^2$, which is more than 100 times that due to the truck!

12-3 Pressure in liquids at rest

The pressure at the bottom of a regular container in which there is some liquid is easy to compute. The volume of liquid is found from the base area A and the height h (Fig. 12-1). If the density of the liquid is d, the weight of the column of liquid is

$$W = mg = (Ahd)g = Ahdg$$

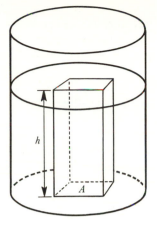

Figure 12-1

The pressure is force per unit area, or $\dfrac{Ahdg}{A}$, whence

$$P = hdg \qquad (12\text{-}1)$$

$$N/m^2 = (m)(kg/m^3)(m/s^2)$$
$$dyn/cm^2 = (cm)(g/cm^3)(cm/s^2)$$

Thus, since g is essentially constant, the pressure at any point in a uniformly dense liquid at rest* depends only on the height and density of the liquid above the point. Suppose two different liquids are used to measure the same pressure. Then $h_1 d_1 g = h_2 d_2 g$, or

$$\frac{h_1}{h_2} = \frac{d_2}{d_1}$$

The height of liquid corresponding to a given pressure is therefore inversely proportional to the density of the liquid. Mercury has a density 13.6 times that of water (Table 2-1); hence 1 cm of mercury is equivalent to 13.6 cm of water.

Equation 12-1 for pressure can be used for the pressure at the bottom of a short column of gas, provided the density of the gas is sufficiently constant. This is *not* true, of course, for the pressure due to the atmosphere, as the following example shows.

---
*If the liquid is moving, then Bernoulli's equation (Sec. 12-8) must be applied.

EXAMPLE 12-3

If the earth's atmosphere were everywhere of the same density as at the surface of the earth (i.e., 1.290 kg/m³), how high would the atmosphere have to extend in order to give rise to the observed atmospheric pressure of 1.013×10^5 N/m²?

From $P = hdg$, we have

$$h = \frac{P}{dg} = \frac{1.013 \times 10^5 \text{ N/m}^2}{(1.290 \text{ kg/m}^3)(9.80 \text{ m/s}^2)}$$

$$= 8.0 \times 10^3 \text{ m} = \boxed{8.0 \text{ km}}$$

An atmosphere only 8 km high would not even cover the top of Mt. Everest, and yet the observed sea-level pressure would be obtained from such an atmosphere if its density were uniform and equal to the sea-level value. We conclude that the density of the atmosphere must taper off at higher altitudes; it is incorrect to use our simple formula $P = hdg$ here.

EXAMPLE 12-4

During a research project, deep-sea underwater photographs were made at a depth of 24 600 feet (7.50 km). (a) What is the pressure at this depth? (b) What is the force on the plane surface of the window of a camera enclosure which measures 12 cm × 15 cm?

(a) We assume that water is sufficiently incompressible so that its density is approximately constant. The specific gravity of sea water is 1.025 (Table 2-1); therefore its density is 1.025 g/cm³.

$$P = hdg$$
$$= (7.50 \times 10^5 \text{ cm})(1.025 \text{ g/cm}^3)(980 \text{ cm/s}^2)$$

$$= \boxed{7.54 \times 10^8 \text{ dyn/cm}^2}$$

This is a pressure of 744 atm.

(b) Force = (pressure)(area)
$$= (7.54 \times 10^8 \text{ dyn/cm}^2)(180 \text{ cm}^2)$$

$$= \boxed{1.36 \times 10^{11} \text{ dyn}}$$

Since 10^5 dyn = 1 N $\approx \frac{1}{4}$ lb, the force on the window is about 10^6 N, or 300 000 lb.

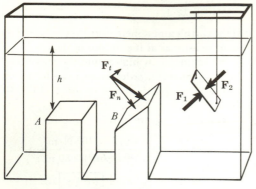

Figure 12-2

Although pressure is a scalar quantity and has no direction, the *force* exerted on a surface immersed in a fluid is a vector, just like any force. If the fluid is not moving, the direction of the force is perpendicular to the surface of the object. This can be demonstrated in an indirect way. Consider two small rectangular sections of the bottom of a container, both at depth h below the surface of a liquid (Fig. 12-2). If the pressure at A were greater than that at B, liquid would flow from A to B, and the liquid would not be motionless. Hence, since the liquid is assumed to be at rest, the pressures at A and B are equal (given by hdg at each point). Next, assume that the force on B is *not* perpendicular to the surface. This would mean that there would be a component $\mathbf{F}_t$ parallel to the surface, as well as a component $\mathbf{F}_n$ normal (i.e., perpendicular) to the surface. Now we have defined a fluid as a substance that has no rigidity, which means that it is unable to withstand a shearing stress. Thus the component $\mathbf{F}_t$ would cause a flow of liquid parallel to the surface. But the liquid was assumed to be at rest; therefore there can be no component of force $\mathbf{F}_t$, and only a normal component $\mathbf{F}_n$ can exist. Since $P = F/A$, the force is given by $F = F_n = PA$, acting perpendicular to the surface in every case. Suppose a small, thin sheet of metal is suspended at an angle under the surface; the fluid exerts two forces on it, $\mathbf{F}_1$ and $\mathbf{F}_2$, each equal in

magnitude to PA. The vector sum of these forces is zero, and the sheet is in equilibrium. These conclusions are equally valid for all fluids at rest, gases as well as liquids.

12-4 Pascal's principle

When the photographic equipment was submerged at great depth in Example 12-4, the tremendous pressure which we calculated was due to the weight of the water above it. In addition, there is a relatively small pressure on a submerged object due to the weight of air above the water. If a change of weather causes an increase in atmospheric pressure, this change, however slight, is transmitted through the water and acts on the submerged camera. This transmission of changes in pressure was studied by Blaise Pascal (1623–1662), a French philosopher and scientist, whose experimental facts may be summarized in *Pascal's principle:*

> **Change of pressure exerted at any point in a confined fluid is transmitted undiminished in all directions to all points in the fluid.**

The hydraulic lift in a garage is a machine (i.e., a device for doing work) based upon Pascal's principle. In the lift (schematically illustrated in Fig. 12-3), the weight of the

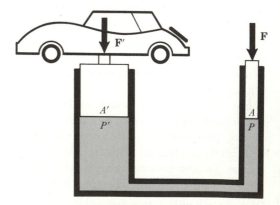

Figure 12-3

car $\mathbf{F}'$ is balanced by a small force $\mathbf{F}$ applied to a piston of small area. The basic idea underlying the operation of such devices is that any increase in pressure at one piston is transmitted undiminished to the other piston. Usually, differences in pressure due to differences in level [which could be found from $\Delta P = (h_2 - h_1)dg$] are negligible compared with the larger pressures built up by the elastic forces of repulsion between the molecules of the fluid. In any case, if the two pistons are at the same level, we can say that $P' = P$; hence

$$\frac{F'}{A'} = \frac{F}{A}$$

and hence

$$F' = \frac{A'}{A} F \qquad (12\text{-}2)$$

If now A' is much greater than A, the force is magnified considerably. Indeed, only the strength and resistance to leakage of the piston chambers limit the forces that can be applied in this manner. P. W. Bridgman (1882–1961) succeeded in designing a leak-proof cylindrical chamber in which the higher the pressure, the greater the efficiency of sealing and the less the leakage. Pressures up to 20 million lb/in.[2] were achieved by Bridgman in this way.[*] We shall use our next example, which deals with the hydraulic press, also as an opportunity to review our earlier study of machines (Sec. 6-14).

EXAMPLE 12-5

In the press shown schematically in Fig. 12-4, the small piston has a radius of 0.500 in., and the large piston has a radius of 8.00 in. The workman operates the press by means of a lever,

---

[*]Magnitude in any field is spectacular, but it is important to note that the 1947 Nobel Prize in physics went to Bridgman not only for exceeding the previous record for pressure; it was the careful planning of experiments and the masterful interpretation of his data in terms of fundamental molecular phenomena that won for the Harvard professor of physics the highest award in his field.

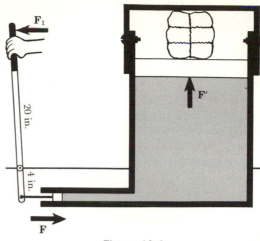

Figure 12-4

with lever arms as shown in the diagram. If he applies a force of 30 lb on the lever, what force is exerted on the bale of scrap paper, assuming 100% efficiency?

The output force of the lever supplies the input force for the press. We have two machines "in series," and the over-all AMA is

$$\text{AMA}_{\text{total}} = \frac{F'}{F_1} = \left(\frac{F'}{F}\right)\left(\frac{F}{F_1}\right)$$
$$= (\text{AMA}_{\text{press}})(\text{AMA}_{\text{lever}})$$

From Eq. 12-2,

$$\text{AMA}_{\text{press}} = \frac{F'}{F} = \frac{A'}{A} = \frac{\pi(8.00 \text{ in.})^2}{\pi(0.500 \text{ in.})^2} = 256$$

Since we are assuming 100% efficiency,

$$\text{AMA}_{\text{lever}} = \text{IMA}_{\text{lever}} = \frac{20 \text{ in.}}{4 \text{ in.}} = 5$$

$$\text{AMA}_{\text{total}} = (256)(5) = 1280$$

$$F' = (\text{AMA})(F_1) = (1280)(30 \text{ lb})$$

$$= \boxed{38\,400 \text{ lb}}$$

Thus a workman can exert more than 19 tons of force by applying a force of only 30 lb. The catch is that the workman will have to move the lever many inches to cause even 0.01 in. of motion of the big piston. Energy is conserved, so that, if the big piston goes up 0.01 in. ($\frac{1}{1200}$ ft), the work done on the bale is

$$(38\,400 \text{ lb}) \left(\frac{1}{1200} \text{ ft}\right) = 32 \text{ ft} \cdot \text{lb}$$

In order to do this work by applying a force of 30 lb, the workman must move his end of the lever 32 ft·lb/30 lb = 1.07 ft = 12.8 in. This disparity of distances is, of course, the essential feature of all machines that magnify force.

As a further illustration of Pascal's principle and the pressure-depth formula (Eq. 12-1), we calculate the pressure at various points in the somewhat complicated-looking apparatus of Fig. 12-5.

EXAMPLE **12-6**

Calculate the pressure at points C and B' in the apparatus of Fig. 12-5. The liquid is alcohol of density 0.816 g/cm³; the enclosed volumes contain air. Atmospheric pressure on the day of the experiment is 1000 millibars.

The pressure at A' is the same as that at A (no change in level). For the liquid column $A'C$ we have

$$\Delta P = hdg = (60 \text{ cm})(0.816 \text{ g/cm}^3)(980 \text{ cm/s}^2)$$
$$= 48 \times 10^3 \text{ dyn/cm}^2 = 48 \text{ mbar}$$

Hence the pressure at C is

$$P_C = P_{A'} + \Delta P$$
$$= 1000 \text{ mbar} + 48 \text{ mbar}$$
$$= \boxed{1048 \text{ mbar}}$$

The pressure at B' is the same as that at B (the difference in level causes a negligible ΔP because d in hdg is negligibly small for air). The

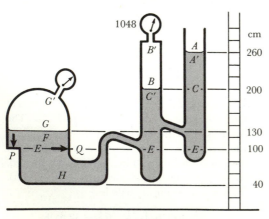

Figure 12-5

pressures at B, C', and C are all equal since these points are at the same level. Hence

$$P_{B'} = P_B = P_{C'} = P_C$$
$$= \boxed{1048 \text{ mbar}}$$

12-5 Measurement of pressure

Many common devices for measurement of pressure make use of a combination of Pascal's principle and the pressure-depth relationship, as described for Fig. 12-5. The liquid-filled *manometer* may take various forms. For instance, in measuring the heating value of fuel gas, it is important to know the pressure at which the gas is passing through the flowmeter. Since the pressure is not very different from atmospheric pressure, an open-tube water manometer is often used. As drawn in Fig. 12-6 the pressure is in excess of atmospheric pressure by an amount equal to hdg. The *sphygmomanometer**—the device the doctor uses to determine your blood pressure—uses two mercury columns to measure the pressure applied to an artery by means of a close-fitting, hollow jacket which can be expanded as air is pumped in. When the external pressure equals or exceeds the maximum pressure in the artery (which occurs during the part of the heartbeat cycle called systole), the doctor hears no rhythmic heartbeat. This pressure is expressed in mm Hg, as read from the difference in level of the two columns of mercury. Maximum pressures usually range from 120 to 180 mm

---
* From the Greek word *sphygmos*, meaning pulse.

Figure 12-6 Open-tube manometer; pressure nearly equal to atmospheric pressure.

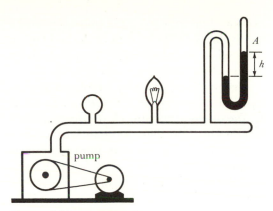

Figure 12-7 Closed-tube manometer; pressure nearly zero. The sealed part of the manometer tube (*A*) is evacuated.

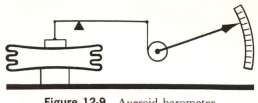

Figure 12-9 Aneroid barometer.

Hg (120 to 180 torr), depending on age, sex, weight, and many other factors.

The following examples of pressure-measuring devices illustrate the great variety of such instruments:

1. For measuring pressures that differ only slightly from an absolute vacuum, the *closed-tube manometer* is used, usually with mercury as the indicating liquid. As drawn in Fig. 12-7, the manometer is indicating a small pressure; the levels would be equal if the experimental system were completely evacuated.

2. The *mercury barometer* (Fig. 12-8) measures atmospheric pressure. Although it is simple in

principle—essentially a closed-tube manometer—a user must take many precautions if his readings are to be precise and repeatable. Temperature increases cause the mercury to expand, so that its density decreases. Thus, according to $P = hdg$, a longer column of mercury is supported by a given atmospheric pressure, and normal atmospheric pressure would seem to be more than 760 mm Hg unless a temperature correction is made. The expansion of the scale used to measure heights (usually of brass) must also be taken into account, as well as the curvature of the top surface of the mercury column due to surface tension (Sec. 12-7).

3. The *aneroid barometer* (Fig. 12-9) depends upon the elastic properties of a sealed evacuated can acted on by external (atmospheric) pressure. A sensitive lever arrangement magnifies the motion and on some instruments moves a pen point up and down a slowly rotating drum. The aneroid barometer can be made so sensitive that a measurable decrease in reading results from merely lifting the instrument from the floor to a desk top.

EXAMPLE **12-7**

What is the fractional change in pressure registered by an aneroid barometer when it is raised vertically 1.00 m?

The density of the air is essentially constant for a change of altitude of only 1 m, so the pressure-depth equation can be used in this problem.

The density of air is 1.29×10^{-3} g/cm³.

$$P = hdg$$
$$= (100 \text{ cm})(1.29 \times 10^{-3} \text{ g/cm}^3)(980 \text{ cm/s}^2)$$
$$= 126 \text{ dyn/cm}^2$$

Since normal atmospheric pressure is about 1.013×10^6 dyn/cm², the fractional change is

$$\frac{126 \text{ dyn/cm}^2}{1.013 \times 10^6 \text{ dyn/cm}^2} = \boxed{0.00012}$$

This is about $\frac{1}{80}$ of 1%!

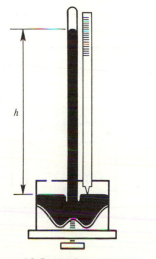

Figure 12-8 Mercury barometer.

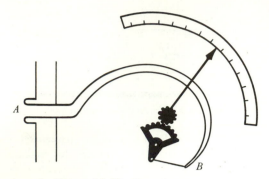

Figure 12-10 Bourdon gauge.

4. Water or steam pressure is usually measured by a *Bourdon gauge* (Fig. 12-10), in which the sealed chamber AB uncoils slightly, actuating a pointer through a system of gears and levers. This is somewhat like a more rugged version of the aneroid barometer.

Any physical property which depends upon pressure could be used for measurement; many properties such as density and electrical resistivity depend slightly upon pressure, and so can be used in instruments for measuring high pressures.

12-6 Archimedes' principle

We are now in a position to make a close study of Archimedes' principle, which has numerous applications. When a body is wholly or partially immersed in a fluid, the body seems to weigh less. Archimedes' principle states that *the loss of weight equals the weight of the fluid displaced by the body.* This apparent loss of weight may be thought of as an upward *buoyant force* (BF) supplied by the fluid. A body *floats* if its weight exactly equals the weight of the displaced fluid, for then the net force on the body is zero. A body *sinks* if the BF on it is less than its weight. However, there is still a BF on a body even if it is not floating. For instance, victims of polio exercise in pools where the BF of the water helps make it possible for weakened muscles to move the limbs. In precise weighing, it is necessary to correct for the buoyancy of the air displaced by the volume of the body being weighed on one pan, and by the brass weights on the other pan (see Prob. 12-C6).

A geometrical proof for Archimedes' principle for a certain special case can be based on the differences between pressures at various depths. Consider the solid block shown in Fig. 12-11, immersed in fluid of uniform density d. The pressure at depth h_1 is $h_1 dg$, and the downward force on the top of the block is $F_1 = Ah_1 dg$. Likewise, the upward force on the bottom of the block is $F_2 = Ah_2 dg$, and, because h_2 is greater than h_1, F_2 is greater than F_1. The net upward force is $F_2 - F_1 = A(h_2 - h_1)dg$, but since $A(h_2 - h_1)$ is the volume of the block, we see that $A(h_2 - h_1)dg$ is the weight of the fluid that would have occupied the space now filled by the block. Hence the BF equals the weight of fluid displaced.

This proof is far from being a general one, but it can be improved and made satisfactory. For instance, we could extend it to a body of irregular shape by dividing the body into a large number of small vertical cylinders each of very small cross-sectional area. Also, we assumed the fluid's density to be uniform; the proof for a fluid of variable density can

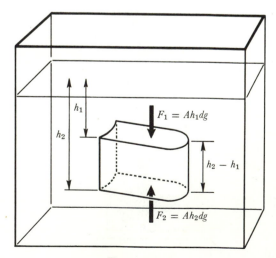

Figure 12-11

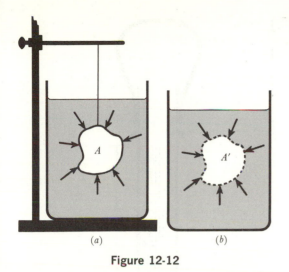

(a)   (b)

Figure 12-12

be carried out using advanced mathematical methods. However, Archimedes himself doubtless was led to his principle by physical reasoning which is elegant in its simplicity and yet correct and quite general. Consider the irregularly shaped body in Fig. 12-12a immersed in a fluid whose density may vary in any way whatever. The forces acting on the surface of the body are due to the surrounding fluid; these forces would be the same whether the body were replaced by an identically shaped one of iron, lead, stone, or whatever. In particular, let us imagine the immersed body to be replaced by a body A' of exactly the same size and shape as A, composed of the same material as the surrounding fluid. The buoyant force is just sufficient to support A', since Fig. 12-12b represents merely a uniform bowl of fluid in equilibrium. Hence the vector sum of all the forces of the fluid on A (the same forces as on A') must be just sufficient to balance the fluid which has been displaced. This is Archimedes' principle.

It is said that Archimedes (who died in 212 B.C.) came upon the principle while in his bath, cogitating on how he might detect a fraudulent crown.[*] Gold is one of the

densest metals (specific gravity 19.3; see Table 2-1), and any substance such as copper or silver which might be used as an alloy would necessarily make the specific gravity of the crown lower than that of pure gold. In modern terminology, the problem resolved itself into finding the density of an irregularly shaped object (a crown) whose volume could not easily be found by measurement.

EXAMPLE 12-8

A crown weighs (4280×980) dyn in air, and when immersed in water it weighs (4000×980) dyn. What is the specific gravity of the material?

Let V = the volume of the crown. By Archimedes' principle,

$$BF = \text{weight of fluid displaced}$$
$$(4280 - 4000)(980) \text{ dyn}$$
$$= V(1.00 \text{ g/cm}^3)(980 \text{ cm/s}^2)$$
$$V = 280 \text{ cm}^3$$

$$d = \frac{\text{mass}}{\text{volume}}$$

$$= \frac{4280 \text{ g}}{280 \text{ cm}^3} = \boxed{15.3 \text{ g/cm}^3}$$

Hence the specific gravity of the material is 15.3. Since pure gold has a specific gravity of 19.3, the crown has been alloyed with a lighter substance such as silver (specific gravity 10.5) or copper (specific gravity 8.9).[*]

EXAMPLE 12-9

An irregular object weighs 200×980 dyn in air, 160×980 dyn when immersed in water, and 170×980 dyn when immersed in oil. Calculate (a) the density of the object; (b) the density of the oil.

(a) The loss of weight in water is 40×980 dyn. By Archimedes' principle this is the weight of the displaced water. The mass of the displaced

---

[*] Perhaps he observed the apparent partial loss of weight of his arms and legs while in the bath. Whether he

actually ran down the streets of Syracuse clad only in a towel and shouting "Eureka!" ("I have found it!") is open to question.

[*] Pure gold is rather soft, and some degree of alloying is necessary for jewelry, coins, and so on. Gold coinage, consisting of 90% gold and 10% copper, has a specific gravity of 17.2.

water is thus 40 g. The volume of displaced water is

$$V_{\text{water}} = \frac{40 \text{ g}}{1.00 \text{ g/cm}^3} = 40 \text{ cm}^3$$

Since this is also the volume of the irregular object, the object's density is

$$d = \frac{m}{V} = \frac{200 \text{ g}}{40 \text{ cm}^3} = \boxed{5.00 \text{ g/cm}^3}$$

(b) The weight of displaced oil is the BF, given by $(200 - 170)980$ dyn; the mass of displaced oil is 30 g.

$$d_{\text{oil}} = \frac{m}{V} = \frac{30 \text{ g}}{40 \text{ cm}^3} = \boxed{0.75 \text{ g/cm}^3}$$

When a body floats in a liquid, it sinks into the liquid just far enough so that the buoyant force equals its own weight. It is then in equilibrium, and net $\mathbf{F} = 0$. If a floating log is pushed down into the water slightly, the buoyant force is increased because more water is displaced. The net force is now upward, and the log accelerates upward (by Newton's second law) and returns to its equilibrium position. In fact, if a floating block has uniform cross section, the restoring force is proportional to the distance below or above the equilibrium position, and hence the motion is SHM (Sec. 9-3). The familiar battery tester (*hydrometer*) measures specific gravity by means of a float whose equilibrium position indicates the specific gravity of the liquid (Fig. 12-13). Since the specific gravity of the acid in a fully charged lead battery is about 1.30, whereas that of the acid in a "dead" battery is only about 1.10, the condition of the battery can be seen at a glance.

EXAMPLE 12-10

The float in Fig. 12-13 is constructed as follows: A portion A having volume 3.00 cm³ is totally immersed, and a stem 10 cm long having cross section 0.200 cm² is partially immersed. The mass of the whole float is 4.80 g. When floating in a liquid of specific gravity 1.20, how much of the stem projects above the liquid surface?

Let $x =$ the length above the surface. The

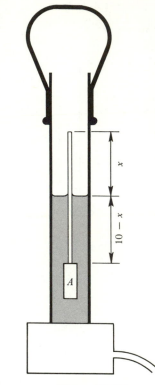

Figure 12-13

volume below the surface in cm³ is $3.00 + (10 - x)(0.200)$. Hence the weight of displaced fluid is found from the volume and the density of the fluid:

$$\{[3.00 + (10 - x)(0.200)]\text{cm}^3\}$$
$$\times \left(1.20 \frac{\text{g}}{\text{cm}^3}\right)\left(980 \frac{\text{cm}}{\text{s}^2}\right)$$

The BF equals the entire weight of the float, since it is floating and is in equilibrium. We now apply Archimedes' principle, expressing both sides of the equation in dynes:

$$\text{BF} = \text{weight of fluid displaced}$$
$$(4.80)(980)$$
$$= [3.00 + (10 - x)(0.200)](1.20)(980)$$
$$x = \boxed{5.00 \text{ cm}}$$

A balloon floating in air presents a slightly different problem. The entire balloon is immersed in the air, so the BF is constant (ignoring slight changes in density of air with altitude). The airship can be kept in equilibrium only by adjusting its weight so that the net force is zero. The downward force consists of the weight of the balloon plus the gas inside it plus the payload plus the ballast, if any. The upward force is the weight of an equal volume of air. In the next example we will use the "weight density," or weight per unit volume.

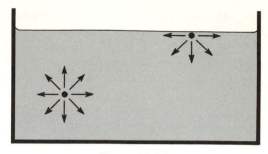

Figure 12-14

EXAMPLE **12-11**

A blimp of volume 100 000 ft³ is filled with helium (weight density 0.0111 lb/ft³). If the blimp bag and framework weigh 800 lb and the payload of personnel, instruments, etc. is 5000 lb, how much ballast should be carried to maintain equilibrium? (The weight density of air is 0.0807 lb/ft³.)

BF = weight of air displaced

= (100 000 ft³)(0.0807 lb/ft³) = 8070 lb

There are three downward forces:

(*a*) Weight of helium in blimp:

(100 000 ft³)(0.0111 lb/ft³) = 1110 lb

(*b*) Weight of blimp: = 800 lb

(*c*) Payload: = 5000 lb

Total downward force = 6910 lb

Hence to secure equilibrium, an additional downward force of

8070 − 6910 = $\boxed{1160 \text{ lb}}$

must be supplied by the ballast.

12-7 Surface tension

Before leaving the subject of fluids at rest, we shall discuss the phenomena that take place at the surface of a liquid. As we have repeatedly pointed out, "cohesive forces" exist between atoms and between molecules; these forces are not gravitational, but are ultimately electrical in nature.

In the liquid state the cohesive forces are insufficient to give the structure rigidity, but

the molecules are close enough to each other so that there is sufficient cohesion to make possible a definite surface. To study surface phenomena more quantitatively, let us consider the energy changes which accompany the formation of a liquid surface. In order to increase the surface area of a given amount of liquid, molecules must be brought from deep[*] inside the liquid to the surface. At the start, the molecule is acted on by no net force, since it is pulled equally in all directions (Fig. 12-14). As it approaches the surface, however, work must be done against a net force, since there are now no molecules (save for a few air molecules, not shown) to counteract the inward force of neighboring molecules of liquid. We see that a molecule at or very near the surface has potential energy, in much the same way that molecules of a stretched rubber band have elastic PE, or a satellite has gravitational PE. The "surface PE" of a liquid therefore increases whenever the total surface area increases. When left to itself, a liquid tends to assume the shape which has the least surface area for a given volume, for in this way the potential energy is least. A falling raindrop would be spherical if it were not for its weight and for air resistance, since a sphere is a surface of minimum surface area. We define the *coefficient of surface tension* of a liquid as potential energy per unit surface area; this constant is denoted

---

[*] The term "deep" here means several molecular diameters, i.e., about 10^{-9} to 10^{-8} m. At greater distances, the short-range electric forces discussed in Sec. 2-1 are of negligible importance.

by γ, the lower-case Greek gamma. A typical value of γ, for a water surface in contact with air, is 0.073 J/m² or, in cgs units, 73 ergs/cm². (This is one of the few places in physics in which the erg is an energy unit of convenient size.)

An alternative way of looking at surface tension is to consider the force that must be applied when a surface is "stretched out" to a larger area. If a force **F**, applied parallel to the surface, moves an imaginary line of length L through a distance Δs perpendicular to the direction of the line, the increase in PE is the work done, equal to $F\,\Delta s$; the area created is $L\,\Delta s$. Hence, using the definition of γ,

$$\gamma = \frac{F\,\Delta s}{L\,\Delta s} = \frac{F}{L} \qquad (12\text{-}3)$$

We therefore interpret γ as a contractile force per unit length—a definition which is equivalent to our earlier definition of γ as surface energy per unit area. Since a joule is a newton·meter, the two units J/m² and N/m are equivalent. The dimensions of surface tension are $[FL^{-1}]$ or $[MT^{-2}]$.

One of the standard methods of measuring the coefficient of surface tension of a liquid is by means of a Du Nouy torsion balance. A light ring of platinum wire, 4.00 cm in circumference, is suspended at one end of a light horizontal rod; the other end of the rod is attached perpendicularly to a fine horizontal steel wire which is fixed at one end and attached to a knob at the other end. As the knob is turned, the end of the rod tends to move upward, but it can be kept stationary by application of a small downward force. Such a torsion balance is capable of measuring a force of a 0.1 dyn and (with a fine quartz fiber instead of a steel wire) can be made sensitive enough to measure forces as low as the weight of a 0.001 microgram object (approximately 10^{-6} dyn). As the ring is pulled out of the liquid, it carries a thin film of liquid with it (Fig. 12-15). When just about to break loose, the liquid film is verti-

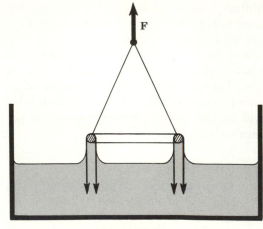

Figure 12-15

cal, and there is a downward force of contraction which is represented by the downward force vectors. The contractile force is proportional to the circumference of the ring; the length of surface along which the force acts is approximately twice the circumference of the ring, since the sheet of liquid which is being stretched has two sides.[*] The working equation for the instrument is therefore $\gamma = F/2L$, where F is the magnitude of the measured force just as the liquid film breaks and L is the circumference of the ring.

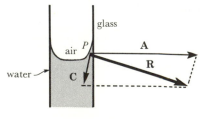

Figure 12-16

The coefficient of surface tension can also be determined by observing the rise in a capillary tube. Let us take water in a clean glass tube as an example (Fig. 12-16). A molecule at P is acted on by three forces: a force of *cohesion* **C** exerted by adjacent water molecules; a force of *adhesion* **A** due to attraction

---

[*] Strictly, the total length is the sum of the inside circumference and the outside circumference of the ring.

by the glass molecules; and a slight force of adhesion caused by the attraction of the air molecules (not shown in the diagram).[*] The reason that water "wets" glass is that the adhesive forces are so much greater than the cohesive forces that the resultant inward force **R** is practically perpendicular to the surface of the glass. The liquid adjusts itself until its surface is perpendicular to the resultant force; if it did not, there would be a force component parallel to the surface, and the liquid, having no rigidity, would flow.[*] Near the glass, the resultant force on a water molecule is directed inward, toward the glass, and hence the liquid rises until the surface is perpendicular to the force. If the inside radius of the capillary tube is small, the top surface of the water in the tube is a hemisphere, with the edge of the liquid surface forming a horizontal circle of radius r. We now consider the equilibrium of the cylinder of water, of height h (Fig. 12-17). The upward force of surface tension (equal to γ dynes for every centimeter of length) is exerted on the circular ring where the surface ends, and the force is γ times the length of the ring. The downward force on the cylinder of water is the weight of the liquid, equal to its volume $\pi r^2 h$ times

its weight per unit volume, dg. For equilibrium,

$$\text{Upward force} = \text{downward force}$$

$$\gamma(2\pi r) = (\pi r^2 h)dg$$

Solving this for h, we get

$$h = \frac{2\gamma}{rdg} \qquad (12\text{-}4)$$

It is evident that the height of rise can be very great if r is small enough.

[*] The term *cohesion* refers to forces between like molecules, and *adhesion* to forces between unlike molecules.
[*] The same phenomenon is illustrated by the fact that the surface of the water in a lake is horizontal, perpendicular to the force of gravity.

EXAMPLE **12-12**

How high does methyl alcohol rise in a glass tube 0.1 mm in diameter? The surface tension is 23 dyn/cm, and the specific gravity is 0.8.

$$h = \frac{2\gamma}{rdg}$$

$$= \frac{2(23 \text{ dyn/cm})}{(5 \times 10^{-3} \text{ cm})(0.8 \text{ g/cm}^3)(980 \text{ cm/s}^2)}$$

$$= \boxed{12 \text{ cm}}$$

If the cohesive forces are sufficiently large in comparison to the adhesive forces, the liquid will not wet the glass or other container, and the rise is less. For any given liquid-solid combination the *angle of contact* (Fig. 12-18) is 0 only if the liquid wets the surface. The upward component of force is $\gamma(2\pi r) \cos \theta$, and Eq. 12-4 is replaced by

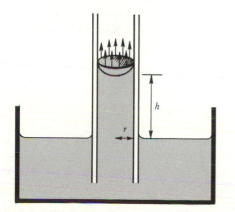

Figure 12-17   Rise of water in a capillary tube.

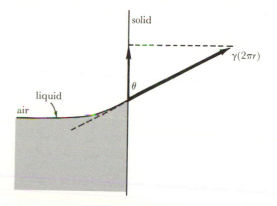

Figure 12-18

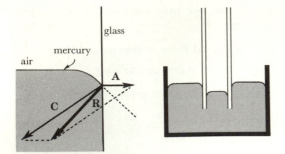

Figure 12-19

$$h = \frac{2\gamma \cos \theta}{rdg}$$

In extreme cases, the level near the glass may actually be depressed, as in the case of mercury in a glass tube (Fig. 12-19). The resultant force **R** can be perpendicular to the surface only if the mercury surface curves downward as shown. Here $\theta = 140°$, $\cos \theta = -0.65$, and h is negative.

Many common phenomena depend on surface tension. For example, absorbency of a blotter or a towel is due to a capillary action in the tiny tubes formed by the crisscross of the fibers of paper or cloth. The dissolving of "instant" powdered coffee depends on the wettability of the granules. Water has a relatively high coefficient of surface tension— about 72 dyn/cm—as compared with other common liquids such as alcohol (23 dyn/cm) or carbon tetrachloride (20 dyn/cm). A commercial (nonalcoholic) antiseptic trademarked ST-37 has a value of only 37 dyn/cm. This minimizes the formation of drops which might block the entrance to tiny cracks in the skin or wound. A not-too-clean needle or razor blade can be "floated" on water due to the combined action of surface tension and Archimedes' principle. The water does not wet the metal (which may have a thin oily film), and, as the needle sinks, the water's surface is deformed, giving rise to an upward component of force which may partially balance the weight if the diameter of the needle is not too large.

TABLE 12-1   Surface constants

Interface	°C	Coefficient of surface tension γ erg/cm^2 or dyn/cm
water--air	0	75.6
water--air	20	72.8
water--air	100	58.9
ethyl alcohol--air	20	22.3
mercury--air	15	487.
NaCl (molten)--N$_2$	803	114.
nitrogen (liq.)--N$_2$	−203	10.5
water--benzene	20	35.0

Interface	Angle of contact
water--clean glass	0°
ethyl alcohol--glass	0°
mercury--glass	140°
water--silver	90°
water--paraffin	107°
kerosene--glass	26°

12-8   Fluids in motion— Bernoulli's equation

Turning now to the subject of fluids in motion, or hydrodynamics, let us first note that the mathematical difficulties are formidable; some of the finest mathematicians and physicists of past and present have found the subject worthy of their best efforts, and only with the advent of electronic computing devices has there been substantial progress in the field. A fundamental equation, by means of which many of the simpler phenomena of fluids can be correlated, is due to Bernoulli.[*] The state of a fluid at any point can be characterized by four quantities: the velocity **v**, the density d, the pressure P, and the height h above some reference level. In *streamline flow,* the molecules of fluid move from point to point without any rotational motion or turbulence. If the fluid in such a flow is incom-

[*]Daniel Bernoulli (1700–1782), whose father and uncle were also famous Swiss mathematicians and physicists.

Figure 12-20 Stream-
line flow and turbulence
around an airfoil.

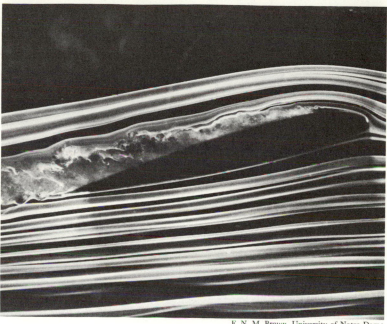

pressible, its speed is large wherever its cross section A is small, and vA remains constant. (For an analogous situation, imagine incompressible cars, moving bumper-to-bumper in both lanes of a highway, coming to a construction area in which one lane is blocked off. If streamline flow is maintained, as assumed, the cars move twice as fast through the constricted area, accelerated perhaps by bumps from behind if the drivers are not alert.) Above a certain critical speed, which depends on the fluid and the shape of the pipe, or the shape of the stationary body past which the fluid is flowing, the motion of fluids is one of *turbulent flow*. Little eddies or whirlpools absorb energy, and the frictional drag of the fluid increases sharply (Fig. 12-20).

With these definitions in mind, we are ready to state *Bernoulli's equation:*

> **If an incompressible fluid is in streamline flow, then the quantity ($\frac{1}{2}dv^2 +$ hdg $+$ P) is constant at every point in the fluid.**

We can express this fact in the form of an equation:

$$\tfrac{1}{2}dv_1{}^2 + h_1dg + P_1 = \tfrac{1}{2}dv_2{}^2 + h_2dg + P_2$$

$$(12\text{-}5)$$

where the subscript 1 refers to an arbitrary point in the fluid and the subscript 2 refers to any other point. A proof of Bernoulli's equation for an incompressible fluid in steady, streamline flow is given in Sec. 12-10; we note here that the proof depends upon two of the conservation laws: the law of conservation of mass and the law of conservation of energy.

Bernoulli's equation can be simplified in many special cases, if P, v, or h is constant, as often happens. A few examples illustrate varied applications of Bernoulli's equation.

Case 1 *Velocity constant.* The equation becomes $hdg + P =$ constant. For instance, the pressure in a tank of water decreases as the height above the bottom level increases. Thus $h_1d_1g + P_1 = h_2d_2g +$

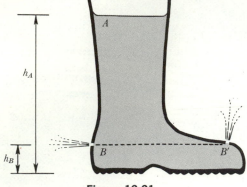

Figure 12-21

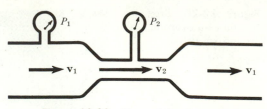

Figure 12-22 Venturi flow meter.

P_2; if the fluid is incompressible $d_1 = d_2 = d$, and the equation can be rearranged to give the familiar formula $P_1 - P_2 = (h_2 - h_1)dg$, relating the magnitudes of the change in pressure and the change in height (consistent with Eq. 12-1).

Case 2 *Pressure constant.* A rubber boot is partially filled with water, and two pinholes are made at the same distance below the surface of the water (Fig. 12-21). (The density is considered to be constant.) The pressure at A is atmospheric pressure, and so is the pressure at hole B, since the water at this point is open to the atmosphere. Hence, canceling out the pressure from each side of Bernoulli's equation, we see that $\frac{1}{2}dv^2 + hdg =$ constant. The speed of the water at A is practically zero, but the speed at B is greater than that at A, since the value of h at B is less than at A. The speeds of the two streams at B and B' are the same, although the water emerges in different directions, since the force due to the pressure is always perpendicular to the surface on which it acts. The term hdg is just the gravitational PE of 1 cm^3 of the liquid, and the term $\frac{1}{2}dv^2$ is the KE of 1 cm^3. The total mechanical energy[*] remains constant, in accordance with the energy principle.

Case 3 *Height constant.* The equation becomes $\frac{1}{2}dv^2 + P =$ constant. The pressure decreases wherever the velocity increases. Among the applications of this aspect of Bernoulli's equation are the Venturi meter for measuring rate of flow of a liquid or gas;

the curve of a spinning baseball; and the lift of the wings of a plane.

In a Venturi meter the rate of flow of a fluid (usually a liquid) is shown by the pressure decrease as the fluid is forced through a constriction in the pipe and consequently speeds up (Fig. 12-22). Such meters are used by water departments which encounter large flow rates.

When a baseball is thrown forward, the air rushes past it at a velocity **v**. If, however, the ball is spinning, it drags some air with it (especially if the surface is rough), with a velocity **u** (Fig. 12-23). Hence the resultant speed of air at B is $v + u$, while at A the resultant speed of the air is $v - u$. By Bernoulli's equation the pressure is greatest at A, where the air speed is least, and the ball is pushed sideways, to the right, as shown. The whole phenomenon is called the *Magnus effect,* and depends on the viscous drag of the fluid as well as on Bernoulli's principle. The ability of a baseball pitcher to throw a curve is by no means an optical illusion. On the same principle, the Flettner rotor ship in the 1920's took advantage of the winds without using sails (Fig. 12-24). A tall, rough cylindrical shaft was kept rotating by a small auxiliary motor; if a wind was blowing, the Bernoulli force acted on the rotor and hence on the ship. The

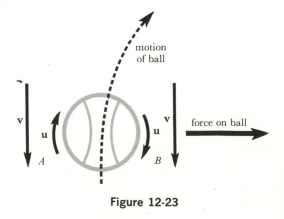

Figure 12-23

---

[*] The energy principle (Sec. 6-4) represents the conservation of *mechanical* energy and applies only when there are no frictional losses. It is for this reason that Bernoulli's principle is restricted to streamline flow, for in turbulent flow energy is dissipated in the form of heat.

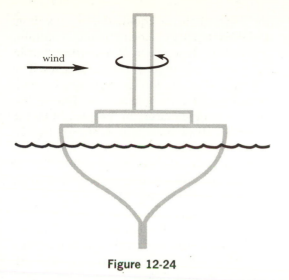

wind

Figure 12-24

power supplied by the motor was needed only to overcome friction in the bearings; the power to drive the ship came from the wind. The ships did not prove a commercial success, however.

As a final example of Bernoulli's equation, we consider the "lift" of an airfoil. If the shape of the leading edge of the airfoil is right, streamline flow is maintained[*] and the air molecules are crowded together above the wing (Fig. 12-25). Since the flow lines are crowded together, the air speed is greater, just as if the air were flowing through a constriction in a pipe. Greater speed means less pressure, and so the downward force caused by the air pressure on the upper surface is less than the upward force on the lower surface. The lift depends on several factors, including the speed and the angle the wing makes with the horizontal. If the undersurface of the airfoil slopes upward, there is a force normal to the surface due simply to the

[*] If streamline flow is not maintained, turbulence sets in and the plane becomes very inefficient due to frictional losses. Hence the designer shapes the airfoil to maintain streamline flow for as high a speed as possible.

impact of the stream of air, and this force has an upward component which supplies some lift in addition to the Bernoulli lift; but this scheme cannot be carried too far, or turbulence results.

12-9　Viscosity

Fluids in motion exhibit a certain resistance to motion which is called *viscosity*—a sort of internal molecular friction. The viscous drag is caused in liquids by short-range molecular cohesive forces, and in gases by collisions between fast-moving molecules. In both liquids and gases, the drag is proportional to the speed, as long as the speed is slow enough so that streamline flow takes place. However, in turbulent flow, the viscous drag increases rapidly, more nearly proportional to the square or cube of the velocity. Streamlining of cars and planes means just what it says—design to permit streamline flow at high velocities and hence with less friction. There is no simple law giving the relationship between force and velocity (see Fig. 12-26 for typical experimental results). Whatever the law, one thing is certain: viscous drag increases as velocity increases.

The dependence of the force of viscosity upon velocity is nicely illustrated by the *terminal velocity* reached by a falling body. As a parachutist falls, his velocity increases, and hence the upward viscous force of the air increases. Sooner or later the upward drag equals the downward force (his weight), and

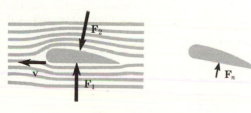

F_2

v

F_1

F_n

Figure 12-25

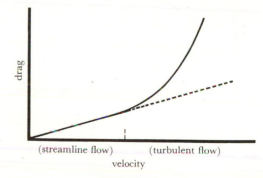

drag

(streamline flow)　　(turbulent flow)

velocity

Figure 12-26

then the net force on the man is zero. The parachutist has reached a constant terminal velocity which depends on his weight and the size of the parachute. A typical value of terminal velocity for a man-parachute combination is about 25 mi/h. While the system is picking up speed, the viscous drag is less than the man's weight. Using Newton's second law, we have

$$\text{net } F = ma$$

or

$$\text{Weight} - \text{viscous drag} = ma$$

The weight is constant, but the viscous drag is increasing, so the net force is not constant, and hence the acceleration is not constant. The velocity increases asymptotically toward the terminal velocity, as shown in Fig. 12-27. Without a parachute, the area exposed to the air stream is much smaller, and the balance of forces occurs at a higher speed. If he falls far enough before opening his chute, the parachutist attains a terminal velocity of about 125 mi/h, regardless of the height from which he falls. In liquids, terminal velocity is illustrated by the motion of a steamship. The forward thrust of the propellers causes the velocity to increase. As the velocity increases, so does the viscous drag of the water. Eventually, equilibrium is reached (a condition of no acceleration) when

the forward thrust equals the viscous drag, and then the ship's velocity remains constant.

Viscosity depends on temperature, but in a different fashion for liquids and gases. The viscosity of most liquids decreases as temperature increases—molasses in June is less viscous than molasses in January. On the other hand, at high temperatures the viscosity of gases increases.[*]

Plastic flow, the permanent distortion of an object by an applied force, may be thought of roughly as an exaggerated case of viscous flow, with a very high coefficient of viscosity. The technical definition of plastic flow includes the provision that the substance flows only after some minimum stress is exceeded; lead, soap, and tallow candles are plastic. On the other hand, tar, sealing wax, some glues, and most glasses are highly viscous liquids. Crystalline quartz generally fractures before it flows appreciably and is not considered a liquid. The detailed study of liquids and plastic solids is complex and difficult. For instance, the apparent elasticity and viscosity of protoplasm depend on many factors, such as the nature of the protein lattice structure within the cells. Every cook has experienced the sudden change in viscosity of a bowl of fudge that is being beaten. These examples show that the terms "liquid" and "solid" are only useful first approximations to the behavior of real substances. Much remains to be understood about the ways in which molecules interact with each other at close quarters, in the no-man's-land between solid and liquid.

[*] It is relatively easy to explain the increased viscosity of gases as being due to the fact that at high temperatures the molecules move faster and collide more often with each other, giving rise to increased internal friction. The decrease of viscosity of a liquid is harder to explain, and it probably depends on the fact that at higher temperatures the molecules are farther apart, and the cohesive forces of internal molecular friction are therefore less effective.

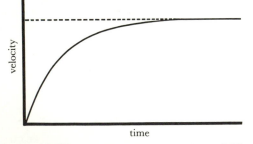

Figure 12-27 Velocity-time graph for a falling body, illustrating terminal velocity.

A fluid is a substance, such as a liquid or gas, which has no rigidity. Liquids are distinguished from gases by the presence of a surface.

Pressure is force per unit area. The difference in pressure between two points in a liquid at rest equals the vertical height separating the two points times the weight per unit volume of the liquid. A similar statement is true for gases, if the gas can be considered to have uniform density. Changes in pressure exerted at any point in a confined fluid are transmitted undiminished in all directions to all parts of the fluid (Pascal's principle).

The loss of weight of a body wholly or partially immersed in a fluid equals the weight of the fluid displaced (Archimedes' principle). The volume of an irregular solid can be measured, using Archimedes' principle, thus leading to a way of determining density.

The molecules at the surface of a liquid are in a state of contractile tension due to cohesive forces between the molecules. The coefficient of surface tension represents the potential energy per unit area of surface; it is numerically equal to the contractile force per unit length of boundary. The angle of contact is the angle between a liquid surface and an adjacent solid surface.

According to Bernoulli's equation, the quantity $(\frac{1}{2}dv^2 + hdg + P)$ is constant throughout an incompressible fluid which is at rest or in streamline flow. At a given level, the pressure is least where the velocity is greatest.

Viscosity is fluid friction, caused (in a liquid) by cohesive forces between molecules and (in a gas) by collisions between molecules. The viscosity of liquids decreases as temperature rises, but the opposite is true for gases.

■ CHECK LIST

fluid	$P = F/A$	turbulent flow
liquid	$P = hdg$	Bernoulli's equation
gas	Archimedes' principle	$\frac{1}{2}dv^2 + hdg + P$
pressure	Pascal's principle	$= $ constant
atmosphere (pressure unit)	coefficient of surface	viscosity
bar	tension	terminal velocity
mm Hg	angle of contact	
torr	streamline flow	

■ QUESTIONS

12-1 At ordinary temperatures, which of the following are not fluids? air; diamond; glass; taffy; kerosene; mercury; iron.

12-2 Suppose the undersea camera of Example 12-4 first is pointed horizontally, then is turned downward to photograph the ocean bottom. Compare the magnitudes of the total force on the glass window in these two orientations.

12-3 Explain why "water seeks its own level."

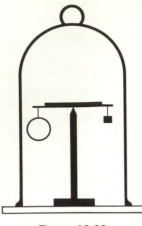

Figure 12-28

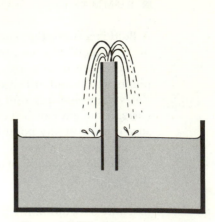
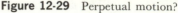

Figure 12-29 Perpetual motion?

12-4 Is the buoyant force on a submerged submarine lying on the floor of the ocean exactly the same as when the submarine is only 20 ft below the surface? Why?

12-5 What factors determine the maximum height to which a balloon can rise?

12-6 In Fig. 12-28 the thin hollow glass ball is empty and is sealed off from the surrounding atmosphere. It is exactly counterbalanced by a small solid brass counterweight. If the air is pumped out of the bell jar, does the glass ball rise or fall?

12-7 A solid cube of ice is floating in a glass of water, with the water level just even with the top of the glass. When the ice melts, will the water level be the same, will it fall below the edge of the glass, or will some water overflow?

12-8 A bucket of water is on a spring balance which reads 40 lb. If a 2 lb live trout is placed in the bucket, and the trout floats, will the reading of the balance be changed?

12-9 Which of the following units could *not* be used for the coefficient of surface tension? erg/cm^2; dyn/cm; N/m^2; N/m; J/m^2; $lb/in.^2$; g/cm.

12-10 A boy sends a Frisbee traveling horizontally through the air, giving it a spin as he throws it forward. As viewed from above, the disk is spinning clockwise. Does the Frisbee veer to the boy's right or left?

12-11 Is the ship of Fig. 12-24 coming toward the observer or moving away?

12-12 What's wrong with the proposed perpetual-motion machine in Fig. 12-29? The capillary tube is of such an inside diameter that water would rise 10 cm in it. The tube is cut off to be only 8 cm long, and it is expected that water will continually rush out in a fountain.

12-13 Collisions have occurred between ships that were steered too close together in parallel paths. Explain.

■ PROBLEMS

12-A1 What is the pressure on the ground beneath one of the spikes of a 240 lb football player whose shoes contain six spikes each, the cross section of the spikes at the ground being 0.1 in.^{2} for each spike?

12-A2 What is the pressure due to wind on the wall of a building 14×3.5 m, if the horizontal force of the wind on the wall is 9800 N? Express your answer in N/m^2 and in atm.

12-A3 What is the total downward force of the atmosphere on a horizontal table top which is 2 ft $\times$ 2$\frac{1}{2}$ ft? What is the total upward force on the underside of the same table top?

12-A4 Make the following conversions: 73.5 lb/in.^{2} = ? atm; 73.5 lb/in.^{2} = ? mm Hg; 20 atm = ? N/m^2; 152 mm Hg = ? torr; 1000 atm = ? lb/in.^{2}.

12-A5 Make the following conversions: 40 atm = ? lb/in.^{2}; 40 mm Hg = ? lb/in.^{2}; 0.2 atm = ? mm Hg; 10^{-10} atm = ? torr.

12-A6 The specific gravity of mercury is 13.6. Calculate the pressure in N/m^2 and in lb/in.^{2} at the bottom of a flask of mercury which is 10 cm deep.

12-A7 Compute the pressure at the lowest point of a tank car full of gasoline (specific gravity = 0.7). The tank is in the form of a closed horizontal cylinder 200 cm in diameter and 10 m long.

12-A8 A student is given a sample of one of the liquids in Table 2-1. He finds that the pressure at the bottom of a 50 cm column of the liquid is 6.17 $\times$ 10^4 dyn/cm^2 greater than atmospheric pressure. Identify the liquid.

12-A9 A stone of volume 2.0 ft^3 weighs 300 lb in air. What fraction of its weight would it lose if it were submerged at the bottom of a fresh-water lake?

12-A10 Calculate the buoyant force on an ice cube of volume 15 cm^3, immersed in alcohol of specific gravity 0.8.

12-A11 Suppose that, as a fog clears, 1000 tiny drops of water coalesce into one larger drop. What is the ratio of the total surface energy of the tiny drops to the surface energy of the large drop?

12-B1 An 8 oz spherical grapefruit is resting on a table. If the material of the fruit yields until the pressure is reduced to 0.5 lb/in.^{2}, what is the radius of the flat spot on the bottom of the grapefruit?

12-B2 A swimming pool measures 8 m $\times$ 25 m $\times$ 3 m deep. (a) What is the force on the bottom due to the water? (b) What is the force on the vertical wall at one end of the pool? (*Hint:* Use the average pressure, which equals the pressure at the average depth.)

12-B3 Seamen attempt to escape from a sunken submarine which is lying on its side in 100 m of water. What force must the men exert on the escape hatch, which is a door 1.2 m $\times$ 0.5 m? The pressure inside the submarine is 1 atm.

12-B4 In Fig. 12-5, what is the pressure at a point in the liquid 80 cm above the level of the table which supports the meter stick?

12-B5 The density of blood is about 1.1 g/cm^3. If there were no valves in a giraffe's neck to give partial support to the blood, what would be the difference in pressure, in torr, between the blood in a giraffe's brain and that in his heart? The brain is 2 m above the heart.

12-B6 In an experiment on the elasticity of living tissues, a researcher used boric acid solution as the indicating liquid in a manometer. He found that, when the pressure changed from 5 cm of boric acid solution to 15 cm of boric acid solution, the volume of a urinary bladder changed from 300 cm^3 to 200 cm^3. The specific gravity of boric acid solution is 1.08. (a) Express the pressure change in dyn/cm^2; in mm Hg. (b) If the bladder were a solid body obeying Hooke's law, what would be the bulk modulus of the organ? (How do you account for the greatly different order of magnitude of your answer from the values listed for other materials in Table 9-2?)

12-B7 A hydraulic lift in a service station is to raise a 8 ton truck by compressed air at a pressure of 20 atm. The large cylinder of the lift has a radius of 4.00 in. Is the air pressure sufficient for the job?

12-B8 In a solid-waste recycling plant, the small piston of a hydraulic press has radius 1 in., and the large piston has radius 10 in. Compressed air is available at 80 lb/in.² above atmospheric pressure. What force can be applied to a load of used metal beverage cans?

12-B9 Calculate the pressure, in millibars, read by the gauge at G' in Fig. 12-5.

12-B10 (a) Calculate the force, in dynes, on a horizontal surface of area 2 cm² at P in Fig. 12-5. (b) Calculate the force on a 2 cm² vertical surface at point Q.

12-B11 A cube of aluminum 2 cm on an edge floats in a beaker of mercury. (a) What volume of aluminum is immersed in the mercury? (b) What volume of gold, placed on top of the cube, would cause the cube to be just immersed in the mercury? (Use densities from Table 2-1.)

12-B12 A geologist, analyzing an irregular piece of ore, finds that the sample is balanced by 88 g when weighed in air and 76 g when weighed in water. What is the specific gravity of the ore?

12-B13 A piece of cork weighs 20 × 980 dyn in air. To find its density, a student hangs the cork from one arm of a balance and hangs an iron sinker from the cork. When the sinker is immersed in water, the apparent weight of sinker plus cork is 145 × 980 dyn; when both sinker and cork are immersed, their total apparent weight is 65 × 980 dyn. From these data, calculate the density of cork.

12-B14 A 60 kg man is just floating in water, with negligible volume in the air. What is his volume?

12-B15 A moon rock weighs 3 × 980 dyn in air, and 2 × 980 dyn when immersed in alcohol of specific gravity 0.8. (a) What is the volume of the rock? (b) What is its specific gravity?

12-B16 The cross section (at the water level) of a sea-going tanker is 700 m². How far will the tanker sink into the water if 1500 m³ of gasoline is pumped into it? (See Table 2-1.)

12-B17 A weather balloon made of light, flexible plastic weighs 10 N when empty; it carries a load of 70 N of apparatus when filled with helium near the earth's surface. What is the volume of the balloon when full?

12-B18 It is desired to measure the specific gravity of olive oil by Archimedes' principle. An irregularly shaped piece of metal, of unknown composition, is available. The following data are taken: weight of metal in air = 108.0 × 980 dyn, weight of metal in water = 68.0 × 980 dyn; weight of metal in oil = 71.2 × 980 dyn. (a) Calculate the specific gravity of the metal. (b) Calculate the specific gravity of the oil.

12-B19 The concept of isostasy in geology implies that mountains "float" on a denser substratum, the earth's mantle, with the tallest mountains penetrating deepest into the mantle. Suppose a cube of side L and density d' floats in a liquid of density d (Fig. 12-30) with a length L_1 beneath the liquid and L_2 above the surface ($L_1 + L_2 = L$). Derive formulas for L_1 and L_2, in terms of L, d, and d', and show that both L_1 and L_2 increase as L increases, thus illustrating isostasy.

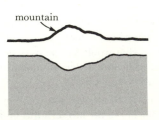
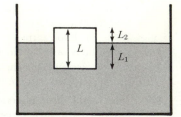

Figure 12-30

12-B20 In Fig. 12-22 pressure gauge P_1 reads 4000 lb/ft² and P_2 reads 3200 lb/ft². If the velocity of the stream at point 1 is 10 ft/s, what is the velocity at point 2? The liquid is sea water. (One cubic foot of sea water weighs 64.0 lb; use W/g for mass.)

12-B21 In Fig. 12-21 the hole at B is 18 in. below the water level and 2 in. above the floor. Assuming no loss of speed due to friction at the hole, how far out from the boot does the stream of water strike the floor? (Use Bernoulli's equation; the density of the liquid cancels out.)

12-B22 A storage tank is filled with water to a depth of 16 ft, and a short hose is connected to a faucet at the bottom of the tank. Neglect friction. (a) With what speed does water leave the nozzle of the hose? (b) If the hose is pointed upward, how high does the stream of water rise?

12-B23 Air flows past the upper surface of a horizontal streamlined airplane wing at 250 m/s and past the lower surface of the wing at 200 m/s. The density of air is 1.0 kg/m³ at the flight altitude, and the area of the wing is 20 m². (a) What is the pressure difference between the two surfaces of the wing? (b) Calculate the net lift on the wing.

12-B24 The coefficient of surface tension of soap solution is 25 erg/cm². What is the increase in total surface energy when a soap bubble is blown up from a radius of 2 cm to a radius of 3 cm? What is the source of this energy?

12-B25 The coefficient of surface tension of a mixture of alcohol and water is measured with the apparatus of Fig. 12-15. It is observed that the film breaks when an upward force of 600 dyn is applied. The average radius of the ring is 1.20 cm. Calculate the coefficient of surface tension of the liquid.

12-B26 What upward force is needed to break the vertical film of water adhering to a thin horizontal wire ring of diameter 3 cm (Fig. 12-15)? The coefficient of surface tension for water is 72 dyn/cm.

12-B27 A pencil 7 mm in diameter is held vertically with its lower end immersed in a beaker of carbon tetrachloride for which the coefficient of surface tension is 20 dyn/cm. The pencil is wet by the liquid. What are the magnitude and direction of the force exerted by the liquid on the pencil?

12-B28 A toy boat of white pine (specific gravity 0.4) is floating on the surface of a calm lake. The wood is 10 cm wide, 40 cm long, and 2 cm thick. (a) If, to the water near the right-hand 10 cm side, is added a detergent making the coefficient of surface tension $\frac{1}{3}$ its normal value, what is the direction of the net force on the boat? (b) Ignoring viscous drag of the water, what will be the velocity of the boat 5 s after the detergent is added to the water?

12-B29 The coefficient of surface tension for water is about 72 dyn/cm. (a) What is the total upward force exerted on the pavement during a rainstorm due to the boundary of a hemispherical water bubble on the pavement? The bubble is 4 cm in diameter. (*Hint:* The water film has two surfaces, each exerting upward contractile force.) (b) The upward force found in part (a), exerted over a circular area of pavement, means that the pressure inside the bubble is greater than atmospheric pressure. Calculate this excess pressure, in dyn/cm². (c) Solve the problem in symbols, showing that for a spherical bubble of radius R the excess pressure is $\Delta P = 4\gamma/R$.

12-B30 How high will water rise in a glass capillary tube 0.01 mm in diameter?

12-B31 In an effort to get an indestructible standard of length depending on the properties of one substance, Sir Humphry Davy (1778–1829) suggested that the natural standard of length be equal to the diameter of a tube, wet by water, in which the capillary rise of water, at the melting point of solid water (0°C), equals the tube's diameter. (a) Derive a formula for this length. (*b*) Check the dimensions of your formula. (*c*) Using 75.6 ergs/cm² for the coefficient of surface tension of water at 0°C, calculate the length of Sir Humphry's unit.

12-B32 Pure water is placed in a clean silver cup. (*a*) Construct a force diagram similar to Figs. 12-16 and 12-17, showing the vectors **A**, **C**, and **R** (ignore the effect of air molecules above the liquid surface). What is the direction of the resultant force **R**? (*b*) For a water-silver interface, what is the ratio C/A of the cohesive force to the adhesive force? (*c*) What would you expect if a capillary tube of silver were placed in a beaker of water?

12-B33 What is the diameter of a capillary tube in which glycerin rises 4 cm? The glycerin wets the glass, and its coefficient of surface tension is 63 dyn/cm.

12-C1 At moderate speeds, air resistance is found to be proportional to the square of the velocity. (*a*) Is such flow streamline or turbulent? (*b*) Show that the power expended against air resistance is proportional to the cube of the velocity. (*c*) Two mile runners, *A* and *B*, run a race timed at 4 min 40 s, as follows: Runner *A* does successive quarters in even time, 70, 70, 70, and 70 s. Runner *B* runs the race in quarters timed at 60, 70, 80, and 70 s. Show that one runner uses up about 3% more energy than the other, in overcoming air resistance.

12-C2 Assume that the left ventricle of the heart is analogous to a simple piston and chamber, and that 70 cm³ of blood is ejected during each stroke of the piston, against an average pressure of 105 torr. (*a*) Calculate the work done by the ventricle in a single contraction. (*b*) Assuming a pulse of 60 beats per minute, calculate the useful power (in horsepower) which the left ventricle develops in the contraction part of the cycle. (*c*) Calculate the total work performed (in foot·pounds) by this organ during a 70 year life span. (*d*) This text asks that the answers to some problems be in British units in order to facilitate comparison with everyday experience. Make such a comparison for the total work done, for instance, in lifting the Great Pyramid of Cheops (Prob. 2-A8) vertically a few feet into the air, or use some comparison of your own choice.

12-C3 In Example 12-4, sea water was assumed to be of uniform density. Check this assumption, using data from Table 9-2 (convert bulk modulus of water to lb/in.²). (*a*) Find the fractional change in volume of a given mass of sea water lowered from the surface to a depth of 24 600 ft, where the pressure is 10 900 lb/in.². (*b*) Calculate the specific gravity of sea water at 24 600 ft, if the sea-level value is 1.03.

12-C4 A 200 kg wooden log is floating with 20% of its volume above the water level. What is the volume of an iron object which must be attached to the underside of the log in order to completely submerge it? What is the mass of the iron object?

12-C5 A skin diver's average specific gravity is 0.98, and his mass is 75 kg. For good maneuverability while walking on a submerged ship's deck in fresh water, the diver desires an apparent weight 0.08 times his weight in air. (*a*) What volume of iron weights should he attach to his feet? (*b*) How much does this amount of iron weigh in air?

12-C6 An aluminum casting has a mass of 400.000 g and a volume of 150.00 cm³. (*a*) What is the buoyant force of the air on the aluminum? (*b*) If the aluminum is weighed on a beam balance, using brass weights as a counterweight, what is the buoyant force of the air on the brass weights? (*Hint:* First find the approximate volume of the brass weights; specific gravity of brass is 8.4.) (*c*) What mass corresponds to the apparent weight of the aluminum casting when weighed in air?

For Further Study

12-10 Derivation of Bernoulli's equation

The derivation uses the law of conservation of mass and the law of conservation of energy. We consider a tube of variable cross section through which an incompressible fluid moves in streamline flow (Fig. 12-31). In a time Δt, the volume of fluid which enters at end 1 is $(v_1 \Delta t)A_1$, and the mass of fluid which enters is $v_1 \Delta t A_1 d$. In this same time, the mass of fluid which leaves end 2 is $v_2 \Delta t A_2 d$. The law of conservation of mass tells us that no mass is created or absorbed within the tube. Since we assume that the flow is steady and the fluid is incompressible, of density d throughout, this means that the mass which enters end 1 in time Δt equals the mass which leaves end 2 during the same interval. That is,

$$v_1 \Delta t A_1 d = v_2 \Delta t A_2 d$$
$$v_1 A_1 = v_2 A_2 \qquad (12\text{-}6)$$

From this we see that the speed v is large where the cross section is small.*

To force the fluid through the tube, some external force $P_1 A_1$ acts at end 1, doing work *on* the fluid, and work is performed *by* the fluid as it exerts a force $P_2 A_2$ at end 2. Thus at end 1, the system gains energy $P_1 A_1 (v_1 \Delta t)$, and at end 2 it loses energy $P_2 A_2 (v_2 \Delta t)$. The

---

*It may be asked what forces act to accelerate the molecules of fluid, which speed up and slow down according to Eq. 12-6. The key word in this connection is that the fluid is "incompressible," which means that short-range repulsive forces act whenever the molecules come too close together. These forces, essentially electric in nature, serve to cause the necessary changes in momentum of the molecules.

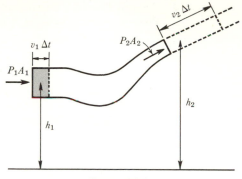

Figure 12-31

net work done on the system is

$$P_1 A_1 (v_1 \Delta t) - P_2 A_2 (v_2 \Delta t)$$

Other energy changes are associated with PE and KE:
The net increase in KE is

$$\tfrac{1}{2}(v_2 \Delta t A_2 d)v_2{}^2 - \tfrac{1}{2}(v_1 \Delta t A_1 d)v_1{}^2$$

and the net increase in PE is

$$(v_2 \Delta t A_2 d)gh_2 - (v_1 \Delta t A_1 d)gh_1$$

Putting all these results together, we apply the law of conservation of energy. The work done on the system equals the increase in KE and PE:

$$P_1 A_1 (v_1 \Delta t) - P_2 A_2 (v_2 \Delta t)$$
$$= \tfrac{1}{2}(v_2 \Delta t A_2 d)v_2{}^2 - \tfrac{1}{2}(v_1 \Delta t A_1 d)v_1{}^2$$
$$+ (v_2 \Delta t A_2 d)gh_2 - (v_1 \Delta t A_1 d)gh_1$$

Remembering that $v_1 A_1 = v_2 A_2$, we simplify and rearrange, obtaining Bernoulli's equation:

$$\tfrac{1}{2}dv_1{}^2 + h_1 dg + P_1 = \tfrac{1}{2}dv_2{}^2 + h_2 dg + P_2$$

■ REFERENCES

1   Greulach, Victor A., "The Rise of Water in Plants," *Sci. American,* Oct., 1952, p. 78.

2   Reiner, Markus, "The Teapot Effect," *Physics Today,* Sept., 1956, p. 16. An interesting and not completely understood example of liquid flow. Reba, I., "Applications of the Coanda Effect," *Sci. American,* June, 1966, p. 84. More about the teapot effect.

3 Webster, David L., "What Shall We Say About Airplanes?" *Am. J. Phys.,* **15,** 228 (1947). Physical problems involved in flight.

4 Hazen, David C., and Rudolf F. Lehnert, "Low Speed Flight," *Sci. American,* Apr., 1956, p. 46.

5 McDonald, James E., "The Shape of Raindrops," *Sci. American,* Feb., 1954, p. 64.

6 Mili, G., "Baseball's Curve Balls—Are They Optical Illusions?" *Life,* Sept. 15, 1941, p. 83. Verwiebe, Frank L., "Does a Baseball Curve?" *Am. J. Phys.,* **10,** 119 (1942). Sutton, R. M., "Baseballs Do Curve and Drag," *Am. J. Phys.,* **10,** 201, (1942).

7 Shapiro, A., *The Magnus Effect* (film).

8 Trefethen, Lloyd, *Examples of Surface Tension; Surface Tension and Contact Angles; Formation of Bubbles; Surface Tension and Curved Surfaces; Breakup of Liquids into Drops* (films).

13

Temperature and Expansion

In our study of mechanics we assumed, without much quantitative evidence, that work done against friction becomes heat energy. We may, if we wish, consider much of the next four chapters as an elaboration of the law of conservation of energy, interpreted to include heat as a form of energy. It is our intention to develop the theory of heat to the point at which we can state two broad generalizations, the first and second laws of thermodynamics. In order to do this, however, we must first find out how to measure temperature and heat energy quantitatively.

13-1 Temperature

The sensations of "hot" and "cold" are among the very first experiences of a newborn baby. Every known primitive language has words for hot, warm, cool, and cold. These descriptive terms are applied to the concept of *temperature,* a concept with which most of us have at least a nodding acquaintance. Yet upon careful consideration, it becomes evident that, if we are to use the idea of temperature in our work, we need to have a way of measuring it. For instance, exactly what is meant when we speak of a tempera-

ture of $37°C$? For that matter, what is meant by a mass of 37 kilograms? Looking back, you will realize that our ability to use the concept of mass depends upon the existence of some fixed or reference mass (the international prototype kilogram), and a means of comparing masses by acceleration or momentum experiments (Sec. 6-3). In a similar fashion we deal with temperature by defining fixed temperatures on a scale, and adopting a *thermometer* as a means of making comparisons between temperatures. Actually there are two kinds of mass: inertial mass (defined by acceleration experiments) and gravitational mass (defined by weighing experiments). The *operation,* or experimental procedure, really defines what is meant by the concept "mass." Just so, the thermometer really defines the concept "temperature," and we shall see that there are as many different kinds of temperature as there are types of thermometers. Fortunately, most common centigrade temperature scales are in close agreement with each other, and for many purposes it is unnecessary to specify the exact type of thermometer used.

It is tempting to think of molecular motion as a fundamental property by which we "ought" to define temperature, for most of

---

Joints in rigid pavements allow for thermal expansion and contraction of roads and sidewalks. This highway buckled when the temperature exceeded expectations of the designer. (The Daily News Photo)

the phenomena of heat are associated with the random motions of atoms or molecules. For instance, when a car remains in direct sunlight for a while, the iron atoms in the fender gain kinetic energy and move faster than before. The faster-moving atoms, acting against the forces of cohesion, will be able to vibrate in slightly larger paths, and so the iron expands. If an observer touches the hot metal, KE is transferred to his finger by a collision process, and the sensation of warmth is produced in his brain. The pressure exerted by the air in a tire is due to the impact of molecules upon the walls of the tire; at higher temperatures, the molecules move faster, and the pressure is greater.

However, we do not define temperature in terms of molecular motion. For one thing, the process of radiation would not easily fit into such a scheme: How is the heat from the sun transmitted through 93 million miles of vacuum to reach the earth? What is the temperature of empty space? And, as we shall see in our discussion of thermodynamics (Sec. 16-6), mere absence of random molecular motion is no guarantee that a body possesses no heat or is at a temperature of absolute zero. There may be other forms of available energy (e.g., potential energy) remaining in the atoms, even though at ordinary temperatures these other forms of energy are of negligible importance in comparison with the random kinetic energies of the molecules of a solid, liquid, or gas.

Let us emphasize once more our reason for bringing up at this time these disquieting suggestions about the definition of temperature. The student will (understandably) want a nice, packaged definition of temperature that will stand up in later work. We are unable to give such a definition until Chap. 16. For the present, we merely point out that many definitions are *possible,* because many types of thermometers have been invented. For ordinary work, it is our good fortune that commonly used thermometers agree well enough with each other (and with

the thermodynamic scale to be discussed later) that we are spared the necessity of choosing any particular thermometer as standard.

13-2   Temperature scales

The human senses are not precise enough to determine small changes in temperature. An anxious mother, feeling her child's brow, can distinguish between high fever and no fever at all, but such measurements are not quantitative. The science of medical diagnosis was tremendously advanced by the invention in about 1593 by Galileo of the "thermoscope," based upon the expansion of gases or liquids, and the later development of thermometers which quantitatively indicated temperature.[*] Before going into constructional details of thermometers, let us first discuss temperature scales and the fixed reference temperatures used to define the various scales.

Of the many temperature scales used in various places and times, only two are still in common use. Both use as fixed points the melting point of ice and the boiling point of water. On the Fahrenheit scale, ice melts at $32°F$ and water boils at $212°F$—a range of 180 Fahrenheit degrees. For scientific work, the Celsius (centigrade) scale is almost universally used. On it, ice melts at $0°C$ and water boils at $100°C$—a range of 100 Celsius degrees.[■] Since an interval of 100 Celsius degrees equals an interval of 180 Fahrenheit degrees, we see that a change of one Fahrenheit degree (1 F°) is only $\frac{100}{180}$, or $\frac{5}{9}$, as large

[*] Early thermometers indicated only *changes* of temperature, and a fever thermometer had to be individually calibrated before each use, by making a mark at the point indicated when the bulb was placed in the mouth of a person known to be in good health.

[■] A minor point may well be explained here. The choice of the ice point and the steam point as fixed points, with the interval divided into 100 degrees, was due to Anders Celsius (1701–1744), a Swedish astronomer. (Celsius originally called the ice point 100°C and the steam point 0°C!) By analogy with the Fahrenheit scale, we designate the boiling point of water as "100 degrees Celsius," abbreviated as 100°C. This is the practice

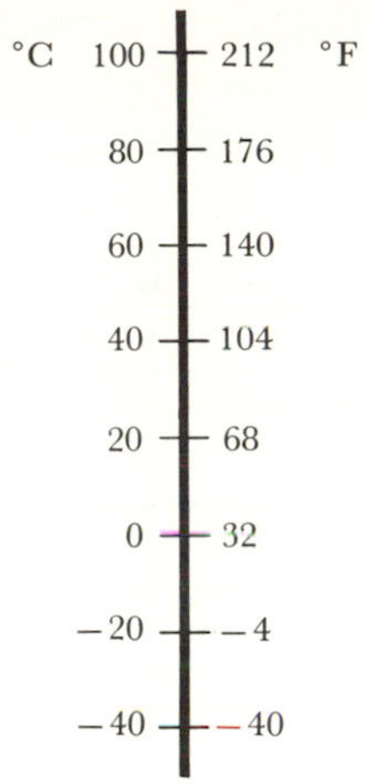

°C		°F
100	—	212
80	—	176
60	—	140
40	—	104
20	—	68
0	—	32
−20	—	−4
−40	—	−40

Figure 13-1

as a change of one Celsius degree (1 C°).[*]
To visualize the relation between temperatures on the Celsius and Fahrenheit scales, a diagram similar to Fig. 13-1 may prove useful. Formulas can be derived which relate °C to °F, but these are actually unnecessary and a burden on the memory; it is preferable to proceed directly from first principles, as in the following examples.

EXAMPLE **13-1**

A mountain climber finds that water boils at 80°C. What is the Fahrenheit temperature of his boiling water?

We consider *changes* in temperature from some standard temperature such as an assumed original temperature of 100°C (which we know to be 212°F). The new temperature is 20 C° less than the normal boiling temperature. This

is a decrease of $\frac{9}{5}(20)$, or 36, F°. (Remember that 1 F° is smaller than 1 C°, so that it takes more of them to represent a given change.) Hence the new temperature is

$$212° - 36° = \boxed{176°F}$$

EXAMPLE **13-2**

Research has indicated that humans can stand an air temperature of 900°F for over a minute, if insulated with clothing about 1 cm thick. Express this air temperature on the Celsius (centigrade) scale.

We can work from a reference temperature of 32°F (which we know to be 0°C). The new air temperature is 868 F° above the melting point of ice. This interval is $\frac{5}{9}(868°) = 482.2$ C°. Since the melting point of ice is at 0°C, the air temperature is

$$0° + 482.2° = \boxed{482.2°C}$$

The Celsius scale we have discussed is the International Practical Scale, which is defined to have 100 degrees between the ice point and the steam point.[*] The thermodynamic or Kelvin scale will be discussed in later chapters.

Any property of a substance which depends uniquely on temperature may be used as a temperature indicator. Such a property is called a *thermometric property* of the substance. For example, a yellow-hot piece of iron is hotter than a red-hot piece; color is therefore a thermometric property, and thermometers can be made which depend on color changes (radiation pyrometers, discussed in Sec. 14-6). The pressure of an enclosed gas also depends on temperature, and gas thermometers can be constructed in which change in temperature is indicated by change in pressure (Sec. 13-5). The electrical

in most European countries. The term "centigrade" refers merely to the division of the fixed interval into 100 degrees and could equally well be applied to a scale based upon the boiling point of sulfur as 0° and the melting point of silver as 100°, for instance. The use of water as standard substance has prevailed, and the scale deserves to be called by the name of Celsius. How fortunate that the initial letter "C" can be used for both Celsius and centigrade!

[*] Note that the symbol °F is used to indicate a temperature, in degrees Fahrenheit, while the symbol F° indicates an interval, in Fahrenheit degrees—that is, in degrees of the size used by Fahrenheit. A similar distinction is made between °C and C°.

[*] It has been found that the triple point of water (at +0.01°C; see Sec. 15-9) is more accurately reproducible than the ice point. On the International Practical Scale the ice point is defined as exactly 0.01 degree below the triple point, and there are exactly 99.99 degrees between the triple point and the steam point.

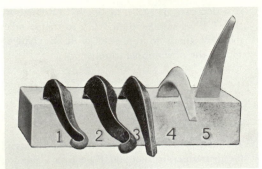

Photo courtesy Edward Orton, Jr. Ceramic Foundation

Figure 13-2 A series of pyrometric cones of different ceramic compositions which have all had the same heat treatment. The degree of softening depends primarily on the kiln temperature.

resistivity of most metals increases as their temperature goes up (Sec. 19-6), and so we have the resistance thermometer, consisting of a coil of platinum wire whose resistance indicates its temperature. An interesting thermometric property is used to determine (roughly) the temperature in a furnace: A series of ceramic cones of different compositions is placed in the furnace; each cone softens at a different temperature, and so the furnace temperature is determined (to within a few degrees) by inspection of the row of cones (Fig. 13-2). Enough has been said to indicate that many properties of substances are usable as thermometric properties. However, thermometers based upon the thermal expansion of solids and liquids are often used for reasons of convenience, ease of reading, and cost.

13-3 Thermal expansion of solids

Almost all solids expand upon heating; no matter which common temperature scale is used, the expansion is sufficiently uniform so that for practical purposes we can say that the length is a linear function of the temperature. Figure 13-3 represents the change in length of a certain brass rod, whose length increases by a small amount Δl for each small temperature increase Δt. However, we

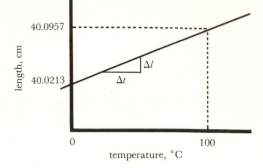

Figure 13-3 Uniform expansion of a certain brass rod.

expect the change in length of a long rod of brass to be more for a given change in temperature than that of a short rod, for the longer rod has more small crystals, each of which expands the same amount. Thus the change in length is proportional not only to the change in temperature, but also to the original length l. We therefore write the basic equation for the expansion of a solid as follows:

$$\Delta l = l\alpha \, \Delta t \qquad (13\text{-}1)$$

where α (the Greek letter alpha) is the *coefficient of linear expansion* for the given material. To find the final length, the amount Δl is added to or subtracted from the original length, according to whether the temperature increases or decreases. Values of α for various solids are shown in Table 13-1.[*]

EXAMPLE **13-3**

Compute the coefficient of linear expansion for brass, using the data given in Fig. 13-3.
  From the diagram

$$\Delta l = 40.0957 \text{ cm} - 40.0213 \text{ cm} = 0.0744 \text{ cm}$$

Solving Eq. 13-1 for α, we get

$$\alpha = \frac{\Delta l}{l \, \Delta t} = \frac{0.0744 \text{ cm}}{(40.0 \text{ cm})(100 \text{ C}°)}$$

$$\boxed{= 18.6 \times 10^{-6} \, (\text{C}°)^{-1}}$$

[*] For a few substances, over a limited range of temperature, α is negative; such substances shrink upon heating.

The unit for α is simply $1/\mathrm{C}°$ or $(\mathrm{C}°)^{-1}$, a fractional change in length per degree. In words, a coefficient of 18.6×10^{-6} $(\mathrm{C}°)^{-1}$ means that any given length of the material changes by 18.6 parts per million of its original length, for every Celsius (centigrade) degree change in temperature.

EXAMPLE **13-4**

A piece of steel railroad track is exactly 20 m long in the winter, at $-10°\mathrm{C}$. What is its length on a hot summer day when the temperature is $35°\mathrm{C}$?

$$\Delta l = l\alpha \, \Delta t$$
$$= (20 \text{ m})[11 \times 10^{-6} \, (\mathrm{C}°)^{-1}](45 \text{ C}°)$$
$$= 0.010 \text{ m}$$

We know that steel expands when it is heated, so we *add* Δl to the original length. The new length is

$$l' = 20.000 \text{ m} + 0.010 \text{ m} = \boxed{20.010 \text{ m}}$$

In railroad construction, rail sections are as much as a mile long; they are prevented from buckling by careful anchoring to the ground. Similarly, concrete roads are poured in slabs to allow for expansion (see photograph on p. 287). The forces generated by thermal expansion are large, as illustrated by the following example.

EXAMPLE **13-5**

A steel rail exactly 20 m long at $-10°\mathrm{C}$ is firmly clamped between two fixed supports exactly 20 m apart. What is the force of compression in the rail if its temperature rises to $35°\mathrm{C}$? The rail's cross-sectional area is 50 cm² $(= 50 \times 10^{-4} \text{ m}^2)$.

The force is the same as would be required to compress the rail back to its original length. For this we use Young's modulus (Sec. 9-1). The change in length Δl is 0.010 m as computed in Example 13-4. According to Table 9-1 the stretch modulus for steel is $20 \times 10^{10} \text{ N/m}^2$.

$$\frac{\Delta F}{A} = \frac{E \, \Delta l}{l}$$
$$\Delta F = \frac{EA \, \Delta l}{l}$$
$$= \frac{(20 \times 10^{10} \text{ N/m}^2)(50 \times 10^{-4} \text{ m}^2)(0.010 \text{ m})}{20 \text{ m}}$$
$$= \boxed{5.0 \times 10^5 \text{ N}}$$

This is more than 100 000 lb of force.

Considered as a thermometric property, linear expansion of solids is exploited in thermostats and inexpensive thermometers. To gain sensitivity, thin strips of different metals are welded or riveted together along

T A B L E **13-1**
Coefficients of expansion

Substance	Coefficient of linear expansion α, $(\mathrm{C}°)^{-1}$	Coefficient of volume expansion β, $(\mathrm{C}°)^{-1}$
Solids		
iron or steel	11 $\times 10^{-6}$ •	33 $\times 10^{-6}$
aluminum	26	77
brass	19	56
ordinary glass	8.5	26
Pyrex glass	3.3	10
fused quartz	0.4	1
platinum	9.0	27
concrete	12	36
lead	29	87
Liquids		
methyl alcohol		1134 $\times 10^{-6}$
carbon tetrachloride		581
glycerin		485
mercury		182
turpentine		900
gasoline		960
Gases •		
air		3670 $\times 10^{-6}$
carbon dioxide		3740
hydrogen		3660
helium		3665

• See p. 199 for a note on powers of 10.
• At constant pressure.

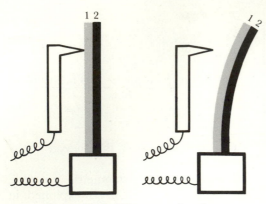

Figure 13-4 Thermostatic relay.

their lengths and clamped at one end. When the temperature changes, the combination must bend in an arc, as shown in the schematic diagram of a thermostat (Fig. 13-4), where metal 1 has a greater coefficient of expansion than does metal 2. As the temperature rises, the electric contact is opened, turning off the furnace. In a thermometer of this type, a more sensitive indication is achieved by forming the bimetallic strip into a compact spiral which would be many inches long if unrolled.

The advantages of low coefficient of expansion are illustrated by the use of Pyrex brand glass for ovenware. Glass is a poor conductor of heat, and so, when placed in an oven, the inner and outer surfaces of a dish become hot before the glass in the interior has a chance to warm up. Because of thermal expansion, stresses are set up which may shatter the glass. One way to avoid this is to heat the glass dish very slowly, to allow time for equalization of temperature throughout. Even rapid heating is possible if the glass has a low coefficient of expansion, as does Pyrex, for then the stresses will be small (see Table 13-1). Fused quartz has an even lower coefficient of linear expansion; a white-hot piece of quartz at 1500°C may be removed from a flame and plunged into liquid air ($-192°C$) without shattering.

Crystalline substances such as calcite (Iceland spar, $CaCO_3$) or blue vitriol (ordinary copper sulfate, $Cu_2SO_4 \cdot 5H_2O$) are held together by cohesive forces between atoms. Unless the crystal is highly symmetrical, the forces are likely to be different along different directions both because of differing interatomic distances such as a, b, and c (Fig. 13-5) and because of different strengths of forces between atoms of different kinds. We expect, therefore, that the coefficients of linear expansion will be different along the various directions in a crystal. Such is actually the case. For example, the linear expansion coefficient of a single crystal of bismuth metal varies from 15.4×10^{-6} $(C°)^{-1}$ in a certain direction to 10.8×10^{-6} $(C°)^{-1}$ in another direction at right angles to the first one. Ordinary bismuth metal expands equally in all directions, however, since it is made up of millions of tiny crystals oriented at random.

Both the elastic properties of a substance and its thermal expansion are intimately dependent upon the same basic phenomenon—cohesive forces between molecules. This is why substances that are easily compressed also usually have high expansion coefficients and low melting points. All these attributes of a metal such as lead are related to the relatively weak cohesive forces between atoms in the crystal.

13-4   Thermal expansion of liquids

In the case of liquids, the cohesive forces are so weak that a rigid crystal is not possible.

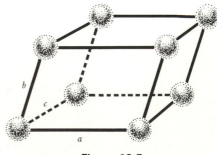

Figure 13-5

The KE of molecular motion is large enough to prevent "locking in" of the molecules. According to these ideas, we can understand why liquids generally are more easily compressible than are solids (Table 9-1) and why their thermal expansion is greater (Table 13-1). The *coefficient of volume expansion* is defined as the fractional change in volume per degree change in temperature. Thus an equation similar to Eq. 13-1 can be used:

$$\Delta V = V\beta\,\Delta t \qquad (13\text{-}2)$$

where ΔV is the change in volume, V the original volume, Δt the change in temperature, and β (Greek letter beta) the coefficient of volume expansion. To find the final volume, the amount ΔV is added or subtracted, according to whether the temperature has increased or decreased.

EXAMPLE **13-6**

A 2000 cm³ quartz beaker is brimful of turpentine at a room temperature of 20°C. If the beaker and contents are put in a refrigerator so that the temperature becomes 5°C, what is the volume of the turpentine?

We can ignore the expansion of the quartz beaker, since its coefficient is so small. In this problem, Δt is 15 C°. From Table 13-1, β is 900×10^{-6} (C°)$^{-1}$. Hence

$$\Delta V = V\beta\,\Delta t$$
$$= (2000 \text{ cm}^3)[900 \times 10^{-6} \text{ (C°)}^{-1}](15 \text{ C°})$$
$$= 27 \text{ cm}^3$$

We subtract this amount from the original volume, since the temperature has decreased. Hence the final volume of the turpentine is

$$2000 \text{ cm}^3 - 27 \text{ cm}^3 = \boxed{1973 \text{ cm}^3}$$

The liquid-in-glass thermometer consists of a bulb and an indicating tube of small inside diameter ("bore"). Mercury is commonly used as a thermometric substance down to −40°C, its freezing point; alcohol or propane can be used down to temperatures below −100°C. If the bulb is of large volume

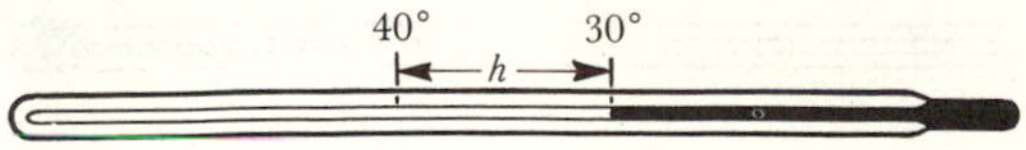

Figure 13-6

and the indicating tube is of very small bore, such a thermometer can be made sensitive enough to indicate changes of 0.001° or less.

EXAMPLE **13-7**

A mercury thermometer (Fig. 13-6) has a quartz bulb of volume 0.300 cm³ and a stem of bore 0.0100 cm. How far does the indicating thread of mercury move when the temperature changes from 30.0°C to 40.0°C? Ignore the expansion of the quartz bulb.

From Table 13-1, the coefficient of volume expansion for mercury is 182×10^{-6} (C°)$^{-1}$. Hence

$$\Delta V = V\beta\,\Delta t$$
$$= (0.300 \text{ cm}^3)[182 \times 10^{-6} \text{ (C°)}^{-1}](10.0 \text{ C°})$$
$$= 5.5 \times 10^{-4} \text{ cm}^3$$

Since the additional volume of the cylinder of mercury in the indicating tube is equal to the change in volume of the mercury, and since volume = (cross-sectional area)(height), we find for the height of rise

$$h = \frac{\text{volume}}{\text{area}} = \frac{5.5 \times 10^{-4} \text{ cm}^3}{\pi(0.005 \text{ cm})^2} = \boxed{7.0 \text{ cm}}$$

In practice, the bulb is made of ordinary glass, and its expansion would be appreciable. The bulb is a solid; since a solid expands in each of three independent directions, its volume coefficient is approximately three times its linear coefficient. (Note the relation between α and β for the solids in Table 13-1.) The volume coefficient of Pyrex is

$$3 \times 3.3 \times 10^{-6} \text{ (C°)}^{-1} = 10 \times 10^{-6} \text{ (C°)}^{-1}$$

The *relative* or *apparent* volume coefficient of expansion for mercury in Pyrex is then

$$(182 - 10) \times 10^{-6} \text{ (C°)}^{-1}$$
$$= 172 \times 10^{-6} \text{ (C°)}^{-1}$$

During its construction, the liquid-in-glass thermometer is sealed off at the end of the

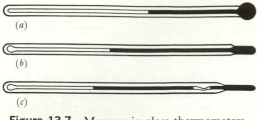

(a)

(b)

(c)

Figure 13-7 Mercury-in-glass thermometers.

capillary tube to prevent loss of liquid by evaporation. The upper part is evacuated (Fig. 13-7*a*) in order to prevent building up a pressure as the mercury expands. Sometimes the space above the liquid is filled with an inert gas, and a little cavity at the very top prevents excessive build-up of pressure (Fig. 13-7*b*). The clinical thermometer (Fig. 13-7*c*) has a constriction just above the bulb so that the mercury thread is broken as the temperature falls. Surface tension keeps the mercury from flowing back through the narrow opening. In this way the patient's temperature can be read after the thermometer has been removed.

Most liquids expand more or less uniformly, as does mercury (Fig. 13-8). However, the expansion of water shows a curious behavior; the volume of 1 g of pure water varies with

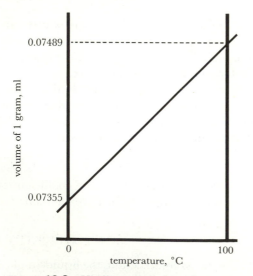

Figure 13-8 Uniform expansion of mercury.

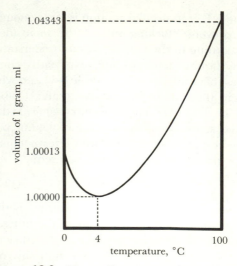

Figure 13-9 Non-uniform expansion of water (scales exaggerated for clearness).

temperature as shown in Fig. 13-9. The cause of this anomalous behavior of water is not fully understood. In the region 0 to 4.0°C, water *contracts* as the temperature increases, and the volume coefficient is negative. At 4.0°C, water has the least volume and maximum density, and the coefficient of expansion is zero. When water is used as a reference material in determining the specific gravity of other substances, the temperature of 4.0°C is often used, since variations in the temperature at which measurements are made will then have only a slight effect on the density of water near this point.

Biological life cycles are profoundly affected by the expansion characteristics of water. In a fresh-water lake, as the surface water becomes chilled to 4°C, it becomes denser and sinks, thereby pushing up warmer water from below which in turn is chilled at the surface. A continual mixing takes place, and the water deep in the lake remains at about 4°C until the colder (and less dense) water at the surface freezes. This is why lakes freeze from the surface down—a fortunate fact for the survival of living organisms in fresh water. The situation is complicated by the fact that the temperature of maximum density decreases with increasing pressure. This leads to incipient instability and allows wind-induced mixing to become effective in cooling deep water below 4°C.

13-5 Thermal expansion of gases

In contrast to those in solids and liquids, the cohesive forces between the widely separated molecules of a gas are so small as to be almost negligible. The thermal expansion of a confined gas is intimately related to its pressure as we shall see in Chap. 15. For the present, let us imagine the pressure on a confined mass of gas to be held constant. Common experience tells us that, as the temperature increases, we must not allow the pressure to build up; otherwise the volume change will be meaningless, or even zero. Therefore we specify that the coefficient of volume expansion for a gas is measured at *constant pressure*. A glance at Table 13-1 shows that the coefficients for gases are much larger than those for liquids or solids. It is also evident that the coefficients for all gases are practically the same. This is to be expected, since the only differences between the various gases would be in their cohesive forces, which are so small as to have only a minor effect, or in the variable sizes of the molecules, which in a gas are tiny compared with the distances between molecules. We shall consider the gas laws in detail in Chap. 15.

In applying the standard expansion equation, $\Delta V = V\beta\,\Delta t$, to gases, we must be cautious in our choice of initial volume V. Since, in contrast to solids or liquids, the change in volume ΔV may be a large fraction of the volume, it is necessary to make a definite statement about the measurement of V itself. The values of β listed in Table 13-1 are *based on the original volume at 0°C.* Thus we may say that hydrogen expands by 0.00366 of its 0° volume for every degree rise in temperature, if the pressure remains constant.

The gas thermometer (Fig. 13-10) is capa-

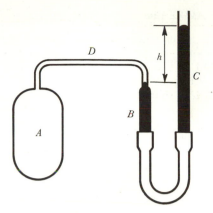

Figure 13-10 Gas thermometer.

ble of high precision, and it is a standard with which other types of thermometers may be compared and calibrated. If the gas container A is heated, the gas expands and pushes the mercury in arm B down and that in arm C upward. Thus the height h changes, and so there is a pressure change as well as a volume change. The flexible connecting tubing allows arm C to be lowered, to bring the pressure back to the original value. When so used, the thermometric property is the *volume* of the gas, and the device is a *constant-pressure gas thermometer*. The connecting neck D is of small volume compared with A, but even so, correction must be made because this portion of the gas is not at the same temperature as that in A. The gas thermometer cannot, of course, be used below the temperature at which the gas liquefies.

The chief advantage of a gas as the thermometric substance is that the readings are relatively independent of the exact composition of the gas. What slight differences remain between the behavior of hydrogen and helium, for instance, can be allowed for, since the necessary corrections are small and well-known. The hydrogen gas scale, so corrected, is called the ideal gas scale and has considerable theoretical importance (Chap. 16). In actual use, the gas thermometer is complicated and bulky and does not respond quickly to changes of temperature.

EXAMPLE **13-8**

In a constant-pressure gas thermometer the volume of the gas is 546 cm³ at 0°C and 746 cm³ at 100°C. What is the temperature when the volume is 628 cm³?

The volume increased by 200 cm³ for a 100 C° change in temperature; each increase of 1 cm³ corresponds to a temperature increase of 0.50 C°. The observed volume is 82 cm³ greater than the volume at 0°C. Hence the temperature is

$$t = (82 \text{ cm}^3)\left(\frac{0.50 \text{ C}^\circ}{1 \text{ cm}^3}\right) = \boxed{41°C}$$

An alternative solution uses a proportion:

$$\frac{100°C - 0°C}{746 \text{ cm}^3 - 546 \text{ cm}^3} = \frac{t - 0°C}{628 \text{ cm}^3 - 546 \text{ cm}^3}$$

$$t = \boxed{41°C}$$

It is also possible and more convenient to use the gas thermometer in such a way that the *volume* remains constant; the thermometric property is then the *pressure* exerted by the gas. If the necessary (small) corrections are made, the calculated temperatures agree with each other for the constant-pressure gas thermometer and the constant-volume gas thermometer.

■ SUMMARY

Temperature is a measure of how hot or cold a body is, as determined by some type of thermometer. The fixed points upon which the Celsius (centigrade) scale is based are the melting point of ice (0°C) and the boiling point of water (100°C). On the Fahrenheit scale these points are at 32°F and 212°F. A change of 1 Celsius degree is the same as a change of $\frac{9}{5}$ Fahrenheit degrees. Various thermometric properties of matter may be used to define temperature scales. Commonly used thermometers make use of the expansion of gases and liquids and also of changes in electrical properties; at ordinary temperatures, such thermometers are in close, though not exact, agreement with each other.

The change in length of a solid is proportional to the original length and to the change in temperature, according to the equation $\Delta l = l\alpha \Delta t$. The factor α is the coefficient of linear expansion. Similarly, the change in volume of a solid, liquid, or gas is given by $\Delta V = V\beta \Delta t$, where β is the coefficient of volume expansion. For a solid, β is approximately 3α.

■ CHECK LIST

thermometric property
Celsius (centigrade) temperature scale
Fahrenheit temperature scale
coefficient of linear expansion

coefficient of volume expansion
constant-pressure gas thermometer
constant-volume gas thermometer

■ QUESTIONS

13-1 Name several thermometric properties of matter.

13-2 In the manufacture of the first light bulbs, the electrical leads were made of platinum, which has a coefficient of expansion nearly equal to that of ordinary glass. Explain how this helps make a firm glass-metal joint.

13-3 If the stem of the thermometer described in Example 13-7 were engraved every 0.2°C, how far apart would the markings be?

13-4 Analyze the act of shaking down a fever thermometer, in terms of mass, forces, and acceleration.

13-5 Describe qualitatively the construction of a mercury-in-glass thermometer which indicates temperatures between 42°C and 46°C on a scale 20 cm long.

13-6 Why are the concrete slabs on a highway separated from each other by narrow cracks filled with pitch?

13-7 In using the equation $\Delta l = l\alpha \,\Delta t$, which length l should be used—the initial length or the final length? Does it really matter which length is used?

13-8 Why is it incorrect to say that hydrogen expands by 0.00366 of its original volume for every Celsius (centigrade) degree rise in temperature?

13-9 If the room temperature goes up, barometric pressure remaining the same, how will the reading of a mercury barometer be affected? Will it seem to indicate a pressure that is too high or too low? (*Hint:* The density of the mercury changes. See Sec. 12-3.)

13-10 Why doesn't an aluminum cup shatter when it is suddenly filled with boiling water?

13-11 A circular metal hoop is sawed through at one point, leaving a gap of 1.4 mm. If the hoop is then put in an oven and heated uniformly, will the air gap become larger or smaller?

13-12 A hole is cut in a flat metal plate. When the plate is heated, will the hole become larger or smaller?

13-13 Explain how a housewife loosens the metal cap on a bottle of catsup by holding it briefly in boiling water.

■ PROBLEMS

13-A1 A typhoid-fever patient's temperature ranged from 96.8°F to 105.8°F during the course of the illness. Express these temperatures in °C.

13-A2 What is the Fahrenheit temperature at which pure water has its maximum density?

13-A3 Express in °C: 68°F; 10°F; −328°F; 98.6°F.

13-A4 Express in °C: 0°F; −40°F; 842°F; 77°F.

13-A5 Express in °F: 20°C; 500°C; −40°C; −273°C.

13-A6 Express in °F: 40°C; −10°C; 6000°C; −78°C.

13-A7 A patient's temperature decreased by 3.6 F° overnight. Express this change in C°.

13-A8 A bridge expands 2 mm for every Fahrenheit degree temperature change. How much does it expand for a change of 1 C°?

13-A9 Steel rails 40 ft long are laid on a winter day when the temperature is −10°C. How much space must be left between the rails to allow for expansion at a summer temperature of 40°C?

13-A10 A metal rod, observed to be 102.302 cm long at 20°C, is 102.433 cm long at 70°C. Out of which of the metals listed in Table 13-1 might the rod be made?

13-A11 A radiator in the attic of a house is connected to the furnace by a vertical iron pipe 10.0 m long. If the radiator rests snugly on the floor when the pipe is at 5°C, how high off the floor will it be lifted when the pipe contains steam at 100°C?

13-A12 (a) Show, directly from Table 13-1, that the coefficient of linear expansion for steel is about 6.1×10^{-6} (F°)$^{-1}$. (b) The steel main span of the Golden Gate bridge is 4200 ft long. What allowance must be made for expansion, if the temperature varies between 10°F and 135°F?

13-A13 What is the volume coefficient of expansion of a metal whose linear coefficient is 15×10^{-6} (C°)$^{-1}$?

13-A14 A block of metal is 5.000 cm $\times$ 8.000 cm $\times$ 10.000 cm at 25°C. The linear coefficient of expansion is 20×10^{-6} (C°)$^{-1}$. What is the volume of the block if it is immersed in an ice bath?

13-A15 A motorist purchases 10 gallons of gasoline on a cold day when the temperature is -20°C. What is the volume of this gasoline after it has warmed up to $+20$°C?

13-B1 Calculate the reading on the original Celsius scale (see second footnote on p. 288) corresponding to 68°F.

13-B2 Newton used a thermometer filled with linseed oil. The ice point was 0°N, and body temperature was 12°N. What is the temperature on this scale of an ice-salt mixture which is at 5°F? Suggest one or more disadvantages of this scale.

13-B3 On the Reaumur temperature scale, still used in some European countries, the ice point is 0°R and the steam point is 80°R. What is the Fahrenheit temperature corresponding to 50°R? What Reaumur temperature corresponds to 68°F?

13-B4 At what temperature is the reading of a Fahrenheit thermometer three times that of a Celsius thermometer?

13-B5 In assembling a steel bridge, a steel rivet 1.613 in. in diameter is to be placed in a hole 1.612 in. in diameter. Would it be possible to do the job by first cooling the rivet in Dry Ice (-78°C)? Air temperature is 22°C.

13-B6 An iron ball 5.000 cm in diameter is 0.001 cm too large to go through a hole in a brass plate when the ball and plate are at 20°C. At what temperature, the same for both ball and plate, will there be a snug fit?

13-B7 An iron rim is to be placed around a wagon wheel whose circumference at 15°C is 270.562 cm. The rim is in the form of a band having cross section 2.50 cm^2 and inside circumference at 15°C of 270.020 cm. (a) To what temperature must the band be heated to fit snugly on the wheel? (b) What tension will exist in the band after it cools to 15°C?

13-B8 In precise surveying, a steel tape is used which is calibrated at 20°C and is at a specified tension. If the tension is maintained constant, but the temperature increases to 35°C, (a) what is the percent error in the reading of the tape? (b) Does the tape read too high or too low? (c) What would be the error, in cm, when measuring a lot 12.00 m wide?

13-B9 A brass cannon has a bore 10.000 cm in diameter at 20°C. A steel liner, in the form of a thin cylinder of outside diameter 10.020 cm at 20°C, is to be inserted. To what temperature should the cannon be heated, so that the liner (still at 20°C) can be inserted?

13-B10 To what temperature should both the cannon and liner of Prob. 13-B9 be heated so that the liner becomes loose and can be extracted?

13-B11 An aluminum rod of length L and cross section 2 cm^2 is cooled from 200°C to 30°C. What force is needed to stretch the rod back to its original length?

13-B12 During a Young's modulus experiment, weights were stretching a brass wire of length L and cross section 2.00×10^{-6} m^2. Room temperature dropped by 10 C°; what added force was needed to restore the wire to its original length?

13-B13 A lead rod and a brass rod, each 80.00 cm long at 0°C, are clamped together at one end with their free ends coinciding. Compute the separation of the free ends of the rods when the system is placed in a steam bath.

13-B14 An aluminum cup is brim-full, containing 200 cm³ of glycerin at 10°C. What volume of glycerin overflows when the cup is placed in an oven which is at 125°C?

13-B15 A 50.00 gallon open steel drum is just filled with turpentine when the temperature is 45°F. The drum and contents are heated to 81°F and then allowed to cool to 45°F again. How much turpentine remains in the drum?

13-B16 An aluminum block of volume V is heated from 10°C to 30°C. What pressure must be applied to keep the block from expanding?

13-B17 What is the apparent coefficient of volume expansion of methyl alcohol in an aluminum container?

13-B18 A fused quartz container is completely filled with mercury and sealed off at 20.00°C. What is the pressure, in atm, inside the container if the temperature becomes 20.30°C?

13-B19 Starting at 0°C, through what temperature change would hydrogen have to be heated in order to cause a 3.66% change in volume?

13-B20 In a constant pressure gas thermometer the volume of the gas is 400 cm³ at 20°C and 440 cm³ at 100°C. What temperature is indicated when the volume is 500 cm³?

13-B21 A rod has original length l_0. When the temperature changes by Δt, its final length is l_t. The coefficient of linear expansion is α. Prove that $l_t = l_0(1 + \alpha \, \Delta t)$.

13-B22 A body has original volume V_0. When the temperature changes by Δt, its new volume is V_t. The coefficient of volume expansion is β. Prove that $V_t = V_0(1 + \beta \, \Delta t)$.

13-C1 A clock has a period of exactly $\frac{1}{2}$ s at 20°C. The pendulum consists of a small, heavy bob on a thin brass rod.   (a) What will the period be if the temperature rises to 30°C? (b) At the end of a day, how many seconds will the clock have gained or lost?

13-C2 At 20°C, a steel rod of length 40.000 cm and a brass rod of length 30.000 cm, both of the same diameter, are placed end to end between two rigid supports 70.000 cm apart, with no initial stress in the rods. The temperature of the rods is now raised to 50°C.   (a) How far from its old position is the new junction between the rods?   (b) What is the stress in each rod, at 50°C? (*Hint:* First find the total expansion, if the rods were free; then find out how much each rod must be compressed to return the system to 70.000 cm length. You have two equations to be solved simultaneously: one for the sum of the two changes, and one for the ratio of the two changes based on the values of the stretch modulus.)

13-C3 At 20°C, the specific gravity of aluminum is 2.699 and that of glycerin is 1.248. A block of aluminum whose volume is 100.0 cm³ at 20°C is immersed in glycerin at 20°C. (a) What is the apparent weight of the aluminum block?   (b) What is the apparent weight of the block if the system is heated to 80°C?

■ REFERENCES

1 Wilson, R. E., "Standards of Temperature," *Physics Today*, Jan., 1953, p. 10.

2 Boyer, Carl B., "Early Principles in the Calibration of Thermometers," *Am. J. Phys.*, **10**, 176 (1942). Two-point calibration; one-point calibration; Fahrenheit, Celsius, Réaumur, and Kelvin scales.

Now that we have studied the measurement of temperature, we are able to discuss the quantitative phenomena of heat which involve the interactions of material bodies.

14-1 The caloric theory and its overthrow

If a kilogram of lead and a kilogram of iron are placed side by side in front of a brisk fire which is burning in a fireplace, and if the blocks of metal are allowed to stand for, say, ten minutes, it is observed that the lead becomes warmer than the equal mass of iron. Experiments of this sort were made in the late 1700's by Black, Rumford, and others, in an effort to test the theory that heat is a fluidlike substance, called "caloric," which flows from a hot body into a cold one. The experimental facts (which are as true today as 200 years ago) are something as follows: Every material object, such as a lead ball or an aluminum saucepan, has a "thermal capacity," which is a measure of the quantity of heat required to make it one degree warmer. Some early measurements were indeed made with fires in fireplaces serving as the source of heat, and it was assumed that a fire, burning steadily, produced in 40 minutes twice as much "quantity of heat" as it did in 20 minutes. It was soon found that different iron objects had different thermal capacities, in proportion to their masses, and so the concept of "specific heat" was introduced to represent the thermal capacity per unit mass. We shall define these quantities in Sec. 14-3 and give units for them.

Nothing in the above paragraph would contradict the idea that heat is a material fluid. As a matter of fact, scientists still speak of the "flow" of heat, but no longer is it supposed that it is an actual physical substance (the caloric of the old theory) that flows. The existence of caloric would mean that heat should have weight and mass, and a body should gain weight and mass when it is heated. That this does not happen was demonstrated about 1760 by Joseph Black (1728–1799) and is evidence against the existence of a caloric having weight.

The investigators could still "save the phenomenon" by imagining caloric to be a *weightless* fluidlike substance, intimately associated with the material particles of a substance. In 1797 Count Rumford (1753–1814) designed a simple but decisive experiment to test whether a fluid (weightless or not) resided in a heated body. One of the most colorful

A black body is a good emitter and absorber of radiation. The word "BLACK" was painted on a porcelain dish and a flame directed at the inside surface, behind the letter A. The heated black surface emits more radiation than does the surrounding porcelain dish, which is at the same temperature.

natural philosophers of his time, Rumford was born Benjamin Thompson in North Woburn, Massachusetts. After a stay in England, he became minister of war in Bavaria, where he received his title. While supervising the boring of cannon at the military arsenal at Munich, he noticed the great rise in temperature of the shavings left in the cannon by the boring tool. Unlike the dozens or hundreds of his predecessors in many countries who must have observed this heat, Rumford had some scientific training and an inquiring mind, which was interested in discovering basic scientific facts, even if they were not immediately useful. Rumford was finally led to press a *blunt* boring tool against the inside of a cannon which was being turned by horses. The stationary tool exerted a considerable force—about 10 000 lb—on the 113 lb cannon, and, after half an hour of turning, the cannon's temperature had risen to 130°F—strong evidence of the production of a large quantity of heat by friction. Only about 0.12 lb of gun metal had been scraped loose by the action of the borer, and Rumford reasoned that it was quite unlikely that so much caloric could come from this minute amount of metallic dust. (He had previously determined that metal powder or scrapings had the same thermal behavior as did bulk metal.) In another experiment, Rumford caused water to boil by the heat generated by friction: "It would be difficult to describe the surprise and astonishment expressed in the countenances of the by-standers, on seeing so large a quantity of cold water heated, and actually made to boil without any fire." And Rumford confessed to "a degree of childish pleasure" at the result of the experiment.

If the cannon were turned for a long enough time, an amount of caloric approaching infinity could be obtained from a finite amount of gun metal which itself suffered no apparent change. Rumford rejected the idea of caloric for this reason, and concluded that heat is a form of motion.

Sir Humphry Davy (1778–1829) also did some crucial experiments at about the same time. In one carefully designed experiment Davy managed to melt some blocks of ice by rubbing them against each other, on a day when the temperature was below 0°C. In another experiment, performed by a clockwork mechanism which rested on a cake of ice in an evacuated chamber, some wax was melted by friction.

The caloric theory was ingenious, and, in its day, explained all the then known phenomena; it was "true" in the same sense that any physical theory is "true" (see Sec. 4-10). The theory was overthrown when well-thought-out experiments gave new results not explainable by the existence of caloric. This is a fascinating chapter in the history of science. Rumford's own description of his motives, his experiments, and his conclusions should be read by everyone interested in the development of scientific methods. (See references at the end of this chapter.)

14-2 Heat as energy— the first law of thermodynamics

We now know that heat is a form of energy and that the molecules of a body gain energy when heat "flows" into it.[*] Heat energy *can* be measured in the usual energy units, such as the joule or foot·pound. However, before these matters were fully understood, special units for "quantity of heat" were used, and they have persisted to this day. A *calorie* was originally defined in thermal terms as the heat energy required to raise the temperature of one gram of water one Celsius degree.[*]

---

[*] Since the molecules of a hot body are, on the average, moving faster than they are when the body is cold, they have slightly greater mass, according to the Einstein equation $\Delta E = (\Delta m)c^2$ (Sec. 7-7). To this extent, heat does have mass. But this increase in mass is so slight as to be, so far, entirely beyond detection.

[*] More precisely, from 14.5°C to 15.5°C. The Calorie, with a capital C, used by dieticians, is the kilocalorie (kcal). It is the heat required to raise the temperature of one kilogram of water one Celsius degree. This Calorie is 1000 calories.

Likewise, the *British thermal unit* (Btu) is the heat energy required to raise the temperature of one pound[•] of water one Fahrenheit degree. It is useful to know that 1 Btu is about equal to 252 cal.

Much effort and ingenious experimentation lie behind the simple statements that *one calorie is equivalent to 4.18 joules of energy*, or *one Btu is equivalent to 778 ft·lb of energy*. Implicit in these statements is the fact that, whenever mechanical energy (KE or PE) is transformed into heat energy, the rate of exchange, under all circumstances, is 4.18 J for each calorie, or 778 ft·lb for each Btu. In other words, the law of conservation of energy is valid, if heat energy is included as one form of energy. When so understood, the law of conservation of energy is called the *first law of thermodynamics.*

The first law of thermodynamics did not spring into being fully formed, but rather it evolved from the speculations and experiments of many workers. As long ago as 1797, Rumford had an inkling of the first law when he reflected on the production of heat in his cannon-boring experiment. He wrote: ". . . in a case of necessity, the Heat thus produced might be used in cooking victuals. But no circumstances can be imagined, in which this method of procuring Heat would not be disadvantageous; for, more Heat might be obtained by using the fodder necessary for the support of a horse, as fuel." Here he anticipated the energy relationships of metabolism, an aspect of biology which he was unequipped to explore. Julius Mayer (1814–1878), a German physician, published in 1842 a paper in which he stated that "the warming of a given weight of water from 0°C to 1°C corresponds to the fall of an equal weight from the height of about 365 meters." We see from this that Mayer had a clear idea

of the transformation of energy between mechanical and thermal forms. James Joule (1818–1889) published a series of papers from 1843 to 1850 in which he measured the mechanical equivalent of heat in many ways: he churned water by a paddle system; he stirred mercury with iron paddles; he heated water by friction, passing it though a thick piston perforated with many small holes; he rubbed iron rings together under mercury; he measured the heat produced electrically in a wire through which electrons flowed. Using all these methods, Joule obtained results agreeing within 5%, truly a remarkable achievement. It is because of the variety of his methods that Joule's work is of such great significance. The German physiologist and physicist Hermann von Helmholtz[•] (1821–1894) presented in 1847 a great paper which consolidated the work of others and gave the first systematic publication of the law of conservation of energy with convincing mathematical arguments.

We see that the decade 1840–1850 was remarkably fruitful, and it is perhaps pointless to try to assign "credit" to any of Mayer, Joule, or Helmholtz for the first law of thermodynamics. Our brief excursion into the history of science shows how a major discovery often develops gradually, with a "break-through" coming as a result of independent but simultaneous efforts on the part of a number of workers. Present-day physics is in this tradition.

14-3 Specific heat

As we noted in Sec. 14-1, substances differ markedly in the temperature rise caused by the absorption of a given amount of heat energy. To describe this property of matter, we define the *specific heat* of a substance as the heat energy absorbed per unit mass, per degree rise in temperature. According to this

---

[•] Strictly speaking, we mean an amount of water which weighs 1 lb on the surface of the earth. On the moon, where the acceleration due to gravity is only $\frac{1}{6}$ that on the earth, 1 lb of water would represent 6 times as many molecules, and would require 6 Btu.

[•] A direct descendant, on his mother's side, of William Penn.

definition, the specific heat c of a substance equals the number of calories required to raise the temperature of one gram of the substance one Celsius degree. Present practice is to define the calorie as exactly 4.184 J; the specific heat of water is then an experimentally determined quantity very close to 1.00 cal/g·C°. As is true of most properties of water, the specific heat varies slightly with temperature. The specific heat of aluminum is 0.22 cal/g·C°, meaning that it requires only 0.22 cal to raise the temperature of 1 g of aluminum 1 C°. We see that water is a "standard substance" with which other substances may be compared easily. In engineering work the British thermal unit is used, and the specific heat of water is 1 Btu/lb·F°. For any given substance, the specific heat in cal/g·C° (or in kcal/kg·C°) is *numerically* equal to the specific heat in Btu/lb·F°. As is seen from Table 14-1, water has by far the largest specific heat of any common substance.

How much heat is required to raise the temperature of 400 g of aluminum from 40°C to 45°C? First we note that if we had 400 g of *water,* the heat required would be 400 cal for every degree, or $400(5) = 2000$ cal in all. But aluminum requires only 0.22 as much heat as does water, since the specific heat of aluminum is 0.22 cal/g·C° (Table 14-1). Therefore the required heat is (2000 cal)(0.22) = 440 cal. If 400 g of aluminum were to cool from 45°C to 40°C, 440 cal of heat would "flow out" of the aluminum. In general, if the temperature of a body changes by an amount Δt, the heat ΔQ that flows in or out of it is given by

$$\Delta Q = mc\,\Delta t \qquad (14\text{-}1)$$

where m is its mass and c is its specific heat. The product mc is called the *thermal capacity,* or *water equivalent,* of the body. The latter term is more descriptive; for instance, in the above discussion, the aluminum could have been replaced by 88 g of water insofar as its thermal behavior is concerned. Then specific

TABLE 14-1
Specific heats of some solids, liquids, and gases

Substance	Temperature or temperature range, °C	Specific heat, cal/g·C° or kcal/kg·C° or Btu/lb·F°
Solids		
iron or steel	20	0.11
copper	15–100	0.093
aluminum	15–100	0.22
glass	20–100	0.20
lead	20	0.0306
sodium chloride	20	0.21
ice	−3	0.50
wood	...	0.4
asbestos	20–98	0.195
silver	20	0.056
Liquids		
water	15	1.00
mercury	20	0.033
glycerin	50	0.60
methyl alcohol	20	0.60
benzene	20	0.41
Gases*		
steam	110	0.48
air	50	0.25
helium	20	1.24
neon	20	0.25

* At constant pressure.

heat would no longer enter into the problem; the heat flow would be given by

(water equivalent)(temperature change)
$$= (88 \text{ cal/C°})(5 \text{ C°}) = 440 \text{ cal}$$

Some of the values of specific heat given in Table 14-1 are average values, calculated from $c = (1/m)(\Delta Q/\Delta t)$ for the temperature intervals Δt shown in the table. In the limit as Δt approaches zero, the specific heat at a given temperature t is found from a derivative:

$$c = \lim_{\Delta t \to 0} \left[\frac{1}{m}\left(\frac{\Delta Q}{\Delta t} \right) \right]$$

or

$$c = \frac{1}{m}\frac{dQ}{dt} \qquad (14\text{-}2)$$

This is the significance of the specific heat values quoted in the table for some definite temperature, such as 20°C.

The variation of specific heat with temperature is ordinarily rather small, but at very low temperatures the specific heat of a solid is usually strongly dependent on the temperature.

EXAMPLE 14-1

A 200 g sample of a chemical compound, originally at $-260°C$, is heated electrically to $-250°C$. The heat absorbed is plotted in Fig. 14-1 as a function of temperature. Calculate the specific heat of this material at $-255°C$.

Since the Q versus T curve is not a straight line, we see that the specific heat is not constant. According to Eq. 14-2, the specific heat is proportional to the derivative of Q with respect to temperature; to evaluate this derivative graphically, we draw a tangent to the curve at $t = -255°C$ and find the slope of the tangent.

$$\frac{dQ}{dt} = \text{slope of tangent}$$

$$= \frac{6.2 \text{ cal} - 0.0 \text{ cal}}{-250.0°C - (-258.1°C)}$$

$$= 0.765 \frac{\text{cal}}{C°}$$

$$c = \frac{1}{m}\frac{dQ}{dt} = \frac{0.765 \text{ cal}/C°}{200 \text{ g}}$$

$$= \boxed{0.0038 \text{ cal}/\text{g}\cdot C°}$$

In many cases, where a system of bodies is isolated and no mechanical work is done on or by the system, conservation of energy means conservation of *heat* energy. The heat gained by one part of the system equals that lost by another part of the system, as illustrated below.

EXAMPLE 14-2

An automobile radiator contains 20 liters of water. If 200 000 cal of heat is given to the

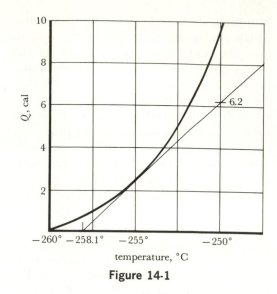

Figure 14-1

cooling system by the motor, what is the rise in temperature?

First we find the mass of the water; 20 liters = 20 000 cm³.

$$\text{Mass} = (\text{density})(\text{volume})$$
$$= (1 \text{ g/cm}^3)(20\,000 \text{ cm}^3) = 20\,000 \text{ g}$$

Now consider the isolated system

$$[(\text{water}) + (\text{cylinder block of motor})]$$

Heat gained by water
$$= \text{heat lost by cylinder block}$$

$$mc\,\Delta t = 200\,000 \text{ cal}$$
$$(20\,000 \text{ g})(1 \text{ cal/g}\cdot C°)(\Delta t) = 200\,000 \text{ cal}$$

$$\Delta t = \frac{200\,000 \text{ cal}}{20\,000 \text{ cal}/C°} = \boxed{10 \text{ C}°}$$

EXAMPLE 14-3

Repeat the previous example, assuming that the radiator is filled with glycerin instead of water.

From Table 2-1, the density of glycerin is 1.26 g/cm³. Twenty liters of glycerin therefore has a mass of 25 200 g.

Heat gained by glycerin
$$= \text{heat lost by cylinder block}$$
$$(25\,200 \text{ g})(0.60 \text{ cal/g}\cdot C°)(\Delta t) = 200\,000 \text{ cal}$$

$$\Delta t = \frac{200\,000 \text{ cal}}{15\,120 \text{ cal}/C°} = \boxed{13.2 \text{ C}°}$$

It is evident that glycerin is not so efficient as is water in cooling a car engine. In fact, if water cost $100 a gallon, it would still be the most desirable coolant, since a given amount of heat would cause less temperature rise in water than in any other common liquid.

The *method of mixtures* is often used to measure specific heat in the laboratory, as in the following example:

EXAMPLE 14-4

400 g of lead BB shot is heated to 100°C and dropped into a 100 g glass beaker containing 200 g of water at 20°C. The specific heat of glass is 0.20 cal/g·C°. The mixture is stirred until temperature equilibrium is reached; the final temperature is 24.2°C. What is the specific heat of lead?

$$\binom{\text{heat gained}}{\text{by water}} + \binom{\text{heat gained}}{\text{by beaker}} = \binom{\text{heat lost}}{\text{by lead}}$$

$$[(200 \text{ g})(1 \text{ cal/g·C°})(24.2° - 20.0°)]$$
$$+ [(100 \text{ g})(0.20 \text{ cal/g·C°})(24.2° - 20.0°)]$$
$$= [(400 \text{ g})(c)(100° - 24.2°)]$$

$$840 \text{ cal} + 84 \text{ cal} = (400 \text{ g})(c)(75.8 \text{ C°})$$

$$c = \frac{924 \text{ cal}}{(400 \text{ g})(75.8 \text{ C°})} = \boxed{0.0305 \text{ cal/g·C°}}$$

The thermal capacity of the beaker is its mass times its specific heat, or

$$(100 \text{ g})(0.20 \text{ cal/g·C°}) = 20 \text{ cal/C°}$$

In other words, it is equivalent to 20 g of water, and we could have added this amount of "water equivalent" to the original 200 g of water for a simpler solution of the problem.

Another example will perhaps clarify the energy principle as exemplified by the first law of thermodynamics. Many mechanical problems are highly simplified by the neglect of friction. Let us now take friction into account, using the first law of thermodynamics (i.e., the law of conservation of energy).

EXAMPLE 14-5

A 30 kg boy on a 5 kg sled takes a running start from the top of a slippery hill, moving at 4.00 m/s. The hill is 60 m long and drops 2.00 m for every 10 m traversed (Fig. 14-2). The boy and sled reach the bottom of the hill, moving at 10.0 m/s. (a) How much mechanical energy has been transformed into thermal energy? (b) If the steel runners of the sled have a mass of 1.500 kg and if we assume that half the heat generated remains in the runners, by how much does the temperature of the steel runners rise during the ride downhill?

(a) The top of the hill is 12 m above the bottom. The weight of the boy and sled together is (35 kg)(9.8 m/s²) = 343 N. As the sled and boy slide down, some (but not all) of their initial PE is transformed into KE. The rest of the energy becomes heat energy. We now apply the first law of thermodynamics:

$$PE_1 + KE_1 = PE_2 + KE_2 + \Delta Q$$

where ΔQ is the heat energy acquired by sled and ground.

$$(343 \text{ N})(12 \text{ m}) + \tfrac{1}{2}(35 \text{ kg})(4.00 \text{ m/s})^2$$
$$= 0 + \tfrac{1}{2}(35 \text{ kg})(10.0 \text{ m/s})^2 + \Delta Q$$

$$4120 \text{ J} + 280 \text{ J} = 0 + 1750 \text{ J} + \Delta Q$$

$$\Delta Q = \boxed{2650 \text{ J}}$$

This is the amount of energy "dissipated" in heat. Of course it has not disappeared; it has merely been transformed into heat energy, a less useful form. To express this heat energy in kilocalories rather than joules we use the fact that 1 kcal = 4184 J.

$$\Delta Q = (2650 \text{ J})(1 \text{ kcal}/4184 \text{ J}) = 0.634 \text{ kcal}$$

(b) It is assumed that half the heat, i.e., 0.317 kcal, remains in the runners. To find the temperature rise of the steel runners is a simple problem involving the specific heat of steel (Table 14-1):

$$\Delta Q = mc \, \Delta t$$

$$0.317 \text{ kcal} = (1.500 \text{ kg})(0.11 \text{ kcal/kg·C°}) \, \Delta t$$

$$\Delta t = \boxed{1.92 \text{ C°}}$$

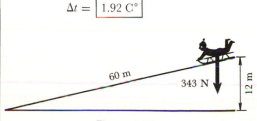

60 m   343 N   12 m

Figure 14-2

Under some circumstances the rise in temperature due to friction can be considerable. A piece of sandpaper becomes noticeably warm while being used. The iron brake shoes may become red-hot when a speeding train is brought to a halt. In other cases the temperature rise might be unobservably small; but it must be present, according to the first law. The stylus and the surface of a phonograph record must become slightly warmer as the record spins. In some cases it is not so easy to see that "disappearing" energy is really being transformed to heat because of friction, as in the case of an object on the end of a spring. As the object vibrates up and down, there is a continual transformation of energy from PE to KE and back to PE (Fig. 9-12). However, the vibration dies out because there are "frictional" forces between atoms which to a slight extent oppose the motion of the spring. To move against these forces requires work, regardless of whether the spring is stretching or contracting. There is no other source for this work, unless the total mechanical energy of the system is to be gradually used up. The spring becomes warmer. As we shall see in Chap. 16, *random* KE (heat energy) of the atoms increases, as the *ordered* KE of the whole body decreases and becomes zero. Energy is conserved.

14-4 Change of phase; latent heats

Matter exists in three states, known as *phases:* solid, liquid, and gaseous. Let us review the behavior of a common substance, H_2O. Let us place 1 g of it in the form of ice at, say, $-20°C$ in a thin metal cup on a hotplate maintained at $130°C$. According to Table 14-1, the specific heat of ice is 0.50 cal/g·C°, so a certain quantity of heat is required to raise it to $0°C$. Further heat is then absorbed, but it is found that, instead of the ice becoming warmer, the temperature remains $0°C$ while the ice melts. Experiment shows that about 80 cal of heat is required to melt

1 g of ice, the temperature remaining constant at $0°C$ the entire time. Only when all the ice has become water at $0°C$ does the temperature again begin to rise. In the early days, before heat was known to be a form of energy, it was expected that *any* absorption of heat would result in a temperature rise, and so the behavior of a melting piece of ice was a puzzle. The 80 cal was called the *latent* [*] *heat of fusion,* since its effect is to cause melting (fusion) rather than even the tiniest rise in temperature.

As we continue to apply heat, the H_2O, in the liquid phase now, warms up until $100°C$ is reached, when a similar behavior is observed. The water remains at $100°C$ until *all* of it is converted to water vapor (steam),[*] and then (if this gas is confined) the temperature begins to rise again. The *latent heat of vaporization* of water at $100°C$ is considerably greater than its latent heat of fusion—about 540 cal/g is absorbed without any rise in temperature. Finally, if water vapor in a sealed box is placed in a furnace, it can be heated as much as desired without any further latent heats, until the molecules (absorbing "heat of dissociation") are separated into hydrogen and oxygen atoms, and H_2O molecules cease to exist as such.

We interpret change of phase as a reorganization of the molecular arrangements. In a solid the molecules are so strongly bound to each other by cohesive forces that a definite crystal structure is possible. When ice melts, the molecules of liquid water are at about the same distance from each other as in the ice crystals, but they are no longer rigidly bound to each other. To disrupt the structure requires work (force times displacement); in much the same way, work must be done to break up a large rock into an equal amount of sand, even though the fragments have about the same total volume as the rock.

---

[*] From the Latin word *latere,* to lie hidden or concealed.
[*] Technically, "steam" is the same as "water vapor" and is invisible. The white visible cloud which is commonly miscalled steam is in reality a collection of many tiny droplets of liquid water, similar to a mist or fog.

TABLE 14-2   Latent heats

Substance	Melting point, °C	Latent heat of fusion Δh_f, cal/g or kcal/kg	Boiling point, °C	Latent heat of vaporization Δh_v, cal/g or kcal/kg
oxygen	−218	3.30	−183	51
ammonia	−75	108	−33	327
water	0	80	100	540
mercury	−39	2.8	357	70
lead	327	5.9	1753	206
aluminum	660	90	2450	2720
copper	1083	32	2300	1211
uranium	1133	20	3900	454
tungsten	3410	44	5900	1150

In the vapor state the molecules are actually separated to a considerable distance, and the required work is larger. The volume of 1 g of H_2O is 1.1 cm^3 in the form of ice, 1.0 cm^3 in the form of water, and 1700 cm^3 in the form of water vapor at 100°C. We know that there are cohesive forces acting to hold the molecules of a liquid together, as shown by the phenomenon of surface tension. Most (about 92%) of the 540 cal/g required to vaporize water at 100°C is used up in separating the molecules, doing work against internal cohesive forces; about 8% of the total energy is used in expanding the gas, doing work on the surrounding molecules which are pushing in with atmospheric pressure.

From what has been said, it is evident that heat energy may be absorbed in two ways: it may become random KE of the molecules, in which case the temperature increases, or it may become PE, in which case the temperature remains constant and the heat is a latent heat, stored in the substance and its surroundings[*] and available for doing work. We may take the gram of water through a reverse cycle, starting with water vapor at 130°C. After the temperature falls to 100°C, liquid drops of water form from the vapor, and the temperature remains at 100°C until all the vapor is condensed. The latent heat—540 cal/g—is *released* during this process. As the liquid H_2O cools down to 0°C, more heat is given off (as the KE of the molecules decreases), and then, while the H_2O changes from liquid to solid, 80 cal/g of heat is given off while the temperature remains at 0°C (the PE decreases and the molecules lock into position in tiny crystals). It is important to note that the latent heats of fusion and of vaporization (denoted by Δh_f and Δh_v) involve no temperature changes, and are expressed in calories per gram rather than calories per gram per degree as are specific heats.[*] Latent heats can be measured by the method of mixtures:

EXAMPLE 14-6

A well-insulated copper calorimeter cup of mass 100 g contains 400 g of water at 5°C. When 40 g of ice at −8°C is added, it is found that

[*]Recall that PE is a joint property of several bodies (Sec. 6-6). When water boils away in an open dish, there are two systems in which PE can be stored. Gravitational PE is stored in the system (earth + H_2O molecules), as the molecules rise. More important by far is the PE stored in the system (liquid molecules + vapor molecules), as work is done against the short-range electric forces of attraction between molecules.

[*]The thoughtful student may wonder why we use ΔQ to represent a flow of heat energy, and introduce an entirely unrelated symbol Δh to represent latent heat. Aside from the fact that ΔQ is measured in calories and Δh in calories per gram, there is a significant difference between the ultimate destinations of these two "heats." ΔQ is change of *internal* energy, either potential or kinetic. The latent heat Δh, however, goes not only into increasing the internal energy but also into doing some *external* work as the substance expands against atmospheric pressure. Thus latent heat is something more than just an increase in internal energy.

23.6 g of the ice melts. What is the latent heat of fusion of ice?

Since not all the ice melts, we know that the final temperature is 0°C, with 16.4 g of ice floating in 423.6 g of water at 0°C. For the specific heats of ice and copper we use the values from Table 14-1, which are 0.50 and 0.093 cal/g·C°, respectively.

$$\begin{pmatrix} \text{heat absorbed to} \\ \text{raise 40 g of ice to} \\ 0°C \end{pmatrix} + \begin{pmatrix} \text{latent heat ab-} \\ \text{sorbed to melt} \\ \text{23.6 g of ice} \end{pmatrix}$$

$$= \begin{pmatrix} \text{heat given off} \\ \text{by water cooling} \\ \text{to } 0°C \end{pmatrix} + \begin{pmatrix} \text{heat given off} \\ \text{by copper cup} \\ \text{cooling to } 0°C \end{pmatrix}$$

$$(40 \text{ g})(0.50 \text{ cal/g·C°})[0° - (-8°)]$$
$$+ (23.6 \text{ g})(\Delta h_f)$$
$$= (400 \text{ g})(1 \text{ cal/g·C°})(5° - 0°)$$
$$+ (100 \text{ g})(0.093 \text{ cal/g·C°})(5° - 0°)$$

$$160 \text{ cal} + (23.6 \text{ g})(\Delta h_f) = 2000 \text{ cal} + 46 \text{ cal}$$

$$\Delta h_f = \boxed{79.9 \text{ cal/g}}$$

EXAMPLE 14-7

What is the result of adding 10 g of steam at 100°C to 50 g of ice at 0°C?

We have to feel our way through this problem in order to find out whether all the ice melts. Then we can set up an equation to find the final temperature. During condensation, 10 g of steam *can* give up as much as

$$(540 \text{ cal/g})(10 \text{ g}) = 5400 \text{ cal}$$

To melt all the ice requires $(80 \text{ cal/g})(50 \text{ g}) = 4000 \text{ cal}$. Hence all the ice *will* melt, heat will be left over to warm up the ice water formed from the ice, and the final temperature will lie between 0°C and 100°C. Call this final temperature t.

Heat given off = heat absorbed

$$\begin{pmatrix} \text{latent heat} \\ \text{of steam} \end{pmatrix} + \begin{pmatrix} \text{cooling of} \\ \text{hot water} \end{pmatrix}$$

$$= \begin{pmatrix} \text{latent heat} \\ \text{of ice} \end{pmatrix} + \begin{pmatrix} \text{warming up of} \\ \text{cold water} \end{pmatrix}$$

$$(10 \text{ g})(540 \text{ cal/g})$$
$$+ (10 \text{ g})(1 \text{ cal/g·C°})(100° - t)$$
$$= (50 \text{ g})(80 \text{ cal/g})$$
$$+ (50 \text{ g})(1 \text{ cal/g·C°})(t - 0°)$$

$$5400 + 1000 - 10t = 4000 + 50t$$

$$t = 40.0°C$$

The result is

$$\boxed{60 \text{ g of water at } 40.0°C}$$

The latent heat given off when water freezes is put to good use on a farm when produce is stored in an unheated shed. Large tubs of water are set in the shed, and, as the outside temperature goes below 0°C, the water freezes. As this happens, the latent heat is given off to the surroundings, keeping the shed and contents at about 0°C until all the water freezes. A similar action takes place when dew forms on the leaves of a plant. As the water vapor in the air condenses and becomes liquid water, the latent heat of vaporization is given out and this heat may well help prevent plant damage by freezing.

14-5 Transfer of heat by convection and conduction

With the understanding that heat is energy and not a material substance, we can talk about the flow of heat, by which we mean transfer of heat energy from one place to another. There are three fundamentally distinct ways in which heat may be transferred: convection, conduction, and radiation.

The *convection* of heat is well illustrated by the blower of a space heater in a large gymnasium. Hot air is moved bodily from the point of origin to the spectators in the bleachers. Kinetic energy of the air molecules is transported by simply moving the molecules *en masse* to the desired point. Convection currents in gases and liquids are often caused by variations in density due to temperature changes. Hot air, which is less dense than cold air, rises from a point above a radiator and strikes the ceiling, spreading out through the room. The force causing this motion arises from pressure exerted by near-by cooler (and denser) air. In this way, a continuous flow takes place. A similar convection phenomenon keeps the temperature of a lake at about 4°C (the temperature of the densest water) before any of it cools to freezing (Sec. 13-4). Most of the changes in air temperature which appear on a weather map are due to convection currents set up as masses of warm or cold air move over land and water.

The *conduction* of heat is a more intimate process, in which KE is handed on from one molecule to an adjacent molecule by collisions. When a spoon is placed in a cup of hot coffee, the vibrating silver molecules in the spoon bump each other, but they do not migrate away from their average positions. The slow-moving molecules in the handle eventually receive KE through a chain of events, and so the handle becomes warm. As one would expect, heat flows by conduction only if there is a temperature difference, for only then (on the average) will fast molecules be striking slow molecules and be able to give up energy.

Substances differ markedly in their *thermal conductivity* and may be classified as insulators (such as asbestos or cork) and conductors (such as aluminum or iron). Metals are especially good conductors of heat, since a metal has free electrons not attached to any particular nucleus. These free electrons can drift through the metal and transfer KE by their collisions with each other and with much more massive ions and nuclei in the crystal lattice. (Such a drift partakes of the nature of convection as well as conduction.) The same free electrons are responsible for the conduction of electric currents; it is no accident that silver and copper, which are the best conductors of heat, are also the best conductors of electricity.

The amount of heat which flows through

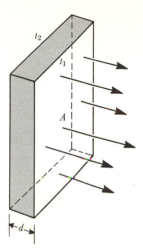

Figure 14-3 Heat flow through a slab of area A. The temperature gradient is $(t_2 - t_1)/d$.

a flat slab of material (Fig. 14-3) in time T is proportional to the time, to the area A of the slab, and to the *temperature gradient* $\Delta t/\Delta x$, where Δt is the temperature difference between the two faces of the slab which are separated by a thickness Δx. In Fig. 14-3 the temperature gradient is $(t_2 - t_1)/d$, where one face is at temperature t_2, the other face is at temperature t_1, and the thickness of the slab is d. The constant of proportionality is called the *thermal conductivity* of the material. Thus, in a time T the heat flow is given by

$$Q = KAT\frac{t_2 - t_1}{d} \qquad (14\text{-}3)$$

$$\text{cal} = \left(\frac{\text{cal}\cdot\text{cm}}{\text{cm}^2\cdot\text{s}\cdot\text{C}°}\right)(\text{cm}^2)(\text{s})\left(\frac{\text{C}°}{\text{cm}}\right)$$

Architects in the U.S. usually express K in Btu·in./ft²·h·F° (see Prob. 14-B27).

14-6   Transfer of heat by radiation

A third method of heat transfer requires no physical medium and is essentially electrical in nature. Whenever electric charge undergoes acceleration, an electromagnetic wave is formed which sends energy through space,

TABLE **14-3**

Thermal conductivity of common materials

Substance	Approximate thermal conductivity K, cal·cm/cm²·s·C°
Masonite	0.00011
white pine	0.0002
building brick	0.0015
glass	0.0025
concrete	0.002
steel	0.11
copper	0.92

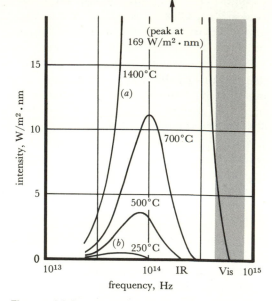

Figure 14-4 Black-body radiation at various temperatures. The shaded region includes the frequencies of visible light. (*a*) White-hot piece of coal, at 1400°C; (*b*) blackberry pie, at 250°C.

in somewhat the same way as waves spread out over the water when a stone is dropped into a pond. Electromagnetic waves, which we will study in Chap. 23, are called radio waves when they are vibrating slowly enough to cause a response in a radio receiver. They may be generated when electrons move up and down in the tower of a transmitting antenna. Electromagnetic waves of various higher frequency ranges are known by other names: heat, infrared radiation, visible light, ultraviolet radiation, x rays, and gamma rays. They all travel through a vacuum at 3×10^8 m/s (186 000 mi/s). Heat waves *are* radio waves and can be generated electrically, as in the diathermy machine or the electronic oven. Far more commonly in solids it is the vibratory motion of electrically charged particles (ions) that gives rise to the electromagnetic radiation which we call "heat radiation" or "infrared radiation." As shown in Table 14-4, the various electromagnetic radiations are characterized by a wide variety of mechanisms for production and reception; the unifying property of all these radiations is their *wave velocity*.

The essential thing for us now is the fact that every body radiates not just one type of radiation, but some of each. Consider two of the graphs in Fig. 14-4. Curve *a* is for a white-hot piece of coal in a fireplace, and it shows that the most intense radiation is in the infrared (IR) region, with some visible light and a small amount of ultraviolet (UV)

radiation (too weak to be seen on the graph). There is even *some* radiation in the radio band, although the amount is relatively so small that we don't have to worry about TV interference from a fireplace! When a blackberry pie (at 250°C) is set outdoors to cool, the total radiation is much less, and also the character of the radiation is different (curve *b*). The peak radiation is now at about 5×10^{13} cycles (in the infrared region), and there are relatively more of the infrared and radio waves. Visible light from the pie is present, but far too weak to detect. The total radiation (all frequencies combined) is, of course, much less for the pie than for the glowing coal.•

The only way heat energy can traverse a

• Precise statement of the laws of radiation makes use of absolute temperature (Sec. 15-5). Stefan's law says that the total radiation, all frequencies included, is proportional to the fourth power of the absolute temperature. Wien's law says that the frequency of the radiation which is most intense is directly proportional to the absolute temperature.

TABLE 14-4 Electromagnetic radiations

Type of radiation	Usual source	Usual detector	Approximate frequency range, Hz (cycles/s)	Approximate wavelength range, meters*
Radio waves	electric circuits	electric circuits		
lowest frequency used in communications....................			10^4	3×10^4
AM broadcast band........................			$0.55 \times 10^6 - 1.60 \times 10^6$	545–188
FM broadcast band........................			$88 \times 10^6 - 108 \times 10^6$	3.40–2.78
TV broadcast bands.......................			$54 \times 10^6 - 890 \times 10^6$	5.55–0.34
radar.......................................			$10^9 - 10^{11}$	$3 \times 10^{-1} - 3 \times 10^{-3}$
highest experimental frequency.......................			10^{12}	3×10^{-4}
Infrared radiation (heat)	hot bodies	thermocouples thermometers skin	$10^{11} - 3.8 \times 10^{14}$	$3 \times 10^{-3} - 8 \times 10^{-7}$
Visible light	hot bodies electric arcs	eye photocell photographic plate	$3.8 \times 10^{14} - 7.5 \times 10^{14}$	$8 \times 10^{-7} - 4 \times 10^{-7}$
Ultraviolet radiation	electric arcs	photographic plate photocell Geiger counter	$7.5 \times 10^{14} - 3 \times 10^{17}$	$4 \times 10^{-7} - 10^{-9}$
X rays	impact of electrons on target	photographic plate Geiger counter ionization chamber	$3 \times 10^{17} - 3 \times 10^{19}$	$10^{-9} - 10^{-11}$
Gamma rays	radioactive nuclei	Geiger counter ionization chamber scintillation counter	$> 3 \times 10^{19}$	$< 10^{-11}$

Note: The boundaries between the various types of electromagnetic radiation are not sharp; for instance, radiation in the neighborhood of 3×10^{17} Hz might equally well be called "extremely high-frequency ultraviolet radiation" or "extremely low-frequency x rays." Likewise, the range from 10^{11} to 10^{12} Hz can be produced either by electrical means (radio) or by thermal means (infrared sources).

*Wavelength = velocity/frequency. Wavelengths, when short, are often expressed in angstrom units, microns, or nanometers (or millimicrons): 1 angstrom (1 Å) = 10^{-10} m; 1 micrometer (1 μm) = 1 micron (1 μ) = 10^{-6} m; 1 nanometer (1 nm) = 10^{-9} m.

vacuum is by electromagnetic radiation. In fact, "empty" space between the stars is crisscrossed by many electromagnetic radiations, and it has been estimated that 4×10^{-14} J of radiant energy is contained in each cubic meter of "empty" space (on the average). Practically all the radiation received by the earth comes from the sun; at a place where the sun is directly overhead, 8.13 J (1.94 cal) would reach each square centimeter of the earth's surface every minute if the atmosphere were perfectly transparent. This has been going on for several billion years; why, then, has the earth's temperature not increased to some fantastic figure? The answer lies in the radiation *from* the earth to the surrounding space. The earth has become just hot enough so that the daily loss of heat by radiation (which depends on the earth's temperature) just balances the daily gain in energy from the sun. The sun itself is a gigantic H-bomb, continually converting rest mass into energy (Sec. 7-7). Most of the radiation from the sun does not reach the earth, but passes on out into space.

Not all surfaces are equally effective in emitting or absorbing radiation, even though their temperatures are the same. A white or shiny body is a poor absorber of radiation, since some of the incoming energy is reflected back away from the interior. By the same token, a white or shiny body is a poor *emitter* of radiation, since radiant energy within the body is reflected back. The surface looks shiny from the inside as well as from the outside! A perfect absorber is called a *black body,* and such a body absorbs all radiations, not merely visible light. A black body is also a perfect emitter. The photograph on p. 300 shows a porcelain dish upon which the word "black" has been applied with black paint and baked in; the central part of the marking is heated by a gas flame. It is evident that the same

region which is a good absorber (when cool) is also a good emitter (when hot). According to these ideas, a living-room "radiator" should be painted black for efficient radiation of heat. A light-colored paint job might be decorative, but hardly in accord with the laws of physics. However, most of the heat transfer from a "radiator" is by direct contact of air molecules with faster-moving iron atoms or paint molecules; such a process is conduction and does not depend greatly on the blackness of the surface.

At a given temperature, all black bodies look much alike and emit the same distribution of frequencies of radiant energy. In a fireplace, the "red-hot" iron poker is almost indistinguishable from the "red-hot" carbon (coal or ash) which is at the same temperature. The *radiation pyrometer* is a device to determine the temperature of a furnace by comparing the color of the contents with that of an electrically heated wire at a known temperature. To the extent to which all substances are "black," it is immaterial just what substance is in the furnace. The same radiation pyrometer can be used to measure the temperature of molten iron or molten glass.

■ SUMMARY

Heat is a form of energy and can be measured in ordinary mechanical units such as joules or foot·pounds. The calorie, an amount of energy equal to 4.184 J, raises the temperature of one gram of water one Celsius degree. The Btu (equal to 778 ft·lb or 252 cal) is the heat energy which raises the temperature of one pound of water one Fahrenheit degree.

The specific heat of a substance is the heat absorbed or given off per unit mass per unit change in temperature. In the cgs system, specific heat is measured in calories per gram per Celsius degree, and in the British system, a similar unit for specific heat is based on the Btu and the pound. The numerical value of specific heat is the same in these two systems. The thermal capacity, or water equivalent, of a body is the heat absorbed or given off per unit change in temperature; thermal capacity equals mass times specific heat. When heat flows into or out of a body, the quantity of heat ΔQ that flows to cause a temperature change Δt is given by $\Delta Q = mc\,\Delta t$, where m is the mass in grams, c the specific heat, and ΔQ the heat that flows, in calories.

The law of conservation of energy, when modified to include heat as a form of energy, is called the first law of thermodynamics. Mechanical energy is not "lost" due to frictional processes; it is merely transformed into heat energy.

When heat flows into a body, the internal energy of the body increases in one or both of two ways. If the random KE of the molecules increases, the temperature increases. If the PE of the molecules increases, as in the separation of the molecules against cohesive forces during vaporization, the temperature does not increase, and the heat absorbed is a latent heat. Both processes are reversible; heat is given out when a body cools down (KE decreases) or when it condenses or solidifies (PE decreases).

Transfer of heat takes place in three ways. Convection is the motion of heated matter, in bulk. Conduction is due to transfer of KE from molecule to neighboring molecule by collision. In metals, heat conduction and electrical conduction are both largely due to a cloud of free electrons within the metal. Radiation includes all types of electromagnetic waves, from radio waves, which are of lowest frequency, to gamma rays, which are of highest frequency. All electromagnetic waves travel with the same velocity $(3 \times 10^8 \text{ m/s})$ through a vacuum. A hot body emits some of each type of radiation, but the frequency of the most intense radiation depends on the temperature of the body. The total radiation, all frequencies combined, increases rapidly with temperature. A good absorber is also a good emitter of radiation. A black body has a surface which absorbs and emits radiation of all frequencies without hindrance; the distribution of frequencies and the total amount depend only on the temperature.

■ CHECK LIST

calorie	latent heat of vaporization
Calorie	convection
Btu	conduction
specific heat	thermal conductivity
thermal capacity	radiation
water equivalent	types of electromagnetic waves
latent heat of fusion	black body

■ QUESTIONS

14-1 What is the difference between "heat" and "temperature"?

14-2 Correct the following statements: (a) "The heat of the sun's surface is 6000°C." (b) "Body heat of a whale is 36°C." (c) "The heat required to melt lead is 5.86 cal/g·C°." (d) "The specific heat of copper is 0.093 cal."

14-3 Which of the following are *not* units of energy: joule, calorie, foot·pound, watt, Calorie, horsepower, erg, Btu?

14-4 If two buckets of equal size are filled, one with water and one with glycerin, at 50°C, which will cool down to room temperature first? Why? Suppose one bucket were painted black and the other white. Would this affect your answer?

14-5 A book lying near the edge of a table is pushed over and allowed to fall to the floor. When the book strikes the floor, its KE disappears. What becomes of this KE?

14-6 How much heat is needed to convert 1 kg of ice at 0°C to 1 kg of water vapor at 100°C?

14-7 An Eskimo places a stone near a fireplace, then takes the warm stone to bed with him and rests his feet on the stone. Explain how convection, conduction, and radiation are involved in this custom.

14-8 Why does a metal seem colder to the touch than wood, outdoors on a cold winter day?

14-9 Is any light of visible frequency emitted by the surface of a soot-covered teakettle filled with boiling water?

14-10 It is known that the interior of the earth is maintained at a temperature of several thousand degrees Celsius by a source of heat due (probably) to radioactive disintegration. This being the case, does the earth absorb the same amount of heat from the sun (Sec. 14-6) as it radiates?

■ PROBLEMS

14-A1 How much heat is required to raise the temperature of 500 g of water from 40°C to 70°C? Express your answer in calories.

14-A2 How much heat is wasted when a steamship engine pours 400 lb of water at 190°F into a lake which is at 40°F? Express your answer in Btu.

14-A3 In a laboratory experiment, 800 J of work was done against friction. How much heat, in calories, was generated?

14-A4 What is the specific heat of water, in $J/kg \cdot C°$?

14-A5 It is found that 640 cal of heat is needed to heat 100 g of a substance from 6°C to 14°C. What is the specific heat of the substance?

14-A6 How much heat, in Btu, is needed to heat 100 lb of asbestos from 40°F to 100°F?

14-A7 What is the water equivalent of 2 kg of lead?

14-A8 What is the water equivalent of 400 g of aluminum?

14-A9 How much heat, in calories, is given off by 200 g of water which cools from 100°C to 20°C?

14-A10 How much heat, in Btu, is needed to bring 500 lb of water from 60°F to within 2°F of the boiling temperature?

14-A11 A metal block of mass 500 g is warmed from 20°C to 30°C when it absorbs 1200 cal of heat energy. Calculate the specific heat of the metal.

14-A12 How much heat is needed to melt 1 kg of lead, originally at 0°C?

14-A13 While condensing from a vapor into a liquid at 61.5°C, 20 g of chloroform liberated 1180 cal of heat. What is the latent heat of vaporization of chloroform?

14-A14 How much heat must be supplied to a 30 g ice cube, originally at 0°C, in order to convert it to 30 g of water at 20°C?

14-A15 The latent heat of vaporization of liquid helium, at its normal boiling point −269°C, is 6 cal/g. How much helium is evaporated when an electric heater coil supplies 30 J of energy to a flask of liquid helium?

14-A16 What is the wavelength of the center frequency of a TV station which sends out programs on Channel 6 (frequency range 82–88 MHz)?

14-A17 A certain type of x-ray radiation from copper atoms has wavelength 1.5×10^{-10} m. What is the frequency of this radiation?

14-A18 A certain infrared radiation has wavelength 10^{-6} m. Express this wavelength in angstrom units; in micrometers; in nanometers.

14-A19 What is the frequency, in GHz, of radar waves whose wavelength is 1.00 cm?

14-A20 What is the wavelength, in angstroms, of the most intense radiation from an electric heater whose temperature is 700°C? (See Fig. 14-4.)

14-B1 When 500 g of lead at 100°C is placed in an aluminum calorimeter cup of mass 100 g which contains 200 g of olive oil at 20°C, the final temperature of the mixture is 29.3°C. What is the specific heat of olive oil?

14-B2 In a laboratory experiment to determine the latent heat of fusion of ice, 46.9 g of ice at 0°C was placed in a calorimeter cup which contained 265 g of warm water. The water equivalent of the cup was 20.1 g. All the ice melted, and the cup and contents were cooled from 32.4°C to 17.1°C. Calculate the latent heat of fusion of ice from these data. (*Note:* The lack of agreement between your answer and the accepted value is typical of the degree of precision commonly obtained in heat measurements in elementary laboratories.)

14-B3 An aluminum pan is placed on a stove at 4:05 P.M. The pan has a mass of 340 g and contains 300 g of water at 20°C. The flame supplies 100 cal of heat to the system each second. (*a*) When will the water begin to boil? (*b*) When will the temperature of the pan begin to rise above 100°C?

14-B4 What is the result of adding 10 g of steam at 100°C to 20 g of ice at 0°C?

14-B5 What is the result of adding 10 g of steam at 100°C to 60 g of ice at 0°C?

14-B6 What is the result of adding 10 g of steam at 100°C to 90 g of ice at 0°C?

14-B7 To measure the temperature of a flask of liquid air, a researcher let a 400 g chunk of aluminum come to thermal equilibrium with the liquid air, then quickly dropped the chilled aluminum into a 200 g copper cup containing 500 g of water. The cup and the water, both originally at 80°C, came to a final temperature of 46°C. What was the temperature of the liquid air? The average specific heat of aluminum for the temperature range used is 0.182 cal/g·C°.

14-B8 A glass beaker whose water equivalent is 60 g contains 200 cm^3 of glycerin at 20°C. What will be the final temperature if 10 cm^3 of copper at 80°C is placed in the beaker?

14-B9 How much steam at 100°C would have to be added to 60 g of ice at 0°C so that the entire system becomes water at 30°C?

14-B10 Steam is allowed to bubble through water contained in an aluminum cup; some of the steam condenses into water, giving off its latent heat of vaporization, and then (as water) cools down. The following data are taken: initial temperature, 15°C; final temperature, 40°C; mass of empty cup, 100 g; mass of cup plus cold water, 246 g; mass of cup plus warm water plus condensed steam, 253 g. Calculate the latent heat of vaporization of steam.

14-B11 A closed, unheated storage room at 1°C contains sacks of vegetables whose total water equivalent is 500 kg. As a result of a sudden temperature drop outside, heat leaves the room at the rate of 400 kcal/h. (*a*) How long will it take for the temperature of the room to go from 1°C to −1°C? (*Note:* The vegetables do not freeze because the starch and sugar in them lower the freezing point to a few degrees below −1°C.) (*b*) How long will be required for the same temperature drop if a tank containing 100 kg of water is in the storage room along with the vegetables?

14-B12 An ice cube of mass 40 g is in an aluminum cup of mass 100 g. The system is originally at $-35°C$. What will be the result of adding 7000 cal to the system?

14-B13 About how much water is evaporated in an hour from a lake 1 km² in area if 0.1% of the sun's radiation gets through the atmosphere and is absorbed by the water? The sun's rays strike the water at an angle of 60° with the vertical.

14-B14 An artisan hammers out a silver ashtray whose mass is 70 g. If the hammer head has a mass of 500 g and it strikes the tray with a speed of 10 m/s, what is the rise in temperature of the tray after 30 strokes of the hammer? (*Note:* Silver is much softer than the iron hammer head. Hence it is reasonable to assume that the hammer rebounds with negligible speed. Assume also that 50% of the heat generated remains in the ashtray.)

14-B15 An automobile weighing 1600 lb is traveling at 16 ft/s. How much heat, in Btu, is developed when it is brought to rest? (Use W/g for mass when calculating KE.)

14-B16 Using the modern value of the mechanical equivalent of heat (4.18 J/cal), calculate the height of fall which Mayer stated to be "about 365 meters" (Sec. 14-2).

14-B17 A lead bullet of mass 20 g is fired at 400 m/s at a tree trunk, and the bullet emerges on the other side with a speed of 300 m/s. (*a*) What is the loss of KE? (*b*) Assuming that 60% of the loss of mechanical energy is stored as heat in the bullet, and also assuming the specific heat to remain equal to the value given in Table 14-1, calculate the bullet's rise in temperature. Will any of the bullet melt?

14-B18 A 40 kg child slides down a playground slide made of steel and arrives at the bottom with a speed of 3 m/s. The slide is 2.5 m long and is inclined at 45°. How much heat, in joules, is shared by the child and the slide?

14-B19 An ice skater of mass 100 kg starts to glide at 10 m/s and comes to rest after traveling 20 m. How much ice is melted? (Assume that half the heat generated by friction is absorbed by the ice, and also assume that the ice beneath the skates is at 0°C.)

14-B20 What is the ratio of the energy needed to evaporate some water in the ocean to the energy needed to lift it to form a cloud 2 km above the ocean?

14-B21 An overweight student attempts to reduce by lifting two 5 kg masses vertically through 0.8 m. How many times must this be done to "work off" the energy equivalent of a 70 kcal slice of bread? Is your answer a reasonable one? Discuss.

14-B22 An office worker prepares for a coffee break by heating a cup of water with a 150 W electric immersion heater placed directly in the cup. The cup is of glass and has mass 200 g; it holds $\frac{1}{4}$ liter and is $\frac{4}{5}$ full of water at 20°C. How much time is required to bring the temperature of the water to 85°C?

14-B23 In a physiological experiment, it was found that when the surrounding air temperature was in the range 19°C to 31°C, a human subject lost about 100 kcal of heat per hour due to all causes. Of this loss, about 20% was due to evaporation from an invisibly thin layer of moisture on the skin. The latent heat of vaporization of water at skin temperature is 580 cal/g. (*a*) Taking the surface area of the subject's body to be 1.0 m², what would be the thickness of a layer of sweat which would be completely evaporated in 10 min? (*b*) Can you explain qualitatively why the latent heat of vaporization of water at body temperature is higher than the value quoted in the text for 100°C? (*c*) Suggest two ways in which the other 80% of the heat loss might take place.

14-B24 A lead ball of mass m kg is dropped onto a pavement from a height of 25 m. Assuming that 60% of the heat generated remains in the lead, calculate the temperature rise of the ball.

14-B25 A frozen water pipe in a house is 5 m long, and its inside diameter is 1 cm. The pipe is iron, and its mass is 2 kg. To thaw the ice, an electric current is set up in the pipe by a transformer which can supply electric energy at the rate of 1500 W, of which 60% is useful in warming up the pipe and ice. The original temperature of pipe and ice is $-20°C$. (*a*) How much heat is needed to melt the ice? (*b*) How much time is required for the job?

14-B26 Estimate graphically the specific heat of the material of Example 14-1 (*a*) at $-260°C$; (*b*) at $-252.5°C$. The mass of the sample is 200 g.

14-B27 Architects in the U.S. often need to know the heat in Btu that flows each hour through a sheet of area 1 ft² due to a temperature gradient of 1 F°/in. By what factor should the coefficient K in the cgs unit $cal \cdot cm/cm^2 \cdot s \cdot C°$ be multiplied to obtain K in $Btu \cdot in./ft^2 \cdot h \cdot F°$?

14-B28 One wall of a brick house is 10 ft × 30 ft × 8 in. thick. How many Btu per day flow through the wall if the inside temperature is 70°F and the outside temperature is 10°F?

14-B29 Because of radioactivity in the crust of the earth (see Sec. 29-1), the temperature gradient in non-volcanic regions is about 3 C° for each 100 m change in depth below the surface. (*a*) Assuming the thermal conductivity of granitic crust material to be about the same as for concrete, calculate the heat flow (in $\mu cal/s$) through 1 cm² of the earth's surface. (*b*) Estimate the heat flow, in J/day, through the entire surface of the earth.

14-B30 A lamp bulb inside a spherical glass shell 20 cm in radius and 0.5 cm thick radiates 100 W of thermal power. What is the difference in temperature between the inner and the outer surface of the glass?

14-B31 An ice-cube container is in the form of a cube of outside edge 30 cm, made of plastic insulating material whose thermal conductivity is $9 \times 10^{-5} \ cal \cdot cm/cm^2 \cdot s \cdot C°$. The base, walls, and lid are 5 cm thick. If 7.5 kg of ice at 0°C is placed in the container and taken on a picnic on a day when the temperature is 33°C, how long will it be before all the ice is melted?

14-C1 The temperature of a sample of molten lead, near its temperature of solidification, is falling at the rate of 6.0 C°/min. If the lead continues to lose heat at the rate indicated by this value and takes 30.0 min to solidify completely, what is the latent heat of fusion for lead indicated by these data? (*Hint:* Let m be the mass of the sample.) What assumption have you made?

14-C2 An iron rod has a diameter 0.50 cm. What will be the change in length of the rod if it absorbs 4000 cal of heat? (*Hint:* Solve in symbols first. Use tables from Chaps. 2, 13, and 14 as necessary.)

14-C3 The ends of an aluminum rod are held clamped between two fixed posts 2 m apart with a tension of 50 N. How many calories of heat must be absorbed by the rod to reduce the tension to zero? (*Hint:* Solve in symbols first; use tables from Chaps. 2, 9, 13, and 14.)

14-C4 A torque of 0.020 m·N is needed to turn the paddles of a cake mixer at 300 rev/min. (*a*) At what rate is heat being generated in the batter? (*b*) What temperature rise would occur in 600 g of batter as a result of mixing it for 5 min? (Assume specific heat of the batter to be 0.9 cal/g·C°.)

14-C5 A "solar house" is designed to store a million kcal of heat. The storage facilities consist of sealed drums in the attic containing Glauber's salt ($Na_2SO_4 \cdot 10H_2O$), which is warmed up from 25°C to 45°C during the day. (*a*) What mass of salt is needed? (*b*) What should be the volume of the drums? [Data for Glauber's salt, in the impure form used: specific heat (solid), 0.46 cal/g·C°; specific heat (liquid), 0.68 cal/g·C°; melting point, 32°C; latent heat of fusion, 58 cal/g; specific gravity, 1.6.]

14-C6 A fossil-fuel power plant which generates 1000 MW (10^9 W) of electric power discharges 1.7×10^6 Btu per second as waste heat into a river.   (a) At what rate (in MW) is thermal energy obtained from the burning of coal?   (b) What is the thermal efficiency (power output/heat input) of the plant?   (c) The river flow is at 4×10^8 kg of water per hour; what is the temperature rise of the river due to this thermal pollution?

14-C7 The waste heat from a nuclear power plant is 6×10^8 cal/s. If it could be recovered and efficiently converted into electric energy, what would be the value of the energy wasted in 1 day? The wholesale cost of 1 kilowatt·hour (kWh)(3.6×10^6 J) is about 1¢.

14-C8 Consider the metals iron, copper, aluminum, and lead. Calculate the specific heat per cubic centimeter for each metal. Also calculate the specific heat per mole for each metal. Arrange your results in the form of a neat table. Discuss any regularities in the table.

14-C9 A factory roof measures 20 m $\times$ 50 m. If the sun's radiation strikes the roof broadside, and the energy could be converted into mechanical work with 10% efficiency, what power, in kilowatts, would be available for use in the factory?

14-C10 Suppose that beginning now only 99% of the solar radiation of 1.94 cal/cm²·min (see p. 311 is reradiated, the balance being trapped by increased CO_2 in the atmosphere arising from man's burning of fossil fuels. Assume also that 10% of the *trapped* radiation goes to melt polar ice caps; how much would the ocean level rise in 10 years? Data: Radius of earth, 6.4×10^6 m; area of ocean surface, 4.0×10^{14} m². (*Hint:* Use the projected earth surface πr^2 in calculating thermal input.)

14-C11 Global average annual rainfall is about 1.0 m. How much energy (in J) is released world-wide by water-vapor condensation during a year? Where did this energy come from?

■ REFERENCES

1 Roller, Duane, *The Early Development of the Concepts of Temperature and Heat: The Rise and Decline of the Caloric Theory* (Harvard Case Histories in Experimental Science No. 3), (Cambridge: Harvard University Press, 1950). See especially Rumford's investigation of weight ascribed to heat (pp. 47–61); his experiments on heat produced by friction (pp. 61–81); and Davy's experiments (pp. 81–89). *Caution:* The early writers sometimes used the words "heat" or "degree of heat" where we would today use "temperature." The context will tell whether temperature or quantity of heat is meant.

2 Magie, W. F., *A Source Book in Physics,* (New York: McGraw-Hill, 1935, also, Cambridge: Harvard University Press, 1963). Selections from the writings of Newton, Amontons, Fahrenheit, Taylor, Black, Rumford, Davy, Gay-Lussac, and others. See especially the work of Black on latent heat (pp. 139–44); Rumford on heat produced by friction (pp. 151–61); Davy's experiments on heat produced by friction (pp. 161–65).

3 Brown, Sanborn C., "Count Rumford's Concept of Heat," *Am. J. Phys.,* **20,** 331 (1952).

4 Brown, Sanborn C., "The Caloric Theory of Heat," *Am. J. Phys.,* **18,** 367 (1950).

5 Powell, Marcy S., "Count Rumford: Soldier, Statesman, and Scientist," *Am. J. Phys.,* **3,** 161 (1935). More about his personal life than about his physics.

6 Wilson, M., "Count Rumford," *Sci. American,* Oct., 1960, p. 42.
Dyson, Freeman J., "What Is Heat?" *Sci. American,* Sept., 1954, p. 58.

15

Thermal Behavior of Gases

Now that we have developed some familiarity with the concepts of temperature and of heat as a form of energy, we proceed to the study of the thermal behavior of gases and vapors. One of the most useful generalizations about gases is the concept of an ideal gas, whose behavior can be described by rather simple equations.

15-1 Ideal gases

In spite of their much greater familiarity in daily life with matter in the solid and liquid state, physicists understand the behavior of gases better than that of either solids or liquids. The gaseous state is actually far simpler than the liquid or solid state, because the molecules in a gas are so far apart that the cohesive forces between them are usually small. Granted that the cohesive forces between molecules of O_2 gas are probably different from those between molecules of CO_2 gas, the practical effect on gas behavior in either case is so small as to be negligible in ordinary work. Again, it is natural to suppose that the behavior of these two gases would be somewhat affected by the sizes of the individual molecules. However, as a practical matter, the volume actually occupied by the

molecules is usually a tiny fraction of the total volume of the container in which the gas is confined. An *ideal* gas is one in which these disturbing effects of cohesive forces and molecular volume are so small as to be negligible. We thus expect the laws of gases to be simple and universal, the same for all ideal gases, regardless of their chemical composition. No real gas is, of course, an ideal gas, but we shall find that under ordinary conditions of temperature and pressure we shall not be very wrong if we consider all gases ideal.

15-2 Boyle's law

How shall we describe the behavior of a gas? Four measurable quantities are of interest: the pressure, volume, mass, and temperature of a given sample of gas. Together, these quantities determine the "state" of the sample of gas. The first experimental results were obtained by the English investigator Robert Boyle (1627–1691), who studied the changes in volume of a gas as the pressure was varied. In order to arrive at a physically significant law, Boyle simplified the problem by doing his experiments under controlled conditions. He kept the mass m of gas constant (i.e., no

---

Condensation of water vapor into water droplets has resulted in the formation of these cumulus clouds. (USDA Photo)

leaks in the container), and he kept the temperature t constant. Under such circumstances, the relation between pressure P and volume V has the simple form known as *Boyle's law:*

$$PV = \text{constant} \quad \left(\begin{smallmatrix}\text{temperature and}\\ \text{mass constant}\end{smallmatrix}\right) \quad (15\text{-}1)$$

or

$$P = \frac{\text{constant}}{V}$$

In other words, the pressure of a given mass of ideal gas is inversely proportional to the volume, if the temperature remains constant. The graph of P versus V is a hyperbola (Fig. 15-1), and, as we increase the pressure, passing from state 1 to state 2, Eq. 15-1 tells us that the product PV is constant. Boyle's law is often written as

$$P_1 V_1 = P_2 V_2 \quad \left(\begin{smallmatrix}\text{temperature and}\\ \text{mass constant}\end{smallmatrix}\right)$$

As elsewhere, the subscript 1 refers to the initial state, and the subscript 2 refers to the final state.

EXAMPLE 15-1

An oxygen tank of volume 2.00 ft³ is filled to a pressure of 3000 lb/in.². What was the original volume of the oxygen, when it was at a pressure of 1 atm? (Assume no temperature change.)

In using Boyle's law, it is necessary to use similar units on each side of the equation. We have seen (Sec. 12-2) that 1 atm is about 15 lb/in.², so the final pressure is

$$P_2 = (3000 \text{ lb/in.}^2)\left(\frac{1 \text{ atm}}{15 \text{ lb/in.}^2}\right) = 200 \text{ atm}$$

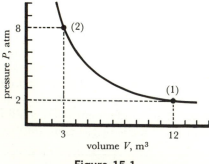

Figure 15-1

Now we apply Boyle's law:

$$P_1 V_1 = P_2 V_2$$
$$(1.00 \text{ atm}) V_1 = (200 \text{ atm})(2.00 \text{ ft}^3)$$
$$V_1 = \boxed{400 \text{ ft}^3}$$

Solving the problem in another way will show that it is unimportant just which units are chosen for P (or V), provided they are the same (for similar quantities) on each side of the equation:

$$P_1 V_1 = P_2 V_2$$
$$(15 \text{ lb/in.}^2) V_1 = (3000 \text{ lb/in.}^2)(2.00 \text{ ft}^3)$$
$$V_1 = \boxed{400 \text{ ft}^3}$$

It is essential to note that the mass of oxygen has not changed. The same mass is involved both at 3000 lb/in.² and at 15 lb/in.².

15-3   The effect of mass

The value of the "constant" in Boyle's law depends on several factors. First, let us imagine that the quantity of gas changes, temperature and volume remaining fixed. The pressure is caused by the impact of molecules as they strike the walls of the container. It is reasonable to expect that doubling the mass (i.e., doubling the number of gas molecules) will cause the number of impacts per second to double, and hence the pressure will be doubled. Experiment verifies that the product PV is proportional to the mass of gas. If m is the mass of gas, we may say that

$$\frac{P_1 V_1}{m_1} = \frac{P_2 V_2}{m_2} \quad \left(\begin{smallmatrix}\text{at constant}\\ \text{temperature}\end{smallmatrix}\right) \quad (15\text{-}2)$$

In other words, if m is increased, then so is PV; and PV/m remains the same.

EXAMPLE 15-2

A compressed-air storage tank whose volume is 112 liters contains 3.00 kg of air at a pressure of 18 atm. How much air would have to be forced into the tank to increase the pressure to 21 atm, assuming no change in temperature?

$$\frac{P_1 V_1}{m_1} = \frac{P_2 V_2}{m_2}$$

$$\frac{(18 \text{ atm})(112 \text{ liters})}{3.00 \text{ kg}} = \frac{(21 \text{ atm})(112 \text{ liters})}{m_2}$$

$$m_2 = 3.50 \text{ kg}$$

The mass of air which must be forced in is

$$3.50 \text{ kg} - 3.00 \text{ kg} = \boxed{0.50 \text{ kg}}$$

When dealing with different kinds of gas, such as hydrogen as compared with oxygen, experiment shows that we must use equal *numbers of molecules* rather than equal masses. For instance, there are actually more molecules in 8 g of H_2 (whose molecular weight is 2.0) than in 80 g of O_2 (whose molecular weight is 32). The 8 g of H_2 represents $(8 \text{ g})(1 \text{ mole}/2 \text{ g}) = 4$ moles, and hence has $4 \times 6.02 \times 10^{23}$ molecules. (One mole of any substance has 6.02×10^{23} molecules, which is Avogadro's number. See Sec. 2-4) Similarly, 80 g of O_2 is only $80/32 = 2.5$ moles, and contains only $2.5 \times 6.02 \times 10^{23}$ molecules. All else being equal, we expect that the pressure exerted by the 8 g of H_2 would be greater, in the ratio of 4 to 2.5. Thus we can say that

$$\frac{PV}{n} = \text{constant} \quad \left(\begin{array}{c}\text{at constant}\\\text{temperature}\end{array}\right) \qquad (15\text{-}3)$$

where n is the number of moles of gas present.

EXAMPLE 15-3

A certain tank contains 6.4 kg of oxygen at a pressure of 4.20 atm. What would be the pressure if the oxygen were pumped out and replaced by 6.6 kg of carbon dioxide (CO_2) at the same temperature?
The molecular weight of O_2 is $16 + 16 = 32$; the molecular weight of CO_2 is $12 + 2(16) = 44$.

$$6.4 \text{ kg of } O_2 = (6400 \text{ g})\left(\frac{1 \text{ mole}}{32 \text{ g}}\right)$$

$$= 200 \text{ moles}$$

$$6.6 \text{ kg of } CO_2 = (6600 \text{ g})\left(\frac{1 \text{ mole}}{44 \text{ g}}\right)$$

$$= 150 \text{ moles}$$

$$\frac{P_1 V_1}{n_1} = \frac{P_2 V_2}{n_2}$$

Here $V_1 = V_2$, since it is the same tank in both cases.

$$\frac{(4.20 \text{ atm})(\cancel{V_1})}{200 \text{ moles}} = \frac{(P_2)(\cancel{V_2})}{150 \text{ moles}}$$

$$P_2 = \boxed{3.15 \text{ atm}}$$

What we are really saying here is that, at a given temperature and volume, the pressure exerted by any single molecule does not depend on the type of molecule involved. This remarkable fact illustrates the essential simplicity of gases as compared with liquids or solids. We shall see reasons for this simplicity in Chap. 16, "The Theory of Heat."

15-4 The effect of temperature

Finally, in our approach to a general gas law, we ask what are the effects of temperature changes upon the pressure exerted by a confined gas, the volume and mass remaining constant. Experiment shows (Fig. 15-2) that the pressure increases *uniformly* as the temperature increases (with certain exceptions to be noted later). The graph of pressure versus temperature is thus a straight line. Note that in Fig. 15-2 the graph of P versus t is shown as a dashed line below a certain temperature. By this we mean to indicate that the pressure would become zero at some temperature *if it continued to decrease at the same rate as it does near*

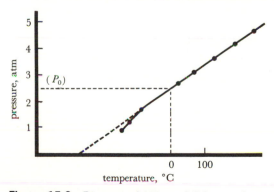

Figure 15-2 Pressure of 100 g of CO_2 gas in a container of volume 20 liters.

room temperature. Any ideal gas is assumed to behave in this ideal way, but in actual practice, before the gas pressure becomes zero the gas liquefies at some temperature, and then, of course, no longer even *is* a gas.

As the temperature of the gas decreases, the pressure deviates somewhat from the straight-line relationship. We expect such a behavior on the basis of what we know about molecules. The slower-moving molecules are more affected by mutual cohesive forces, and the pressure exerted by them drops off a little. Finally, when the temperature is low enough and the molecules slow enough, the cohesive forces cause the molecules to stick together as a liquid.

15-5   The absolute temperature scale

The mathematical equation of the straight line in Fig. 15-2 may be written as

$$P_t = P_0(1 + bt) \qquad (15\text{-}4)$$

where P_t is the pressure at a temperature t, P_0 is the pressure at $0°C$, and b is the *pressure coefficient* for the gas. Experiment shows that the constant b is practically the same for all gases, being about $0.00366 \ (C°)^{-1}$ for such gases as hydrogen which are most nearly "ideal" (Table 15-1). The differences that do exist are minor and show that no gas is strictly ideal. This being so, we can rework Eq. 15-4 into a form that makes calculations

TABLE **15-1**
Pressure coefficients* for some gases, in $(C°)^{-1}$

air	0.00366
carbon dioxide	0.00371
hydrogen	0.00366
oxygen	0.00367
helium	0.00367
ethane	0.00375
nitrogen	0.00367

* Measured at room temperature with original pressure about 1 atm.

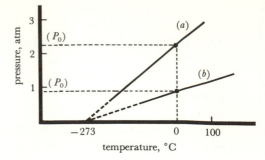

Figure 15-3   At absolute zero, the pressure exerted by any ideal gas would be zero. (*a*) 4 kg of H_2 in 20 m³; (*b*) 36 kg of O_2 in 30 m³.

very simple. Using the numerical value of $0.00366 \ (C°)^{-1}$ for b (the same for all ideal gases),

$$P_t = P_0(1 + 0.00366t) \qquad (15\text{-}5)$$

At what temperature would the pressure be zero, if the straight line continued to be a true description of the course of events? From Eq. 15-5, it is evident that to make $P_t = 0$, we must have $(1 + 0.00366t) = 0$. That is, the required temperature is

$$t = -\frac{1}{0.00366 \ (C°)^{-1}} = -273°C$$

Thus there is a certain temperature, *the same for all ideal gases,* at which the pressure would become zero *if* the gas behaved "ideally" all the way down. We call this temperature *absolute zero.* In Fig. 15-3 the different masses of different gases (*a* and *b*) are enclosed in containers of different volume, but the graphs each extend toward the same point, zero pressure at $-273°C$. By measuring the pressure of a gas at two temperatures (which may be $0°C$ and $100°C$ if desired) and extrapolating the graph down to zero pressure, it is possible to determine absolute zero in the laboratory without any risk of frostbite. Precise experiments have given $-273.15°C$ as the value of this important constant, but we shall usually round the value off to $-273°C$. *Absolute temperature* is defined as the Celsius (centigrade) temperature (*t*) plus $273°$ and is denoted by T. Absolute temperature, measured in degrees Kelvin (°K), represents the number of degrees that a body is above absolute zero. On the absolute scale the size of the degree is the same as on the Celsius scale, and on both scales there are 100 degrees between the ice point and the steam point.

The concept of absolute zero has deep significance, as we shall see in Chap. 16. However, for the present we are defining absolute zero merely in order to get a simpler equation describing the effect of temperature on the pressure of an ideal gas. A little algebra is needed here. Since $b = 1/273$, Eq. 15-5 can be written as

$$P_t = P_0\left(1 + \frac{1}{273}t\right) \tag{15-6}$$

$$P_t = P_0\left(\frac{273 + t}{273}\right) = \frac{P_0}{273}(273 + t)$$

Calling $0°C$ ($= 273°K$) equal to T_0, and remembering that $273° + t = T$, we can write

$$P_t = \left(\frac{P_0}{T_0}\right)T \tag{15-7}$$

or

$$\frac{P_t}{T} = \frac{P_0}{T_0} \tag{15-8}$$

We see that, while the pressure *depends on* the Celsius (centigrade) temperature t (Eq. 15-6), it is actually *proportional to* the absolute temperature T (Eq. 15-7). This is the reason for using the absolute temperature in the gas laws—it allows a simple proportion.

The proportionality between P_t and T can be clearly seen directly from Fig. 15-4. Since the pressure versus temperature graph for any particular amount of gas in a given volume is a straight line, triangles ABC and AED are similar. Hence $P_t/T = P_0/T_0$. This proportion is not true unless absolute temperatures are used.

EXAMPLE **15-4**

The pressure gauge of a truck tire reads 30 lb/in.² on a cold day when the thermometer

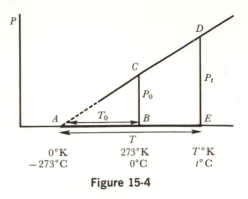

Figure 15-4

stands at $-3°C$. What will the pressure reading be after the tire has stood for a while in a heated garage at 27°C?

First, we take account of atmospheric pressure, which is about 15 lb/in.² If the gauge read zero, the air in the tire would still be at atmospheric pressure; at a gauge reading of 30 lb/in.² the total (absolute) pressure in the tire is 30 lb/in.² + 15 lb/in.² = 45 lb/in.² Next, we convert both Celsius temperatures to absolute temperatures by adding 273°: $-3°C$ is $270°K$, and $27°C$ is $300°K$.

Since the mass and volume of the air remain constant, the pressure is proportional to the absolute temperature.

$$\frac{P_2}{T_2} = \frac{P_1}{T_1}$$

$$\frac{P_2}{300°K} = \frac{45 \text{ lb/in.}^2}{270°K}$$

$$P_2 = 50 \text{ lb/in.}^2$$

This is the total pressure; the gauge will read

$$50 \text{ lb/in.}^2 - 15 \text{ lb/in.}^2 = \boxed{35 \text{ lb/in.}^2}$$

We have used Eq. 15-6 to describe the effect of temperature on the pressure of a gas at constant volume. A very similar equation describes the effect of temperature on the *volume* of a gas at constant pressure. As the temperature decreases, both the pressure and the volume of an ideal gas must tend toward zero at the *same* temperature (absolute zero). Thus

$$V_t = V_0(1 + \tfrac{1}{273}t) \tag{15-9}$$

The same factor, $(1 + \frac{1}{273}t)$, applies both to pressure changes and to volume changes of an ideal gas. We see now that it is no accident that the coefficients of volume expansion for most gases (Table 13-1) are almost equal to $0.00366\ (C°)^{-1}\ [= \frac{1}{273}(C°)^{-1}]$, the same as the pressure coefficients in Table 15-1.

15-6   The general gas law

We have seen (Eq. 15-3) that PV/n is a constant for any given temperature, but the constant depends on the temperature. We have also seen (Eqs. 15-7 and 15-8) that the pressure is proportional to the absolute temperature, other things being equal. Combining these ideas, we can write the *general gas law*

$$\frac{PV}{nT} = R \qquad (15\text{-}10)$$

where P is the pressure, V is the volume, n is the number of moles, T is the absolute temperature, and R is a new constant. Here at last we have a constant that is really constant —not dependent on the mass of gas, its chemical formula, or the temperature. We denote this *universal gas constant* by the symbol R. Its value[*] depends on the units for P, V, and T, but for our purposes we will not actually have to use *any* number for R, if we remember one combination of P, n, T, and the corresponding experimental value of V. The following set of values is called *standard temperature and pressure (STP)* for 1 atm pressure and 0°C (273°K):

$$\boxed{\begin{aligned} P_0 &= 1\ \text{atm} \\ n_0 &= 1\ \text{mole} \\ T_0 &= 273°\text{K} \\ V_0 &= 22.4\ \text{liters} \end{aligned}}$$

---

[*] If P is in atmospheres, V in liters, and T in degrees Kelvin, then R has the value 0.0821 liter·atm/mole·K°. If P is in newtons per square meter, V in cubic meters, and T in degrees Kelvin, then R is 8.314 J/mole·K°. The latter value is the mks value for R.

Several examples will make clear how to use the general gas law in various systems of units. In all cases, *absolute* temperatures are used.

EXAMPLE **15-5**

If 4 moles of gas exerts a pressure of P_1 lb/in.^{2} when confined in a volume of 40 m^3 at 300°C, what would be the pressure of 200 moles of the same gas when placed in a 100 m^3 tank at 600°C?

$$\frac{P_1 V_1}{n_1 T_1} = \frac{P_2 V_2}{n_2 T_2}$$

$$\frac{(P_1\ \text{lb/in.}^2)(40\ \text{m}^3)}{(4\ \text{moles})(573°\text{K})} = \frac{(P_2)(100\ \text{m}^3)}{(200\ \text{moles})(873°\text{K})}$$

$$P_2 = \boxed{30.5\,P_1\ \text{lb/in.}^2}$$

It is not necessary to worry about combining fps units (lb/in.^{2}) and mks units (m^3) as long as similar units appear on each side of the equation.

EXAMPLE **15-6**

How much helium gas (molecular weight 4.00) is contained in a high-altitude balloon whose volume is 5.00 m^3, if the temperature is $-23°$C and the pressure is 30 cm Hg?

Here we use the set of values for standard conditions to fill in one side of the gas law equation, remembering that 1 atm is 76 cm Hg pressure and 22.4 liters $= 0.0224$ m^3. The unknown is n_1, the number of moles of helium.

$$\frac{P_0 V_0}{n_0 T_0} = \frac{P_1 V_1}{n_1 T_1}$$

$$\frac{(76\ \text{cm Hg})(0.0224\ \text{m}^3)}{(1\ \text{mole})(273°\text{K})} = \frac{(30\ \text{cm Hg})(5.00\ \text{m}^3)}{(n_1)(250°\text{K})}$$

$$n_1 = 96.2\ \text{moles}$$

The mass of helium is

$$(96.2\ \text{moles})\left(\frac{4.00\ \text{g}}{1\ \text{mole}}\right) = \boxed{385\ \text{g}}$$

EXAMPLE **15-7**

Solve the preceding example, assuming the balloon to be filled with hydrogen instead of helium. The molecular weight of hydrogen (H$_2$) is 2.02.

Just as before, we calculate that there are 96.2 moles of gas in the balloon. However, since the molecular weight is different, the mass of gas is also different.

The mass of hydrogen is

$$(96.2 \text{ moles})\left(\frac{2.02 \text{ g}}{1 \text{ mole}}\right) = \boxed{194 \text{ g}}$$

The *ratio method* of using the gas law is easily applied and appeals to "common sense." Starting with

$$\frac{P_1 V_1}{n_1 T_1} = \frac{P_2 V_2}{n_2 T_2} \qquad (15\text{-}11)$$

we solve algebraically for the desired unknown. For instance, if we wish to compute the new pressure, the equation becomes

$$P_2 = P_1 \left(\frac{V_1}{V_2}\right)\left(\frac{T_2}{T_1}\right)\left(\frac{n_2}{n_1}\right) \qquad (15\text{-}12)$$

We think of the new pressure P_2 being related to the old pressure P_1 through a series of "ratio factors" such as V_1/V_2, etc. Each ratio factor is dimensionless if similar units are used in numerator and denominator. It is unnecessary to derive the formula each time; common sense will tell you whether to use V_1/V_2 or V_2/V_1, and similarly for the other ratio factors, as the following examples show.

EXAMPLE **15-8**

A balloon containing 6 ft³ of hydrogen at 10°C and 1 atm pressure rises to an altitude where the temperature is −30°C and the pressure is 0.2 atm. Assuming that the balloon bag is free to expand, what is its new volume? We first write $V_2 = \left(6 \text{ ft}^3\right)\left(—\right)\left(—\right)$ where the blank spaces are to be filled in from the given data. We know that decreasing the temperature makes the volume *less*, so the temperature ratio factor must be less than 1. Therefore we write $\left(\frac{243°\text{K}}{283°\text{K}}\right)$ rather than $\left(\frac{283°\text{K}}{243°\text{K}}\right)$. Similarly, we know that decreasing the pressure causes the volume to *increase*, so the pressure ratio is greater than 1. Hence we use $\left(\frac{1 \text{ atm}}{0.2 \text{ atm}}\right)$ for

the pressure ratio. We now fill in the ratios and solve:

$$V_2 = (6 \text{ ft}^3)\left(\frac{243°\text{K}}{283°\text{K}}\right)\left(\frac{1 \text{ atm}}{0.2 \text{ atm}}\right) = \boxed{25.8 \text{ ft}^3}$$

EXAMPLE **15-9**

What pressure is exerted by 64 g of oxygen (O_2), confined at 40°C in a flask of volume 5 liters?

If we had 32 g of O_2 (1 mole) in 22.4 liters at 0°C, the pressure would be 1 atm, or 14.7 lb/in.². However, we actually have 64 g of O_2, so this mass-ratio factor would itself double the pressure. Also, the temperature and volume are not standard, and each affects the pressure in its own way. We use a succession of ratio factors:

$$P = \left(14.7 \frac{\text{lb}}{\text{in.}^2}\right)\left(\frac{64 \text{ g}}{32 \text{ g}}\right)\left(\frac{313°\text{K}}{273°\text{K}}\right)\left(\frac{22.4 \text{ liters}}{5 \text{ liters}}\right)$$

$$= \boxed{151 \text{ lb/in.}^2}$$

It is a matter of personal preference whether you use the gas law in the form of a proportion (Eq. 15-11) or in some ratio factor form (such as Eq. 15-12). The two methods are completely equivalent. In neither case is it necessary to use a numerical value for the gas constant R if you remember the standard volume, 22.4 liters for 1 mole of ideal gas at 0°C and 1 atm pressure.

When two or more different gases are in the same container, a *partial pressure* can be calculated for each gas which is the pressure that each gas would exert if it alone occupied the whole container. Dalton's *law of partial pressures* states that the total pressure is the sum of the partial pressures of the component gases. This behavior of a mixture of gases is in accord with the model of an ideal gas discussed in Sec. 15-1.

15-7 Saturated vapor pressure

Suppose we have an open dish of water in a room at 20°C. As we shall see in Chap. 16, the average KE of the molecules depends on the temperature, being greater at 20°C than at 10°C, for instance. Because of incessant collisions, head-on as well as glancing, the molecules continually exchange energy. At *any* temperature, however, not all molecules have the same speed. If the motions are in random directions, there may be chance collisions resulting in very high or very low speeds. For instance, two molecules might "gang up on" a third molecule to give it a larger-than-average speed. At any instant there must be some molecules which are moving slowly or not at all and others which have acquired exceptionally high speeds. These "average-raisers" are able to escape from the liquid, overcoming the cohesive forces pulling them back to the surface, in much the same way that a spaceship leaves the earth when it is fired upward at sufficient speed. In each case the initial KE must be sufficient to supply the PE gained by the molecule (or spaceship) as it moves against the cohesive molecular force (or gravitational force, in the case of the spaceship). In this way we account for the evaporation that takes place when a bowl of water is left standing in the open. We can also explain why evaporation is a cooling process: the molecules that leave are the faster ones, and the *average* KE of those left behind decreases. In arid regions, desert travelers carry drinking water in porous bags so that it will be cooled by evaporation. Looking at it from another viewpoint, we may say that the latent heat of vaporization required to change the liquid to a vapor is supplied by the water in the bag, and, since heat flows out of the water, it is cooled. Evaporation can also take place from a solid substance, although the cohesive forces are much stronger than in liquids, and fewer molecules leave the surface.

Let us next imagine uncovering our bowl of water inside a cylinder which has been sealed off after all air molecules have been removed by a vacuum pump (Fig. 15-5a). After a short while, which may be less than 0.001 s, the empty space becomes filled

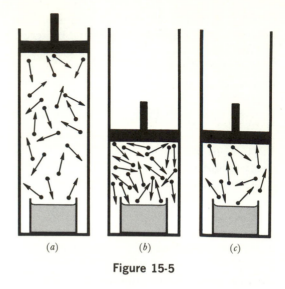

(a)   (b)   (c)

Figure 15-5

with water molecules, crisscrossing and bouncing off the walls of the enclosure. Some of these molecules will find their way back into the liquid; the process builds up until the number leaving per second just equals the number bouncing back in. The space is then said to be *saturated* with water vapor. The molecules in the cylinder exert a pressure which is called the *saturated vapor pressure*.

If the volume of the cylinder is now suddenly reduced (Fig. 15-5b), the pressure exerted by the gas is momentarily increased. The vapor is now *supersaturated*, but this is not a stable condition. The number of molecules leaving the surface per second is the same as before (since the temperature has not changed, and the average molecular KE is the same). However, the number coming back into the liquid, per second, has been increased because of the increased density of vapor. There is a net flow of molecules back into the liquid, and the vapor pressure returns to the original saturated value, at which there is a balance between molecules leaving and molecules returning (Fig. 15-5c). Our conclusion is that saturated vapor pressure depends only on the temperature of the liquid, and not at all on the volume of the container. In Table 15-2 are listed values of the saturated

TABLE 15-2
Saturated vapor pressure and density of water vapor

Temperature, °C	Saturated vapor pressure, torr	Saturated density of water vapor, g/m³
0	4.58	4.85
10	9.21	9.40
20	17.55	17.30
30	31.86	30.37
.	.	.
.	.	.
.	.	.
90	526	424
99	733	579
100	760	598
101	788	618
110	1 074	827
200	11 650	7 840
300	64 400	45 600
374	166 000	318 000

vapor pressure for water at various temperatures; pressures are expressed in torr (1 torr = 1 mm Hg).

In actual practice, we are more likely to have to deal with a cup of water on a dining-room table, with the added complication that air molecules fill up most of the space in the room. Collisions with air molecules make the diffusion and return of water molecules to the cup much slower; equilibrium is reached only after several hours, rather than practically instantaneously as in the previous illustration. If a cup of water at 20°C is placed on the table in a sealed room full of dry air, and if the pressure of the air is originally 760.0 torr (= 760.0 mm Hg), then in due course of time the total pressure in the room will rise to

$$760.0 + 17.6 = 777.6 \text{ torr}$$

What is more probable is that the room is not airtight. If outside atmospheric pressure is 760 torr, there will be a slight excess of pressure in the room, which causes a flow of air and water vapor through chinks and cracks around the doors and windows. Eventually, in the steady condition, the pressure in the room will be a total of 760 torr, of which about 742 torr is due to air molecules and 18 torr is due to water molecules (see Table 15-2). The air in the room will be practically* saturated with water vapor, as long as any water at all remains in the dish.

15-8   Relative humidity

Relative humidity is the ratio of the density of the water vapor in the air at any temperature to the density that would exist if the air were saturated with water vapor at that temperature. Relative humidity has great biological importance. A warm summer day when the relative humidity is 80% or more is likely to be uncomfortable. The human skin is normally cooled by evaporation, and if the surrounding air is nearly saturated, there will be only a slight net evaporation, and hence the skin will remain warm. On the other hand, if the air is not saturated, moist tissues will tend to lose more water molecules than they receive. Generally, a relative humidity of at least 40% is desirable. Some simple calculations will show that many gallons of water might have to be evaporated to "humidify" a large room in winter.

EXAMPLE **15-10**

An auditorium measures 10 m × 20 m × 30 m and is heated with air which has been drawn in from the outside at 0°C. Assuming the most favorable case, that is, that the outside air is completely saturated, what is the relative humidity in the auditorium?

At 0°C the incoming air, which is saturated, contains 4.85 g of water in each cubic meter (Table 15-2). After it has been heated to 20°C,

---
*Not quite saturated; after all, slightly more water molecules must be leaving the cup than returning, since a few of those leaving the cup are being lost through the chinks and crevices to the outside world.

we have the same 4.85 g in a slightly larger volume. This volume of the heated air, which was 1 m³ when at 0°C, is found from the gas law:

$$\frac{P_1 V_1}{n_1 T_1} = \frac{P_2 V_2}{n_2 T_2}$$

$$\frac{1 \text{ m}^3}{273°\text{K}} = \frac{V_2}{293°\text{K}}$$

$$V_2 = (1 \text{ m}^3)\left(\frac{293°\text{K}}{273°\text{K}}\right) = 1.07 \text{ m}^3$$

Therefore the density of water vapor in the auditorium is 4.85 g/1.07 m³ = 4.52 g/m³. At 20°C, the air *can* hold 17.30 g/m³. Hence the relative humidity is

$$\frac{\text{Density of water vapor}}{\text{Density if saturated}} = \frac{4.52 \text{ g/m}^3}{17.30 \text{ g/m}^3}$$

$$= \boxed{26.1\%}$$

EXAMPLE **15-11**

How much water would have to be evaporated in the auditorium of Example 15-10 in order to bring the relative humidity up to 40%?

At 20°C, a 40% relative humidity would correspond to a water-vapor density 40% of the saturated value, and each cubic meter would contain 0.40(17.30 g) = 6.92 g. We already have 4.52 g of water vapor in each cubic meter brought in from the outside, so we must add 2.40 g to each cubic meter. The volume of the room is 6000 m³, so the total amount of water that would have to be evaporated into the air is

$$\text{Mass of water} = (6000 \text{ m}^3)(2.40 \text{ g/m}^3)$$

$$= 14\,400 \text{ g} = \boxed{14.4 \text{ kg}}$$

This is almost 4 gallons of water!

To add such amounts of water to the air and to replace it several times an hour as the air is circulated is a formidable engineering problem. There is also the practical difficulty of dew formation ("sweating"). The humidified air that finds its way to a cold windowpane or a cold wall might at the lower temperature be supersaturated, and dew would form; i.e., water vapor would leave the air and become liquid.[*]

[*] Relative humidity is often defined as the ratio of the pressure of water vapor present to the pressure of satu-

15-9 The vapor pressure curve and change of phase

We have discussed rather fully one type of change of phase—that from liquid to vapor (*evaporation*) and vice versa (*condensation*). Much the same sort of process is involved when a solid changes to a liquid (*melting*) or a liquid changes to a solid (*freezing*). A third process, known as *sublimation,* takes place when molecules of a solid go directly into vapor form. The "wasting away" of Dry Ice or a moth ball are examples of sublimation. The opposite process, when molecules go from vapor phase to solid phase, may be called *deposition.*

To get a unified view of changes of phase, let us plot a graph of saturated vapor pressure versus temperature, using the substance H_2O as an example (Fig. 15-6). The curve AB represents the same data on vapor pressure as those which are presented in Table 15-2; the other segments of the graph, AC and AD, are also based on experiment. At any point such as q, e, b, or m on curve AB, liquid and saturated vapor are in dynamical equilibrium with each other and can exist together, with a continual exchange of molecules. Likewise, curve AC gives the pressure and temperature at which solid ice and liquid water can exist together (in equilibrium), and AD gives the conditions for steady coexistence of solid ice and saturated water vapor. Point A is the *triple point,* which for H_2O is at $t = +0.01°\text{C}$ and $P = 4.58$ torr. At the triple point, solid, liquid, and gaseous H_2O can exist together, freely exchanging molecules but with no tendency to accumulate in any one of the three phases.

rated water vapor at the same temperature. This is entirely equivalent to our definition which involves densities, since water vapor obeys the gas laws closely. The definition in terms of pressures makes possible an easy calculation; for instance, in Example 15-10 we can read the pressures directly from Table 15-2 and obtain the relative humidity as (4.58 torr)/(17.55 torr), which equals 26.1%, as previously obtained. In spite of this computational ease, we prefer to emphasize the density of water vapor rather than its pressure.

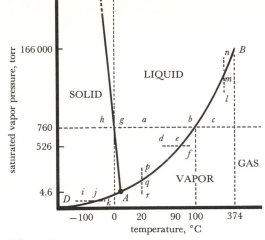

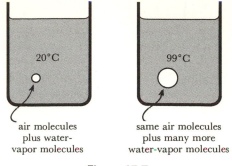

Figure 15-6 Phase diagram for H_2O; saturated vapor pressure as a function of temperature (scales exaggerated for clearness).

To illustrate the use of the graph, let us make some simple changes in temperature or pressure (one at a time).

1. At a pressure of 760 torr (= 760 mm Hg, standard atmospheric pressure) we start at room temperature, say, 20°C. Point *a* on the graph corresponds to these conditions, and the substance is liquid. As we increase the temperature (keeping pressure constant all the while) we move along the line *abc*. The saturated vapor pressure increases until finally it reaches 760 torr when the temperature reaches 100°C. Any small bubble in the water at 20°C has by now expanded greatly (Fig. 15-7).* Bubbles rise rapidly and break

---

*The space within the bubble is saturated with water vapor at any time, since it is a closed volume in contact with liquid. At 20°C the water molecules inside a bubble exert a pressure of 17.55 torr (Table 15-2), and so by subtraction we find that the air molecules in the bubble exert a pressure of 742.45 torr, to make up the total pressure of 760 torr. As the temperature rises, the saturated vapor pressure of the water molecules increases greatly, as more water molecules leave the liquid and enter the bubble. The pressure of the *air* molecules must decrease greatly, and according to the gas law this can be done only if the volume of the bubble expands greatly. The total pressure inside the bubble remains 760 torr. The pressures at the two temperatures are divided as follows: at 20°C: (17.55 torr due to water vapor) + (742.45 torr due to air); at 99°C: (733 torr due to water vapor) + (27 torr due to air).

Figure 15-7

the surface, and the water boils. In general, the *boiling point* of a liquid is the temperature at which the saturated vapor pressure of the liquid equals the surrounding atmospheric pressure. At point *b* we have a change of state from liquid to vapor. Continuing to increase the temperature, we reach a point (such as *c*) when the water vapor is superheated.

2. If we heat some water on a mountain-top where atmospheric pressure is 526 torr, we will be on a line such as *def*. The water will boil at 90°C (point *e*), at which temperature the saturated vapor pressure equals atmospheric pressure. The curve *AB* therefore represents a boiling-point curve; the boiling point at any pressure can be read off from the curve, or taken from Table 15-2. Pressure cookers are a necessity at high altitudes, and even at sea level cooking can be hastened in a pressure cooker in which the boiling point may be as high as 125°C.

3. If we move from point *a* toward lower temperatures, the water freezes when we reach 0°C (point *g*), and we enter the region marked "solid." At point *h* the ice is at a temperature of, say, −10°C, and is unable to remain in equilibrium with liquid water if the pressure is 760 torr. Point *g* is the *melting point* of water at standard atmospheric pressure of 760 torr (= 760 mm Hg); in general, the melting point is that temperature at which the solid and liquid can exist together, with molecules leaving the solid and returning in equal numbers. The curve *AC* (practically a straight line) is the melting-point

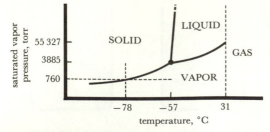

Figure 15-8 Saturated vapor pressure of CO_2 as a function of temperature (scales exaggerated for clearness).

curve for H_2O. According to the figure, the melting point of ice *decreases* as the pressure increases, since the line AC slopes upward to the left. This characteristic is related to the fact that water expands when it freezes.

4. Finally, sublimation is possible along a line such as *ijk*, provided the temperature is low enough and the surrounding vapor pressure is low. This is not usually the case for H_2O. Carbon dioxide (CO_2) sublimes at ordinary atmospheric pressure. It is this fact that makes Dry Ice "dry." Study of the vapor pressure diagram for CO_2 (Fig. 15-8) shows that solid CO_2 becomes vapor at $-78°C$ if the pressure is 1 atm. In fact, CO_2 can be a liquid only if the pressure is 3885 torr or more. In general, the *sublimation point* is that temperature at which the solid and vapor can exist together, with molecules leaving the solid and returning in equal numbers.

The liquid-vapor curve AB does not continue indefinitely; at a certain temperature called the *critical temperature* ($374°C$ for water) the curve ends. Below $374°C$, application of sufficient pressure will cause the vapor to liquefy, the change taking place along a line such as *lmn*. However, if the temperature is above $374°C$, no amount of pressure is suffi-

cient to cause formation of liquid. This is understandable, since at high temperatures the molecules move so fast that the cohesive forces are unable to form a surface. *The critical temperature is the temperature at or above which no amount of pressure, however great, will cause a gas to liquefy.* At $400°C$, it would be possible to compress 1 g of water vapor to a volume so small that the density would be greater than that of lead, but it would still be a gas, and no liquid surface would form. Sometimes a substance is called a *gas* above the critical temperature, and a *vapor* below the critical temperature. From this point of view, nitrogen and oxygen are gases at ordinary temperatures. Before the discovery of critical temperature, scientists tried (and failed) to make liquid oxygen by use of tremendous pressures. If oxygen is first cooled to a temperature below $-119°C$, its critical temperature, then a very moderate pressure of 50 atm is sufficient to liquefy it.[*]

In our discussion of vapors, we have dealt mostly with a familiar substance, H_2O, in solid, liquid, and gaseous form. All of these ideas are equally true for other substances, such as CO_2 or oxygen, but their behaviors may appear entirely different because the numerical values of critical temperature and triple point are so different. However, these are differences in degree only.

[*] If highly compressed oxygen is allowed to expand suddenly, the cohesive forces are no longer negligible, since the molecules are close together. To expand the gas requires that work be done against the cohesive forces, and as a result the gas may well be cooled below its critical temperature as its internal random KE is used up to do work. The same phenomenon occurs on a lesser scale if a tank of compressed air is opened. As the air expands and rushes out, it is often noted that the valve becomes cold enough so that water vapor from the air around the tank condenses on the valve and freezes. (See photo on p. 336.)

■ SUMMARY

If the molecules of a gas exert no appreciable cohesive forces on one another, and if the molecules are small compared with the space between them, then the gas is an ideal gas. To a large extent these conditions are fulfilled by ordinary gases at ordinary temperatures, with the result that all gases obey the same gas law regardless of chemical composition. The form of the gas law is simplified by using absolute temperature, which is equal to the Celsius temperature plus 273°. The pressure, volume, and temperature of an ideal gas are related by the general gas law: $PV/nT = R$, where the amount of gas is given by n, the number of moles, and R is the general gas constant. Boyle's law is a special case of the general gas law, with temperature held fixed.

Any gas which is below its critical temperature is called a vapor and can exist in liquid form if the pressure is sufficiently great. When the number of molecules that leave a liquid surface each second equals the number returning to the surface, the space adjacent to the surface is said to be filled with saturated vapor. Under these conditions, the pressure exerted by the molecules is called the saturated vapor pressure; the value of this pressure does not depend on the volume of the container, but only upon the temperature. Saturated vapor pressure increases rapidly as temperature increases, and when the saturated vapor pressure equals the surrounding atmospheric pressure the liquid boils.

Relative humidity is given by the ratio of the density of water vapor in the air to the density of water vapor that could be contained if the air were saturated.

The saturated vapor pressure diagram for a substance gives much information about changes of phase from liquid to vapor (boiling), solid to liquid (melting), and solid to vapor (sublimation). At the triple point the three phases can co-exist and remain in equilibrium with each other. If the temperature is above the critical temperature, no amount of pressure will cause a gas to form a liquid surface.

■ CHECK LIST

ideal gas	relative humidity	critical temperature
Boyle's law	boiling point	vapor
mole	melting point	gas
absolute temperature	sublimation point	$T = 273 + t$
partial pressure	triple point	$PV/nT = R$
saturated vapor		

■ QUESTIONS

15-1 Molecules of different gases have different sizes. Explain why this fact is of little practical importance as far as the gas laws are concerned.

15-2 Starting with the general gas law (Eq. 15-10), write a simpler law for the relation of volume and absolute temperature, when mass and pressure remain constant. (This is called the law of Charles and Gay-Lussac.) The same formula can be derived starting with Eq. 15-9.

15-3 Explain why the saturated vapor pressure of water at 100°C is so much greater than at 20°C (about a 43:1 ratio—see Table 15-2), whereas the ratio of absolute temperatures is 373:293 (only 1.27:1). In your explanation, solve the general gas law for P, and tell which factor has increased so much.

15-4 Why is evaporation a cooling process?

15-5 How is it possible to have boiling water at room temperature (20°C)? Would a flask containing such boiling water be hot to the touch?

15-6 Use Fig. 15-6 to explain what would happen to ice at -5°C if sufficient pressure were applied. (This is what makes ice skating possible.)

15-7 Explain in words what happens when water at 20°C undergoes a change from a state p to a state r along a line such as pqr in Fig. 15-6.

15-8 Two thermometers are hung side by side in a room which is not at 100% relative humidity. One thermometer has a moist cloth wrapped around its bulb. Are the readings of the two thermometers identical?

15-9 Carbon dioxide is usually shipped in tanks at a pressure of about 1000 lb/in.^{2} at room temperature. Is the material in the tank solid, liquid, or gas?

15-10 What is the difference between a gas and a vapor?

15-11 What is meant by the statement: "The boiling point of liquid hydrogen is -253°C"?

15-12 A biological preparation is to be dried by boiling off the water in a tissue sample, but to avoid damaging the specimen the temperature should not be above 30°C. How can this be done?

15-13 Air is (principally) a mixture of nitrogen (boiling point -196°C) and oxygen (boiling point -183°C). When first made, liquid air contains both of these substances. Is a flask of liquid air at the temperature of liquid nitrogen or at the temperature of liquid oxygen (assume that it is boiling)? After standing for a while, one of the materials boils away, leaving the other substance in almost a pure state. Which of the two substances is contained in a flask of liquid air that has stood for a while?

15-14 What is the condition (i.e., solid, liquid, vapor, or gas) of CO_2 at   (*a*) 0°C and 760 torr; (*b*) 20°C and 3000 torr;   (*c*) 20°C and 50 000 torr.   (*d*) Is there any temperature at which it would be possible to have liquid CO_2 in a beaker open to the atmosphere?

■ PROBLEMS

(*Note:* Assume atmospheric pressure to be 15 lb/in.^{2} unless otherwise stated.)

15-A1 The cork of a popgun is inserted so tightly that a pressure of 3 atm is required to dislodge it. The air is admitted through a hole at A which is 1.00 ft from the cork at B (Fig. 15-9). How far from A is the piston when the cork pops out, assuming no temperature change?

15-A2 Water is pumped into the bottom of a vertical storage tank whose volume is 2000 ft^3. When the pressure gauge on the tank shows that the air pressure in the tank is 45 lb/in.^{2} above atmospheric pressure, what is the volume of water in the tank?

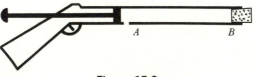

Figure 15-9

15-A3 A large research balloon whose volume is 20 000 ft³ is to be filled with helium at atmospheric pressure. The helium is stored in cylinders of volume 2 ft³ at a pressure of 300 lb/in.² How many cylinders are required?

15-A4 How many moles of H_2O are there in a drinking glass which contains 360 g of water?

15-A5 How many moles of H_2O are there in a piece of ice which has a mass of 360 g?

15-A6 A tank containing 0.1 m³ of nitrogen at room temperature and at 8×10^6 N/m² is connected through a valve to an empty tank of volume 0.4 m³. The valve is opened and the nitrogen allowed to expand. After the system returns to room temperature, what is the pressure in the tank?

15-A7 Convert to °K: 40°C; −80°C; 68°F; 212°F.

15-A8 Given that a certain mass of ideal gas occupies a volume of 30 liters at 300°K and 1 atm, compute the volume of the same mass of the same gas at 1500°K and 3 atm.

15-A9 A sealed tank contains air at 27°C and a pressure of 6×10^5 N/m². What will be the pressure if the temperature goes up to 127°C?

15-A10 A truck tire contains air at 10°C. A gauge reads 24 lb/in.² above atmospheric pressure. What will the gauge read if the temperature rises to 30°C?

15-A11 A certain amount of an ideal gas occupies 20 ft³ at 200°K and 30 lb/in.². What pressure will the same mass of gas exert if confined in 5 ft³ at 500°K?

15-A12 On a hot, steamy day, the temperature is 30°C and there is 26 g of water vapor in each cubic meter of air. What is the relative humidity?

15-A13 How much water vapor is contained in each cubic meter of air if the temperature is 20°C and the relative humidity is 40%?

15-A14 At what temperature does water boil if atmospheric pressure is 788 torr?

15-A15 What is the atmospheric pressure on a mountaintop where water boils at 90°C?

15-B1 Writing Boyle's law in the form $PV = k$, show by dimensional analysis that the constant k has the same dimensions as does work.

15-B2 A vertical cylindrical tank is 42 cm tall and 2.00 cm in radius and is open at the top. Atmospheric pressure is 1.013×10^5 N/m². A close-fitting cylindrical plug of mass 5.00 kg is inserted at the top and falls inside. If the temperature of the trapped air does not change, how far from the top of the cylinder is the base of the plug when it comes to rest?

15-B3 A mercury barometer column has an air bubble of radius 1.00 mm, 60 cm below the top of the column. What will the depth of the bubble be when it has risen to a point where its radius is 2.00 mm? Assume constant temperature.

15-B4 (a) How many moles of methane gas (CH_4) is contained in a tank of volume 50 liters, if the temperature is 40°C and the pressure is 10 atm? (b) What is the mass of this gas? (c) How many molecules of methane are in the tank?

15-B5 A scientist stores 18 g of hydrocyanic acid (a highly poisonous gas whose chemical formula is HCN) at a pressure of 900 torr. Overnight, the container develops a slight leak and the pressure drops to 860 torr. What mass of the gas has escaped?

15-B6 What pressure is exerted by 1200 g of sulfur dioxide (SO_2) confined in a tank of volume 90 liters at 50°C?

15-B7 Calculate the pressure exerted by 10 moles of carbon dioxide gas (CO_2) in a tank of volume 400 liters at a temperature of 80°C.

15-B8 (a) How many kilomoles of hydrogen gas (H_2) at $-60°C$ occupy a volume of 6 m^3 at a pressure of 4.6 atm?   (b) What is the mass of this gas?

15-B9 Calculate the density of ozone (O_3) at 650 torr and 60°C.

15-B10 A flask of volume 100 liters contains nitroxyl fluoride (NO_2F) at 760 torr and 31°C. What will the pressure become if 13 g of the gas leaks out?

15-B11 Calculate the density of neon (a monatomic gas whose atomic weight equals its molecular weight; see Periodic Table in the Appendix), when the pressure is 30 atm and the temperature is 300°C.

15-B12 One of the few gaseous compounds of uranium is uranium hexafluoride (UF_6). Calculate the ratio of the density of the gas UF_6 at 1000°K to that of N_2 gas at 300°K, both gases being under the same pressure.

15-B13 A balloon is filled with helium at 0°C and at 1 atm. The weight of the balloon is 0.50 N, and it contains 15 moles of helium.   (a) What is the volume of the balloon? (b) What is the useful payload that the balloon can lift at the earth's surface? (Mol. wt. of helium, 4.00; average mol. wt. of air, 29.)

15-B14 A soccer ball of constant volume 2.24 liters is pumped up with air at 20°C so that a pressure gauge reads 13 lb/in.^{2} above atmospheric pressure.   (a) How much air is in the ball?   (b) During the game, the temperature rises to 30°C. How much air must be allowed to escape, to bring the gauge pressure back to its original value? (The molecular weight of air may be taken to be approximately 29.)

15-B15 In a vessel of 1 m^3 capacity are placed the following: (a) hydrogen (H_2), which occupies 1 m^3 at atmospheric pressure,   (b) nitrogen (N_2), which occupies 4 m^3 at 0.5 atm pressure, and   (c) oxygen (O_2), which occupies 3 m^3 at 2 atm pressure. Calculate the partial pressures and the total pressure of the mixture. Calculate the mass of the mixture if the above figures refer to a temperature of 0°C.

15-B16 A tank contains 5 liters of oxygen under a pressure of 5520 torr. A second tank, cubical in shape and 40 cm on an edge, contains hydrogen at 690 torr pressure. The two tanks are connected and their contents allowed to mix. Assuming that the temperature does not change, calculate the resultant pressure, the same in each tank.

15-B17 On a day when atmospheric pressure is 15 lb/in.^{2}, a cylindrical diving bell 12 ft tall is lowered to the bottom of a lake. It is observed that water rises inside the bell to within 5 ft of the top.   (a) What is the pressure of the air trapped in the diving bell?   (b) How deep is the lake? (Assume constant temperature.)

15-B18 On a day when atmospheric pressure is 760 torr ($= 760$ mm Hg), a defective barometer tube of uniform inside diameter contains mercury to a height of 740 mm (h in Fig. 12-8); the 60 mm of clear space above the mercury contains a small amount of air. What is the true atmospheric pressure on another day when this barometer reads 700 mm?

15-B19 At what temperature are the readings on the Kelvin scale and the Fahrenheit scale the same? (*Hint:* Use the precise value of absolute zero.)

15-B20 From the data given in Example 15-5, calculate the value of the pressure P_1.

15-B21 Dew has formed on the outside of a glass of cold water which is on a table in a well-ventilated room at 30°C. As the glass and its contents warm up, it is observed that the dew disappears when the temperature of the water reaches 10°C. What is the relative humidity in the room?

15-B22 In a damp basement, the relative humidity is 70% at 10°C. How many grams of water could be evaporated into each cubic meter of the basement before moisture would start to condense on the walls?

15-B23 Consider water vapor at its critical temperature and pressure. Express the critical pressure in atm; in lb/in.^{2}. Express the density of water vapor at its critical point, in g/cm^3.

15-B24 What is the vapor pressure of water in a greenhouse where the temperature is 30°C and the relative humidity is 70%?

15-B25 A classroom measures 2 m × 4 m × 5 m and is supplied with air drawn from the outside at 10°C and heated to 20°C. The relative humidity outdoors is 80%. What is the relative humidity in the classroom?

15-C1 What is the weight, in pounds, of the oxygen in the tank of Example 15-1, if the temperature is 77°F? (This is as much an exercise in conversion of units as it is an exercise in physics!)

15-C2 (*a*) Do you expect water vapor at 100°C and 1 atm to be an ideal gas? (*b*) To test whether the ideal gas law is even approximately true for water vapor at 100°C and 1 atm, calculate the density of this substance from its molecular weight, the temperature, and the pressure, assuming the ideal gas volume of 22.4 liters for 1 mole at STP. Compare with the experimental density listed in Table 15-2.

15-C3 If the volume at 20°C of the bubble in Fig. 15-7 is 1.00 mm^3, calculate the volume at 99°C.

15-C4 In a damp basement 5 m × 10 m × 3 m, the relative humidity, at 30°C, is 90%. How much water would have to be removed by a dehumidifier in order to bring the relative humidity down to 20%? Would a gallon jug hold this much water?

15-C5 In a chemistry experiment the measured volume of hydrogen gas collected over water is 380 cm^3 at 20°C; the barometer stands at 750 torr, and the pressure inside the collecting chamber is 6.8 cm of water greater than atmospheric pressure. What dry volume would the gas have if no water vapor were present?

15-C6 Calculate the density of water vapor at the critical temperature and pressure, assuming the vapor to obey the ideal gas law. Why is your calculated value less than the experimental value given in Table 15-2?

■ REFERENCES

1 Conant, J. B., *Robert Boyle's Experiments in Pneumatics* (Harvard Case Histories in Experimental Science No. 1), (Cambridge, Mass.: Harvard University Press, 1950). See especially the Foreword; the Introduction (pp. 11–19); and Boyle's law (pp. 57–67).

2 Bridgman, P. W., "Synthetic Diamonds," *Sci. American,* Nov., 1955, p. 42. See phase diagram for the substance carbon (p. 46). Is the diamond phase stable at room temperature, according to the phase diagram?

3 Gaines, J. L., "The Dunking Duck," *Am. J. Phys.,* **27,** 190 (1959).

4 "A Robot Shadoof—New Waterbird for Egypt," *Sat. Rev.,* June 3, 1967. Introduction by John Lear (p. 49) and technical article by R. B. Murrow (p. 51). Proposes a "dunking bird" for river-valley irrigation.

5 Miller, Franklin, Jr., *Critical Temperature* (film).

16

The Theory of Heat

In this chapter, we shall study the basic theory of heat. To be sure, we have already used many of the concepts about to be discussed, such as temperature, molecular motions, and specific heat. It is time now to look at certain aspects of heat more closely and to form a unified theory tying together various fundamental phenomena of heat. We shall build upon Newton's second law for some of our theory (the kinetic theory of gases, Sec. 16-2), but we shall also see how the non-Newtonian quantum theory is indicated by measurements of specific heats (Sec. 16-3). The theory of heat has been, in more than one way, a steppingstone for the advance of physical knowledge.

16-1 Heat and temperature

First of all, let us make a sharp distinction between heat and temperature. From the molecular point of view, *heat* is internal energy. When a gram of ice melts at 0°C, the internal energy of the molecules increases. About 80 cal of heat energy goes into PE, as the molecules are pulled apart against the

cohesive forces binding them together. Heat energy may also become KE when molecules are speeded up. On the other hand, *absolute temperature in a gas or liquid is proportional to the average translational KE per molecule due to random motions.*[*] As an illustration, consider a swimming pool full of cold water at 5°C and a teacup full of hot water at 90°C. There is much more heat energy (total KE) in the pool, since there are many molecules, but the average KE per molecule is less than the average per molecule in the cup.

Each word in our statement about absolute temperature has definite significance: "proportional," since the factor relating absolute temperature to average KE depends on the choice of temperature scale (we shall see later what the factor of proportionality is); "average," because not all the molecules are moving at the same speed, and, if they were, collisions would soon destroy equality; "translational," since rotational or vibrational types of KE are without effect in causing collisions

---

[*] In a solid the molecules are in contact but are not free to "drift" and can have no random translational KE. In solids the *vibrational energy* is proportional to the absolute temperature.

Small cohesive forces between molecules must be overcome as air expands from an initial pressure of 2600 lb/in.² The work required to separate the molecules comes from the kinetic energy of the gas molecules, and therefore the gas cools during expansion. In this experiment the initial temperature was 54°F and the final temperature −5°F; atmospheric water vapor has condensed as frost on the low-pressure side of the system. (Cooper Industries)

in a gas and hence cannot cause heat transfer (a basic effect of temperature difference); "kinetic" energy, since increases of PE such as a latent heat take place without rise in temperature; "per molecule," since we are distinguishing between temperature and heat; due to "random" motions, since the "ordered" motions of all the molecules of a body are considered to be mechanical energy. The molecules of a baseball are given ordered motions when it is thrown. These "organized" motions of translation and rotation are not in themselves indicative of a temperature rise, but are classed as KE of the ball as a whole. When the ball strikes a catcher's mitt and is stopped, its KE is transformed into an increase in random KE of molecules of ball and mitt, and the ball's temperature rises.

16-2   The kinetic theory of gases

The interpretation of temperature we have just given is based upon our understanding of nature, as summarized by Newton's laws of motion and the general gas law. The first step in the analysis is to use Newton's second law to derive a formula for the pressure P exerted by the molecules in a closed box of volume V. We assume that there are N molecules, each of mass m, moving with equal speeds[*] v in the x-direction toward the face bc of the box (Fig. 16-1). The problem is to find the force on the box face bc due to impacts of molecules as they strike and bounce back. This problem is similar to the problem of the fire hose sending water against a burning building (Example 6-1). Now, as then, we find it convenient to use Newton's second law in the form net force = rate of change of momentum. A molecule has a momentum $+mv$ before collision with the wall and a momentum $-mv$ after collision, assuming that it bounces back without loss of

[*] The assumption of equal speeds is for simplicity; a derivation of Eq. 16-2 not making this assumption is given in Sec. 16-7.

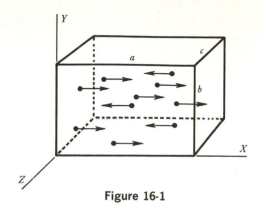

Figure 16-1

energy. The change in momentum is

$$(+mv) - (-mv) = 2mv$$

at each collision. In a time t, a molecule moves vt meters, and, since the molecule has to travel a distance of $2a$ meters between successive collisions at the same surface, the number of collisions in time t is

$$\frac{(vt \text{ meters})}{(2a \text{ meters/collision})} = \frac{vt}{2a} \text{ collisions}$$

Now we apply the second law:

$$\text{net force} = \frac{\begin{pmatrix} \text{change in} \\ \text{momentum} \\ \text{per collision} \end{pmatrix}\begin{pmatrix} \text{total} \\ \text{number of} \\ \text{collisions} \end{pmatrix}}{\text{time}}$$

$$= \frac{(2mv)(vt/2a)}{t} = \frac{mv^2}{a}$$

Thus the force exerted by each molecule is mv^2/a. There are N molecules, so the total force is Nmv^2/a, and the pressure on face bc is force/area:

$$P = \frac{Nmv^2}{(a)(bc)} = \frac{Nmv^2}{V}$$

where $V = abc$ = volume of the box. To make our derivation more realistic, we must take account of the fact that the molecules are moving in random directions, not just parallel to the x axis. This introduces a factor of $\frac{1}{3}$, since on the average a molecule has a component of velocity along any one of three axes. Only $\frac{1}{3}$ of the momentum changes are "effective" in any one direction. Our final result is, therefore,

$$P = \frac{1}{3}\frac{Nmv^2}{V} \qquad (16\text{-}1)$$

or

$$PV = \frac{1}{3}Nmv^2 \qquad (16\text{-}2)$$

We can use this equation to compute the speeds of molecules in gases.

EXAMPLE **16-1**

What is the speed of the molecules in a flask of oxygen (O_2) at 0°C and 1 atm pressure?

If there is 1 mole of oxygen present, the volume is 22.4 liters = 2.24×10^{-2} m³. The pressure of 1 atm is equal to 1.01×10^5 N/m² (see Sec. 12-1). First, we solve Eq. 16-2 for the velocity:

$$3PV = Nmv^2$$

$$v = \sqrt{\frac{3PV}{Nm}}$$

Since N is the number of molecules and m is the mass of each, the product Nm is the total mass of gas, 32 g = 3.2×10^{-2} kg in this case, since we are assuming that 1 mole of oxygen is present.

$$v = \sqrt{\frac{3(1.01 \times 10^5 \text{ N/m}^2)(2.24 \times 10^{-2} \text{ m}^3)}{3.2 \times 10^{-2} \text{ kg}}}$$

$$= \sqrt{21.2 \times 10^4 \text{ m}^2/\text{s}^2} = 4.60 \times 10^2 \text{ m/s}$$

$$= \boxed{460 \text{ m/s}}$$

This is an average speed, about $\frac{1}{4}$ mi/s in the example. Individual O_2 molecules may have speeds much greater or smaller than this value, which is called the root-mean-square (rms) speed.[*]

Let us look at Eq. 16-2 more closely. We recognize mv^2 as just twice the translational

KE per molecule. Therefore, replacing mv^2 by 2KE, we get

$$PV = (\tfrac{1}{3}N)(2\text{KE per molecule})$$

or

$$PV = (\tfrac{2}{3}N)(\text{KE per molecule}) \qquad (16\text{-}3)$$

This, then, is the result of applying Newton's second law to calculate the pressure due to moving molecules.

We now observe that the general gas law, based upon experiments with gases at various temperatures and pressures, also makes a statement about the product PV:

$$PV = (nR)(T) \qquad (16\text{-}4)$$

Since n, the number of moles, is proportional to N, the number of molecules (each mole has 6.02×10^{23} molecules), we can write Eqs. 16-3 and 16-4 in words as follows:

16-3: pressure times volume is proportional to number of molecules times KE per molecule

16-4: pressure times volume is proportional to number of molecules times absolute temperature

The conclusion is immediate and far-reaching: *Absolute temperature is proportional to KE per molecule.* This is what we set out to prove. In so doing, we have not only strengthened our understanding of temperature; we have also strengthened our belief in the adequacy of Newton's laws, and we have made more plausible our model of molecules as hard, elastic particles like billiard balls or machine-gun bullets. The derivation has been made for an ideal gas. The fact that most real gases deviate slightly from the general gas law can be accounted for by a detailed analysis including factors which we have neglected, such as the weak cohesive forces between molecules and the volume of the molecules themselves.

Although our derivation has been made for gases, the final conclusion is also true for liquids and solids. Imagine an aluminum

---

[*] There are several ways of averaging molecular speeds. The rms speed computed by use of Eq. 16-2 is about 8% greater than the mean speed. (See Sec. 16-7.)

cup full of water, sitting on a table at 20°C. The water molecules are being continually bombarded by air molecules, and there would be a tendency for an exchange of energy (heat flow) unless the average KE's were the same for air and water molecules. We know that if the temperatures of air and water are the same there is no heat flow. Hence we conclude that the average KE's are the same in air and water at the same temperature. A similar argument shows that the average KE of the vibrating aluminum molecules has the same value as for the air and the water molecules.

One illustration will show the power of the kinetic theory of gases in correlating and illuminating earlier observations and hypotheses. In 1811, Avogadro was led to suggest that, at any temperature and pressure, the number of molecules per unit volume is the same for all gases. The kinetic theory, which came later, gives a direct proof of Avogadro's brilliant hypothesis, as follows: Solving Eq. 16-1 for the number of molecules divided by the volume, we get for the number per unit volume $N/V = 3P/mv^2$. Now the product mv^2 is proportional to the absolute temperature T. Thus at any given temperature and pressure, both P and mv^2 are fixed. Therefore N/V is completely determined by the temperature and the pressure, and this ratio is independent of anything else, such as the identity of the gas. We now know that if the temperature is 0°C and the pressure is 1 atm, the ratio N/V equals 6.02×10^{23} molecules/22.4 liters for any ideal gas.

16-3   Theory of specific heats

We are already familiar with specific heat, measured in calories per gram per degree—a property of matter which tells how the temperature of a substance changes upon the addition of heat energy (Sec. 14-3). In this section we shall see that close study of specific heat will help us understand better the nature

of temperature and heat. We shall find in this study some discrepancies, of fundamental significance, which are related to the quantum theory, previously discussed in Sec. 9-6.

First, we attempt to predict the specific heat of a gas from the kinetic theory, using helium as an example. Equating Eqs. 16-3 and 16-4 gives

$$(\tfrac{2}{3}N)(\text{KE per molecule}) = nRT$$

$$\text{Total KE of } N \text{ molecules} = \tfrac{3}{2}nRT$$

$$\text{KE} = \tfrac{3}{2}nRT \quad (16\text{-}5)$$

Suppose we have 1 mole of helium (4.00 g) at 0°C ($= 273°$K), and heat it up by 1° to a temperature of 1°C ($= 274°$K). How much heat energy is required to give the additional KE to the molecules? Here we use the value of the gas constant R, assumed to be exactly 8.315 J/mole·K°.

$$\text{KE}_{274°\text{K}}$$
$$= \tfrac{3}{2}(1 \text{ mole})(8.315 \text{ J/mole} \cdot \text{K}°)(274.00 \text{ K}°)$$
$$= 3417.46 \text{ J}$$

$$\text{KE}_{273°\text{K}}$$
$$= \tfrac{3}{2}(1 \text{ mole})(8.315 \text{ J/mole} \cdot \text{K}°)(273.00 \text{ K}°)$$
$$= 3404.99 \text{ J}$$

$$\Delta\text{KE} = 3417.46 \text{ J} - 3404.99 \text{ J}$$
$$= (12.47 \text{ J})\left(\frac{1 \text{ cal}}{4.18 \text{ J}}\right) = 2.98 \text{ cal}$$

Hence the specific heat of helium is expected to be

$$\frac{2.98 \text{ cal/C}°}{4.00 \text{ g}} = 0.745 \text{ cal/g} \cdot \text{C}°$$

Measurement of the specific heat of helium gives a value of 0.75 cal/g·C°, in excellent agreement with our predicted value.

A more sophisticated way of arriving at the specific heat of helium is to use the derivative. The size of the C°, Δt, is the same as that of the K°, ΔT, and therefore from Eq. 14-2 the specific heat is

$$c = \lim_{\Delta T \to 0} \left[\frac{1}{m}\left(\frac{\Delta Q}{\Delta T}\right) \right] = \frac{1}{m}\frac{dQ}{dT}$$

where Q is the heat content of m grams of the gas, considered here to be entirely KE. We now use Eq. 16-5 to find the specific heat:

$$Q = \frac{3}{2}nRT = \left(\frac{3}{2}\frac{m}{W}R\right)(T)$$

where n, the number of moles, is given by (mass)/(molecular weight), or m/W.

$$\frac{dQ}{dT} = \left(\frac{3}{2}\frac{m}{W}R\right)\left(\frac{dT}{dT}\right)$$

The derivative of T with respect to T is 1, hence

$$c = \frac{1}{m}\frac{dQ}{dT} = \frac{3}{2}\frac{R}{W}$$

Since $R = 8.315$ J/mole·K° $= 1.99$ cal/mole·K° $= 1.99$ cal/mole·C°, and the molecular weight W of helium is 4.00 g, this gives the same result as before:

$$c = \frac{3}{2}\frac{1.99 \text{ cal/mole·C°}}{4.00 \text{ g/mole}} = 0.745 \text{ cal/g·C°}$$

Calculation of the specific heat of neon (molecular weight 20.2) proceeds in much the same way. Just as for helium, we expect 2.98 cal to be required to heat 1 mole of neon 1 C°; then the specific heat (per gram) would be (2.98 cal/C°)/20.2 g $= 0.148$ cal/g·C°. The experimental value is 0.15 cal/g·C°. So far, so good.

Things are different if we calculate the specific heat of H_2, O_2, or N_2. Experiment shows that to heat 1 mole of one of these diatomic gases 1 C° requires not 2.98 cal, but about 4.9 cal. However, we understand well enough where the extra energy goes (see Sec. 16-8). The hydrogen molecule can rotate on its axis and have rotational KE ($= \frac{1}{2}I\omega^2$) (Sec. 8-8) in addition to translational KE ($= \frac{1}{2}mv^2$) (Fig. 16-2). The helium and neon molecules, on the other hand, consist of single atoms, have no appreciable rotational inertia, and hence can have no appreciable rotational energy. To raise the temperature of hydrogen 1 C° requires more heat than for

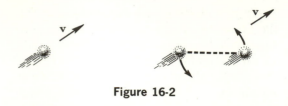

Figure 16-2

helium, since that portion of the internal energy which is stored as rotational energy is not effective in raising the temperature. This illustrates our statement that temperature is proportional to the *translational* KE.

Another prediction of the theory is borne out by experiment. It is found that if a gas is allowed to expand while being heated (thereby preventing a build-up of pressure), the specific heat is greater than if the volume is kept constant. Suppose a mole of any gas is confined within a container having a piston (Fig. 16-3). As heat is supplied and the piston is allowed to move, mechanical work is done against an opposing force PA, where P is the pressure both inside and outside the container, and A is the area of the piston. This energy has to come from somewhere.

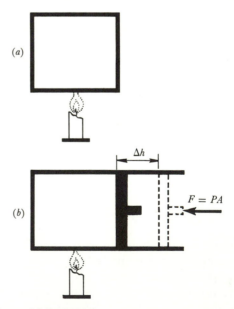

Figure 16-3 Which gas becomes warmer? Container a is at constant volume, and container b is at constant pressure.

The source of heat must supply not only the increased KE of the molecules but also the external work done as the piston moves. The specific heat is greater, by an amount which is almost the same for all gases, since all gases obey the same gas law. Experiment confirms this fully. At constant pressure the specific heat c_p of H_2 or O_2 is 40% greater than the specific heat c_v at constant volume.[*] For further discussion of the ratio of the specific heats of a gas see Sec. 16-8.

In view of all these successes, it came as a distinct shock, in the early 1900's, to find that some predictions of the kinetic theory of gases were not at all true. Consider the case of hydrogen, for which we expect the specific heat at constant volume to be 4.9 cal/mole· C°. This is not far from the observed value, if we observe a 1° change near room temperature, say from 293°K to 294°K. However, to change 1 mole of H_2 from 93°K to 94°K, only about 2.98 cal is required, the same as for 1 mole of helium. Evidently at low temperatures the hydrogen molecules are unable to absorb rotational energy! In Sec. 9-6 we discussed molecular rotations and vibrations, as revealed by the spectroscope. Not all rotational speeds are possible; a molecule may possess 0, 1, 2, 3, . . . quanta of rotational KE, but it is impossible for a molecule to rotate very slowly, with just $\frac{1}{2}$ or $\frac{1}{5}$ or $\frac{1}{1000}$ of a quantum of KE. Here is an explanation of the curious behavior of H_2 gas at low temperatures. At 93°K, the average translational KE of the molecules is much less than even one quantum of rotational KE. As a result, the collisions take place without angular recoil, as in Fig. 16-4b. To all intents and purposes, the molecules cannot rotate, and all the heat energy is effective in raising temperature, just as in the case of helium. At room temperature the collisions are energetic enough to allow more or less free rota-

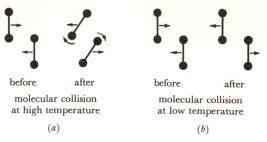

before   after   before   after

molecular collision   molecular collision
at high temperature   at low temperature

(a)   (b)

Figure 16-4 Collisions between molecules of a diatomic gas.

tion, with 5 or 10 quanta, and the specific heat has its expected higher value because extra energy may now be stored as rotational KE.

It is interesting to note that this "freezing out" of rotational motion at low temperatures provided some early evidence for the quantum of mechanical energy. The kinetic theory has served both to give us a model for the cause of pressure as due to molecular impacts, and to give us an indication that the model is insufficient to fully describe reality.

16-4   The first law of thermodynamics

The science of thermodynamics provides powerful methods of analyzing changes in pressure, temperature, and volume (for solids and liquids as well as gases), and the relations between these variables and quantity of heat and mechanical work. We have already used the first law of thermodynamics (Sec. 14-2); we repeat it for the sake of completeness:

All the heat energy added to a closed system can be accounted for as mechanical work, increase in internal energy, or both.

The first law is merely a statement of the law of conservation of energy, with heat energy being included along with mechanical energy or work.

EXAMPLE **16-2**

A hailstone of mass 200 g is falling at 20 m/s when it strikes the ground. If $\frac{1}{5}$ of the heat gen-

[*] The ratio of specific heats of a gas, c_p/c_v, is denoted by γ (gamma); this constant enters into the formula for the speed of sound in a gas (Sec. 16-10) and it is usually measured by acoustical experiments.

erated remains in the ground and $\frac{4}{5}$ remains in the hailstone, how much ice is melted?

The "closed system" here is (earth + hailstone). The original KE of the hailstone is

$$KE = \tfrac{1}{2}mv^2 = \tfrac{1}{2}(0.200 \text{ kg})(20 \text{ m/s})^2 = 40 \text{ J}$$

Of this energy, $\frac{1}{5}$, or 8 J, remains in the ground, and 32 J is available to melt some ice. The latent heat of fusion of ice is 80 cal/g, and hence the amount melted

$$m = \frac{(32 \text{ J})(1 \text{ cal/}4.18 \text{ J})}{80 \text{ cal/g}} = \boxed{0.0957 \text{ g}}$$

The investigations of Joule and others (Sec. 14-2) showed that the mechanical equivalent of heat energy is the same whether the heat is generated by friction between solid bodies, by internal molecular friction (as when water or mercury is stirred), by electrical heating in wires, or in some other way.

16-5 The second law of thermodynamics

It is more difficult to give a brief and all-inclusive statement of the second law, partly because there are so many ways of stating it, all of them equivalent. The second law expresses a *trend* in nature. To put it crudely, heat tends to flow from a hot place to a cold place. If we mix 1 kg of boiling water at 100°C with 1 kg of ice-cold water at 0°C, the result is 2 kg of water at 50°C; such calculations were made in Chap. 14. Now the total energy of the system has remained constant—as much heat was lost by the hot water as was gained by the cold water (first law of thermodynamics). Nothing in the first law would prohibit the reverse process from happening: heat *might* leave the cold water (thereby freezing it) and flow into the hot water (thereby causing it to boil). For that matter, the first law would not be violated if the molecules of the 2 kg of 50° water were to sort themselves out into two groups, with the faster ones in one corner of the bucket and the slower ones in another corner, setting up a temperature difference again. All these

unlikely happenings would violate the second law, for we would have heat flowing spontaneously "from cold to hot." The second law says that it is highly unlikely that heat will flow from cold to hot—so unlikely that we can safely assert it to be an impossibility in everyday life.[*]

If we are willing to expend some energy and do some work, it *is* possible to cause heat to flow "uphill." The refrigerator is a good example: heat is caused to flow from a cold place (the ice cubes) to a warm place (the kitchen). But this is done only by virtue of the fact that a motor is turning, expending electrical energy as the price to be paid for the backward transfer of heat. Some homes are built with "heat pumps" which are reversible. In summer they are air conditioners, moving heat from a cool place (the rooms of the house) to a warm place (the outside). In winter the heat is moved from outside to inside—once more from a cool place to a warm place. Such a "furnace" burns no fuel and has no chimney. The source of heat is free—usually from pipes buried in the ground—but the owner pays for his comfort by paying for electric power to keep the motor of the heat pump turning.

The key problem here is the unavailability of heat. There is a tremendous amount of heat energy in the oceans, for instance, but it is not available for doing work unless there is a region of lower temperature to which the heat can flow. Energy is "available" when it can be converted into mechanical or "useful" work.

A *heat engine* is a device for getting useful mechanical work out of the heat energy of the fuel. The question is, how much of this heat energy is "available"? The science of

---

[*] On a very small scale, say a dozen molecules in a small region of a container of gas, random collisions might act together to make little "pockets" of hot spots (faster molecules) and cold spots (slower molecules). Averaged out, there is no lasting observable effect. Ordinarily we deal not with a dozen, but with 10^{23} or so molecules. Under such circumstances, "highly improbable" means "practically impossible." Such considerations of probability are at the very heart of the second law.

thermodynamics had its origin in the studies of N. L. Sadi Carnot (1796–1832), a young French engineer and physicist who tried to improve the efficiency of steam engines. His conclusions were general and apply to steam engines, diesel engines, gas turbines, and all other devices which convert heat into work. We may represent a heat engine by the schematic diagram of Fig. 16-5. There is a source of heat A which, according to the second law, must be at a higher temperature than the reservoir B. Heat therefore can flow from A to B. The heat Q_2 rejected by the engine to the reservoir is less than the heat input Q_1 received by the engine from the source; the remainder of the heat energy is converted into useful work W. By the first law, $Q_1 = Q_2 + W$, hence

$$W = Q_1 - Q_2$$

$$\text{Efficiency} = \frac{\text{useful work output}}{\text{heat input}} = \frac{W}{Q_1}$$

$$= \frac{Q_1 - Q_2}{Q_1}$$

Carnot proved that, for an engine of *maximum efficiency*, the efficiency depends only on T_2/T_1,

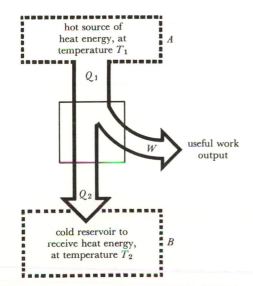

hot source of
heat energy, at
temperature T_1 A

Q_1

W useful work
output

Q_2

cold reservoir to
receive heat energy,
at temperature T_2 B

Figure 16-5 Schematic representation of a heat engine.

and Kelvin defined absolute temperature in such a way that

$$\frac{Q_1 - Q_2}{Q_1} = \frac{T_1 - T_2}{T_1}$$

where these temperatures are absolute temperatures of source and reservoir. Not all engines have this efficiency, but Carnot's analysis showed that at best the efficiency of an engine has a theoretical limit. The only way to get 100% efficiency would be to have the cold reservoir at absolute zero; if $T_2 = 0$, then the efficiency would be T_1/T_1, or 100%. This would mean that $Q_2 = 0$; an engine that was 100% efficient would reject no heat, and would convert *all* the heat input into useful work.

Actual heat engines are thus less than 100% efficient, even in the absence of frictional losses. An engine which has the maximum possible efficiency, $(T_1 - T_2)/T_1$, is called a *Carnot engine*. Most real engines, such as a steam engine or a gas turbine, cannot be designed to have the Carnot efficiency, even in theory. Frictional losses and conduction of heat through the metal parts of the engine lower the efficiency still more. To illustrate these ideas: In a simple type of steam engine, the heat input might be at 200°C, the temperature of superheated steam at a pressure of 11 650 torr (Table 15-2). The steam is allowed to expand, pushing a piston, and the working substance, H_2O, condenses into water at 100°C. According to Carnot's theorem, the greatest possible efficiency for this operation is

$$\frac{473°K - 373°K}{473°K} = 21\%$$

In other words, at least 79% of the input heat energy is unavailable for doing work, and this 79% is carried away in the hot water which is the end result of the cycle. In practice, frictional and other losses cut down the over-all efficiency of a piston type of steam engine from 21% to about 10% or 12%. The efficiency can be improved by operating at higher temperatures and by improved mechanical design, as illustrated in the next example.

EXAMPLE 16-3

The most efficient engine yet constructed is a supercharged, spark-ignited engine of the piston type, developed by the Cooper Industries. The working substance, a mixture of natural gas and air, operates between temperatures of 1870°C in the firing chamber and 430°C in the exhaust. The over-all efficiency, including all frictional losses, is such that in 1 hour, 6.85×10^9 cal of heat energy produce 1.20×10^{10} J of useful mechanical energy. Calculate (a) the Carnot efficiency, (b) the actual efficiency, and (c) the horsepower of this engine.

(a) For this engine, with $T_1 = 2143°$K and $T_2 = 703°$K, the Carnot efficiency is

$$\frac{2143°\text{K} - 703°\text{K}}{2143°\text{K}} = \boxed{67\%}$$

(b) The efficiency of the engine as actually constructed is

$$\frac{\text{Work output}}{\text{Work input}} = \frac{1.20 \times 10^{10} \text{ J}}{(6.85 \times 10^9 \text{ cal})(4.18 \text{ J/1 cal})}$$

$$= \boxed{42\%}$$

(c) The power output is

$$\frac{\text{Work output}}{\text{Time}} = \frac{1.20 \times 10^{10} \text{ J}}{3600 \text{ s}} = 3.33 \times 10^6 \text{ W}$$

$$= 3.33 \times 10^6 \text{ W}\left(\frac{1 \text{ hp}}{746 \text{ W}}\right)$$

$$= \boxed{4460 \text{ hp}}$$

The term *entropy* is used to describe the *unavailability of energy*. Entropy is a measurable property of a body, just as is its temperature or its internal energy. In the example of mixing 1 kg of hot water and 1 kg of cold water, *some* energy was available at the start of the operation, since a heat engine could (conceivably) have been operated to make use of the temperature difference between the two bodies of water. However, after the mixing, this energy is no longer available,

since there is no temperature difference. The energy of the system has remained constant (first law), but the entropy of the system has increased (second law). More energy is unavailable than was unavailable at the start of the process. Entropy is such a fundamental property that the second law may be stated in terms of it:

The entropy of the universe never decreases; during any process, the entropy either remains constant or else increases.

Eventually, when the entropy of the universe has reached its maximum value, everything will be at the same temperature (estimated to be a few degrees above absolute zero), and there will be no way to convert the heat energy of the universe into useful mechanical work. This "heat death" is far in the future.

16-6 Absolute zero

So far, we have used absolute zero simply as a convenient tool in the description of the behavior of ideal gases. The kinetic theory showed that for an ideal gas the KE of the molecules is zero at absolute zero. At temperatures of 5 or 10 degrees above absolute zero, however, most gases are far from ideal, if indeed they have not yet condensed into liquid or solid. It is questionable, therefore, whether our definition of absolute zero has much meaning, since the gas laws which were used to define absolute zero no longer are working well near absolute zero. We now have a better way of defining absolute zero, based on Carnot's work with heat engines.

We have seen (Sec. 16-5) that the efficiency of a Carnot engine would be 100% if the temperature of the cold reservoir were at absolute zero. For 100% efficiency, *all* the heat energy Q_1 would be converted into useful work, so that no heat would be rejected to the reservoir. This leads to Lord Kelvin's definition of absolute zero:

Absolute zero is the temperature of a reservoir to which a Carnot engine would reject no heat.

This is the only definition which will stand up under close scrutiny. A Carnot engine* is not restricted to any particular substance. Lord Kelvin, extending these ideas, developed the *absolute temperature scale,* which does not depend on the properties of any substance, and so there is no problem of deciding which thermometric property and which substance should be used. (This was the problem raised in Sec. 13-1.) Kelvin also showed that the ideal gas thermometer scale is identical with the absolute thermodynamic scale, which is one reason we have stressed the properties of ideal gases in our work. Unlike the Celsius (centigrade) scale, the Kelvin scale makes use of just *one* reproducible temperature. The size of the Kelvin degree is fixed by *defining* the triple point of H_2O (Sec. 15-9) to be exactly $273.16°K$. Since $0°C$ is $0.01°C$ below the triple point of water, this makes absolute zero equal to $-273.15°C$.

Absolute zero is sometimes defined (erroneously) as the temperature at which all random molecular motion stops. If there *is* random molecular motion which can be utilized, then surely we are not at absolute zero. But this is not the whole story. Even if the molecules were devoid of random linear, rotational, or vibrational KE, there might

---
* Remember that a "heat engine" is any device for transforming heat energy into mechanical energy. This process may be carried out entirely without moving parts, and the working substance may be solid, liquid, or gas.

still be sources of available energy, such as magnetic energy or electric energy or some as yet unknown form of internal energy. Such molecules (in theory, at least) could run an engine and so would *not* be at absolute zero, since there would then be a lower temperature to serve as a reservoir.

Of course, the residual magnetic and other energies are so small that they are usually of no significance whatever. At ordinary temperatures, the molecular KE so far overshadows these smaller energies that it may well be said that absolute temperature is proportional to the KE of the molecular motion. At very low temperatures, below $1°K$, magnetic and other energies are the major source of available energy, since molecular KE is, by comparison, already zero. The quest for ever lower temperatures is one of the active fields of research in physics. At present, the lowest attainable temperature for bulk matter in the solid or liquid state is about $10^{-3}\ °K$. On an atomic scale, temperatures of $10^{-6}\ °K$ have been achieved for spinning nuclei; but such temperatures are not useful since no larger body such as a crystal lattice can be put into thermal equilibrium with the nuclei for which the temperature is defined. The properties of matter are radically different in some respects at very low temperatures (for instance, note the phenomenon of superconductivity in some metals—Sec. 19-6), and low-temperature research is bringing us a better understanding of the properties of matter and the basic laws of physics.

■ SUMMARY

Absolute temperature of an ideal gas is proportional to the average random translational KE per molecule. Heat energy is total internal energy, including PE and KE. The pressure exerted by the molecules of a gas can be computed from Newton's second law, and this allows determination of molecular speeds from observed data on temperature, pressure, and volume. At any given temperature, molecular speeds are less for molecules of greater mass, since the average KE's of molecules of all gases are the same at any temperature.

Study of specific heats of gases shows that ordinarily a diatomic molecule such as H_2 absorbs some heat energy as rotational KE in addition to random translational KE. At low temperatures, rotation is no longer possible, since the average energy of a colliding molecule is much less than a single quantum of rotational KE. At such temperatures, gases with complex molecules behave like monatomic gases.

The first law of thermodynamics is essentially an extension of the law of conservation of energy, including heat energy along with mechanical energy. The second law of thermodynamics may be stated in several ways: (*a*) Heat tends to flow from a hot place to a cold place; (*b*) energy tends to become more and more unavailable for conversion from heat to mechanical work; (*c*) the entropy of the universe tends toward a maximum. Heat engines are devices for extracting useful work from a source of heat; it is impossible to convert all the heat from a source with 100% efficiency unless the cold reservoir is at absolute zero. A heat engine operated in reverse is a refrigerator, in which heat flows from a cold place to a hot place; an external source of mechanical work is required to perform this feat.

At absolute zero, random molecular motions would cease; in addition, all other forms of available energy would have to be removed from the molecules.

■ CHECK LIST

$PV = \frac{1}{3}Nmv^2$

translational KE $= \frac{3}{2}nRT$

specific heat per mole of gas

specific heat at constant volume

specific heat at constant pressure

first law of thermodynamics

second law of thermodynamics

heat engine

refrigerator

efficiency of a heat engine

$$\text{efficiency} = \frac{Q_1 - Q_2}{Q_1}$$

$$\text{maximum efficiency} = \frac{T_1 - T_2}{T_1}$$

Carnot engine

available heat energy

entropy

absolute zero

■ QUESTIONS

16-1 What is the difference between "heat" and "temperature"?

16-2 The molecules in a flask of helium have a total random translational KE of 10 ergs; a flask of oxygen contains twice as many molecules and has a total random translational KE of 14 ergs, and 8 ergs of internal rotational energy. Which flask is hotter?

16-3 Isotopes of gases are sometimes separated by diffusion, depending upon the differences of the speeds of the molecules. Neon consists of a mixture of two stable isotopes of molecular weight 20 and 22, respectively. Which neon molecules are moving faster, if both isotopes have the same temperature? Which would you expect to diffuse more readily through a porous membrane?

16-4 Give two reasons, having to do with the molecules of a gas, why the general gas law might fail to represent the behavior of that gas.

16-5 If the molecules of a gas were not tiny points, but had a finite (though small) diameter, would they make more or fewer collisions per second with the walls of the box in Fig. 16-1? (Assume the same temperature, i.e., the same speed.) Would the pressure be greater or less than that computed from Eq. 16-1?

16-6 In deriving Eq. 16-1, we ignored the collisions that must be taking place between molecules in the box. Explain why this is justified. (*Hint:* See Sec. 6-3.)

16-7 A wire wrapped around a long bar of iron carries electric current for a certain length of time, supplying a definite amount of heat to the bar. Would the temperature rise of the bar be at all affected if the ends were rigidly clamped, so that the bar were prevented from expanding as the temperature increases?

16-8 A complex molecule such as alcohol or ether is able to absorb energy in the form of internal vibrations in many ways, as well as by rotation of the molecule as a whole and by linear motion of the center of mass of the molecule itself. Would you expect the specific heat per molecule of alcohol vapor to be greater or less than that of oxygen, which does not have so many ways in which to vibrate?

16-9 Mercury can be made to boil (under pressure) at a temperature as high as 473°C. Why would mercury be a more efficient working substance for a heat engine than water? (Mercury boilers are actually used in some high-efficiency power plant installations.)

16-10 The plunger of a bicycle pump is suddenly pushed down, compressing the air within the cylinder. Explain why the air becomes warm.

16-11 In a real gas there are small cohesive forces between molecules. If the valve of a tank of compressed air is opened and the gas allowed to expand suddenly, the metal parts of the valve become cold. Explain, on the basis of the kinetic theory of gases.

16-12 In some parts of the Indian Ocean, the surface water is 23°C warmer than the water 100 m below the surface. Would it be possible to operate a steam engine continuously, using this temperature difference? Would some working substance in the engine other than H_2O make this possible, or is the whole idea theoretically unsound?

16-13 Would it be possible to cool a room by placing an electric refrigerator with its door open in the room and allowing a fan to blow cool air from the ice cube compartment out into the room?

16-14 Quantum theory predicts that a vibrating molecule can have only certain specified values of the vibrational energy, and that there is a least amount of vibrational energy that is possible. In other words, it is impossible for a molecule to have zero vibrational energy. The molecule has this "zero-point" energy even at absolute zero and is presumably vibrating with a certain amplitude even at absolute zero (insofar as a model is possible). Is such energy "available"? Does possession of this energy conflict with Kelvin's definition of absolute zero?

■ PROBLEMS

16-A1 The total random translational KE of the water molecules in a 5 gallon can is 4.4×10^6 J; the total random translational KE of the water molecules in a 100 000 gallon swimming pool is 8×10^{10} J. Which body of water—can or pool—is at the higher temperature?

16-A2 A 1 liter bottle contains liquid mercury at a temperature such that the total random translational KE of the molecules is 3.2×10^5 J. What is the total random translational KE of the molecules in a beaker containing 200 cm³ of the same material, at an absolute temperature 1.2 times that of the bottle?

16-A3 An oxygen molecule (O_2) bounces back and forth between two walls of a cubical container 20 cm on an edge and makes 1500 round trips each second. What is its speed?

16-A4 A heat engine absorbs 10 cal from the source of heat and does 4.18 J of mechanical work. What is its efficiency?

16-A5 What is the maximum efficiency of an engine which operates between fixed temperatures of 500°K and 400°K?

16-A6 A heat engine performs 1500 J of work, and at the same time rejects 7500 J of heat energy to the cold reservoir. What is the efficiency of the engine?

16-A7 An engine which has efficiency 10% does 180 ft·lb of work. How much heat energy does it take in from the hot reservoir?

16-A8 A heat engine performs 400 ft·lb of work while rejecting 1200 ft·lb of heat energy to the cold reservoir. Calculate the efficiency of the engine.

16-A9 An engine whose maximum (Carnot) efficiency is 40% obtains energy from a hot reservoir at 500°K. What is the temperature of the cold reservoir?

16-B1 A machine gun fires 20 bullets per second. Each bullet has mass of 100 g, is moving at 250 m/s and bounces back from the target with its speed unaltered. What is the average force on the target?

16-B2 Calculate the rms speed of the molecules in a tank of N_2 gas (mol. wt. 28) if the temperature is 819°C and the pressure is 3 atm.

16-B3 A total mass of 2.0 kg of a certain gas is confined in a tank of volume 6 m³, and the pressure is 1.6×10^4 N/m². What is the rms speed of the molecules?

16-B4 Suppose, in Example 16-1, that the pressure were P instead of 1 atm, the temperature still being 0°C. Show that Eq. 16-2 will give the same velocity. Thus show that the molecular speed for a given gas depends only on temperature and not at all on pressure.

16-B5 (*a*) Calculate the average KE of a single O_2 molecule in Example 16-1.   (*b*) Calculate the gravitational PE increase when a molecule moves from bottom to top of a tank 2 m on an edge.   (*c*) Are gravitational effects significant in this tank?

16-B6 Make a dimensional check for Eq. 16-2.

16-B7 (*a*) Calculate the rms speed of the molecules in a tank of volume 4 m³ which contains 100 kg of gas at a pressure of 5 atm.   (*b*) If the temperature of the gas is 20°C, what is the molecular weight?

16-B8 At 100°C the rms speed of certain molecules is v. What will the rms speed of these same molecules become if the temperature rises to 400°C?

16-B9 Calculate the temperatures of the water in the can and in the pool in Prob. 16-A1. (*Hint:* 1 gallon of H_2O contains 210 moles.)

16-B10 Calculate the temperature of the bottle of Prob. 16-A2.

16-B11 What is the temperature of the molecule in Prob. 16-A3?

16-B12 Calculate the total translational KE of 7 moles of an ideal gas at 500°K.

16-B13 The best modern vacuum pumps are able to reduce the pressure in a container to about 1 picotorr (10^{-12} torr).   (*a*) At such a pressure, how many molecules are in 1 cm³ at 0°C?   (*b*) Assuming each molecule to be at the corner of a cube x cm on an edge, calculate x, which is the approximate average distance between molecules in the gas.

16-B14 "Absolute temperature is proportional to KE per molecule." This means that KE per molecule = (constant)(absolute temperature). Find the mks value of this constant.

16-B15 A Carnot engine performs work at the rate of 1200 kW. The source of heat is at 627°C, and there is a difference of temperature of 300°C between the input heat source and the output heat reservoir. How much heat energy is being "lost" each hour?

16-B16 A steam power plant has a boiler which can safely withstand a pressure of 1240 lb/in.². The steam is rejected at 100°C, and 40% of the mechanical output is used to overcome friction. (*a*) What is the maximum temperature of the boiler? (See Table 15-2.) (*b*) What is the Carnot efficiency of the engine? (*c*) What is the maximum possible over-all efficiency of the power plant?

16-B17 The useful power output of an engine is 20 kW, and the frictional losses within the engine are 4 kW. The engine operates between temperatures of 227°C and 727°C. (*a*) What is the Carnot efficiency of the engine? (*b*) How many joules of thermal energy are taken in per hour by the engine? (*c*) What is the over-all efficiency of the engine?

16-B18 An engine, working at maximum efficiency between temperatures of 300°K and 500°K, rejects 1800 cal of heat during a certain time. How much work is performed during this period?

16-B19 A Carnot engine takes in 50 kcal of heat from a reservoir at 900°K and performs 3×10^4 J of work. What is the temperature of the cold reservoir?

16-C1 Starting with Eq. 16-2, prove that the rms speed of molecules of a gas of molecular weight W at a temperature T°K is given by the equation $v = \sqrt{3RT/W}$. Use the equation to check the result of Example 16-1.

16-C2 Prove that Eq. 16-2 is also equivalent to $P = \frac{1}{3}dv^2$, where d is the density of the gas.

16-C3 Make dimensional checks for the equations derived in Probs. 16-C1 and 16-C2.

16-C4 The rms speed of atoms in a certain star's atmosphere at 10 000°K is v. What is the rms speed of the same kind of atoms, higher up in the star's atmosphere where the temperature is 2500°K?

16-C5 Measurements on sodium atoms (Na) are made using a beam that escapes through a small hole in an oven heated to 500°K. With what speed do atoms leave the oven?

16-C6 Isotopes of uranium are separated by the thermal diffusion process which depends on slight difference of rms speeds for two gaseous UF_6 molecules. Calculate the ratio of rms speeds of $^{235}U^{19}F_6$ to that of $^{238}U^{19}F_6$. (*Hint:* See Prob. 16-C1.)

16-C7 A run-down watch is dropped into a beaker of acid which dissolves it completely, releasing a considerable amount of heat in a chemical reaction. (*a*) Would the amount of heat released be less, or would it be greater, if the watch were first wound before it was dropped into the acid? (*b*) Make a rough calculation of the temperature difference to be expected due to the winding of the watch. Assume that the acid and watch have a total water equivalent of 140 g, and that the owner of the watch, while winding it, turned the stem through 10 revolutions, exerting a force of 0.8 N at a lever arm of 3 mm.

16-C8 A household refrigerator can be considered to be a heat engine operating in reverse, with the hot source at room temperature (say 27°C) and the cold reservoir at the temperature of the freezing compartment (−21°C). (*a*) What is the maximum efficiency of such a device, if operating as a heat engine? (*b*) Operating as a maximum-efficiency engine, how much work output would be obtained if enough heat flowed into the cold reservoir to transform 1 kg of ice cubes at −21°C into water at 27°C? (*c*) If this maximum-efficiency engine were reversed to act as a refrigerator, how much work input would be needed to move enough heat from the ice cube compartment to the hot source in order to change 1 kg of water at 27°C into ice at −21°C? (*d*) With the cost of mechanical energy at 3¢ per kWh (see Chap. 6), what would be the cost of making 1 kg of ice cubes if the refrigerator used 4 times as much energy as calculated in part (*c*)?

For Further Study

16-7 The distribution of molecular speeds

We derived in Sec. 16-2 an expression for the pressure exerted by molecules moving in a box: $P = \frac{1}{3}Nmv^2/V$. The derivation is marred by several weaknesses. For one thing, we assumed that all molecules have the same speed v. Also, the factor $\frac{1}{3}$ was introduced in a way which appealed to intuition but which was not rigorous. We now proceed to a more detailed study of molecular motions, leading to the same result which we earlier derived with the aid of simplifying assumptions.

We assume identical molecules, each of mass m, but suppose that they have different speeds $v_1, v_2, \ldots, v_N$. We also suppose that the molecules are moving through the box in various directions, not all parallel to the x axis. Four such molecules are shown in Fig. 16-6. Molecule 1 has velocity $\mathbf{v_1}$ (a vector), and the components of its velocity have magnitudes denoted by $v_{1x}, v_{1y},$ and v_{1z}. Molecule 2 has components v_{2x}, v_{2y}, v_{2z} (for clearness, only v_{2x} is shown), and so forth for molecules 3, 4, $\ldots$, N. In particular, molecule 3 is moving along the y axis, so $v_{3x} = 0$.

Consider the impacts of molecule 1 on the face bc of the box. Molecule 1 approaches this face with a velocity v_{1x}, and the x-component of the change of momentum at each collision

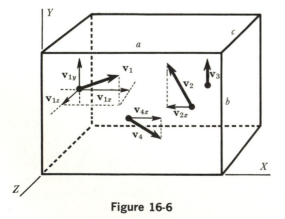

Figure 16-6

is $2mv_{1x}$. Molecule 1 makes $v_{1x}/2a$ collisions per second with face bc. By Newton's second law, the force on bc due to repeated impacts of molecule 1 is the rate of change of momentum:

$$F_1 = (2mv_{1x})\left(\frac{v_{1x}}{2a}\right)$$

$$= \frac{mv_{1x}^2}{a}$$

(The components of velocity v_{1y} and v_{1z} do not contribute at all to the force on bc.) Similarly, molecule 2 causes a force on bc equal to

$$F_2 = \frac{mv_{2x}^2}{a}$$

The total force of all N molecules on face bc is the sum of the forces caused by the separate molecules.

$$F = \frac{mv_{1x}^2}{a} + \frac{mv_{2x}^2}{a} + \cdots + \frac{mv_{Nx}^2}{a}$$

$$= \frac{m}{a}(v_{1x}^2 + v_{2x}^2 + \cdots + v_{Nx}^2)$$

$$= \frac{m}{a}(\Sigma v_x^2) \qquad (16\text{-}6)$$

To evaluate the sum, we recall that the magnitude of a vector is found by applying the Pythagorean theorem to the perpendicular components. This theorem is general and applies to vectors in three dimensions as well as two. Thus $v_1^2 = v_{1x}^2 + v_{1y}^2 + v_{1z}^2$, and similarly for v_2, etc. Let us make a table and form the sum of each column:

$$v_1^2 = v_{1x}^2 + v_{1y}^2 + v_{1z}^2$$
$$v_2^2 = v_{2x}^2 + v_{2y}^2 + v_{2z}^2$$
$$v_3^2 = v_{3x}^2 + v_{3y}^2 + v_{3z}^2$$
$$\cdot \qquad \cdot \qquad \cdot \qquad \cdot$$
$$\cdot \qquad \cdot \qquad \cdot \qquad \cdot$$
$$\cdot \qquad \cdot \qquad \cdot \qquad \cdot$$
$$\underline{v_N^2 = v_{Nx}^2 + v_{Ny}^2 + v_{Nz}^2}$$
$$\Sigma v^2 = \Sigma v_x^2 + \Sigma v_y^2 + \Sigma v_z^2 \qquad (16\text{-}7)$$

For example, suppose molecule 1 has a large velocity, more or less diagonally through the box, with components (3, 2, 3) in some system of units. Suppose molecule 2 is moving upward and to the left, with components $(-1, 2, 0)$. If molecule 3 is moving upward, its components might be $(0, 3, 0)$. The table of values would look like this:

$$v_1^2 = 9 + 4 + 9$$
$$v_2^2 = 1 + 4 + 0$$
$$\frac{v_3^2 = 0 + 9 + 0}{\Sigma v^2 = 10 + 17 + 9}$$

We now make the assumption that the molecules are moving in *random directions*. That is, there is no breeze blowing through the box, and there is no preferred direction of motion for the molecules. In our example with only 3 molecules, the y-direction is distinctly favored; but in the case of a measurable quantity of gas* we can surely say that, if the molecules are in random motion,

$$\Sigma v_x^2 = \Sigma v_y^2 = \Sigma v_z^2$$

Therefore Eq. 16-7 becomes

$$\Sigma v^2 = \Sigma v_x^2 + \Sigma v_x^2 + \Sigma v_x^2$$

or

$$\Sigma v_x^2 = \tfrac{1}{3}\Sigma v^2 \qquad (16\text{-}8)$$

Substituting into Eq. 16-6, we obtain

$$F = \frac{m}{a}\left(\frac{1}{3}\Sigma v^2\right) \qquad (16\text{-}9)$$

Since there is no speed common to all the molecules, our next step is to introduce some sort of average speed. Guided by Eq. 16-9, we define the mean-square speed as

$$\overline{v^2} = \frac{v_1^2 + v_2^2 + \cdots + v_N^2}{N} = \frac{\Sigma v^2}{N}$$

It is the average (or "mean") of the squares of the speeds of the N molecules. Multiplying by N, we see that

$$\Sigma v^2 = N\overline{v^2}$$

and hence

$$F = \frac{1}{3}\frac{m}{a}N\overline{v^2}$$

---

*Even a millionth of a mole of gas contains 6×10^{17} molecules!

The rest of the proof goes much as in Sec. 16-2:

$$P = \frac{\text{force}}{\text{area}} = \frac{1}{3}\frac{m}{abc}Nv^2$$
$$PV = \tfrac{1}{3}Nm\overline{v^2} \qquad (16\text{-}10)$$

This means that PV is proportional to the average of the squares of all the speeds of the molecules, rather than to the square of any one speed.

In Example 16-1 we found that $v = \sqrt{3PV/Nm}$. How shall we interpret speed computed in this way? From Eq. 16-10 we see that $\overline{v^2} = 3PV/Nm$, so $\sqrt{\overline{v^2}} = \sqrt{3PV/Nm}$. Thus we are calculating the *root-mean-square speed*—the square root of the mean of the squares of the speeds. Without further specification, it is the rms speed ($\sqrt{\overline{v^2}}$ or v_{rms}) that is meant when we speak of the "speed" of the molecules in a gas. There are several other ways of describing the distribution of molecular speeds. For instance, we could simply average the speeds. This would result in the *mean speed* (v_m). Another index of molecular speeds which is sometimes useful is the *most probable speed* (v_p), that speed which the greatest number of molecules have. The distribution of speeds in a gas has been studied by James Clerk Maxwell in Cambridge, Ludwig Boltzmann in Vienna, and others. The *Maxwell distribution curve* (Fig. 16-7) shows that at any instant some molecules have very small speeds such as v_1 (a few are nearly at rest), and others have speeds such as v_2, far above the average value. The rms speed is about 8% greater than the mean speed, for a large number of molecules in random motion.*

In view of these results, it ordinarily makes little difference which of the various molecular average speeds are used to describe a gas, as long as we remember that there are a considerable number of molecules in the assemblage whose speeds differ by a substantial amount from the average value. Since the

---

*This is to be expected, since the rms method of averaging gives greater weight, through squaring, to the larger speeds.

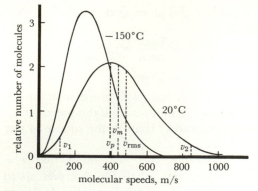

Figure 16-7 Maxwell's distribution curve for molecular speeds at two temperatures. The ordinate is in per cent, for a 10 m/s range of speeds. Example: At 20°C, about 2% of the molecules have speeds between 395 and 405 m/s.

total KE of a collection of molecules equals $\frac{1}{2}M\overline{v^2}$ (Prob. 16-C10), and since KE is so importantly related to temperature, the rms speed is usually preferred as an all-around indicator of molecular speeds.

16-8 The ratio of the specific heats of a gas

In Sec. 16-3 we saw that the specific heat at constant pressure should be greater than that at constant volume, since at constant pressure some extra heat energy must be supplied in order to do work as the gas expands against external pressure.

At *constant volume,* for an ideal monatomic gas, we can carry through a calculation in symbols similar to that in Sec. 16-3 for helium. For n moles,

At $T_1 + 1°$:   KE $= \frac{3}{2}nR(T_1 + 1°)$

At T_1:   KE $= \frac{3}{2}nRT_1$

Since the volume does not change and no heat is absorbed as rotational or vibrational energy, all the heat required to raise the gas 1 K° becomes extra translational KE of the molecules. Subtracting, we obtain

$$\Delta KE = \frac{3}{2}nR$$

and specific heat per mole $= \frac{3}{2}R$ ($= 3.0$ cal/mole·K°, since $R = 2.0$ cal/mole·K°).

At *constant pressure,* the gas must expand against the surroundings. As seen from Fig. 16-3, the work required is

$$\text{Work} = (\text{force})(\text{distance})$$
$$= (PA)(\Delta h)$$
$$= P\,\Delta V$$

since $A\,\Delta h$ is ΔV, the change in volume. From the gas law,

$$PV = nRT$$

hence

$$\Delta W = P\,\Delta V = nR\,\Delta T$$

and the extra work, for a 1 K° rise in temperature ($\Delta T = 1$ K°) is

$$\Delta W = nR \text{ (for } n \text{ moles)}$$

or

$$\Delta W = R \text{ per mole} = 2.0 \text{ cal/mole·K°}$$

Thus the *difference* between the specific heat[*] at constant pressure C_p and the specific heat at constant volume C_v is expected to be equal to R—that is, 2.0 cal/mole·K° for all ideal gases, regardless of composition. Table 16-1 shows this difference to be approximately 2.0 cal/mole·K° for many gases.

The *ratio* of the specific heats is denoted by the symbol γ (gamma) and is a dimensionless number. Measurement of the speed of sound in the gas (Sec. 16-10) allows calculation of γ. For helium and other monatomic gases,

$$\gamma = \frac{C_p}{C_v} = \frac{\frac{3}{2}R + R}{\frac{3}{2}R} = \frac{\frac{3}{2} + 1}{\frac{3}{2}}$$
$$= \frac{5}{3} = 1.67 \quad \text{(monatomic gas)}$$

For H_2, O_2, N_2, and other diatomic gases, the possibility exists that energy may be absorbed by molecular rotation. These molecules can rotate in two significant ways, not three (rotation about the molecular axis absorbs no appreciable energy, since the

---

[*] We use C_p and C_v for specific heat per mole, and c_p and c_v for specific heat per gram.

moment of inertia about this axis is assumed to be zero); we say that a diatomic molecule has two *degrees of freedom*, as far as rotation is concerned. On the other hand, these molecules (as well as all other molecules) have three translational degrees of freedom, as the molecule can move along the x, y, or z axis. It seems very reasonable that any available energy absorbed by the molecules is shared among these various degrees of freedom; for instance, energy would not be absorbed in such a way that *all* the molecules would gain translational energy along the y axis only. This share-the-energy behavior is a fundamental prediction of classical mechanics, and is called the *law of equipartition of energy*. In the case of H_2, there is a total of 5 degrees of freedom; $\frac{3}{5}$ of the energy absorbed goes into translation, and $\frac{2}{5}$ goes into rotation. Translational energy is to rotational energy as $3:2$, and our ratio of specific heats becomes

$$\frac{C_p}{C_v} = \frac{\left(\genfrac{}{}{0pt}{}{\text{trans-}}{\text{lational}}\right) + \left(\genfrac{}{}{0pt}{}{\text{rota-}}{\text{tional}}\right) + \left(\text{work}\right)}{\left(\genfrac{}{}{0pt}{}{\text{trans-}}{\text{lational}}\right) + \left(\genfrac{}{}{0pt}{}{\text{rota-}}{\text{tional}}\right)}$$

$$\gamma = \frac{\frac{3}{2}R + \frac{2}{2}R + R}{\frac{3}{2}R + \frac{2}{2}R} \qquad (16\text{-}11)$$

$$= \tfrac{7}{5} = 1.40 \quad \left(\genfrac{}{}{0pt}{}{\text{diatomic gas with molecules}}{\text{free to rotate}}\right)$$

It is evident that if other ways of absorbing energy, such as vibration, exist, both the numerator and the denominator of the fraction increase by the same amount, and the ratio becomes closer to 1. Table 16-1 shows how these ideas are verified by experiment. Note particularly how the complex vibrations of ethane (C_2H_6) and ether ($C_4H_{10}O$) are indicated by the large values of their specific heats, and by values of γ approaching 1. The fact that $C_p - C_v$ for ether departs from the predicted value of 2.0 cal/mole·K° merely indicates that ether is not an ideal gas at the temperature of these measurements.

Other departures are attributed to quantum effects. At very low temperatures, collisions are not energetic enough to give the molecules even one quantum of rotational energy (Sec. 16-3). As a result, the rotational terms ($\frac{2}{2}R$) are missing from both numerator and denominator in Eq. 16-11, and γ for H_2, O_2, and N_2 becomes $\frac{5}{3}(= 1.67)$, the same as for a monatomic gas. This "breakdown of the law of equipartition of energy," as it was called, or the "freezing out of degrees of freedom" was an unexplained symptom of the breakdown of classical physics itself. The difficulty was resolved only when quantum ideas were applied, in about 1910, to molecular rotation and vibration.

TABLE 16-1   Measured specific heats of gases at room temperature*

Type of gas	Chemical formula	C_p	C_v	$C_p - C_v$	$\gamma = C_p/C_v$
Monatomic	He	4.97	2.98	1.99	1.67
	Ne	4.97	2.98	1.99	1.67
Diatomic	H_2	6.87	4.88	1.99	1.41
	O_2	7.03	5.03	2.00	1.40
	N_2	6.95	4.95	2.00	1.40
Polyatomic	CO_2	8.83	6.80	2.03	1.32
	H_2O	8.20	6.20	2.00	1.32
	C_2H_6	12.35	10.30	2.05	1.20
	$C_4H_{10}O$	32.5	31.	1.5	1.05

* In cal/mole·K°

16-9   The adiabatic gas law

An *isothermal*[*] process is one in which the temperature does not change. Boyle's law, $P_1V_1 = P_2V_2$, describes such a process (Fig. 16-8a). An *adiabatic*[*] process is one in which no heat is transferred. For example, a piston is suddenly pushed down, doubling the pressure on an enclosed mass of gas (Fig. 16-8b). The process is adiabatic, since there is insufficient time for the heat which is generated to flow out of the gas. In effect, heat has been created in the gas, and the temperature rises.

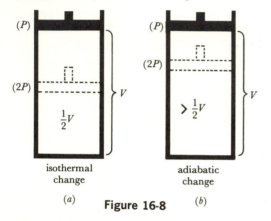

isothermal   adiabatic
change   change
(a)   **Figure 16-8**   (b)

This temperature rise causes an expansion of the gas, so to speak; the pressure is doubled, but the volume is still more than half its original value. It is reasonable to expect that the specific heats of the gas will be involved in this process, since we must relate the heat production to the change in temperature. The result of analysis is

$$P_1V_1^\gamma = P_2V_2^\gamma \qquad \left(\begin{array}{c}\text{adiabatic process}\\ \text{for ideal gas}\end{array}\right)$$

The final temperature can be found from the gas law, as usual, since $PV = nRT$ at all times.

EXAMPLE 16-4

A cylinder contains 10 liters of air at 3 atm and 300°K. (a) If the pressure is suddenly doubled, what are the new volume and temperature? (b) If the pressure is slowly doubled, what are the new volume and temperature?

[*] From Greek roots meaning "equal temperatures."
[*] From Greek roots meaning "not going through."

(a) For the sudden change, the adiabatic gas law is used. Air consists mostly of N_2 and O_2, both diatomic molecules, and so we may take γ for air as 1.40.

$$P_1V_1^\gamma = P_2V_2^\gamma$$
$$(3\ \text{atm})(10\ \text{liters})^{1.40} = (6\ \text{atm})(V_2)^{1.40}$$
$$\left(\frac{10\ \text{liters}}{V_2}\right)^{1.40} = \frac{6\ \text{atm}}{3\ \text{atm}} = 2$$
$$\frac{10\ \text{liters}}{V_2} = 2^{1/1.40} = 2^{0.714} = 1.64\ ▲$$

$$V_2 = \frac{10\ \text{liters}}{1.64} = \boxed{6.10\ \text{liters}}$$

Now to find the new temperature. The mass of gas has not changed, and so we use

$$\frac{P_1V_1}{T_1} = \frac{P_2V_2}{T_2}$$
$$T_2 = (300°\text{K})\left(\frac{6.10\ \text{liters}}{10\ \text{liters}}\right)\left(\frac{6\ \text{atm}}{3\ \text{atm}}\right)$$
$$= \boxed{366°\text{K}}$$

(b) For a slow change, heat flows out of the cylinder as the gas is compressed, and the temperature remains constant. Boyle's law applies, and we easily obtain

$$V_2 = \boxed{5\ \text{liters}}$$
$$T_2 = T_1 = \boxed{300°\text{K}}$$

If a gas is well insulated, no heat can flow, and even a slow change is adiabatic under such circumstances. We should point out that it is perfectly possible to have changes which are neither isothermal nor adiabatic, so that the division we have made is to some extent arbitrary, dictated by ease of solving problems.

16-10   The speed of sound in a gas

The ratio of specific heats of a gas appears in the formula for the speed of sound, and values

▲ The use of logarithms to find $2^{0.714}$ goes as follows (the logarithms can be looked up on a slide rule or in a log table): $\log 2 = 0.301$; $(0.714)\log 2 = 0.215$; $2^{0.714} = $ antilog $0.215 = 1.64$.

of γ are usually found by measuring the speed of sound. In Sec. 10-3 we saw that the speed of sound in a fluid is

$$v_w = \sqrt{\frac{B}{d}}$$

where B is the bulk modulus and d is the density. We will use the gas laws to calculate d and B for an ideal gas (i.e., for a gas which obeys the gas laws). The density is

$$d = \left(\frac{W}{V_0}\right)\left(\frac{P}{P_0}\right)\left(\frac{T_0}{T}\right)$$

where W is the molecular "weight" (actually, the mass of 1 mole) and V_0, P_0, T_0 refer to standard temperature and pressure. For 1 mole of any ideal gas at any temperature and pressure,

$$\frac{PV}{T} = \frac{P_0V_0}{T_0} = R$$

This gives

$$d = \frac{WP}{RT} \qquad (16\text{-}12)$$

The bulk modulus was defined in Sec. 9-1 as

$$B = \frac{\Delta P}{-\Delta V/V} = -\frac{V\,\Delta P}{\Delta V}$$

or, in the limit of a small pressure change,

$$B = \lim_{\Delta V \to 0}\left(-\frac{V\,\Delta P}{\Delta V}\right) = -V\frac{dP}{dV}$$

(The $-$ sign indicates that P decreases when V increases.)

To calculate dP/dV we use the gas law, but which shall we use: the isothermal law $P_1V_1 = P_2V_2$, or the adiabatic law $P_1V_1{}^\gamma = P_2V_2{}^\gamma$? We must recognize that in a sound wave the molecules of the gas undergo temperature changes; the gas becomes momentarily warmer wherever a condensation exists. A gas is a poor conductor of heat. At audible frequencies of, say, 1000 Hz, there is insufficient time for heat to flow out of the compressed region before the pressure relaxes and the condensation becomes a rarefaction,

cooling the gas by expansion. The tendency now is for heat to flow into the region, but before this can happen the region becomes a condensation and warms up again. The whole process is an adiabatic one (no flow of heat) rather than an isothermal one (no temperature change). Therefore, to find the bulk modulus, we differentiate the adiabatic gas law.

$$PV^\gamma = P_0V_0{}^\gamma$$
$$P = (P_0V_0{}^\gamma)V^{-\gamma}$$
$$\frac{dP}{dV} = (P_0V_0{}^\gamma)(-\gamma V^{-\gamma-1})$$
$$= (PV^\gamma)(-\gamma V^{-\gamma-1})$$
$$= -\gamma PV^{-1}$$

The bulk modulus is, then,

$$B = -V\frac{dP}{dV} = -V(-\gamma PV^{-1})$$
$$= \gamma P \qquad (16\text{-}13)$$

We have proved that the adiabatic volume modulus of an ideal gas is γ times the pressure.

Finally, we calculate the speed of sound:

$$v_w = \sqrt{\frac{B}{d}} = \sqrt{\frac{\gamma P}{d}} = \sqrt{\frac{\gamma P}{WP/RT}}$$

or

$$v_w = \sqrt{\frac{\gamma RT}{W}} \qquad (16\text{-}14)$$

The velocity is proportional to the square root of the absolute temperature and is independent of pressure.

EXAMPLE 16-5

Calculate the speed of sound in CO_2 gas at 50°C, if the speed at 0°C is 258 m/s.

$$\frac{(v_w)_2}{(v_w)_1} = \sqrt{\frac{T_2}{T_1}}$$

$$(v_w)_2 = (v_w)_1\sqrt{\frac{T_2}{T_1}} = (258 \text{ m/s})\sqrt{\frac{323°\text{K}}{273°\text{K}}}$$

$$= \boxed{281 \text{ m/s}}$$

■ PROBLEMS

16-C9 Ten molecules have speeds as follows, in arbitrary units: 1, 3, 0, 2, 2, 4, 3, 6, 1, 2. Calculate (a) the most probable speed; (b) the mean speed; (c) the rms speed.

16-C10 Prove that the total translational KE of N molecules is given by $\frac{1}{2}$(total mass)(v_{rms}^2), and hence, insofar as KE and absolute temperature are concerned, the molecules behave as if all had the same speed, equal to the rms value.

16-C11 The specific heat at constant pressure of a certain gas is 0.250 cal/g·C°. The molecular weight of the gas is 28. (a) What is the specific heat at constant volume, in cal/g·C°? (b) What is γ for this gas? (c) What can you tell about the molecular structure of this gas?

16-C12 For SO_2 gas, $\gamma = 1.29$ and $C_p - C_v = 2.15$ cal/mole·K°. Calculate C_p and C_v.

16-C13 Reconsider Prob. 15-A1, taking account of the fact that the temperature of the air would inevitably be increased during the sudden compression while the popgun is being fired. The process is described by the adiabatic gas law instead of by Boyle's law. (a) How far does the piston move before the cork pops out? (b) What is the temperature of the air just before the cork pops out (assumed to be at an initial temperature of 20°C)?

16-C14 A cylinder contains 400 cm³ of neon at 17°C and 1 atm pressure. The gas is suddenly compressed to a pressure of 900 torr. Calculate (a) the final volume, and (b) the final temperature of the neon.

16-C15 How much heat is required to increase the temperature of 2 moles of ethane (C_2H_6) by 10 K° at constant volume? (See Table 16-1.)

16-C16 Starting at 0°C, how much heat must be added to 3 moles of oxygen (a) to double the volume, at constant pressure? (b) to double the pressure, at constant volume?

16-C17 Explain in words why the pressure has no effect on the speed of sound in a gas.

16-C18 Calculate the speed of sound in CO_2 gas at 400°C. (*Hint:* See Table 16-1.)

16-C19 If the speed of sound in a certain gas is 360 m/s at 250°C, what is the speed in the same gas at 500°C?

16-C20 At what temperature would the speed of sound in air be 1000 ft/s as assumed in Prob. 10-C4?

16-C21 A closed pipe 80.0 cm long resonates in its fundamental mode with frequency 105.0 Hz when filled with ethylene vapor (C_2H_4) at 27°C. Calculate the value of γ for ethylene at this temperature.

16-C22 If the temperature in Prob. 11-B16a was 333°K, at what temperature was Prxxg successful?

■ REFERENCES

1 Summers, C. M., "The Conversion of Energy," *Sci. American,* Sept., 1971, p. 148.

2 Angrist, S. W., "Perpetual Motion Machines," *Sci. American,* Jan., 1968, p. 115.

3 Ford, N. C., and J. W. Kane, "Solar Power," *Bull. Atomic Scientists,* Oct., 1971, p. 27.

4 Meinel, A. B., and M. P. Meinel, "Is it Time for a New Look at Solar Energy?" *Bull. Atomic Scientists,* Oct., 1971, p. 32.

5 Rubber when under tension contracts when thermal energy is absorbed. Simple rubber-band engines are described in C. L. Stong's "Amateur Scientist" department of *Scientific American* by R. Hayward, May, 1956, p. 149, and by P. B. Archibald, Apr., 1971, p. 118.

17

Electric Charge

As we have seen in Chap. 2, one of the distinguishing attributes of matter is the existence of electric forces of several kinds. We are now prepared to study these forces in more detail in the next six chapters, and to learn how the application of fundamental ideas has led to so many useful developments in science, industry, and the home. With very few exceptions, we shall deal only with the long-range electric forces. These are complex enough, to be sure. One of the difficulties in making an orderly presentation of the subject is the high degree of interrelationship between the concepts. At the start, you will be asked to take certain facts for granted and even to use freely some terms that have not yet been precisely defined. However, before you finish the course you will (we hope) gain for yourself an inclusive view of electric phenomena.

17-1 Electric and magnetic forces

Our study of the electrical structure of matter (Sec. 2-3) showed that each atom has a heavy, positively charged central nucleus, surrounded by a cloud of small, negatively charged electrons. A "charged body" can be produced by creating an excess or a de-ficiency of electrons, so that a body can contain a net negative charge or a net positive charge.

One kind of force between *stationary* charged bodies is called *Coulomb force* and will be discussed in Sec. 17-5. In addition to the Coulomb forces which are always present, there are forces which arise only when two charged bodies are in motion relative to each other. These forces due to *motion* of charges are called *magnetic forces;* they are discussed in detail in Chap. 21. Magnetism is thus a subdivision of electricity, dealing with the effects of charges in motion.

Another way of looking at this division is to note the distinction between charge and current. The rate of flow of charge is called *current*. Using this terminology, we can say that Coulomb forces are caused by electric charges, as such, while magnetic forces are caused by electric currents.

17-2 Electrification of bodies

It has long been known that small bits of paper, fluff, and so on, can be "picked up" by a rod of glass, hard rubber, amber, plastic, etc., which has been rubbed with fur, silk, or cloth. The phenomenon was observed by the

---

Negative charge on the rubbed fountain pen has repelled an excess of negative charge (electrons) to the leaves of the aluminum-foil electroscope. (Photo by Kenneth Jewell)

Greeks, whose word for amber was "elektron." In Elizabethan times a piece of amber, glass, or other material which could exert a force on a small test body such as a scrap of paper or a silk thread was said to be "electrified."• These long-range electric forces could be observed at distances up to several inches. Not until the early nineteenth century did investigators come to the conclusion that there are two, and only two, kinds of electricity. For instance, let body A be a glass rod which has been rubbed with silk, and let B be a rubber rod which has been rubbed with fur. A force of attraction is observed between body A and body B. However, if C is a glass rod rubbed with silk (i.e., C is identical with A), then it is found that A is repelled by C and B is attracted to C. If A and B are oppositely charged, it is impossible to find a third kind of charge which will be attracted by both A and B, or repelled by both A and B.

Historically, a great deal of needless confusion has arisen, related to the naming of the two kinds of charge. The kind of electricity usually found on rubbed glass was called "vitreous," and the kind usually found on rubbed amber was called "resinous." Charles Du Fay (1698–1739) advocated a "two-fluid" concept of electricity, embracing vitreous and resinous electricity on equal footing. Benjamin Franklin (1706–1790) proposed a "one-fluid" theory in which vitreous electricity was an excess of positive electricity, and resinous electricity a deficiency of positive electricity—hence called negative electricity by Franklin. As we now know, there *are* two kinds of electricity, and so the two-fluid theory is actually "correct"; yet the one-fluid theory has some merit, since in most large-scale applications only one type of electricity is free to "flow." Franklin's choice of the terms "positive" and "negative" was unfortunate, since the free electrons

which move through wires and light our lamps and turn our motors are of the negative sort. This negative electricity actually exists and is not merely a deficiency of positive electricity.

Let us look briefly at those features of these two electric theories which made them so difficult to accept. In the one-fluid theory, the fluid was "electricity," and it repelled itself. Since oppositely charged bodies attract each other, and since a negative body was characterized by absence of the electric fluid, it became necessary to assume that electric fluid attracts ordinary matter. Now, however, arises the problem of repulsion between two negatively charged bodies, which were supposed to have no electric particles of any sort. Franklin himself worried about this; his follower the German Aepinus (1724–1802) boldly assumed that particles of ordinary matter repel each other! In the ordinary, unelectrified state, all matter was assumed to have just enough electric fluid so that the force of attraction on the particles of matter exerted by the electric particles was exactly balanced by the repulsion exerted by the particles of matter on each other. A cumbersome, artificial system of this kind could not endure long except as a last resort.

The proponents of the two-fluid theory had their troubles, too. A major problem was posed by the idea that uncharged matter contained equal quantities of positive and negative electric particles, ready to be separated by friction and yet not producing any observable effect. The production of "something" out of "nothing" is always open to question. From the vantage point of the twentieth century we see no difficulty, for we are accustomed to the idea of a nuclear atom with electrons and positive charges already in existence, separated by a small distance. Let us realize, however, that the efforts of Du Fay, Franklin, and others in the eighteenth century to construct models are in the best modern tradition. Our own models of atomic structure and the nature

---

• The earliest use of the term "electrification" included magnetic as well as electrostatic effects.

of electricity will no doubt seem equally confused to a generation of physicists 200 years hence; we can only hope that in the present generation we do not suffer unduly from modelitis (see footnote, p. 88).

Our belief in the atomic nature of electricity rests on a firm experimental foundation which we shall study in due course. Among such experiments are the Rutherford scattering experiment (Sec. 28-1), which showed the existence of a positively charged nucleus, and the Millikan oil-drop experiment (Sec. 27-1), which showed that charges on bodies are multiples of a unit charge, the charge of one electron. In a solid body, the nuclei are fixed in position except for the slight to-and-fro vibrations which are the manifestation of heat energy. At higher temperatures, the nuclei vibrate with greater amplitude. However, unless the solid melts, nuclei so rarely leave their places in the crystal structure that they take little part in observable electric phenomena in solids. On the other hand, in certain substances some of the orbital electrons are rather easily detached from the electron cloud surrounding a nucleus.

Electrification by friction is really a process of electrification by contact. If a glass rod is touched momentarily to a piece of silk, electrons have a tendency to leave the glass and attach themselves to the silk. This leaves an excess positive charge on the glass at the point of contact. It would be most surprising if such a tendency did not exist when dissimilar substances are brought into intimate contact. Rubbing the glass with silk is a more effective way of electrifying it than merely touching it with silk, because more parts of the surfaces are brought into contact in this way. The primary mechanism is still one of contact, and surface layers of excess charge are created—a negative layer on the silk (excess of electrons) and a positive layer on the glass (deficiency of electrons). The process of electrification by friction is, as yet, very imperfectly understood.

17-3 Conductors and insulators

Metals, unlike other solids, contain *free electrons*. For instance, in iron each nucleus has 26 protons, and there are 26 electrons per atom, making iron as a whole electrically neutral (see Fig. 2-2). However, not all the 26 electrons are bound to one nucleus; roughly 1 to 3 of them are free to move through the crystal, attached to no particular nucleus. In a small piece of iron of mass 56 g (1 mole) there are 6×10^{23} atoms[*] arranged in a crystal structure shown schematically in Fig. 2-1a. There are $26 \times 6 \times 10^{23}$ electrons in the piece of iron, of which perhaps $24 \times 6 \times 10^{23}$ are bound to their respective nuclei, and $2 \times 6 \times 10^{23}$ are free to move throughout the metal. The large number of free electrons in iron make it a good *conductor* of electricity. In an *insulator,* such as glass, all electrons are bound to one or another of the nuclei, and it is only the relatively few electrons right at the surface which can move under the influence of a closely approaching dissimilar atom or molecule.

We have described two extreme cases, the good conductor and the good insulator. Intermediate behavior is possible, and all substances may be classified in a sequence ranging from "good conductors" to "poor conductors" (which are also called "good insulators"). Silver and copper are the best conductors of electricity, but all metals are good conductors. Glass, quartz, sulfur, and paraffin are typical insulators. In Chap. 19 we shall see how the conductivity of a substance can be measured quantitatively.

17-4 Electrostatic experiments— conservation of charge

Very simple instrumentation will suffice for the qualitative study of electric forces and the motions of electric charges. The *electroscope*

[*]This is Avogadro's number. See Sec. 2-4.

(p. 357) has a central rod and a flexible leaf of gold or aluminum foil which is free to move. The outer case may be connected to the earth, although this is not necessary for all experiments. The central rod assembly is insulated from the case by a support made of rubber, sulfur, or some other insulating material. Sometimes an electroscope is constructed with two movable leaves. In one modern form, rugged enough to be used by mineral prospectors in the field, the moving conductor is a fine quartz fiber coated with gold to make it a good conductor of electricity. With the aid of an electroscope several experiments can be performed which give important information about charges and charge distribution.

1. *Like charges repel each other and unlike charges attract each other.* The repulsion of like charges is shown by the fact that the leaves of the electroscope diverge when a charged rod of either sign is touched to the knob (Fig. 17-1). Some of the surface charge on the

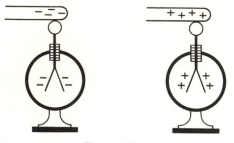

Figure 17-1

rod is shared among the rod, the knob, and the leaf structure. If the rod is removed, the leaves remain somewhat spread apart, indicating that some charge has remained on the leaves. The repulsion of like charges spreads the leaves apart until mechanical equilibrium is reached when Coulomb forces are balanced by gravitational forces.

2. *Substances differ markedly in their electrical conductivity.* In Fig. 17-2, a rod R is placed in contact with a charged sphere and an electroscope which was originally uncharged. The leaves begin to diverge, but at a rate which

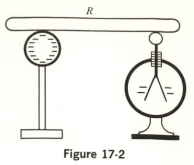

Figure 17-2

depends on the rod. If R is of copper, iron, aluminum, or some other metal, the leaves diverge almost instantly; but if R is of glass, rubber, etc., the leaves diverge slowly—perhaps imperceptibly. We can distinguish between conductors and insulators in this fashion and get a rough idea of their relative conductivities. The electroscope itself makes good use of these differences. The leaves must be good conductors so that charges can flow; hence the use of a metal foil. The supporting plug at the top of the container must be a good insulator so that charge applied to the knob will not "leak" to the surrounding case.

3. *Electric charges are of two kinds.* This can be shown by first charging the electroscope negatively (Fig. 17-3). If, now, a negatively charged rod is brought near the knob, not touching it, the leaves diverge a bit. We interpret this as a repulsion of electrons driving additional excess negative charge to the leaves. If a positively charged rod is brought up, the leaves collapse a little. We interpret this as an attraction of unlike charges; the rod pulls up some of the excess electrons that were on the leaves.

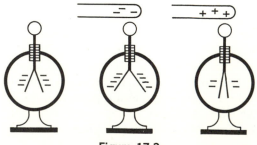

Figure 17-3

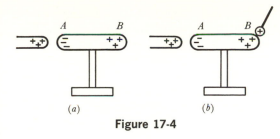

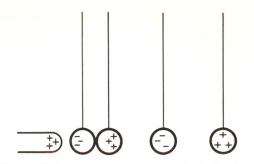

(a)   (b)

Figure 17-4

4. *Electric forces act at a distance.* This fact is shown by the experiments just described to illustrate item 3. The quantitative measurement of how electric forces depend upon distance will be discussed in Sec. 17-5.

5. *A body can be charged by induction.* If a positively charged rod is brought near a neutral metal body mounted on an insulating stand (Fig. 17-4a), free electrons in the metal are attracted toward the rod, leaving an excess positive charge at the opposite end of the body. We say that a charge has been *induced* at each end of the metal body. Of course, no charge has been created; it has merely been separated into two equal and opposite amounts. We can test this with a "proof plane," which might be a small metal disk or ball on a long insulating handle. Touching the proof plane to end B allows a sharing of charge (Fig. 17-4b), and the proof plane receives a small positive charge, a sample of the charge at B. The sign of the charge of the proof plane can be tested with the aid of an electroscope, as in item 3. If two metal spheres, initially uncharged, are hung from silk threads and are touching each other, they can be charged by induction as in Fig. 17-5. The spheres can be separated

Figure 17-5 Separation of charge by induction.

while the charge is held bound by the rod, and a considerable amount of positive and negative charge can be isolated in this way.[*] Still another induction experiment consists of "grounding" one end of the body while the negative charge is held bound at the other end by the charged rod's action-at-a-distance (Fig. 17-6). Electrons flow from the earth to end B and neutralize the positive charge at this end. Upon removing the earth connection, the negative charge is still bound at end A. Finally, when the charged rod R is removed, the negative charge spreads out, and we have a body with a net excess of negative charge. Induced charges are produced even in good insulators, but the separation of such charges is impossible since no flow can take place. Nevertheless, the effect is noticeable, as when a fountain pen is rubbed on the clothing and held near a small piece of paper. The molecules of the paper are dis-

_____

[*]This sort of operation, done in a mass-production fashion, is the basis for electrostatic generators delivering large quantities of charge at high voltage suitable for making artificial lightning or doing experiments in high-energy physics (see Sec. 30-2).

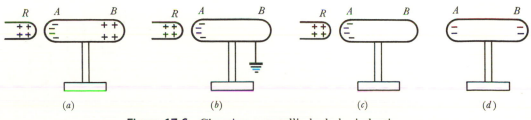

(a)   (b)   (c)   (d)

Figure 17-6 Charging a metallic body by induction.

torted, and a layer of positive charges appears at the surface. If a positively charged rod had been used, the induced surface layer would have been negative. In either case, a force of attraction between unlike charges would cause the pen or rod to "pick up" the paper. The formation of thunderclouds and lightning strokes illustrates both electrification by friction (water droplets move through air) and induction (charged cloud attracts opposite charges from earth). Similarly, a gasoline truck moving on the highway becomes charged as the tires separate from the pavement. To prevent hazardous "lightning strokes" to the earth, a dangling chain provides a conduction path to the ground and prevents an accumulation of charge.

6. *When charges are separated by contact, equal amounts of positive and negative charge are produced.* A small piece of fur glued to an insulating handle is stroked by a piece of hard rubber. It is easy to show qualitatively that the rubber gains a net negative charge and the fur gains a net positive charge. For a quantitative test of the equality of these charges, we use an induction method: A metal cup[*] on an insulating stand is connected by a wire to an electroscope (Fig. 17-7). Originally, the cup and electroscope are uncharged. If the negatively charged piece of rubber is held in the cup, not touching the metal, a separation of charge takes

place by induction, and the outer surface of the cup becomes negatively charged. The electroscope leaves share this charge, and they diverge a certain amount. If the rubber rod is removed and the fur is inserted, the leaves diverge the same distance as before, but this time the electroscope is positively charged. As a final test of the equality of the charges, the rubber and the fur are both inserted, not touching each other or the metal cup. No induction takes place, and the electroscope leaves do not diverge. By experiments of this sort, Faraday was led to the idea of the indestructibility of electric charge.

7. *The charge on a solid or hollow conducting body resides entirely on the surface of the body.* The electrons in the body of a metal conductor are free to move, and, since like charges repel, any excess charge moves to the surface.[*] This can be tested by a proof plane which acquires a charge when touched to point X (Fig. 17-8) but acquires no charge when touched to an interior point such as Y.

In summary, we can express the facts revealed by these and other experiments in the form of a *law of conservation of electric charge:*

The algebraic sum of the electric charge in any closed system remains constant.

In other words, charge can be neither created

---

[*] Michael Faraday (1791–1867) used an ordinary pewter ice pail for his historic researches in 1843. Faraday was an English physicist noted for his experimental investigations in many fields of electricity and magnetism.

[*] This would be true only for an inverse-square law of electric force, and the absence of charge on the inner surface of a hollow conductor is a sensitive proof of Coulomb's inverse-square law. See Sec. 17-7.

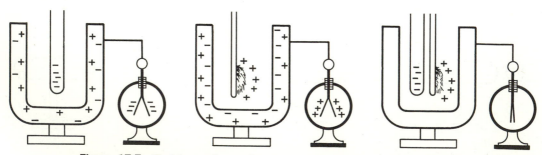

Figure 17-7 Positive and negative charge is produced in equal amounts.

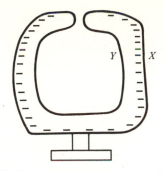

Figure 17-8 Any excess charge is distributed over the *outside* surface of a hollow conductor.

nor destroyed, though it can be moved from place to place. We accept this as a fundamental conservation law, with the same validity as the other conservation laws dealing with mass, energy, and momentum. We recall how the facts of relativity required us to qualify the laws of conservation of energy and mass, and are understandably cautious about proclaiming any conservation law as absolute and final. The validity of the law of conservation of charge in the realm of nuclear particles is still under discussion, and it may well be that some modification may eventually have to be made.[*] This possibility need not concern us, for whatever law is developed in the future must include the familiar conservation law as a special case. This is the way in which scientific advances come about.

17-5 Coulomb's law

The law of force between two charged bodies[■] was investigated by the French engineer and scientist Charles-Augustin Coulomb (1736–

1806), who also made important studies of sliding friction and the twisting of wires. To measure the small forces involved, Coulomb in 1788 used a torsion balance very similar in principle to that used 10 years later by Cavendish in his measurement of the force of gravitation described in Sec. 4-8. A small charged metal sphere was placed at a known distance from another similar small charged sphere which was part of the moving system, and the force was calculated from the observed twist of the fiber. Although he experienced some difficulty with forces due to undesired induced charges, Coulomb was able to conclude that an inverse-square law of electric force accounted for his results. *Coulomb's law* states that

> The force between two charges at rest is directly proportional to the product of the magnitudes of the charges and inversely proportional to the square of the distance between them.

In symbols,

$$F = k\frac{Q\,Q'}{r^2} \qquad (17\text{-}1)$$

where F is the magnitude of the force, Q and Q' are the magnitudes of the charges, r is the distance[▲] between them, and k is an as yet undefined constant, to be determined by experiment.

Note that Coulomb's law contains an implicit definition of magnitude of charge: the ratio of the magnitudes of two charges equals F_1/F_2, the ratio of the magnitudes of the forces they would exert on a third "test charge" at the same distance r. This definition is made plausible by the experimental

---

[*] For example, when a positive electron (Sec. 29-2) and a negative electron come together, their rest masses are converted into radiant energy, according to the equation $\Delta E = (\Delta m_0)c^2$. The equal and opposite charges cancel out, to be sure, so in a sense the law of conservation of charge is valid. In another sense, we are not merely rearranging charges; we are *destroying* charges which originally added up to zero.

[■] The law was surmised by Daniel Bernoulli in 1760. Joseph Priestley in 1767, John Robison in 1769, and

Henry Cavendish in about 1775 all made experiments to test an inverse-square law. See Sec. 17-7 for a discussion of scientific priority, as illustrated by the investigations of the law of force for electric charges.

[▲] The charges in Coulomb's law are supposed to be point charges, which is an idealization. The distance between the centers of *spherically symmetric* distributions of charge can also be used as r in Coulomb's law; we omit proof of this fact. A similar situation holds for gravitational force, which is also an inverse-square force (cf. the first footnote on p. 70).

fact that the total force exerted by n equal charges all at one point is n times the force exerted by any one of them placed at the same point. The situation is quite analogous to the way the ratio of two gravitational masses can be defined (F_1/F_2 for attraction by a given test body such as the earth). Similar operational definitions have been made in Sec. 4-3 for the ratio of two forces (a_1/a_2 for acceleration of a given test body), and the ratio of two inertial masses (F_1/F_2 for the force necessary to cause a given acceleration).

You will immediately be impressed by the similarity between Coulomb's inverse-square law and Newton's inverse-square law for gravitational forces, $F = Gmm'/r^2$. Many scientists believe that this similarity is more than accidental, but the precise connection, if any, has eluded theoretical physicists of the highest caliber. We shall consider Coulomb's law as an independent law of physics, based upon experimental evidence.

The constant k in Coulomb's law can be determined as soon as we adopt a system of units for charge, distance, and force. In this book, only *mks units are used when dealing with electrical quantities*. Indeed, one reason for introducing mks units such as the newton and joule in our study of mechanics was to lay the foundation for our use of these same units in electricity. The mks unit of charge is the *coulomb* (C), which is based on the mks unit of current called the *ampere* (A). We shall define the ampere in Chap. 21, in terms of the magnetic force between two currents. In other words, in defining the ampere, we place emphasis on the forces between moving charges. For the present we need only say that current I measures the rate of flow of charge. A coulomb is the quantity of charge carried past a given point in one second by a current of one ampere. Symbolically, we can say

$$Q = It \qquad (17\text{-}2)$$

coulombs = (amperes)(seconds)

We could write $1\ \text{A} = 1\ \text{C}/1\ \text{s} = 1\ \text{C/s}$, but

this statement, though true, is not a definition. Equation 17-2 emphasizes our choice of the ampere as the *basic* electric unit for this and future chapters.[*] The coulomb is a derived unit, in the same sense that the joule is a derived unit in mechanics.

EXAMPLE **17-1**

How much charge flows through the bulb in a student's desk lamp in 1 h if the current through the bulb is 0.5 A?

$$Q = It$$
$$= (0.5\ \text{C/s})(3600\ \text{s}) = \boxed{1800\ \text{C}}$$

A problem in language arises; strictly speaking it is charge that flows, and current *exists* in a wire. "Current" means "rate of flow of charge." Thus the sentence "A current of 0.5 A flows through the bulb" really says (ungrammatically) "A rate of flow of charge of 0.5 C per second flows through the bulb." We shall avoid such inexact statements as "Current flows through a wire."

On the atomic scale, a "natural" unit of charge immediately suggests itself—the charge e of the electron. The most fundamental of all atomic constants, the charge of the electron has been the subject of many brilliant investigations since about 1897, with the result that all electrons are now known to be identical, having a charge equal to 1.6020×10^{-19} C (Sec. 27-1). In symbols, accurate to three significant figures, the charge of the electron is

$$e = 1.60 \times 10^{-19}\ \text{C/electron}$$

We could therefore define one coulomb in terms of the charge of the electron. The reciprocal of e is

$$\frac{1}{1.6020 \times 10^{-19}\ \text{C/electron}}$$
$$= 6.2422 \times 10^{18}\ \text{electrons/C}$$

In other words a coulomb might be *defined* as the charge carried by this number of elec-

[*] The ampere is one of the six fundamental units of the Système International (SI) units (p. 6). The dimensions of charge are $[Q] = [IT]$.

trons. Such a definition would be a permanent one, based on atomic standards. However, the present-day precision of such a standard of charge would fall far short of the precision of the atomic standards of length and time mentioned in Sec. 1-4.

EXAMPLE 17-2

How many electrons flow through the wire filament in a flashlight bulb in 1 millisecond ($= 1$ ms $= 10^{-3}$ s), if the current is 50 milliamperes ($= 50$ mA $= 50 \times 10^{-3}$ A)?

$$Q = It$$
$$= (50 \times 10^{-3} \text{ A})(10^{-3} \text{ s}) = 50 \times 10^{-6} \text{ C}$$

Number of electrons
$$= (50 \times 10^{-6} \text{ C})\left(\frac{1 \text{ electron}}{1.60 \times 10^{-19} \text{ C}}\right)$$
$$= \boxed{3.12 \times 10^{14} \text{ electrons}}$$

In 0.001 s, 312 trillion electrons flow through the filament, carrying a total charge of 50 μC.

Having defined the coulomb as a unit of charge, we are able to ask what is the value of the constant k in Coulomb's law. In principle, k can be found by experiments such as Coulomb performed, in which known charges are separated by a measured distance. However, the most precise values for k are found indirectly, by analysis of experiments which are in turn based upon Coulomb's law. Such experiments give the value of k as

$$k = 9.0 \times 10^9 \text{ N} \cdot \text{m}^2/\text{C}^2 \qquad (17\text{-}3)$$

It is significant that k is numerically equal to 10^{-7} times the square of the speed of light; this is a consequence of the electrical nature of light and the other electromagnetic waves. We use k to avoid fractions and a factor 4π. Coulomb's law is often written in the form $F = (1/4\pi\epsilon_0)QQ'/r^2$, where $1/4\pi\epsilon_0 \equiv k$.

EXAMPLE 17-3

What is the force of attraction between a positive charge of 4 microcoulombs (μC) and a negative charge of 5 μC, separated by 30 cm?

$$F = \frac{kQ\,Q'}{r^2}$$
$$= \frac{\left(9 \times 10^9 \frac{\text{N} \cdot \text{m}^2}{\text{C}^2}\right)(4 \times 10^{-6} \text{ C})(5 \times 10^{-6} \text{ C})}{(0.30 \text{ m})^2}$$
$$= \boxed{2.00 \text{ N}}$$

17-6 Electrolysis

Many liquids such as oil, alcohol, or pure water are poor conductors of electricity. However, water solutions of many salts are good conductors. Such solutions, known as *electrolytes*, differ from nonconducting solutions (such as sugar in water) in that the molecules of solute are dissociated into charged fragments called *ions*. In a typical case, silver nitrate ($AgNO_3$) breaks up spontaneously into silver ions, Ag^+, and nitrate ions, NO_3^-. Each silver atom has given one electron to an NO_3^- ion, and hence each Ag^+ ion consists of a nucleus with 47 positive charges (its atomic number) surrounded by only 46 electrons. Neutral NO_3 would have 31 electrons: 7 electrons for the N atom (of atomic number 7) and 24 electrons for the three O atoms (each of atomic number 8). After receiving one electron from the Ag atom, each NO_3^- ion has 32 electrons in its cloud instead of the 31 which would make it neutral.

If a solution of silver nitrate is placed in a cup with two metallic electrodes A and B and a charge is caused to flow, a movement of ions takes place in the liquid. We do not at this time have to specify just why the charge flows; it would be sufficient to connect the electrodes to spheres (Fig. 17-9) which had been charged by electrostatic means.[*] For a continuous effect, we could connect the electrodes to a battery or a generator which could cause a steady flow of charge through the circuit. We are interested

---

[*] Electrolysis was first observed in this way by Giovanni Beccaria in 1758, some 22 years before the discovery of electric current and the invention of electric batteries.

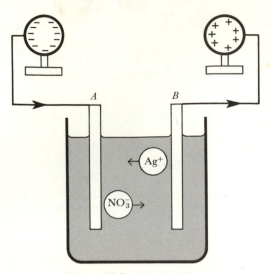

Figure 17-9 Electrolysis.

at present in what takes place in the electrolytic cell. The positive Ag^+ ions are attracted to the negative electrode A, and, when they touch it, each Ag^+ ion receives a free electron from the metal electrode, thus increasing the number of electrons in the cloud from 46 to the normal 47. The silver ion thus becomes an atom of silver metal and sticks to the electrode. In this way electrode A becomes silver-plated, and gains weight. At electrode B, electrons leave the solution and become free electrons in the metal. However, the prediction of just where these electrons come from requires considerable knowledge of chemistry. It turns out that, if B is an inert metal such as platinum, electrons leave the OH^- ions that are in the solution, and water and oxygen gas are produced at B. If B is silver metal, silver atoms leave the metal and go into solution as Ag^+ ions, leaving free electrons behind in the metal electrode; these free electrons flow through the external metallic circuit. To summarize: in the solution, the carriers of charge are positive and negative ions moving in opposite directions; in the metal, the carriers of charge are free electrons. The reactions at the electrodes cannot always be easily predicted, although it is usually safe to say that the positive ions

will accept free electrons from a negative metal electrode and will become neutralized and "plate out" at the negative electrode.

It is an easy matter to calculate the mass of substance which is deposited during electrolysis. We need to know the atomic weight of the ion, its excess charge,* and the total quantity of charge that flows. Many ions can exist in several states of charge. For instance, copper has an atomic number 29, and the neutral atom has an electron cloud of 29 electrons. The Cu^+ ion has 28 electrons in the cloud, and the Cu^{++} ion has only 27 electrons in the cloud. To plate out each Cu^{++} ion requires that two electrons flow through the external circuit in order to neutralize the doubly charged ion.

EXAMPLE **17-4**

A current of 8 A is maintained for 10 min through an electrolytic cell containing Cu^{++} ions. How much copper is deposited?

$$Q = It = (8 \text{ A})(600 \text{ s})$$
$$= 4800 \text{ C}$$

If n electrons flow through the cell,

$$n = (4800 \text{ C})\frac{1 \text{ electron}}{1.60 \times 10^{-19} \text{ C}}$$
$$= 3.00 \times 10^{22} \text{ electrons}$$

Since each atom requires 2 electrons for neutralization, the number of atoms of Cu is

$$\frac{3.00 \times 10^{22} \text{ electrons}}{2 \text{ electrons/atom}} = 1.50 \times 10^{22} \text{ atoms}$$

Finally, we use Avogadro's number and the atomic weight of copper:

Mass of Cu
$$= (1.50 \times 10^{22} \text{ atoms})\frac{63.6 \text{ g}}{6.02 \times 10^{23} \text{ atoms}}$$
$$= \boxed{1.58 \text{ g}}$$

Although the calculation of Example 17-4 is simple and straightforward, you can save

---

* In chemistry, the excess charge on an ion is often called its valence; Cu^{++} is a divalent copper ion.

time and trouble by using the concepts of the gram-equivalent weight and the faraday. The *gram-equivalent weight* (g.e.w.) of an atomic ion is its atomic weight divided by n, the number of electrons needed to neutralize the ion. The *faraday* of charge, Q_F, is defined as the charge necessary to deposit or liberate 1 gram-equivalent weight of any ion. We can easily calculate the value of the faraday as follows: let Avogadro's number be N_A, and the charge of the electron be e.

1 mole has N_A atoms.

1 gram-equivalent weight has $\dfrac{N_A}{n}$ ions, each with charge $\pm ne$.

The charge needed to deposit 1 g.e.w. is, therefore,

$$Q_F = \left(\frac{N_A}{n} \frac{\text{ions}}{\text{g.e.w.}}\right)\left(n \frac{\text{electrons}}{\text{ion}}\right)\left(e \frac{\text{C}}{\text{electron}}\right)$$

$$= N_A e \frac{\text{C}}{\text{g.e.w.}} \tag{17-4}$$

$$= (6.023 \times 10^{23})(1.602 \times 10^{-19})$$

$$= 96\,500 \text{ C/g.e.w.}$$

In view of Eq. 17-4, we see that the faraday is the total charge of a mole of electrons.

Faraday formulated the laws of electrolysis as follows: (1) the mass of substance deposited or liberated from a solution is directly proportional to the quantity of charge which flows through the circuit; (2) when the same quantity of charge passes through a number of cells the masses of substances deposited or liberated are in the ratio of their respective gram-equivalent weights. We can summarize these laws in one statement:

In electrolysis, one faraday of charge (96 500 coulombs) is required to deposit or liberate one gram-equivalent weight of any substance.

According to this statement, a proportion can be set up:

$$\frac{m}{Q} = \frac{\text{g.e.w.}}{Q_F} \tag{17-5}$$

where m is the mass deposited or liberated by a charge Q, and g.e.w., the gram-equivalent weight, is the mass in grams liberated by a faraday, Q_F. Solving for m, we obtain

$$m = \frac{\text{g.e.w.}}{Q_F} Q \tag{17-6}$$

or

$$m = zQ \tag{17-7}$$

where z, defined as the *electrochemical equivalent*, is the mass deposited by 1 C of charge; it is given by the g.e.w. divided by the faraday.

EXAMPLE **17-5**

Solve Example 17-4, using the faraday and the gram-equivalent weight.

As before, the charge passing through the cell is 4800 C.

The g.e.w. of Cu^{++} is 63.6 g/2 $= 31.8$ g. We use a proportion to find the unknown mass m which is deposited.

$$\frac{m}{4800 \text{ C}} = \frac{31.8 \text{ g}}{96\,500 \text{ C}}$$

$$m = (31.8 \text{ g})\left(\frac{4800 \text{ C}}{96\,500 \text{ C}}\right)$$

$$= \boxed{1.58 \text{ g}}$$

The study of electrolysis brings out the close connection between the atomic nature of matter and the atomic nature of electricity. Several decades before the discovery of the electron, it was clearly realized that, if matter is atomic, the facts of electrolysis require the existence of "atoms of electricity" having a charge of the order of magnitude of 10^{-19} coulomb. The equation $Q_F = N_A e$ has been used in several ways. The faraday is experimentally known with great precision (the latest value is written as $96\,487 \pm 1$ coulombs); hence N_A can be found if e is known, and vice versa. The first rough estimates of e were made in this way, using approximate values for Avogadro's number. It is customary now to invert the process and use the precise modern value of e (Sec. 27-1) to find Avogadro's number.

Electric forces between charges at rest are called Coulomb forces, and electric forces between charges in relative motion are called magnetic forces. When two dissimilar substances are placed in close contact, electrons tend to leave one substance and go to the other. If the substances are insulators, the electrons are bound to the nuclei, and transfer of charge takes place only at the points of contact. Electrification of insulators by friction is effective because contact is made at many points. In metallic conductors, free electrons can flow through the metal and take part in the conduction process.

There are two kinds of charge, positive and negative, and the law of conservation of charge states that the algebraic sum of the electric charge in any closed system remains constant. A neutral, or uncharged, body has equal amounts of positive and negative charge. Like charges repel each other, and unlike charges attract each other. The law of force between stationary electric charges is Coulomb's law, which states that the force is directly proportional to the product of the magnitudes of the charges and inversely proportional to the square of the distance between them. The mks unit of charge is the coulomb, which is the charge carried past a given point in one second due to a current of one ampere.

In electrolytes, electric current is carried by ions which are produced by dissociation of molecules. An ion has an excess or deficiency of electrons, and the net charge of an ion is a small multiple of the electron charge. When an ion reaches a metal electrode, its charge may be neutralized by the flow of free electrons to or from the electrode. In electrolysis, the mass of substance deposited or liberated is proportional to the charge that flows and to the gram equivalent weight of the substance. One faraday is that amount of charge which will deposit or liberate one gram-equivalent weight of a substance. The numerical value of the faraday is 96 500 C, which is the charge of one mole of electrons.

■ CHECK LIST

Coulomb forces	insulator	ion
magnetic forces	electric induction	gram-equivalent weight
positive electricity	Coulomb's law	faraday
negative electricity	coulomb	electrochemical equivalent
conductor	electrolyte	

■ QUESTIONS

17-1 Why is magnetism considered a subdivision of electricity?

17-2 In what respects is modern electric theory a two-fluid theory? In what respects is it a one-fluid theory?

17-3 Do protons contribute to the flow of electricity through a metal?

17-4 Why is a plastic pen electrified more strongly when it is rubbed vigorously on a coat sleeve than when it is merely touched to the cloth?

17-5 Reword item 3 of Sec. 17-4 to apply to a situation in which the electroscope was originally charged positively (i.e., had a deficiency of electrons), and rods of either sign are brought up.

17-6 Explain the action of a rubbed comb which picks up bits of paper. Why do the pieces eventually fly off the comb with an initial velocity? Why is it difficult to pick up pieces of paper on a damp day?

17-7 Suggest a method by which Coulomb might have placed on one of two identical metal balls exactly half as much charge as on the other.

17-8 Correct the following statement: "The charge on the raindrop was 463 μA."

17-9 Why should an airplane be connected to ground before a refueling operation begins?

17-10 A small battery has fallen out of a transistor radio, and it is found that the $+$ and $-$ labels on the terminals are illegible. Describe how the principles of electrostatics can be used to determine the polarities of the battery terminals.

17-11 How many electrons are required to neutralize the charge of a gram-equivalent weight of a substance?

17-12 How many electrons are in the electron cloud of a Pb^{4+} ion? (*Hint:* Use the Periodic Table in Appendix to find the atomic number of lead.)

■ PROBLEMS

17-A1 In 5 min, 3000 C of free electrons enter one end of a conductor, and 3000 C move out the other end. What is the current through the conductor?

17-A2 A service station charges a battery for 8 h, using a current of 25 A. How much charge passes through the battery?

17-A3 How many electrons are needed to carry a picocoulomb of charge (10^{-12} C)?

17-A4 What quantity of charge passes through the wire filament in a lamp bulb in 15 min if the current is 0.4 A?

17-A5 What current in picoamperes is equivalent to the flow of a million electrons per second?

17-A6 What force does a charge of 10^{-10} C exert on a charge of 4×10^{-8} C which is 2 m away?

17-A7 Two charged bodies exert a force on each other of 24 mN. What will be the force between the same two bodies if the distance between them is doubled?

17-A8 How far apart must two charges, each of 1 μC, be placed so that the force between them is 1 μN?

17-A9 What is the gram-equivalent weight of the antimony ion Sb^{3+}? (*Hint:* Use the Periodic Table in Appendix to find the atomic weight.)

17-A10 Calculate the gram-equivalent weight of Pb^{4+}. (*Hint:* Use the Periodic Table in Appendix to find the atomic weight.)

17-A11 (*a*) What charge is needed to neutralize 10 gram-equivalent weights of a substance? (*b*) How many electrons does this require?

17-A12 How many electrons are needed to neutralize the charge of 3 gram-equivalent weights of a substance?

17-A13 A certain ion has 3 excess electrons. How many moles of this ion will be deposited by 12 faradays of charge?

17-B1 Assuming that the electron in a hydrogen atom is 0.53×10^{-10} m $(= 0.53 \text{ Å})$ from the nucleus (which consists of a proton), (*a*) compute the force of electrical attraction between the electron and the proton. (*b*) Compute the gravitational attraction between the electron and proton. (*c*) Show that the ratio of these two forces is approximately 10^{39} (Use Appendix for necessary data.)

17-B2 What would be the force of repulsion between a gram of electrons on the moon separated by 3.70×10^{5} km from another gram of electrons on the earth? (Use Appendix for necessary data.)

17-B3 How far apart are two electrons if the Coulomb force between them equals the weight (on earth) of an electron? (Use Appendix for necessary data.)

17-B4 Two helium nuclei (alpha particles), each of charge $+2e$, approach each other to within a distance of 10^{-10} m in vacuum. What is the force of repulsion?

17-B5 A cosmic-ray particle as shown in Fig. 30-12 consists of a bare $_{26}$Fe nucleus (an iron atom stripped of all its orbital electrons). Calculate the Coulomb force exerted on an electron 1 nm from the particle in a photographic emulsion.

17-B6 Calculate the repulsive Coulomb force between two protons 5×10^{-15} m apart inside a nucleus. Why doesn't a force of such magnitude, acting on a proton of mass about 10^{-24} kg, cause the nucleus to fly apart? (See Sec. 2-1.)

17-B7 Calculate the force of repulsion between a proton, of charge $+e$, and a nitrogen nucleus, of charge $+7e$, at a separation of 10^{-11} m.

17-B8 A small plastic sphere coated with a thin metallized surface has mass 0.08 g and carries a charge of $+10$ nC. It is suspended by a light insulating thread at a point 3 cm below the center of a fixed conducting sphere carrying -5 nC. When the thread is cut, with what acceleration does the plastic sphere start to fall?

17-B9 A charge $Q_1 = +10$ μC is on the x axis at $x = 0$, and a second charge $Q_2 = -2$ μC is on the x axis at $x = 5$ m. A third charge $Q_3 = -10$ μC is placed on the x axis at $x = 15$ m. Calculate the magnitude and direction of the force on Q_2.

17-B10 A triangle ABC is marked out in a flat surface, having sides of length as follows: $AB = 4$ m, $BC = 5$ m, and $AC = 3$ m. At the corners are the following charges: -30 μC at A, -160 μC at B, and $+90$ μC at C. What are the magnitude and direction of the net force on the charge at A?

17-B11 Three charges, each of $+30$ μC, are equally spaced along a straight line, successive charges being 3 m apart. Calculate (*a*) the force on one of the end charges; (*b*) the force on the central charge.

17-B12 Equal charges of $+10$ μC are placed at the four corners of a square 0.3 m on a side. Calculate the magnitude and direction of the force on one of the charges.

17-B13 Equal charges of $+2$ μC are placed at the three corners of an equilateral triangle 3 m on a side. Calculate the magnitude and direction of the force on one of the charges.

17-B14 In 1 μs, how many electrons flow through a sensitive research galvanometer which reads 80 pA?

17-B15 Iron articles are sometimes cadmium-plated to increase rust resistance. Calculate the mass of cadmium deposited by a current of 50 A during an 8 h working day, if the cadmium is present in the form of Cd^{++} ions.

17-B16 The electrochemical equivalent is defined as the mass of substance deposited by 1 C of charge. Calculate the electrochemical equivalent for Cu^{++}.

17-B17 A student performs a laboratory experiment in which he plates copper from a solution containing copper ions (which may be Cu^+ or Cu^{++}). The initial mass of the electrode was 22.043 g. After a current of 0.400 A was passed through the solution for 25 min, the plate was removed and dried, and its mass was found to be 22.441 g. What can you conclude about the charge of the copper ions in the solution?

17-B18 In the production of hydrogen gas, a current of 90 A is used. Charge flows for 8 h through a cell containing water. H_2 gas is released at one electrode, and O_2 gas at the other electrode. The ions involved are H^+ and O^{--}. (*a*) What mass of hydrogen is produced? (*b*) What volume is occupied by the released hydrogen, measured at 27°C and 760 torr? (*c*) What volume of oxygen is produced?

17-B19 Chlorine is often produced by electrolysis from a solution containing Cl^- ions. (*a*) What is the current through an electrolytic cell in which 10 kg of Cl_2 gas is produced every 24 h? (*b*) What is the volume of this gas, stored as Cl_2 at 0°C and 100 atm?

17-C1 Two ping-pong balls painted with aluminum paint are suspended from the same point by threads 60 cm long. The mass of each ball is 10 g. When equal charges are given to the two balls, they come to rest in an equilibrium position in which their centers are 60 cm apart. Calculate the charge on each ball.

17-C2 Each of two small spheres is charged positively, the combined charge totaling 8×10^{-9} C. When they are placed 3 cm apart, the force of repulsion is 7×10^{-5} N. What is the magnitude of each charge?

17-C3 Two small charges Q_1 and Q_2 are situated on the x axis at points having the following coordinates: $Q_1 = -4\ \mu C$ at $x = -3$ m; $Q_2 = +1\ \mu C$ at $x = 2$ m. Find a point on the x axis at which a third charge Q_3 could be placed and experience no net force.

17-C4 Three charges are located as follows: $+6\ \mu C$ at the origin of coordinates; $+2\ \mu C$ on the x axis at $x = +2$ m; and $-3\ \mu C$ on the y axis at $y = +2$ m. Calculate the magnitude and direction of the net force on the $+2\ \mu C$ charge.

17-C5 A small ball of mass 0.02 kg carries charge $+2 \times 10^{-6}$ C and is suspended by a light thread 50 cm long which is attached at the intersection of a wall and the ceiling. What positive charge, fixed on the wall 40 cm below the ceiling, will cause the ball to be in equilibrium 30 cm out from the wall?

17-C6 It is desired to gold-plate a metal object to a thickness of 0.015 mm. The object has a total surface area (all sides) of 100 cm^2 and is in an electrolytic bath containing Au^{3+} ions. If a current of 4 A is used, how much time will be required for the job?

17-C7 A lead-covered telephone cable is buried in moist earth, and the leakage current from a 4 m section of the cable into the earth is expected to be 20 mA. The reaction at the sheath involves Pb^{4+} ions. The lead sheath is 5 cm in diameter and 2 mm thick. What is the useful lifetime of the cable?

17-C8 A spoon has surface area 30 cm^2 and is to be silver-plated by passing a current of 0.2 A through a solution containing Ag^+ ions. How long is required to deposit a silver coating 10^{-3} cm thick? (The density of silver is 10.5 g/cm^3.)

17-C9 What current should be used in an electrolytic bath containing Ag^+ ions to silver-plate in 1 min both sides of a coin 3 cm in diameter with silver 100 μm thick? (The density of silver is 10.5 g/cm^3.)

17-C10 Each heartbeat of a long-distance runner causes 0.0041 C of charge to flow through a small electrolytic cell strapped to his chest. When he ran a 5000 m race in 13 min 20 s, a total of 5.73 mg of silver was deposited on the cathode from a solution containing Ag^+ ions. Calculate the average rate, in min^{-1}, of the athlete's heartbeat during the race.

For Further Study

17-7 Scientific priority and Coulomb's law

Over the years, undignified squabbles over priority have taken place between workers, each of whom claimed to have first discovered some physical law or invented some device. Robert Hooke (1635–1703), a versatile English scientist who was a bit more sensitive to these matters than some of his contemporaries, found it expedient to announce his law of elasticity in anagram form as "ceiiinosssttuu" in 1676. He thus guarded his priority, and unraveled the anagram two years later as *Ut tensio sic uis,*[*] which can be translated "As the extension, so the force." Christian Huygens (1629–1695) used a similar Latin anagram to "pin down" his priority in the matter of the true shape of Saturn's rings.

It is always a legitimate question to ask who is to be credited for a major discovery. To illuminate this matter a little, let us look closely at the unfolding of the inverse-square law for electric forces, known as Coulomb's law. Action at a distance had been introduced to physics in the shape of Newton's law of universal gravitation, published in 1687. When it became clear that electric forces decrease rapidly as distance increases, a natural assumption was made that $F \propto 1/r^n$, where n is some exponent. This assumption that the force depends on some *power* of r was justified by experiment. It was generally known by 1760 that the charge on the surface of an irregularly shaped body (Fig. 17-10) is distributed irregularly, but for similarly shaped bodies the pattern of charge distribution is the same. This observation, plus some geometrical reasoning, is sufficient to show that the force law must be of the form $F \propto 1/r^n$, and not, for instance, of the form

$$F \propto a^{-kr} \quad \text{or} \quad F \propto 1/r^n + 1/r^m$$

The problem, then, was to determine the value of the exponent. Daniel Bernoulli, in 1760, guessed that $n = 2$, but had no proof for his guess.

Joseph Priestley (1733–1804), the discoverer of oxygen, was informed in 1766 by his friend Benjamin Franklin that cork balls inside a metal cup were totally unaffected by the electrification of the cup. Now from Newton's work it was known that if the earth were a hollow spherical shell, a body anywhere within the shell would experience no force of gravitation. This is a consequence of the fact that gravitation follows an inverse *square* law. Hence Priestley drew the correct conclusion that since the cork balls experienced no force, the law of electric force must also be an inverse-square law.

To prove this statement, consider a point P somewhere inside a uniformly charged hollow sphere (Fig. 17-11), and construct a small double cone through P, as shown, intersecting the surface in areas A_1 and A_2. If ϕ is small, the surface segment is almost plane, and the projection of the segment s_1 (perpendicular to the line of sight) is given by $s_1 \cos \theta_1$. Similarly, the projection of the other arc is $s_2 \cos \theta_2$. By similar triangles,

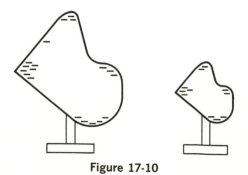

Figure 17-10

---

[*] In the twentieth century a Latin phrase might still seem to many to be a "hidden message," but in 1678 Latin was the accepted language for communicating learned discoveries.

$$\frac{s_1 \cos \theta_1}{r_1} = \frac{s_2 \cos \theta_2}{r_2}$$

But $\theta_1 = \theta_2$ (they are exterior base angles of an isosceles triangle whose vertex is at O). Hence, after squaring, we get

$$\frac{s_1{}^2}{r_1{}^2} = \frac{s_2{}^2}{r_2{}^2}$$

The area of each portion of the surface is proportional to the square of its linear dimension, so we have

$$\frac{\text{Area } A_1}{r_1{}^2} = \frac{\text{Area } A_2}{r_2{}^2} \qquad (17\text{-}8)$$

Now the shell is uniformly charged; therefore the charges on areas A_1 and A_2 are proportional to the areas; thus

$$\frac{Q_1}{r_1{}^2} = \frac{Q_2}{r_2{}^2} \qquad (17\text{-}9)$$

If the inverse-square law is true, a test charge Q_0 placed at P will experience two opposing forces $kQ_1Q_0/r_1{}^2$ and $kQ_2Q_0/r_2{}^2$, and these forces are equal by Eq. 17-9. The net force on the test charge due to both charged areas is zero at any arbitrary point P; the greater charge on area A_1 has been exactly compensated for by the greater distance r_1. The entire sphere can be divided up in this way by small cones passing through P, and so the net force on a charge anywhere inside the

Figure 17-12 Cavendish's experiment to test the inverse-square law for electrostatic force.

spherical shell is zero. To complete the proof, one must also show that no exponent other than 2 will give this result. Priestley apparently did not take account of the fact that the electrified cup was not a sphere.[*]

Priestley published his conclusion in London in 1767. For some reason the paper seems to have escaped the notice of other scientists of his day, or perhaps he himself failed to present his case forcefully enough. At any rate, in 1769 John Robison of Edinburgh attempted a direct experimental proof of the law, measuring the forces between small charged bodies. He arrived at the exponent 2.06—satisfactory enough in view of the difficulty of the experiment.

A few years later Henry Cavendish (1731–1810), who was to become known for his work on Newton's law of gravitation a decade later, turned his attention to the problem and chose to expand upon Priestley's indirect method. He placed a metal sphere of diameter 12.1 in. within two close-fitting hemispheres 13.3 in. in diameter. The inner sphere was completely enclosed but nowhere touched the outer hemispheres except by means of a removable wire (Fig. 17-12). Cavendish now electrified the system as strongly as possible, allowing every opportunity for charge to flow to the inner sphere by means of the connecting wire. He then withdrew the wire, isolating

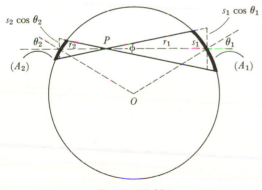

Figure 17-11

---

[*] Analysis by Karl Friedrich Gauss (1777–1855) extended the proof to include any *closed* metallic cup of whatever shape.

the inner sphere, and finally separated and removed the hemispheres. He tested the inner sphere by touching it with a pith ball, and found it to be electrically neutral. His conclusion was similar to Priestley's—that the absence of any flow of charge to the inner sphere showed that electric forces obey an inverse-square law.

Cavendish went further: he tested the sensitivity of his apparatus by deliberately charging the inner sphere to the least detectable amount. In this way he placed limits of accuracy upon his conclusion, and stated that the exponent must lie between 1.96 and 2.04. This important experiment of Cavendish's, as well as his gravitational experiment (Sec. 4-8), deserves to be studied carefully by students interested in the development of scientific methods. Cavendish's results were not published during his lifetime.

The next investigator of the law of electric force was, of course, Coulomb, who published his researches in a series of papers during the years 1788 and 1789. Coulomb no doubt expected to get an inverse-square law, as did all the others. He had already invented the torsion balance, independently of the Rev. John Michell (1724–1793) (whose torsion balance was applied by Cavendish to the gravitational problem), and he had studied the torsion of wires. It was a natural step for Coulomb to apply the torsion balance to the measurement of the exceedingly small forces of electrical repulsion and attraction.* Using figures from his first paper, we can construct a table (Table 17-1) relating distance between charges and the force of repulsion (expressed in arbitrary units). Within experimental error, the data indicate an inverse-square law of force.

Following Coulomb, other workers repeated Cavendish's experiment with more sensitive detectors of charge. Faraday, in 1838, actually put himself inside a 12 ft wire

---

* Forces as small as 0.0001 grain (less than a millionth of an ounce) could be determined with considerable accuracy by Coulomb.

TABLE 17-1   Coulomb's experiment

| Observed force | DISTANCE | |
	Observed	Calculated from inverse-square law
36 units	36 units	36 units
144	18	18
576	8½	9

cage and had the cage charged to an extent that artificial lightning would jump to the walls of the room. He felt no effect upon himself, and a sensitive electroscope inside the cage showed not the slightest indication of charge inside the cage. A century after Cavendish, Maxwell in 1870 used a delicate electrometer and found no effect, leading to a value of the exponent between 1.9999 and 2.0001. In 1971 it was shown by E. R. Williams, J. E. Faller, and H. A. Hill that if $F \propto 1/r^{2+x}$, then the deviation x from an exact inverse-square force law is less than 3×10^{-16}.

In view of all this activity, why then has the law become known as Coulomb's law when Coulomb's experiment was neither the first nor the most precise? Scientific priority is based upon *publication* of results in a form suitable for the use of others. For this reason, Cavendish lost out because he never published his method or his results, even though he kept an excellent record of his experiments in his private notebooks. It was Maxwell who brought Cavendish's notes to light and edited and published them for the first time in 1879, a century too late. It is hard to explain Cavendish's eccentricity in this respect, since he did publish some of his researches in the usual channels, including the gravitational experiment for which he is famous. He was best known in his lifetime as a chemist, and is one of those to whom the discovery of the composition of water is attributed. He inherited great wealth and retired from much of public life and from society to devote himself to scientific work. As Maxwell said,

Cavendish cared more for investigation than for publication. He would undertake the most laborious researches in order to clear up a difficulty which no one but himself could appreciate, or was even aware of, and we cannot doubt that the result of his enquiries, when successful, gave him a certain degree of satisfaction. But it did not excite in him that desire to communicate the discovery to others which, in the case of ordinary men of science, generally ensures publication of their results. How completely these researches of Cavendish remained unknown . . . is shown by the external history of electricity.[*]

Coulomb, on the other hand, contributed to his subject through a series of well-organized and comprehensive papers. He described his apparatus in detail, discussed the possible sources of error, and, most important, checked his result by an entirely different ex-

periment. His first method used the torsion balance to measure repulsive forces between two small, similarly charged spheres. His second method involved attraction instead of repulsion, and was dynamic instead of static (he timed oscillations of a small charged body placed near a charged sphere several feet in diameter). Having obtained an inverse-square law by two different methods, he felt able to announce the law as proved.

The subsequent experiments of Faraday, Maxwell, and others contribute nothing new but merely extend the precision of the value of the exponent. Such experiments have to be made, of course, for if they fail to come out exactly 2 (within experimental error) the consequences would be disastrous for much related electrical theory. But for its scientific importance, which includes publication and comprehensiveness, Coulomb's work stands out, and we justifiably speak of the law as Coulomb's law.

[*] *The Electrical Researches of the Hon. Henry Cavendish,* ed. by J. Clerk Maxwell, 1879. Quoted in W. C. D. Whetham, *The Theory of Experimental Electricity,* Cambridge Univ. Press, 1912, p. 15.

■ REFERENCES

1 Magie, W. F., *A Source Book in Physics* (New York: McGraw-Hill, 1935), pp. 400–03, 408–20. Selections from original papers of Franklin and Coulomb.

2 Roller, Duane, and Duane H. D. Roller, *The Development of the Concept of Electric Charge* (Harvard Case History in Experimental Science No. 8), (Cambridge: Harvard University Press, 1954). See especially Priestley's work on the inverse-square law (pp. 69–72) and Coulomb's experiments (pp. 72–80).

3 Cohen, I. Bernard, "In Defense of Benjamin Franklin," *Sci. American,* Aug., 1948, p. 36.

4 Whetham, W. C. D., *The Theory of Experimental Electricity* (Cambridge University Press), 1912, pp. 1–22. Cavendish's experiments are described on pp. 15–20.

5 Jefimenko, O., and D. K. Walker, "Electrostatic Motors," *The Physics Teacher,* **9,** 121 (1971).

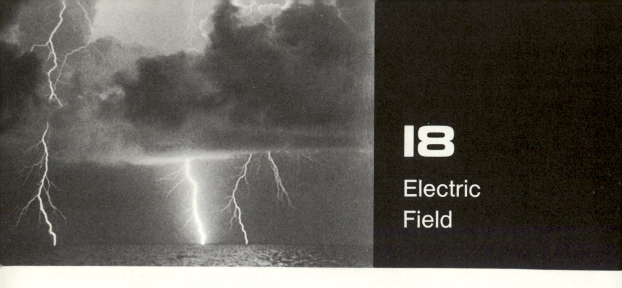

18

Electric Field

18-1   The concept of electric field

What can we say about the mechanism of
electric force? The theory of action at a
distance described in Chap. 17 is sufficiently
noncommittal to be accepted without much
critical thought. Electric forces exist; these
forces act at a distance. Yet this is almost no
theory at all. To *describe* what happens is
often useful, but most of us wish to *explain* a
phenomenon in terms of familiar laws and
facts of physics. In other words, we intuitively
seek a model for electric forces. To do so is
permissible, but only if we have been inocu-
lated against the scourge of modelitis. If we
keep in mind the limitations of any model
and remember that a model is not the same
thing as reality, we can gain much help
from a well-constructed model.

Let us discuss several models which have
been proposed as descriptions of electric force.
A useful concept, that of electric field, was
introduced by Michael Faraday (1791–1867).
Consider the space surrounding a positively
charged sphere (Fig. 18-1a). We explore the
region, using an imaginary test charge, and
we find that, no matter where we put the test
charge, it experiences a force such as $\mathbf{F}_1$, $\mathbf{F}_2$,
etc. The force $\mathbf{F}_3$ is larger than $\mathbf{F}_2$, since at

m the test charge is closer to the sphere than
at n. At any point in space a force vector can
be drawn; such a collection of infinitely many
vectors is called a *vector field*, or simply a *field*.
The direction of the field is defined as the
direction of the force on a *positive* test charge.
The strength or intensity $\mathbf{E}$ of the electric
field is defined as the force on a test charge
divided by the magnitude of the test charge.
Thus

$$\mathbf{E} = \frac{\mathbf{F}}{Q} \qquad (18\text{-}1)$$

In this definition, $\mathbf{E}$ and $\mathbf{F}$ are parallel vectors. The magnitude of $\mathbf{E}$ is given by

$$E = \frac{F}{Q} \qquad (18\text{-}2)$$

Electric field strength equals force per unit
charge, and E is measured in newtons per
coulomb. It is always assumed that the test
charge itself is so small that it does not appreciably alter the distribution of the charges
which cause the field. For instance, if a very
small test charge Q is placed at a, b, or c near
a uniformly charged metal sphere (Fig.
18-1a), the force vectors labeled $\mathbf{F}_1$ are all
equal in magnitude if a, b, and c are equi-
distant from the sphere's center. However,

During a thunderstorm, large electric fields are built up in the region between clouds and the earth, as well as in the region
between the clouds themselves. (Westinghouse)

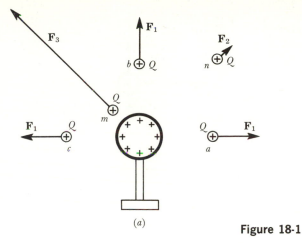

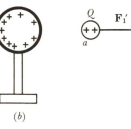

(a)   (b)

Figure 18-1

if the test charge is larger, the charges on the sphere will be redistributed, as in Fig. 18-1b, and the force on the (larger) test charge at a will be $\mathbf{F}_1'$. The ratio $\mathbf{F}_1'/Q$ will be equal to the magnitude of the field strength at point a, to be sure; but this ratio is the field strength due to the redistributed charge and is not the same as the original field strength we wished to measure. It is not unusual in physics for a measuring device (the test charge, in this case) to have an effect on what is being measured. It is like trying to explore the surface of the moon on foot—the explorer's footprints are part of what he sees. To overcome this difficulty, we imagine using a test charge so small that making it still smaller would cause no appreciable change in the ratio F/Q. In other words, the ratio F/Q approaches a limit as Q (and F) approach zero, and this limit is E.

EXAMPLE **18-1**

Calculate the magnitude and direction of the electric field at a point 50 cm directly above a charge $Q' = -2 \times 10^{-6}$ C.

Since by definition the field strength equals the force on a unit *positive* charge, and a positive charge would be attracted by the given negative charge, we see at once that the direction of the field at the point in question is downward. The magnitude of the force on a test charge Q is found from Coulomb's law:

$$E = \frac{\text{force}}{\text{charge}} = \frac{kQ\,Q'/r^2}{Q} = \frac{kQ'}{r^2}$$

$$= \frac{(9 \times 10^9 \ \text{N} \cdot \text{m}^2/\text{C}^2)(2 \times 10^{-6} \ \text{C})}{(0.50 \ \text{m})^2}$$

$$= \boxed{7.2 \times 10^4 \ \text{N/C}}$$

It is not always as easy to calculate the magnitude and direction of the electric field vector $\mathbf{E}$ as it is in the example just given. If there are many charges, perhaps on several conductors, it might be a difficult task. We would have to use vector addition, with one vector contributed by each of the charges (Fig. 18-2). The point we wish to emphasize is that, regardless of the origin of a field, we can *measure* the field by use of a test charge. Here, as elsewhere in physics, to specify a measurement procedure is to define a con-

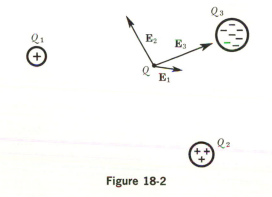

Figure 18-2

cept.[*] We have set up a procedure (measuring the force per unit charge) which *defines* what we mean by electric field. In general, Example 18-1 shows the magnitude of the electric field of a point charge Q to be given by

$$E = kQ/r^2$$

We have met with a vector field before—in Sec. 4-9 where we found $\gamma = GM/r^2$ for the magnitude of the gravitational field of a point mass M. In principle, we can explore the gravitational field in any region of space by measuring the magnitude and direction of the force on a small test mass m. The nature of any force field is defined by the kind of test body we use: a small test charge Q to explore an electric field, or a small test mass m to explore a gravitational field. In our study of the magnetic field, we will use still another kind of test body to explore the field.

The concept of field is not limited to force fields such as we have been discussing. For an example of a vector field which is not a force field, consider a river of varying depth, flowing in a curved path, perhaps having here and there some swirling eddies or stagnant places. The flow pattern can be defined as a velocity field, for at every point in the river a velocity could be measured in both magnitude and direction. For an example of a scalar field, we could measure the density of the sea at various latitudes, longitudes, and depths. Density is a scalar quantity, and so this collection of density values, one for every point in space, is a scalar field.

18-2   Lines of force

As an aid in visualizing electric fields, Faraday and his successors drew *lines of force* which everywhere have the direction of the

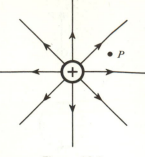

Figure 18-3

electric field. In Fig. 18-3 we show lines of force which represent the electric field surrounding a positive point of charge. The lines are directed away from the + charge, since like charges repel each other and the imaginary test charge is a + one. The lines extend all the way (!) to infinity, since there is *some* Coulomb force on the test charge no matter how far away it is. Only a representative number of lines are drawn; a line could equally well be drawn through *any* point, such as P. For this special case of a single isolated point charge, the lines of force are straight lines; in general, when the direction of **E** is calculated at various points the field lines turn out to be curved. The field near a pair of equal but opposite charges is shown in Fig. 18-4*a*, and the field near a pair of equal and like charges is shown in Fig. 18-4*b*.

Lines of force originate on + charges and terminate on − charges; no line starts in midair, so to speak. Another descriptive property of the lines of force is their "density"; the field is strongest where the lines are closest together.

18-3   Models for electric field

Nowadays, we consider that lines of force have no objective reality, but are merely a convenient and useful *representation* of a field. Faraday went much farther; he made a model of electric forces according to which the lines

---

[*] For other examples of this "operational" viewpoint, see Secs. 6-3 and 9-3, where operational definitions of inertial mass are discussed. Also, in Chap. 16, our definition of absolute zero was an operational one made in terms of a procedure involving Carnot engines.

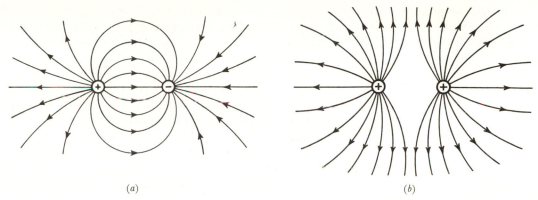

(a)   (b)

Figure 18-4 (a) Electric field in the neighborhood of a pair of unlike charges of equal magnitude; (b) electric field in the neighborhood of a pair of like charges of equal magnitude.

of force were considered to be the bounding edges of what he called tubes of force (Fig. 18-5). Faraday, like most scientists of his time, imagined all space to be filled with an invisible weightless "ether" which had certain elastic properties. The tubes of force, being made of ether, were like stretched rubber bands, or—better yet—like the springy wires used to give shape to the Victorian era's bustle and corset. In addition, Faraday advanced the hypothesis that adjacent tubes of force repelled each other. Thus in Fig. 18-4a, the repulsion between adjacent tubes causes the lines of force to bulge outward, and, since the lines of force tend to straighten out, the force of attraction between unlike charges is explained. This ingenious model was refined and developed by Maxwell and others of the "Cambridge school" of English physicists to include a description of electromagnetic forces as well as electrostatic forces.

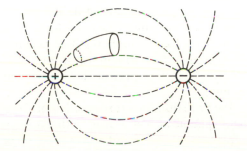

Figure 18-5 A short section of a tube of force.

The difference between the action-at-a-distance model and the lines-of-force model can be illustrated by considering the nature of charge according to the two models. The action-at-a-distance model considers charges to be located at certain places, attracting or repelling each other much as the earth attracts the moon. The electric field is then just a mathematical or geometrical description of the Coulomb forces. In the Faraday-Maxwell model, the emphasis is on the field, that is, on the space between the charges. This space is filled with elastic tubes of force (of varying cross sections), and the charges are thought of as having existence only as the representation of a sliced-off "raw end" of a tube of force.

No present-day physicists still adhere in detail to Faraday's concept of the electric field, in which the forces are a sort of by-product of properties of a medium. For one thing, the medium (ether) would have to be present even in a vacuum, and its mechanical properties would have to be miraculous indeed. The Michelson-Morley experiment (Sec. 7-4) in 1887 showed that the velocity of light is the same in a north-south direction as in an east-west direction, even though the rotation of the earth is continually taking us in an easterly direction, and other motions are surely present as the earth revolves around the sun, the sun moves through our galaxy, and our galaxy moves through space. This experiment showed that there is no "ether wind" blowing past a moving observer. All other attempts to prove the existence of the ether have also failed. The Faraday-Maxwell model also lost ground with the discovery of the free electron about 1900. We are accustomed now to think of charged particles moving through cyclotrons and radio tubes, and even bombarding us from outer space in the form of cosmic rays. For such "free" particles, the cumbersome apparatus of tubes of force seems to be of little help.

To Faraday goes the credit for introducing and emphasizing the concept of field; even if we reject the idea of an ether, we can still think of electric field as a property of space itself, requiring no medium of any kind. This viewpoint pervades scientific thought even now, and we recognize Faraday's contribution as a major one in the history of physics.

A third model for describing forces which act at a distance has been developed in the last several decades. Imagine two boys standing on frictionless carts (Fig. 18-6), playing a game of catch. Boy A recoils to the left when he throws the ball; boy B receives an impulse to the right when he catches the ball. We can look upon this as an example of action at a distance, with the force being "transmitted" by the ball. If the boys used a heavier (i.e., "massier") ball, or if they exchanged it more often or threw it with greater speed, there would be a greater force of repulsion between the boys. The simple

model we have considered is incapable of describing an attractive force. (See Ques. 18-9 at the end of this chapter.) However, working along these lines, theoretical physicists have constructed a satisfactory model for electromagnetic forces of attraction as well as repulsion, including Coulomb forces, magnetic forces, and the interatomic forces which are responsible for elasticity and molecular cohesion. The theory is complex and highly mathematical. The "particles" which are, in a sense, bandied back and forth are called *photons*. Under suitable circumstances, photons can have a "real" existence and are emitted as light from an atom which is electrically excited, as in a neon sign. We see here how intimately related are the theories of electricity and light; the very existence of electric forces is postulated to be due to particles of light.

In Chap. 2 we stated that there are three types of force between matter—electric, nuclear, and gravitational. For each type of force, there is a field theory according to which particles are emitted and absorbed at a very rapid rate. In the model of strong nuclear forces, the field particles are called *mesons;* it is for this reason that mesons are sometimes spoken of as "nuclear glue." For example, the continual exchange of mesons between the proton and the neutron is thought to keep the two nucleons bound together in the $^2_1$H nucleus. Even gravitational forces have their field particles, called *gravitons;* but we should point out that this model of gravitation has not yet been fully accepted. Under suitable circumstances, free photons can be formed (emission of light), and free mesons can be formed (the particles materialize in large atom-smashers). The possibility of detecting free gravitons, if they exist, seems to be beyond present-day experimental technique.

We have discussed three models of electric field: action at a distance, with no further explanation; the Faraday-Maxwell theory, in which lines and tubes of force represent a

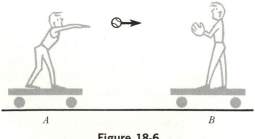

A   B

Figure 18-6

strain in the ether; and the photon-field theory, which is action at a distance with an added model for transmission of force by impact of particles. In our future work, we shall use the first model of field, drawing lines of force as a representation but without giving them any mechanical properties.

18-4 Potential difference

When we explore the electric field in the region of space around a charged body or near the terminals of a battery or generator, we imagine that a small, positive test charge is moved from point to point. For simplicity, we consider first a uniform electric field directed from point A to point B. If we move the test charge from B to A and electric forces act on it, we must do work on the charge, and its PE will increase. The change in PE per unit charge is called the *electric potential difference* between the points in question; the mks unit for potential is the *volt* (V). *Two points differ in potential by 1 volt if 1 joule of work is required to move 1 coulomb of charge from one point to the other.* In symbols,

$$V_{AB} = \frac{W_{B \to A}}{Q} \qquad (18\text{-}3)$$

$$\text{volts} = \frac{\text{joules}}{\text{coulombs}}$$

where V_{AB} is the difference in potential of point A relative to point B, and $W_{B \to A}$ is the work done against electric forces to move a test charge Q from B to A.

A gravitational analogy will prove helpful in our study of electric potential.* When a test mass m, such as the book in Fig. 18-7a, is moved around on a horizontal table, no work is done against gravity, and there is no change of gravitational potential. All points on the table top are at the same potential; simi-

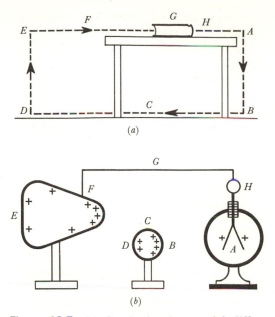

Figure 18-7 (a) Gravitational potential difference; (b) electric potential difference.

larly, all points on the floor are at the same (but lower) potential. If the height of the table is h, the difference of gravitational PE is mgh, and the gravitational potential difference between the two levels is

$$(V_{AB})_{\text{grav}} = \frac{W_{B \to A}}{m} = \frac{mgh}{m} = gh$$

Gravitational potential is expressed in joules per kilogram, a unit for which no special name has been invented. In mechanical problems a body may be assigned zero PE at any convenient level—when it is at sea level, or at the level of the floor, or on the table top, etc. There is thus no absolute value of gravitational potential for any given location. It is potential *difference* between two levels that is useful. The energy principle is stated in terms of *changes* of energy, and the choice of "reference level" of zero PE is arbitrary.

These ideas are carried over into our discussion of electric potential. *Potential difference* is the useful quantity. Just as in the gravitational case, "the potential at a point" has meaning only if some point of zero po-

---

*The analogy is not perfect, however. There are two kinds of electric charges, positive and negative; but, as far as we know, negative masses do not exist outside the pages of science fiction.

tential is specified. Often the earth (the "ground" of a circuit) is taken to be at zero electric potential, but other choices are possible.

Examples of potential difference (PD) are not hard to find. Let us discuss a few representative situations in detail.

(*a*) Two metal spheres are charged, with *A* having an excess of positive charge and *B* an excess of negative charge (Fig. 18-8). Since the test charge is defined as a positive one, we see that work must be done on it to bring the unit positive test charge from *Q* to *P*. Therefore *P* is at higher potential than *Q*.

(*b*) Water tends to flow downhill. The usual explanation is from the point of view that the force of gravity has a component parallel to the hill, and this force component "pulls the water down the hill." A more sophisticated way of looking at this same fact is to consider energy changes. Water flows downhill because its gravitational PE is less in the lower position. A heavy particle, subjected to no other force than its own weight, always seeks the lowest attainable level, where its gravitational PE is least. Similarly, *positive* electric charge tends to flow from points of high potential to points of low potential—"downhill," so to speak. We do not have to consider electric forces in detail to know that their net result is to push or pull charges into a position of least PE. This means that, if the charges on a metallic conductor are at rest, all points of

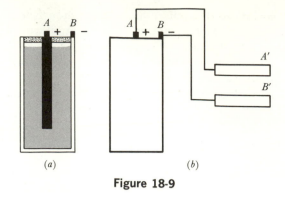

Figure 18-9

the conductor are at the same potential. In Fig. 18-7*b*, the pear-shaped object, the connecting wire, and the leaves of the electroscope are all at the same potential, for, if this were not so, the free electrons in the metal would readjust their distribution, flowing away from any region in which they had a higher PE into a surrounding region in which they had a lower PE. All parts of the small metallic sphere are also at the same potential; indeed, the distribution of its charge is self-adjusting to make this so. Points *E*, *F*, *G*, *H*, and *A* are all at one potential; points *B*, *C*, and *D* are at another potential. The analogy between parts (*a*) and (*b*) of Fig. 18-7 is complete.

(*c*) An ordinary dry cell (Fig. 18-9*a*) consists essentially of a carbon rod and a zinc can, separated by a moist paste containing ammonium chloride and manganese dioxide, which can be considered an electrolytic solution. When the cell is on the shelf or on a laboratory table, with nothing connected to it, experiment shows that there is a slight excess of positive charge on the carbon rod[*] and a slight excess of negative charge on the zinc can. To move a charge of $+1$ C from *B* to *A* through the air requires 1.5 J of work, and the PD is 1.5 J/C, or 1.5 V. If the terminals are connected by wires to a pair of plates (Fig. 18-9*b*), the plates become charged (this process takes a small fraction of a

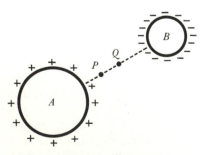

Figure 18-8 Potential difference; *P* is at a higher potential than *Q*.

---

[*] Remember that a deficiency of electrons is entirely equivalent to an excess of positive charge.

second), and then the PD between A' and B' is also 1.5 V.

EXAMPLE 18-2

What is the PD between two points if 200 J of work is required to move 40 C from one point to the other?

$$V_{AB} = \frac{W_{B \to A}}{Q} = \frac{200 \text{ J}}{40 \text{ C}} = 5 \text{ J/C}$$

$$= \boxed{5 \text{ V}}$$

EXAMPLE 18-3

How much work is required to move an electron between the two terminals of an atom-smasher whose PD is 4 million volts?

$$W_{B \to A} = V_{AB}Q$$
$$= (4 \times 10^6 \text{ V})(1.60 \times 10^{-19} \text{ C})$$
$$= \boxed{6.40 \times 10^{-13} \text{ J}}$$

EXAMPLE 18-4

An electron leaves the heated cathode K of a radio tube (Fig. 18-10) with negligible initial velocity and is accelerated through an applied PD of 500 V. What is the velocity of the electron as it approaches the accelerating electrode A?

While the electron is being repelled by the cathode K and attracted by the anode A, it loses PE, just as does a baseball which falls toward the earth. The change in PE equals the change in potential times the charge. Now we apply the energy principle:

Loss of PE = gain of KE

$$(1.60 \times 10^{-19} \text{ C})(500 \text{ V})$$
$$= \tfrac{1}{2}(9.11 \times 10^{-31} \text{ kg})(v^2)$$

$$v = \sqrt{\frac{2(1.60 \times 10^{-19} \text{ C})(500 \text{ V})}{9.11 \times 10^{-31} \text{ kg}}}$$

$$= \sqrt{1.77 \times 10^{14} \text{ C} \cdot \text{V/kg}}$$

$$= \boxed{1.33 \times 10^7 \text{ m/s}}$$

A note on units: In this example, we arrived at velocity $= \sqrt{\text{C} \cdot \text{V/kg}}$. This fine collection of units can be shown to be equal to m/s, as follows:

$$\sqrt{\frac{\text{C} \cdot \text{V}}{\text{kg}}} = \sqrt{\frac{\text{C} \cdot (\text{J/C})}{\text{kg}}} = \sqrt{\frac{\text{J}}{\text{kg}}}$$
$$= \sqrt{\frac{\text{N} \cdot \text{m}}{\text{kg}}} = \sqrt{\frac{(\text{kg} \cdot \text{m/s}^2)\text{m}}{\text{kg}}}$$
$$= \sqrt{\frac{\text{m}^2}{\text{s}^2}} = \text{m/s}$$

Whenever a large velocity turns up in a problem, as in Example 18-4, we must consider the possibility of relativity effects. The answer for Example 18-4 is about $\frac{1}{20}$ the speed of light, and so we realize that a small error has been made in using the simple expression $\tfrac{1}{2}mv^2$ for kinetic energy. A correct (relativistic) solution of a similar problem is outlined in Prob. 18-C1. However, relativity effects depend on v^2/c^2, which in this problem is

$$\frac{v^2}{c^2} = \frac{(1.33 \times 10^7 \text{ m/s})^2}{(3.00 \times 10^8 \text{ m/s})^2} = 0.0020$$

Since in this case v^2/c^2 is small compared to 1, the error in our "classical" or "non-relativistic" treatment is not a serious one.

18-5 Equipotential surfaces

An *equipotential surface* is defined as a surface all points of which are at the same potential. The PD between any two points on the surface is zero. Since PD is work per unit charge, our definition implies that no work is done when a test charge is moved from one point to another on an equipotential surface. As an example, consider a point charge $+Q$ from which lines of force radiate outward (Fig. 18-11). We can easily prove that the equipotential surface designated by V_1 is a

Figure 18-10

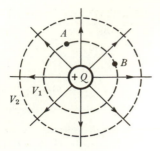

Figure 18-11 Lines of force (solid lines) and equipotential surfaces (dashed curves) for an isolated point charge.

sphere surrounding the charge. To take a test charge from A to B along a path which lies in the spherical surface would require no work, since the motion is at all times perpendicular to the direction of the force. Hence A and B are at the same potential, and the sphere is an equipotential surface. The larger sphere, labeled V_2, is also an equipotential surface, and there is a PD equal to $(V_1 - V_2)$, or ΔV, between any point on the first sphere and any point on the second sphere. A more complicated example is shown in Fig. 18-12, which represents the field surrounding a pair of equal and like charges. This is the same as Fig. 18-4b, with the addition of dashed lines to represent some of the equipotential surfaces; they are no longer spherical in shape.

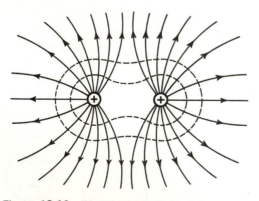

Figure 18-12 Electric field in the neighborhood of a pair of like charges. The equipotential surfaces are shown as dashed curves.

An important relationship between lines of force and equipotential surfaces is seen in both figures. The equipotential surfaces are everywhere perpendicular to the lines of force. This must be true in general, for if the lines of force were *not* perpendicular to an equipotential surface, then a test charge moving along the equipotential surface would experience a component of force parallel to its motion, and work would be done. This means that a PD would exist between two points on the same equipotential surface—which is contrary to definition.

The close relationship between field and potential leads to a new interpretation of electric field strength. Since

$$\text{Work} = (\text{force})(\text{displacement})$$

where the displacement is assumed to be parallel to the force, we can write

$$\frac{\text{Work per unit charge}}{\text{Distance moved}} = \text{force per unit charge}$$

or

$$\frac{\text{Potential difference}}{\text{Distance moved}} = \text{field strength}$$

$$-\frac{\Delta V}{\Delta s} = E \qquad (18\text{-}4)$$

A positive test charge tends to move from a region of high potential to a region of low potential—"downhill," in the direction of *decreasing* potential. The $-$ sign is used in Eq. 18-4 so that E is positive when $\Delta V/\Delta s$ is negative. The change in potential per unit distance* is called the *potential gradient;* we see that the magnitude of the field strength equals the negative of the potential gradient. Potential gradients are measured in volts per meter.

---

*The distance moved must be measured along the direction of the lines of force, i.e., one must proceed from one equipotential surface to a neighboring one along the shortest (perpendicular) path. To obtain the field strength "at a point," we must take the limit of Eq. 18-4 as $\Delta s \to 0$; in other words, $E = -dV/ds$.

EXAMPLE 18-5

A fine iron wire 25 cm long is connected between the terminals of a battery whose terminal voltage is 6 V. What are the magnitude and direction of the electric field inside the wire?

$$E = -\frac{\Delta V}{\Delta s} = -\frac{6\ \text{V}}{0.25\ \text{m}} = \boxed{-24\ \text{V/m}}$$

The direction of the field is toward the negative terminal of the battery, since a positive test charge inside the wire would move in this direction. The direction of field is always in the direction of *decreasing* potential.

EXAMPLE 18-6

In Example 18-5, what is the PD between point P of the wire, which is 5 cm from the $+$ terminal of the battery, and point Q, which is 15 cm from the $+$ terminal?

The field inside the wire is uniform, equal to -24 V/m. Hence

$$\Delta V = -E\,\Delta s = -(-24\ \text{V/m})(0.10\ \text{m})$$

$$= \boxed{2.4\ \text{V}}$$

Point P is 2.4 V higher in potential than is point Q, since P is closer to the $+$ terminal.

We now have two mks units for field strength: newtons per coulomb (force per unit charge) and volts per meter (potential gradient). We can show that these units are equivalent, using the definition of the volt:

$$E = \frac{\text{V}}{\text{m}} = \frac{\text{J/C}}{\text{m}} = \frac{\text{J}}{\text{C}\cdot\text{m}} = \frac{\text{N}\cdot\text{m}}{\text{C}\cdot\text{m}} = \frac{\text{N}}{\text{C}}$$

In most engineering applications, the volt per meter or the volt per centimeter is used to measure electric field. For instance, tables show that the field strength required to cause a spark in dry air is about 10^6 V/m. The electric field near the surface of the earth (due to an excess of negative charge on the earth) is normally about 100 V/m, but this figure may increase to 600 V/m or more during a thunderstorm. Since a volt per meter is equal to a newton per coulomb, this means that if a coulomb of positive charge could be collected together and held at one point near the earth's surface, it would experience a downward force of about 100 N (approximately 23 lb).

The potential, relative to infinity, of an isolated point charge or a uniformly charged sphere can be found using integral calculus. This is done in Sec. 28-7, where the result is used in our study of the Bohr theory for the hydrogen atom.

18-6 Capacitance

If two conductors are placed near each other, the combination is called a *capacitor*. Usually, but not necessarily, the conductors are in the form of parallel plates. In Fig. 18-13 we show two arbitrarily shaped conductors, with some $+$ charge on one body and an equal $-$ charge on the other. Of one thing we can be sure: If the charges are at rest, there is no electric field at any point within the metal of either conductor. If this were not so, the free electrons in the metal would move under the influence of the field, and so would not be at rest. By similar reasoning, at any point on the surface of a conductor there can be no component of electric field parallel to the surface, for if there were, free electrons would flow along the surface. All points of conductor A must therefore be at the same potential, since the potential gradient along the surface is zero. We conclude that *if the charges are at rest, the surface of a conductor is an equipotential surface,* and all points in the interior have the same potential as the surface. The piling up of

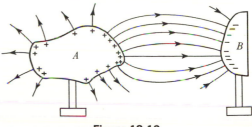

Figure 18-13

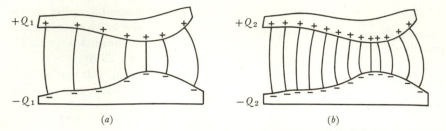

Figure 18-14 A capacitor.

charge at the pointed end of conductor A is explained by saying that if the charges were *not* so distributed, the conductor would not be an equipotential surface.[*] The charges automatically adjust themselves so that the surface of A is an equipotential surface. The surface of B is also an equipotential surface, at some different potential. There is a single PD between body A and body B. In Fig. 18-13, A is at a higher potential than B, since it would require work to move a unit test $+$ charge from B to A.

We now prove that the PD between the plates of any capacitor is directly proportional to the charge on either plate. We assume equal and opposite charges on the plates.[*] If the charge on the plates is increased, the strength of the field between the plates is increased in proportion to the charge. The relative distribution of lines of force remains the same, but, if there are twice as many charges on the plates, the lines are twice as close together, and the field is twice as great. This is illustrated in Fig. 18-14, where the same capacitor is charged with a small charge Q_1 in (*a*) and with a large charge Q_2 in (*b*). The PD is defined as work per unit test charge; and, if a test charge is moved from

the bottom plate to the top plate along any of the lines of force, the work is greater in (*b*), since the force is greater while the displacement is the same. Therefore the potential difference V is proportional to the charge. For any pair of conductors, the ratio Q/V would be expected to be a constant. This constant is called the *capacitance* C of the capacitor.

$$\text{Capacitance} = \frac{\text{charge on either plate}}{\text{PD between the plates}}$$

$$C = \frac{Q}{V} \qquad (18\text{-}5)$$

$$\text{farads} = \frac{\text{coulombs}}{\text{volts}}$$

The mks unit for capacitance is the *farad* (F); a capacitor has a capacitance of one farad if one coulomb of charge causes a PD of one volt. The farad is an inconveniently large unit. For most applications, the microfarad (1 μF $= 10^{-6}$ F) or the picofarad (1 pF $= 10^{-12}$ F) is used. Circuit-diagram symbols for a capacitor are —||— or —|(— or —|/(—, the last one representing a variable capacitor.

For the special case of a pair of parallel plates close together, it can be shown that

$$C = \frac{A}{4\pi k d} \qquad (18\text{-}6)$$

where A is the area of each plate, d is their separation, and k is the constant ($= 9 \times 10^9$ N·m²/C²) which enters into Coulomb's law (Sec. 17-5). Putting in the numerical value of k, we get

$$C = 8.85 \times 10^{-12} \frac{A}{d}$$

---

[*] Incidentally, as we have seen in Sec. 17-4, the charges all reside on the surface of a metallic conductor, whether the body is solid or hollow.

[*] It is possible, of course, to place unequal charges on two metallic plates. However, if the plates are close together and of large area, the charges on the facing portions of the plates will be equal, and the excess charge of the more highly charged plate will lie on the outer surface of that plate. Lines of force from these excess charges will terminate on opposite charges on some neighboring conductor or conductors.

C is in farads if A is in square meters and d is in meters. A check of the units of this equation for capacitance is instructive.

$$C = \frac{A}{4\pi k d} = \frac{m^2}{(N \cdot m^2/C^2)(m)} = \frac{C^2}{N \cdot m}$$

$$= \frac{C^2}{J} = \frac{C}{J/C} = \frac{C}{V} = F$$

The check of units is successful.

One way of charging a capacitor is to connect the two plates to a source of PD such as a battery. In Fig. 18-15 the battery (represented by the symbol ⊣∣⊢) causes a momentary flow of electrons from the top plate through the battery to the bottom plate. If we arbitrarily "ground" the bottom plate and call its potential zero, then the top plate will be at a potential of $+45$ V. The PD between A and B is 45 V. Just enough charge (given by $Q = CV$) flows so that the top plate is an equipotential surface at a potential of $+45$ V and the bottom plate is an equipotential surface at a potential of 0 V. The lines of force are perpendicular to the equipotential surfaces, as must always be true; for this special case of a parallel-plate capacitor with plates close together, the lines of force are parallel, and the field between the plates is *uniform* in magnitude and direction. Work is done when a capacitor is charged; see Sec. 18-9 for a discussion of energy storage in capacitors.

The process of "grounding" a conductor or a point in a circuit, as in Fig. 18-15, deserves a brief comment. The earth is such a large body that charges can flow into or out of it without appreciably changing its potential relative to infinity (or relative to the moon or the sun). This means that the capacitance of the earth is so large that ΔV (given by $\Delta V = \Delta Q/C$) is negligible for any charge flow ΔQ encountered in the laboratory or even in large-scale engineering.[*] The potential of the earth is, therefore, a useful reference potential, and grounding a circuit by connecting a point to earth gives that point a fixed potential which is designated 0 volts. In circuit diagrams "ground" is represented by ⏚ or ⏛.

18-7 Dielectrics

So far, in our discussion of capacitors we have tacitly assumed the two plates to be separated by a vacuum. If this is not so, the capacitance is increased by a factor which depends on the electrical nature of the insulating medium between the plates. Such a medium is called a *dielectric medium,* or simply a *dielectric.* We define the *dielectric constant K* of a medium as a ratio of capacitances: if C_{med} is the capacitance of a capacitor whose plates are separated by the medium, and C_{vac} is the capacitance of the same pair of plates separated by a vacuum, then

$$K = \frac{C_{med}}{C_{vac}} \qquad (18\text{-}7)$$

Equation 18-7 is an operational definition of dielectric constant, since it specifies a way of measuring K. Let us now try to interpret the dielectric behavior of matter by a molecular model.

The molecules of many substances are said to be *polar,* which means that the positive and negative charges of the molecules are separated by a small distance. For instance, research has shown that H_2O is a bent molecule (see Fig. 18-16), with the negative oxy-

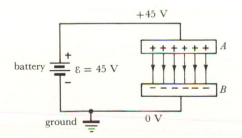

+45 V

battery $\varepsilon = 45$ V

+ + + + + A

- - - - - B

ground 0 V

Figure 18-15 A parallel-plate capacitor, showing uniform field between the plates.

[*] A good analogy is the water in the oceans; sea level doesn't change appreciably even if many gallons of water are added or removed from the ocean.

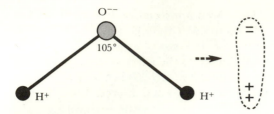

Figure 18-16 A water molecule is equivalent to a dipole.

gen ion not coinciding with the point midway between the two positive hydrogen ions. This permanent polar structure of the water molecule is equivalent to a pair of equal and opposite charges separated by a small distance. Such an arrangement of charges is called a *dipole*. The dipoles in water give rise to electric forces within the liquid which partially cancel the Coulomb forces between other charged bodies immersed in the water.

Coulomb's law is always true, whether or not there is a dielectric medium. That is, there is only one force law, namely,

$$F = \frac{kQ\,Q'}{r^2}$$

for the force between two stationary electric charges, wherever they may be. No new law of force is needed to explain the reduced force

between charges immersed in a dielectric. To illustrate this point for a liquid, let us analyze the forces acting on the sodium ion when sodium chloride is dissolved in water. The water molecules, which are polar, cluster around the Na^+ and Cl^- ions, as shown in Fig. 18-17. The net electric force on the Na^+ ion is the vector sum of three forces caused by (a) the Cl^- ion, (b) the ring of negative charges such as that at B, and (c) the ring of positive charges such as that at C. All other charges, such as those at D and D', have no effect, since they occur in equal and opposite pairs close together. Each of the three forces (a), (b), and (c) can be computed (in principle, at least) by application of Coulomb's law: (a) is the force on the Na^+ ion, directed toward the right, exerted by the Cl^- ion; (b) is zero, since the forces exerted on Na^+ by this symmetrical ring of negative charges have a vector sum equal to zero; (c) is directed toward the left and tends to oppose (a). The *net* force on Na^+ is therefore less than it would have been if the charges had been in a vacuum. It is for this reason that water is such a good solvent for NaCl and so many other diverse chemical compounds; the attraction between the component parts of the molecules is weakened. Even symmetrical

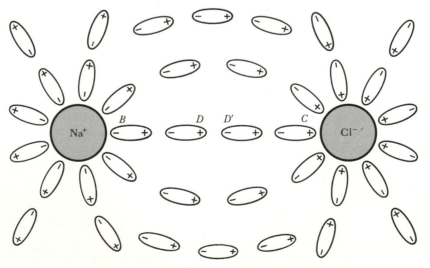

Figure 18-17

nonpolar molecules such as those of benzene are distorted slightly by neighboring charged bodies, and so there is a slight dielectric effect, but not nearly so strong an effect as in the case of polar molecules, where dipoles are already permanently present.

Consider now a charged capacitor with plates separated first by a vacuum, then (Q remaining the same) with the plates immersed in a dielectric fluid such as water, benzene, or air. According to the argument of the preceding paragraphs, the effect of the dipoles in the fluid is to reduce the net force on a test charge immersed in the medium, and hence less work is required to take a test charge from one plate to the other. Since potential difference is work per unit charge, we see that V has been decreased by the presence of the dielectric medium, and therefore $C (= Q/V)$ has been increased. The ratio of the two capacitances is defined as the dielectric constant of the medium, as in Eq. 18-7.

It is difficult to imagine an experiment to measure the force on a test charge immersed in a solid dielectric medium; for this reason we have limited our discussion of dielectric constant to fluids. Nevertheless, it should be evident that dipoles *are* formed in a solid insulator when an electric field is applied, and the capacitance increases when a solid slab of glass, mica, etc., is inserted between a pair of plates. Even though the molecular interpretation of solid dielectrics is complex, Eq. 18-7 still serves as a definition of dielectric constant for a solid and provides a procedure for its measurement. Note that K is the ratio of two capacitances, hence the dielectric constant of a substance is a dimensionless quantity. Dielectric constants for some substances are given in Table 18-1.

For any parallel-plate capacitor, the definition of K requires that Eq. 18-6 be modified to read

$$C = \frac{KA}{4\pi k d} \qquad (18\text{-}8)$$

In the mks system, $C = 8.85 \times 10^{-12} \dfrac{KA}{d}$.

TABLE 18-1   Dielectric constants*

vacuum	1
dry air at 1 atm	1.0006
water	80
carbon tetrachloride	2.24
benzene	2.28
castor oil	4.67
methyl alcohol	33.1
glass	4–7
amber	2.65
wax	2.25
mica	2.5–7

* All at 20°C.

18-8   The use of capacitors in circuits

In electric circuits, capacitors are used to store charge, somewhat as a bucket can be used to store a quantity of water. Do not imagine, however, that the capacitance of a capacitor is the "amount of charge it can hold." Since $Q = CV$, the charge Q can be increased to any desired value, with a corresponding increase in the potential difference V. The limit is reached only when the insulation between the plates breaks down and a spark jumps across the gap. The water analogy would be closer if we were to think of a cylindrical storage tank of very great height. As water is pumped in, the pressure rises, and the "capacity" of the tank is limited only by the ability of the seams and joints at the bottom of the tank to withstand the pressure of the water. At any given time, the amount of water stored in the tank is proportional to the pressure P, just as the charge stored in a capacitor is proportional to the potential difference V.

EXAMPLE 18-7

A capacitor whose capacitance is 20 μF is momentarily connected to a dry cell which maintains a potential difference of 1.5 V between its terminals and is then disconnected. How much charge is stored in the capacitor?

The resistance of the connecting wires and the

internal resistance of the cell are of no consequence except to determine how rapidly the capacitor becomes charged. In a typical case, after 10^{-6} s or 10^{-8} s or some similar very short time, the capacitor is fully charged and the electrons have ceased flowing. The plates have reached a PD equal to the PD of the cell.

Using Eq. 18-5,

$$Q = CV = (20 \times 10^{-6} \text{ F})(1.5 \text{ V})$$
$$= 30 \times 10^{-6} \text{ C}$$
$$= \boxed{30 \ \mu\text{C}}$$

EXAMPLE **18-8**

In measuring the charge of the electron by the Millikan oil-drop method (Sec. 27-1), a uniform electric field is obtained by connecting a pair of parallel circular plates to a source of PD, as in Fig. 18-15. If the plates are 10 cm in radius and are separated by 3 cm, what charge on the plates will give a field of 400 V/cm?

If the field is to be 400 V/cm, the PD between the plates must be

$$V = (400 \text{ V/cm})(3 \text{ cm}) = 1200 \text{ V}$$

We next compute the capacitance of the capacitor formed by the plates. They are separated by air, whose dielectric constant is very nearly 1.

$$C = 8.85 \times 10^{-12} \frac{KA}{d}$$
$$= \left(8.85 \times 10^{-12} \frac{\text{C}^2}{\text{N}\cdot\text{m}^2}\right) \frac{(1)(\pi)(1 \times 10^{-1} \text{ m})^2}{3 \times 10^{-2} \text{ m}}$$
$$= 9.27 \times 10^{-12} \text{ F}$$

Now we can compute the charge on the plates:

$$Q = CV = (9.27 \times 10^{-12} \text{ F})(1.20 \times 10^3 \text{ V})$$
$$= \boxed{1.11 \times 10^{-8} \text{ C}}$$

The physical construction of capacitors depends upon the desired capacitance and the PD that must be sustained without rupture of the dielectric. In the formula

$$C = \frac{KA}{4\pi kd}$$

either K, A, or d can be varied. For low values of C (such as those needed in the tuning stage of a radio receiver), parallel plates separated by air are used (Fig. 18-18a). The capacitance is changed by rotating one set of plates to give a variable effective area of overlapping A. Small "trimmer" capacitors used for fine adjustment (b) have mica dielectric, and C is changed by changing the separation d. Capacitors of fixed value are often constructed of two rolled-up aluminum foils separated by waxed paper (c). An essential compromise must be made between high capacitance (which requires small separation d) and the ability to withstand high voltages without rupture (which requires a large d). Capacitors are therefore rated by "working voltage" as well as capacitance. For about a dollar, one can buy 1000 μF rated at 6 V, 40 μF rated at 450 V, or 0.0005 μF rated at

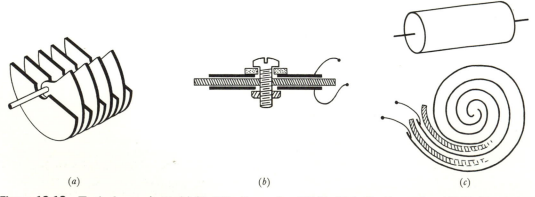

| (a) | (b) | (c) |

Figure 18-18 Typical capacitors: (a) 30–300 pF, rated at 300 V; (b) 5–40 pF, rated at 100 V; (c) tubular fixed capacitor, 0.05 μF, rated at 600 V.

10 000 V. Some large oil-filled capacitors are designed to store a small charge at a PD of a million volts or more, for use in artificial lightning experiments.

As circuit elements, capacitors are notable for the fact that no steady flow of electrons can take place. Instead, a charge builds up, and the dielectric material, or vacuum, between the plates is subject to an electric field.

Capacitors can be connected together in several ways, and the equivalent capacitance of the combination can be calculated by certain formulas. When capacitors are connected in parallel as in Fig. 18-19, charge flows from the battery or other source and is shared among the various plates that are connected together. If a PD of V is applied, each capacitor receives a different quantity of charge, given by $Q = CV$ in each case. The total charge that leaves the battery is

$$Q = Q_1 + Q_2 + Q_3 + \cdots$$

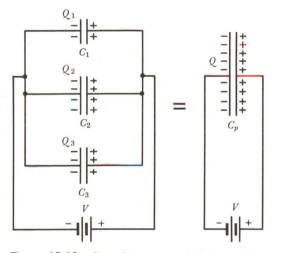

Figure 18-19 Capacitors connected in parallel.

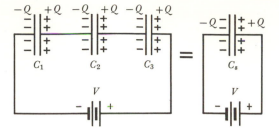

Figure 18-20 Capacitors connected in series.

Hence

$$C_p V = C_1 V + C_2 V + C_3 V + \cdots$$

Canceling V, we get

$$C_p = C_1 + C_2 + C_3 + \cdots$$
$$\text{(capacitors in parallel)} \quad (18\text{-}9)$$

If a number of capacitors are connected in series as in Fig. 18-20, a certain charge Q leaves the battery and flows on one plate of C_1. This repels an equal charge $-Q$ to the left-hand plate of C_2, leaving a charge $+Q$ on the right-hand plate of C_1. Similarly, the plates of C_3 acquire charges $-Q$ and $+Q$. The total PD from A to B is the sum of the PD's of the individual capacitors, but the charges on all the capacitors are the same, equal to the charge that flows from the battery.

$$V = V_1 + V_2 + V_3 + \cdots$$
$$\frac{Q}{C_s} = \frac{Q}{C_1} + \frac{Q}{C_2} + \frac{Q}{C_3} + \cdots$$

Canceling Q, which is the same for each capacitor, gives

$$\frac{1}{C_s} = \frac{1}{C_1} + \frac{1}{C_2} + \frac{1}{C_3} + \cdots$$
$$\text{(capacitors in series)} \quad (18\text{-}10)$$

■ SUMMARY

An electric field exists in any region of space where a small test charge would experience an electric force. The magnitude, or intensity, of the field equals the force per unit charge, and the direction of the field is the direction of the force on a small positive test charge. Another example of a force field is gravitational field, whose magnitude is defined as force per unit mass. In general, a field may be thought of as a collection of vectors or scalars which are defined throughout a region of space. Lines of force are used to represent the direction of an electric field. Various models have been proposed for electric, nuclear, and gravitational fields.

Electric potential is the electric potential energy per unit charge. The potential difference between two points is measured in joules per coulomb, or volts. The PD is 1 V if 1 C of charge gains or loses 1 J of electric PE when moved from one point to another. An equipotential surface is one all points of which have the same potential. If the free charges in a conductor are at rest, they reside on the surface and are distributed in such a way that the surface is an equipotential surface. Lines of force "originate" on positive charges and "terminate" on equal negative charges, and equipotential surfaces are always perpendicular to the lines of force. The electric field strength can also be measured in terms of potential gradient, which is the rate of change of potential with respect to distance, measured along a line of force.

Capacitance is defined as the stored charge per unit PD between the plates of a capacitor. For a parallel-plate capacitor, the capacitance depends on the area and separation of the plates, and on the dielectric material between the plates. The dielectric constant of a medium is the factor by which the capacitance of a capacitor is increased when immersed in the medium. Formulas exist for the combined capacitance of capacitors in series and in parallel.

■ CHECK LIST

electric field strength
gravitational field
   strength
lines of force
photons, mesons,
   gravitons
potential difference

volt
equipotential
   surface
potential gradient
capacitance
farad
dielectric constant

$$V_{AB} = \frac{W_{B \to A}}{Q}$$

$$C = KA/4\pi kd$$

$$C_p = C_1 + C_2 + C_3 + \cdots$$

$$\frac{1}{C_s} = \frac{1}{C_1} + \frac{1}{C_2} + \frac{1}{C_3} + \cdots$$

■ QUESTIONS

18-1 Is electric field strength a vector quantity or a scalar quantity? Is electric potential a vector quantity or a scalar quantity?

18-2 Why must a *small* test charge be used when measuring an electric field?

18-3 Draw in some representative equipotential lines in Fig. 18-4a.

18-4 Figure 18-21a represents a small part of the surface of the moon, assumed to be horizontal, and the space above the moon. Draw in a few representative lines of force (solid) and a few equipotential surfaces (dotted) for the gravitational field above the lunar surface. Repeat, for the *mascon* (mass concentration) of Fig. 18-21b, in which a small deposit of heavy material is buried below the surface. How can the existence of a mascon be detected?

(a) Normal lunar surface

(b) Lunar surface near a mascon

Figure 18-21

18-5 Why are the lines of force in Fig. 18-13 perpendicular to the surfaces where the lines and surfaces meet?

18-6 What is the gravitational field at the center of the earth?

18-7 A spaceship moves around Venus in an orbit which is the intersection of a gravitational equipotential surface and a plane. Describe the orbit. (Ignore the gravitational attraction of the sun and the other planets.)

18-8 On the weather map of Fig. 18-22, wind velocities and temperatures are shown at various points. Explain how the map illustrates the following concepts: (a) vector field; (b) scalar field; (c) temperature gradient. Where is the temperature gradient the greatest?

18-9 Construct a model by which two boys on frictionless carts, as in Fig. 18-6, can exert an attractive force upon each other by throwing and catching a boomerang rather than a ball. (See also Reference 40 at the end of Chap. 30.)

18-10 What is the difference between potential difference and difference of potential energy?

18-11 In Fig. 18-7b, suppose the charge on the small sphere to be distributed evenly, instead of concentrated at B, and all other charges in the diagram to remain as shown. Would the potential energy of the sphere be increased or decreased?

18-12 Explain why all parts of a conductor are at the same potential, if there is no flow of electrons.

18-13 Two plates of metal which have fixed charges $+Q$ and $-Q$ are immersed in a tank of oil. If the oil is pumped out, does the electric field at a point midway between the plates increase, decrease, or remain the same?

18-14 Ten equal capacitors are connected in series. Is the resultant capacitance greater than, less than, or equal to the capacitance of each capacitor? Repeat, for ten equal capacitors connected in parallel.

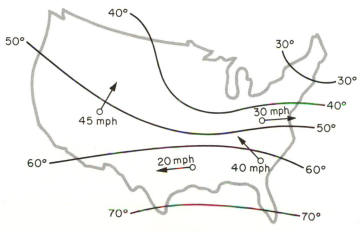

Figure 18-22

■ PROBLEMS

18-A1 The force on a charge of -3×10^{-7} C is measured to be 0.24 N, in a downward direction. What are the magnitude and direction of the electric field at this point?

18-A2 What electric force is exerted on a charge of 6×10^{-8} C which is placed in a field whose strength is 5×10^4 N/C?

18-A3 What are the magnitude and direction of the electric field at a point where an electron experiences an upward force of 3.2×10^{-17} N?

18-A4 Calculate the force on a charge of 20 μC which is in a field whose strength is 300 N/C.

18-A5 What is the PD between two points if 4 J of work is required to move 20 C from one point to the other?

18-A6 How much work is needed to move a charge of $+6$ C from A to B, if the potential of A relative to ground is $+100$ V and the potential of B relative to ground is $+112$ V?

18-A7 Three points A, B, and C are on a straight line, at intervals of 25 cm. The potentials of the points are $+30$ V, $+80$ V, and $+130$ V, respectively. What is the magnitude of the potential gradient?

18-A8 (a) What is the potential gradient in the space between the parallel plates of a capacitor which are separated by 2 mm if the PD between the plates is 2400 V? (b) What is the electric field in the space between the plates?

18-A9 A charge of $+5000$ μC is on one plate of a capacitor, and -5000 μC is on the other plate. The PD between the plates is 200 V. What is the capacitance of the capacitor?

18-A10 What charge flows from a 12 V battery when a 100 μF capacitor is connected to it?

18-A11 What charge can be placed on the plates of a 40 μF capacitor whose maximum (breakdown) PD is 250 V?

18-A12 Calculate the capacitance of a capacitor which acquires a PD of 400 V when 10^{12} electrons are taken from one plate and placed on the other plate.

18-B1 Two charges of value $Q_1 = +40$ μC and $Q_2 = -200$ μC are 7 m apart in air. At a point P, which is on the line between the charges, 2 m from Q_1, calculate   (a) the field due to Q_1;   (b) the field due to Q_2;   (c) the resultant field.

18-B2 Two charges of $+20$ μC and $+60$ μC are 4 m apart. Calculate the magnitude and direction of the electric field at a point midway between the two charges.

18-B3 Calculate the magnitude and direction of the electric field at the center of a square 2 m on an edge, if three charges of $+6$ μC each are at three of the corners of the square.

18-B4 The moon has, in round numbers, $\frac{1}{81}$ as much mass as the earth. Show that, on a journey from moon to earth, the resultant gravitational field is zero when a spaceship is $\frac{1}{10}$ of the way to the earth. (*Hint:* Let M = mass of earth, x = distance from earth, and y = distance from moon.)

18-B5 An electron is located at a point at which the electric field is 5000 V/m.   (a) What is the force on the electron?   (b) What is the acceleration of the electron?

18-B6 A drop of water in a fog has a net charge of 100 electrons. What is the radius of the drop if it is suspended motionless on a day when the earth's electric field is 200 V/m?

18-B7 It is found that 1.2 J of work is needed to move a charge of 0.3 mC from one insulated conductor to another insulated conductor which is 20 cm away. What is the average field strength in the region between the conductors?

18-B8 Starting from rest, a charge of -3×10^{-8} C is moved from P to Q in Fig. 18-8 by a muscle which does 11×10^{-6} J of work. The test particle arrives at Q with 2×10^{-6} J of KE. What is the PD between P and Q?

18-B9 If a proton gains 8×10^{-13} J of electric PE while being moved from X to Y, what is the PD between X and Y, and which point is at the higher potential?

18-B10 A proton moving horizontally southward at 3×10^5 m/s enters a region of electric field. What is the magnitude and direction of the field that will cause the proton to come to rest while traveling 2 cm? (See Appendix for mass of proton.)

18-B11 What speed will an electron acquire if it is accelerated through a PD of 90 V?

18-B12 A storm cloud may have a PD relative to a tree of, say, 800 MV. If, during a lightning stroke, 50 C of charge is transferred through this PD and 1% of the energy is absorbed by the tree, how much water (sap in the tree) can be boiled away, starting at 30°C?

18-B13 A parallel-plate capacitor has square plates 6 cm $\times$ 6 cm, separated by 1 mm of glass for which the dielectric constant is 5; calculate the capacitance of the capacitor, in pF.

18-B14 A parallel-plate capacitor of capacitance 0.04 μF has a charge of 36 μC. The plates are separated by 0.5 mm of air. What is the magnitude of the electric field at a point between the plates?

18-B15 A 200 μF capacitor is charged up to a PD of 25 V, and then connected to a circuit consisting of a switch and an electrolytic cell in series. The electrolytic cell contains Zn^{++} ions. When the switch is closed, the capacitor discharges through the cell. The operation is repeated 1000 times. What mass of zinc is plated out?

18-B16 A parallel-plate capacitor has a slab of glass (of dielectric constant 4) separating the plates. The capacitor is first connected to a 6 V battery, and then disconnected from the battery. The PD between the plates is observed by connecting them to a meter, which reads 6 V. The slab of glass is now pulled out. What is the new reading of the meter?

18-B17 A parallel-plate capacitor with plates separated by air acquires 1 μC of charge when connected to a battery of 1000 V. The plates, still connected to the battery, are then immersed in benzene. How much charge flows from the battery?

18-B18 It requires 8×10^{-17} J to move an electron from one plate of a charged capacitor to the other. The plates are 2.5 mm apart. Calculate the field strength at a point between the plates.

18-B19 Three capacitors having capacitances 5 μF, 10 μF, and 30 μF are available. (*a*) What is the maximum value of the capacitance that can be formed from these three? (*b*) What is the minimum value? Draw circuit diagrams showing the connections for each case.

18-B20 A capacitor of 15 μF and one of 10 μF are connected in parallel and charged by connecting the combination to a battery of 60 V. Calculate (*a*) the PD of each capacitor; (*b*) the charge on each capacitor.

18-B21 Capacitors of 15 μF and 10 μF are connected in series, and the combination charged by connecting to a battery of 60 V. Calculate (*a*) the charge on each capacitor; (*b*) the PD of each capacitor.

18-C1 (*a*) Show that a PD of only 1.02×10^6 V (1.02 megavolts) would be sufficient to give an electron a speed equal to twice the speed of light, if Newtonian mechanics remained valid at high speeds. (*b*) What speed would an electron actually acquire in falling through a PD of 1.02×10^6 V? (*Hint:* Find the mass of the electron from its total mc^2, and then find its speed from $m = m_0 / \sqrt{1 - v^2/c^2}$.)

18-C2 Solve Prob. 18-B3 if one charge next to the vacant corner is changed to $-6\ \mu C$.

18-C3 In a small television picture tube an electron moves horizontally at 5×10^7 m/s through the space between two 4 cm $\times$ 4 cm horizontal deflection plates separated by 1 cm. The PD between the plates is 80 V. (a) How long is the electron in the deflecting electric field? (b) Calculate the vertical impulse (see Sec. 6-2) given to the electron. (c) What is the vertical component of the electron's velocity after it leaves the deflection plates?

18-C4 A capacitor of 5 μF is connected in parallel with the combination of a 6 μF and a 3 μF capacitor which are in series. Find (a) the net capacitance of the entire combination; (b) the PD across the 6 μF capacitor when 12 V is maintained across the 5 μF capacitor.

18-C5 A capacitor is designed to have a capacitance of 0.1 μF and is to operate safely at a PD of 1000 V. The dielectric is mica, for which the dielectric constant is 3 and the breakdown field is 100 kV/mm. What must be the area of each plate, allowing a safety factor of 5?

18-C6 If the breakdown field strength for dry air is 3×10^6 V/m, what is the maximum charge that can be stored in a parallel-plate air capacitor whose plates are each 20 cm $\times$ 50 cm? (*Hint:* The spacing of the plates is not given. Call it d, carry d through your calculation, and trust that it will cancel out of your final answer.)

18-C7 (a) Derive a formula correlating the charge density (Q/A) on the plates of an air-filled parallel plate capacitor with the field strength (E) in the space between the plates. (b) Assuming (correctly) that your formula is also valid for the space near the surface of a charged metal sphere, calculate the radius of the smallest sphere which can hold a charge of 0.1 C in air if the breakdown field strength of dry air is 3×10^6 V/m.

For Further Study

18-9 Energy storage in capacitors

The act of forcing charge into a capacitor is analogous to the stretching of a spring. At the start, when the capacitor is "empty," it is easy to move a few electrons from one plate to the other through an external circuit. However, as the capacitor becomes charged, an electron approaching the negative plate is repelled by the charges already on that plate; and an electron leaving the positive plate is attracted back by the excess of positive charge on that plate. Thus the greater the charge on a capacitor, the more work is required to place additional charge on the plates. In a similar fashion, the more a spring is already stretched, the more work is required to stretch it an additional small amount. In Sec. 9-3 we found that the PE stored in a stretched spring is the area under the force-displacement graph; this PE turned out to be $\frac{1}{2}kA^2$ where k is the force constant

or stiffness of the spring and A is the final (maximum) displacement. The area under the curve was found both graphically and by an integration process.

Exactly similar reasoning is used to find the work done in charging a capacitor to some final charge Q_f. The work ΔW required to move a charge of ΔQ coulombs against an opposing PD of V volts is given by $\Delta W = V\Delta Q$ joules. While the capacitor is being charged, its PD varies from 0 to its final value V_f in such a way that V is at all times proportional to Q (Fig. 18-23)—just as F is always proportional to x as a spring is stretched. (The factor of proportionality between V and Q is $1/C$, since Eq. 18-5 can be rearranged to read $V = Q/C$.) Hence the total work done in charging the capacitor is the area under the curve, and from Fig. 18-23 we obtain

$$PE = \tfrac{1}{2}V_f Q_f$$

We can drop the subscripts, which have now

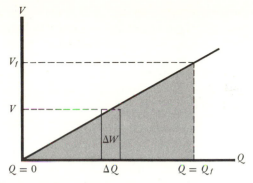

Figure 18-23 The PE of a charged capacitor is $\frac{1}{2}Q_fV_f$.

served their purpose, using Q and V for the final charge and PD, and we can also use the relation $C = Q/V$ to obtain three equivalent equations for the PE of a charged capacitor:

$$\text{PE} = \frac{1}{2}VQ = \frac{1}{2}CV^2 = \frac{1}{2}\left(\frac{1}{C}\right)Q^2 \quad (18\text{-}11)$$

The same result can be obtained by an integration:

$$\text{PE} = W = \lim_{\Delta Q \to 0} \sum \Delta W = \lim_{\Delta Q \to 0} \sum \frac{1}{C}Q\,\Delta Q$$

$$= \frac{1}{C}\int_{Q=0}^{Q=Q_f} Q^1\,dQ = \frac{1}{C}\left[\frac{Q^2}{2}\right]_0^{Q_f}$$

$$= \frac{1}{C}\left(\frac{Q_f^2}{2}\right) - \frac{1}{C}\left(\frac{0^2}{2}\right)$$

$$= \frac{1}{2}\left(\frac{1}{C}\right)Q_f^2$$

Let us note the close analogy with the formula for the PE of a spring.

$$\text{PE} = \frac{1}{2}\left(\frac{1}{C}\right)Q^2 \quad \text{(PE of a charged capacitor)}$$

$$\text{PE} = \frac{1}{2}(k)A^2 \quad \text{(PE of a stretched spring)}$$

The reciprocal of the capacitance is a measure of the "stiffness" of a capacitor, just as the force constant k is a measure of the stiffness of a spring. If C is very large, the capacitor absorbs a great deal of charge with only a little rise in its PD, and it might be said to be not very stiff. Mechanical analogies of

this sort are useful in discussing electrical oscillations (Sec. 23-3).

Just *where* is the energy stored in a charged capacitor? The answer to this question depends to a large extent upon which model of electric field we adopt. If we like action-at-a-distance, we probably think of the PE as stored at or on the charges themselves. After all, we did mechanical work on the charges while we were charging up the capacitor. If we like the Faraday-Maxwell field theory, we think of the energy as stored in the space between the plates, somehow related to an elastic stress of the lines of force. This is an attractive model, for we know that the molecules of a dielectric medium such as glass or mica *are* distorted and under stress while in a field. The model fails to explain how the energy is stored in a vacuum, where there are no molecules to be polarized. It is easy enough to write down an expression for the energy density in the space between the plates of a capacitor. Using symbols already defined, we have

$$\text{Stored energy} = \frac{1}{2}CV^2 = \frac{1}{2}\left(\frac{KA}{4\pi kd}\right)V^2$$

$$\text{Volume of dielectric} = Ad$$

$$\text{Energy density} = \frac{\text{energy}}{\text{volume}}$$

$$= \frac{\frac{1}{2}(KA/4\pi kd)V^2}{Ad}$$

$$= \frac{K}{8\pi k}\left(\frac{V}{d}\right)^2$$

But the field strength E is the potential gradient, or V/d. Hence our final result is

$$\text{Energy density} = \frac{K}{8\pi k}E^2 \quad (18\text{-}12)$$

This formula says that, even in a vacuum, where $K = 1$, energy is stored in any region of space where a field exists. Indeed, energy density may well be the most fundamental aspect of electric fields.

The present-day photon model gives a simple interpretation of energy storage in a field.

It will be recalled that electric forces are considered to be caused by many photons (particles of light) being exchanged between the charges. Energy storage in the space between the charges is a natural result of this model, since each photon has a certain amount of energy. In an exactly similar fashion, a juggler stores energy "in space" if he maintains a group of oranges moving from hand to hand. The concept of energy storage can be extended to the gravitational field, for which the model postulates the exchange of gravitons (Sec. 18-3) between two attracting masses.

■ PROBLEMS

18-C8   A 30 mF capacitor is charged to a PD of 40 V. How much electric PE is stored in this capacitor?

18-C9   A capacitor has parallel plates separated by 2 mm of air. When 5×10^{15} electrons are moved from one plate to the other, a field of 80 kV/m is set up between the plates. (a) What is the PE of the charged capacitor? (b) How much work would be required to move one additional electron from one plate to the other?

18-C10   A photographer's flash-bulb outfit consists of a capacitor which is slowly charged by five 90 V batteries in series, and then discharged quickly through a xenon-filled glow tube. The tube requires 100 J of energy for a successful discharge. What must be the capacitance of the capacitor?

18-C11   The electric field surrounding the earth is about 100 V/m in the air near the surface. How much electric energy is stored in a layer of air 2 m thick over a football field which measures 91.4 m × 48.8 m?

18-C12   A 3 μF capacitor is charged to 15 V, and a 5 μF capacitor is charged to 7 V. The batteries are removed, and the capacitors are then connected together, + to + and − to −. (a) Compute the new PD, the same for each capacitor. (b) Compute the new charge on each capacitor. (*Hint:* Total charge is conserved, algebraically, during the redistribution process.) (c) What is the loss of electric PE during this process? What becomes of this PE?

■ REFERENCES

1   Kondo, Herbert, "Michael Faraday," *Sci. American,* Oct., 1953, p. 90.

2   Loeb, Leonard B., "The Mechanism of Lightning," *Sci. American,* Jan., 1949, p. 22.

3   Gemant, Andrew, "Electrets," *Physics Today,* March, 1949, p. 8. Discusses permanently "frozen in" electric dipoles.

4   Methods of making electrets are described in C. L. Stong's "Amateur Scientist" department of *Scientific American* by G. O. Smith, Nov., 1960, p. 202, and by S. L. Khanna, July, 1968, p. 122.

5   McDonald, James E., "The Earth's Electricity," *Sci. American,* Apr., 1953, p. 32.

19-1   The energy method in electricity

Our preliminary survey of electric charge in Chap. 17 revealed a general law of force —Coulomb's law. One would expect that the motions of electrons, protons, and other charged particles could be predicted by a straightforward use of Newton's laws of motion in conjunction with Coulomb's law. In like fashion, one would expect to be able to use Newton's laws to discuss the complex motion of a speeding airplane (and it *is* a complex motion, if we seek to take account of turbulence, air resistance, production of shock waves, and so on). Electric forces are even more complex, if we wish a detailed picture of what happens. In mechanical problems, we have several times used the energy principle for an easy solution of problems for which the use of Newton's laws would have been too difficult. One example of this procedure is our derivation of the period of a SHM (Sec. 9-3) by the energy principle. Another example is afforded by our use of Bernoulli's equation (Sec. 12-8), which is based on the energy principle. In applying Bernoulli's equation we talk about pressure, velocity, and height at "point A" and "point B," but we have no detailed picture of forces and accelerations, and we do not ask what happens between points A and B, or how it happens. The energy method is easier than a complete dynamical solution, but, as payment for this ease, we get less information. For instance, the energy method is unable to give us an equation for the velocity as a function of time, in SHM; for this information we used the reference circle to derive Eq. 9-4.

Fortunately, energy considerations are sufficient to give us much useful information about the behavior of electric charge. In Chap. 18 we introduced the concept of potential difference, which is work per unit charge. In this chapter we will continue to place emphasis on energy changes of charges, rather than on the electric fields which would give a more detailed description than we usually need. We do not concern ourselves with how the force on a charge varies as it is moved from point to point; when using the energy method, we pass over the details and deal only with the total work done.

19-2   Electromotive force

Much of the usefulness of electricity in science and technology and in our daily lives arises

The electric eel, *Electrophorus electricus*, is able to transform chemical energy into electric energy. Such a device is called a seat of electromotive force. (New York Zoological Society Photo)

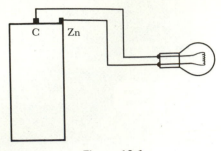

Figure 19-1

from the possibility of continuous flow of charge through an electric circuit. A simple circuit is shown in Fig. 19-1, in which a dry cell is connected to a flashlight bulb. We know that the bulb becomes warm and lights up, and hence we conclude that some agency is at work forcing charges to move in an endless path around the circuit. This agency is the cell; we say that the cell is the *seat* of an *electromotive force*.

Let us look more closely at the action of the cell which we discussed in Sec. 18-4(*c*). Before they are connected to the lamp, the carbon rod has an excess positive charge and the zinc can has an excess negative charge (Fig. 18-9). A PD of 1.5 V exists between Zn and C. How did the charge become separated in this way? We are dealing here with chemical forces* as well as electric forces. When the cell is first constructed, electrons flow from carbon to zinc, being impelled by chemical forces of some sort. However, electric forces of the Coulomb sort come into play as soon as charge is separated; unlike charges attract each other. The charging-up process continues until just enough charge is separated so that the electrostatic forces balance the chemical forces. The charges are then in equilibrium, as shown in Fig. 18-9a. The statement "A

---

*Chemical forces are short-range electric forces between atoms; their origin is related to the wave nature of matter (Sec. 27-6). They can be studied by methods of quantum mechanics, using mathematical techniques beyond the scope of this book. The reactions at the electrodes of an electrolytic cell (Sec. 17-6) are also governed by chemical forces of this type.

PD of 1.5 V exists between Zn and C" means that, for every coulomb of + charge that is moved through the solution (a moist paste) from Zn to C, 1.5 J of work must be done *against electric forces*. The charge gains energy from chemical sources, and this energy is exactly transformed into electric PE by the time the charge reaches C. The situation is much simpler if the test charge is moved from Zn to C through the air. Now no chemical forces do any work, and the increased PE of the charge must be supplied from outside—perhaps by the muscles of the person who moves the charge from Zn to C. Whether the test charge is moved from Zn to C through the solution or through the air, 1.5 J/C of work is done against *electric* forces.

If, now, the dry cell's terminals are connected by a wire and a lamp, free electrons will be repelled by the − charges on the zinc can and will flow through the lamp to the + terminal of the cell. In so doing, the electrons lose electric PE; this energy is somehow transferred to the nuclei of the tungsten filament of the bulb, and the nuclei vibrate with greater amplitude and hence become warmer. If we wish to make a crude model for this process, we can think of free electrons rubbing the ions as they drift through the metal of the filament on their way toward the positive terminal of the cell. According to this model, the heat is produced as a sort of frictional effect. After an electron reaches the + terminal of the cell, it moves through the cell from + to −, gaining electric PE at the expense of some of the cell's chemical PE. When our wandering electron finally reaches the zinc cup again, it is ready to start on another round trip. We have here a mechanism for the steady flow of charge, as long as the chemical PE of the cell holds out. The essential action of the cell is to transform energy from chemical PE to electric PE. The reverse process is also possible; if charges are forced backward through the cell, electric energy is transformed into chemical energy. This is what happens when a car's storage

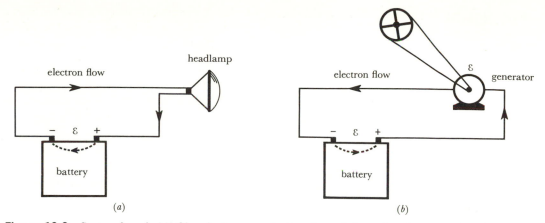

Figure 19-2 Seats of emf. (*a*) Chemical energy is transformed into electric energy in the battery. (*b*) Electric energy, which was transformed from mechanical energy in the generator, is transformed into chemical energy in the battery.

battery is "charged up" (Fig. 19-2*b*). The transformation of energy is said to be *reversible*. Enough energy can be stored in recently developed types of reversible cells to make electric automobiles an economic possibility.

Our discussion of the electric cell can be paralleled by a discussion of other seats of electromotive force. In an electric generator, it is mechanical energy which is transformed into electric energy. A thermocouple transforms thermal energy into electric energy. These devices will be discussed in Chap. 22.

With this background, we now define a seat of electromotive force.

A seat of electromotive force is a device within which nonelectric energy can be reversibly transformed into electric energy.

The numerical value of an electromotive force (emf) is denoted by $\mathcal{E}$, and is measured in joules per coulomb, or volts. Thus in a battery whose emf is 6 V, each coulomb of charge passing through the battery gains or loses 6 J of electric PE, depending on the direction of flow. The transformation of energy takes place at the rate of 6 J for every coulomb involved. The emf of a battery depends only on its physical and chemical construction and the temperature; it is not de-

pendent on the amount of current through it.

The concepts of potential difference and electromotive force are easily confused, since both are measured in terms of work per unit charge. The distinguishing feature of a seat of emf is the *transformation* of nonelectric energy into electric energy. To illustrate this point, let us imagine connecting a wire between two oppositely charged bodies such as *A* and *B* in Fig. 18-8. There is a PD between the bodies, but no emf. When the connection is made, electrons will momentarily flow from *B* to *A*, to be sure; but no continuous current is possible unless another wire is added to provide a return path from *A* to *B*. However, negatively charged electrons would not flow along this return wire back to the negatively charged conductor *B*, since like charges repel each other. Therefore some source of electric energy, a seat of emf, would have to be added to the circuit in order to obtain a continuous flow. It is not hard to extend this type of reasoning to the mechanical system of Fig. 18-7*a*. The book would fall to the floor "on its own," and its KE would be dissipated as heat energy during the inelastic collision with the floor (Sec. 6-11). Even if the table and floor were frictionless, continuous motion around the path could take place only if a source of mechanical energy were "in the circuit" (for instance, a muscle helping the book back up from D to E on each trip).

19-3   The electric circuit—Joule's law

There must be at least one seat of emf in any circuit in which continuous current exists. The electric energy supplied by the seat of emf can, of course, be converted into mechanical or chemical energy. In addition, a certain amount of the electric energy is transformed into heat, as we noted for the light bulb in the last section. There is a fundamental difference in the way in which electric energy is utilized in a motor and in a lamp bulb. The motor is (electrically speaking) reversible, and if it were turned by a crank powered by a muscle or a waterfall, it would act as a generator to send electric energy into a circuit (see Sec. 22-2). On the other hand, the lamp bulb is not reversible; it is impossible to generate electric current in a wire by building a fire under a lamp bulb which is connected to the two ends of a wire. This is why we say that a lamp bulb "dissipates" electric energy; it is a one-way street for electric energy. Such a conductor, in which electric energy is dissipated no matter in which direction the charges flow, is called a *resistor*. Every conductor is a resistor, in the sense that the flow of electrons is impeded to a greater or lesser extent.

During his classical investigations of the mechanical and electrical equivalent of heat, Joule studied the production of heat in resistors. He found that the rate of production of heat energy in a metallic resistor is proportional to the square of the current. Now the rate of production of energy is power (P), measured in the mks system as joules per second (= watts), and so we state the experimental facts as Joule's law: P/I^2 = constant for a given metallic resistor at a given temperature. We define the constant in Joule's law as the resistance R of a resistor:

$$R = \frac{P}{I^2} \qquad (19\text{-}1)$$

The mks unit of resistance is the *ohm;* a resistor has a resistance of 1 ohm if it dissipates energy at the rate of 1 watt when the current is 1 ampere. Rearranging Eq. 19-1 gives another form of Joule's law:

$$P = RI^2 \qquad (19\text{-}2)$$
$$\text{watts} = (\text{ohms})(\text{amperes})^2$$

The symbol Ω, the Greek capital letter omega, is often used to represent ohms.

EXAMPLE **19-1**

What is the resistance of a 40 W lamp bulb which is lit at full brilliance by a current of $\frac{1}{3}$ A?

$$R = \frac{P}{I^2} = \frac{40 \text{ W}}{(\frac{1}{3} \text{ A})^2} = \boxed{360 \ \Omega}$$

EXAMPLE **19-2**

An electric iron has a resistance of 10 Ω. How much heat energy is produced in 10 min if the current is 10 A?

$$P = RI^2 = (10 \ \Omega)(10 \text{ A})^2 = 1000 \text{ W}$$

Heat energy is being produced at the rate of 1000 J/s.

$$\text{Energy} = Pt = (1000 \text{ J/s})(10 \text{ min})\left(\frac{60 \text{ s}}{1 \text{ min}}\right)$$
$$= 6 \times 10^5 \text{ J}$$

Converting to heat units, we obtain

$$6 \times 10^5 \text{ J} \left(\frac{1 \text{ cal}}{4.18 \text{ J}}\right) = \boxed{1.44 \times 10^5 \text{ cal}}$$

Joule's law is based on experiment, and it tells us something about the behavior of matter. In this respect it has about the same status as Hooke's law, which can be derived from an exact knowledge of short-range electric forces between atoms. Similarly, Joule's law can be derived from knowledge of how electrons behave as they pass through matter. Neither Hooke's law nor Joule's law is of the same logical importance as Newton's laws of motion or Coulomb's law. Nevertheless, Joule's law is an important generalization which is the basis for much circuit theory, including Ohm's law, which we discuss next. First, however, we must decide on a convention for representing the direction of an electric current.

19-4 Conventional current

There are two equivalent ways of describing the same electric current: as a flow of negative charge in one direction (e.g., toward the right), or as an equal flow of positive charge in the opposite direction (i.e., toward the left). In metals and in vacuum tubes the carriers of charge are electrons, which are negative. The flow of positive charge is not entirely unknown, however. For example, charge carriers of both signs are involved in electrolysis (see Fig. 17-9, where Ag^+ ions move toward the left while NO_3^- ions move toward the right). Also, the current in many high-energy nuclear accelerators consists of a flow of positive charges (protons); and on a smaller scale transistors of some types make use of the flow of "holes" which behave like positive charges. Since both positive and negative carriers of charge are known to be able to flow, our choice of words to describe the direction of electric current can be made quite arbitrarily.

In this book we define the direction of current to be the direction in which *positive* charges would have to move in order to give the same electric and magnetic effects as does the flow of actual charges (+, −, or both) through the circuit. This *conventional current*, the flow of an equivalent positive charge, is what we mean by the term "current." Our definition of the direction of current, which is in widespread use, fits in well with the use of a *positive* test charge for measuring electric potential difference. A positive test charge loses energy when it moves from a region of high potential to a

region of low potential, much as a test mass loses gravitational PE when it flows downhill. Thus in Fig. 19-2a the conventional current is directed away from the + terminal of the battery, and the hypothetical positive charge flows "downhill" from the + terminal through the headlamp and returns to the battery at the − terminal. The direction of this conventional current is opposite to the direction of electron flow.

19-5 Ohm's law

In most practical cases we are interested in how a resistor limits the current, rather than in the rate at which heat is being developed. For such calculations, Ohm's law is used. We can make a simple derivation of this very useful law, based on Joule's law.

Consider a resistor AB (Fig. 19-3), whose resistance is R, through which the current is I. We are not concerned with *why* the charges flow; it is enough to know that they *do* flow, for some reason connected with one or more emf's somewhere in the circuit. We assume that the resistor itself is free from sources of energy. Since there is a flow of charge through the resistor and heat is produced, this charge is losing electric PE. In other words, there must be a potential difference between the terminals of the resistor; A is positive relative to B. *The mere existence of a current through a resistor causes the terminal at which conventional (positive) charge enters to become positive relative to the other terminal.* To compute the PD requires only a little algebra and the use of definitions already made.

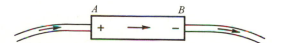

Figure 19-3 When charge flows through a resistor, the end at which positive charge enters the resistor becomes positive relative to the end at which the charge leaves. The arrows represent the direction of the conventional current through the resistor.

PD ≡ work per unit charge: $V = \dfrac{W}{Q}$

Charge ≡ (current)(time): $V = \dfrac{W}{It}$

Power ≡ work per unit time: $V = \dfrac{P}{I}$

Joule's law: Power = I^2R: $V = \dfrac{I^2R}{I}$
(experiment, defines R)

Hence, $V = IR$, or

$$R = \frac{V}{I} \qquad (19\text{-}3)$$

This is *Ohm's law* for a part of the circuit.

The ratio of potential difference between any two points in a circuit to the current is constant and equals the resistance between the two points.

In 1827, several decades before Joule's investigations, German physicist Georg Simon Ohm (1787–1854) discovered the law which bears his name. Ohm's law, like Joule's law, is valid only for metallic resistors at a definite temperature and for steady currents, but within these limitations it is remarkably precise. The law is approximately valid for other types of resistors, but fails completely for semiconductors and for gases. Unless specifically stated otherwise, all resistors in circuits discussed in this book are assumed to obey Ohm's law.

We now have two equivalent definitions of resistance: $R = P/I^2$ (Joule's law) and $R = V/I$ (Ohm's law). For any current I the ratio V/I always equals the ratio P/I^2. But this ratio (the resistance) may not be constant; it may depend on the current I or the potential difference V. The constancy of R is a definite statement about the behavior of a certain class of substances (metals), and it is possible to derive Ohm's law from fundamental considerations about the drift of free electrons through metals.

In Chap. 20 we will study electric circuits in detail; we will derive formulas for resistors in series and in parallel, and we will consider emf's in series and in parallel. For the pres-

Figure 19-4

ent, we consider a very simple circuit (Fig. 19-4) in which a device D is connected to the two terminals of a "black box" B. Without knowing anything about the nature of B and D, we can say that if conventional charge flows from the positive terminal of B at a rate of I amperes, and if the PD across the terminals of B (and D) is V volts, then energy is being supplied by B, and absorbed by D, at the rate of P watts, where

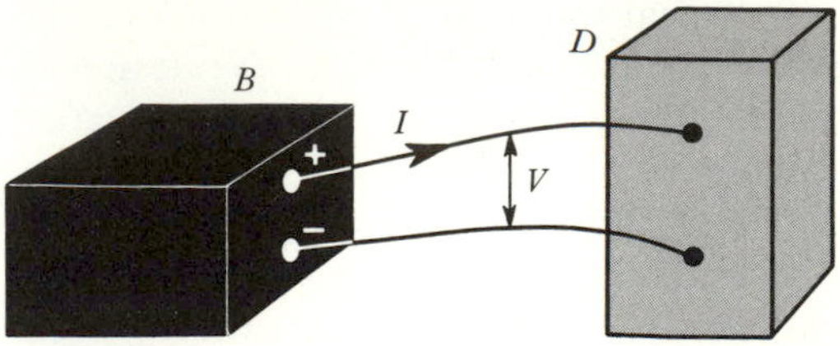

$$P = VI \qquad (19\text{-}4)$$

This follows directly from the definition of the volt as a joule per coulomb:

$$\text{watts} = \left(\frac{\text{joules}}{\text{coulomb}}\right)\left(\frac{\text{coulombs}}{\text{second}}\right)$$

In Eq. 19-4, we use V, the symbol for PD, since we do not know, or need to know, the *reason* for the potential difference between the terminals of the boxes. Similarly, if the box B contains an emf $\mathcal{E}$ and the current through the seat of emf is I, the rate at which nonelectric energy is being converted into electric energy by the seat of emf is

$$P = \mathcal{E}I \qquad (19\text{-}5)$$

These power equations are general and do not depend on Ohm's law. $\mathcal{E}$ is used in Eq. 19-5, since we are concerned here with the transformation of energy in a seat of emf.

EXAMPLE **19-3**

Suppose that in Fig. 19-4, the device D consists of a motor which uses 400 W, and a lamp bulb. If $V = 100$ V and $I = 5$ A, how much electric power is received by the bulb?

Power input to D:

$$VI = (100 \text{ V})(5 \text{ A})$$
$$= 500 \text{ W}$$

Power to bulb:

$$500 \text{ W} - 400 \text{ W} = \boxed{100 \text{ W}}$$

EXAMPLE 19-4

If the black box B of the preceding problem dissipates heat at the rate of 40 W, calculate the value of the emf inside the box.

The box is delivering electric energy at the rate of 500 W to device D, and electric energy is being transformed into heat energy inside the box at the rate of 40 W. Hence the total rate of production of electric energy from nonelectric forms must be 540 W.

$$P = \mathcal{E}I$$

$$\mathcal{E} = \frac{P}{I} = \frac{540 \text{ W}}{5 \text{ A}} = \boxed{108 \text{ V}}$$

Electric energy may be measured in joules ($=$ watt·seconds). Another common unit is the kilowatt·hour (kWh) which is the energy produced, transformed, or dissipated in 1 h if the power is 1000 W. One kilowatt·hour equals $(1000 \text{ J/s})(3600 \text{ s}) = 3.6 \times 10^6$ J. The commercial cost of electric "power" (really energy) is from $\frac{1}{2}$¢ to 5¢ per kWh, depending upon the amount purchased. Of course, the unit kilowatt·hour is not reserved for electric energy; it can equally well be used for thermal, mechanical, or other forms of energy.

19-6   Resistivity

In practice, it is difficult to apply Joule's law, and the precise measurement of metallic resistance is always made by means of Ohm's law, either directly or indirectly. To compare the resistances of various substances, we need to define a property which does not depend on the size and shape of the conductor. Other things being equal, we would expect the resistance to be directly proportional to the length of the conductor. We also expect that the electron flow will be easier if the conductor has a larger cross section, since the free electrons can "spread out" more. Thus we define the *resistivity* ρ (Greek letter rho) as the constant in the equation

$$R = \frac{\rho L}{A} \qquad (19\text{-}6)$$

where R is the resistance, L is the length, and A is the cross section. The dimensions of resistivity are ohm·meters, as seen by solving for ρ:

$$\rho = \frac{RA}{L} = \frac{\Omega \cdot \text{m}^2}{\text{m}} = \Omega \cdot \text{m}$$

Table 19-1 shows that silver has the smallest resistivity of any metal, with that of copper

TABLE 19-1

Approximate* resistivities ρ and temperature coefficients of resistance α, all at 0°C

Substance	ρ, $\Omega \cdot$m	α, (C°)$^{-1}$
silver	1.47×10^{-8}	0.0038
copper	1.59×10^{-8}	0.0039
gold	2.27×10^{-8}	0.0034
aluminum	2.60×10^{-8}	0.0040
tungsten	5.0×10^{-8}	0.0046
iron	11.0×10^{-8}	0.0052
platinum	11.0×10^{-8}	0.00392
constantan	49×10^{-8}	0.00001
(60 Cu, 40 Ni)		
mercury	94×10^{-8}	0.0009
Nichrome	100×10^{-8}	0.0002
(60 Ni, 24 Fe, 16 Cr)		
carbon	4×10^{-5}	-0.0005
germanium	2×10^{0}	
silicon	3×10^{4}	
boron	1×10^{6}	
wood (maple)	3×10^{8}	
Celluloid	4×10^{12}	
glass	$10^{11} - 10^{13}$	
amber	5×10^{14}	
sulfur	1×10^{15}	
mica (colorless)	2×10^{15}	
fused quartz	5×10^{17}	

*Exact values depend on purity, heat treatment, etc.

not much greater. Of all physical properties of matter, electrical resistivity perhaps shows the greatest range of values. Fused quartz, an excellent insulator, has a resistivity which is more than 10^{25} times that of silver. Our next examples show that the resistance of a conductor does not depend solely on the amount of metal used.

EXAMPLE 19-5

In a power station, a "bus bar" designed to carry many amperes of current is in the form of a slab of copper 2 m long and 1 cm × 10 cm in cross section. (a) What is the resistance of the bar at 0°C? (b) What PD is needed to cause a current of 5000 A through the bar?

(a) The cross section is $(10^{-1}$ m$)(10^{-2}$ m$) = 10^{-3}$ m^2. To find the resistance, we look up the resistivity of copper in the table and substitute into the formula:

$$R = \frac{\rho L}{A} = \frac{(1.59 \times 10^{-8}\ \Omega \cdot \text{m})(2\ \text{m})}{10^{-3}\ \text{m}^2}$$

$$= \boxed{3.18 \times 10^{-5}\ \Omega}$$

(b) To find the necessary PD, we use Ohm's law:

$$V = IR = (5 \times 10^3\ \text{A})(3.18 \times 10^{-5}\ \Omega)$$

$$= 15.9 \times 10^{-2}\ \text{V} = \boxed{0.159\ \text{V}}$$

Such a small PD would be negligible in a power station, where the PD between the terminals of the generator might be several hundred volts.

EXAMPLE 19-6

What will the resistance of the bar of Example 19-5 become if it is stretched out to form a long wire which is 1 mm × 1 mm in cross section?

The length of the wire equals the volume divided by the area of the cross section:

$$L = \frac{(2\ \text{m})(10^{-1}\ \text{m})(10^{-2}\ \text{m})}{(10^{-3}\ \text{m})(10^{-3}\ \text{m})} = 2 \times 10^3\ \text{m}$$

The wire has the same amount of copper as the bar had, but it is now in the form of a wire 2000 m long (over a mile long), and its cross section is now only 10^{-6} m^2.

$$R = \frac{\rho L}{A} = \frac{(1.59 \times 10^{-8}\ \Omega \cdot \text{m})(2 \times 10^3\ \text{m})}{10^{-6}\ \text{m}^2}$$

$$= \boxed{31.8\ \Omega}$$

Like most physical properties, resistivity depends on temperature. The resistivity of a pure metal increases with temperature, except for certain special alloys, and we describe this increase by use of a *temperature coefficient of resistance* α (alpha) which is analogous to the temperature coefficient of linear expansion (Sec. 13-3). The defining equation, for temperature changes that are not too great, is

$$\rho_t = \rho_0(1 + \alpha t) \qquad (19\text{-}7)$$

where t is the temperature in degrees Celsius (centigrade), ρ_t is the resistivity at $t°$C, ρ_0 is the resistivity at 0°C, and α is the temperature coefficient of resistance, in $(\text{C}°)^{-1}$, based on a reference temperature of 0°C. Values for α are tabulated in Table 19-1; these temperature coefficients are several hundred times as large as the linear coefficients of expansion for solids and are comparable to the volume coefficients of expansion for gases. It is noteworthy that for most common metals α is about 0.004 $(\text{C}°)^{-1}$, which means that the resistivity would become zero at about $-1/0.004°$C $= -250°$C if the linear relationship of Eq. 19-7 remained valid at low temperatures. Actually, the resistivities of all pure metals tend toward zero as absolute zero $(-273.2°$C$)$ is approached. If impurities are present, the resistivity tends to "level off" at some finite value as the temperature is lowered.

For any given conductor, the resistance R is proportional to the resistivity ρ, so

$$R_t = R_0(1 + \alpha t) \qquad (19\text{-}8)$$

The variation of resistance with temperature affords one way of making precise temperature measurements. Platinum is often used for this purpose because of its high melting point and its freedom from corrosion and oxidation effects.

EXAMPLE 19-7

The resistance of a platinum resistance thermometer is 200.0 Ω at 0°C, and is 257.6 Ω when immersed in a crucible containing melting $SbCl_3$. What is the melting point of this compound?

Solving Eq. 19-8 for the temperature t, and substituting, we find

$$t = \frac{R_t - R_0}{\alpha R_0} = \frac{257.6\ \Omega - 200.0\ \Omega}{[0.00392\ (\text{C}°)^{-1}](200.0\ \Omega)}$$

$$= \frac{57.6\ \Omega}{0.784\ \Omega \cdot (\text{C}°)^{-1}} = \boxed{73.5°\text{C}}$$

In 1911, the Dutch physicist Kamerlingh Onnes discovered that below 4.2°K the resistivity of mercury becomes essentially zero. Some (but not all) metals and their compounds show an abrupt decrease of resistivity at some finite temperature, called the *transition temperature,* and the resistivity then remains exactly zero[*] for all lower temperatures. This phenomenon of *superconductivity* is an active subject of research. A current in a loop of superconducting material, once started, persists for hours or days, with no emf in the circuit, and can be detected by its

[*] A more precise statement would be that the resistivity becomes unmeasurably small. Present techniques show that the resistivities of metals in this state are less than $4 \times 10^{-25}\ \Omega \cdot \text{m}$.

magnetic field. Transition temperatures vary from 0.0002°K (for rhodium) to as high as 21°K (for a niobium-aluminum-germanium compound $Nb_3(Al_{0.75}Ge_{0.25})$. Placing a superconducting material in a magnetic field lowers its transition temperature; in a sufficiently strong field, characteristic of the substance, superconductivity is not possible even at absolute zero.

Semiconductors are characterized by moderate resistivities (Table 19-1) and have negative temperature coefficients; germanium and silicon are typical examples. The model of a semiconductor attributes poor conduction either to a relatively few free electrons (*n*-type semiconductors) or to a relatively few vacant spots ("holes") in the crystal for which a large cloud of free electrons compete (*p*-type semiconductors). The model successfully explains the negative temperature coefficient. (At higher temperatures, there are more free electrons, but still far fewer than in a metal. Hence the conductivity goes up, and the resistivity goes down.) The model also explains the strong dependence on a minute concentration of impurity atoms. (The impurities may bring in the free electrons necessary to increase conduction.) The transistor (Sec. 22-10) is a semiconductor device, as is the "crystal" of the early crystal-set radio.

■ SUMMARY

A seat of electromotive force is a device which is able to transform mechanical, chemical, thermal, or other nonelectric energy into electric energy, and vice versa. Emf's are measured in joules per coulomb, or volts. A device has an emf of 1 V if 1 J of energy is transformed during the passage of 1 C of charge. Common seats of emf are electric generators, cells, and thermocouples.

The energy principle in electricity focuses our attention upon how a unit positive test charge gains or loses energy. Conventional current is therefore defined as the flow of positive charge. In metals the direction of the conventional current is opposite to the direction of flow of the free electrons which are the charge carriers.

Joule's law states that the rate of production of heat energy in a conductor is proportional to the square of the current. The proportionality constant is called the resistance of the conductor; it equals the power dissipated in heat divided by the square of the current.

Ohm's law states that for a metallic conductor at any given temperature, the ratio of potential difference to current is a constant, and this constant is the resistance as defined by Joule's law. A conductor's resistance is 1 Ω if a current of 1 A generates heat at the rate of 1 W, or (an equivalent statement) if there is a current of 1 A when the PD is 1 V.

Resistivity is a property which depends only on the substance and its temperature, and is defined as ρ in the formula $R = \rho L/A$. Resistivity is measured in ohm·meters. There is a wide variation in resistivities of substances, ranging from metallic conductors through semiconductors to insulators. Resistivity and resistance depend on temperature, and the temperature coefficient of resistivity is defined as α in the formulas $\rho_t = \rho_0(1 + \alpha t)$ and $R_t = R_0(1 + \alpha t)$. For most metals, α is positive and resistance increases with temperature, but for semiconductors α is usually negative.

■ CHECK LIST

seat of electromotive force	Ohm's law	temperature coefficient of
Joule's law	$V = IR$	resistance
resistance	$P = VI$	$R_t = R_0(1 + \alpha t)$
ohm	$P = \mathcal{E}I$	semiconductor
$P = I^2R$	resistivity	superconductivity
conventional current	$R = \rho L/A$	transition temperature

■ QUESTIONS

19-1 Electrons leave a dry cell and flow through a lamp bulb back to the cell. Which terminal, the + or the − terminal, is the one from which electrons leave the cell? In which direction does the conventional current flow?

19-2 When a dry cell is furnishing current through an external circuit, electrons inside the cell are moving toward the − terminal. Explain how they are able to do this.

19-3 Both PD and emf are measured in volts. What is the difference between these concepts?

19-4 Suppose a research worker invented a device capable of converting mass energy (Sec. 6-6) into electric energy and vice versa. Would he be justified in calling his invention a seat of emf?

19-5 A piece of wire is cut into four equal parts, and the pieces are bundled together side by side to form a thicker wire. How does the resistance of the bundle compare with that of the original wire?

19-6 Could you construct two wires of the same length, one of copper and one of iron, which would have the same resistance at the same temperature?

19-7 What is "constant" about the alloy constantan (Table 19-1)?

19-8 State two ways in which p-type semiconductors differ from ordinary metallic conductors such as copper.

19-A1 What is the current through a 1250 W electric heater if the resistance is 50 Ω?

19-A2 Calculate the resistance of a 600 W heater in which the current is 6 A.

19-A3 What is the current through a 25 Ω resistor if the PD between its terminals is 150 V?

19-A4 What PD is needed to cause a current of 5 mA through a resistance of 2 MΩ?

19-A5 Calculate the resistance of the filament of a radio tube if the current is 0.06 A when the PD is 6 V.

19-A6 Calculate the energy in kWh expended in lighting a 100 W lamp bulb for 30 min.

19-A7 A certain wire has a resistance 100 Ω. Find the resistance of a wire of the same material, at the same temperature, which is 3 times as long and has a cross section 4 times as large.

19-B1 A current of 4 A through a battery is maintained for 20 s, and in this time 480 J of chemical energy is transformed into electric energy. (*a*) What is the emf of the battery? (*b*) How much electric power is available for Joule heating and other uses?

19-B2 If energy costs 3¢ per kWh, what does it cost to leave a 40 W porch light burning all night (average 10 h) every day for a year?

19-B3 In some television sets the picture tube filament is kept warm at all times, using 40 W of power to provide "instant-on" viewing. If the set is idle 16 h/day: (*a*) calculate the dollar cost per year, at 3¢ per kWh, for this convenience. (*b*) calculate the ecological cost, in kg/y of coal (fuel value 7000 kcal/kg) burned in a 38% efficient power plant.

19-B4 At 2¢ per kWh, what does it cost to heat a 500 liter tank of water from 20°C to 70°C?

19-B5 What is the efficiency of a $\frac{1}{4}$ hp motor which draws a steady current of 2.00 A from a 115 V line? (See Appendix for conversion of horsepower into watts.)

19-B6 An electric iron draws 10 A from a 120 V line. The iron's mass is 0.8 kg, and it is originally at 20°C. Forty per cent of the heat is lost to the room by radiation. What will the temperature of the iron be 2 min after it is connected? Assume constant resistance.

19-B7 A car battery of emf 12 V is used to run a radio receiver through which the current is 3 A, and at the same time to operate the motor of a 20 W electric shaver. How much chemical energy is transformed into electrical energy in 10 min?

19-B8 What is the current from a 12 V automobile battery while it supplies 0.8 hp to a starter motor?

19-B9 An average electron beam current of 15 μA strikes a 1 kg copper target in a linear accelerator for which the equivalent PD is 20 GV (Sec. 30-2). If the target were not cooled, how long would be required for its temperature to rise from 20°C to the melting point? For this range, the specific heat of copper is about 0.11 cal/g·C°.

19-B10 A coil of wire is immersed in a cup of water and is connected to a source which delivers 5 A. After the water starts to boil, it is found that 90 g of water is vaporized in 10 min. (*a*) What is the resistance of the coil? (*b*) Explain why the specific heat of the cup does not enter into this problem.

19-B11 The 1980 annual consumption of electric energy in the U. S. is predicted to be 11 500 kWh per person. (*a*) How much coal (fuel value 6930 kcal/kg) will be burned during 1980 in a power plant of 40% efficiency to generate electric energy for a family of 4? (*b*) How much heat will be rejected to the environment?

19-B12 A bird whose feet are 5 cm apart perches on a power line which carries 900 A. If the resistance of the wire is 40 μΩ/m, what is the PD between the bird's feet?

19-B13   A wire of resistance 10 Ω is drawn out to four times its original length. What is the new resistance of the wire?

19-B14   A telephone cable contains many insulated 22-gauge copper wires, each of diameter 0.64 mm. Calculate the total resistance at 0°C of one circuit pair from the central office to a subscriber 2.00 km away and back to the central office.

19-B15   The resistance of a coil of copper wire is measured when immersed in an ice bath; then the coil is placed in an oil bath and heated. At what temperature will the resistance be double the original value?

19-B16   A tungsten filament in a lamp bulb is rated at 60 W when connected to a 120 V source. When cold, the filament's resistance is 14 Ω.   (a) Calculate the hot resistance of the filament.   (b) Estimate the temperature of the filament, assuming an average temperature coefficient of resistance, for the temperature range considered, to be 0.009 $(C°)^{-1}$.

19-B17   A resistor is made by winding on a spool a 30 m length of constantan wire of diameter 0.8 mm. Calculate the resistance of the wire at   (a) 0°C;   (b) 50°C.

19-B18   A wire of uniform cross section is stretched along a meter stick, and a PD of 0.400 V is maintained between the 0 cm mark and the 100 cm mark. How far apart on the wire are two points which differ in potential by 1 mV?

19-C1   A researcher has available 2 g of gold and wishes to form it into a wire having resistance 80 Ω at 0°C. How long should the wire be?

19-C2   (a) Derive a formula for the resistance of a wire in terms of its radius, density, total mass, and resistivity. Check your formula for dimensional consistency.   (b) Calculate the resistance at 0°C of a copper wire 1 mm in radius whose mass is 30 g.

19-C3   The copper wire of a magnet winding has resistance 10.0 Ω at 20°C. What is the resistance when the current is large enough to raise the temperature to 80°C? (*Hint:* First find the resistance at 0°C.)

19-C4   A motor connected to 120 V direct-current mains develops 400 W of useful mechanical power, and 32 W is lost through friction. The current from the mains is 4 A.   (a) What electric power is being supplied to the motor?   (b) What is the motor's efficiency?   (c) What is the resistance of the motor?

■ REFERENCES

1   Magie, W. F., *A Source Book in Physics* (New York: McGraw-Hill, 1935), pp. 420–31, early discoveries of Galvani and Volta about electric currents; pp. 465–72, Ohm's work.

2   Austin, L. G., "Fuel Cells," *Sci. American,* Oct., 1959, p. 72.

3   Steinbach, H. B., "Animal Electricity," *Sci. American,* Feb., 1950, p. 40.

4   Cox, R. T., "Electric Fish," *Am. J. Phys.,* **11,** 13 (1943).

5   Grundfest, H., "Electric Fishes," *Sci. American,* Oct., 1960, p. 115.

6   Stong, C. L., in the "Amateur Scientist" department, *Sci. American,* Jan., 1962, p. 145. Report on an apparatus to demonstrate nerve and muscle potentials of aquatic animals.

7   Wheeler, R. F., in C. L. Stong's "Amateur Scientist" department, *Sci. American,* Feb., 1966, p. 120. Electrical signals from microscopic animals.

8   Steel, Earl L., "Descriptive Theory of Semiconductors," *Am. J. Phys.,* **25,** 174 (1957).

9   Matthias, B. T., "Superconductivity," *Sci. American,* Nov., 1957, p. 92.

20
Electric Circuits

So far, we have discussed three circuit elements—capacitors, seats of emf, and resistors. A surprisingly large number of simple circuits involving only these circuit elements are in common use, and can be discussed with the help of Ohm's law. Among such instruments are the Wheatstone bridge and the potentiometer, which are used routinely in the physics laboratory and in medical research. We shall discuss these and other circuits in this chapter, but we shall introduce no new theory, relying upon our understanding of PD, emf, and Ohm's law to see us through.

In drawings of electric circuits, certain symbols are standard. A capacitor is represented by ⊣⊢ or ⊣⊢. A resistor is represented by a zigzag line —⋁⋁⋁— ; a variable (adjustable) resistor, also called a rheostat, is represented by —⋁⋏⋁— or ⋀⋁⋀ . A cell or a battery is represented by —⊣⊢, where the long line is the positive terminal and the short line is the negative terminal. A battery (several cells connected together in series) is also sometimes represented thus: —⊣⏐⏐⏐⊢. Unless specifically stated otherwise, we shall assume all connecting wires to have zero resistance. The junction between two connected wires is indicated by

a thick spot: ——•—. A grounded point, connected to earth, is represented by ⏚ or ⏛.

20-1 Kirchhoff's laws

For the systematic application of Ohm's laws to circuits, we use Kirchhoff's laws, which in turn are based upon two of the conservation laws. Kirchhoff's first law is based upon the law of conservation of charge. In words:

The algebraic sum of the currents entering any junction point in a circuit is zero.

That is, in a given time as much charge flows into any point as flows away from that point. If this were not true, there would be a net accumulation or diminution of charge at the point. Kirchhoff's first law is illustrated at each of the junctions in Fig. 20-1. At A, $(12 + 3)$ A enters and 15 A leaves. At C, $(6 + 4 + 2)$ A enters and 12 A leaves. You should verify Kirchhoff's first law at the other junctions B and D.

Kirchhoff's second law (the loop theorem) is based on the law of conservation of energy:

The printed circuit is well adapted to mass production. The extremely small size of this microcircuit is shown by comparison with cubic crystals of ordinary table salt. (IBM)

The algebraic sum of the changes in potential around any closed path is zero.

If we remember that potential changes are measures of work done per unit charge, it should be evident that Kirchhoff's second law is a statement of the energy principle, in a form useful for electric circuits. A mechanical analogy may prove helpful here: Suppose a student slides his physics book to the edge of the table, lets it fall to the floor, slides it along the floor, then lifts it up and returns it to the starting point (Fig. 20-2; see also the discussion of Fig. 18-7). The total work done on the book against gravitational forces is zero, for the round trip $GHABCDEFG$. If this were not so, and the gravitational PE of the book at G after the trip were greater than the gravitational PE before the trip, then a perpetual-motion machine could be constructed (letting the book fall the "hard" way and lifting it up the "easy" way). The impossibility of constructing such a machine is asserted by the law of conservation of energy (more generally, by the first law of thermodynamics). Similarly, for electric charge, conservation of energy requires that no net work be done *against electric forces* in taking a test charge around any complete

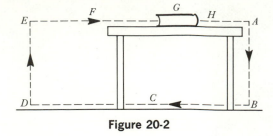

Figure 20-2

circuit. Therefore there can be no net change in potential after going completely around a circuit.

In a circuit, changes of potential can arise in several ways. If there is a current through a resistor, the PD across the resistor is given by Ohm's law, $V = IR$. The product IR is called the "IR drop" in the resistor. Another way in which potential may change is in a cell or battery. In a cell there is a PD equal to the emf; in addition, if the cell has an internal resistance r, there is a PD equal to Ir in the cell. Let us apply Kirchhoff's second law to several simple circuits.

(*a*) A dry cell of negligible internal resistance is connected to a load of resistance $0.5\ \Omega$; the emf of the cell is 1.5 V. In Fig. 20-3 the arrows represent the direction of the conventional current away from the positive terminal of the cell and through the load resistor from A to B. Whenever there is a current through a resistor, the end of the resistor at which charge enters becomes positive relative to the end from which it leaves. In this case, A is positive relative to B, and we indicate this fact by drawing a small + sign near A and a small − sign near B. We also label the two terminals of the cell + and −. Note well the difference in method here.

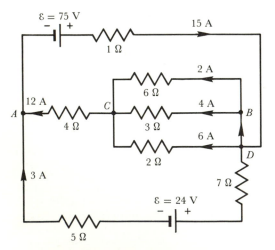

Figure 20-1 Kirchhoff's first law can be verified at each junction.

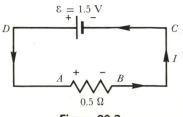

Figure 20-3

At the seat of emf, the $+$ and $-$ signs attached to ε are determined by the physical structure of the cell, battery, generator, etc. For a dry cell, the $-$ sign is at the zinc can, regardless of the direction of the current. On the other hand, the $+$ and $-$ signs by which we label the resistor depend entirely on the direction of the current (if any). A resistor has no inherent $+$ or $-$ terminals, as does a seat of emf. Having labeled both the emf and the resistor in Fig. 20-3, we now apply Kirchhoff's second law. Traversing the circuit in the direction $ABCDA$, we have a *fall* of potential ($+$ to $-$) in AB equal to $-I(0.5\ \Omega)$, followed by a *rise* of potential ($-$ to $+$) in CD equal to $+1.5$ V. Kirchhoff's second law says that the net change of potential is zero:

$$-I(0.5\ \Omega) + 1.5\ \text{V} = 0$$

whence

$$I = \frac{1.5\ \text{V}}{0.5\ \Omega} = \boxed{3\ \text{A}}$$

We can with equal success traverse the circuit in the opposite direction, $ADCBA$: we have a fall ($+$ to $-$) in the emf followed by a rise ($-$ to $+$) as we go through the resistor from B to A:

$$-1.5\ \text{V} + I(0.5) = 0$$

whence

$$I = \frac{-1.5\ \text{V}}{-0.5\ \Omega} = \boxed{3\ \text{A}}$$

which is the same answer as before.

(b) To study a more complex circuit, let us redraw Fig. 20-1, this time filling in $+$ and $-$ signs for each resistor and each emf (assumed to have no internal resistance). We also fill in the values of the IR drops in the resistors, obtained in each case by Ohm's law (Fig. 20-4). In this diagram the currents are assumed known, and we wish to verify Kirchhoff's second law. Around path ACM-$NBDGHKA$:

$$+48\ \text{V} + 12\ \text{V} + 15\ \text{V} - 75\ \text{V} \stackrel{?}{=} 0$$
$$0 = 0$$

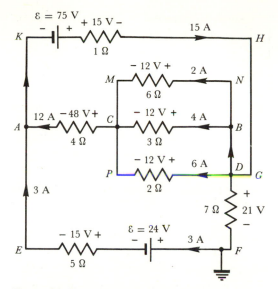

Figure 20-4 Kirchhoff's second law can be verified around any closed path.

Also, around path $AEFDPCA$:

$$+15\ \text{V} + 24\ \text{V} + 21\ \text{V}$$
$$- 12\ \text{V} - 48\ \text{V} \stackrel{?}{=} 0$$
$$0 = 0$$

Similarly, the net change of potential is zero around *any* closed path in Fig. 20-4.

EXAMPLE **20-1**

What is the PD between points A and D in Fig. 20-4?

Starting at A, we go to D along the path $ACMNBD$:

$$\Delta V = +48\ \text{V} + 12\ \text{V} = \boxed{+60\ \text{V}}$$

Hence D is 60 V higher in potential than is A. As a check, we can use the path $AKHGD$:

$$\Delta V = +75\ \text{V} - 15\ \text{V} = +60\ \text{V}$$

When some point of a circuit such as F in Fig. 20-4 is "grounded," the potential of such a point may be considered to be zero.

EXAMPLE **20-2**

If point F (Fig. 20-4) is grounded, what is the potential of point K?

Here F is assumed to be at a potential of 0 V. Around path $FDGHK$:

$$\Delta V = +21\text{ V} + 15\text{ V} - 75\text{ V} = -39\text{ V}$$

Hence if F is grounded, the potential of K is

$$\boxed{-39\text{ V}}$$

20-2 Terminal voltage of a cell

A cell or battery is no exception to the rule that irreversible "Joule" heat is produced by a current. This means that a cell has *internal resistance*, denoted by r. The internal resistance of a fresh "dry" cell is about $0.05\ \Omega$, but, as the moist electrolyte dries out, the internal resistance may increase to as much as $100\ \Omega$ or more. Also, hydrogen gas released by electrolysis during the passage of charge through the cell collects around the carbon rod; this too increases the resistance, since a gas is a poor conductor of electricity.[*] The cell is said to become *polarized* during use. The addition of the compound MnO_2 to the cell is helpful in two ways: the MnO_2 combines with the H_2 gas to form water, thereby not only removing the H_2 but also replenishing moisture which may have evaporated. From what has just been said, one would expect the internal resistance of a cell not to be constant but to depend upon the age and past history of the cell and upon the amount of current through it. Nevertheless, in problem solving we usually can consider the internal resistance to be a constant, at least to a first approximation.

Taking into account the internal resistance, we find that the PD between the terminals of a cell, the *terminal voltage* of the cell, is not always equal to the cell's emf. Consider a 6.08 V storage battery which is delivering 4 A to a headlamp of resistance $1.5\ \Omega$. In

---

[*]The emf of a cell depends on the nature of the electrodes and the chemical composition of the electrolyte. Thus the emf of a dry cell also changes when it becomes polarized, since it is now to some extent a hydrogen-zinc cell instead of a carbon-zinc cell.

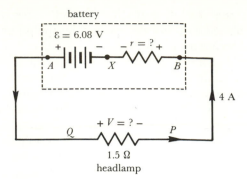

Figure 20-5 A car battery AB gives rise to a current through a headlamp PQ.

Fig. 20-5 we represent the internal resistance of the battery as a resistor r drawn next to the ε ⊣|ı|⊢ symbol; the battery as a whole is represented by a box which includes ε and r, with terminals A and B. In actuality, ε, r ⊣Wı⊢ might be a better symbol for a battery of emf ε and internal resistance r, but by drawing these symbols separately we make it easier to apply Kirchhoff's second law. We must not forget, however, that the junction point such as X in Fig. 20-5 is never available to us and we could never connect a voltmeter between A and X to measure the emf ε directly.

EXAMPLE 20-3

For the circuit in Fig. 20-5, calculate (*a*) the PD across the headlamp terminals PQ, (*b*) the terminal voltage of the battery, and (*c*) the internal resistance of the battery.

(*a*) By Ohm's law, across the headlamp

$$V = IR = (4.00\text{ A})(1.50\ \Omega) = \boxed{6.00\text{ V}}$$

(*b*) The terminal voltage of the battery is the PD between its terminals. We calculate this PD by going from B to A. Taking the bottom path $BPQA$, we see that the terminal voltage is

$$(4.00\text{ A})(1.50\ \Omega) = \boxed{6.00\text{ V}}$$

the same as the PD across the headlamp which is connected across the terminals of the battery.

(c) Since the current through r is from B to X, the point B is positive relative to X and is so marked in the figure. The internal Ir drop is $(4.00 \text{ A})(r)$. Now we apply Kirchhoff's second law around the path $AQPBXA$:

$$-6.00 \text{ V} - (4.00 \text{ A})(r) + 6.08 \text{ V} = 0$$

$$r = \frac{0.08 \text{ V}}{4.00 \text{ A}} = \boxed{0.02 \text{ }\Omega}$$

Calling the terminal voltage V, we see that in Example 20-3 the terminal voltage is less than the emf of the battery; in fact,

$$V = \mathcal{E} - Ir$$

where $\mathcal{E}$ is the emf and r is the internal resistance of the battery. It is also possible for the terminal voltage of a battery to be greater than its emf, if electrons are being forced through the battery in a direction opposite to that in which they would normally flow. This is what happens to a string of storage batteries being charged in a service station.

EXAMPLE **20-4**

Two batteries, of emf 6.20 V and 12.45 V, are being charged by a battery charger which sends 20 A through the cells (Fig. 20-6). The internal resistances of the batteries are 0.01 Ω and 0.03 Ω, respectively. What is the terminal voltage of each battery?

We use Ohm's law to calculate the Ir drop inside each battery. In the 6.20 V battery, $Ir = (20 \text{ A})(0.01 \text{ }\Omega) = 0.20 \text{ V}$. In the 12.45 V battery, $Ir = (20 \text{ A})(0.03 \text{ }\Omega) = 0.60 \text{ V}$. These internal Ir drops are labeled $+$ to $-$ as shown, since cur-

rent is being forced through the internal resistances of the batteries and the $+$ end of a resistor is always the one at which the current enters. Now we find the PD from A to B as an algebraic sum of two potential changes, the emf and the internal Ir drop:

$$V = +6.20 \text{ V} + 0.20 \text{ V} = \boxed{6.40 \text{ V}}$$

Similarly, for the other battery, the PD from C to D is

$$V = +12.45 \text{ V} + 0.60 \text{ V} = \boxed{13.05 \text{ V}}$$

These terminal voltages are *greater* than the emf's of the batteries. In this case, $V = \mathcal{E} + Ir$.

In general, the terminal voltage of a battery or other seat of emf is given by

$$V = \mathcal{E} \pm Ir \qquad (20\text{-}1)$$

Whether to use the $+$ or the $-$ sign depends on the direction of the current through the battery. As a special case, we see that, if the current through a cell or battery or other seat of emf is zero, then $V = \mathcal{E}$, and the terminal voltage equals the emf. This points out one way to measure emf; we need only measure the terminal voltage under circumstances in which there is no current through the seat of emf. The potentiometer circuit (Sec. 20-8) is designed to do just this. An ordinary voltmeter is definitely ruled out for this purpose, since there must be *some* current through the voltmeter to produce a deflection of the needle by magnetic force.

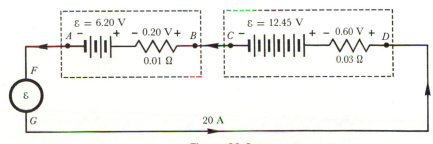

Figure 20-6

20-3   Resistors in series and in parallel

Many (but not all) complex circuits can be simplified by mentally replacing a group of resistors by a single resistor equivalent to the group.

The resistors R_1, R_2, and R_3 in Fig. 20-7 are said to be connected *in series,* because the charges move through the resistors one after the other, with no short cuts. The three resistors as a unit have resistance R_s and carry a current I. The PD from A to D is V_s, so by Ohm's law $R_s = V_s/I$. Now in a series circuit, the PD from A to D equals the sum of the PD's across the individual resistors (the total work required to take a unit test charge from A to D equals the sum of the work from A to B plus that from B to C plus that from C to D).

$$V_s = V_1 + V_2 + V_3 + \cdots$$

$$R_s = \frac{V_s}{I} = \frac{V_1 + V_2 + V_3 + \cdots}{I}$$

$$= \frac{IR_1 + IR_2 + IR_3 + \cdots}{I}$$

Hence, for resistors in series,

$$R_s = R_1 + R_2 + R_3 + \cdots \quad (20\text{-}2)$$

Our proof is general, for any number of resistors in series, although for simplicity we have shown only three resistors in Fig. 20-7. The essential features of the proof are (*a*) that the PD's are added for resistors in series, and (*b*) that the current is the same through each resistor in a series group.

The resistors in Fig. 20-8 are said to be connected *in parallel.* In a parallel connection, the incoming current divides, some going through each resistor. On the other hand, the PD's across each resistor are the same, and are equal to the PD across the equivalent resistor R_p. (The work required to take a unit

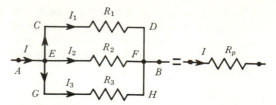

Figure 20-8   Resistors connected in parallel.

test charge from A to B is the same by any path, whether CD, EF, or GH.) By Kirchhoff's first law, the total current entering point E equals that leaving point E.

$$I = I_1 + I_2 + I_3 + \cdots$$

By Ohm's law, the equivalent resistance

$$R_p = \frac{V}{I}$$

and therefore

$$\frac{1}{R_p} = \frac{I}{V} = \frac{I_1 + I_2 + I_3 + \cdots}{V}$$

$$\frac{1}{R_p} = \frac{I_1}{V} + \frac{I_2}{V} + \frac{I_3}{V} + \cdots$$

Hence, for resistors in parallel,

$$\frac{1}{R_p} = \frac{1}{R_1} + \frac{1}{R_2} + \frac{1}{R_3} + \cdots \quad (20\text{-}3)$$

The essential features of this proof are (*a*) that the currents are added for resistors in parallel, and (*b*) that the same PD exists across each resistor in a parallel group.

EXAMPLE **20-5**

Resistors of resistance 6 Ω, 12 Ω, and 4 Ω are connected in parallel. What is the equivalent resistance of the combination?

$$\frac{1}{R_p} = \frac{1}{6} + \frac{1}{12} + \frac{1}{4} = \frac{2}{12} + \frac{1}{12} + \frac{3}{12} = \frac{6}{12}$$

$$R_p = \frac{12}{6} = \boxed{2\ \Omega}$$

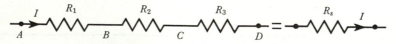

Figure 20-7   Resistors connected in series.

It will be helpful to remember that the combined resistance of any number of resistors in parallel is always less than the smallest resistance in the combination. As an analogy from highway engineering, the opening of a parallel bypass, no matter how narrow, will always allow a greater flow of cars, and hence the "resistance" of the combination is less than that of either the road or the bypass taken separately.

A complicated circuit can often (but not always) be simplified by repeated application of the formulas for resistors in series and in parallel.

EXAMPLE **20-6**

What is the net resistance between A and F in Fig. 20-9a?

This circuit is a combination of series and parallel units. We start from the "inside" and use our two basic formulas.

Step 1: Replace the parallel combination BC by a single resistor.

$$\frac{1}{R} = \frac{1}{30} + \frac{1}{15} = \frac{1}{30} + \frac{2}{30} = \frac{3}{30}$$

$$R = \frac{30}{3} = 10 \ \Omega$$

Substituting 10 Ω for the combination, we can redraw our circuit as in Fig. 20-9b.

Step 2: Replace the series combination BD by a single resistor.

$$R = 10 + 2 = 12 \ \Omega$$

We are now able to redraw our circuit as in Fig. 20-9c.

Step 3: Replace the parallel combination AE by a single resistor.

$$\frac{1}{R} = \frac{1}{12} + \frac{1}{60} = \frac{5}{60} + \frac{1}{60} = \frac{6}{60}$$

$$R = \frac{60}{6} = 10 \ \Omega$$

This allows us to redraw the circuit as in Fig. 20-9d.

Step 4: Finally, using the series formula, we see that the resistance from A to F is

$$R = 10 \ \Omega + 7 \ \Omega = \boxed{17 \ \Omega}$$

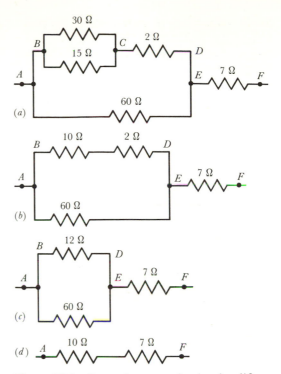

Figure 20-9 Successive stages in the simplification of a circuit diagram.

Not all circuits can be simplified by this technique. Figure 20-10a shows a "simple" circuit whose net (equivalent) resistance A to K can be found by repeated use of the series and parallel formulas (the answer turns out to be 10 Ω). The circuit of Fig. 20-10b, however, is actually a "non-simple" circuit which cannot be further simplified; it is impossible to use the method outlined above to find the resistance from A to B. The difficulty is that the 5 Ω resistor is neither in series nor in parallel with any other resistor. In this book we shall usually deal with simple circuits, except in special cases; see Sec. 20-9 for a method of finding currents in non-simple circuits.

20-4 Emf's in series and in parallel

When several emf's are connected in series, they can be replaced by a single emf equal to the algebraic sum of the individual emf's.

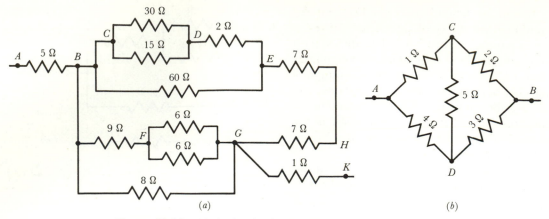

Figure 20-10 (*a*) A simple circuit; (*b*) a non-simple circuit.

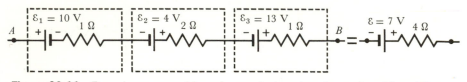

Figure 20-11 Batteries connected in series; equivalent ε and r are found by addition.

This follows from the energy principle; the total rate of transformation of nonelectric energy into electric energy equals the algebraic sum of the rates of transformation in the various individual seats of emf. Thus in Fig. 20-11 three batteries are shown; the net emf between A and B is $-10\,\text{V} + 4\,\text{V} + 13\,\text{V} = +7\,\text{V}$, with B positive relative to A. Since the cells are in series, their internal resistances can also be added up, to give $4\,\Omega$.

The situation is more complicated when batteries are connected in parallel. In Fig. 20-12, it is impossible to compute easily the effective ε and r of the combination. Certain currents exist within the batteries, and all we know is that the terminal voltages of all three batteries will be equal since they are connected in parallel. In the specific example of Fig. 20-12, the 4 V battery is probably being charged by current from the other batteries, but the magnitude of the current through it cannot be easily found.[*] There is only one way in which we can make a

simple treatment of cells in parallel: if they are identical in emf *and in internal resistance*, then the net emf equals the emf of each cell, and the net internal resistance equals the parallel resistance of all the cells (and hence is $1/n$ times the internal resistance of any one cell). Common dry cells and batteries illustrate these points (Fig. 20-13). The tall

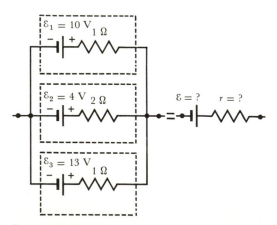

Figure 20-12 Batteries connected in parallel; equivalent ε and r cannot be found in any simple way.

---

[*] Such questions can be answered by the systematic use of Kirchhoff's laws, as discussed in Sec. 20-9.

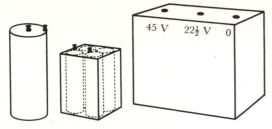

Figure 20-13 Cells and batteries.

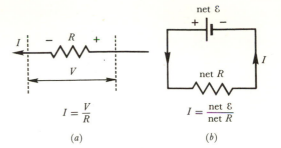

Figure 20-14 Two aspects of Ohm's law: (a) Ohm's law for a part of a circuit; (b) Ohm's law for a complete simple circuit.

cylindrical cell is a true single cell of emf $1\frac{1}{2}$ V. The short stubby $1\frac{1}{2}$ V "cell" is actually a battery of four flashlight cells, connected in parallel. The 45 V battery consists of thirty $1\frac{1}{2}$ V cells, connected in series to give a total emf of 45 V. A 12 V automobile storage battery consists of six 2 V batteries in series, and each 2 V battery may have up to several dozen individual cells in parallel with each other. The internal resistance of a car battery is thus very low (not only because of the use of cells in parallel but also because a liquid electrolyte is used instead of a paste). This allows large currents; in some cars the starter motor may require 50 A or more for a brief time.

20-5 Ohm's law for a complete circuit

We have now discussed the conditions under which a circuit can be replaced by the simplified equivalent form of a seat of emf connected to a single resistor. Ohm's law takes a simple form in such a situation: the current through the circuit equals the net emf divided by the net resistance of the circuit. Thus we see that there are two rather different aspects of Ohm's law:

Ohm's law for a part of a circuit: for any resistor or group of resistors, anywhere in a circuit:

$$I = \frac{V}{R} \qquad (20\text{-}4)$$

Ohm's law for a complete simple circuit:

$$I = \frac{net\ \mathcal{E}}{net\ R} \qquad (20\text{-}5)$$

We emphasize that these equations are to be applied in different situations (Fig. 20-14), even though both may be abbreviated as "amperes = volts/ohms." Equation 20-4 is always true (for a metallic resistor in which there is a steady current); Eq. 20-5 applies only to certain "simple" circuits in which there *is* a net emf[*] and a net resistance. In the solution of problems, we use both forms of Ohm's law to calculate currents and PD's.

EXAMPLE **20-7**

What is the current in the 5 Ω resistor of Fig. 20-15a?

First we simplify the circuit and find the total current. For the combination of the 6 Ω and the 3 Ω resistors, $1/R = \frac{1}{6} + \frac{1}{3}$, whence $R = 2\ \Omega$. We now redraw the original circuit as in 20-15b, remembering to draw in the internal resistance of the batteries. We now have a series combination, and the net resistance is given by $R = 1\ \Omega + 5\ \Omega + 1\ \Omega + 2\ \Omega = 9\ \Omega$. To find the net emf, we observe that the two batteries tend to send charges around the circuit in opposite directions, so that the net emf is 60 V − 6 V = 54 V. Now we apply Ohm's law to the entire circuit:

$$I = \frac{net\ \mathcal{E}}{net\ R} = \frac{54\ V}{9\ \Omega} = \boxed{6\ A}$$

This is the current through each battery and through each 5 Ω resistor.

---
[*]Sometimes the symbol Σ𝓔 is used for "net 𝓔," just as we use Σ*F* for "net *F*" in Newton's second law.

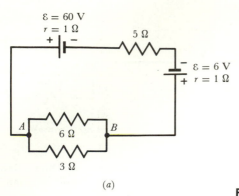

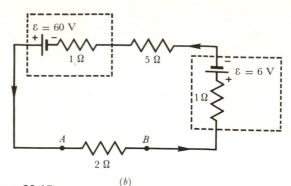

(a)

(b)

Figure 20-15

EXAMPLE 20-8

What are the currents through the 3 Ω and 6 Ω resistors in Fig. 20-15a?

Here we apply Ohm's law to a portion of the circuit. The section AB is redrawn in Fig. 20-15b. Since 6 A flows between A and B, and the net resistance from A to B is 2 Ω, Ohm's law tells us that $V = IR = (6 \text{ A})(2 \text{ Ω}) = 12$ V. This is the PD across each of the two resistors. The current through each resistor is found by further application of Ohm's law:

$$I_1 = \frac{12 \text{ V}}{6 \text{ Ω}} = 2 \text{ A}$$

$$I_2 = \frac{12 \text{ V}}{3 \text{ Ω}} = 4 \text{ A}$$

As a check, we find that I_1 and I_2 add up to 6 A, which is the total current entering point A. Thus Kirchhoff's first law is satisfied.

EXAMPLE 20-9

Calculate the terminal voltage of each battery in Fig. 20-15b.

To find the terminal voltages, we use the previous solution, in which we calculated the current through each battery to be 6 A. In Fig. 20-16 we fill in the Ir drops in the internal resistances of the batteries, paying due regard to signs. (Remember that the current is from + to − through a resistor.) The terminal voltage of the 60 V battery is found by going from E to F through the battery; a drop of 6 V is followed by a rise of 60 V, so the net PD is −6 V + 60 V = +54 V, with F positive relative to E. Similarly, in going through the other battery from D to C, we find a rise of 6 V followed by a

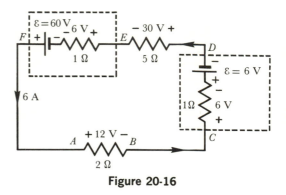

Figure 20-16

rise of 6 V; the net PD is +6 V + 6 V = +12 V, and the terminal voltage of this battery is 12 V, with C positive relative to D.

Suppose that we connect ideal voltmeters• across the various parts of our circuit. According to the calculations of the preceding examples, the voltmeters would read as shown in Fig. 20-17. Kirchhoff's second law is illustrated here, and the net change of potential around the circuit is zero, as it should be. For instance, around the path $ABCDEFA$, the potential changes indicated by the voltmeters add up to zero as follows:

$$-12 \text{ V} - 12 \text{ V} - 30 \text{ V} + 54 \text{ V} = 0$$

We have not yet extracted all possible nourishment from our study of the circuit of Fig. 20-15. Let us make an energy check,

---

• That is, voltmeters of very high resistance, which would not divert any appreciable current from the circuit (see Sec. 20-6).

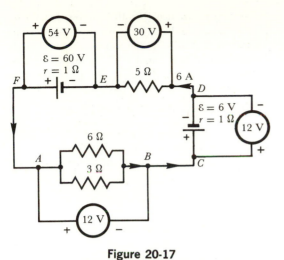

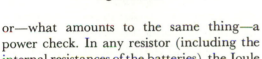

Figure 20-17

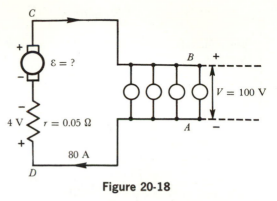

Figure 20-18

or—what amounts to the same thing—a power check. In any resistor (including the internal resistances of the batteries), the Joule heating is at the rate of $P = I^2R$, where the power P is measured in watts if I is in amperes and R in ohms. The table below shows that heat is produced at a total rate of 324 W. We can account for this by the action of the two seats of emf. The 60 V battery is delivering current in the "normal" direction, and chemical PE is being transformed into electric energy at the rate of $P = \mathcal{E}I = (60 \text{ V})(6 \text{ A}) = 360$ W. The 6 V battery is being charged as charge carriers are forced "backward" through it. Here electric energy is being transformed into chemical PE at the rate of $P = \mathcal{E}I = (6 \text{ V})(6 \text{ A}) = 36$ W. The *net* rate of production of electric energy is $360 \text{ W} - 36 \text{ W} = 324$ W; and we see from the table that this electric energy is transformed into heat in the various resistors and in the batteries themselves at a total rate of 324 W.

R, ohms	I, amperes	V, volts	P, watts
3	4	12	48
6	2	12	24
1	6	6	36
5	6	30	180
1	6	6	36
			324 W

EXAMPLE 20-10

A portable generator is delivering 80 A to a string of lamps at a carnival (Fig. 20-18). The terminal voltage of the generator is 100 V, and its internal resistance is 0.05 Ω.

(*a*) What is the emf of the generator?   (*b*) Ignoring friction, what mechanical power, in horsepower, must be supplied by the gasoline engine which turns the generator?

We represent the generator schematically as a seat of emf in series with 0.05 Ω.

(*a*) The Ir drop inside the generator is $(80 \text{ A})(0.05 \text{ Ω}) = 4$ V. Applying Kirchhoff's second law around the circuit $ABCDA$:

$$+100 \text{ V} - \mathcal{E} + 4 \text{ V} = 0$$

$$\mathcal{E} = \boxed{104 \text{ V}}$$

(*b*) The rate of transformation of energy from mechanical to electric form[*] is

$$P = \mathcal{E}I = (104 \text{ V})(80 \text{ A}) = 8320 \text{ W}$$

$$= (8320 \text{ W})\left(\frac{1 \text{ hp}}{746 \text{ W}}\right) = \boxed{11.2 \text{ hp}}$$

This is the rate at which the gasoline engine does work to supply the electrical energy.

EXAMPLE 20-11

In the preceding example, the load consists of a number of 50 W lamps connected in parallel. How many lamps are used?

In each lamp, $P = VI$; hence the current in each lamp

---

[*] We use $\mathcal{E}I$ rather than VI, since $\mathcal{E}$, an emf, represents the reversible transformation of energy.

$$I = \frac{P}{V} = \frac{50 \text{ W}}{100 \text{ V}} = 0.5 \text{ A}$$

Since the total current is 80 A, the number of lamps is

$$\frac{80 \text{ A}}{0.5 \text{ A/lamp}} = \boxed{160 \text{ lamps}}$$

Check: The resistance of each lamp bulb is 100 V/0.5 A = 200 Ω. Connecting 160 of these lamps in parallel gives a total resistance R_p which we may calculate as follows:

$$\frac{1}{R_p} = \frac{1}{200} + \frac{1}{200} + \cdots + \frac{1}{200} = \frac{160}{200}$$

$$R_p = \frac{200}{160} = 1.25 \text{ Ω}$$

The total current through the system is

$$\frac{V}{\text{net } R} = \frac{100 \text{ V}}{1.25 \text{ Ω}} = 80 \text{ A}$$

which checks with the given data.

Kirchhoff's second law can be used to compute the PD across a capacitor, as in the following example.

EXAMPLE **20-12**

Calculate the charge on the plates of the capacitor in the circuit of Fig. 20-19.

We ignore the capacitor while we first calculate the steady current, using Ohm's law for a complete circuit.

$$I = \frac{\text{net } \mathcal{E}}{\text{net } R} = \frac{16 \text{ V}}{(2 + 5 + 1) \text{ Ω}} = 2 \text{ A}$$

The PD across C is found by taking a Kirchhoff loop around the path $ABYXA$: we have $+10$ V

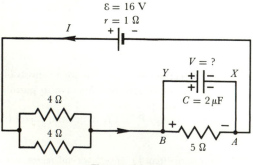

$$\mathcal{E} = 16 \text{ V}$$
$$r = 1 \text{ Ω}$$
$$V = ?$$
$$C = 2 \mu\text{F}$$
$$4 \text{ Ω}$$
$$4 \text{ Ω}$$
$$5 \text{ Ω}$$

Figure 20-19

$-V = 0$, whence $V = 10$ V across the capacitor. (This could also be found by noting that the PD from X to Y is the same as from A to B, and the latter is equal to IR, or 10 V.) Finally, we calculate the charge on the capacitor from C and V:

$$Q = CV = (2 \times 10^{-6} \text{ F})(10 \text{ V})$$

$$= \boxed{20 \times 10^{-6} \text{ C}}$$

There is $+20 \ \mu\text{C}$ on plate Y, and $-20 \ \mu\text{C}$ on plate X.

20-6   The ammeter and the voltmeter

As their names indicate, the ammeter measures current (expressed in some multiple of the ampere), and the voltmeter measures potential difference (expressed in some multiple of the volt). The instruments in common use are both based on the same moving-coil galvanometer, which is fundamentally a current-indicating device depending on magnetic forces. We shall discuss the magnetic force on a wire carrying current in Chap. 21, and we shall study the moving-coil galvanometer in detail in Chap. 22; at this time we accept the galvanometer as a working instrument and inquire how to use it in practical measurement of current and potential difference.

Small currents can be measured by a galvanometer without further modification. For instance, a basic galvanometer movement might have a coil of resistance 100 Ω, and it might require 200×10^{-6} A (200 microamperes, or 200 μA) for a full-scale deflection. A current of 160×10^{-6} A would give a reading 80% of full scale, and so forth. Usually, however, we need to measure larger currents—for instance, the current of 2 A through a lamp bulb or a small motor. The unmodified galvanometer is far too sensitive for such a use.

To make a galvanometer less sensitive, we provide a low-resistance bypass to carry most of the current. Such a resistor, called a *shunt,*

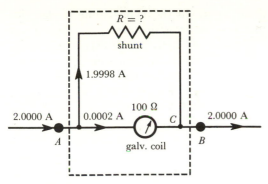

Figure 20-20 Conversion of a galvanometer into an ammeter, using a low-resistance shunt.

is connected in parallel with the galvanometer coil (Fig. 20-20). Let us compute the value of the shunt resistor needed to convert our galvanometer into an *ammeter* reading 2 A at full scale deflection. Using a path ACB, we find that the PD from A to B is $(2 \times 10^{-4} \text{ A})(100 \,\Omega) = 2 \times 10^{-2}$ V. Applying Kirchhoff's first law at A we find that there is 1.9998 A through the shunt when the needle is deflected full-scale by the current of 0.0002 A through the coil. By Ohm's law,

$$R_{\text{shunt}} = \frac{2 \times 10^{-2} \text{ V}}{1.9998 \text{ A}} = 0.010 \,\Omega$$

The shunt has a resistance of about 0.01 Ω, a very small resistance indeed. Another way of looking at the action of the shunt is to consider how the current divides. In Fig. 20-20, the shunt has a resistance which is about $\frac{1}{10\,000}$ that of the coil. If the wire at A is carrying 2 A, the current splits in the ratio of 10 000 to 1; that is, $\frac{1}{10\,001} \times 2$ A through the galvanometer coil, and $\frac{10\,000}{10\,001} \times 2$ A through the shunt. To three-figure accuracy, $\frac{1}{10\,001} \times 2 = 2.00 \times 10^{-4}$ A, and the instrument reads full-scale, as desired. After adjustment of the shunt, the final step in the conversion of our galvanometer into an ammeter is to mark off the scale with values ranging from 0 to 2 A, so that it will be direct-reading. The shunt is usually mounted inside the case, out of sight.

To construct a *voltmeter*, we modify a gal-

vanometer in a somewhat different way. The original instrument could be used to measure a small PD: full-scale deflection requires 200×10^{-6} A through 100 Ω, and the required PD is given by

$$V = IR = (200 \times 10^{-6} \text{ A})(100 \,\Omega) = 0.020 \text{ V}$$

The meter reads 20 mV full-scale, and its resistance is 100 Ω.

To measure a dry cell's terminal voltage, however, would require a larger range. This is made possible by a high-resistance resistor called a *multiplier,* connected in series with the galvanometer coil. In Fig. 20-21 we show the construction of a voltmeter which reads 2 V full-scale, starting with the same galvanometer used for the ammeter of Fig. 20-20. The galvanometer coil and the multiplier are in series, and at full-scale deflection each carries the current 2×10^{-4} A. By Ohm's law, the resistance of the circuit from A to B is

$$\Sigma R = \frac{2 \text{ V}}{2 \times 10^{-4} \text{ A}} = 10\,000 \,\Omega$$

Hence

$$100 \,\Omega + R_{\text{multiplier}} = 10\,000 \,\Omega$$
$$R_{\text{multiplier}} = 9900 \,\Omega$$

Thus we need to connect a resistor of 9900 Ω in series with the coil, in order to obtain a voltmeter of range 0 to 2 V. If the applied potential difference is reduced from 2 V to 1.6 V, the current through both coil and multiplier is reduced, and the deflection is 80% of full scale. We mark off the scale with

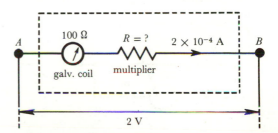

Figure 20-21 Conversion of a galvanometer into a voltmeter, using a high-resistance multiplier.

values ranging from 0 to 2 V, so that the voltmeter is direct-reading. The multiplier is usually mounted inside the case, out of sight.

The voltmeter we have described has a high resistance (10 000 Ω total), and it diverts no more than 0.0002 A from the device to which it is connected. An *ideal voltmeter* is one whose resistance is infinite, and which therefore draws no current from the circuit whose PD is being measured. Voltmeters in common use in elementary physics laboratories have resistances such that they require about 0.005 to 0.01 A for full-scale deflection. Voltmeters used for research may need as little as 10^{-5} A; if even less disturbance of the circuit is required, vacuum-tube voltmeters are available which require currents of 10^{-9} A or less.

To summarize: ammeters and voltmeters are modifications of the same moving-coil galvanometer. Ammeters have very low resistance and are connected in series with the circuit element for which the current is to be measured. Voltmeters have high resistance and are connected in parallel with the circuit element for which the PD is to be measured.

20-7 The Wheatstone bridge

As final illustrations of electric circuits, we describe two instruments in which the galvanometer is used only as a "null" indicator, to show when the current is zero. The first of these devices is the Wheatstone bridge[*] for measurement of resistance.

First let us closely examine the "ammeter-voltmeter" method of measuring resistance. If in Fig. 20-22a the voltmeter reads 10.0 V and the ammeter reads 2.00 A, it would seem that $R = (10.0 \text{ V})/(2.00 \text{ A}) = 5.00 \text{ Ω}$. This would be true if the voltmeter were ideal. If, however, the voltmeter resistance is 1000 Ω, we know that the current through the voltmeter is

$$I = \frac{V}{R} = \frac{10.0 \text{ V}}{1000 \text{ Ω}} = 0.01 \text{ A}$$

and hence only 1.99 A goes through the unknown resistor R. The correct value of R is given by

$$R = \frac{V}{I} = \frac{10.0 \text{ V}}{1.99 \text{ A}} = 5.03 \text{ Ω}$$

We can obtain the correct value of R only by allowing for the effect of the voltmeter, i.e., by knowing the resistance of the voltmeter.

A similar error arises in the connection of Fig. 20-22b; this time the current is correctly measured, but the 10.0 V reading of the voltmeter includes the IR drop in the ammeter, as well as that in R. To find the correct value of R, we would have to know the resistance of the ammeter.

The Wheatstone bridge avoids this difficulty. The unknown resistance is determined

---

[*] Invented by Sir Charles Wheatstone (1802–1875), who used the bridge to locate breaks and short circuits in telegraph lines.

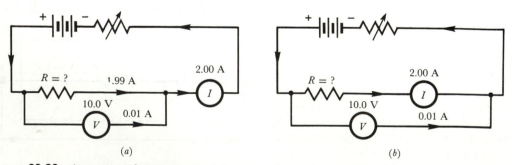

(a)   (b)

Figure 20-22 Ammeter-voltmeter method for measuring resistance. For either circuit, the unknown resistance is not given exactly by $R = V/I$.

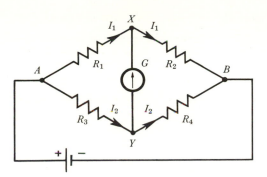

Figure 20-23 Wheatstone bridge circuit.

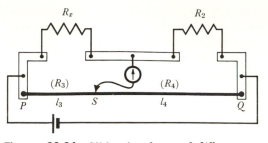

Figure 20-24 Slide-wire form of Wheatstone bridge.

in relation to three other resistances which are assumed known. In Fig. 20-23 the current entering at A splits in some proportion, with I_1 going through the top branch and I_2 through the bottom branch. If there is no current through the galvanometer G, what must be the relation between R_1, R_2, R_3, and R_4? Since (by hypothesis) there is no current between X and Y, all of I_1 goes on from R_1 to R_2; likewise, I_2 passes through both R_3 and R_4. Since X and Y are at the same potential,* the IR change from A to X must equal the change from A to Y.

Therefore,

$$I_1 R_1 = I_2 R_3$$

Similarly,

$$I_1 R_2 = I_2 R_4$$

Dividing equals by equals gives

$$\frac{I_1 R_1}{I_1 R_2} = \frac{I_2 R_3}{I_2 R_4}$$

whence

$$\frac{R_1}{R_2} = \frac{R_3}{R_4} \qquad (20\text{-}6)$$

This is the equation of balance for a Wheatstone bridge. Any unknown "arm" of the bridge can be found if the other three arms are known.

In a common form of Wheatstone bridge, R_3 and R_4 are segments of a uniform wire stretched along a meter stick (Fig. 20-24).

*If there were a PD between X and Y, there would be a current through the galvanometer.

According to Eq. 19-6, $R = \rho l / A$ and, if the cross section is uniform, R is proportional to l, and so R_3/R_4 equals l_3/l_4. This takes care of two of the resistances in the equation of balance—only the ratio need be known. R_2 is a standard resistance of known value, and R_x is the unknown resistance. The equation becomes

$$\frac{R_x}{R_2} = \frac{l_3}{l_4}$$

from which R_x is easily found by solving a slide-rule proportion.

Since the resistivity of a metal depends on temperature (Sec. 19-6), a Wheatstone bridge can be used to measure temperature. In this case R_x is a coil of wire, usually platinum or nickel, whose resistance is found by balancing the bridge. The temperature is then calculated from Eq. 19-7. One test for the purity of distilled water involves measurement of the resistance of a standard-sized sample by means of a Wheatstone bridge. The temperatures encountered in spacecraft are transmitted to earth by radio signals; the sensing device can be some form of Wheatstone bridge. Some lie detectors measure skin resistance by a Wheatstone bridge; it is presumed that a moist skin (indicated by lower electrical resistance) can be correlated with the subject's psychological state.

20-8 The potentiometer

We have already pointed out that an ordinary voltmeter cannot be used to measure the emf of a cell. Such a voltmeter measures the terminal voltage correctly enough, but, since the cell must supply some slight current through the voltmeter, the reading is given by $V = \varepsilon - Ir$, which is less than the emf ε.

EXAMPLE **20-13**

A voltmeter of resistance 1000 Ω is connected to an "old" dry cell whose emf is 1.500 V and whose internal resistance is 100 Ω (Fig. 20-25). What is the reading of the voltmeter?

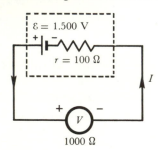

Figure 20-25 The voltmeter reads less than the emf of the cell.

The net resistance for the entire circuit is 1000 Ω + 100 Ω = 1100 Ω. Hence

$$I = \frac{\text{net } \varepsilon}{\text{net } R} = \frac{1.500 \text{ V}}{1100 \text{ }\Omega} = 0.001364 \text{ A}$$

Now applying Ohm's law to the voltmeter alone,

$$V = IR = (0.001364 \text{ A})(1000 \text{ }\Omega)$$

$$= \boxed{1.364 \text{ V}}$$

Compared with the emf of 1.500 V, this represents a considerable error, due entirely to current through the voltmeter.

The potentiometer circuit[*] measures an emf by a null method which draws no current from the cell or other seat of emf. In the basic potentiometer circuit of Fig. 20-26, a "working battery" sets up a steady current through a slide wire *BCD* in the closed loop *ABCDEA*. The sliding contact *C* is adjusted so that there is no current through the galvanometer. Suppose that the unknown cell has an emf ε

[*]Invented by the German physicist Johann Poggendorff (1796–1877).

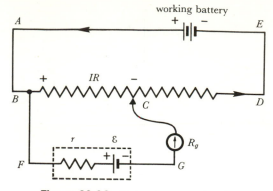

Figure 20-26 Potentiometer circuit.

and internal resistance r, and let the galvanometer's resistance be R_g. There is no current through r or R_g when the potentiometer is in balance; therefore there are no IR drops in r and R_g, and no little + and − signs are shown for them. We now apply Kirchhoff's second law to the path *BCGFB*:

$$-IR + (0 \text{ A})(R_g) + \varepsilon + (0 \text{ A})(r) = 0$$

$$\varepsilon = IR$$

When the potentiometer is in balance, the emf of the cell equals the IR drop in a portion of the slide wire. The internal resistance of the cell is of no consequence, since there is no current through it.

In practice, the unknown cell is compared with a *standard cell* whose emf ε_s is known (Fig. 20-27). The potentiometer is first standardized by throwing the switch to position 1 and adjusting the slide wire for balance,

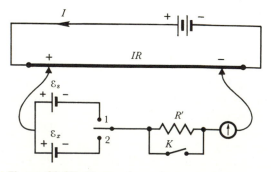

Figure 20-27 Comparison of two emf's, using a slide-wire potentiometer.

giving a length l_1 (proportional to R_1). Then the potentiometer is balanced with the switch thrown to position 2, giving a length l_2 (proportional to R_2).

$$\frac{\mathcal{E}_x}{\mathcal{E}_s} = \frac{IR_2}{IR_1} = \frac{R_2}{R_1} = \frac{l_2}{l_1}$$

$$\mathcal{E}_x = \mathcal{E}_s\left(\frac{l_2}{l_1}\right)$$

Several forms of potentiometer have been developed for special purposes; many are made direct-reading (in volts) by proper adjustment of the working current I. One common type of standard cell is the Weston cell, which uses as its negative electrode a cadmium-mercury amalgam instead of the more common zinc. Such a cell has an emf which is reproducible and is relatively independent of temperature. At 20°C, $\mathcal{E}_s = 1.0183$ V for the saturated form of Weston cell (Fig. 20-28). These cells have high internal resistance, but this is allowable because the current is zero when the potentiometer is balanced. To prevent undesired electrolysis within the cell, care must be taken not to draw more than about 0.0001 A from a standard cell. The protective resistor R' (which may be 5000 Ω or more) is in series with the standard cell while initial adjustments are being made. When the potentiometer is almost balanced, R' is shorted out by closing the key K, and the final critical adjustment of the slide wire is made to bring the galvanometer current to zero.

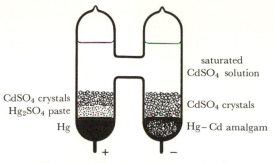

saturated CdSO$_4$ solution

CdSO$_4$ crystals
Hg$_2$SO$_4$ paste

Hg

CdSO$_4$ crystals

Hg–Cd amalgam

Figure 20-28   Weston standard cell.

The potentiometer is a most valuable tool for chemists, biologists, and medical researchers. Since the emf of a cell depends on its electrodes and on the nature and concentration of the electrolyte, measurement of emf gives valuable information to the chemist. Also, numerous potentials exist in living tissue. For instance, there is a steady emf of about 50 mV between the inside and the outside of the giant axon nerve cell[*] of the squid *Loligo forbesi,* when no nerve impulse is being transmitted. This emf rises sharply to about 95 or 100 mV during nerve action. These potential differences can be measured best with a potentiometer, since the potentiometer requires essentially no current from the source and the high internal resistance of biological tissue has no effect on the measurements.

_____

[*] We must distinguish between a biological "cell" and an electric "cell." In this case, the nerve cell is also an electric cell in the same sense that a dry cell is a "cell."

■ SUMMARY

Kirchhoff's first law is based on the law of conservation of charge, and it states that the algebraic sum of currents entering any point in a circuit is zero. Kirchhoff's second law is based on the law of conservation of energy, and it states that the algebraic sum of the changes in potential around any closed path is zero. These laws offer a systematic way of applying Ohm's law to a circuit.

The internal resistance of a cell depends on its physical construction and its past history. Because of internal resistance, the terminal voltage of a cell differs from its emf when there is current through the cell. If the cell is discharging, $V = \mathcal{E} - Ir$, and, if the cell is being charged, $V = \mathcal{E} + Ir$.

Formulas for the equivalent resistance of resistors in series and in parallel can be used to reduce many circuits to a simple circuit. Not all circuits can be studied by such elementary means, however. Emf's in series can be added up, but there is no simple way to obtain the equivalent emf to represent a number of emf's connected in parallel, unless they are identical in emf and internal resistance. For a simple circuit for which a net emf and net resistance exist, Ohm's law states that $I = $ net $\mathcal{E}/$net R.

The deflection of a moving-coil galvanometer is proportional to the current through its coil. An ammeter is a galvanometer with a low-resistance shunt in parallel with the coil. A voltmeter is a galvanometer with a high-resistance multiplier in series with the coil. A voltmeter will usually have some effect on the circuit to which it is connected unless the resistance of the voltmeter is sufficiently large.

A Wheatstone bridge is used to make precise measurement of resistance. The method requires one standard resistor whose resistance is known, and two other resistors for which the ratio of resistances is known. This pair of resistors may be two segments of a uniform slide wire.

The potentiometer is used for measuring the value of an emf without drawing current from the seat of emf. The disturbing effect of a cell's internal resistance is thus eliminated.

■ CHECK LIST

Kirchhoff's first law
Kirchhoff's second law
terminal voltage
polarization of a cell
$V = \mathcal{E} \pm Ir$
$R_s = R_1 + R_2 + R_3 + \cdots$
$\dfrac{1}{R_p} = \dfrac{1}{R_1} + \dfrac{1}{R_2} + \dfrac{1}{R_3} + \cdots$
simple circuit
non-simple circuit
Ohm's law for any
  part of a circuit
$I = V/R$

Ohm's law for a
  complete simple circuit
$I = $ net $\mathcal{E}/$net R
galvanometer
ammeter
voltmeter
shunt
multiplier
Wheatstone bridge
$\dfrac{R_1}{R_2} = \dfrac{R_3}{R_4}$
potentiometer

■ QUESTIONS

20-1 In Fig. 20-29, place $+$ and $-$ signs at the ends of the resistors A, B, C, D, and E, to indicate the polarity of the PD across each resistor.

20-2 Check Kirchhoff's second law around the path *EFDGHKAE* of Fig. 20-4.

20-3 Why is the internal resistance of a cell not a constant?

20-4 Under what circumstances is the terminal voltage of a battery greater than its emf?

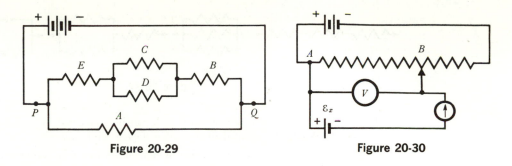

Figure 20-29   **Figure 20-30**

20-5   Can the terminal voltage of a battery be zero?

20-6   Is the mechanical construction of a voltmeter essentially different from that of an ammeter?

20-7   How does the electrical circuit of a voltmeter differ from that of an ammeter?

20-8   If the working battery of a Wheatstone bridge runs down a little so that its terminal voltage decreases, will this affect the value of R_x calculated from Eq. 20-6?

20-9   In what way is a potentiometer superior to a voltmeter for measuring a PD? In what way is a voltmeter superior?

20-10   The circuit of Fig. 20-30 is a potentiometer which requires no standard cell. Show that if the sliding contact B is adjusted so that the galvanometer reads zero, the voltmeter reading equals the emf of the unknown cell $\mathcal{E}_x$.

20-11   Why would it ruin a standard cell if you tried to measure its emf with a low-resistance voltmeter?

■ PROBLEMS

20-A1   What is the PD between points F and P of Fig. 20-4?

20-A2   What is the PD between points E and N in Fig. 20-4?

20-A3   If point H in Fig. 20-4 is grounded instead of point F, what is the potential of point C?

20-A4   In Fig. 20-31, what is the PD across each of the three resistors?

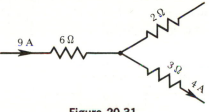

Figure 20-31

20-A5   A 30 Ω resistor is connected in series with a parallel combination of 20 resistors, each of 100 Ω. What is the net resistance of the circuit?

20-A6   Five 10 Ω resistors are connected   (a) in series and   (b) in parallel. Compute the equivalent resistance in each case.

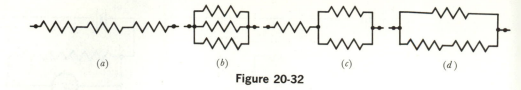

Figure 20-32

20-A7 Three equal resistors, each of resistance 6 Ω, can be connected in four different ways (Fig. 20-32). What is the equivalent resistance of each combination?

20-A8 A string of Christmas tree lights consists of eight bulbs in series, connected to a 120 V source of PD. (a) What is the PD across each bulb? (b) If one bulb is removed from its socket while the string is plugged in, what is now the PD across the socket terminals? (c) While this bulb is removed, what is the PD across the remaining seven bulbs?

20-A9 The emf of an automobile storage battery is 12.00 V, and its internal resistance is 0.010 Ω. What is the terminal voltage of the battery when the starter motor is drawing 140 A from the battery?

20-A10 A test for a fresh dry cell is to connect the terminals directly to the terminals of a heavy-duty ammeter. [*Warning:* Don't try this with an ammeter of insufficient range. (Why not?)] If the ammeter reads at least 30 A the cell is considered fresh. What is the internal resistance of a fresh dry cell?

20-A11 In Fig. 20-23, $R_1 = 200$ Ω, $R_2 = 100$ Ω, $R_3 = 46.2$ Ω. For what value of R_4 will the bridge be balanced?

20-A12 In Fig. 20-24, $R_x = 30.0$ Ω and $R_2 = 20.0$ Ω. The slide wire PQ is 100 cm long with the 0 cm mark at P and the 100 cm mark at Q. When the bridge is balanced, what is the reading of the slider S, in centimeters?

20-A13 In Fig. 20-24, the slide wire PQ is 100 cm long, and the distance PS is 20 cm. The value of R_2 is 10 Ω; what is the value of R_x?

20-B1 What is the resistance between P and Q in Fig. 20-29, if $A = 8$ Ω, $B = 3$ Ω, $C = 6$ Ω, $D = 6$ Ω, and $E = 2$ Ω?

20-B2 What is the resistance between P and Q in Fig. 20-29, if $A = 15$ Ω, $B = 18$ Ω, $C = 6$ Ω, $D = 3$ Ω, and $E = 10$ Ω?

20-B3 Show that the resistance between A and K in Fig. 20-10a is 10 Ω as stated in the text.

20-B4 What is the equivalent resistance between C and D in Fig. 20-10b?

20-B5 The battery in Fig. 20-29 has terminal voltage 12 V. What is the current through the battery, if $A = 8$ Ω, $B = 4$ Ω, $C = 6$ Ω, $D = 3$ Ω, and $E = 2$ Ω?

20-B6 (a) In Fig. 20-33, calculate the current through each resistor. (b) At what rate, in watts,

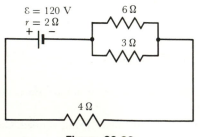

Figure 20-33

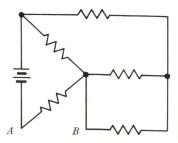

Figure 20-34

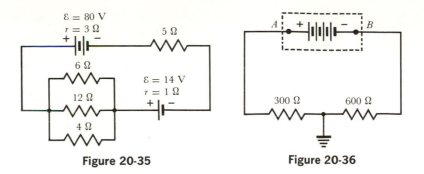

Figure 20-35   **Figure 20-36**

is energy being transformed from chemical to electrical form in the battery? (*c*) Make a table of power dissipated in heat in the various circuit elements and show that the sum of these powers agrees with your answer to part (*b*).

20-B7 Solve Prob. 20-B6, with the battery changed to one with ε 36 V and $r = 3\ \Omega$.

20-B8 In Fig. 20-34 the battery has $\varepsilon = 20$ V and $r = 0.4\ \Omega$, and each resistor is 6 Ω. (*a*) Calculate the current through the battery. (*b*) If *A* and *B* are connected by a wire of zero resistance, what will the current through this wire be?

20-B9 In Fig. 20-35, (*a*) what power is dissipated in the 5 Ω resistor? (*b*) What is the current in the 6 Ω resistor? (*c*) What is the terminal voltage of each battery?

20-B10 In Fig. 20-36, *AB* represents a radio battery of emf 9 V and negligible internal resistance. Calculate the potential of each terminal of the battery relative to the ground.

20-B11 In Fig. 20-36, suppose the battery has emf 15 V and internal resistance 100 Ω. (*a*) What is the terminal voltage of the battery? (*b*) What is the potential of point *A* relative to the ground?

20-B12 In the amplifier circuit of Fig. 20-37, electrons flow from cathode *K* through the tube to the plate *P* and around through the resistor and the battery back to *K*. No electrons flow to or from the grid *G*. The electron current is 3 mA. Compute the potentials relative to the ground of (*a*) the cathode, (*b*) the plate, and (*c*) the grid.

20-B13 Four batteries are connected in series, as shown in Fig. 20-38. Compute the terminal voltage of each battery, and verify Kirchhoff's second law by showing that the algebraic sum of the terminal voltages equals zero.

20-B14 Solve Prob. 20-B13, with the 8 V battery's polarity reversed.

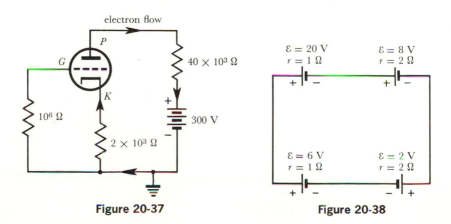

Figure 20-37   **Figure 20-38**

20-B15 Two cells are connected in parallel, + to + and − to −. Cell 1 has $\mathcal{E} = 1.5$ V and $r = 1\ \Omega$; cell 2 has $\mathcal{E} = 2.1$ V and $r = 2\ \Omega$. What is the common terminal voltage of the cells when connected together in this way, with no external load?

20-B16 Two lamp bulbs, each designed to use 100 W when connected to a 120 V source, are connected in series across a 120 V source of PD. Assume that at the new (cooler) temperature each bulb's resistance is half of its value at the normal temperature. Calculate the total power used by the two bulbs when connected in series.

20-B17 In Fig. 20-6, (*a*) what is the PD between points *F* and *G*? (*b*) The battery charger has an emf of 20 V. What is its internal resistance?

20-B18 A milliammeter reads 1 mA full scale and has a resistance of 50 Ω. How would you convert this to a voltmeter reading 10 V full scale, by use of a single additional resistor? Draw the circuit and compute the value of the resistor that is needed.

20-B19 A biophysicist needs a voltmeter which reads 20 V full scale and has available a milliammeter of resistance 100 Ω which requires 5 mA for full-scale deflection. Draw a circuit showing how a single additional resistor can be used to convert the milliammeter into the desired voltmeter. Calculate the value of the resistor that is needed.

20-B20 What is the value of a single resistor which would convert the milliammeter of Prob. 20-B18 into an ammeter reading 5 A full scale? Draw the circuit and compute the value of the resistor that is needed.

20-B21 A milliammeter of resistance 100 Ω requires 5 mA for full-scale deflection. Draw a circuit showing how a single additional resistor can be used to convert the milliammeter into an ammeter reading 2 A full scale. Calculate the value of the resistor that is needed.

20-B22 (*a*) In Fig. 20-22*a*, the voltmeter (of resistance 2000 Ω) reads 50.0 V and the ammeter (of resistance 0.010 Ω) reads 1.00 A. What is the value of *R*? (*b*) Calculate *R* for the same meter readings, using the circuit of Fig. 20-22*b*.

20-B23 Figure 20-39 shows a Wheatstone bridge that is almost balanced, with point *C* grounded. (*a*) Calculate the potential of point *A*. (*b*) Calculate the potential of point *B*. (*c*) If a galvanometer is connected between *B* and *A*, what is the direction of the current through it? (*d*) For what value of the resistor *BC* would the bridge be in balance?

20-B24 To what value should the 3 Ω resistor in Fig. 20-39 be changed, in order to make the potential of point *B* equal to that of point *A*?

20-B25 A coil of iron wire is wound on an insulating support and immersed in an oil bath. The coil is made one arm of a Wheatstone bridge, and when the oil is at 0°C the bridge is balanced as shown in Fig. 20-40. If the oil bath is heated to 70°C, how far and in what direction must the sliding contact be moved? (*Hint:* See Table 19-1.)

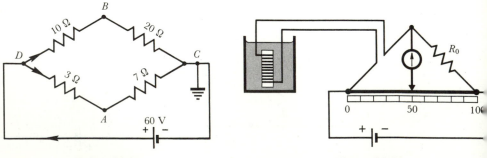

Figure 20-39   **Figure 20-40**

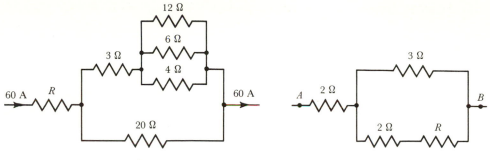

Figure 20-41   Figure 20-42

20-B26 The slide wire BD of the potentiometer of Fig. 20-26 is 11.000 m long and its resistance is 20 Ω. The working battery's terminal voltage is 1.540 V. What length of wire BC is needed to balance the potentiometer when the cell FG is a standard cell of emf 1.018 V and internal resistance 100 Ω?

20-C1 A 100 W, 120 V lamp bulb is connected in parallel with a 40 W, 120 V lamp bulb. What is their combined resistance?

20-C2 Compute the current through the 6 Ω resistor in Fig. 20-41.

20-C3 Compute the power dissipated in the 12 Ω resistor of Fig. 20-41.

20-C4 In Fig. 20-42, calculate the value of the unknown resistor R which makes the total resistance of the circuit from A to B also equal to R.

20-C5 The terminal voltage of a generator is 111 V when it delivers 3 A, and is 105 V when it delivers 5 A. Calculate the emf and the internal resistance of the generator.

20-C6 A voltmeter of range 0 to 10 V requires 0.010 A for full-scale deflection. When connected to a battery of emf 6.09 V, the voltmeter reads 6.00 V. Calculate the internal resistance of the battery.

20-C7 When a voltmeter of resistance 100 Ω is connected to a dry cell, it reads 1.5200 V. If a 100 Ω resistor is connected in parallel with the cell and the voltmeter, the reading falls to 1.5100 V. Compute the emf and internal resistance of the cell. (*Hint:* Use simultaneous equations.)

20-C8 Two identical resistors (R) and a cell ($\mathcal{E}$, r) are available. Currents through the cell are measured for various loads: I_1 for one resistor; I_2 for two resistors in series; I_3 for two resistors in parallel. Show that $2I_1I_2 + I_1I_3 = 3I_2I_3$.

20-C9 The reciprocal of resistance is conductance, denoted by G. The mks unit for conductance is the mho (read it backward!). (*a*) What is the conductance of an electric iron which develops 1000 W when connected to a PD of 100 V? (*b*) Derive a formula for the conductance G_p of a group of conductors G_1, G_2, G_3, ... connected in parallel.

20-C10 An automobile battery has an emf of 12.50 V and internal resistance of 0.05 Ω. The headlights have total resistance of 2.00 Ω (assumed constant). What is the PD across the lamp bulbs (*a*) when they are the only load on the battery, and (*b*) when the starter motor is operated, taking an additional 30 A from the battery?

20-C11 A galvanometer reads 0.01 A full scale and has resistance 20 Ω. It is desired to construct an ammeter reading 50 A full scale, by attaching a constantan shunt between the terminals of the meter. The terminals are 8 cm apart. What should be the cross section of the shunt, which is in the form of a straight bar?

For Further Study

20-9 Kirchhoff's laws applied to non-simple circuits

We have seen that not all circuits can be simplified by use of the formulas for resistances in series and in parallel. To solve problems involving non-simple circuits, we use Kirchhoff's systematic formulation of Ohm's law. The method can best be explained by an example. We wish to find the current in each resistor of Fig. 20-43. Simple methods fail, since the two seats of emf are not in series (they do not carry the same current), nor are they exactly in parallel (their terminal voltages are not the same). There are three unknown currents which we call I_1, I_2, and I_3; therefore we need three equations in this example. The general procedure is as follows:

1. Assume arbitrary current directions in each conductor; if you choose wrongly, this will become apparent by a $-$ sign in the answer.

2. Label each resistor and emf with little $+$ and $-$ signs, the signs for the resistors being determined by the assumed directions of the currents.

3. Write Kirchhoff's second law (the loop theorem, p. 412) for as many loops as possible, making sure that each loop contains at

least one new circuit element not already used in another loop.

4. Write Kirchhoff's first law for all but one of the junction points (nodes).

5. You will find that you have exactly enough equations to solve simultaneously for the unknown currents.

In this example, we use two loops and one node.

$ABEFA$: $-8I_3 - 2I_1 - 1I_1 + 7 = 0$
$BCDEB$: $+1I_2 - 36 + 3I_2 + 8I_3 = 0$
Node at B: $I_1 + I_2 - I_3 = 0$

Upon simplification the equations become

$$\begin{cases} 3I_1 \qquad\quad + 8I_3 = 7 \\ \qquad 4I_2 + 8I_3 = 36 \\ I_1 + I_2 - I_3 = 0 \end{cases}$$

Especially if a computer is available, these equations can easily be solved for the three unknown currents. The result is

$$I_1 = -3 \text{ A}, \quad I_2 = 5 \text{ A}, \quad I_3 = 2 \text{ A}$$

Since I_1 turned out to be negative, we know that our choice for its direction was wrong and I_1 actually is from A to F. Our final solution is shown in Fig. 20-44. Note how Kirchhoff's first law is satisfied at B and at E.

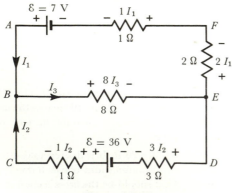

Figure 20-43

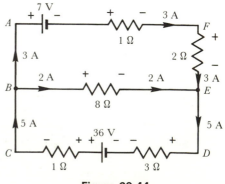

Figure 20-44

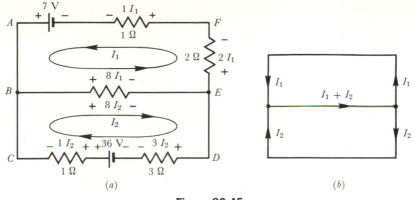

Figure 20-45

As a final check, the PD between E and B works out to be the same for any of the three paths, $EDCB$, EB, or $EFAB$ (Prob. 20-C12).

You should guard against writing down equations that are superfluous. Thus the loop $ACDFA$ correctly leads to

$$+1I_2 - 36 + 3I_2 - 2I_1 - 1I_1 + 7 = 0$$

but this equation is nothing new; it is simply the sum of the equations for the two loops already used. This is why each loop equation must have at least one new circuit element. Likewise, a node equation could be written for point E, but it would be algebraically equivalent to the node equation already written for B.

If a computer is not available, algebraic complexity can be considerably reduced by using the *loop* currents as unknown. In Fig. 20-45*a* we assume arbitrary directions for I_1 in the upper loop and I_2 in the lower loop. According to our diagram, the current in the central 8 Ω resistor is the algebraic sum of I_1 and I_2. This technique automatically

satisfies Kirchhoff's first law at the junction points (Fig. 20-45*b*). We next label the emf's and the resistors with little $+$ and $-$ signs; the 8 Ω resistor is traversed simultaneously by two currents, so we draw two IR drops in it, one for each current.

Now we apply Kirchhoff's second law to each loop, getting a pair of simultaneous equations:

$ABEFA:$ $-8I_1 - 8I_2 - 2I_1 - 1I_1 + 7 = 0$
$BCDEB:$ $+1I_2 - 36 + 3I_2 + 8I_2 + 8I_1$
$$= 0$$

Upon simplification, the equations become

$$\left\{ \begin{matrix} 11I_1 + 8I_2 = 7 \\ 8I_1 + 12I_2 = 36 \end{matrix} \right\}$$

whence $I_1 = -3$ A and $I_2 = +5$ A. The current through the 8 Ω resistor is $I_1 + I_2 = -3$ A $+ 5$ A $= +2$ A, in the direction of the assumed I_2. Our final solution is the same as obtained by the more direct method using three equations; the currents are shown in Fig. 20-44.

■ PROBLEMS

20-C12 Compute the PD from B to E in Fig. 20-44 by each of the three possible paths.

20-C13 Solve the circuit of Sec. 20-9, assuming other directions for I_1 and I_2.

20-C14 Make a power check (see p. 407) for the circuit of Fig. 20-44. Show that the net rate of heat production is equal to the net rate of conversion of chemical energy into electric energy.

20-C15 (a) Compute the current through a 155 V battery, of negligible internal resistance, connected between points A and B of Fig. 20-10b. (*Hint:* With the battery $\mathcal{E}$ drawn in the circuit, there are three loops: $ACDA$; $CBDC$; $B\mathcal{E}ADB$. Use six simultaneous equations, or three equations if the loop currents are used as unknowns.) (b) From the current found in part (a), calculate the resistance between A and B.

20-C16 (a) Calculate the current through each battery in the circuit of Fig. 20-12. (b) Calculate the terminal voltage (the same for each battery).

20-C17 (a) What is the current through the 2 Ω resistor in Fig. 20-46? (b) What is the terminal voltage of each battery?

20-C18 (a) Compute the PD between X and Y in Fig. 20-47, and tell which is at the higher potential. (*Hint:* This part of the problem can be solved without use of simultaneous equations.) (b) Points X and Y are now connected by a 7 Ω resistor. What is the current through this resistor?

20-C19 A battery of emf $\mathcal{E}$ and internal resistance r is connected to a load R. (a) Show that the power P dissipated in the load resistor is $\mathcal{E}^2 R(R + r)^{-2}$. (b) Set $dP/dR = 0$ to find the relation between R and r for maximum transfer of power from the battery to the load. (c) Show that at most 50% of the chemical energy in the cell can be delivered as thermal energy in the load resistor.

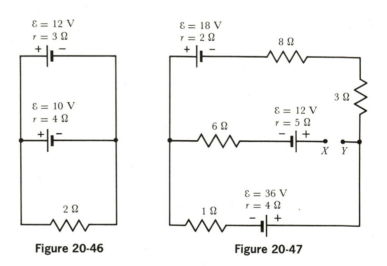

Figure 20-46   **Figure 20-47**

■ REFERENCES

1 Smith, O. H., "An Inexpensive High Resistance Voltmeter," *Am. J. Phys.,* **19**, 224 (1951).

2 Busse, J., "Thermometry," in Glasser, O., *Medical Physics* (Year Book Publishers, 1944), pp. 1561–64. A discussion of the Wheatstone bridge and potentiometer.

3 Baker, P. H., "The Nerve Axon," *Sci. American,* Mar., 1966, p. 74.

21-1   Magnetic field

The ancients knew that a certain naturally occurring ore called magnetite (impure iron oxide, Fe_3O_4) could serve as a navigational aid, for a piece of the material pointed in a north-south direction when suspended by a thread. These "lodestones" were first found near Magnesia, an ancient city in Asia Minor, and became known as *magnets*. As early as A.D. 121 the Chinese knew that a piece of ordinary iron metal could be "magnetized" by bringing it near a lodestone, and navigation with the aid of the magnetic compass dates back ten centuries into the past. Without knowing it, the early natural philosophers who studied magnetism were dealing with electric forces of a kind that is different from the electrostatic (Coulomb) forces exerted by a rubbed piece of amber which is at rest. The early investigators studied *magnetic forces,* and, as we stated in Chap. 2, these forces are exerted by charges in motion. The lodestone fits into this scheme; the magnetic force is that between certain spinning orbital electrons in the Fe_3O_4 and other spinning orbital electrons in the iron which it attracts. When a compass needle points north, it does so be-

cause of a magnetic force between some spinning orbital electrons in the iron atoms of the needle, and (presumably) electrons, protons, or ions which move in a circular path deep within the earth. Our study of magnetism will consist of the study of forces exerted by moving charges.[*]

From what has been said, it would seem that the magnetic properties of matter are complex. The basic ideas come to the fore when we consider isolated charges moving in a vacuum—an electron on its way to a television picture tube's front surface, or a proton moving in a cyclotron. The magnetic forces are additional forces due to motion, over and above any Coulomb forces which would act on the charge even if it were stationary. To discuss magnetic forces in detail, we shall find it convenient to introduce the concept of magnetic field. In Chaps. 4 and 18 we described gravitational field and electric field; in each case the field is explored by measuring the force on a suitable test object. The magnitude of a gravitational field is force per unit mass; the magnitude of an electric field

---

[*] The term "electromagnetic force" is used to describe the totality of all electric forces, including electrostatic and magnetic forces.

Electrons flowing in flat, vertical coils give rise to a horizontal magnetic field directed away from the reader. At the center of the apparatus a beam of electrons is emitted upward from an electron gun in a partially evacuated tube. The magnetic force on the moving electrons bends the beam into a circle.

Figure 21-1

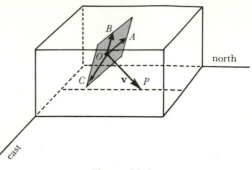

Figure 21-2

is force per unit charge. In similar fashion, we say that a *magnetic field* exists in any region of space where magnetic force would be exerted on a suitable test object. For our test object, we use a moving charge; an electron moving down the axis of a TV picture tube would serve (Fig. 21-1). As one might expect, in a given field the magnetic force on a test charge depends both on the magnitude of the charge and on its velocity. Thus $F \propto Qv$, and we can use the ratio F/Qv (force per unit charge-velocity product) as a measure of the strength of the magnetic field at the point in question.

Now a complicating factor appears. Experiment shows that, at a given point in space (say, the geometrical center of the room in which you are studying), the force on a moving charge depends on the direction of motion as well as on the product Qv. If the only magnetic field is that of the earth and if the TV tube is aimed so that the electrons move downward and to the north (*OP* in Fig. 21-2), then the moving electrons experience no force; but if the stream of electrons is directed at right angles to the line *OP* (direction *OA*, *OB*, *OC*, or any other line in the tilted plane), then the force on the moving charge is a maximum.[*] The only unique

direction in this situation is the direction *OP*, so we call this the direction of the magnetic field at *O*.

> **A moving charge experiences no force when moving parallel to a magnetic field; the force on a moving charge is a maximum when the motion is perpendicular to the magnetic field.**

Keeping these facts in mind, we represent a magnetic field at any point by a vector **B**, which is known as the *magnetic induction* or, more simply, the *magnetic field strength*. The *magnitude* of **B** is defined as

$$B = \frac{F}{Qv} \qquad (21\cdot1)$$

where F is the magnetic force on a test charge Q moving with velocity v in such a direction as to give rise to the maximum force at the point in question. The *direction* of **B** is the direction in which a moving charge would have to move in order to experience zero force. Like gravitational and electric fields, a magnetic field can be represented by field lines drawn through representative points, always in the direction of the vector **B**. The mks unit for B is the *tesla* (T); Eq. 21-1 gives

$$\text{tesla} = \text{T} = \frac{\text{N}}{\text{C}\cdot\text{m/s}} = \frac{\text{N}\cdot\text{s}}{\text{C}\cdot\text{m}}$$

EXAMPLE **21-1**

An electron moving at 10^6 m/s experiences a force of 8×10^{-13} N when moving vertically upward ($\mathbf{v}_1$ in Fig. 21-3), and it experiences a

---

[*] This is an idealized experiment; in actual practice even the maximum force on an electron moving through a TV tube would be too small to observe if the earth's magnetic field were the only field present. Much larger magnetic forces are exerted by the "deflection coils" in the set.

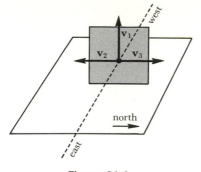

Figure 21-3

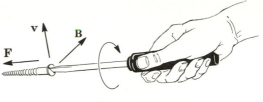

Figure 21-4

The direction of magnetic force on a moving positive charge is in the direction that an ordinary (right-hand) screw would advance when rotated from the direction of v toward the direction of B.

force of 8×10^{-13} N when moving horizontally from north to south ($\mathbf{v}_2$) or from south to north ($\mathbf{v}_3$). For no other direction of motion is the force larger than 8×10^{-13} N. (*a*) What can you say about the direction of the magnetic field? (*b*) What is the magnetic induction?

(*a*) The plane in which the electron moves to receive the maximum force is evidently a vertical north-south plane. The electron would experience no force if moving perpendicular to this plane; hence **B** is directed horizontally, either toward the east or toward the west. (Our present definitions do not allow us to decide between these two possibilities.)

(*b*) The magnitude of **B** is

$$B = \frac{F}{Qv} = \frac{8 \times 10^{-13} \text{ N}}{(1.6 \times 10^{-19} \text{ C})(10^6 \text{ m/s})}$$

$$= 5 \text{ N per C·m/s} = \boxed{5 \text{ teslas}}$$

This is a strong field; the magnetic induction B for the earth's field is about 6×10^{-5} T, and B is about 10 or 20 T for the field produced in the air gap between the highly magnetized iron pole pieces of a very strong electromagnet.

21-2   Direction of magnetic force on a moving charge

To specify the direction of magnetic force on a moving charge, we agree to define the direction of **B** so that the following *right-hand screw rule* is true:

(For example, when the screw in Fig. 21-4 is turned so that vector **v** is rotated toward the direction of vector **B**, the screw advances in the direction of vector **F**.) This procedure is summarized by a vector symbolism called the *cross product:*

$$\mathbf{F} = Q\mathbf{v} \times \mathbf{B} \qquad (21\text{-}2)$$

The magnitude of **F** is:

$$F = QvB \sin \theta \qquad (21\text{-}3)$$

The cross not only indicates the direction of **F**, but also includes the factor $\sin \theta$, where θ is the angle between the directions of **v** and **B**.[*] The magnitude of the cross product is $QvB \sin \theta$, and the direction of $Q\mathbf{v} \times \mathbf{B}$ is perpendicular to **v** and **B**, according to the right-hand screw rule. [Note that for any two vectors, $\mathbf{M} \times \mathbf{N} = -(\mathbf{N} \times \mathbf{M})$, so it is important to preserve the order of the vectors in a cross product.] In Eq. 21-3 we can think of $B \sin \theta$ as the component of **B** perpendicular to **v**. Often **v** and **B** are perpendicular to each other, $\sin \theta = 1$, and then we have

$$F = QvB \qquad (21\text{-}4)$$

This equation, which is equivalent to Eq. 21-1, gives the magnitude of the maximum magnetic force on a charge Q moving with

---

[*] Recall that the dot in the dot product (Sec. 6-4) symbolizes the *cosine* of the angle between two vectors. The dot product of two vectors gives a scalar, whereas the cross product of two vectors gives another vector.

velocity **v** through a magnetic field whose induction is **B**. The direction of the force on a moving charge is perpendicular both to **v** and to **B**. This three-way perpendicularity of **F**, **v**, and **B** is illustrated in cyclotrons (Fig. 30-2). In circular accelerators such as those shown in Figs. 8-11, 30-3, and 30-6 protons move horizontally, in a vertical magnetic field supplied by currents in iron-cored electromagnets. From $F = q\mathbf{v} \times \mathbf{B}$ the electromagnetic force is perpendicular to both **v** and **B** and is directed horizontally, toward the center of the orbit. This force has magnitude QvB and serves as the centripetal force necessary to keep the protons moving in a circle. We shall discuss this and other particle accelerators more fully in Chap. 30.

To repeat: The magnetic force on a moving charge is perpendicular to the direction of motion and to the direction of the magnetic field; and, for maximum effect, the motion should be perpendicular to the direction of the magnetic field.

EXAMPLE **21-2**

In a particle accelerator a proton moves horizontally toward the south through a region in which the magnetic induction B is 10 T in an upward direction. The proton's speed is 0.1 that of light. Find the magnitude and direction of the force on the proton.

The speed is

$$v = (0.1)(3 \times 10^8 \text{ m/s}) = (3 \times 10^7 \text{ m/s})$$

Since **v** is perpendicular to **B**, we use Eq. 21-4 to calculate the magnitude of the force on the proton.

$$F = QvB$$

$$= (1.60 \times 10^{-19} \text{ C})(3 \times 10^7 \text{ m/s})(10 \text{ T})$$

$$= \boxed{4.80 \times 10^{-11} \text{ N}}$$

Although this may seem to be a small force, the mass of a proton is so small that a very large acceleration is produced. The acceleration can be found by Newton's second law.

To find the direction of the magnetic force on the proton, we note that **v** is horizontal, to the

south, and **B** is upward. By the right-hand rule, a screw rotating from the direction of **v** toward that of **B** would advance toward the west.

Hence **F** is $\boxed{\text{horizontal, to the west.}}$

EXAMPLE **21-3**

In Fig. 21-3, assume a positive charge to be moving upward with velocity $\mathbf{v}_1$, experiencing a force horizontally to the north. What is the direction of **B**?

We already found in Example 21-1 that **B** is horizontal, either toward the east or toward the west. We test both assumptions, and find that only for **B** toward the east does the right-hand screw rule give a force toward the north as specified in the problem.

21-3 Force on a current segment in a magnetic field

Aside from charges in accelerators and in TV and radio tubes, most of the moving charges with which we are familiar are the free electrons which drift through a wire carrying a current. A simple geometrical proof allows us to modify the equations of Sec. 21-2 to forms useful for currents. (Corresponding equations are numbered similarly.) Consider a cylindrical segment of wire of length Δl (Fig. 21-5). Suppose there are N free electrons, each of charge q, in this segment of wire, and suppose each electron has a drift velocity **v** toward the left, equivalent to a drift of N positive (conventional) charges to the right. If the segment is placed in a magnetic field of induction B, directed at right angles to the wire, the total magnetic force on the N charges is

$$F = QvB = NqvB$$

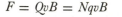

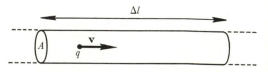

Figure 21-5 A current segment.

Now current = charge/time, and the time required for any charge to travel a distance Δl is $\Delta l/v$. During this time, the cylinder will be emptied of the charge that is in it, and an entirely new set of charges will be in the cylinder. The charge that crosses one face of the cylinder in this time is Nq; hence

$$I = \frac{\text{charge}}{\text{time}} = \frac{Nq}{\Delta l/v} = \frac{Nqv}{\Delta l}$$

$$I\,\Delta l = Nqv$$

and the magnitude of the maximum force on the cylinder is given by

$$F = I\,\Delta l\,B \qquad (21\text{-}4')$$

In a given magnetic field the magnetic force on a wire is not determined solely by the current. The force on a long wire is greater than that on a short wire, and the essential factor is the product $I\,\Delta l$, which is called a *current segment*. In view of Eq. 21-4', we can write

$$B = \frac{F}{I\,\Delta l} \qquad (21\text{-}1')$$

which means that a current segment $I\,\Delta l$, instead of a moving charge Qv, could be used as a test body, and B could be measured as the force per unit current segment. According to Eq. 21-1', B in teslas can be measured in $N/A \cdot m$ as well as in N per $C \cdot m/s$. These units are equivalent, since amperes equal coulombs per second. (In Sec. 21-9 we will introduce still another unit equivalent to the tesla, the weber/m^2.) Just as for a single moving charge, directions are important. The force on a current segment is greatest if the current is perpendicular to the magnetic field. Also, the force on a current segment is perpendicular to the direction of the current and perpendicular to the direction of the magnetic field.

In general, $\mathbf{I}$ and $\mathbf{B}$ are not perpendicular. The vector equation analogous to Eq. 21-2 is

$$\mathbf{F} = I\,\Delta l \times \mathbf{B} \qquad (21\text{-}2')$$

and the magnitude of $\mathbf{F}$ is given by

$$F = I\,\Delta l\,B \sin\theta \qquad (21\text{-}3')$$

where θ is the angle between the directions of $\mathbf{I}$ and $\mathbf{B}$.

EXAMPLE 21-3

A wire carrying 400 A is stretched horizontally in an east-west direction between two towers 60 m apart. The earth's magnetic field is downward and to the north, and the value of B at that location is 7×10^{-5} N/A·m. What is the magnitude of the magnetic force exerted on the wire?

After convincing ourselves that $\mathbf{B}$ is perpendicular to the current $\mathbf{I}$, we use Eq. 21-4'.

$$F = I\,\Delta l\,B = (400\text{ A})(60\text{ m})(7 \times 10^{-5}\text{ N/A}\cdot\text{m})$$

$$= \boxed{1.68\text{ N}}$$

It may seem strange to consider a 60 m wire as a short segment. However, since the wire is straight and all parts of it are subjected to the same field, our procedure can be justified. Consider, for instance, dividing the wire into 6000 parts, each 0.01 m long. The force on each little segment would then be $\frac{1}{6000}$ of 1.68 N, and the total of all the forces on all 6000 segments would be 1.68 N, as we have already obtained by our short cut. If the wire were curved, or if $\mathbf{B}$ were not constant over the length of the wire, then our simplified procedure would be incorrect, and we would use integral calculus to find the total force.

The magnetic force on a current segment is what makes an electric motor work. The moving-coil galvanometer and its modifications, the ammeter and the voltmeter, also depend on magnetic forces acting upon current segments.

EXAMPLE 21-4

A single wire in a household motor may be 10 cm long and carry a current of 2 A. If the magnetic induction B is 0.6 N/A·m, what is the force on the wire?

$$F = I\,\Delta l B = (2\text{ A})(0.1\text{ m})(0.6\text{ N/A}\cdot\text{m})$$

$$= \boxed{0.12\text{ N}}$$

Since this is only about half an ounce, it is evident that in practical motors many current segments must be acted upon simultaneously in order to obtain a useful force.

21-4 Current loops

We now have described two ways in which we might, in principle, determine the magnitude and direction of a magnetic field at any point. We can measure force per unit Qv, or force per unit $I\Delta l$. Both methods are based on the fact that a charge experiences a force when it moves in a suitable direction in a magnetic field. However, the idealized current segment of strength $I\Delta l$ is hard to imagine, for there must be *some* conducting path by which the current enters and leaves the segment. For a more practical indicator of the direction of a magnetic field, we can use a circular or rectangular current loop (Fig. 21-6). Here we have a complete path for the current, but the net magnetic force on the two lead-in wires M and N is zero, since they are close together and carry equal currents in opposite directions. Such a loop tends to orient itself broadside to a magnetic field, for if it were at an angle (Fig. 21-7a) the two forces $\mathbf{F}_1$ and $\mathbf{F}_2$ would give rise to a torque, and the loop would rotate. When in the broadside position of Fig. 21-7b, the loop is in equilibrium, and the net force and net torque are both zero.* To obtain a larger effect, we can form a stiff wire into a long narrow coil (Fig. 21-8). When an outside source sets up a current through the wire, the coil tends to rotate until the axis AA' is parallel to the field $\mathbf{B}$, with each loop broadside to the field. The coil points in the direction of the field, like a compass needle.

---

*Figure 21-7 illustrates a method for drawing vectors which are perpendicular to the plane of the paper. The current in the bottom wire is toward the reader and is represented by $\odot$, which may be thought of as the head of an arrow just bursting through the paper. The symbol $\otimes$ represents the tail feathers of an arrow just disappearing into the paper. We shall use these symbols many times in the future.

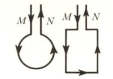

Figure 21-6   Current loops.

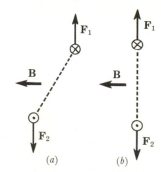

Figure 21-7   Top view of current loop; the torque is zero when the loop is broadside to the magnetic field.

The most ancient device for indicating the direction of a magnetic field is, of course, the compass needle itself. There is more than accidental similarity between our long narrow coil and a magnetized iron nail hung by a thread (Fig. 21-9). In the model of an atom which we are using, electrons move in orbits around the nucleus. The electrons also spin on their own axes. Both orbital motion and spin can cause a magnetic field, but in many atoms, such as neon and copper, the combined effects of orbital and spin magnetism cancel out; orbits (and spins) are randomly oriented. In other atoms, including iron and aluminum, there are unpaired spins that do

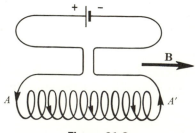

Figure 21-8

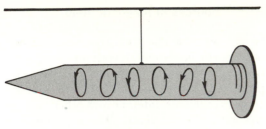

Figure 21-9

not cancel, and an external magnetic field can to some extent align the spin magnets in spite of thermal agitation.

Iron is very much more magnetic than aluminum because of a "cooperative" effect. Groups of 10^{15} or so iron atoms are locked together with all their electron-spin axes parallel. These groups, called *domains*, are roughly 0.01 to 0.1 mm on an edge in unmagnetized iron. During the magnetization process, some domains grow at the expense of others (Fig. 21-10). In union there is strength, and each domain exerts a force on its neighbor, helping to keep it oriented under the influence of the field. This is what is meant by the "cooperative" effect. In aluminum, where there are no domains, it is every atom for itself; aluminum is much less magnetic than iron. An aluminum rod will set itself parallel to a magnetic field, but the effect is noticeable only in the strongest fields.

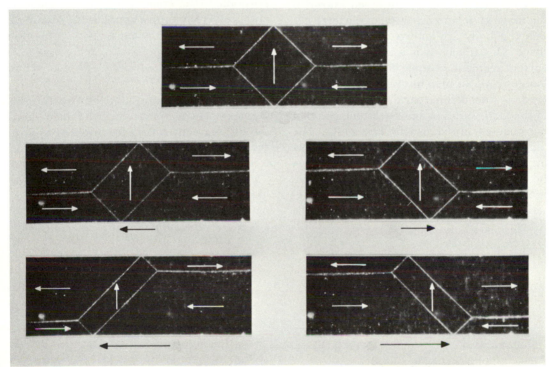

Courtesy of R. W. De Blois and C. D. Graham, Jr.

Figure 21-10 Growth of magnetic domains in a section of a single-crystal iron whisker 0.1 mm thick. The top photograph is for no applied magnetic field; domains oriented to the right and the left are of equal size. In the other photographs, arrows drawn below the whisker show the magnitude and direction of the applied field. It is seen that the process of magnetization consists of growth of some domains and shrinking of others. Domain boundaries are made visible by spreading a colloidal suspension of magnetite over the surface; strong magnetic fields in the region of the boundaries attract the magnetite to form the white lines visible in the photomicrographs.

A magnetized compass needle is, therefore, basically a collection of current loops. The needle points northward because of the earth's magnetic force on the atomic currents of the domains. Another way of indicating the direction of a magnetic field is familiar to you from your high school general science. You have undoubtedly sprinkled iron filings on a piece of paper which has been laid on top of one or more permanent magnets. Each little sliver of iron contains domains, and the magnetic force on the current loops which make up the domains acts to turn the sliver parallel to the field. This is an easy way of "mapping" the field. In the final analysis, we are still making use of magnetic forces on moving charges.

A magnetic field is considered known if the magnitude and direction of the force per unit test body are known at every point in space. Our test body is a moving charge, a current segment, or some equivalent current loop or coil. To help in visualizing the field, we draw lines of force parallel to the direction of the field. Lines of force near one end of a bar magnet are shown in Fig. 21-11. The diagrams show three different ways of exploring the field. A line of force could, of course, be drawn through any point at which there is a field, such as P. However, only a representative number of lines can actually be drawn.

21-5 Sources of magnetic field

So far, we have discussed the measurement of magnetic field strength without saying anything about how a magnetic field might be produced, except to intimate that moving charges (currents) interact with other moving charges (other currents). The experimental facts may be summarized by use of current segments $I \, \Delta l$. We use one current segment $I_1 \, \Delta l_1$ as a test body to explore the magnetic field caused by a second current segment $I_2 \, \Delta l_2$. The magnetic induction is the force per unit current segment of the test body ($B = F/I_1 \, \Delta l_1$). Experiment shows that at any point the magnitude of the induction caused by a current segment is directly proportional to the strength $I_2 \, \Delta l_2$ of the segment and is inversely proportional to the square of the distance from it. That is,

$$B = \frac{F}{I_1 \, \Delta l_1} = k' \frac{I_2 \, \Delta l_2}{r^2} \qquad (21\text{-}5)$$

where k' is a constant, and the two current segments ($I_2 \, \Delta l_2$ producing the field, and the test segment $I_1 \, \Delta l_1$) are oriented to give the maximum force. A current segment $\mathbf{I} \, \Delta l$ is a vector, having magnitude $I \, \Delta l$. Only the "broadside" component of $\mathbf{I_2} \, \Delta l_2$ is effective in producing a magnetic field. Thus the magnetic induction at P (Fig. 21-12) is given by $k' I_2 \, \Delta l_2 / r^2$; the induction at P'' is zero (head-

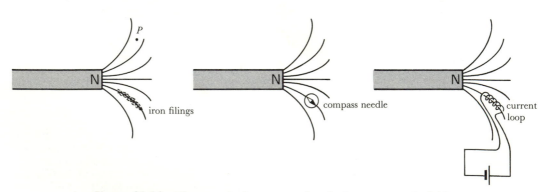

Figure 21-11 Three equivalent ways of exploring a magnetic field.

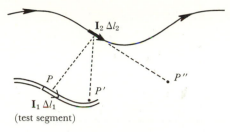

(test segment)

Figure 21-12

on view of $I_2 \Delta l_2$); and the induction at P' could be computed by first resolving the vector $I_2 \Delta l_2$ into a broadside component and a head-on component, using only the broadside component to compute B. Orientation is usually not a problem, since practical devices such as motors and galvanometers are (understandably) constructed to take advantage of proper orientation for maximum force.

The magnetic field surrounding a long straight wire is represented by lines of force which are concentric circles (Fig. 21-13). A simple *right-hand rule* is used to determine the direction of the field surrounding a wire which carries a current. Imagine grasping the wire with your right hand, your thumb pointing in the direction of the conventional (positive) current. The fingers are curved around the wire and indicate the direction

of the magnetic field.[*] This thumb rule for the direction of **B** and the right-hand screw rule for the direction of **F** will take care of all our future needs, including the study of motors, generators, and cyclotrons.

For our first application of Eq. 21-5, we calculate the magnitude of the field at the center of a circular current loop of radius R (Fig. 21-14). All the little segments $I \Delta l$ are at a distance R from the center, and each segment $I \Delta l$ is oriented for maximum effect, broadside to the radius. The total field is

$$k' \frac{I \Delta l_1}{R^2} + k' \frac{I \Delta l_2}{R^2} + k' \frac{I \Delta l_3}{R^2} + \cdots$$

$$= k' \frac{I}{R^2} (\Delta l_1 + \Delta l_2 + \Delta l_3 + \cdots)$$

For a complete circular loop, the total of all the Δl's is the circumference, or $2\pi R$; hence

$$B = \frac{k'I}{R^2} (2\pi R)$$

$$= \frac{2\pi k'I}{R} \qquad (21\text{-}6)$$

[*] A current of negative charges, such as a beam of electrons in a TV tube, creates a magnetic field just opposite to that of a positive current such as a beam of protons in a particle accelerator; on such occasions we can use a left-hand rule to find the direction of **B**, or we can replace the negative current by a conventional (positive) current in the opposite direction.

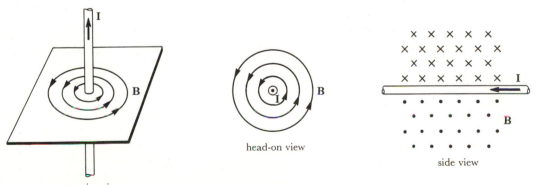

perspective view

head-on view

side view

Figure 21-13 Three views of the magnetic field surrounding a long straight wire. In each case, the fingers of the right hand would represent the direction of the field if the thumb pointed in the direction of the conventional (positive) current.

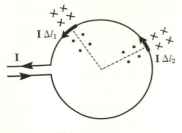

Figure 21-14

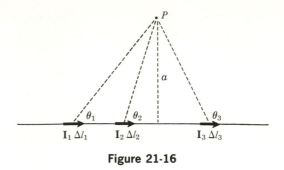

Figure 21-16

for the field at the center of a circular loop. At the center of the loop, the direction of **B** is out of the paper in Fig. 21-14. Here we notice an interesting inversion of our right-hand rule. For a straight wire, the right thumb represents the current and the fingers represent the field. For a circular loop, we can let the fingers of the right hand represent the current and the thumb represent the field. (Try it in Fig. 21-14.) In either case (Fig. 21-15a and b), the curved fingers represent whatever is inherently curved—be it field or current—and the thumb represents the other quantity.

Considering next the magnitude of the field at a point near an infinitely long wire,

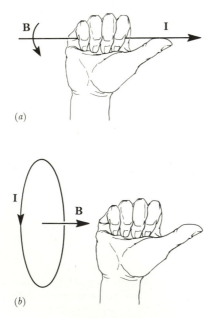

Figure 21-15

we must find the resultant of infinitely many fields due to the infinitely many tiny current segments **I** Δl, of which three are shown in Fig. 21-16. The segments differ in distance from P, and differ also in orientation relative to P. The result of a summation, performed by the methods of integral calculus, is a rather simple equation for the magnitude of the magnetic induction at a point which is a distance a from an infinitely long straight current (see Sec. 21-10 for details):

$$B = k' \frac{2I}{a} \qquad (21\text{-}7)$$

Logically, Eqs. 21-6 and 21-7 are equivalent to Eq. 21-5, but they are easier to apply.

We are now able to define the ampere (and, therefore, the coulomb)—something we have put off for more than four chapters. The discussion revolves around the constant k'. If we had already defined units of current, force, and length, then we would use Eq. 21-7 to find k' experimentally by substituting measured values of B ($= F/I_1 \Delta l_1$), I, and a. You will recall that this is what we did when we found the constant k in Coulomb's law (Sec. 17-5). However, since we have not yet defined the size of the ampere, we are free to select any numerical value for k' that we please. We choose the value of the constant k' to be 10^{-7} N/A², for historical reasons,[*] and then Eq. 21-7 serves to define the unit of current which we call the ampere. To see

---

[*] This choice of k' makes our ampere equal to the ampere which earlier workers had defined before mks units became popular.

446 Electromagnetism

how the choice of k' leads to a definition of the ampere, let us compute the force per unit length on a long straight wire carrying current $I_1 = 1$ A, which is 1 m from a parallel wire also carrying a current $I_2 = 1$ A.

$$B = k' \frac{2I_2}{a}$$

$$\frac{F}{I_1 \Delta l_1} = k' \frac{2I_2}{a}$$

$$\frac{F}{\Delta l_1} = k' \frac{2I_1 I_2}{a}$$

$$= (10^{-7}\ \text{N/A}^2) \left[\frac{2(1\ \text{A})(1\ \text{A})}{1\ \text{m}} \right]$$

$$= 2 \times 10^{-7}\ \text{N/m}$$

We see that making $k' = 10^{-7}$ N/A^2 leads to the following definition:

One ampere is the current in a long straight wire which exerts a force per unit length of exactly 2×10^{-7} N/m on a neighboring long parallel wire, 1 m distant, which carries an equal current.

We use k' to avoid fractions and a factor 4π. Eq. 21-7 is often written in the form $B = (\mu_0/4\pi)2I/a$, where $\mu_0/4\pi$ is the same as k'.

The National Bureau of Standards maintains a "current balance" which is based on Eq. 21-7, but which uses coils instead of long wires. The ammeters you use in the laboratory are (indirectly) calibrated relative to this current balance by use of Eq. 21-7. In this way, our standards of current, charge, and PD can be reproduced in terms of the meter, the kilogram, and the second.

EXAMPLE 21-5

At an electroplating plant, a long horizontal wire connecting two buildings carries 500 A toward the east. What are the magnitude and direction of the magnetic induction at a point on the ground 10 m directly below the wire?

The *direction* of the field is found from the right-hand rule (Fig. 21-17). The field at the ground level is toward the north.

The *magnitude* of the induction is found from Eq. 21-7:

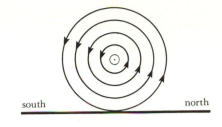

south   north

Figure 21-17

$$B = k' \frac{2I}{a} = (10^{-7}\ \text{N/A}^2) \left[\frac{2(500\ \text{A})}{10\ \text{m}} \right]$$

$$= 10^{-5}\ \text{N/A} \cdot \text{m} = 10^{-5}\ \text{T}$$

Thus

$$\mathbf{B} = \boxed{10^{-5}\ \text{T, toward the north}}$$

Two parallel wires which carry currents in the same direction attract each other. To see this, consider Fig. 21-18a, which shows the two wires in cross section. Wire Y is in the field caused by wire X; applying the right-hand screw rule $\mathbf{F} = \mathbf{I}\,\Delta l \times \mathbf{B}$ to wire Y shows that $\mathbf{F}$ is to the left. A similar argument shows that the force on X is toward the right. (We would expect this by Newton's third law—the force of X on Y is opposite to

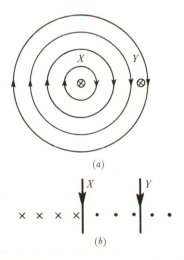

(a)

(b)

Figure 21-18 Force of attraction between two wires carrying current in the same direction.

the force of Y on X.) Figure 21-18b shows the same situation as viewed from the side. In general, like currents attract each other, and unlike currents repel each other.

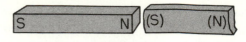

Figure 21-19

21-6 Magnets and poles

In the early days, it was thought that a lodestone or a magnetized iron rod had "poles," and that the force of attraction or repulsion between two magnets was due to a sort of Coulomb law between magnetic poles of two kinds. The "north" pole of a bar magnet is the "north-seeking pole," which points northward in the earth's magnetic field. It was assumed that a compass needle had an excess of N poles at one end and an excess of S poles at the other end. This idea was reinforced by the phenomenon of magnetic induction (Fig. 21-19). If a piece of iron is brought near a permanent magnet, poles are induced in the iron in a way which reminds one of the separation of electric charge by induction (Sec. 17-4). Sometimes, if the right kind of iron is used, the piece becomes permanently magnetized. The effect is more pronounced if the iron bar is stroked or hammered while in the magnetizing field. One might say that magnetic poles have been separated "by induction."

Magnetic poles were once believed to have an objective existence of their own, separate from the atoms and molecules of the substance in which they were found. This concept of magnetic poles had to be abandoned, how-

ever, as additional facts were brought to light. For one thing, no *free* poles have ever been observed.* If you try to cut off a north pole from one end of a magnet (Fig. 21-20a), all you get is another short magnet, with a north pole at one end and a south pole at the other. No magnetic current has ever been observed, and there are no "conductors" of magnetic poles. The magnetic properties of a few substances such as iron, cobalt, nickel, and certain alloys are radically different from those of other metals, but the electrical properties of these same metals differ only slightly.

It cannot be denied that the north end of a bar magnet attracts the south end of another bar magnet, and so we still speak of north poles and south poles of magnets, and we say that like poles repel and unlike poles attract one another. We now know that these magnetic forces are essentially electrical in nature; they arise from the interactions between the moving and spinning orbital electrons which are bound to the nuclei. We know that the radical differences between iron and copper are due to the possibility of cooperation between oriented domains in the iron. We presume that *all* magnetic phenomena are

---

* Free poles are not absolutely prohibited by present-day complete electromagnetic theory. The failure to observe them is somewhat of a mystery.

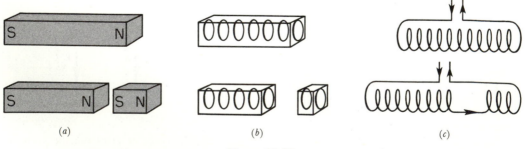

(a) (b) (c)

Figure 21-20

ultimately to be described in terms of moving charges, and the free magnetic pole is only a useful fiction, a model that has probably outlived its usefulness. All statements containing the phrase "magnetic pole" can be interpreted in this way, the pole actually being the result of electronic motions. Figures 21-20b and c, which represent a magnet in terms of atomic current loops or equivalent long coils, show how impossible it would be to produce free poles by cutting a bar magnet.

A long, thin bar magnet offers the closest approximation to isolated poles. The magnetic field near one end of such a needle-like bar is shown in Fig. 21-21; almost all the lines of force leave the immediate vicinity of the end of the needle. Under these conditions, it makes sense to think of poles located at the ends of the needle. It can even be shown that, if a second long needle is brought near the first one, the magnetic force between the current loops represented by the poles should vary inversely as the square of the distance between the "poles." Only in this limited sense is there a Coulomb's law for magnetic poles.[*] Even for a bar magnet which is not a needle, the general direction of the field is away from the region designated as a north pole and toward the region designated as a south pole. Permanent magnets are usually marked with polarities in accordance with this scheme; this does not mean that the poles are "real." The two situations in Fig. 21-22 are equivalent.

The net magnetic field caused by a coil or other current-carrying conductor is greatly increased if a piece of iron or magnetic alloy is brought nearby. The oriented domains are equivalent to current loops inside the metal, and the magnetic field caused by these "atomic current loops" is added to the original field caused by the current in the external coil or wire. Three electromagnets are shown in Fig. 21-23; the magnet b can sup-

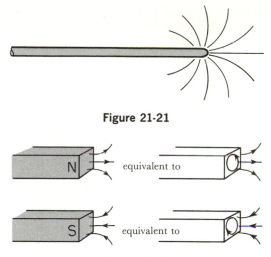

Figure 21-21

Figure 21-22

port a much greater load than a, since much of the force is supplied by the domains within the iron core. The relatively weak magnetic field of the current in the winding has been sufficient to rotate or enlarge many domains in the iron. The magnet c is still stronger, for the domains are head to tail for complete paths in the iron, thus helping keep the domains in alignment.

In most large electromagnets, the coil current causes Joule heating which must be dissipated by a cooling system; the cost of the

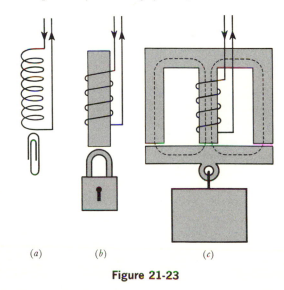

(a)   (b)   (c)

Figure 21-23

---

[*]Coulomb made some measurements of this sort with his torsion balance, and established an inverse-square law for long, needle-shaped magnets.

electric power (I^2R) wasted in heat is also a consideration. Modern efficient magnet coils use wires of a superconducting alloy (which is cooled below its transition temperature). For example, a commercially available magnet uses an alloy of 75% Nb, 25% Zr cooled to 4.2°K. Such a wire of only 0.01 in. diameter carries 25 A with no Joule heating whatever.

21-7 The earth's magnetism

As long ago as 1570, William Gilbert, court physician to Elizabeth I of England, constructed a permanent magnet in the form of a large lodestone sphere. Gilbert used a small compass needle to survey the magnetic field near the surface of the sphere and found that his model successfully represented the main features of the earth's magnetic field. At a point in the Northern Hemisphere, such as in England or the United States, the field is directed downward and to the north, as shown in Fig. 21-24a. These observations can be accounted for by assuming that a relatively short magnet, several hundred miles long, is buried deep inside the earth (Fig. 21-24b). Since the lines of force are directed downward for an observer in the Northern Hemisphere, we must assume that the "earth's magnet" has an S pole on the end which is beneath the north magnetic pole.

We know that permanent magnets are collections of current loops, and so we draw Fig. 21-24c as a modern equivalent of the outmoded permanent-magnet hypothesis. If the axis of the assumed current loop is more or less along the direction of the earth's axis of rotation, the magnetic North Pole will approximately coincide with the geographical North Pole, as it actually does. The magnetic North Pole is in northern Canada; this means that a compass needle does not point exactly north except at certain places. The difference between geographic north and magnetic north is called *magnetic declination;* this quantity varies from about 25°E to 20°W for different places in the United States, and also varies slowly from year to year. (See the map, Fig. 21-25.)

The ordinary compass needle responds only to the horizontal component of the earth's magnetic field, since it is pivoted in such a way as to prevent up-and-down motion. If a magnetized needle is mounted so as to be free to swing in a vertical north-south plane,* it is free to point in the direction of the field, and in the Northern Hemisphere it will assume a position in which it points downward and to the north. Such an instrument is called a *dip needle* (Fig. 21-26); the angle of dip is measured from the horizontal. Values of the horizontal and

---

* Magnetic north, not geographic north.

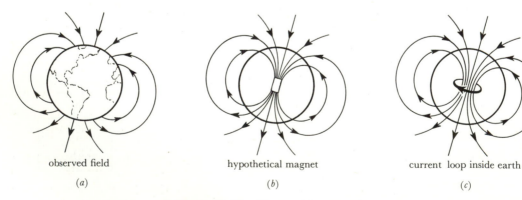

observed field   hypothetical magnet   current loop inside earth

(a)   (b)   (c)

Figure 21-24 The earth's magnetism.

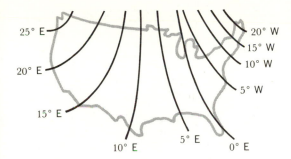

Figure 21-25 Lines of equal magnetic declination.

vertical components of **B** for the earth's field at Washington, D.C., are shown in Fig. 21-26; the vertical component is about three times the horizontal component at this location.

The ultimate explanation of the earth's magnetic field must somehow be found in circulating currents deep within the earth, or in the upper atmosphere, or both. The approximate coincidence of the magnetic poles and the geographic poles is surely significant, but the exact mechanism is still unknown. Perhaps a stream of charged particles (electrons, ions, or both) has been created by the high temperatures in the earth's core, and the motion of these charges is affected by the earth's rotation. The strength and location, or both, of the internal currents change gradually, for the angles of declination and dip change. For example, in 1580 the declination at London was measured by Gilbert to be 11°E; it was 0° in 1657, reached 25°W by 1820, and is decreasing again, with

a value of about 7°W in 1972. A second type of change is sudden and short-lived; such "magnetic storms" are correlated with sunspot outbreaks and are a result of temporary currents of ions in the upper atmosphere. Finally, evidence points to radical changes in the magnetic axis of the earth in the geological past; during the last several million years the magnetic axis reversed direction several times, in addition to undergoing irregular wandering. In the period from 2.5 My ago to 0.7 My ago, the "north" magnetic pole of the earth was in the Southern Hemisphere. This implies a change in the direction of the field-causing currents inside the earth.

The phenomena of terrestrial magnetism are complicated and variable, and not yet well understood. So far, however, physicists are confident that the earth's field orginates in the motion of charges, as do all other magnetic fields.

21-8   Induced electromotive force

We have seen in Sec. 21-1 that in general a charged particle moving through a magnetic field experiences a magnetic force. If, then, a conductor moves in a suitable direction relative to a magnetic field, a magnetic force acts on the free electrons in the conductor, the electrons are accelerated, and work is done. We have here a mechanism for the transformation of mechanical energy into electric energy; in other words, the conductor becomes a seat of emf. We say that an *induced emf* is generated in a conductor which moves in a suitable direction through a magnetic field.

Until the discovery in the 1830's of electromagnetic induction, the only practical way of obtaining electric current was from cells and batteries of various types. It is instructive to contemplate the tremendous advance of technology which resulted from the discovery of induced emf in about 1832 by Joseph Henry in the United States and

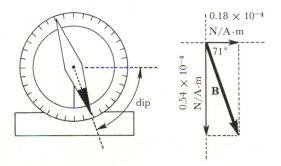

Figure 21-26 Dip needle.

Michael Faraday in England. The generators of the world's power stations transform mechanical energy into electric energy with high efficiency; more than 99% of the electric energy used per year in the United States is "generated" in this way, and less than 1% comes from energy transformations in other seats of emf, such as chemical cells.

The original investigators found that an emf appears in a wire which moves relative to a magnetic field. The magnitude of the induced emf can be derived by a "thought experiment." Imagine a horizontal rod (Fig. 21-27) which rolls along rails with velocity v. A charge q which is carried along by the moving rod experiences a force $F = q\mathbf{v} \times \mathbf{B}$; this force is perpendicular to both $\mathbf{B}$ and $\mathbf{v}$ and hence is parallel to the rod. When the charge moves along the rod a distance l, the work done by the magnetic force is $W = Fl = (qvB \sin \theta)l$. The emf generated in the rod is work/charge, or

$$\mathcal{E} = Blv \sin \theta \qquad (21\text{-}8)$$

For the maximum induced emf this equation simplifies to

$$\mathcal{E} = Blv$$

The induced emf equals the product of the only factors that might be expected to be significant: B is the magnitude of the magnetic induction, and l is the length of the

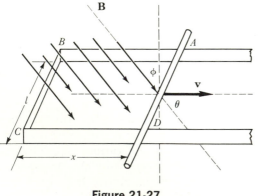

Figure 21-27

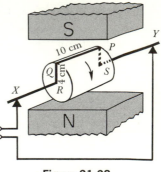

Figure 21-28

wire which is moving with velocity v at an angle θ with the field. We interpret $v \sin \theta$ as the component of velocity perpendicular to $\mathbf{B}$. To check units, note that

$$\mathcal{E} = Blv \sin \theta = (\text{N/A} \cdot \text{m})(\text{m})(\text{m/s})$$
$$= \text{N} \cdot \text{m/A} \cdot \text{s} = \text{J/C} = \text{V}$$

For maximum effect, the wire should be at right angles to the field, and the velocity should be at right angles to both the field and the wire. Such a motion can be described as that of a conductor which is "cutting" lines of force, as in Fig. 21-28.

EXAMPLE **21-6**

The rotating armature of a simple generator consists of a rectangular loop $SPQR$, to which connections are made by means of sliding contacts. The armature is turned at 1200 rev/min, and the magnetic induction is 0.5 N/A · m. What is the magnitude of the induced emf between the terminals of the generator at the instant shown in Fig. 21-28?

The wire PQ is moving to the right, with a velocity given by the circumference of its circular path, multiplied by the number of revolutions per second.

$$v = 2\pi r \nu = 2\pi (0.04 \text{ m})\left(1200 \frac{\text{rev}}{\text{min}}\right)\left(\frac{1 \text{ min}}{60 \text{ s}}\right)$$
$$= 5.03 \text{ m/s}$$

The magnetic field is directed vertically upward, from N pole to S pole. Since the wire is cutting perpendicularly across the lines of force, the induced emf is a maximum. Here $\sin \theta = 1$, and

$$\mathcal{E} = Blv = (0.5 \text{ N/A} \cdot \text{m})(0.10 \text{ m})(5.03 \text{ m/s})(1)$$

$$= \boxed{0.25 \text{ V}}$$

No emf's are induced in wires SP and RQ because the magnetic force on free electrons in these wires (given by Eq. 21-2) has no component parallel to the wires. Also, there is no emf in the lead-in wires, which are not in motion. Therefore the total emf between the terminals is just that due to the segment PQ, and $\mathcal{E} = 0.25$ V at this particular instant.

The *direction* of an induced emf is given by Lenz's law:[*]

An induced emf tends to set up a current whose action opposes the change that caused it.

Let us apply Lenz's law to the generator of the preceding example, in which a conductor moves to the right (Fig. 21-29a). The best way to proceed is to *assume* a direction for $\mathcal{E}$, and then apply Lenz's law to see if the assumption is correct. Let us assume that, if there were a complete circuit, positive (conventional) charge would flow toward us from P to Q and on through the circuit. Using the right-hand rule to get the force on this assumed current, we find that the force on the wire PQ would be toward the left (Fig. 21-29b). This force would indeed oppose the "change that caused it"; in this case the change is a motion to the right. Hence our assumed direction is correct. Charge would tend to flow from P to Q, out at X, through an external circuit, and back in at Y. The generator's positive terminal is at X, and its negative terminal is at Y at the instant shown.

Lenz's law is a deduction from the law of conservation of energy. Thus in Fig. 21-29b, the force $\mathbf{F}$ is opposite to the velocity $\mathbf{v}$, and work must be done by an outside agent to move the wire to the right against the op-

[*] Developed by the German physicist Heinrich Lenz (1804–1865).

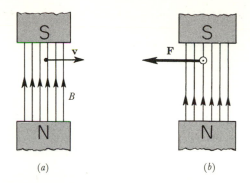

Figure 21-29 (a) Conductor moving to the right, cutting across a magnetic field. (b) Lenz's law: a conductor is moving to the right, giving rise to an induced current which in turn causes a force to the left, opposing the motion.

posing force $\mathbf{F}$. Suppose the generator is connected by wires to a lamp. Mechanical energy is put into the system and transformed into Joule heat (I^2R) in the wires of the generator and in the lamp. However, if the induced current were in the opposite direction ($\otimes$ instead of $\odot$), the wire would be doing mechanical work ($\mathbf{F}$ and $\mathbf{v}$ in the same direction), and *also* heat energy would be produced in the wires of the generator and in the lamp. This would clearly violate the law of conservation of energy, since there would be an expenditure of energy without any source of energy. We see that Lenz's law is a necessary consequence of the law of conservation of energy.[*]

21-9 Magnetic flux

We have discussed one way in which induced emf's can arise—from the motion of a con-

[*] In chemistry, Le Chatelier's principle is similarly related to the law of conservation of energy: "if the equilibrium of a system is disturbed by a change in one or more of the determining factors (as temperature, pressure, or concentration) the system tends to adjust itself to a new equilibrium by counteracting as far as possible the effect of the change" (Webster's New International Dictionary, 3rd ed.)

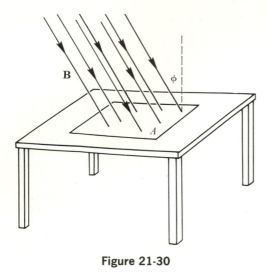

Figure 21-30

ductor across a magnetic field. It is also possible for an emf to be induced without physical motion of any conductor. An important example of such an emf is that induced in the secondary coil of a transformer.

To formulate a general law for induced emf, we first define *magnetic flux* Φ as a product:

$$\Phi = B \cos \phi \, A \qquad (21\text{-}9)$$

where $B \cos \phi$ is the component of **B** perpendicular to the plane of some complete circuit whose area is A (Fig. 21-30) and ϕ is the angle between **B** and the perpendicular to A.* In many cases **B** is broadside to the circuit, making $\phi = 0°$ and $\cos \phi = 1$, in which case

$$\Phi = BA \qquad (21\text{-}10)$$

The unit for flux is the *weber* (Wb).

---

* We assume that the circuit lies in a plane and that **B** is uniform in magnitude and direction at all points in the plane of the circuit. In a more general treatment, Φ is calculated as a limit of a sum of terms:

$$\Phi = B_1 \cos \phi_1 \, \Delta A_1 + B_2 \cos \phi_2 \, \Delta A_2 + \cdots .$$

The limit of the sum is, of course, a definite integral, and hence $\Phi = \int B \cos \phi \, dA$. The vector notation for flux uses the dot product (Sec. 6-4): $\Phi = \int \mathbf{B} \cdot d\mathbf{A}$.

From this definition of Φ, we can derive a new formulation of Faraday's law of induced emf. The flux through the area $ABCD$ (Fig. 21-27) is

$$\Phi = B \cos \phi \, (lx)$$

If B, Φ, and l are constant, the only change in flux is caused by a change in x. Thus the rate of change of flux is

$$\frac{d\Phi}{dt} = B \cos \phi \, l \, \frac{dx}{dt}$$

But $\cos \phi = \sin \theta$ (they are complementary angles), and dx/dt is the velocity v. Hence

$$\frac{d\Phi}{dt} = B \sin \theta \, lv$$

From Eq. 21-8 we then conclude that the induced emf equals the rate of change of magnetic flux:

$$\mathcal{E} = \lim_{\Delta t \to 0} \frac{\Delta \Phi}{\Delta t}$$

or

$$\mathcal{E} = \frac{d\Phi}{dt} \qquad (21\text{-}11)$$

The direction of $\mathcal{E}$ is found, as before, from Lenz's law, according to which the emf's direction tends to oppose the change that causes it. For this reason, a minus sign is sometimes used in writing Faraday's law, indicating that the emf induced in a circuit equals the time rate of *decrease* of magnetic flux through the circuit. If a circuit or coil has N turns and the rate of change of flux is the same through each turn, then the magnitude of the total induced emf in the coil is found from

$$\mathcal{E} = N \frac{d\Phi}{dt} \qquad (21\text{-}12)$$

An induced emf arises whenever Φ changes; it is not necessary that the area change. Thus an emf can also arise because of a change in B (transformers) or a change in $\cos \phi$ (generators). These devices will be discussed in Chap. 22.

EXAMPLE 21-7

A square loop of wire 4 cm on an edge is lying on a horizontal table. An electromagnet above and to one side of the loop is turned on, causing a uniform magnetic field which is downward at an angle of 30° from the vertical as in Fig. 21-30. The magnetic induction is 0.500 T, or 0.500 Wb/m². Calculate the average induced emf in the loop if the field increases from 0 to its final value in 200 ms.

The component of magnetic induction perpendicular to the plane of the loop is

$B \cos \phi = (0.500 \text{ Wb/m}^2)(\cos 30°)$
$\qquad\qquad = 0.433 \text{ Wb/m}^2$

The area of the loop is 1.6×10^{-3} m². Hence the flux is

$\Phi = B \cos \phi \, A$
$\quad = (0.433 \text{ Wb/m}^2)(1.6 \times 10^{-3} \text{ m}^2)$
$\quad = 6.93 \times 10^{-4} \text{ Wb}$

In finding the average induced emf we do not take a limit as $\Delta t \to 0$; hence we calculate $\Delta\Phi/\Delta t$ and we denote the average emf by a bar over the $\mathcal{E}$, as is usual for average values (Sec. 3-3).

The magnitude of the average induced emf is

$$\bar{\mathcal{E}} = \frac{\Delta\Phi}{\Delta t} = \frac{6.93 \times 10^{-4} \text{ Wb}}{200 \times 10^{-3} \text{ s}}$$

$$= \boxed{3.47 \times 10^{-3} \text{ V}}$$

The direction of $\mathcal{E}$ is found from Lenz's law. Assume that $\mathcal{E}$ is clockwise as viewed from above. By the right-hand rule, the induced magnetic field caused by this assumed $\mathcal{E}$ would be downward; this would *aid* the downward field which is being established by the magnet. The induced emf must *oppose* the change that caused it; hence the assumed direction is wrong. We conclude that $\mathcal{E}$ is directed counterclockwise around the loop during the time that the downward magnetic field is increasing.

Magnetic units in the cgs system. Some research workers in physics, chemistry, and biology still use cgs magnetic units which may have unfamiliar names. The cgs unit for magnetic induction B is the gauss (G), where 1 gauss = 10^{-4} Wb/m² = 10^{-4} tesla. The cgs unit for magnetic flux Φ is the maxwell, where one maxwell = 10^{-8} Wb.

■ SUMMARY

A magnetic field exists in any region of space where a moving electric charge would experience an electric force other than Coulomb force. Magnetic forces are caused by moving charges acting on other moving charges; we therefore speak of magnetism as a subdivision of electricity. A magnetic field can be explored by using a moving charge as a test body. The magnitude of the vector representing magnetic field is given by the force per unit Qv product. The force on a moving charge is always perpendicular to its motion and to the direction of the field. A moving charge experiences maximum force if its motion is perpendicular to the field. A current segment $I \, \Delta l$ can be used as a test body instead of a moving charge. If current segments are laid end to end, forming a loop or a coil, the coil tends to align its axis with the direction of magnetic field. The force on a current segment in a magnetic field equals the product of the magnetic induction and the strength of the current segment.

Permanent magnets such as a compass needle are essentially collections of current loops arising from the spins of the electrons in the electron clouds of the atoms. In iron and other strongly magnetic substances, the magnetic effect is strong because atoms are grouped in domains in which cooperation between neighboring atoms causes many orbits and spins to be in parallel alignment.

Magnetic fields are mapped by drawing representative lines of force. In the space surrounding a wire through which electrons flow, the magnetic field is represented by concentric circles. The right-hand rule gives the direction of the concentric lines of force caused by a conventional current.

The earth's magnetic field is caused by circulating electric currents inside the earth, but the exact mechanism is not known. In the United States, the field is directed downward and to the north. Three quantities are used to describe the earth's field at any given point: the magnitude of the field, the declination, and the dip.

When a wire moves across a magnetic field, an emf is induced in the wire. The direction of the induced emf is given by Lenz's law, which states that an induced emf tends to oppose the change that caused it. Magnetic flux is the product of the perpendicular component of magnetic field strength and the area of a circuit. Induced emf in a circuit equals the time rate of change of magnetic flux.

■ CHECK LIST

magnetic induction
test bodies for measurement of field strength
$B = F/Qv$
$\mathbf{F} = Q\mathbf{v} \times \mathbf{B}$
$B = F/I\,\Delta l$
tesla
current segment
$\mathbf{F} = \mathbf{I}\,\Delta l \times \mathbf{B}$
domain
right-hand rule
ampere

magnetic declination
angle of dip
induced emf
Lenz's law
$\mathcal{E} = Blv \sin \theta$
magnetic flux
weber
$\Phi = BA \cos \phi$
$\mathcal{E} = N\dfrac{d\Phi}{dt}$

■ QUESTIONS

21-1 Upon what factors does the magnetic force on a moving charge depend?

21-2 How would you determine the direction of magnetic field at a given point, (a) using a compass, and (b) using a moving test charge?

21-3 Give two mks units for magnetic field strength, both equivalent to the tesla, and show that they are equivalent to each other.

21-4 What characteristic of iron causes it to concentrate magnetic lines of force?

21-5 Is there any known way in which a magnetic field can be caused other than by the motion of electric charge?

21-6 When a bar magnet is cut in half along its length, what is the magnetic state of the two pieces? What is the result of cutting the same bar magnet in half by a perpendicular cut?

21-7 Lightning strikes a vertical metal flagpole, and a stream of electrons momentarily flows up the pole. What is the direction of the magnetic field at a point in the air just east of the center of the pole?

21-8 Suppose that a plastic phonograph record is rubbed along its edge with fur, charging the edge with an excess of electrons. When the charged record is turning in the usual direction, in which direction is the magnetic field at a point directly above the center of the record?

21-9 There is an upward current in a vertical wire which is in a magnetic field directed horizontally to the south. In what direction is the force on the wire?

21-10 (*a*) In what direction (to the right or to the left) is the magnetic field at a point between the pole pieces of the magnet in Fig. 21-31? (*b*) In what direction (up or down) is the magnetic field at a point above the N pole? (*c*) In which direction (right or left) does the wire *AB* move when the switch is closed?

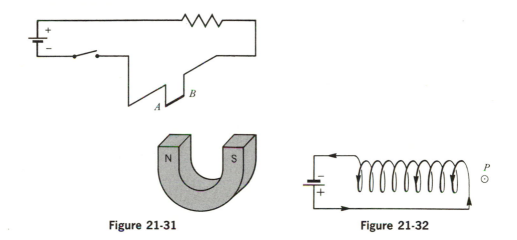

Figure 21-31   Figure 21-32

21-11 What is the direction of the force (upward or downward) on the current-carrying wire at *P* in Fig. 21-32?

21-12 The general equation for the force on a charge q is $\mathbf{F} = q\mathbf{E} + q\mathbf{v} \times \mathbf{B}$. Interpret each symbol in this equation.

21-13 Which is more important for a navigator to know—the dip or the declination?

21-14 Describe, in three dimensions, the general direction of the earth's magnetic field in Australia.

21-15 Without using iron or any other magnetic material, how could you construct a device which would serve as a compass?

21-16 Explain how magnetic storms are caused by eruption of charged particles from the sun.

21-17 Upon what factors does the induced emf in a conductor depend?

21-18 An iron girder in a railway bridge is in a vertical north-south plane, at an angle of 70° from the horizontal, with the bottom end of the girder north of its top end. During the year, the girder becomes magnetized as trains pound over the bridge. Explain.

21-19 What is the experimental evidence against the existence of free magnetic poles?

21-20 What is the difference between magnetic induction and magnetic flux?

21-21 A horizontal rod is falling through a gap between the poles of the permanent magnet of Fig. 21-31. Does the induced emf cause electrons to move toward or away from the reader? Explain your answer, using Lenz's law.

21-22 An automobile is traveling south along a level highway in the United States, and the front axle is therefore cutting the earth's magnetic field. (*a*) Upon which of the front hub caps do electrons tend to accumulate because of the induced emf? (*b*) Would the effect be greater or less if the car were coasting southward down a hill 20° below the horizontal?

■ PROBLEMS

21-A1 A charge of 2×10^{-10} C is moving at 1 km/s through a magnetic field in such a direction that the magnetic force on it is a maximum. What is the magnetic induction, if the maximum force on the charge is 6×10^{-8} N?

21-A2 What is the magnitude of the force on an electron which moves at 1% of the speed of light horizontally through a vertical magnetic field of 0.12 T?

21-A3 A rifle bullet having a net charge of 10^{-12} C moves at 300 m/s perpendicular to the earth's magnetic field. The intensity of the earth's field is 6×10^{-5} T. What is the magnitude of the magnetic force on the bullet? Is this force observable?

21-A4 Calculate the magnetic force on a current segment 3 cm long, placed broadside to a magnetic field of strength 1.2 T. The current in the wire is 2000 A. Could a strong man supply enough force to hold the wire in equilibrium?

21-A5 What is the magnetic induction at a point 10 cm away from a long wire in which the current is 20 A?

21-A6 A vertical wire 40 cm long is in a horizontal magnetic field of 800 T. What current through the wire will cause a force of 200 N on the 40 cm segment?

21-A7 The earth's magnetic field at a certain place is 0.60×10^{-4} T, and the angle of dip is 60°. Calculate the value of the horizontal component of the induction.

21-A8 Calculate the current in a circular loop 20 cm in radius which would cause a magnetic induction at the center of the loop equal to that of the earth's magnetic field, which is about 5×10^{-5} T.

21-A9 What emf is induced in a horizontal rod 2 cm long which is falling at 4 m/s in a region where there is a horizontal magnetic induction of 0.3 T?

21-A10 A horizontal wire 8 cm long moves with a speed of 3 m/s across a magnetic field of induction 0.04 T. What emf is induced in the wire?

21-A11 What is the magnetic flux through a coil 20 cm tall and 30 cm wide, if the coil is placed with its plane horizontal in the earth's magnetic field shown in Fig. 21-26?

21-A12 Calculate the emf induced in a coil of wire through which the magnetic flux changes from $+40$ Wb to -20 Wb in 0.500 s.

21-A13 Prove: $1 \text{ T} = 1 \text{ V·s/m}^2$; $1 \text{ T} = 1 \text{ }\Omega\text{·C/m}^2$; $1 \text{ Wb} = 1 \text{ }\Omega\text{·C}$; $1 \text{ Wb} = 1 \text{ J·s/C}$.

21-B1 (*a*) What is the magnetic force on a proton moving horizontally at (almost) the speed of light through a synchrotron (Fig. 8-11) where there is a vertical magnetic induction of 1.9 T? (*b*) What is the ratio of the magnetic force calculated in part (*a*) to the weight of the proton when it is at rest?

21-B2 Calculate the strength of the magnetic field in the cyclotron of Example 8-8 (p. 169).

21-B3 A deuterium nucleus ^2_1H (i.e., a deuteron; see Sec. 2-3) has a speed of 2×10^7 m/s perpendicular to a magnetic field of induction 0.3 T. (*a*) What is the force on the deuteron? (*b*) What is the acceleration of the deuteron? (*c*) Calculate the radius of curvature of the orbit of the deuteron. (*Hint:* The magnetic force causes a centripetal acceleration.)

21-B4 An electron in a small cathode-ray tube in an oscilloscope is moving horizontally toward the viewer at 5×10^7 m/s and is deflected upward by a horizontal magnetic field of 10^{-3} T. Relativity effects can be neglected (why?). (*a*) What is the magnetic force on the electron? (*b*) What is the radius of the circular path of the electron while in the magnetic field? (*Hint:* The magnetic force causes a centripetal acceleration.)

21-B5 An electron moves horizontally toward the east through a region where two fields exist: upward electric field $E = 1000$ V/m, and horizontal magnetic field $B = 0.5$ T. (*a*) What is the speed of an electron which passes through this "velocity selector" with no resultant deflection? (*b*) In which direction is **B**?

21-B6 A current in a circular path through a horizontal coil is clockwise as viewed from above. The coil, which has 10 turns, has a radius of 5 cm, and the current is 300 mA. (*a*) What is the direction of the magnetic field at the center of the loop? (*b*) What is the magnitude of the magnetic induction at the center? (*c*) Is this field stronger or weaker than the earth's magnetic field?

21-B7 Calculate the magnetic induction at the nucleus of an atom caused by the revolution of an electron in a circle of radius 0.5×10^{-10} m, if the moving orbital electron is equivalent to a current of 10^{-4} A. (*Note:* These values are approximately correct for the motion of the electron in a hydrogen atom in its normal state.)

21-B8 In a long vertical wire, 4×10^{20} electrons per second are flowing upward; 40 cm west of the wire is a vertical current segment 10 cm long carrying 3 A, with electrons flowing downward. What are the magnitude and direction of the force on the segment?

21-B9 (*a*) In the United States, what should be the direction of a horizontal current-carrying wire, if it is to experience a maximum force due to the earth's magnetic field (assume declination to be 0°)? (*b*) Approximately what would this force be on a properly oriented current segment 1 cm long which carries a current of 25 A?

21-B10 Two long wires, 2 cm apart, carry currents of 50 A in opposite directions. (*a*) Is the magnetic force between them one of attraction or repulsion (explain)? (*b*) What is the magnetic force on a 1 cm length of one wire due to the field of the (very long) other wire?

21-B11 A heart surgeon monitors the rate of flow of blood through an artery using an electromagnetic flow meter (Fig. 21-33) in which electrodes A and B are attached to the outer surface of a blood vessel of inside diameter 4 mm. Blood and vessel both have good electrical conductivity. (*a*) For a magnetic induction 0.02 T an emf of 100 μV is

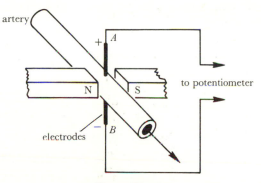

Figure 21-33

developed. Calculate the velocity of the blood. (b) Verify that electrode A is +, as shown. Does the sign of the emf depend on the sign of the ions in the blood?

21-B12 During a space walk, two astronauts are separated by a connecting wire 10 m long which is moving horizontally, broadside to its length, at an orbital speed of 8.0 km/s over the north magnetic pole of the earth. The earth's magnetic induction at this point is 6×10^{-5} T. Calculate the induced emf, which is also the PD between the two astronauts.

21-B13 A flat circular wire loop of radius 4 cm and resistance 3 Ω is in a region of space where the magnetic field strength is 0.06 Wb/m², perpendicular to the plane of the loop. Calculate the current in the loop if the field decreases at a steady rate becoming 0 after 200 μs.

21-B14 A flat coil has 100 turns of fine wire, each of radius 20 cm. The coil is first oriented broadside to a uniform magnetic field and then, in 0.050 s, it is turned through 90° so that the plane of the coil is parallel to the field. The average induced emf is found to be 0.040 V. Calculate the magnetic induction at the coil.

21-B15 What emf is induced in a flat coil of 40 turns, each of radius 10 cm, which is broadside to a uniform magnetic field for which B is changing at the rate of 90 T/s?

21-C1 A coil of diameter 5 cm having 80 turns and resistance 4.0 mΩ is flat on a table, and a magnetic induction perpendicular to its plane changes at a steady rate from 0 to 0.6 T. How much charge flowed through the coil during the time required for the field to change?

21-C2 Suppose the apparatus of Fig. 21-27 is tilted so the rails make an angle α with the horizontal, and **B** is vertical ($\phi = 0$). Show that the frictionless rod (of resistance R) slides down the rails at a constant (terminal) speed given by $v = mgR \sin \alpha / B^2 l^2 \cos^2 \alpha$. (*Hint:* Equate the magnetic force to the component of weight parallel to the rails.) What if $\alpha = 0$? What if **B** is reversed in direction, still vertical?

21-C3 A flat square coil 50 cm on an edge has 200 turns and is in a magnetic field perpendicular to its plane which changes with time t according to the following equation: $B = 0.5$ T $+ (0.01t)$ T·s^{-1} $- (0.006t^2)$ T·s^{-2} $+ (0.0001t^3)$ T·s^{-3}. (a) What is the emf in the coil at $t = 0$? (b) At what time is $\mathcal{E}$ a maximum?

For Further Study

21-10 Magnetic field due to a long straight wire

The magnetic induction $d\mathbf{B}$ at point P (Fig. 21-34) due to a small current element $I\,dx$ is out of the plane of the paper, perpendicular to both the current **I** and the radius vector **r**. The field is caused by the component of $\mathbf{I}\,dx$ perpendicular to **r**, which is given by $I\,dx \sin \phi$. Thus, from Eq. 21-5,

$$dB = k' \frac{I\,dx \sin \phi}{r^2}$$

The total field at P due to half the wire is

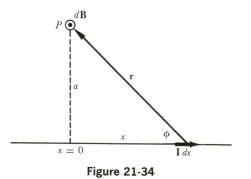

Figure 21-34

given by an integral between suitable limits:

$$B = \int_{x=0}^{x=\infty} dB \quad \text{or} \quad \int_{\phi=\pi/2}^{\phi=0} dB \quad \text{or} \quad \int_{r=a}^{r=\infty} dB$$

To carry out the integration, dB must be expressed in terms of one variable—x, ϕ, or r. If you like trigonometry, it is perhaps simplest to express everything in terms of ϕ. From the diagram, $\tan \phi = a/x$, or $x = a(\tan \phi)^{-1}$. Using the chain rule (Appendix E) we find

$$\frac{dx}{d\phi} = a(-1)(\tan \phi)^{-2}(\sec^2 \phi)$$

$$= -a\frac{\cos^2 \phi}{\sin^2 \phi}\frac{1}{\cos^2 \phi} = -\frac{a}{\sin^2 \phi}$$

Thus

$$dx = -\frac{a \, d\phi}{\sin^2 \phi}$$

Also, $\sin \phi = a/r$, whence $1/r^2 = (\sin^2 \phi)/a^2$.

Finally, we put the pieces together to get dB in terms of ϕ and $d\phi$:

$$dB = k'I\left(-\frac{a \, d\phi}{\sin^2 \phi}\right)(\sin \phi)\left(\frac{\sin^2 \phi}{a^2}\right)$$

The limits on ϕ for the positive half of the wire are $\phi = \pi/2$ (at $x = 0$) and $\phi = 0$ (at $x = \infty$). Thus the total B, due to both halves of the wire, is

$$B = 2\int_{\phi=\pi/2}^{\phi=0}\left(-\frac{k'I}{a}\right)\sin \phi \, d\phi$$

$$= -\frac{2k'I}{a}\left[(-\cos 0) - \left(-\cos \frac{\pi}{2}\right)\right]$$

$$= -\frac{2k'I}{a}[(-1) - (0)]$$

or

$$B = k'\frac{2I}{a}$$

This is Eq. 21-7.

■ PROBLEMS

21-C4 A flat square loop of wire of length a on each side carries a current I. (a) Show that the magnetic induction B at the center of the square is $11.31k'I/a$. (Hint: First calculate the induction caused by half of one side. The limits of integration are from $\phi = \pi/2$ to $\phi = \pi/4$.) (b) Check the answer to part (a) by showing that it lies between the value $12.56k'I/a$ for an inscribed circular loop (tangent to the square) and the value $8.88k'I/a$ for a circumscribed circular loop (touching the four corners of the square).

21-C5 Obtain Eq. 21-7 for the field due to a long straight wire by integrating with respect to x. (Hint: Express $\sin \theta$ and r in terms of a and x. The resulting integral can be evaluated between $x = 0$ and $x = \infty$ with the aid of a table of integrals more extensive than that in this book.)

■ REFERENCES

Note: In some references, the cgs unit for magnetic induction is used: 10^4 gauss = 1 Wb/m² = T. For a vacuum, 1 oersted = 1 gauss.

1 Magie, W. F., *A Source Book in Physics* (New York: McGraw-Hill, 1935), pp. 437–44, Faraday; 473–89, Henry and Faraday on magnetic induction; 511–13, Lenz's law; 513–19, Henry's investigations.

2 Wilson, Mitchell, "Joseph Henry," *Sci. American*, July, 1954, p. 72.

3 Nielsen, J. Rud, "Hans Christian Oersted—Scientist, Humanist and Teacher," *Am. J. Phys.*, **7**, 10 (1939).

4   Runcorn, S. K., "The Earth's Magnetism," *Sci. American,* Sept., 1956, p. 152.

5   Benfield, A. E., "The Earth's Magnetism," *Sci. American,* June, 1950, p. 20.

6   Elsasser, Walter M., "The Earth as a Dynamo," *Sci. American,* May, 1958, p. 44.

7   Doell, R. R. and G. B. Dalrymple, "Geomagnetic Polarity Epochs: A New Polarity Event and the Age of the Brunhes–Matuyama Boundary," *Science,* **152,** 1060 (1966).

8   Cox, A., C. B. Dalrymple, and R. R. Doell, "Reversals of the Earth's Magnetic Field," *Sci. American,* Feb., 1967, p. 44. Evidence for 9 reversals in 3.6 million years.

9   Dunn, J. P., M. Fuller, H. Ito, and V. A. Schmidt, "Paleomagnetic Study of a Reversal of the Earth's Magnetic Field," *Science,* **172,** 840 (1971).

10 Furth, H. P., M. A. Levine, and R. W. Waniek, "Strong Magnetic Fields," *Sci. American,* Feb., 1958, p. 54.

11 Kunzler, J. E., and M. Tannenbaum, "Superconducting Magnets," *Sci. American,* June, 1962, p. 60.

12 Shiers, G., "The Induction Coil," *Sci. American,* May, 1971, p. 80.

13 Lee, E. W., *Magnetism* (Baltimore, Md.: Penguin Books, Inc., 1963).

14 Miller, F., Jr., *Paramagnetism of Liquid Oxygen; Ferromagnetic Domain Wall Motion* (films).

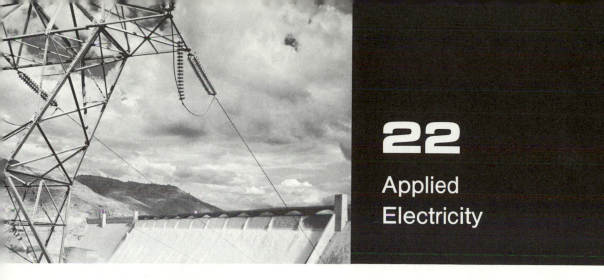

22

Applied Electricity

In this chapter we touch briefly on a representative few of the electrical machines and devices which play such a large part in the technical aspects of present-day culture. Our aim is to show these devices as logical applications of the basic principles we have already studied. First, we see that meters, motors, and generators have much in common.

22-1 Meters

At the heart of applied electricity is the direct-current meter. In Sec. 20-6 we saw how a galvanometer can be modified by a series resistor to make a voltmeter, or by a parallel shunt to make an ammeter. What factors determine the sensitivity of a galvanometer? In other words, what factors determine the deflection caused by a given current in the coil of the galvanometer? This is a problem in design. Let us study the galvanometer in detail, as an illustration of how physical principles are used in the design of apparatus.

The moving-coil galvanometer consists of a flat coil supported in the magnetic field of a permanent iron-alloy magnet (Fig. 22-1). As electrons move through the coil, they experience a magnetic force, which causes the

coil to turn against the resisting torque of the springs S and S'. The "forward" torque is due to the magnetic force on the vertical segments of the coil, PQ and XY. We calculate this torque τ, which equals force × lever arm, as follows: Using Eq. 21-4', we find the force to be $F = I \Delta l \, B$; the total Δl for each segment is l, and the total force on each vertical wire is IBl. The lever arm is $a/2$, hence

$$\tau = IBl\left(\frac{a}{2}\right)$$

The total torque acting on both wires is

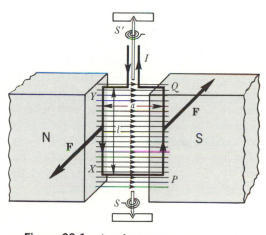

Figure 22-1 A galvanometer movement.

At this power-generating station, gravitational potential energy of water is transformed into electric energy which is sent over high-voltage transmission lines. (Photo for Allis-Chalmers by Robert Yarnall Richie)

463

therefore $2(IBl)(a/2)$. Finally, if the coil consists of N turns of wire, the total torque τ is $NIBla$, or $NIBA$, since la is the area A of the coil.[*] This torque causes the coil to start to turn; as it does so, the springs wind up and give rise to a restoring torque. By Hooke's law, this restoring torque τ' is proportional to the angle of twist—say, $\tau' = k\theta$, where k is the "torsion constant" of the spring system. The coil comes to rest at such an angle that $\tau' = \tau$, for then the net torque $\tau' - \tau$ is zero and there is no angular acceleration of the coil. Thus

$$\tau' - \tau = 0$$
$$k\theta - NIBA = 0$$
$$\theta = \left(\frac{NBA}{k}\right)I \qquad (22\text{-}1)$$

If the design problem is to get the most deflection θ for a given current I, our "working equation" shows that we should strive for a strong magnetic induction B and a coil of large area A having many turns N. It also shows that we should use a spring system that is easily twisted, in other words, that has a small k.

Closer study reveals a defect in the design of our galvanometer movement. In Fig. 22-2a, where the galvanometer movement has turned through an angle θ, the force **F** is no longer fully effective, since the lever arm is less than $a/2$. In fact, if θ reached 90°, the magnetic force would produce no torque at all (Fig. 22-2b). To overcome this defect, the pole pieces are curved, and a fixed cylindrical iron core is placed inside the coil to concentrate the lines of force along radii (Fig. 22-2c). The magnetic force on the current-carrying segments is the same as before, but now the lever arm is the full $a/2$, and the torque caused by the current does indeed depend only on N, B, A, and I, and not upon the coil's position.

We have discussed one electrical device at

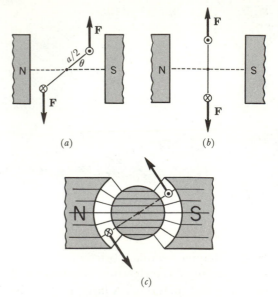

Figure 22-2 Top view of galvanometer coil. (a) Torque is reduced as coil starts to move. (b) Torque is zero when coil is broadside to the magnetic field. (c) Shaping the pole pieces and inserting iron core restore lever arm to its full value.

some length, as an illustration of a typical engineering procedure. Its successful design follows from a full understanding of the underlying physical principles, as expressed in a working equation such as Eq. 22-1. The designer uses the equation to predict the effect of changes in the design or construction. As often happens, compromises must be made. For instance, for high sensitivity we need a large coil (large A), but then the large moment of inertia of the coil will make the period of oscillation long,[*] and perhaps the meter will be too sluggish for convenient use. Again, high sensitivity can be obtained by using a very delicate spring system (small k), but this would also lengthen the period, and in addition it would make the meter fragile and easily damaged. Looking at other factors in our working equation, we see that a strong magnetic induction is desirable

---

[*] Although our derivation is for a rectangular coil, the result is valid if A is the area of a flat coil of any shape.

[*] The period is given by $T = 2\pi\sqrt{I/k}$, where k is the torsion constant τ'/θ, and I is the moment of inertia (see Secs. 9-4 and 9-7).

(large B). Modern alloys such as Alnico V (51% Fe, 24% Co, 14% Ni, 8% Al, 3% Cu) provide strong fields with small pole pieces; hence modern meters are considerably smaller than those of seventy years ago.

Permanent magnets are not "permanent," strictly speaking, for the domains always have a tendency to become disoriented. As a meter ages, the magnitude of the magnetic induction B tends to weaken, and so the sensitivity drops off. For this reason, in precise work a meter is calibrated at regular intervals. Finally, we must compromise on the number of turns of wire (N): for utmost sensitivity a coil should have many turns, but this would be physically awkward unless fine wire is used. In that event, the many turns of fine wire would have a relatively high resistance (Eq. 19-6), which is undesirable for a meter that is to measure small potential differences such as the emf of a thermocouple.

Our analysis of the galvanometer movement is similar to the analysis that could be made for every instrument, machine, or device. This general procedure is typical of applied science, or engineering.

22-2 Motors and generators

A direct-current motor is essentially a galvanometer movement which is allowed to turn continuously. In cross section (Fig. 22-3) we see the similarity: a motor has a coil (called the *armature*) wound on an iron core, which concentrates the magnetic induction just as in the galvanometer. To make continuous rotation possible, the armature leads are brought out to a pair of sliding contacts called a *commutator*. While the armature coasts through the "dead center" position of Fig. 22-4*b*, the direction of current is reversed as the brushes make contact with the other segment of the commutator. Note the reversal of current direction in the armature. Thus whichever conductor is near the north pole is at that time carrying current in the proper

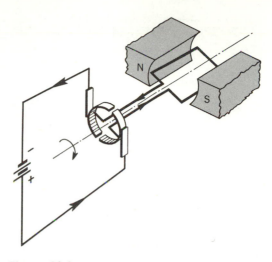

Figure 22-3 A simple motor (schematic; iron core of armature not shown).

direction to give rise to an upward force. In actual practice, an electromagnet is usually used to supply the magnetic field. In order to obtain a smooth-running motor, the armature may have a number of partial windings brought out to a many-segmented commutator.

The construction of a generator is basically the same as that of a motor. The armature is turned by some external torque (Fig. 22-5), and as the wires cut across the magnetic lines of force an emf is induced according to $\mathcal{E} = Blv$ (Eq. 21-8); for a given generator B and l are constant, and the output emf is proportional to the speed with which the armature turns. Let us fix our attention on one conductor of the armature. In Fig. 22-6*a* it is moving upward across the lines of force, and there is a maximum induced emf. When the armature reaches position *b*, the conductor is moving parallel to the field, and there is no induced emf. At *c*, this same conductor moves down across the lines of force, and the induced current is directed opposite to that in *a*. If we plot a graph of the current in the lamp as a function of time, we get a sinusoidally varying current known as an *alternating current* (Fig. 22-6*d*). We see that it

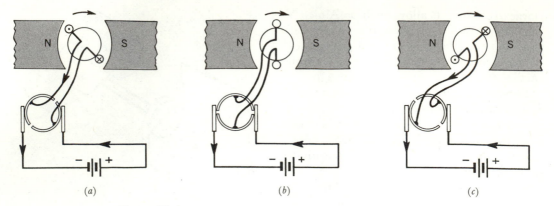

(a)   (b)   (c)

Figure 22-4   Commutator action in a direct-current motor.

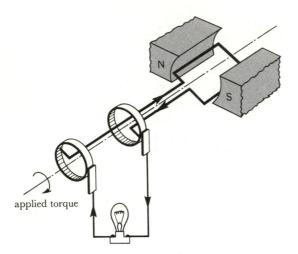

applied torque

Figure 22-5   A simple generator. The armature is turned by a torque supplied by some source of mechanical power.

is not difficult to produce alternating currents. The frequency of an alternating current equals the number of times per second that the current changes from + to − and back again. Ordinary house current in the United States fluctuates at a frequency of 60 cycles/s. The study of alternating currents is highly technical and of great importance in engineering.

It is sometimes necessary to generate a steady, or *direct, current*—for instance, for the purpose of charging a storage battery in a car. The reversing action of a commutator can be

used to make pulsating direct current out of alternating current (Fig. 22-7). The commutator is a quick-acting switch which reverses the connections to the armature at just the right times to compensate for the reversals in current which would otherwise take place. In most cities, however, the direct-current motor and generator have disappeared from the scene, as far as public distribution of power is concerned. We shall see a reason for this when we study transformers in Sec. 22-5. Modern alternating-current motors do not need a commutator, and are more compact and rugged than the direct-current motor we

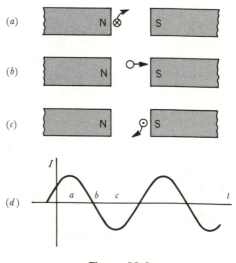

Figure 22-6

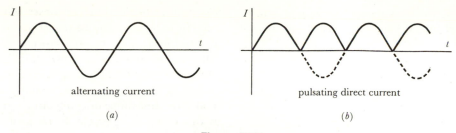

alternating current

(a)

pulsating direct current

(b)

Figure 22-7

have described. Here, too, the full story is technical, and we shall not pursue the matter further.

22-3 Back emf and back torque

Since the physical construction of a motor is practically the same as that of a generator, it follows that if a dc generator is connected to a battery, it will run as a motor; and if a motor is turned by any external means, it will generate an emf.[*] This dualism is with us willy-nilly: whenever a motor is running, its generator action cannot be turned off. By Lenz's law (Sec. 21-8) the induced emf tends to oppose the change which causes it. In this case, the "cause" is the current through the armature, and so the induced emf tends to

_____
[*] In some early automobiles, as in the 1924 Dodge, after the car was started the starter motor was used as a generator (turned by the car's engine) to keep the storage battery charged.

reduce this armature current. This induced emf, which is the unavoidable result of generator action in a motor, is called *back emf.* When the motor is just starting up, its armature is not turning, and hence it is not generating a back emf. As the motor picks up speed, the back emf increases, hence the armature current decreases. This accounts for the fact that the current through a motor is larger while it is starting than while it is running at full speed.

EXAMPLE **22-1**

A motor has an armature of resistance 5 Ω. The magnetic field is supplied by field coils, which form an electromagnet of resistance 55 Ω connected in parallel with the armature. The motor and field coils are both connected to a 110 V dc line, and at full speed the back emf is 100 V. (*a*) What is the current through the motor when it first starts up?   (*b*) What is the current when the motor is running at full speed?

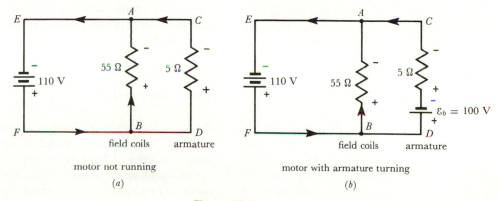

motor not running

(a)

motor with armature turning

(b)

Figure 22-8

(*a*) When starting, the circuit is a simple parallel circuit (Fig. 22-8*a*). The PD across *AB* is 110 V, and the PD across *CD* is also 110 V. We use Ohm's law to find the current in each branch.

Field coils:

$$I = \frac{V}{R} = \frac{110 \text{ V}}{55 \text{ }\Omega} = 2 \text{ A}$$

Armature:

$$I = \frac{V}{R} = \frac{110 \text{ V}}{5 \text{ }\Omega} = 22 \text{ A}$$

Total current when starting is

$$2 \text{ A} + 22 \text{ A} = \boxed{24 \text{ A}}$$

(*b*) As the motor picks up speed, the back emf $\mathcal{E}_b$ grows. We represent this emf by the symbol for a battery; it is in every way equivalent to a battery connected in series with the armature (Fig. 22-8*b*).

Field coils:

$$I = \frac{V}{R} = \frac{110 \text{ V}}{55 \text{ }\Omega} = 2 \text{ A}$$

which is the same current as before.

Armature: We apply Kirchhoff's second law around the loop *CDBFEAC*, letting *I* represent the unknown current through the armature. The PD across the 5 Ω resistor is 5*I*, directed as shown. Hence, starting at *C* and going around the loop, we have

$$+5I + 100 \text{ V} - 110 \text{ V} = 0$$

$$5I = 10 \text{ V}$$

$$I = 2 \text{ A} \quad \text{(instead of the previous 22 A)}$$

The total current through the motor, while it is running, is

$$2 \text{ A} + 2 \text{ A} = \boxed{4 \text{ A}}$$

The large starting current of a motor is familiar in everyday life. A car's headlights may dim when the starter is operated; the house lights may momentarily dim when a refrigerator or furnace motor starts up. As we saw in Sec. 20-2, the terminal voltage of any source of current decreases when the current increases, due to the *IR* drop in the internal resistance of the source. When a refrigerator motor starts up, the "source" consists of everything behind the floor socket, including the house wiring. Usually we can ignore the resistance of connecting wires, but in this case, an appreciable *IR* drop occurs in the house wiring because of the large value of *I* while the motor is starting. A lamp plugged into the same floor socket as the refrigerator becomes dim, since its terminal voltage always equals that of the refrigerator with which it is connected in parallel.

Not only does a motor display generator action, but a generator behaves to a certain extent like a motor. If a generator is causing charges to flow through an external circuit, the charges must, of course, also flow through the armature. But these charges flowing in an armature which is in a magnetic field must experience a torque, and according to Lenz's law this torque must oppose the applied torque by which the generator is being turned. The *back torque* exists only when there is a complete circuit for the flow of charges; a big generator turning at full speed could be wrecked by a sudden attempt to obtain a large current from it. We can consider the generator as a link in a chain by which mechanical energy is transformed into electric energy, and eventually into heat energy in a load such as a lamp bulb. The back torque is a necessary part of this chain, for it is only by pushing against *something* (the back torque) that the source of mechanical energy is able to do work.

22-4 Eddy currents

We have seen (Sec. 21-8) that an emf is induced in any conductor which moves across magnetic lines of force or in any conductor across which lines of force move. The *relative* motion of conductor and magnetic field gives rise to an emf in a conductor. It often happens

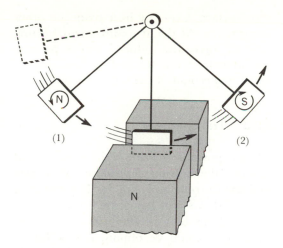

Figure 22-9 Eddy currents induced in the swinging plate are in such a direction as to produce magnetic poles which oppose the motion.

that there is no well-defined path for the induced current. For instance, a flat metal plate on the end of a rod is allowed to swing as a pendulum (Fig. 22-9), passing through a strong magnetic field at the lowest point of its swing. As the plate starts to enter the region of magnetic field (position 1), the induced emf causes free electrons in the metal to swirl around. Such currents, distributed throughout the body of a piece of metal, are called *eddy currents*. By Lenz's law, the direction of an eddy current must be such as to oppose the change that causes it. As the plate swings down, electrons flow clockwise, conventional current is counterclock-

wise, and by the right-hand rule the front face of the plate becomes a north pole and the rear face becomes a south pole. These induced poles are repelled by the poles of the magnet which they are approaching. The eddy currents reverse direction at the instant when the plate is passing through its lowest point. As the plate swings away (position 2), the eddy currents create poles which are attracted to the magnet—again opposing the motion. The currents reverse again when the plate is at its highest position and starts to move down again.

The result of these "educated" eddy currents is always to oppose the motion of the plate, and so the plate soon comes to rest. Such "electromagnetic damping" is put to practical use in some analytical balances in which solid plates of aluminum hang between the poles of small permanent magnets (Fig. 22-10). The balance arm would normally oscillate to and fro when the load or the weights on the pans are changed; and the more sensitive the balance, the more time would be required to reach a steady "zero." With the damping plates, however, the induced eddy currents react with the magnetic field to provide a force which opposes any swinging motion in either direction. Much time can be saved in a chemical or biological laboratory by use of electromagnetic damping.

On a larger scale, some subway and rapid-transit cars have electromagnetic brakes

William Ainsworth and Sons

Figure 22-10 Electromagnetic damper for an analytical balance. Thin aluminum plates move vertically between the poles of permanent magnets.

which depend on eddy currents. An electro-magnet hangs from the car near the steel rails (Fig. 22-11), and, when the operator wishes to stop the car, he sets up a large current through the coils of the electromagnet. If the car is moving, eddy currents are induced in the rails, and the direction of the currents must be such as to oppose the motion of the car. Such brakes are free from sudden jerks, for, as the car slows down, the eddy currents automatically die away. Thus the force de-creases steadily to zero, and the car "eases in" to a quick but gentle stop. The current may be partially supplied by the car's motor; the operator throws a switch to disconnect the motor from the third rail or trolley line and connect it to the magnet. The motor now functions as a generator, and the back torque of the generator also helps stop the car. From an energy viewpoint, the over-all result is that the KE of the car is transformed mostly into heat in the rails. This is Joule heat, given by I^2R, as the eddy currents circulate in the rails.

Sometimes eddy currents represent a wasteful dissipation of energy. For instance, if the armature of a motor were wound on a solid iron core, eddy currents would be in-duced in the rotating iron, the armature would become warm, and the motor would not be very efficient. To reduce eddy currents, the core is "laminated," that is, constructed of many thin iron sheets rather than of one solid piece. The sheets are electrically insu-lated from one another by lacquer, or by

thin surface layers of iron oxide or other compounds. As the armature turns, cutting lines of force, the induced emf is present, just as before, but the induced *currents* are small on account of the high resistance.

22-5 Transformers

Electric power is the rate of doing work, and this rate depends on the product of the emf and the current in a circuit. The equation $P = \mathcal{E}I$ shows that the same power can be obtained in a 12 V headlight bulb operating at 2 A as could be obtained in a 120 V house lighting circuit at a current of 0.2 A. For practical reasons, however, one or the other arrangement may be more suitable. The low-voltage circuit in the automobile is dictated by the low emf of the battery, free-dom from shock hazard, and reliability of insulation. The high-voltage circuit in the home is desirable so that the current to any bulb or appliance will be relatively low; this avoids the expense of installing the heavy copper wire which would be necessary if a low-voltage, high-current circuit were used. The *transformer* is a device for changing the effective value of the emf in a circuit. Indeed, the ease with which this can be done with alternating currents has led to the almost universal present-day use of ac power for home and industry.

A common type of transformer consists of an iron core upon which are wound a primary coil and a secondary coil. Let us start by imagining the primary connected to a battery through a switch S (Fig. 22-12). When the switch is closed, a current is estab-

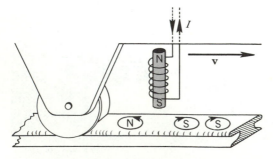

Figure 22-11 Eddy current brake of a streetcar (schematic).

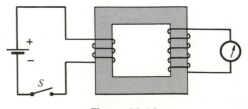

Figure 22-12

lished in the primary, causing an increasing magnetic field which orients the magnetic domains in the entire iron ring. While the domains inside the secondary are being oriented, their lines of force are cutting the wires of the secondary, and therefore a momentary emf is induced in the secondary. An equivalent way of describing the process would be to consider that the magnetic flux lines arising in the primary are confined to the iron core and also pass through the secondary coil; when the flux changes, the induced emf $\mathcal{E}$ is $d\Phi/dt$. As soon as the primary current reaches its steady value,[*] given by Ohm's law, there is no further change in the magnetization of the iron core, and the emf in the secondary becomes and remains zero. In other words, an emf *pulse* of short duration appears across the terminals of the secondary coil. If a galvanometer is connected to the secondary, there is a complete circuit, the induced emf is able to send a pulse of charge through the galvanometer, and a momentary deflection is observed. If at some later time the switch S is opened, the primary current becomes zero, the domains return to their random orientation, and while this is happening lines of force cut the secondary coil in a direction opposite to the cutting that previously took place while the domains were being oriented. Now the flux is decreasing, $d\Phi/dt$ is negative, and a momentary pulse is observed in the opposite sense to that previously observed (Fig. 22-13). Note that a *steady* current in the primary produces no effect in the secondary.

It should now be clear that a continually changing current in the primary will produce a continually changing current in the secondary. It can be shown that the ratio of primary emf $\mathcal{E}_p$ to secondary emf $\mathcal{E}_s$ equals the ratio of the number N_p of turns in the primary to the number N_s of turns in the secondary (Fig. 22-14):

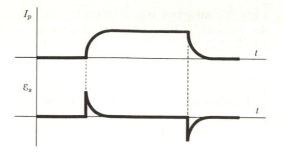

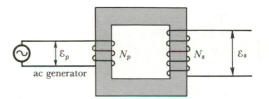

Figure 22-13 An emf $\mathcal{E}_s$ is induced in the secondary only when the current in the primary changes.

Figure 22-14

$$\frac{\mathcal{E}_s}{\mathcal{E}_p} = \frac{N_s}{N_p} \qquad (22\text{-}2)$$

This equation follows in part from the fact that, if a certain emf is induced in each turn, the total emf is the sum of the emf's induced in the individual turns, and hence $\mathcal{E}_s$ is expected to be proportional to N_s. Thus to get a large induced emf, we need many turns in the secondary coil.

Transformers can be either of the "step-up" or "step-down" variety, depending on whether the secondary emf is greater or less than the primary emf. The law of conservation of energy implies that the *power* in the primary equals the *power* in the secondary.[*] Hence, since $P = \mathcal{E}I$,

$$\mathcal{E}_p I_p = \mathcal{E}_s I_s \qquad (22\text{-}3)$$

or

$$\frac{I_s}{I_p} = \frac{\mathcal{E}_p}{\mathcal{E}_s} \qquad (22\text{-}4)$$

---

[*] This may take only a fraction of a second, or it may take several seconds, depending on the number of turns and amount of iron.

[*] This statement is true for an ideal transformer of 100% efficiency.

Thus for an ideal transformer, the ratio of the currents is the reciprocal of the ratio of the emf's.

EXAMPLE **22-2**

A certain ideal transformer's primary has 2000 turns, and its secondary has 50 turns. The primary is connected to a source having emf 120 V; the secondary is connected to a lamp bulb having resistance (when lit) of 0.6 Ω. (a) Calculate the emf of the secondary; (b) calculate the current through the bulb; (c) calculate the current in the primary; (d) calculate the power in the primary and in the secondary.

(a) This is a step-down transformer, since the secondary has fewer turns than the primary. The turns ratio is 2000 turns/50 turns = 40. Hence

$$\mathcal{E}_s = (\tfrac{1}{40})(120 \text{ V}) = \boxed{3 \text{ V}}$$

(b)
$$I_s = \frac{\mathcal{E}_s}{R_s} = \frac{3 \text{ V}}{0.6 \text{ Ω}} = \boxed{5 \text{ A}}$$

(c)
$$\frac{I_s}{I_p} = \frac{\mathcal{E}_p}{\mathcal{E}_s}; \quad \frac{5 \text{ A}}{I_p} = \frac{120 \text{ V}}{3 \text{ V}}$$

$$I_p = \boxed{0.125 \text{ A}}$$

(d) Power in primary:
$$\mathcal{E}_p I_p = (120 \text{ V})(0.125 \text{ A}) = \boxed{15 \text{ W}}$$

Power in secondary:
$$\mathcal{E}_s I_s = (3 \text{ V})(5 \text{ A}) = \boxed{15 \text{ W}}$$

Modern transformers can be constructed to have efficiencies greater than 99%, and so our assumption of 100% efficiency in Example 22-2 is not far out of line.

Electric energy is transmitted through long distances at high voltage, in order that, for a given power, the current in the line will be as small as possible. Any transmission line has some resistance, and the energy dissipated as Joule heat in the line, proportional to the square of the current, is wasted. Our next examples illustrate the advantage of using high voltages for transmitting electric energy.

EXAMPLE **22-3**

A transmission line between a power station and a factory has a resistance of 0.1 Ω in each of two wires. If 200 A is "delivered" at 100 V, (a) what useful power is delivered into the load? (b) What power is wasted in heating the line? (c) What total power must be supplied by the generator?

(a) Power into load:
$$\mathcal{E}I = (100 \text{ V})(200 \text{ A}) = \boxed{20\,000 \text{ W}}$$

(b) Power used in heating line:
$$I^2 R = (200 \text{ A})^2(0.2 \text{ Ω}) = \boxed{8000 \text{ W}}$$

(c) Power supplied by generator:
$$20\,000 \text{ W} + 8000 \text{ W} = \boxed{28\,000 \text{ W}}$$

EXAMPLE **22-4**

Repeat Example 22-3 for the same line, assuming that the same useful power is delivered, but at a voltage of 1000 V instead of 100 V. Find (a) the current in the line, (b) the power used in heating the line, (c) the total power supplied by the generator.

(a) Power into load: 20 000 W, as before.
$$I = \frac{P}{\mathcal{E}} = \frac{20\,000 \text{ W}}{1000 \text{ V}} = \boxed{20 \text{ A}}$$

(This is $\tfrac{1}{10}$ as much current as before.)

(b) Power used in heating line:
$$I^2 R = (20 \text{ A})^2(0.2 \text{ Ω}) = \boxed{80 \text{ W}}$$

(c) Power supplied by generator:
$$20\,000 \text{ W} + 80 \text{ W} = \boxed{20\,080 \text{ W}}$$

By using higher voltage, perhaps by employing a different generator design, the power used in heating the wires is reduced from 8000 W to only 80 W. This energy is lost, unless it possibly helps keep warm some birds perched on the wires. Of course, if the power is delivered to the city at high voltage, it must be transformed to a useful (lower) emf by use of a step-down transformer at the place where it is to be used. Distribution of electric energy has played a major role in the expanding economy of the past 50 years, and we should give the transformer full credit as the device which has made this possible. Its basic principles were discovered by Joseph Henry and Michael Faraday in the 1830's, and the engineering genius of Nikola Tesla and Charles P. Steinmetz in the late nineteenth century implemented the change from the dc system of Edison to our modern flexible and economical ac distribution system.[*]

In each step-up or step-down transformer, no direct connection exists between primary and secondary, and the insulation between the windings must be sufficient to withstand high voltages—up to hundreds of thousands of volts in some cases. The lines in the alleys and along the streets of a city are usually at 2300 V, serving areas of many blocks, and each house or group of several houses is served by its own step-down transformer whose secondary has a grounded center tap. This makes available 115 V for lamps and small appliances, and also 230 V for large appliances such as stoves, furnaces, and dryers, in which the current would be too large if power had to be supplied at 115 V.

The primary winding of a pole transformer is of heavy wire, and its resistance may be only 0.1 Ω or less. This winding is permanently connected across a 2300 V line. At first glance, Ohm's law leads us to expect that this winding would continuously carry some 23 000 A. Actually, it carries only a very small fraction of an ampere, unless some lamp, motor, or other load is turned on in the secondary circuit inside the house. To explain this paradox, we need to study more closely just how a coil limits, or "impedes," alternating current.

[*] Direct-current power distribution has staged a comeback recently, with the advent of high-voltage, high-current mercury-arc rectifiers. See Prob. 22-C4 for an example.

22-6   Impedance of a coil

Suppose we measure the current through a certain coil connected to a source of direct current (Fig. 22-15a). Using Ohm's law, we find the resistance of the coil to be

$$R = \frac{6 \text{ V}}{6 \text{ A}} = 1 \text{ } \Omega$$

However, repeating the experiment with a 6 V source of 60 Hz ($=$ 60 cycle/s) alternating emf, represented by —Ⓐ— , we find a much smaller alternating current. The "apparent resistance" in Fig. 22-15b is

$$\left(\frac{6 \text{ V}}{0.006 \text{ A}} \right) = 1000 \text{ } \Omega$$

This ratio is not constant for alternating current of different frequencies, for, if we repeated the experiment using a 120 Hz source, we would find that I was only 0.003 A, so that the ratio V/I would be 2000 Ω. The coil "impedes" the flow of charge, and we define *impedance Z* in the same way that resistance can be defined for a dc circuit:

$$\text{Impedance} = \frac{\text{alternating PD}}{\text{alternating current}}$$

$$Z = \frac{V}{I} \qquad (22\text{-}5)$$

$$\text{ohms} = \frac{\text{volts}}{\text{amperes}}$$

For the 60 Hz experiment, the coil of Fig. 22-15 has a resistance of 1 Ω and an impedance of 1000 Ω.

The reason the current in Fig. 22-15b is so small is that the coil acts as a transformer,

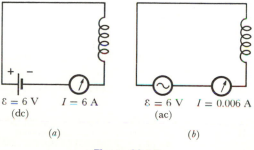

$\varepsilon = 6 \text{ V} \qquad I = 6 \text{ A}$
(dc)

$\varepsilon = 6 \text{ V} \qquad I = 0.006 \text{ A}$
(ac)

(a)   (b)

Figure 22-15

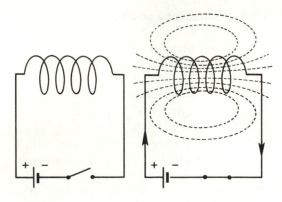

Figure 22-16

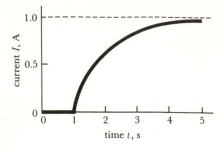

Figure 22-17 Rise of current through a coil; switch is closed at $t = 1$ s.

with the same winding serving as both primary and secondary. When there is a current through a coil, a magnetic field proportional to the current exists within the coil (Fig. 22-16). While the current is building up, lines of force are cutting the wires of the coil, and so an emf is induced in the coil by its own changing magnetic field. This phenomenon is called *self-induction*. The induced emf is a back emf, and by Lenz's law it tends to oppose the change. Thus the back emf keeps the current in a large iron-cored magnet from rising instantaneously to its final value (Fig. 22-17). For any coil, we can define a constant called its *self-inductance L,* measured in *henries* (H). The larger the self-inductance, the greater the back emf for a given rate of change of current.* As shown in Sec. 22-12, the impedance of a coil is given by the formula

$$Z = 2\pi\nu L \qquad (22\text{-}6)$$

where ν is the frequency in Hz (cycles/s); it is assumed that the ordinary resistance of the coil is negligible.

We see now how the impedance of a coil can be so much larger than its resistance. If the induced back emf counteracts to a large extent the applied PD, then the net emf in the circuit is small and the current is small. As with any transformer action, the induced emf is greater if the alternating current reverses more rapidly. Also, the induced back emf in a coil will be greater if the coil has many turns, especially if it is wound on an iron core in which magnetic lines of force are concentrated. The primary coil of a transformer fulfills all these conditions admirably, and so we expect the primary current to be small because of the large back emf.

EXAMPLE **22-5**

Calculate the self-inductance of a transformer primary winding, if a 60 Hz generator whose emf is 115 V sends 0.1 A through the coil.

$$Z = \frac{V}{I} = \frac{115 \text{ V}}{0.1 \text{ A}} = 1150 \ \Omega$$

To find L, we use Eq. 22-6:

$$Z = 2\pi\nu L$$

$$L = \frac{Z}{2\pi\nu} = \frac{1150 \ \Omega}{2\pi(60 \text{ Hz})}$$

$$= \boxed{3.05 \text{ H}}$$

Finally, we seek a mechanism by which a current *does* exist in the primary, when a circuit is closed in the secondary. For instance, when a housewife plugs in an electric iron, how does this cause the *primary* current to increase from its previously negligible value? The interaction is, of course, through the iron core of the transformer. The secondary current creates its own changing mag-

---

* The *self-inductance* of a coil is defined as back emf divided by rate of change of current; a coil's self-inductance is 1 H if a back emf of 1 V is induced in the coil when the current through it changes at the rate of 1 A/s.

netic field, which in turn induces an emf in the primary—a sort of "back-back emf." The primary current increases because the original back emf (which limited the current) is partially canceled by this third emf; the cancellation takes place only when there is current in the secondary. In this way, an increased load in the secondary circuit results in an increased current in the primary —all without direct connection between the two coils.

22-7   Impedance of a capacitor

The behavior of a capacitor in an ac circuit deserves a brief discussion, for there is a sharp contrast between its response to alternating and to direct currents. That is, the impedance of a capacitor is quite different from its resistance. When a capacitor is connected to a dc source of potential, such as a battery, in series with an ammeter and a lamp, there is a momentary rush of charge as the capacitor becomes charged. The ammeter needle swings over for an instant if its inertia is not too great, but it soon returns to zero and stays there. The lamp flashes momentarily. The situation is different if the source is an ac generator. The capacitor is alternately charged in one sense and then in the other, and charges are rushing to and fro through the meter and the lamp (Fig. 22-18a and b). No charge actually moves

across the gap between the plates of the capacitor,* but the lamp and meter don't know the difference. From the lamp's point of view, just as much heat and light are produced as if the charge really did flow in a complete circuit, as in Fig. 22-18c. To all intents and purposes, there is an alternating current *through* the capacitor as far as the rest of the circuit is concerned.

What limits the current through the capacitor in Fig. 22-18? In other words, what factors determine the impedance of a capacitor? The larger the capacitance, the more charge flows to and fro, since more charge can be momentarily stored on the plates (according to the equation $Q = CV$). The higher the frequency of the alternating current, the greater the rate of change of charge, and hence the greater the current. Thus the current is directly proportional both to C and to ν; and the impedance, which is V/I, is inversely proportional to C and ν. As shown in Sec. 22-12, the impedance of a capacitor is given by the formula

$$Z = \frac{1}{2\pi\nu C} \tag{22-7}$$

Especially in audio amplifiers and radio circuits, where ν is large, even a small capacitor's impedance is often low enough to be negli-

---

* We assume that the dielectric material between the plates is a perfect insulator; that is, the dc resistance of the capacitor is assumed to be infinite. This is a valid assumption for most practical capacitors.

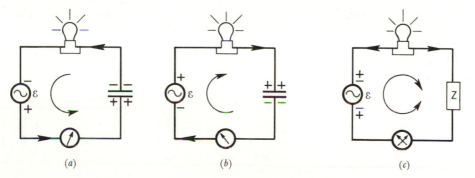

(a)   (b)   (c)

Figure 22-18   (a) and (b) Flow of charge in an alternating-current circuit containing a capacitor. (c) Equivalent circuit, replacing the capacitor by an inductance Z.

gible. To this extent, a capacitor is a short circuit for alternating current but an open circuit for direct current.

See Prob. 22-C5 for an illustration of the opposite situation, in which a lamp would be practically as bright as if the capacitor were not present.

EXAMPLE 22-6

Calculate the impedance of a 0.02 μF capacitor in a circuit for which the frequency is (a) 60 Hz; (b) 6 MHz.

(a) $Z = \dfrac{1}{2\pi(60\ \text{Hz})(0.02 \times 10^{-6}\ \text{F})}$

$= \boxed{1.33 \times 10^5\ \Omega}$ (at 60 Hz)

(b) $Z = \dfrac{1}{2\pi(6 \times 10^6\ \text{Hz})(0.02 \times 10^{-6}\ \text{F})}$

$= \boxed{1.33\ \Omega}$ (at 6 MHz)

At 6 MHz (a radio frequency), the impedance is about 1 Ω, which is negligible in comparison with other impedances in a radio circuit.

EXAMPLE 22-7

A 110 V lamp bulb rated at 50 W is connected to a 110 V, 60 Hz power line through a capacitor of 1μF. Will the lamp light?

This problem is stated in such a way that some preliminary spadework must be done. The normal current in a 110 V, 50 W bulb is given by

$$I = \frac{P}{V} = \frac{50\ \text{W}}{110\ \text{V}} = 0.455\ \text{A}$$

If the current is much less than this value, the lamp will not light. To find the current, we next compute the impedance of the capacitor:

$$Z = \frac{1}{2\pi\nu C} = \frac{1}{2\pi(60\ \text{Hz})(1 \times 10^{-6}\ \text{F})}$$

$$= 2650\ \Omega$$

Even if the bulb were replaced by a short circuit, the capacitor would limit the current to

$$I' = \frac{110\ \text{V}}{2650\ \Omega} = 0.0415\ \text{A}$$

and, if the bulb is also in series with the capacitor, the current would surely be even less than 0.0415 A. Since the normal current for the bulb when fully lit is 0.455 A, we conclude that the current through the capacitor would be insufficient to cause a visible glow.

22-8 Thermoelectricity

One way of measuring temperature electrically has already been discussed in Sec. 20-7. The resistance of a coil of wire depends on its temperature, and, if the resistance is measured with a Wheatstone bridge, the temperature can be computed. Another common way of measuring temperature is by means of a *thermocouple,* which consists of a pair of junctions between two dissimilar metals. A simple thermocouple circuit, using iron and copper, is shown in Fig. 22-19. The two iron-copper junctions are X and Y, with X held at some reference temperature ($0°$C in the diagram). If X and Y are at different temperatures, there is a net emf in the circuit. For small temperature changes, the emf is proportional to the temperature *difference* between X and Y. Tables in handbooks give the value of this *thermal emf* for various pairs of metals; for iron-copper junctions near room temperature, the thermal emf is about 14×10^{-6} V for a temperature difference of 1 C° (14 μV/C°). Such small emf's can be measured by observing the deflection of the galvanometer of Fig. 22-19. Ohm's law for a complete circuit says that $I = \text{net } \mathcal{E}/\text{net } R$; in this circuit the net emf is the thermal emf due to the temperature difference, and the net re-

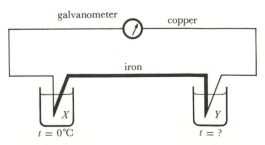

Figure 22-19 Thermocouple circuit; transformation of thermal energy into electric energy.

sistance is the resistance of the junctions, the connecting wires, and the galvanometer coil. For the most precise work, thermal emf's are often measured with the help of a potentiometer (Sec. 20-8).

In order to discuss the theory of thermoelectricity, let us first consider two insulated blocks of metal which are initially uncharged (Fig. 22-20*a*). Iron and copper differ in crystal structure, density, conductivity, thermal expansion, and other properties, and so we expect that the number of free electrons per unit volume will also be different in these dissimilar metals. If now we bring the blocks into contact, an electron which finds itself just at the boundary between iron and copper will experience a net force in one direction or the other, and electrons will tend to flow from one metal block to the other. Experiment shows that in the situation in Fig. 22-20*b*, an excess of electrons flows from iron to copper. A PD is set up between copper and iron merely because the metals are in contact; we call this the *contact potential difference* between iron and copper. If the blocks are separated again, as in (*c*), the excess charges become redistributed over their surfaces, as can be shown by using an electroscope and a proof plane (see Sec. 17-4). All this takes place with the two blocks at the same temperature.

In a complete circuit such as Fig. 22-21, if *X* and *Y* are at the same temperature, the contact PD at *X* exactly opposes that at *Y*, and there is no current. However, the contact PD between any pair of metals depends on temperature, just as almost every physical property does. If, then, *X* and *Y* are at

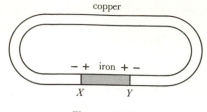

Figure 22-21

different temperatures, the contact PD's will be different, a net PD will exist, and a steady current will exist. From this point of view, thermoelectricity is simply a manifestation of the fact that contact potential difference varies with temperature.

Why do we speak of thermal emf rather than thermal PD? Let us recall that a seat of emf is a place where some form of nonelectric energy is *reversibly* transformed into electric energy. The pair of junctions *X* and *Y* illustrate this transformation of energy. Heat energy is being converted into electric energy; in a similar fashion, chemical energy is transformed into electric energy in a battery, and mechanical energy is transformed into electric energy in a generator. The reversibility of the thermocouple (considered as a seat of emf) is illustrated by the following experiment. A battery sets up a steady electron current through an iron-copper circuit, as shown in Fig. 22-22. If both water baths are originally at 20°C, junction *X* becomes slightly warmer and junction *Y* becomes slightly cooler. While the electrons flow, electric energy is being transformed at junction *X* into heat energy in a way that is en-

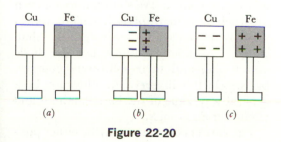

(*a*)   (*b*)   (*c*)

Figure 22-20

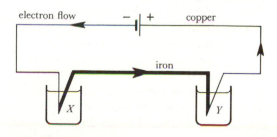

Figure 22-22 Thermoelectric heater (at *X*) and refrigerator (at *Y*).

tirely different from the Joule heating in all resistors. At the same time, heat energy is being absorbed from the water at junction Y. If electrons are forced through the junctions in the other direction, the roles of X and Y are interchanged, and X absorbs heat instead of liberating heat. At a given junction, the transformation can go in either direction; this is what is meant by reversibility, a characteristic of all seats of emf. "Thermoelectric refrigerators" are based upon this principle; they have no moving parts and can be operated as heaters simply by reversing the direction of the current.

Although thermal emf's are small, it is often possible to get a larger net emf by connecting several junctions in series, to form a *thermopile*. Of course, the internal resistance of a thermopile is greater than that of a single pair of junctions. Thus the terminal voltage (given by $V = \varepsilon - Ir$) cannot be increased indefinitely. In practice, 10 to 50 thermocouples in series are about as many as can be used effectively.

Since a thermopile transforms heat energy into electric energy, it is tempting to consider the possible utilization of solar energy by such means. The conversion of solar energy to useful forms is an intriguing research problem. As we pointed out in Sec. 14-6, about 1.94 cal of heat energy falls on each square centimeter of the earth's surface every minute when the sun is directly overhead, if absorption in the atmosphere is neglected. This works out to be a surprisingly large and potentially useful source of energy; about 840 hp falls* on a factory roof which measures 50 ft $\times$ 100 ft. If this heat can be utilized, it is a source of energy which may well prove to be a greater economic boon than nuclear energy sources. One promising avenue of approach is the use of thermopiles. The problem is to obtain a reasonably high efficiency at a cost which is economically feasible. The electrons must flow through the internal re-

sistance of the junctions, and in so doing produce Joule heat, which is wasted. Efficiencies as high as 20% of the theoretical Carnot efficiency have been obtained by judicious choice of junction materials. In a general sense, a thermocouple is a heat engine, since heat energy is converted to a useful (electric) form of energy, and the maximum efficiency is given by Carnot's relation (Sec. 16-5)

$$\text{Efficiency} = \frac{T_1 - T_2}{T_1}$$

It is advantageous therefore to focus the sun's rays to obtain as large a temperature difference as possible. A major problem in all solar energy devices is the storage of energy during hours of sunlight so that useful heat, light, and power can be obtained during the night and on cloudy days. An all-electrical system could use storage batteries which are charged by the current from a thermopile, but batteries are expensive and require careful maintenance. The full exploitation of solar energy awaits development of convenient, inexpensive, and reliable means for storage of energy.

22-9 Vacuum tube devices— diode and triode

No aspect of applied electricity is more characteristic of mid-twentieth–century technology than is the field of electronics. The "miracle of electronics" has become a byword of advertising, and, to be sure, some of the end results such as television are spectacular and are deeply ingrained in contemporary culture. We shall not touch even superficially on many of the ingenious devices and circuits of electronics. Instead, we shall study two simple vacuum tubes, the diode and triode, and their solid-state counterparts, which will serve to illustrate in a general way some of the basic methods and circuits of electronics.

The carriers of charge may be either posi-

---

*On a sunny day if the sun is directly overhead.

tive or negative in certain semiconductors. In vacuum tubes, the carriers are negative (electrons). First we consider some basic principles underlying the operation of vacuum tubes.

The essential parts of an electron-tube *diode*, shown in Fig. 22-23, are the cathode K and the plate P. These electrodes are sealed in an evacuated tube, with connections brought out to two of the base pins. In addition, a heater coil (HH) inside the cathode is provided so that the surface of the cathode can be brought to a dull red heat. There is no direct electrical connection between the heater and the cathode.

We have seen that a distinguishing feature of a metal is the presence of a large number of free electrons which wander about inside the metal, attached to no particular nucleus. They are usually confined to the metal, but at any given temperature a certain number of electrons per second will be able to escape from the surface and fly off into space. This process is very similar to the evaporation of liquid molecules (Sec. 15-7); the analogy can be pushed far enough to include similarities to latent heat of vaporization and vapor

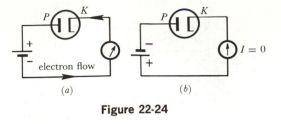

(a) (b)

Figure 22-24

pressure. The important thing for us is that this tendency for electrons to leave the surface increases greatly as the temperature of the metal goes up (compare the vapor-pressure curve of Fig. 15-6). When a diode's cathode is heated to a visible glow, electrons leave the surface at such a rate that the electron current from cathode to plate may easily become measurable. To obtain this current, the plate must be positive relative to the cathode so that electrons are attracted (Fig. 22-24a). If the plate is negative relative to the cathode, electrons are repelled and cannot reach the plate, and the current is zero (Fig. 22-24b). The diode is thus a one-way valve, allowing electrons to flow only if the plate is positive.[*]

The diode is often used to make pulsating direct current when an ac source of potential is available, perhaps from a step-up transformer. Thus in Fig. 22-25, the secondary of the transformer makes AB a source of alternating PD. During half of the cycle, when A is + relative to B, electrons flow in the direction $KPXYABK$; but during the other half of the cycle, when A is − relative to B, electrons cannot leave the cathode and the current is zero. The graph of electron current I vs. time shows a pulsating direct current, whose average value is given by the dashed line. The conventional current through the load resistor R is opposite to the electron current, hence Y is + relative to X. To

Figure 22-23 Diode: P, plate; K, cathode; HH, heater.

---

[*]It is interesting to note that Thomas Edison in 1883 observed a current through the vacuum to an electrode sealed into the side of an incandescent lamp bulb. This was before the discovery of the electron, and Edison missed the significance of his observation that there was no current unless the collector wire was positive.

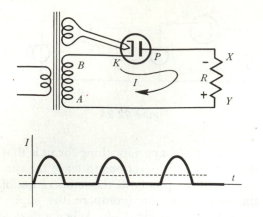

Figure 22-25 Rectifier circuit. The arrow represents the direction of electron flow.

smooth out this current, a capacitor is used in parallel with the load resistor (Fig. 22-26). During the "rainy season" (a), while electrons are flowing through the diode, some electrons go through R, and in addition some flow onto the top plate of the capacitor, charging it as shown. During the "dry season"

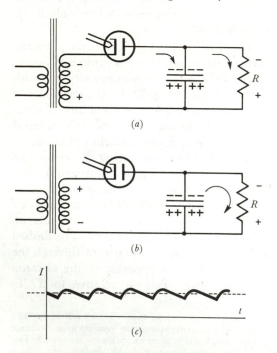

Figure 22-26 Rectifier circuit with filter. The arrows represent the direction of electron flow.

(b), when no electrons flow through the diode, the capacitor discharges, sending electrons through R in the same direction as before. The capacitor acts as a reservoir to smooth out the variations in current. With such an arrangement, called a *filter*, the current can be made practically constant (Fig. 22-26c). Also, the average value of the current is much larger than without the filter, and the PD between X and Y is practically constant. The whole circuit, called a *rectifier circuit*, is used in place of batteries to provide a steady PD. Every ac operated radio or television set has a rectifier circuit of some sort, although other one-way devices such as the solid-state rectifier are often used instead of the vacuum-tube diode described.

As its name implies, the *triode* has three electrodes. In the form invented by Lee De Forest in 1906, an open-mesh *grid* is placed between the cathode and the plate (Fig. 22-27). The grid (G) is made of fine wire so that it does not directly obstruct electrons from reaching the plate, but the electron flow can be controlled by changing the potential of the grid. In a suitable circuit a relatively small change of grid potential can produce a large change in plate current, and

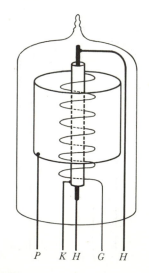

Figure 22-27 Triode: *P*, plate; *K*, cathode; *G*, grid; *HH*, heater.

therefore a large change in plate potential. In this way a triode is used to amplify small changes of potential. We work this out in detail for a typical amplifier circuit, shown in Fig. 22-28, where a 10 000 Ω resistor is connected between a 250 V battery and the plate. The + terminal of the battery is toward the plate in order to attract the (negative) electrons from the cathode. When the grid potential is changed from −8 V to −7 V, it becomes less negative, repels fewer electrons, and allows more electrons to flow to the plate. The plate current increases from 0.005 A (= 5 mA) to 0.007 A (= 7 mA). In (a), the IR drop in the 10 000 Ω resistor is (0.005 A) (10 000 Ω) = 50 V, and hence the potential of the plate P is 250 V − 50 V = +200 V. Similarly, in (b), the potential of the plate works out to be +180 V. What this means is that a 1 V change of the grid causes a 20 V change of the plate; this amplifier has a "voltage gain" of 20 to 1. In practice, the 250 V battery would usually be replaced by a rectifier circuit using a diode and a filter.

Let us contrast the usefulness of a 20:1 step-up transformer and that of an amplifier whose voltage gain is 20:1. Both devices magnify the input voltage the same amount. The transformer, if 100% efficient, delivers as much power as it receives; certainly no increase of power can be obtained from a transformer. Such a result would violate the law of conservation of energy. The amplifier, on the other hand, can deliver much more power than it receives. The ultimate source of the output power is the battery; the grid acts only to *control* the flow of electrons. In much the same way, an engineer at a hydro-electric installation can, by his own unaided muscular efforts, open or shut a valve which may control the flow of many thousands of tons of water.

22-10 Solid-state devices— diode and transistor

To be able to discuss the solid-state diode and the transistor (which is the solid-state analog of the triode), we must consider semiconductor materials. Let us consider a crystal of germanium, which is a typical semiconductor (see Table 19-1). Although a full understanding of semiconductors requires a knowledge of quantum mechanics which is beyond the scope of this book, it turns out that for pure germanium an electron can only gain energy if it receives more than some fixed amount of energy.* There is an *energy gap*,

---

*We have already seen an example of a quantum phenomenon such as this in our study of molecular rotation and vibration in Sec. 9-6.

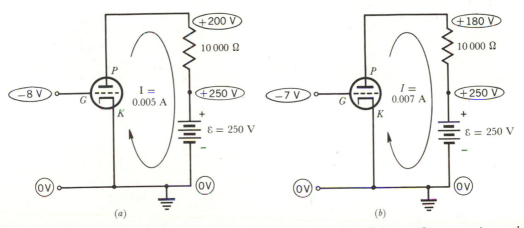

Figure 22-28 A triode amplifier. The arrows represent the direction of electron flow, opposite to the direction of the conventional current.

and small increases of an electron's energy in a germanium crystal are ruled out. It is this energy gap that makes germanium crystals semiconductors; in a copper crystal there is no energy gap and certain electrons are "free" to receive any amount of energy, no matter how small, and thus take part in the conduction process. The semiconductor's conductivity increases as temperature rises, as was mentioned in Sec. 19-6, because thermal agitation can help supply some of the energy required to get an electron across the gap. This is true for pure germanium.

To make a diode, we create two different types of germanium by adding very small, carefully controlled amounts of impurity, measured in parts per million (Fig. 22-29). Germanium has valence 4, and in the crystal each Ge^{4+} ion is surrounded by 4 electrons (shared with other Ge^{4+} ions). If a few germanium atoms are replaced by donor atoms of valence 5, such as arsenic, only 4 of the arsenic's valence electrons are tightly bound in the crystal, and one electron is loosely bound. The "doped" material has an excess of "almost free" negative electrons and is called n-type germanium. Similarly, p-type germanium can be made by replacing some germanium atoms by "accepter" atoms of valence 3, such as aluminum; there are empty places to which electrons can move.

These empty places, called *positive holes,* are equivalent to positive charges which are "almost free" to move. Another commonly used semiconductor material is silicon, with phosphorus or boron as impurities.

A *p-n junction* is a tiny single-crystal piece of germanium, silicon, or some other semiconductor which has been treated to make one side p-type and the other side n-type. The magnitude of the current through the semiconductor depends on the polarity of the battery or other source of emf in the external circuit; the determining factor is what happens at the junction. In Fig. 22-30a, the electrons in the n-side are pushed across the junction by the − terminal of the battery (and are also attracted by the + terminal of the battery). Also, positive holes in the p-side are aided in their travel across the junction by the battery's PD. There is a large current, and the diode is said to be *forward biased.* If the battery is connected as in Fig. 22-30b, the PD of the battery tends to assist electrons in the p-side to cross the junction; but there are very few free electrons in this material, so the current is very small. Likewise the PD would help positive holes to flow from the n-side to the p-side, but there are very few positive holes in the n-type material. Under these circumstances the current is very small, and the diode is said to be *reverse biased.* We

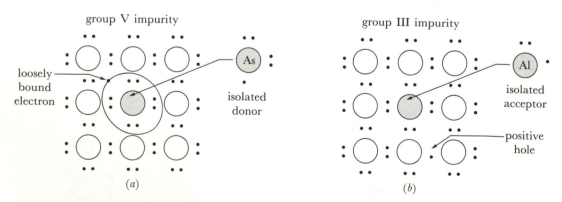

group V impurity

loosely bound electron

As

isolated donor

group III impurity

Al

isolated acceptor

positive hole

(a)

(b)

Figure 22-29 Doped semiconductor materials (schematic). The open circles represent Ge^{4+} ions. (a) n-type germanium; (b) p-type germanium. (Adapted from Fig. 2-12 of *Principles of Solid-State Microelectronics* by S. N. Levine, © 1963 by Holt, Rinehart and Winston.)

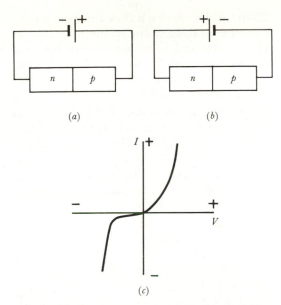

(a)

(b)

(c)

Figure 22-30 Junction diode. (*a*) Forward bias. (*b*) Reverse bias. (*c*) Graph of current vs. bias voltage.

see that the *p-n* junction can serve as a recti-fier, just as does the vacuum diode. There is a large forward current and a very small reverse current (Fig. 22-30*c*). Solid-state di-odes are commonly used in rectifier circuits (Fig. 22-31) which are similar to the vacuum-tube diode rectifier circuit of Fig. 22-25.

The *transistor* is the solid-state analog of the vacuum triode. A common type of tran-sistor is formed of a single-crystal fragment doped with impurity atoms to make two *p-n* junctions back to back. In the *n-p-n* transistors shown in Fig. 22-32 the emitter and the

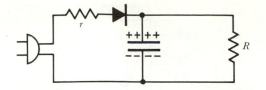

Figure 22-31 Solid-state power supply. If the diode accidentally becomes a short circuit, the small resistor *r* will overheat and burn out, serving as a fuse to protect the capacitor and load.

collector regions are *n*-type with a heavy concentration of impurity atoms; the base is a very thin region with only a small concen-tration of *p*-type impurity atoms.

The operation of a simple transistor am-plifier (Fig. 22-33) can best be explained by an example. The circuit is very similar to that of the triode amplifier of Fig. 22-28. The 10 V battery gives a forward bias to the junction formed by the emitter and the base (the "emitter-base junction"), so a relatively large number of electrons flow into the base region. This flow is strongly dependent on the bias voltage. If the input voltage changes from 0.2 V to 0.3 V, the emitter-base junc-tion is biased more strongly than before, and many more electrons find their way from the emitter into the base region. The "base-collector junction" is reverse biased, but electrons injected from the emitter into the base can diffuse across the very narrow base region; when they reach the base-collector junction they are attracted into the collector and flow through the external circuit. In a

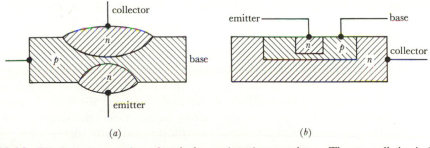

(a)

(b)

Figure 22-32 Physical construction of typical *n-p-n* junction transistors. The over-all size is 1 mm or less; the space between emitter and collector is about 0.01 mm.

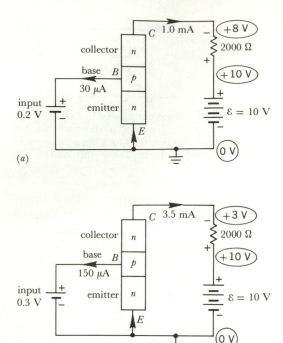

(a)

(b)

Figure 22-33 A transistor amplifier. The arrows represent the direction of electron flow.

22-33a and b; there is a close correspondence with the analysis of the triode amplifier in Fig. 22-28a and b. An increase of the base voltage from $+0.2$ V to $+0.3$ V (this is the signal input) causes an increase in collector current from 1.0 mA to 3.5 mA; the IR drop in the 2000 Ω resistor increases from 2.0 V to 7.0 V, and the potential of C changes from $+8.0$ V to $+3.0$ V. The voltage gain is $(8 - 3)\text{V}/(0.3 - 0.2)\text{V} = (5 \text{ V})/(0.1 \text{ V})$; in this circuit a gain of 50:1 is achieved. You should make a point-by-point comparison of this analysis with that of the triode amplifier —in each case it is the change in current through an external resistor that gives rise to a change in the potential of the plate or the collector, as the case may be. A p-n-p transistor would function in a similar fashion in a circuit for which the polarity of each battery is reversed.

We have discussed two electron devices which have the ability to amplify. Compared with vacuum tubes, transistors have the advantages of small size, no heater power, no warm-up time, and they operate on low voltages such as flashlight cells. On the other hand, transistors are sensitive to temperature changes, their operation at very high frequencies is difficult, and their power output is less than that of vacuum tubes.

typical case, some 95% of the electrons entering the base make it across to the collector without being captured by the p-material's positive holes.

Analysis of the amplifier is made in Fig.

■ SUMMARY

The sensitivity of a galvanometer depends upon several design factors, and the solution of the design problem is one of compromise aided by study of the working equation for the instrument. A direct-current motor is basically a galvanometer with a commutator that automatically reverses the current at the proper times.

Alternating currents are produced by a generator which has no commutator. The chief advantage of alternating currents is that transformers can be used to increase or decrease the emf in a circuit. No power is gained; in an ideal transformer the product of emf and current is the same in the primary as in the secondary. A dc motor is also a dc generator; this duality is illustrated by the phenomena of back emf in a motor and back torque in a generator.

Coils and capacitors behave differently in ac circuits and in dc circuits. Impedance, a generalization of resistance, is defined as the ratio of PD to current, for an ac circuit. The impedance of a coil is largely due to the back emf induced in its windings by its own changing current. The impedance of a capacitor is infinite for direct current, since there can be no steady current through a capacitor. Alternating current through a capacitor *is* possible, since charges flow back and forth in an external circuit as the plates are charged and discharged many times per second. The impedance of a coil goes up as frequency increases, while the impedance of a capacitor goes down as frequency increases.

A thermoelectric emf exists in a circuit where two junctions of dissimilar metals are at different temperatures. The effect is reversible, and a flow of charge through a junction generates or absorbs thermal energy.

Diodes are used to change alternating current into pulsating direct current. A filter capacitor can be used to smooth out the pulses. Triodes are used to amplify small potential changes; the electron flow is controlled by the potential of the grid. In both types of vacuum tube, electrons are emitted by the heated cathode and flow through the tube to the positively charged plate.

Solid-state devices depend on the electrical properties of semiconductors with controlled amounts of substituted impurity atoms. The transistor is a three-terminal, solid-state device in which the collector current is controlled by changes in the base current.

■ CHECK LIST

$$\theta = \left(\frac{NBA}{k}\right)I \qquad Z = \frac{V}{I}$$

		cathode
		plate
armature	self-inductance	grid
commutator	henry	n-type semiconductor
back emf	$Z = 2\pi\nu L$	p-type semiconductor
back torque	$Z = 1/2\pi\nu C$	p-n junction
eddy currents	thermocouple	junction transistor
	thermopile	emitter
$\dfrac{\mathcal{E}_s}{\mathcal{E}_p} = \dfrac{N_s}{N_p}$	diode	base
	triode	collector
impedance		

■ QUESTIONS

22-1 Why is a fixed cylindrical iron core placed inside the moving coil of a galvanometer?

22-2 The output of a generator is, to start with, an alternating current. Describe two ways (one mechanical, one electrical) to convert alternating current to pulsating direct current.

22-3 Why has alternating current been so widely adopted in place of the earlier direct current system prevalent some 80 years ago?

22-4 When a refrigerator motor starts up, the house lights dim momentarily. When an electric iron is plugged in, the lights stay dim for as long as the iron is turned on. Discuss the similarities and differences between these two situations.

22-5 Through a system of gears and belts, the armature of a car's generator is turned at a speed proportional to the engine speed. The headlights are connected directly to the armature. (a) Explain why the lights become brighter when the engine is "gunned" at high speed. (b) Considering the system (engine + generator + lamp filament), discuss the energy transformations and tell which are reversible and which are not reversible.

22-6 Would the eddy-current brake (Fig. 22-11) operate on aluminum rails (which are non-magnetic) as well as on steel rails?

22-7 Why is the core of a transformer made of many thin sheets of iron instead of one solid piece?

22-8 What can you say about the construction of a transformer for which the secondary emf equals the primary emf?

22-9 What would happen if the primary of the pole transformer were supplied with 2300 V dc instead of 2300 V ac?

22-10 Delegate a member of the class to find out from your local electric power company at what voltage the power is generated which eventually supplies the lamps of your classroom. Find out also how many transformers of all sorts there are between the generator and the lamps and at what voltages the primary and secondary of each transformer operate.

22-11 Water flows over a dam, electric energy is generated, energy is transmitted over a line and through various step-up and step-down transformers, and eventually an elevator rises in an office building. Did any *mass* flow through the transformers?

22-12 Can a diode be used for amplifying a weak signal?

22-13 What would happen if the battery of Fig. 22-28a were reversed so that the − terminal was connected to the 10 000 Ω resistor, and the grid was made +8 V instead of −8 V? Would the circuit still function as an amplifier?

22-14 What is the difference between *n*-type germanium and *p*-type germanium?

22-15 Is the resistance of a *p-n* junction the same for forward bias as for reverse bias? If not, in which direction is the resistance greater?

■ PROBLEMS

22-A1 A certain galvanometer needle deflects 30° when the current through the coil is 6 μA. What deflection would be produced by the same current if the magnetic field were decreased to $\frac{4}{5}$ of its original value?

22-A2 The strength of the magnetic field of a "permanent" magnet of a certain moving coil ammeter decreased to 90% of its original value in several decades. The meter was originally calibrated to read 2 A full scale. What is now the reading caused by a current of 2 A?

22-A3 A generator furnishes an emf of 48 V when its armature turns at 400 rev/min. All other factors remaining the same, what will the emf be if the armature turns at 300 rev/min?

22-A4 The back emf in a motor is 100 V when the motor is turning at 2500 rev/min. What is it when the motor turns at 3000 rev/min, if the magnetic field remains the same?

22-A5 A model-train transformer has a 600-turn primary and a 100-turn secondary. When the primary is connected to a 60 Hz, 120 V line, what is the secondary emf?

22-A6 The transformer of Prob. 22-A5 delivers a current of 9 A to the engine; what is the current from the 120 V supply circuit?

22-A7 A step-up transformer in a small radio has 200 turns in the primary and 600 turns in the secondary. When the primary is connected to a 120 V ac source, the current through it is 30 mA. What are the emf and current in the secondary?

22-B1 A dc motor has field coils which are connected in parallel with the armature and with the supply line. The field coils have resistance 100 Ω, and the armature has resistance 4 Ω. When connected to a 100 V source of PD, the motor turns at such a speed that the back emf in the armature is 108 V. What is the total current through the motor?

22-B2 The armature of the motor of Prob. 22-B1 becomes "frozen" owing to improper lubrication and can no longer turn. If no fuses blow, what will the total current become?

22-B3 A motor connected as in Fig. 22-8 draws a total of 6 A from a 120 V dc line. The resistance of the armature is 3 Ω, and the back emf generated in it is 108 V. Calculate   (a) the current through the armature;   (b) the current through the field coil;   (c) the resistance of the field coil.

22-B4 Two flat coils of wire are lying on a table, with the small one (the "secondary") completely inside the larger coil (the "primary") (Fig. 22-34).   (a) If the switch is closed, in which direction, clockwise or counterclockwise, will conventional charge flow in the primary? (b) What will be the direction of the momentary induced current in the secondary, just after the switch is closed?

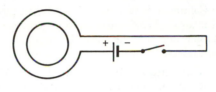

Figure 22-34

22-B5 If the switch in Fig. 22-34 has been closed for some time and is then opened, what will be the direction of the momentary current in the secondary?

22-B6 A transmission line between a city and a suburb consists of two wires, each of resistance 0.5 Ω; the current is 1000 A; the PD at the input end of the line is 2 kV. Calculate   (a) the power input;   (b) the power loss in the line;   (c) the power delivered to the load; (d) the efficiency of the line.

22-B7 Repeat Prob. 22-B6, for the same input power to the same line, but with the PD at the input end of the line 20 kV instead of 2 kV. (*Hint:* First find the current.)

22-B8 Power is delivered to a factory at the rate of 50 000 W at a terminal voltage of 5000 V measured at the factory. The total resistance of both wires connecting the factory and the power station is 10 Ω.   (a) What power is lost in the line?   (b) What is the efficiency of the line?   (c) What is the PD at the terminals of the generator?

22-B9 What is the impedance of a coil of self-inductance 15 mH, at a frequency of 10 kHz, assuming the coil to have a negligible dc resistance?

22-B10 What back emf is induced in a coil of self-inductance 400 mH if the current changes at a steady rate from 12 A to 0 in 3 ms?

22-B11 At what frequency is the impedance of a 0.01 H coil equal to 200 Ω?

22-B12 The self-inductance of an iron-cored magnet is 5 H. What is the current through the magnet when it is connected to a 60 Hz alternating emf of 120 V?

22-B13 At what frequency does a capacitor of value 0.02 μF have an impedance of 20 kΩ?

22-B14 What is the alternating current through a 250 pF capacitor which is connected to a radio-frequency source whose frequency is 400 kHz and whose terminal voltage is 30 V?

22-B15 In a triode circuit similar to Fig. 22-28, the emf of the battery is 110 V, and the load resistor is 12 000 Ω. When the grid potential is −5.0 V, the ammeter in the plate circuit reads 2.5 mA. When the grid potential is changed to -5.4 V, the ammeter reads 2.0 mA. (a) What is the plate potential in each case? (b) What is the voltage gain of the amplifier?

22-B16 In a transistor circuit similar to Fig. 22-33, the emf of the battery is 9 V, and the load resistor is 1500 Ω. When the input potential is 0.25 V, the collector current is 2 mA; when the input potential is changed to 0.35 V, the collector current is 5 mA. (a) What is the collector potential in each case? (b) What is the voltage gain of the amplifier?

22-B17 Use Kirchhoff's first law to calculate the emitter current in each part of Fig. 22-33.

22-C1 What will be the total current through the motor of Example 22-1 if the motor is loaded so as to run at only half speed? (*Hint:* The back emf is proportional to the speed.)

22-C2 What total current will the motor of Prob. 22-B3 use if it is loaded down so that its speed is $\frac{3}{4}$ the original speed?

22-C3 What is the back emf in a motor which draws 13.1 A from a 220 V line, if the field coils have a resistance of 200 Ω and the armature has a resistance of 1 Ω?

22-C4 The output of a generator is stepped up and rectified to give 800 kV at a power station on the Columbia River, and 2700 MW of power is delivered to Los Angeles, about 1350 km away. What must the total cross-sectional area of aluminum wires in the transmission line be if the power loss in the line is to be only 1% (i.e., 27 MW)? (*Hint:* First find the current.)

22-C5 A 110 V bulb rated at 5 W is connected to a 110 V, 60 Hz source through a 40 μF capacitor. Show that the bulb is lit to practically full brilliance. (*Hint:* Compare the resistance of the bulb and the impedance of the capacitor.)

22-C6 An iron-cored coil has self-inductance 5 H and resistance 0.8 Ω. What is the PD across the coil when the current is 20 A, changing at the rate 80 A/s?

For Further Study

22-11 The emf of a generator

An equation for the emf of a generator can be derived with the help of differential calculus. According to Eq. 21-11, the magnitude of an induced emf is equal to the rate of change of magnetic flux: $\varepsilon = -d\Phi/dt$.[*] In a generator the flux changes because the *orientation* of the plane of the loop changes. In Fig. 22-5, the component of **B** perpendicular to the plane of the armature is $B \cos \phi$, where ϕ measures the rotation of the loop from its horizontal position, and the flux through the loop is $B \cos \phi\, A$. If the generator is turning steadily at an angular speed ω rad/s, the angle ϕ increases accord-

[*]The $-$ sign is used with $d\Phi/dt$ for the reasons discussed just after Eq. 21-11.

ing to $\phi = \omega t = 2\pi\nu t$, where ν is the frequency in rev/s and t is the time. Hence

$$\Phi = B \cos 2\pi\nu t\, A$$

$$\varepsilon = -\frac{d\Phi}{dt} = -BA\,\frac{d}{dt}(\cos 2\pi\nu t)$$

To find the derivative of $\cos 2\pi\nu t$ we use the table of derivatives (Appendix, p. 708, where we find that if $y = \cos ax$, then $dy/dx = -a \sin ax$. Here a is $2\pi\nu$, x is t, and so we obtain

$$\varepsilon = 2\pi\nu BA \sin 2\pi\nu t$$

The generator's emf is a sinusoidal function of time. The magnitude of the emf is proportional to the frequency of rotation of the armature, as well as to the area of the loop and the strength of the magnetic induction.

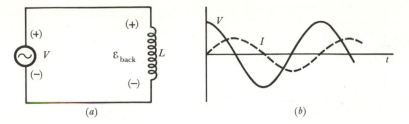

Figure 22-35 (a) Circuit containing an ac source and a coil. The (+) and (−) signs indicate the polarities of coil and generator at an instant when $I = 0$ and is increasing. (b) The current through the coil lags $\frac{1}{4}$ cycle behind the PD.

22-12 Formulas for impedance

Consider the current through a coil of self-inductance L connected to an ac generator whose terminal voltage is V (Fig. 22-35). The frequency is ν cycles/s, and the graph of I versus t is a sinusoidal curve whose equation may be written as $I = I_0 \sin 2\pi\nu t$. As we saw in Sec. 22-6, the magnitude of the back emf in the coil depends on the rate of change of current and is given by $\mathcal{E}_{\text{back}} = L(dI/dt)$. The slope of the current graph is greatest when $t = 0$, so at this instant, when the current is *changing* most rapidly, the rate of change of flux is greatest, and the back emf is greatest. The generator and the coil are in parallel, so their terminal voltages are equal. Hence

$$V = \mathcal{E}_{\text{back}} = L\frac{dI}{dt}$$

Qualitatively we see from the graphs that the current through the coil is *out of phase* with the PD across the coil. In fact, the current reaches its maximum value $\frac{1}{4}$ cycle *after* the applied PD is at its peak. This delay is, of course, caused by the back emf which by Lenz's law opposes the change (an increase, in this case) of current. If the frequency is increased, the necessary terminal voltage of the generator must be larger, for then the time Δt is less for a given ΔI, the slope dI/dt is larger, and the back emf is larger.

To obtain a quantitative formula for the impedance of a coil of negligible resistance we use a formula for the derivative of the sine function (Appendix p. 708) which states that if $y = \sin\ ax$, then $dy/dx = a \cos\ ax$. Here x is t and the constant a is $2\pi\nu$. Hence

$$V = \mathcal{E}_{\text{back}} = L\frac{dI}{dt} = L\frac{d}{dt}(I_0 \sin 2\pi\nu t)$$

$$= LI_0\frac{d}{dt}(\sin 2\pi\nu t)$$

$$= LI_0\, 2\pi\nu(\cos 2\pi\nu t)$$

Thus

$$V = V_0 \cos 2\pi\nu t$$

where V_0, the maximum value of V, is $2\pi\nu LI_0$. The impedance is

$$Z = \frac{\text{max. value of applied PD}}{\text{max. value of current}} = \frac{2\pi\nu LI_0}{I_0}$$

or

$$Z_{\text{coil}} = 2\pi\nu L$$

This is the formula stated without proof in Sec. 22-6. The $\frac{1}{4}$-cycle lag of the current through a coil is apparent from the fact that I is a sine function if V is a cosine function.

The impedance of a capacitor can be calculated in much the same way. Consider the PD across the capacitor in Fig. 22-36, equal at all times to the generator's terminal voltage. If the origin of the time axis is chosen so that the graph of the PD is a sine curve, then the charge on the capacitor is also a sine curve, since $Q = CV$ at all times. The current is the rate of flow of charge, and the charges are able to flow into the capacitor most easily when the capacitor is uncharged and presents no opposing PD. This makes the current $\frac{1}{4}$ cycle out of phase with the ap-

Formulas for impedance **489**

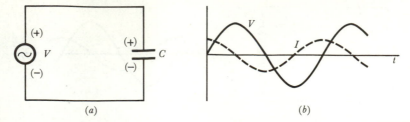

Figure 22-36 (*a*) Circuit containing an ac source and a capacitor. (*b*) The current through the capacitor leads the PD by $\frac{1}{4}$ cycle.

plied PD; but in contrast to the behavior of the coil, the current in a capacitor reaches its peak *before* the PD does.

$$I = \frac{dQ}{dt} = \frac{d}{dt}(CV) = C\frac{d}{dt}(V_0 \sin 2\pi\nu t)$$

$$= CV_0 2\pi\nu \cos 2\pi\nu t$$

or

$$I = I_0 \cos 2\pi\nu t$$

where I_0, the maximum value of I, is $2\pi\nu CV_0$.

The impedance is

$$Z = \frac{\text{max. value of applied PD}}{\text{max. value of current}} = \frac{V_0}{2\pi\nu CV_0}$$

or

$$Z_{\text{capacitor}} = \frac{1}{2\pi\nu C}$$

This is the formula stated without proof in Sec. 22-7. This $\frac{1}{4}$-cycle lead of the current through a capacitor is apparent from the fact that I is a cosine function if V is a sine function (Fig. 22-36).

In advanced work, the property of a coil or a capacitor which gives rise to a $\frac{1}{4}$-cycle phase difference between current and PD is called its reactance. For any circuit, reactance (in ohms) and resistance (in ohms) can be combined, in a certain way, to obtain the total impedance of the circuit (in ohms). For circuits containing resistance as well as inductance and capacitance, the phase relationships between current and PD make the study of ac circuits rather technical, and we will not pursue these matters further.

■ PROBLEM

22-C7 A flat coil of resistance 5 Ω has 500 turns each 5 cm × 20 cm and rotates at 1800 rev/min in a field of 0.04 T. Calculate (*a*) the generated emf; (*b*) the maximum current through a 20 Ω load resistor connected to the coil.

■ REFERENCES

1 Pierce, J. R., "Electronics," *Sci. American,* Oct., 1950, p. 30. Basic processes.

2 Shiers, G., "The First Electron Tube," *Sci. American,* Mar., 1969, p. 104. The Edison effect; the Fleming valve (diode).

3 Shockley, W. S., "Transistor Physics," *Am. Scientist,* **42,** 41 (1954).

4 David, E. E., Jr., "The Reproduction of Sound," *Sci. American,* Aug., 1961, p. 72.

5 Chapman, R. A., "De Forest and the Triode Detector," *Sci. American,* Mar., 1965, p. 92.

6 Mackay, R. S., in C. L. Stong's "Amateur Scientist" department, describes some interesting cyclic effects (very low frequency oscillations) related to variations of impedance with current. *Sci. American,* Aug., 1961, p. 143.

23

Electromagnetic Waves

23-1　The radiation spectrum

The type of electromagnetic radiation known as "light" is, of course, familiar to all of us, but the identification of light as an electromagnetic wave is a comparatively recent development. The wave nature of light has been known only since about 1800, and just over a century has elapsed since Maxwell proved (in 1864) that electromagnetic (e-m) waves are a necessary consequence of the laws of electricity and magnetism. As we pointed out in Chap. 14, not all electromagnetic radiation is visible, and in Table 14-4 we listed various types of e-m radiation, which range from the low-frequency radio waves of communications to the high-frequency gamma rays of nuclear physics.

How shall we describe electromagnetic radiation? At the outset, you should be warned that the description is a strange one, full of seeming contradictions. For instance, it is a known fact that energy moves from place to place by e-m radiation. The earth is warmed by infrared radiation from the sun; a photographic emulsion is "exposed" when it absorbs light energy; radio waves transmit enough energy to control the flow of electrons in a television receiver. It is not

hard to think of illustrations of energy transfer by other e-m radiations such as ultraviolet radiation, x rays, or gamma rays. Certainly, our description (or "model") of e-m radiation must include a means by which energy can be transported from place to place.

One simple model for radiation would be a stream of particles, or corpuscles; this view was held in the earliest times. By the time of Newton it was realized that energy can also be transmitted by means of waves. The pounding of the surf upon the shore, the transmission of audible speech by sound waves—these illustrate the transmission of energy by wave motions. To decide between a corpuscular model and a wave model of light (and other e-m radiation), we must look for experiments which can be described by one model but not the other.

Here we come upon one of the most baffling chapters in the history of physics. Some experiments confirm the corpuscular model and seemingly cannot be explained by the wave model; other experiments are equally definite in causing us to reject the corpuscular model in favor of the wave model. In the end, we shall be led to accept a new model, which combines features of both. We cannot describe this dual model of radiation at this

The navigator of this small craft rotates the loop antenna until the radio receiver indicates a minimum signal. The plane of the loop is then broadside to the source of electromagnetic radiation. (Radiomarine Corporation of America)

point; first we must learn about the wave model and the corpuscular model separately before attempting to fuse them into a single acceptable model.

In Chaps. 23 through 26 we shall be concerned with those aspects of radiation that are best described by a wave model, chiefly the phenomena associated with the propagation of light, including reflection, refraction, interference, and diffraction. In Chaps. 27 through 30 we shall study those aspects of radiation that are best described by a corpuscular model, chiefly the phenomena associated with the emission and absorption of radiation. We cannot say that either model is "correct," nor can we say that it is "incorrect." Rest assured that we are not going to use either model unless it is "correct enough" to allow us to correlate observations and predict the results of new experiments. After all, such correlation and prediction are about all that one can ask of any model, however good.

23-2   Electric oscillations

If energy is to be transmitted by some sort of electric wave, we must look for some type of electric oscillation that can be analogous to the vibration of a bell or string which sends out sound waves. We can learn a great deal by exploring the analogies between electric circuits and mechanical systems. Let us look at the characteristics of a vibrating body—say, an object of mass m on the end of a spring whose force constant is k (see Fig. 9-12). There is a continual interchange of energy from one form to another. The total energy of the system is sometimes stored as KE, sometimes as PE. In Sec. 9-3 we were able to use the energy principle to derive a formula for the period of oscillation T of the mass. The result was $T = 2\pi \sqrt{m/k}$. During this time the oscillating body goes through one cycle of its vibration. The period, which is independent of the amplitude of vibration, depends on the stiffness of the spring (a fac-

tor related to the force constant k, and hence to the PE) and also on the inertia of the body (a factor related to the mass m, and hence to the KE). Such a vibrating mass has a *natural frequency,* and it can set into motion a medium such as air or water, thus giving rise to a wave. The frequency of the wave will be the frequency of the oscillating body; since frequency is the reciprocal of the period, we can write

$$\nu = \frac{1}{2\pi} \sqrt{\frac{k}{m}} \qquad (23\text{-}1)$$

In searching for an electrical analogy to mechanical oscillation, we recall that electric PE can be stored in a capacitor. In Sec. 18-9 the PE of a charged capacitor was derived:

$$\text{PE} = \tfrac{1}{2}(1/C)Q^2 \qquad (23\text{-}2)$$

where $1/C$ is analogous to the force constant k. (The PE of a stretched spring is given by $\tfrac{1}{2}kA^2$; see Sec. 9-3.) The electrical analog of KE must be sought for among magnetic effects, for KE is energy due to motion, and magnetism is caused by the motion of charges. Consider a coil whose self-inductance is L, through which a battery sends a steady current. If the resistance of the coil and battery are very small, only a negligible amount of Joule heat is being dissipated in the coil, and hence the seat of emf is transforming chemical energy into electric energy at a negligible rate. Not so during the time just after the switch was closed; while the magnetic field was being established, the magnetic flux through the coil was changing, a back emf was generated in the coil, and work had to be done to force charges through the coil against the opposing PD of the back emf. After the current reached its final value, the flux was constant and the back emf disappeared; but while the current (and field) were building up, a back emf existed and work had to be done. We can say that energy is stored in the magnetic field, and our analysis involving back emf gives us a mechanism by which work is done in establishing the field.

The energy stored in a current-carrying coil may be likened to KE, for it is due to the motion of charges. The situation is reminiscent of the storage of energy in a moving body such as an automobile. Disregarding friction, once the car gets up to speed no further work need be done, and the KE remains constant. But while it is being accelerated, there is a continuous flow of energy from the fuel, and work is done. We shall not here derive the formula for energy stored in a coil carrying a steady current, except to note that the back emf is proportional to the self-inductance L; hence we would expect the stored energy to be proportional to L. The result of a calculation given in Sec. 23-9 is that

$$KE = \tfrac{1}{2}LI^2 \qquad (23\text{-}3)$$

where L is the self-inductance (or simply the "inductance") of the coil, in henries, and I is the current, in amperes. The stored energy is measured in joules. There is a good analogy between this formula and the form-ula $KE = \tfrac{1}{2}mv^2$ for mechanical KE; L is analogous to m, and I^2 (related to motion of charge) is analogous to v^2 (related to motion of the body).

We are ready now to consider electric oscillations. To produce an oscillation we must arrange for the exchange of energy from KE to PE and vice versa. This is accomplished in the circuit of Fig. 23-1. Suppose that the capacitor is initially charged (a), and then the switch is closed. Charge starts to flow through the coil (b), and some of the PE stored in the capacitor becomes KE as the magnetic field of the coil builds up. In (c), the capacitor is empty of charge, but charges are in motion in the coil, and all the energy of the system is in the form of KE. The process doesn't stop, however. The charges move on through the coil and begin to accumulate (d) on the bottom plate of the capacitor. Eventually (e) the capacitor is fully charged again (opposite to its original condition), and all the energy is again stored as PE, since the current is momentarily zero.

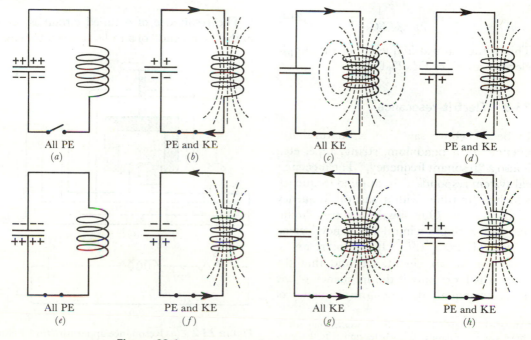

Figure 23-1 Energy transformations in an L-C circuit.

The process could go on indefinitely except for the fact that the wires have *some* small resistance, and the energy is dissipated in Joule heat and the oscillation dies out.

Note the point-by-point similarity of this process to the mechanical oscillation of a mass on a spring. There, too, the oscillation eventually dies out as work is done against internal friction in the spring and against the external air resistance. The ultimate fate of the mechanical energy is heat energy, just as in the electric oscillation. We can use the analogy between mechanical and electric oscillations to write down a formula for the frequency of an electric oscillation. If the force constant k is replaced by $1/C$, and if the mass m is replaced by L, the formula

$$\nu = \frac{1}{2\pi}\sqrt{\frac{k}{m}}$$

becomes

$$\nu = \frac{1}{2\pi}\sqrt{\frac{(1/C)}{L}}$$

Simplifying, we obtain

$$\nu = \frac{1}{2\pi}\sqrt{\frac{1}{LC}} \qquad (23\text{-}4)$$

This is the natural frequency of the simple electric circuit of Fig. 23-1.

23-3 Electric resonance

In Sec. 11-4 we saw that a "natural frequency" of a pendulum, string, pipe, etc., is also a "resonant frequency." The mechanical system responds to an applied frequency which is "in tune" with a natural frequency of oscillation. Our simple electric circuit has only a single natural frequency,[*] given by $\nu = (1/2\pi)\sqrt{1/LC}$. If we insert an ac generator in our circuit, we find that the current is large only if the frequency of the generator equals the resonant frequency of

---

[*] More complicated circuits can be worked out which have a series of natural frequencies similar to the series of frequencies of a string or a pipe.

the circuit. The resonance apparatus of Fig. 23-2 demonstrates this nicely. A resistance R (the lamp bulb) is included in the circuit; it does not alter the resonant frequency, and the brilliance of the bulb is an indication of the current. If the value of L is changed by raising or lowering the iron yoke, the lamp is brightest for the setting which makes L have just the right value in Eq. 23-4. In other words, we *tune* the circuit by adjusting its resonant frequency to match that of the source of energy.

EXAMPLE 23-1

For what value of L does the circuit of Fig. 23-2 resonate at 60 Hz, if $C = 10\ \mu\text{F}$ and $R = 100\ \Omega$?

The resistance of the bulb has no effect except to determine the value of the current at resonance. The condition for resonance is

$$\nu = \frac{1}{2\pi}\sqrt{\frac{1}{LC}}$$

$$L = \frac{1}{4\pi^2\nu^2 C} = \frac{1}{4\pi^2(60\ \text{s}^{-1})^2(10 \times 10^{-6}\ \text{F})}$$

$$= \boxed{0.704\ \text{H}}$$

The resonance of a tuned circuit is used in the "front end" of a radio receiver to select

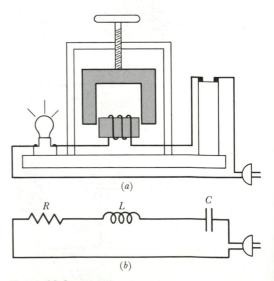

Figure 23-2 (*a*) Resonance apparatus. (*b*) Circuit diagram for resonance apparatus.

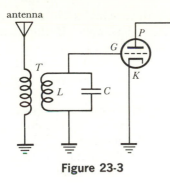

Figure 23-3

which of several stations will be heard. Schematically, Fig. 23-3 shows an antenna which receives e-m radiation of all frequencies—regular broadcast, television, aircraft, satellites, police calls, and so on. All these signals give rise to currents through the primary of the antenna transformer T. Emf's of many frequencies are induced in the secondary, but since the secondary is part of a resonant LC circuit,* the circulating current in the tank circuit is large only for that incoming signal whose frequency just matches that of the circuit. It is this resonance phenomenon that allows a listener to select the desired station. In radios, the tank circuits are usually tuned by changing the capacitors; however, it is also possible to tune a receiver by varying L by means of ferrite slugs which are moved in and out of the coils. Radio frequencies are high enough so that the required values of L and C are very small.

EXAMPLE **23-2**

Calculate the resonant frequency of a tank circuit in a television receiver, for which the capacitance is 9 pF and the inductance is 4 μH.

$$\nu = \frac{1}{2\pi} \sqrt{\frac{1}{LC}}$$

$$= \frac{1}{2\pi} \sqrt{\frac{1}{(4 \times 10^{-6}\ \text{H})(9 \times 10^{-12}\ \text{F})}}$$

$$= \frac{10^9}{2\pi(6)} = 26.5 \times 10^6\ \text{Hz}$$

$$= \boxed{26.5\ \text{MHz}}$$

---

*Picturesquely called a "tank circuit," on account of its energy storage.

23-4 Radiation

Electric oscillations differ from mechanical oscillations in one important respect: the electric oscillator loses energy not only because of Joule heat in the connecting wires but also by radiation. The close interaction between electric fields and magnetic fields is responsible for the electromagnetic waves (radio waves, light waves, x rays, etc.) which are broadcast by an oscillator. We have seen that an induced emf appears when a magnetic field changes; for example, the induced emf in a generator or a transformer is caused by a changing magnetic induction as magnetic lines of force "cut across" a conductor or as magnetic flux through a circuit changes. Now when an emf exists between two points in a wire, or in empty space, there must be an electric field (force per unit charge) in the region between the points. Thus when a wire is connected to the two terminals of a dry cell, the electric field set up in the wire accelerates the free electrons in the wire and causes a current. Even if there is no wire between the terminals, the electric field exists in the space between the terminals. Thus an emf implies an electric field. We can therefore restate Faraday's law of magnetic induction by saying that *a changing magnetic field causes an induced electric field*. Maxwell showed that the opposite is also true: *a changing electric field causes an induced magnetic field*. This symmetrical relationship between electric and magnetic fields provides the interaction needed to propagate a wave. When an e-m wave is passing through some point in space, both the electric field and the magnetic field at that point are changing. Maxwell showed that the electric field **E** and the magnetic induction **B** fluctuate. **E** and **B** both are zero at the same time, and they reverse direction (together) twice each cycle. Another prediction of the Maxwell theory is that **E** and **B** are perpendicular to each other and that both are perpendicular to the direction of propagation of the wave. The velocity of the wave (in a vacuum) is given by $\sqrt{k/k'}$, where k is the constant in Coulomb's law for charges (Sec. 17-5), and k' is the constant in the law of magnetic force (Sec. 21-5). It is natural that these electric and magnetic constants would enter into an expression for the speed of an electromagnetic wave. One of the early triumphs of the Maxwell theory was the numerical value of the predicted velocity of electromagnetic waves:

$$c = \sqrt{\frac{k}{k'}} = \sqrt{\frac{8.99 \times 10^9 \ \mathrm{N \cdot m^2/C^2}}{10^{-7} \ \mathrm{N/A^2}}}$$
$$= \sqrt{8.99 \times 10^{16} \ \mathrm{m^2 \cdot A^2/C^2}}$$
$$= 3.00 \times 10^8 \ \mathrm{m/sec}$$

Thus the constant $\sqrt{k/k'}$, an electrical quantity which can be measured by electrical means, works out to be equal to c, the speed of light in a vacuum. This is strong evidence for the fact that light waves are *electromagnetic* waves; the evidence becomes even stronger when it is realized that all matter, including sources of light, is composed of electric charges which might be expected to oscillate and hence broadcast e-m waves. Before we describe the production of an e-m wave by an oscillating charge, let us first take a more detailed look at the description of an e-m wave.

23-5 Description of an electromagnetic wave

We can describe an electromagnetic (e-m) wave in the same way that we described a water wave in Sec. 10-5. We use the technique of a series of instantaneous snapshots taken at equal intervals of time. Suppose that the antenna tower of a radio station is located at the center of town. We station a number of observers at 75 m intervals along a street leading north from the station and equip each observer with an inertialess positively charged test body which can (in our imagination) respond to very rapid changes in electric field. For a test body we might

use a pith ball attached to the end of a light, thin, springy glass fiber. If the charge on the pith ball is Q and it experiences a force $\mathbf{F}$, then the direction of the field is the direction of that force. At twelve noon (which we call $t = 0$) the observers' pith balls all indicate electric fields which are vertical or zero; but, as shown in Fig. 23-4, some balls are urged upward, others are urged downward, and still others indicate the field to be zero. A quarter of a microsecond later, the situation has changed, and A's pith ball registers zero field, B's is urged upward, etc. At $t = \frac{1}{2} \ \mu s$, the force on A's pith ball is downward, B's has returned to zero, etc. We have here all the attributes of a wave: at any given place, the electric field fluctuates ($\mathbf{E}$ is a function of t); and at any given time, the electric field varies along the street ($\mathbf{E}$ is a function of x). The physically measurable quantity which varies is electric field; hence we could call our wave an "electric" wave. It is a transverse wave, since the quantity which varies is a vector perpendicular to the direction of propagation. The wave we have described is plane-polarized, although other kinds of polarization can also be obtained. The wavelength λ of our wave is 300 m, the distance from A to

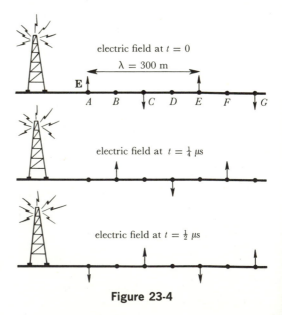

electric field at $t = 0$

$\lambda = 300$ m

E

A B C D E F G

electric field at $t = \frac{1}{4} \ \mu s$

electric field at $t = \frac{1}{2} \ \mu s$

Figure 23-4

E, B to *F*, etc.—always the shortest distance between two points in the same phase of vibration. The period is 1 μs, which is the time for one complete cycle of changes in **E** at any given point; the frequency *ν* is $1/T = 10^6 \text{ s}^{-1} = 10^6 \text{ Hz}$. The velocity of the wave is given as usual by the product of frequency and wavelength:

$$c = \nu\lambda = (10^6 \text{ Hz})(300 \text{ m})$$
$$= 3.00 \times 10^8 \text{ m/s}$$

While all this is going on, a second group of observers stationed at the same points are measuring the *magnetic* field. Each of these observers is equipped with an ideal, inertialess compass needle which points in the direction of the magnetic induction. The results of a series of rapid-fire measurements are shown in Fig. 23-5. At *t* = 0, the magnetic induction at point *A* is directed horizontally to the east (out of the plane of the paper); but at this same instant, the magnetic induction at *B* is zero, and at *C* the magnetic induction is directed to the west, into the plane of the paper. Just as before, we interpret these measurements by means of a "magnetic" wave, transverse because the variable quantity (magnetic induction **B**) is a vector per-

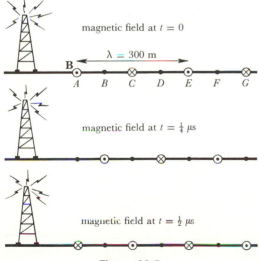

Figure 23-5

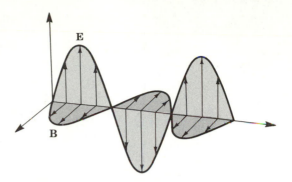

Figure 23-6 The **E** and **B** vectors in a plane electromagnetic wave, traveling along the positive *x* axis. The field vectors are mutually perpendicular and are in phase.

pendicular to the direction of propagation.

We see that a radio wave is really two simultaneous waves (Fig. 23-6). This is the significance of the term "electromagnetic." A purely electric wave, as shown in Fig. 23-4, is impossible; likewise, a purely magnetic wave, shown in Fig. 23-5, is impossible. In a sense, the two waves keep each other going by mutual interaction.

Although our illustration was based upon the long-wavelength e-m waves which we call radio waves, the wave model for visible light is exactly similar. Except for their short wavelengths (and high frequencies), light waves are in no way different from radio waves. They *are* radio waves, and their velocity, in a vacuum, is given by the electrical quantity $\sqrt{k/k'}$, which equals 3.00×10^8 m/s.

23-6 Production of electromagnetic waves

A simple generator of e-m waves is shown in Fig. 23-7, where an antenna is formed by two metal rods connected to some sort of electric oscillator. The exact nature of the oscillator need not concern us here; a tuned *LC* circuit might be used, but other circuits are also possible. The oscillator causes charges to flow back and forth along the antenna. The system is shown at several times. In

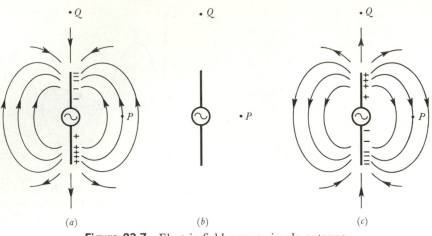

Figure 23-7 Electric field near a simple antenna.

(a), negative charge has been pushed to the end of the upper rod, and an excess of positive charge remains on the bottom rod. At a time which is $\frac{1}{4}$ of a cycle later (b), the charges have come together again, and the rods are momentarily uncharged. Still later, at (c), the oscillator has reversed polarity and the charges are reversed, with the top rod positively charged. Considering the situation at any one instant, say (a), we see that the electric field surrounding the antenna is similar to that due to a pair of equal and opposite charges. At point P in (a), for instance, the electric field is upward; but half a cycle later, at (c), the field at P is downward. During the time between (a) and (c), positive charge is flowing up the antenna from bottom end to top end. This upward current produces a magnetic field, and the right-hand rule tells us that the lines of force of the magnetic field are in the form of circles (Fig. 23-8). At point P, the magnetic induction is perpendicular to the plane of the paper, directed horizontally away from us. Later, while charges are flowing back down the antenna, the magnetic induction **B** reverses direction, but is still horizontal.

We have had to overlook some details, but it is evident from our discussion that at point P *two* fields exist, and that the **E**-field and the **B**-field are perpendicular to each

other. At P, the wave is propagated horizontally, broadside to the rods; at a point such as Q, the magnetic induction is zero (head-on view of the current; see Sec. 21-5), and hence no wave is propagated along a direction parallel to the length of the antenna. There is a smooth variation in the intensity of the wave, from a maximum value if the antenna is viewed broadside, to a zero value if viewed head-on. If, as is usually the case, the observer is far from the antenna (compared with its length), **E** and **B** are "in phase," and both fields pass through their zero values at the same time.

At any instant, energy is stored in the space surrounding the antenna. Figure 23-9 shows the electric field (for simplicity, the accom-

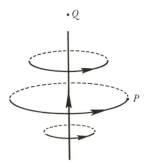

Figure 23-8 Magnetic field near a simple antenna.

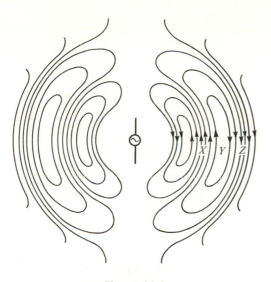

Figure 23-9

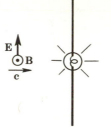

Figure 23-10 An e-m wave moving toward the right is approaching an antenna. The electric vector of the wave is parallel to the antenna wire.

panying magnetic field is not shown) for a given instant; the field in region X is upward, at Y it is zero, and in Z it is downward. Electric PE is stored wherever there is an electric field; the diagram shows that the energy is (at this instant) localized in regions such as X and Z, separated by regions such as Y where no energy is stored. As the wave spreads out, the regions of energy concentration spread out. This is how the wave model explains the propagation of energy by radiation. The situation is analogous to the spreading out of a water wave; gravitational PE is stored wherever there is a crest, and energy is propagated as the crest moves. One basic difference should be noted: an electric or magnetic field can exist in a vacuum, and hence no medium is needed to transmit energy by an e-m wave. In a water wave, gravitational PE can exist only where there is water, and hence such a wave requires a medium (the water).

23-7 Radio

If a radio receiving antenna is placed parallel to the electric field vector of an e-m wave (Fig. 23-10), the free electrons in the metal will be urged back and forth as the field fluctuates. At any instant, the force on an electron (whose charge is Q) could be calculated from the equation $E = F/Q$. That is, $F = QE$. If the current in the antenna is great enough, a lamp bulb at the center can be lit to full brilliance by the charges which pulse back and forth. The effect is greatest if the antenna is parallel to the electric vector of the wave, since otherwise only a component of the force QE would be effective. An antenna whose length is equal to about half the wavelength of the radiation is much more efficient than a shorter or a longer one; this applies both to the transmitting antenna and to the receiving antenna.[*] Still another way to receive radio signals is to use a loop antenna (p. 491). As the e-m wave progresses through the loop, the changing magnetic field induces a current if the wires of the loop are properly oriented. A loop can be used as a direction-finder on ship or plane to locate the direction of a source of e-m radiation, such as a commercial radio station or a radio beacon.

A radio transmitter and receiver together constitute a system by which information can be sent from one place to another. As we use the term, *information* is the opposite of randomness. For instance, if a 1000 Hz sound is being emitted by a tuning fork, "information" is being transmitted to the ear of the listener.

---

[*] This is a case of electric resonance; recall that an open pipe which is $\frac{1}{2}\lambda$ long responds well to an incident sound wave (Sec. 11-4).

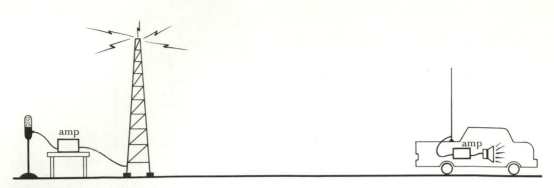

Figure 23-11 Hypothetical system of communication using unmodulated e-m waves, impractical because of long wavelengths required.

A more complex type of information is transmitted by a speaking person, or by a symphony orchestra. To transmit the information of a 1000 Hz fork, it would seem natural to use a system somewhat like that in Fig. 23-11. A microphone would pick up the sound, producing feeble voltages which would be amplified and cause electrons to flow up and down in a vertical transmitting tower. The tower would radiate an e-m wave whose frequency is 1000 Hz, which would be received by a distant antenna. After further amplification, a pulsating current would be caused in the loudspeaker. Unfortunately, the system of Fig. 23-11 is entirely impractical because of the long wavelength of the necessary e-m wave. Since the velocity is 3.00×10^8 m/s, we have

$$\lambda = \frac{\text{velocity}}{\text{frequency}} = \frac{3.00 \times 10^8 \text{ m/s}}{10^3 \text{ Hz}}$$

$$= 3.00 \times 10^5 \text{ m} = 300 \text{ km}$$

The wavelength of our 1000 Hz e-m wave would be 300 km (about 186 mi), and to obtain a useful amount of radiation both the transmitting and receiving antenna would have to be at least $\frac{1}{4}\lambda$ in height—some 46 mi high! While it is true that the antennas could be stretched out horizontally, they would have to be supported many miles above the surface of the earth in order to radiate such a long wave with useful intensity. Such dimensions are fantastically large. Radio waves of frequency 10^6 Hz ($\lambda = 300$ m) are practical enough—but the frequency of 1 000 000 Hz (1 MHz) is far above the audible range.

The problem of conveying audible information by radio waves was solved by the technique of *modulation.* To modulate a wave, some property of the wave is varied at a slow (or "audio") rate. Two types of modulation are in common use: *amplitude modulation* (AM) and *frequency modulation* (FM). If the amplitude of a wave is periodically changed, then information can be sent at a relatively slow rate, using the rapid, easily transmitted high-frequency wave as a *carrier.* For instance, if

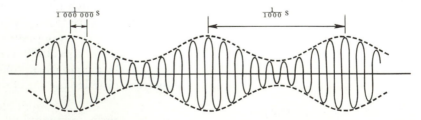

Figure 23-12 Amplitude modulation: carrier wave of frequency 1 MHz modulated by tone of frequency 1000 Hz.

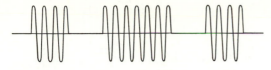

Figure 23-13 A simple form of amplitude modulation.

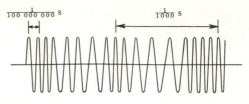

Figure 23-15 Frequency modulation: carrier wave of frequency approximately 100 MHz modulated by tone of frequency 1000 Hz.

the amplitude of a 1 MHz wave varies as shown in Fig. 23-12, information is sent at 1000 Hz by a carrier wave whose frequency is 1 000 000 Hz. The very earliest communication by radio waves was by the use of dots and dashes (Fig. 23-13), a form of amplitude modulation. An on-off switch (telegrapher's key) at the transmitter is sufficient to modulate a carrier wave in this simple fashion. We shall not discuss the circuits by which a microphone produces a smoothly varying amplitude modulation of a carrier as in Fig. 23-12.

In the receiver the wave must be demodulated (or "detected") in order to recover the information. If the wave of Fig. 23-14a were applied directly to a loudspeaker, nothing would be heard. The speaker has far too much inertia to respond to variations as rapid as 10^6 Hz, and so it responds only to the average value of the current (heavy line), which is zero. A simple way to demodulate an amplitude-modulated signal is to use a diode which passes current in only one direction (assumed to be the positive direction in Fig. 23-14). During the positive half of each cycle the diode conducts easily, but during the other half of each cycle, when the diode is reverse biased, it no longer conducts. Either a vacuum-tube or a solid-state diode can be used. After the negative portions of Fig.

23-14a are thus chopped off, the resulting wave (Fig. 23-14b) has a slowly varying average value (heavy line), and the loudspeaker can follow these variations to produce a sound wave which duplicates the original sound wave in the studio at the transmitter.

In frequency modulation the amplitude of the carrier wave remains constant while the frequency changes. In Fig. 23-15, which is typical of a commercial FM station whose "center frequency" is 99.9 MHz, the carrier frequency varies from 99.8 MHz to 100.0 MHz, and this variation takes place 1000 times per second if a flutist is playing a tone having audio frequency 1000 Hz. We shall not discuss the means by which frequency modulation is produced at the transmitter, nor shall we attempt to describe the circuits in a receiver which are used to demodulate an FM wave and thus recover the original (audio) information. One advantage of FM reception is that it is practically free from "static." The e-m radiation caused by lightning, ignition, etc., would show up as an amplitude change, and hence the FM receiver, which responds only to frequency changes, is unaffected.

A television receiver receives two kinds of information: picture information ("video")

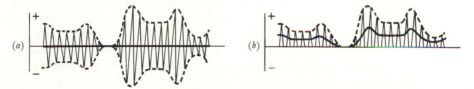

Figure 23-14 Demodulation of an AM signal. The heavy line represents the average loudspeaker current.

and sound information ("audio"). The video information is transmitted by AM, and the net result is to control the intensity of a beam of electrons. The picture is built up as the electron beam is caused to sweep back and forth, striking the face of the picture tube in a succession of spots. The brightness of the spot of light changes rapidly, in proportion to the beam current, which is controlled by a grid whose potential in turn is determined by the demodulated video signal. The sound part of the information is received by FM, with a demodulating circuit and an audio amplifier which feeds a loudspeaker. Typical carrier frequencies are: 85 MHz for the center of TV channel 6, which extends from 82 to 88 MHz; and 98 MHz for the middle of the FM radio band, which extends from 88 to 108 MHz. Working out the wavelengths, we find that λ is about 3.3 m for the TV carrier and also for the FM radio carrier. This has the practical advantage that it is possible to use antennas tuned for maximum efficiency—a structure of dimension $\frac{1}{2}\lambda$ is not too bulky to be placed on a rooftop or chimney. If you look carefully at TV or FM antennas you will find that most of them have rods about a meter in length; what is more, you will note that the rods are horizontal, which shows that carrier waves for TV are polarized horizontally (i.e., the electric vector is horizontal and the magnetic vector is vertical). The antennas are set broadside to the direction of propagation of the wave. The carrier waves of FM stations are also polarized horizontally; hence a TV antenna serves admirably to receive FM broadcasts, especially if it is necessary to receive distant stations.

Standard-band AM stations require tall transmitting antennas, since $\frac{1}{2}\lambda$ is of the order of several hundred meters. One way to economize is to build the tower only $\frac{1}{4}\lambda$ high. Charges are continually rushing up and down the tower. The oscillator tends to cause charges to flow from the top, down the tower, through the oscillator, and out into the earth.

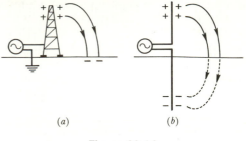

(a)   (b)

Figure 23-16

Thus, when there is a net $+$ charge at the top of the tower, there is a net $-$ charge on the earth's surface, and lines of force can terminate on these charges, as shown in Fig. 23-16a. Half a cycle later, the signs of the net charges, both on tower and ground, are reversed. At a distance which is approximately $\frac{1}{2}\lambda$ from the tower, some of the lines of force become detached and energy is radiated. The tower and ground together are equivalent to the longer doublet of Fig. 23-16b. Usually a set of heavy copper wires is buried in the ground, radiating out from the base of the tower, to facilitate the necessary to-and-fro flow of charges in the earth near the tower.

23-8   Sources of radiation

If all e-m radiation is alike, except for wavelength and frequency, we naturally look for some similarity in the means of production of radiations. What general statement can we make which will apply to such diverse sources of radiation as radio transmitters, incandescent filaments in light bulbs, and x-ray tubes? The unifying concept here is that *electromagnetic radiation is produced when electric charge is accelerated.* We have seen this illustrated for radio waves. As charges rush back and forth along the antenna, they are undergoing rather rapid accelerations, especially at the two ends. (Recall that in SHM the acceleration is greatest at the turning points.) A stream of charges moving with a steady velocity emits no e-m wave, for then there is no acceleration of charge. True enough, a steady magnetic field surrounds a wire carrying a steady current, but an induced electric field is caused only by a *changing* magnetic field. Thus we see that the interaction between electric and magnetic fields requires a changing current, i.e., an acceleration of charge, in order for an e-m wave to arise.

The sources of radio waves quite obviously involve acceleration of charge; free electrons drift back and forth through the metal of the wires and towers, driven by an electric oscillator. Going down the e-m spectrum to shorter wavelengths and higher frequencies, we next consider infrared radiation of typical wavelength 10^5 angstroms ($1 \text{ Å} = 10^{-10}$ m). Infrared radiation (often called heat waves) is emitted by the atoms of a solid or liquid which are in thermal agitation. Unless a solid is at absolute zero, the atoms are not rigidly locked in a crystal structure, but vibrate to and fro about their equilibrium positions. Since all atoms contain electric charges, it is reasonable to expect that the acceleration of these charges during thermal vibration should give rise to e-m waves.

Progressing now to still shorter wavelengths, we identify visible light and ultraviolet radiation as waves often related to accelerations of orbital electrons. In a general way, we would expect the orbital electrons in atoms to vibrate with high frequency because of the small mass of an electron as compared with that of a nucleus or an atom or molecule. Still higher frequencies are represented by x rays. When electrons are given a high speed by application of a large potential difference (say, 40 000 V), they strike a target and are suddenly stopped. This is a negative acceleration, and an e-m wave is emitted from the target. Here, too, we see that the acceleration of charge gives rise to electromagnetic radiation.

The wave model of e-m radiation is best for low-frequency radio waves, and the model becomes progressively worse as we go to higher and higher frequencies. As a sample of the trouble we get into, consider the motion of an orbital electron as it moves around the nucleus. We know that a particle in circular motion is accelerated toward the center of the circle; why, then, does not the electron in a hydrogen atom, for instance, continually radiate an e-m wave? The energy of such a wave would have to come from the electric PE of the electron and its KE. The situation is analogous to the frictional loss of energy by an artificial satellite, which spirals down to the earth as it loses energy. Calculation leads us to expect that all electrons would lose energy by radiation and spiral into the nucleus, with an increasing angular speed. The whole universe should go up in a blinding flash of radiation—the so-called "violet death." Such a catastrophe obviously has not taken place, and so we conclude that in this respect our wave model is inadequate. There is no doubt, however, that visible light *does* have wave properties, and so we shall continue to use the wave model whenever it is proper and fruitful to do so.

■ SUMMARY

A changing magnetic field gives rise to an induced electric field, and a changing electric field gives rise to an induced magnetic field. This interaction between electric and magnetic fields allows the propagation of electromagnetic waves in which the electric and magnetic fields are perpendicular to each other, and both are perpendicular to the direction of propagation. Electromagnetic waves are transverse and are often plane-polarized; their velocity in a vacuum is $c = \sqrt{k/k'} = 3.00 \times 10^8$ m/s.

In general, electromagnetic waves are radiated whenever electric charge is accelerated. The complete electromagnetic spectrum extends from radio waves to gamma rays, but only for the relatively slow vibrations of radio waves can charges be set into continuous oscillation by purely electrical means. Many electric oscillators use tuned circuits whose frequency is given by $\nu = (1/2\pi) \sqrt{1/LC}$. Infrared, visible, ultraviolet, x-ray, and gamma-ray radiations are also connected with acceleration of charge, but the wave model is not entirely adequate to describe the emission of radiation of high frequency.

In order to communicate by use of radio waves, information is conveyed by some form of modulation. Either the amplitude or the frequency of the carrier wave can be varied.

■ CHECK LIST

$\nu = (1/2\pi) \sqrt{1/LC}$	modulation	carrier wave
electromagnetic wave	amplitude modulation	demodulation
information	frequency modulation	angstrom unit

■ QUESTIONS

23-1 Are radio waves a form of "light"?

23-2 Can e-m waves be propagated through a perfect vacuum?

23-3 What is the velocity of x rays in a vacuum?

23-4 Two radio stations having different frequencies are broadcasting in the same city at the same time. Why does a listener's radio receiver respond to only one of these stations at a time?

23-5 In an L-C circuit, why doesn't the capacitor simply lose its charge and remain in a discharged condition, as in Fig. 23-1c?

23-6 A police radio transmitter has a vertical antenna. Should the receiving antenna on a squad car be horizontal or vertical for the most efficient reception?

23-7 Explain, in terms of Eq. 21-11, why the loop antenna shown on p. 491 picks up no induced emf when its plane is broadside to the transmitting radio beacon. (*Hint:* Imagine the antenna wire of Fig. 23-10 replaced by the loop.)

23-8 What is the difference between amplitude modulation and frequency modulation?

23-9 A Boy Scout sends a message by blinking his flashlight on and off, using Morse code. This is an example of a modulated e-m wave. (*a*) What is the approximate frequency of the carrier? (*b*) Is the Boy Scout using amplitude modulation or frequency modulation?

23-10 A traffic signal changes from red to green. What sort of modulation is used to convey the information that it is safe to proceed?

23-11 Discuss the concept of information in relation to that of entropy (Sec. 16-5).

■ PROBLEMS

23-A1 What is the wavelength of a radar wave of frequency (*a*) 10 000 MHz? (*b*) 24 GHz?

23-A2 What is the wavelength of a radio wave of frequency 108 MHz emitted by a transmitter in an artificial satellite? About how long should each section of the antenna be? (See Fig. 30-17 on p. 680.)

23-A3 Calculate the frequency of a radar wave whose wavelength is 1 cm.

23-A4 What is the wavelength of the "interstellar" hydrogen spectrum line whose frequency is 1.42×10^9 Hz?

23-A5 Compute the frequency of green light, whose wavelength in vacuum is 5000 Å.

23-A6 Compute the wavelength of a standard-band AM station which broadcasts on an assigned frequency of 750 kHz.

23-A7 Express in angstroms (*a*) 4.5×10^{-5} cm; (*b*) 6×10^{-10} km; (*c*) 550 nanometers; (*d*) 12 microns. (*Note:* 1 micron = 1 μ = 1 μm = 10^{-6} m).

23-A8 Express in meters: (*a*) 5893 Å; (*b*) 1.54 Å; (*c*) 12 μm.

23-A9 Make the following conversions: (*a*) 4000 Å = ? cm; (*b*) 0.71 Å = ? m; (*c*) 10^{-10} m = ? μm; (*d*) 5461 Å = ? nm; (*e*) 30 μm = ? Å.

23-B1 The current through the coil of a certain electromagnet is 3 A, and 100 J of energy is stored in the magnetic field. How much energy will be stored if the current is 15 A?

23-B2 (*a*) How much energy is stored in a coil of inductance 25 H when the current through the coil is 400 mA? (*b*) How far would a 1 kg object have to fall to gain this much KE?

23-B3 The frequency of a tuned circuit is determined by a fixed inductance connected in parallel with a variable capacitor. The frequency is 500 kHz when the capacitor is set at 100 pF; what is the frequency when the capacitor is increased to 400 pF?

23-B4 Calculate the resonant frequency of a tank circuit for which L is 40 mH, and C is 900 pF.

23-B5 What is the resonant frequency of a tank circuit in which $C = 0.2$ μF and $L = 5$ H?

23-B6 Calculate the wavelength of an e-m wave whose frequency is determined by a tank circuit inductance 100 μH and capacitance 100 pF.

23-B7 In physiotherapy, a diathermy machine generates e-m waves of wavelength 6 m which give "deep heat" when absorbed in tissue. What effective capacitance is needed for the tank circuit, if the inductance is 1.4 μH?

23-B8 A standard broadcast band AM radio can be tuned by a variable capacitor through a range 500 kHz to 1600 kHz. What is the ratio of the maximum capacitance of the tuning capacitor to the minimum capacitance? How is this change usually effected?

23-B9 A coil of inductance L and two capacitors, each of capacitance C, are available to form a tank circuit. What is the ratio of the maximum possible frequency to the smallest possible frequency?

23-B10 An amateur radio operator wishes to construct a receiver to operate in the "15 meter band" (i.e., to receive waves of wavelength in the neighborhood of 15 m). He has a variable capacitor whose average value is 50 pF, and he wishes to wind a coil to use in a tank circuit. What should be the inductance of the coil?

23-B11 In some radios, the tuned circuit consists of a fixed capacitor connected in parallel with a variable inductance. The radio is tuned to 88 MHz (low end of the FM band) when the inductance is 6.33×10^{-8} H. (*a*) What must be the value of the inductance if it is desired to receive 108 MHz (high end of the FM band)? (*b*) What is the value of the fixed capacitance?

23-B12 In an oscillating tank circuit, prove that the maximum current through the coil is related to Q, the maximum charge on the capacitor, by the equation $I = 2\pi\nu Q$. (*Hint:* Use the energy method.)

23-C1 (*a*) Show that in an *L-C* circuit, at the resonant frequency the impedance of the coil equals that of the capacitor. Derive a formula for this impedance, in terms of L and C; check your answer dimensionally. (*b*) What is the value, in Ω, of the impedance of the coil and of the capacitor in the circuit of Prob. 23-B6?

23-C2 The charge of the capacitor in a resonant circuit varies with time according to the equation $Q = Q_0 \sin 2\pi\nu t$, where Q_0 is the maximum value of the charge. By differentiation, derive the result stated in Prob. 23-B12.

23-C3 (*a*) What capacitance will resonate with a small one-turn loop of inductance 1 nH to give a radar wave of wavelength 3 cm (see Table 14-4)? (*b*) If the capacitor has square parallel plates separated by 1 mm of air, what should the edge length of the plates be? (*c*) What is the common value, in Ω, of the impedance of the loop and of the capacitor, at resonance?

For Further Study

23-9   Energy storage in coils

A coil of self-inductance L is connected to a battery of emf ε through a switch as in Fig. 22-16. After the switch is closed, charges leave the battery and move through the coil; work is done on the charges because they move against the back emf $L(dI/dt)$ of the coil. This back emf is not constant (it becomes zero when the current through the coil reaches its final steady value), so we must use integration to find the total work done on the charges.

If in a time Δt a charge ΔQ moves through the coil, the average current is $\Delta Q / \Delta t$, and the back emf is $L(\Delta I / \Delta t)$. Hence

$$\text{joules} = \left(\frac{\text{joules}}{\text{coulomb}}\right)(\text{coulombs})$$

$$\Delta W = \varepsilon_{\text{back}} \Delta Q = L\frac{\Delta I}{\Delta t} \Delta Q$$

Rearranging gives

$$\Delta W = L\frac{\Delta Q}{\Delta t} \Delta I$$

The total work done while the current is in-

creasing from 0 to its final value I is found from a definite integral (Appendix, p. 709):

$$W = \lim_{\substack{\Delta I \to 0 \\ \Delta t \to 0}} \sum L \frac{\Delta Q}{\Delta t} \Delta I$$

$$= \int_{I=0}^{I=I} LI \, dI = L\int_0^I I \, dI$$

$$= L\left(\frac{I^2}{2} - \frac{0^2}{2}\right) = \frac{1}{2}LI^2$$

This work may be considered to be stored in the magnetic field which has been set up in the coil. It can also be interpreted as kinetic energy of the charges whose motion gives rise to the current in the coil.

23-10   Gravity waves

Recently discovered *gravity waves* have many similarities to e-m waves, but as far as is known there is no direct connection. A simplified view of the production of gravity waves involves the propagation of gravitational field. Granted that the gravitational force between two laboratory-sized objects is

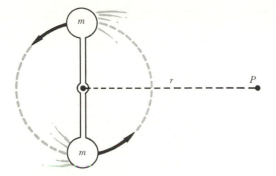

Figure 23-17 A possible source of gravitational radiation is the centripetal acceleration of the mass particles of a rotating body.

extremely small (Prob. 17-B1), nevertheless we can imagine generating a fluctuating gravitational field at a point if a distant "source mass" moves back and forth. In Fig. 23-17, a gravitational wave arises from the centripetal acceleration of the masses m; a test mass at P would experience a small periodic variation in gravitational force in synchronism with the rotation of the dumbbell. Note that it is the *acceleration* of mass that generates a gravitational wave, just as acceleration of charge generates an e-m wave. The rapid rotation of a heavy unsymmetrical body such as a distorted star would also give rise to gravitational radiation.

The theory of gravitation is essentially based on Einstein's general theory of relativity, with mathematical and conceptual complexities such that theorists are not yet in complete agreement. All theories agree in showing that gravitational waves propagate through a vacuum at the speed of light (c), as do e-m waves. (Recall our discussion of c in Sec. 7-4 where this number is most properly considered simply as the speed of any particle whose rest mass is zero.) For the propagation of gravity waves the massless particles which are exchanged (Fig. 18-3) are not photons, but are gravitons, to be discussed in Sec. 30-9. The gravitational interaction is the weakest of all (Table 30-3 on p. 674), so it is not surprising that gravity

waves have escaped notice for so many years.

In 1969 Joseph Weber, working at the University of Maryland, succeeded in observing the effect of gravity waves, using a resonance phenomenon. A 3000 kg solid aluminum cylinder (Fig. 23-18) is suspended in a vacuum by fine wires so as to be practically free of external influences such as seismic waves (earthquakes), traffic vibrations, or impact from air molecules. In our study of resonance (Sec. 11-4) we saw how small impulses, timed properly, can set a large system into vibration. So, also, the tank circuit (Fig. 23-3) of a radio receiver resonates with weak incoming e-m waves to build up a large response. In Weber's experiment, the aluminum cylinder has various modes of vibration; the fundamental frequency for longitudinal vibration is 1660 Hz. If an incoming burst of gravitational radiation has a frequency spectrum which includes this frequency, the cylinder will be set into oscilla-

Figure 23-18 Gravitational-radiation detector. Solid aluminum cylinder of mass about 3000 kg is suspended freely on a steel wire between two iron piers, one of which is shown. When excited by gravitational waves, the cylinder is set into internal vibration. The small blocks bonded to the surface of the cylinder are strain gauges which can detect axial deformations as small as 10^{-16} m. (Photo courtesy J. Weber)

tion. In a similar way a small stroke of a clapper causes a large bell to ring, vibrating with its own set of natural frequencies. Delicate strain gauges monitor the distortion of Weber's aluminum cylinder; the apparatus is sensitive enough to detect a change in length of 10^{-16} m, less than the size of a single nucleus!

Most of the changes observed were due to local mechanical, electric, or magnetic causes. To separate the gravity-wave events from the much more numerous local disturbances, identical cylinders located 1000 km apart were monitored—one in Maryland, the other in Illinois. During a 1-year period several hundred events occurred simultaneously at the two locations, far more coincidences than would be observed by chance. A time-of-day analysis showed a significant increase in the number of coincident events when the center of the galaxy was nearly overhead or was nearly in the opposite direction. (The earth is essentially transparent to gravity waves.) It is concluded that bursts of gravitational waves —non-recurrent waves, or pulses—reach the earth from outside the solar system (not from the sun, since there was no correlation with the sun's position in the sky). By using a properly oriented large disk instead of a cylinder, Weber showed that most, if not all, of the gravitational radiation reaching the earth from the galactic center is in the form of "tensor waves"—somewhat more complex than the simple transverse waves of e-m radiation, and (gratifyingly) predicted by the Einstein theory of gravitation.

Further work in this exciting new field may make use of the earth itself as a resonant detector. Possibly a search will be made for correlations between minute, but simultaneous, deformations of the earth and of the moon.

■ REFERENCES

1 Magie, W. F., *A Source Book in Physics* (New York: McGraw-Hill, 1935), pp. 528–38. Brief extracts from the writings of Maxwell and Rowland.

2 Newman, James R., "James Clerk Maxwell," *Sci. American,* June, 1955, p. 58.

3 Morrison, Philip and Emily, "Heinrich Hertz," *Sci. American,* Dec., 1957, p. 98. An account of the work of the man who first obtained experimental evidence for the electromagnetic waves predicted by Maxwell's theory.

4 Weber, J., "Gravitational Waves," *Phys. Today,* Apr., 1968, p. 34; "The Detection of Gravitational Waves," *Sci. American,* May, 1971, p. 71.

Geometrical Optics

The study of optics is the study of a small part of the electromagnetic spectrum. Visible light waves range in wavelength from about 3800 angstroms (Å) for the violet to about 7600 Å for the red (3800 × 10⁻¹⁰ m to 7600 × 10⁻¹⁰ m). The corresponding frequencies are high—in the neighborhood of 6×10^{14} Hz for green light of wavelength 5000 Å. The approximate wavelength ranges of various colors are shown in Table 24-1. The nanometer (1 nm = 10^{-9} m = 10 Å), also called the millimicron (mμ), is a useful unit for wavelength, since the human eye can just about detect the difference in color of two sources that differ by 1 nm. In spite of the fact that visible light comprises only about one "octave" of the complete e-m spectrum, it is worthwhile to devote a considerable time to the study of optical phenomena. For one thing, of course, much of our knowledge of the physical world comes to us through optical instruments such as the eye, the telescope, and the microscope. Then, too, our study of visible light is to a certain extent the study of all e-m radiation. The wave-particle duality of radiation is perhaps best shown by visible light. Radio radiations (low frequencies) are practically "all wave" and show few or no particle properties; nu-clear gamma rays (high frequencies) are practically "all particle," with wave properties observable only with extreme difficulty. Visible light occupies a favored position in the e-m spectrum where wave and particle aspects are about equally prominent. Since our aim is to explore and understand nature, we welcome the challenges as well as the practical uses of visible light. We shall find the wave model entirely adequate to explain how light often travels in a straight line, how it is reflected by a mirror and bent as it enters a prism or a lens, and how it spreads out into a diffraction pattern when it passes through a small opening.

24-1 Huygens' principle

Christian Huygens[*] (1629–1695) was a versatile Dutch physicist and natural philosopher, born 14 years before Newton. His work in mechanics did much to pave the way for Newton's achievement, and he, along with Galileo, is credited with developing the pendulum clock, an invention of prime importance for the science of mechanics, where

[*] The English pronunciation is Hy'genz.

Refraction at the faces of a parallel slab of glass; the emerging rays are parallel to the incoming rays. Also to be noted are the weak partial reflections at the top and bottom faces. (Bausch & Lomb)

TABLE 24-1   **Wavelengths of visible light**

Color	Approximate wavelength range	
	(Å)	(nm)
violet	3800–4500	380–450
blue	4500–4900	450–490
green	4900–5600	490–560
yellow	5600–5900	560–590
orange	5900–6300	590–630
red	6300–7600	630–760

precise measurement of time intervals is necessary. Huygens was a firm believer in the wave theory of light, and he used a geometrical principle (without proof) to explain many of the properties of the transmission of light and other waves.

The concentric circles formed when a stone is dropped into a lake and water waves spread out (as in Fig. 24-1) are called wave fronts. The circle *a* represents a crest, as does the circle *b*. Between these circles is another circle *t*, which is a trough. In general, a *wave front* is defined as the locus of points having the same phase of vibration. All the points on *a* have the same phase, since all the molecules on this particular wave front are at their maximum displacement *above* the normal level. Likewise, *t* is a wave front on which all particles are at their maximum displacements *below* the normal level. Halfway between *t* and *b* is another wave front where the particles are (momentarily) at the normal level, but moving upward. This technical use of the word "front" is not unfamiliar; in meteorology a cold front is said to connect cities *X*, *Y*, *Z*, which are simultaneously experiencing a sudden drop in temperature. As a wave spreads out, the wave fronts advance, and in the case of our water wave, the radius of each circular wave front continuously increases. *Huygens' principle* tells us how to predict a new wave front when we know the position of an earlier one. The construction is as follows: Let every point on the wave front be considered the source of a small *wavelet* which spreads out in the forward direction (Fig. 24-2). The new wave front is the envelope of all the wavelets; that is, the line or surface tangent to all the little wavelets. If we now consider light waves spreading out in three dimensions, it is evident that the wavelets are small hemispheres, provided

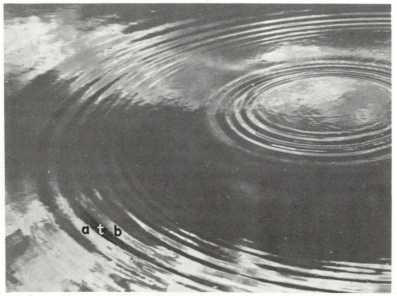

a t b

Helen Faye

Figure 24-1 Wave fronts on the surface of water.

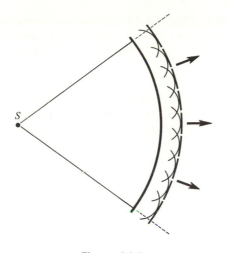

Figure 24-2

Figure 24-3

the speed of light is the same in all directions; hence we see that a spherical wave front will remain spherical and that the energy of the wave is carried away equally in all directions. Such directions of energy flow are called *rays;* the short arrows in Fig. 24-2 represent rays of light which diverge from the source *S*. Unless specifically stated otherwise, we shall always assume that light travels through a medium equally fast in all directions; such a medium is called an *isotropic* medium.[*] Glass, water, air, Lucite, and most other common substances are isotropic, but certain transparent crystals have enough structure to give rise to "easy" and "hard" directions for the propagation of light. Such a crystal is said to be *anisotropic.* In an isotropic medium, the wavelets are spherical, and the rays are always perpendicular to the wave fronts, as in Fig. 24-3. In an anisotropic medium the wave fronts are ellipsoidal, as discussed in Sec. 25-9.

24-2 Straight-line propagation of light

We are ready now to consider some of the well-known properties of light. Huygens'

---

[*] Greek: *iso* = equal, *tropic* = turning or changing.

principle describes the propagation of light in a straight line, since wavelets from a plane wave front give rise to a new plane wave front (Fig. 24-4), and the corresponding rays are parallel to each other. Such a plane wave front can be thought of as a section of a very large spherical wave front; for instance, sunlight strikes the earth in wave fronts of radius 93 million miles. For all practical purposes, such a wave front is "plane" and the rays are "parallel" to each other.

Newton never did accept the wave model, for he felt (quite justifiably) that if light waves existed they would naturally "bend around corners" and would not travel in straight lines through openings or around obstacles, as they are observed to do. Newton did not realize that the extreme smallness of the wavelength of light compared with the size of the usual opening or obstacle could account for this behavior. Let us use water waves to illustrate Newton's point. When plane waves strike a gap in a long breakwater (Fig.

Figure 24-4

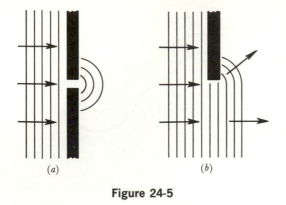

Figure 24-5

24-5*a*), a disturbance spreads out in all directions in the water on the other side; and if the wave train strikes the end of the break-water (as in *b*), the waves diffuse around the corner. On the other hand, if light waves fall on an opening a millimeter or so in width or strike the end of an obstacle, the rays go straight through, with only a very slight "fuzziness" around the edges. We conclude that the straight-line propagation of a wave is a relative thing. We shall prove in Sec. 25-5 that the determining factor is the size of the opening or obstacle *as compared with the wavelength*. Water waves have wavelengths of the order of several centimeters to several meters; for such waves, the "shadow" cast by a floating log, for instance, is not sharp. But visible light waves have wavelengths of about 5×10^{-5} cm, which is small compared with the diameter of a visible object or the pupil of the eye; thus we anticipate that in the usual situations light waves are propagated in practically straight lines.

These ideas are illustrated by sound waves. Many sounds have wavelengths of several meters or more (a 300 Hz blast from an automobile horn has wavelength of approximately 1.1 m), hence street noises diffuse through an open window without any obvious straight-line propagation. Nevertheless, sound shadows and reflection (such as echoes) are possible; it is only necessary to have an obstacle that is large compared to a wave-

length. This can be done in two ways: an echo can be received when an ordinary-sized (audible) wave is reflected from a mountainside or the wall of a large building, or very short-wavelength sound waves of ultrasonic frequency can be reflected by obstacles of ordinary size. The latter method is used in underwater range finding (sonar) and in navigation by bats. Light and other e-m waves also show a gradual transition from diffuse nature to straight-line propagation as wavelength is decreased. Standard-band radio waves ($\nu = 10^6$ Hz, $\lambda = 300$ m) spread out diffusely; TV broadcasts ($\nu = 10^8$ Hz, $\lambda = 3$ m) are easily blocked out by buildings or by the curvature of the earth; radar waves ($\nu = 10^{10}$ Hz, $\lambda = 0.03$ m) are small compared with a plane or even a man, and thus cast sharp shadows.

24-3 Reflection

When a ray of light strikes a smooth surface such as a mirror, the reflected ray leaves the mirror in a definite direction which is determined by two rules: (1) the incident ray, the reflected ray, and the normal to the surface all lie in the same plane; (2) the angle of incidence equals the angle of reflection. These angles are measured from the normal and are denoted by i and r in Fig. 24-6*a*. To prove that $i = r$, we use Huygens' prin-

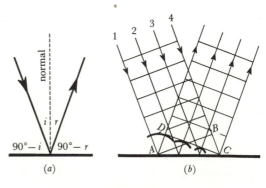

Figure 24-6

ciple. In Fig. 24-6*b* we show a wave front *AB* which is perpendicular to the incoming rays 1, 2, 3, 4. While ray 4 is traveling from *B* to *C*, ray 1 spreads out from *A* in a sphere whose radius is *AD*. Hence *AD* = *BC*. Rays 2 and 3 strike the surface later than ray 1, and therefore their wavelets are correspondingly smaller. According to Huygens' principle, the reflected wave front is *CD*, which is tangent to all the wavelets. We now use plane geometry to prove that *i* = *r*. The right triangles *ADC* and *ABC* are congruent, since they have the same hypotenuse, and the leg *AD* of one triangle equals the leg *CB* of the other triangle. Hence $\angle BCA = \angle DAC$:

$$90° - i = 90° - r$$

$$i = r \qquad (24\text{-}1)$$

The *specular reflection** from a smooth surface such as we have just described is often a nuisance. If a surface is not perfectly smooth, *diffuse reflection* takes place, with many reflected bundles of rays coming from small flat spots. Much of what we see is made visible by such a process. If the surface is so rough that there are no flat spots which are much larger than a wavelength, the light is *scattered* diffusely. We can think of scattering as the absorption and re-emission of light by particles of a medium. The light of the sky reaches us after being scattered by air molecules, water droplets, or dust particles.

24-4   Refraction

The bending of light at the interface between two mediums is called *refraction*. Just as for reflection, the angles are measured from the normal (Fig. 24-7), and, as before, the incident ray, the refracted ray, and the normal all lie in the same plane. Refraction is caused by the fact that the speed of light in a medium is less than in a vacuum and is different in different mediums. This is in accordance

*Latin: *speculum* = mirror.

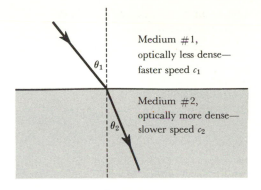

Medium #1, optically less dense— faster speed c_1

Medium #2, optically more dense— slower speed c_2

Figure 24-7

with Maxwell's theory of electromagnetic waves, and advanced theory shows how to correlate the wave velocity with the electrical and magnetic properties of the medium. Most transparent substances are nonmagnetic, and the velocity of an e-m wave then works out to be $c/\sqrt{K}$, where *K* is the dielectric constant. For the very high frequencies of visible light, the dielectric constant is not strictly a constant but depends on frequency. Hence we expect the speed of light in a medium to depend somewhat upon frequency, i.e., upon color. However, this effect is at most only a few per cent over the whole range of visible light and can often be ignored.

The speed of light in a medium is denoted by c_n, and the speed of light in a vacuum is denoted by *c* (= 3.00×10^8 m/s). We now define a constant *n*, called the *index of refraction:*

$$n = \frac{\text{speed of light in vacuum}}{\text{speed of light in medium}} = \frac{c}{c_n} \qquad (24\text{-}2)$$

The index of refraction is a measure of *optical density;* for instance, the speed of light in glass is less than in air, and so glass is said to be optically denser than air. Some values of index of refraction are shown in Table 24-2; these are average values for visible light. The variation with color is minor; typical values of *n* for window glass might be 1.51 for red light, 1.52 for green light, and 1.53 for violet light. This means of course

TABLE 24-2   Index of refraction

Substance	n
crown glass (67% SiO$_2$, 12% Na$_2$O, 11% BaO, misc. 10%)	1.52
flint glass (39% SiO$_2$, 3% Na$_2$O, 49% PbO, misc. 9%)	1.66
diamond	2.42
ice	1.31
water	1.333
benzene	1.50
air at 0°C and 1 atm	1.00029
hydrogen at 0°C and 1 atm	1.00013

that in glass red light travels faster than does violet light. In a medium the dependence of the speed of light on color (i.e., on frequency) is called *dispersion*. Some of the consequences of dispersion, both useful and annoying, will be discussed more fully in Chap. 26, where we study optical instruments.

EXAMPLE 24-1

Calculate the speed of light in diamond.
Since $n = c/c_n$,

$$c_n = \frac{c}{n} = \frac{3.00 \times 10^8 \text{ m/s}}{2.42}$$

$$= \boxed{1.24 \times 10^8 \text{ m/s}}$$

EXAMPLE 24-2

The index of refraction of a certain glass is 1.50 for light whose wavelength in vacuum is 6000 Å. What is the wavelength of this light as it passes through glass?

The frequency remains the same when light enters the glass. This frequency is

$$\nu = \frac{c}{\lambda} = \frac{3 \times 10^8 \text{ m/s}}{6000 \times 10^{-10} \text{ m}}$$

$$= 5 \times 10^{14} \text{ Hz}$$

The speed in glass is given by

$$c_n = \frac{c}{n} = \frac{3 \times 10^8 \text{ m/s}}{1.50}$$

$$= 2 \times 10^8 \text{ m/s}$$

Hence the wavelength in glass is

$$\lambda_n = \frac{c_n}{\nu} = \frac{2 \times 10^8 \text{ m/s}}{5 \times 10^{14} \text{ Hz}}$$

$$= 4 \times 10^{-7} \text{ m}$$

$$= 4000 \times 10^{-10} \text{ m}$$

$$= \boxed{4000 \text{ Å}}$$

The wavelength in glass is only two-thirds as much as it is in vacuum.

We can generalize Example 24-2, working in symbols instead of with numbers. If λ_{vac} is the wavelength in vacuum and λ_n is the wavelength in the medium, we have

$$\nu = \frac{c}{\lambda_{\text{vac}}} = \frac{c_n}{\lambda_n}; \quad \frac{c}{\lambda_{\text{vac}}} = \frac{c/n}{\lambda_n}; \quad \lambda_n = \frac{\lambda_{\text{vac}}}{n}$$

Thus the wavelength is reduced by a factor equal to the index of refraction.

Returning now to refraction, we apply Huygens' principle to derive a relationship between θ_1 and θ_2 (Fig. 24-7). The construction in Fig. 24-8 is much the same as for reflection; during the time interval Δt, ray 1 travels from A to C while ray 4 travels from B to D. The radius AC of the wavelet through C is given by (velocity)(time) $= c_2 \Delta t$, whereas the distance BD is given by $c_1 \Delta t$. Using plane geometry, we identify $\angle BAD$ as equal to θ_1, and $\angle ADC$ as equal to θ_2. We now use the definition of the sine of an angle.

$$\sin \theta_1 = \frac{c_1 \Delta t}{AD}$$

$$\sin \theta_2 = \frac{c_2 \Delta t}{AD}$$

Dividing one equation by the other and simplifying, we get

$$\frac{\sin \theta_1}{\sin \theta_2} = \frac{c_1}{c_2}$$

But since $c_1 = c/n_1$ and $c_2 = c/n_2$, we can write

$$\frac{\sin \theta_1}{\sin \theta_2} = \frac{c/n_1}{c/n_2} = \frac{n_2}{n_1}$$

whence, finally,

$$n_1 \sin \theta_1 = n_2 \sin \theta_2 \qquad (24\text{-}3)$$

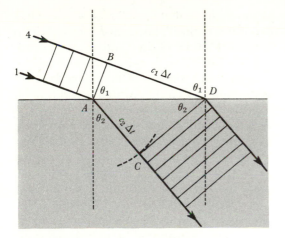

Figure 24-8 Derivation of Snell's law using Huygens' principle.

This is Snell's law, discovered experimentally in 1621 by the Dutch scientist Willebrord Snell. In words, the product of the index of refraction for a given frequency of light and the sine of the angle measured from the normal is a constant, the same in each medium. Thus the angle is less in the medium whose index of refraction is greater. That is, light bends toward the normal when it enters an optically denser medium. Conversely, light bends away from the normal as it enters an optically less dense medium.

EXAMPLE **24-3**

A ray of light in air strikes a slab of glass, as shown in Fig. 24-9. Calculate the angle of refraction in the glass, which has an index of refraction of 1.5.

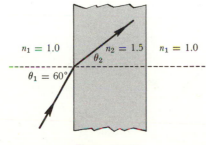

Figure 24-9

The index of refraction of air is 1.0003, and for most problems it is sufficiently accurate to call it exactly 1, the same as for a vacuum. Using Snell's law, we have

$$n_1 \sin \theta_1 = n_2 \sin \theta_2$$
$$(1)(\sin 60°) = (1.5)(\sin \theta_2)$$
$$\sin \theta_2 = \frac{(1)(0.866)}{1.5} = 0.577$$
$$\theta_2 = \boxed{35°}$$

EXAMPLE **24-4**

An insulated lamp bulb is on the bottom of a swimming pool at a point 2.5 m from a wall. The pool is 2.5 m deep and filled with water to the top (Fig. 24-10). At what angle does the light leave the water at the edge of the pool?

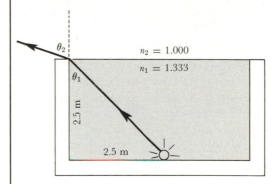

Figure 24-10

First we find θ_1, using simple trigonometry.

$$\tan \theta_1 = \frac{2.5 \text{ m}}{2.5 \text{ m}} = 1.00$$
$$\theta_1 = 45°$$

Next we apply Snell's law:

$$n_1 \sin \theta_1 = n_2 \sin \theta_2$$
$$(1.333)(\sin 45°) = (1.000)(\sin \theta_2)$$
$$(1.333)(0.707) = \sin \theta_2$$
$$\sin \theta_2 = 0.942$$
$$\theta_2 = \boxed{70°}$$

As expected, θ_2 is greater than θ_1, and the ray bends away from the normal as it enters the less dense medium (air).

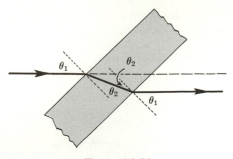

Figure 24-11

One common application of Snell's law is shown in Fig. 24-11. Light strikes a slab of glass having parallel surfaces (e.g., a window pane), and the rays emerge without change of direction. Each ray is displaced slightly, parallel to itself.

24-5 Total reflection

It is always possible for an oblique ray to bend toward the normal as it enters a denser medium. If the ray travels exactly along the normal, it continues to do so in the second medium. However, a ray in a dense medium does not always emerge into a less dense me-

dium. Suppose that $n_1 = 2$ and $n_2 = 1$. Then according to Snell's law $2 \sin \theta_1 = \sin \theta_2$. This is fine for values of $\sin \theta_1$ up to and including 0.5. But no angle exists for which $\sin \theta$ is greater than 1; the sine of 90° is 1, and $\sin 91°$ is the same as $\sin 89°$, which is less than 1. We then ask what happens when θ_1, n_1, and n_2 have values which would make $\sin \theta_2$, as calculated from Snell's law, greater than one. We find experimentally that no refraction takes place; the ray is completely reflected back into the first medium. This is called *total reflection,* and it occurs if the angle θ_1 in the denser medium is greater than a certain *critical angle.* We also find that *some* reflection takes place at any angle; as the angle in the denser medium is gradually increased (Fig. 24-12a, b, c, d),* the reflected ray becomes stronger and the refracted ray becomes weaker. In the limit, at the critical angle, the refracted ray just grazes the surface (e), but its intensity is zero. Beyond the critical angle, there is no refracted ray and the re-

---

*The intensities have been calculated for a glass-air interface with $n = 1.50$, using the Maxwell theory of e-m waves.

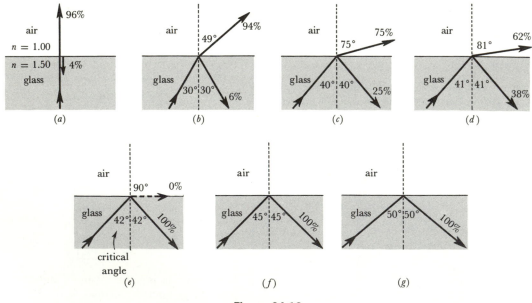

Figure 24-12

flected ray's intensity exactly equals that of the incident ray (*f*, *g*). This is one of the few examples in physics of a process that is truly 100% efficient; absolutely no energy is absorbed at the surface. The critical angle can be easily computed, using Snell's law; for glass surrounded by air, the angle works out to be about 42°, as calculated in the following example.

EXAMPLE **24-5**

Calculate the critical angle for a glass-air interface. The glass is a light crown glass of index of refraction 1.50 (Fig. 24-13).

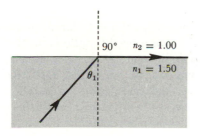

Figure 24-13

When θ_1 just equals the critical angle, the emerging ray just grazes the surface, and $\theta_2 = 90°$.

$$n_1 \sin \theta_1 = n_2 \sin \theta_2$$
$$(1.50)(\sin \theta_1) = (1)(\sin 90°)$$
$$1.50 \sin \theta_1 = (1)(1)$$
$$\sin \theta_1 = 1/1.50 = 0.667$$
$$\theta_1 = \boxed{42°}$$

Total reflection is used in many optical instruments, including binoculars (Fig. 26-25). In order to erect the image, as well as to obtain a longer path length between the objective lens and the eye lens, two 45° prisms are used. Silvered mirrors might be used, but the prisms serve the same purpose, and give 100% reflection, whereas a conventional mirror might become tarnished. Note that the angle in the glass is 45°, which is a few degrees greater than the critical angle

for a glass-air interface. Total reflection takes place, and it is not necessary to silver the reflecting surface at all.

24-6 Thin lenses

Lenses are used to concentrate or to disperse light and to form images. In this section we shall discuss the image-forming properties of simple lenses, with emphasis on the wave fronts and how they change. A rudimentary lens could be formed by placing two prisms back to back, as in Fig. 24-14. Each of the rays is refracted toward the normal as it enters

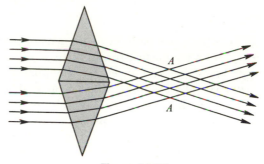

Figure 24-14

the glass, and is refracted away from the normal as it leaves the glass. Both refractions bend the ray away from the original (horizontal) direction. After passing through the "lens," the rays intersect in a zone *AA*, which might be called an image. A much better lens is shown in Fig. 24-15*a*, where the surfaces are curved. The various rays strike the surface at different angles, are refracted differently, and are brought to a much sharper focus *F*. If the lens surfaces are spherical (an easy shape to manufacture), the image is almost, but not quite, a mathematical point. For the time being, we shall consider that *thin* spherical lenses make perfectly sharp images, and we put off until Chap. 26 a discussion of the aberrations which cause fuzzy or color-fringed images.[*] In Fig. 24-15*a*, the

---

[*] Our treatment of lenses is limited to thin lenses and to rays which do not make too large an angle with the axis of the lens. In the figures, the thickness of the lenses have been exaggerated for clarity.

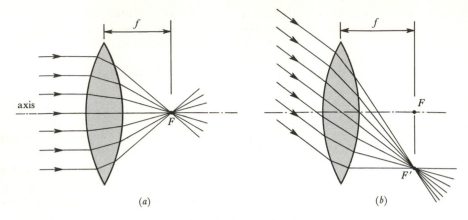

Figure 24-15 Converging lens.

incident rays are parallel to the axis of the lens, and after passing through the lens the rays converge and meet at the *principal focus F.* If the rays are parallel to each other but not to the axis, as in Fig. 24-15*b*, the image *F'* is formed at an off-axis point. The plane *FF'* is called the *focal plane,* and the distance from the lens to the focal plane is the *focal length f.* The focal length is usually measured from the center of the lens, but, since the lens is assumed to be thin, the thickness is negligible compared with *f*, and the focal length can be measured from either surface without serious error.

We have drawn a double-convex lens— one in which both surfaces bulge outward. Parallel rays striking such a lens produce a *real image,* defined as a point at which the

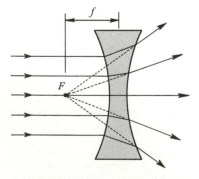

Figure 24-16 Diverging lens.

rays actually intersect. A real image can be observed as a bright spot on a screen, photographic film, etc. Another useful lens is the double-concave lens of Fig. 24-16. Refraction at the surfaces causes the parallel rays to diverge, giving rise to a *virtual image* at *F*, which is defined as the point from which the diverging rays seem to come.

24-7 Objects and images—ray tracing

We usually represent an object by a small vertical arrow (*AB* in Fig. 24-17) to the left of the lens. To a good approximation, *all* the rays leaving *B* pass through the image point *B'*. Our symbols are: *p* = object distance, *q* = image distance, *f* = focal length (all measured from the center of the lens); *h* = object size, *h'* = image size. The magnification *m* is defined as

$$ m = -\frac{h'}{h} $$

where the − sign is used so that an *inverted* image as in Fig. 24-17 will give a *negative* ratio. Three rays are particularly easy to trace; any two of them would suffice to locate the image point *B'*. Ray 1, parallel to the axis, goes through the focal point *F'* (this follows from the definition of focal point).

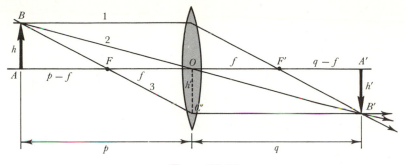

Figure 24-17

Ray 2, headed for the center of the lens, passes through without change of direction (Fig. 24-18) since the lens here is like a thin parallel-faced slab. Ray 3 passes through the other focal point F and emerges parallel to the axis. Any other ray leaving B will also pass through B' if the thin-lens approximation is valid.

Triangle ABO is similar to $A'B'O$, from which we obtain the magnification:

$$m = -\frac{h'}{h} = -\frac{q}{p} \qquad (24\text{-}4)$$

or, in words,

$$\frac{\text{Image size}}{\text{Object size}} = -\frac{\text{image distance}}{\text{object distance}}$$

Also, triangle FOC' is similar to FAB, which yields

$$\frac{h'}{f} = \frac{h}{p - f}$$

From Eq. 24-4, $h' = qh/p$, so

$$\frac{qh}{pf} = \frac{h}{p - f}$$

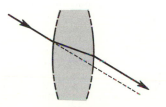

Figure 24-18

After canceling h, this equation can be rearranged to give the *lens equation*

$$\frac{1}{p} + \frac{1}{q} = \frac{1}{f} \qquad (24\text{-}5)$$

Note that the lens equation is consistent with the definition of focal length: for incident parallel rays, the object is at infinity. Then $1/\infty + 1/q = 1/f$, which gives $q = f$.

The same graphical method can be used to find the location and size of a virtual image formed by a diverging lens (Fig. 24-19a). Here, too,

$$m = -\frac{q}{p}$$

but since p is positive, we agree to call q *negative* for a virtual image, to preserve the form of the equation for m and still obtain a positive m for the erect image. A virtual image can also be formed by a converging lens (Fig. 24-19b). These diagrams also show that for any single lens a real image is inverted, whereas a virtual image is erect.

The following sign convention can be used: p, q, and f are each positive for the "typical case" of a converging lens forming a real image of a real object. Any change from this situation requires a minus sign. Thus f is negative for a diverging lens; q is negative for a virtual image (formed to the left of the lens); p is negative for a virtual object (located to the right of the lens). In Fig. 24-20, A is a virtual object since rays striking the lens are coming toward A instead of

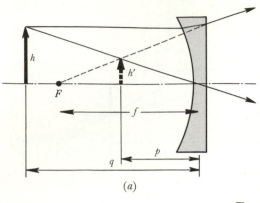

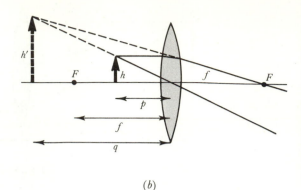

(a)   (b)

Figure 24-19

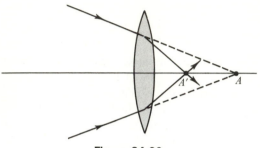

Figure 24-20

diverging, as is the usual case, from a real object to the left of the lens.

EXAMPLE **24-6**

An object 2 mm high is placed 100 cm from a converging lens of focal length 20 cm. Where and how large is the image?

The lens is a converging one, so its focal length is $+100$ cm. The lens equation gives

$$\frac{1}{p} + \frac{1}{q} = \frac{1}{f}$$

$$\frac{1}{100 \text{ cm}} + \frac{1}{q} = \frac{1}{20 \text{ cm}}$$

from which

$$q = \boxed{+25 \text{ cm}}$$

The magnification is

$$m = -\frac{q}{p} = -\frac{25 \text{ cm}}{100 \text{ cm}} = -0.25$$

The image is real, it is inverted, and its size is

$$(0.25)(2 \text{ mm}) = \boxed{0.5 \text{ mm}}$$

EXAMPLE **24-7**

How far from the lens of the previous example should a 3 mm object be placed, in order to form an image on a screen that is 80 cm from the lens? How large will the image be?

Since the image is formed on a screen, we know that it is a real image.

$$\frac{1}{p} + \frac{1}{q} = \frac{1}{f}$$

$$\frac{1}{p} + \frac{1}{80 \text{ cm}} = \frac{1}{20 \text{ cm}}$$

from which

$$p = \boxed{26.7 \text{ cm}}$$

The magnification is

$$m = -\frac{q}{p} = -\frac{80 \text{ cm}}{26.7 \text{ cm}} = -3.00$$

The real image is inverted, and its height is

$$(3.00)(3 \text{ mm}) = \boxed{9 \text{ mm}}$$

EXAMPLE **24-8**

An object is placed 15 cm from the same lens used in the previous two examples. Compute the position and the magnification of the image.

$$\frac{1}{15 \text{ cm}} + \frac{1}{q} = \frac{1}{20 \text{ cm}}$$

$$q = \boxed{-60 \text{ cm}}$$

$$m = -\frac{q}{p} = -\frac{-60 \text{ cm}}{15 \text{ cm}} = \boxed{+4}$$

The image is virtual, erect, and 4 times as large

as the object. It is 60 cm from the lens, and, since q is negative, the image is on the same side of the lens as the object.

24-8   Types of lenses—the diopter

To gain an intuitive understanding of what happens as light passes through a lens, let us consider wave fronts instead of rays. The *curvature* of a wave front is defined as the reciprocal of its radius; thus a sphere of radius 0.2 m has a curvature of $1/(0.2$ m$)$ $= 5$ m^{-1}. Wave fronts reaching the earth from the sun have practically no curvature ($\frac{1}{93\,000\,000}$ mi^{-1} or 0.000000011 mi^{-1}); a plane wave, which is represented by parallel rays and plane wave fronts, has zero curvature. Spherical wave fronts are classified according to the behavior of the rays which are perpendicular to them (see Fig. 24-2). A point source of light, which might serve as an object for a lens, emits rays which travel outward in all directions; the wave fronts continually grow larger in radius. Such a wave front is called a *diverging wave front* and is said to have *negative curvature*. Similarly, if the rays associated with a spherical wave front are all heading for some point, called a real image, such a wave front is called a *converging wave front* and is said to have *positive curvature*.

We can redraw Fig. 24-15a, replacing the parallel rays by a plane wave AC, which strikes the double-convex lens (Fig. 24-21).

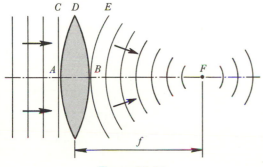

Figure 24-21

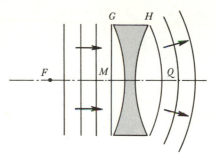

Figure 24-22

During a short time interval Δt while the center of the wave front is moving through glass from A to B, the outer edge of the wave front moves through air from C to E. Since the light travels faster in air than in glass, the wave front must curve around and become converging, as shown. The radius of curvature of the wave front BE is f,[*] but the wave front rapidly shrinks down to a point and then diverges. Fixing our attention on the two wave fronts AC and BE, we see that the lens has had a converging effect. The lens has acted on a wave front which had no curvature to start with, and it has produced a wave front which is converging with a positive curvature equal to $1/f$. For this reason a converging lens is called a *positive lens*. The exact shape of the lens is immaterial as far as the converging effect is concerned; any lens which is thicker in the center than at the edges is a converging, or positive, lens. Of course, a lens with a thicker bulge produces a greater effect and has a shorter focal length.

A *diverging lens* changes a plane wave front into a diverging one. Thus, in Fig. 24-22, the top of the wave front moves from G to H while the center of the same wave front (which has less glass to go through) moves from M to Q. After leaving the lens, the wave front continues to spread out, seeming to come from some virtual image at F. Any lens

---

[*] In the thin lens approximation which we are using, $BF \approx f$.

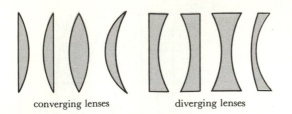

converging lenses diverging lenses

Figure 24-23

which is thinner in the middle than at the edges is a diverging, or negative, lens. Several converging and diverging lenses are shown in Fig. 24-23.

The effect of a lens on the curvature of a wave front is called its *power*. That is, a powerful lens, one which has a short focal length, changes the curvature a great deal. We define the power of a lens as the reciprocal of its focal length. The power of a lens is often measured in *diopters:*

$$\text{Power in diopters} = \frac{1}{\text{focal length in meters}}$$

For any lens, the power depends on the curvatures of the surfaces and on the index of refraction of the glass according to the *lens-maker's equation* which is derived in Sec. 24-12:

$$\frac{1}{f} = (n - 1)\left(\frac{1}{R_1} + \frac{1}{R_2}\right) \qquad (24\text{-}6)$$

In this equation R_1 and R_2 are considered positive for convex surfaces; for a plane surface, $R = \infty$ and $1/R = 0$.

EXAMPLE **24-9**

What is the power of a spectacle lens made of glass for which $n = 1.50$, if the outer surface is convex with radius 200 cm and the surface nearest the eye is concave with radius 100 cm? This is a diverging lens, similar to the right-hand one in Fig. 24-23. We use Eq. 24-6 to calculate the power, first expressing the radii in meters.

$$\frac{1}{f} = (1.50 - 1)\left(\frac{1}{2.0 \text{ m}} + \frac{1}{-1.0 \text{ m}}\right)$$

$$= \boxed{-0.25 \text{ m}^{-1}}$$

The power is -0.25 diopter. The $-$ sign indicates a diverging or negative lens. The shape of this spectacle lens, both surfaces bulging outward from the eye, allows room for eyelashes and the closing of the eyelids.

The power of the human eye (including both the lens and the cornea) is about $+60$ diopters, and the lenses ordinarily used for eyeglasses have powers of from $+5$ diopters to about -5 diopters, in steps of $\frac{1}{8}$ diopter. (See Sec. 26-2 for a more detailed study of the eye.)

24-9 Objects and images— method of curvatures

A simple rearrangement of terms in the lens equation (24-5) gives us a new outlook which in some ways is more intuitive and allows us to solve complicated problems using a cause-and-effect procedure. Equation 24-5 can be written as

$$-\frac{1}{p} + \frac{1}{f} = \frac{1}{q} \qquad (24\text{-}7)$$

Each term in this equation has a simple physical significance. (Remember that the curvature of a wave front is the reciprocal of its radius.) Let us interpret Fig. 24-24 with the help of our new form of the lens equation. We see that $-1/p$ is the curvature of the wave front just as it reaches the lens; the curvature is negative since the waves diverge from the object A. The power of the lens is $1/f$ (Eq. 24-6), and $1/q$ is the curvature of

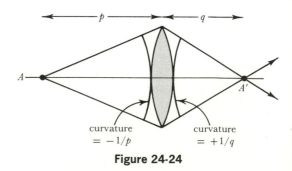

Figure 24-24

the wave front just as it leaves the lens, before it shrinks down to form the image at A'. Our equation tells us how the lens changes the curvature of the wave front; the lens is an *operator*. All the action represented by the equation takes place at the lens; p and q are involved in the problem only insofar as they are related to the curvatures of the wave fronts that are entering and leaving the lens. To illustrate the method of curvatures, we will solve some simple problems.[*]

EXAMPLE 24-10

An object (Fig. 24-25) is placed 100 cm from a lens of focal length $+20$ cm. Where is the image?

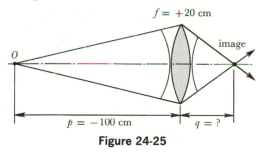

$$f = +20 \text{ cm}$$

$$O$$

$$\text{image}$$

$$p = -100 \text{ cm} \qquad q = ?$$

Figure 24-25

The lens is a converging one, since it is thicker in the center than at the edges, and so it has positive power. The rays diverge from the object at O, and, by the time the wave front reaches the lens, the curvature is $-\frac{1}{100}$ cm^{-1}.

$$\begin{matrix} \text{original} \\ \text{curvature} \end{matrix} + \begin{matrix} \text{power of} \\ \text{lens} \end{matrix} = \begin{matrix} \text{final} \\ \text{curvature} \end{matrix}$$

$$-\frac{1}{100 \text{ cm}} + \frac{1}{20 \text{ cm}} = \frac{1}{q}$$

$$\frac{1}{q} = \frac{1}{20} - \frac{1}{100} = \frac{5}{100} - \frac{1}{100}$$

$$= +\frac{4}{100} = +\frac{1}{25}$$

Hence

$$q = \boxed{+25 \text{ cm}}$$

Note that, if the original wave front comes from a point source of light, its curvature is negative, since light *diverges* from a source.

---

[*] These are also solved by the ray method in Sec. 24-7. Which form of the lens equation you use is a matter of personal preference.

EXAMPLE 24-11

How far from the lens of the previous example should an object be placed, in order to form an image on a screen that is 80 cm from the lens? (See Fig. 24-26.)

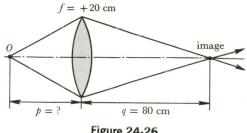

$$f = +20 \text{ cm}$$

$$O$$

$$\text{image}$$

$$p = ? \qquad q = 80 \text{ cm}$$

Figure 24-26

Since the image is formed on a screen, we know that it is formed by converging light; hence the curvature of the beam as it leaves the lens is positive, and equal to $+\frac{1}{80}$.

$$\begin{matrix} \text{original} \\ \text{curvature} \end{matrix} + \begin{matrix} \text{power of} \\ \text{lens} \end{matrix} = \begin{matrix} \text{final} \\ \text{curvature} \end{matrix}$$

$$-\frac{1}{p} + \frac{1}{20 \text{ cm}} = \frac{1}{80 \text{ cm}}$$

$$-\frac{1}{p} = \frac{1}{80} - \frac{1}{20} = -\frac{3}{80}$$

$$p = \frac{80}{3} \text{ cm}$$

$$= \boxed{26.7 \text{ cm}}$$

The object must be 26.7 cm to the left of the lens. Note that p is positive, and the object is in the "typical" position discussed in Sec. 24-7.

EXAMPLE 24-12

An object is placed 15 cm from the same lens used in the previous two examples. Compute the position of the image (Fig. 24-27).

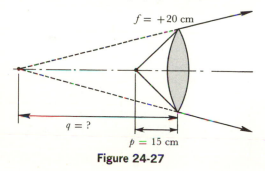

$$f = +20 \text{ cm}$$

$$q = ?$$

$$p = 15 \text{ cm}$$

Figure 24-27

$$\text{original curvature} + \text{power of lens} = \text{final curvature}$$

$$-\frac{1}{15\text{ cm}} + \frac{1}{20\text{ cm}} = \frac{1}{q}$$

$$\frac{1}{q} = -\frac{1}{15} + \frac{1}{20} = -\frac{4}{60} + \frac{3}{60} = -\frac{1}{60}$$

$$q = \boxed{-60\text{ cm}}$$

The $-$ sign for the image distance indicates a virtual image. The lens is not strong enough to remove all the divergence of the original wave front; the rays still diverge (but to a lesser extent) after they leave the lens.

A negative lens always gives a virtual image of an object, since the original diverging wave front can only be made more diverging by the action of a negative lens.

EXAMPLE 24-13

A negative lens of focal length 10 in. is placed 30 in. from an object. Where is the image? (See Fig. 24-28.)

$$\text{original curvature} + \text{power of lens} = \text{final curvature}$$

$$-\frac{1}{30\text{ in.}} + \frac{1}{-10\text{ in.}} = \frac{1}{q}$$

$$\frac{1}{q} = -\frac{1}{30} - \frac{1}{10} = -\frac{4}{30}$$

$$q = \boxed{-7.5\text{ in.}}$$

The $-$ sign for q indicates a virtual image formed by diverging rays.

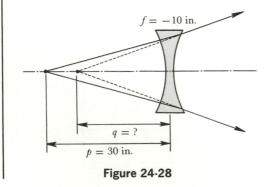

$f = -10$ in.

$q = ?$

$p = 30$ in.

Figure 24-28

EXAMPLE 24-14

A slide projector has a converging lens of focal length 2 in., and the screen is 10 ft from the lens. (a) How far from the lens should the slide be placed? (b) What is the height (on the screen) of the image of a portion of the slide which is 4 mm high?

In this problem, the object distance is unknown, and the image distance is 10 ft ($= 120$ in.)

$$(a) \quad -\frac{1}{p} + \frac{1}{2\text{ in.}} = \frac{1}{120\text{ in.}}$$

$$\frac{1}{p} = \frac{1}{2} - \frac{1}{120} = \frac{118}{240}$$

$$p = \frac{240}{118}\text{ in.} = \boxed{2.03\text{ in.}}$$

The slide should be placed just outside the principal focus of the lens.

$$(b) \quad \text{Magnification} = -\frac{\text{image distance}}{\text{object distance}}$$

$$m = -\frac{120\text{ in.}}{\frac{240}{118}\text{ in.}} = -\frac{(120)(118)}{240} = -59$$

The image size is -59 times the object size (the negative sign indicates an inverted image). Since the object is 4 mm high, the height of the image on the screen is

$$(4\text{ mm})(59) = \boxed{236\text{ mm}}$$

Our next example requires us to find the size of a virtual image.

EXAMPLE 24-15

A magnifying glass of power $+5$ diopters is held 16 cm from a page of fine print. How large is the image of a letter which is $\frac{1}{20}$ in. high?

First we locate the *position* of the image, using the method of curvatures. The power of the lens is $+5$ diopters, so we know that it is a converging lens, with a focal length equal to $\frac{1}{5}$ m, or 20 cm. The object distance is 16 cm, and the image distance is unknown.

$$-\frac{1}{16\text{ cm}} + \frac{1}{20\text{ cm}} = \frac{1}{q}$$

$$q = -80\text{ cm}$$

The negative sign indicates a virtual image,

which is expected, since the object is inside the principal focus of the lens. We can now compute the magnification:

$$\text{Magnification} = -\frac{\text{image distance}}{\text{object distance}}$$

$$m = -\frac{-80 \text{ cm}}{16 \text{ cm}} = +5$$

The positive magnification indicates an erect image (see Fig. 24-19b). Finally, the image size is given by

$$(\tfrac{1}{20} \text{ in.})(5) = \tfrac{5}{20} \text{ in.} = \boxed{\tfrac{1}{4} \text{ in.}}$$

Since the simple converging lens is used so often, it is worthwhile summarizing the image-object relationships for various cases. The results shown in Fig. 24-29 can be obtained by applying the method of curvatures

as we have done in the examples. Two principal foci are shown, each at a distance f from the lens.

24-10 Mirrors

We have already used Huygens' principle to derive the law of reflection for plane mirrors. From a general point of view, a mirror is just another surface which can modify the curvature of a wave front. The same ideas that we used for lenses can be applied to mirrors, with the added understanding that the wave front reverses its direction of motion while being reflected. First, consider a plane mirror. It is evident from Fig. 24-30 that the image distance q equals the object distance p, and that $h' = h$. When you look at your-

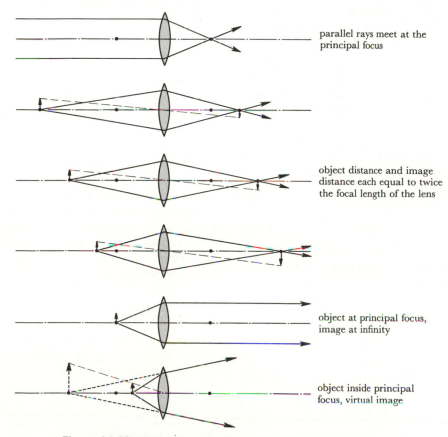

parallel rays meet at the principal focus

object distance and image distance each equal to twice the focal length of the lens

object at principal focus, image at infinity

object inside principal focus, virtual image

Figure 24-29 Object-image relationships for a converging lens.

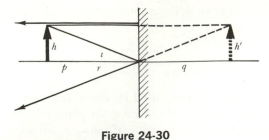

Figure 24-30

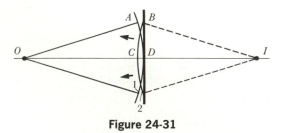

Figure 24-31

Figure 24-32

self in a plane mirror, the virtual image of your face is as far back of the surface as you are in front of it. The same result is found from the method of curvatures.

In Fig. 24-31, the wave front's curvature is not changed by reflection, and the original diverging wave front 1 becomes wave front 2, which is also diverging, seemingly from a virtual image I. Since $AB = CD$, the wave fronts 1 and 2 have the same radius of curvature; hence the image distance equals the object distance.

When parallel rays strike a concave mirror, they converge to a focal point F (Fig. 24-32). We see that a concave mirror is a converging, or positive, mirror. It is fortunate that the same equation, $1/p + 1/q = 1/f$, that we used for lenses can be applied to mirrors. First, however, we need to derive the mirror-maker's formula, which is much simpler than the lens-maker's formula. The focal length of a mirror is given by

$$f = \tfrac{1}{2}R \qquad (24\text{-}8)$$

where R is the radius of curvature of the mirror surface. The proof follows easily from Fig. 24-32, where C is the center of curvature. The angles i, r, and θ are all equal.

Hence, if θ is small,* the two equal sides of the isosceles triangle CFP are approximately equal to half the base. This means that $R - f \approx \tfrac{1}{2}R$, whence $f \approx \tfrac{1}{2}R$. As for a lens, the power of a mirror is $1/f$, in diopters if f is in meters.

For a mirror of focal length $f\ (=\tfrac{1}{2}R)$, we use

$$\frac{1}{p} + \frac{1}{q} = \frac{1}{f}$$

with the following sign conventions: p, q, and f are all positive for the "typical case" of a concave (converging) mirror forming a real image of a real object. Thus f is negative for a convex (diverging) mirror; q is negative for a virtual image formed to the right of the mirror (behind the surface); p is negative for a virtual object (located to the right of the mirror, behind the surface). Just as for a lens, a real image is inverted and a virtual image is erect. The magnification is given, as for a lens, by $m = -q/p$. A converging mirror, like a converging lens, forms a real, inverted image if the object is outside the principal focus, and it forms a virtual, erect image if the object is inside the principal focus. The six typical situations shown in Fig. 24-33 are analogous to those of Fig. 24-29, which are drawn for a converging lens.

---

*This is the sort of approximation made for all of our work with lenses and mirrors. See the footnote on page 517.

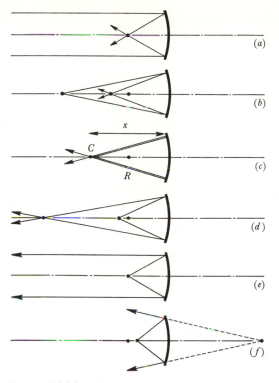

Figure 24-33 Object-image relationships for a converging mirror.

EXAMPLE 24-16

An object of height 6 mm is placed 60 cm from a concave mirror of radius of curvature 30 cm. Where and how large is the image?

The focal length is half the radius, hence $f = 15$ cm. The mirror is stated to be a concave one, so we know that it is a positive mirror.

$$\frac{1}{p} + \frac{1}{q} = \frac{1}{f}$$

$$\frac{1}{60 \text{ cm}} + \frac{1}{q} = \frac{1}{15 \text{ cm}}; \quad q = \boxed{+20 \text{ cm}}$$

Since q is positive, the image is real. The magnification is

$$m = -\frac{q}{p} = -\frac{20 \text{ cm}}{60 \text{ cm}} = -\frac{1}{3}$$

The negative sign indicates an inverted image, and its size is

$$\tfrac{1}{3} \times 6 \text{ mm} = \boxed{2 \text{ mm}}$$

The image is 20 cm to the left of the mirror, real, inverted, and 2 mm tall.

The next example illustrates the solution of a mirror problem by the method of curvatures.

EXAMPLE 24-17

A dentist holds a concave mirror of radius of curvature 50 mm at a distance of 20 mm from a cavity in a tooth (Fig. 24-34). Where and how large is the image of the cavity?

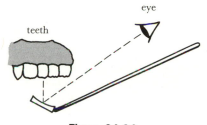

Figure 24-34

The mirror's focal length is $f = \tfrac{1}{2}R = \tfrac{1}{2}(50 \text{ mm}) = 25 \text{ mm}$; the power of this mirror is $\frac{1}{25}$ mm^{-1}.

$$\frac{\text{original}}{\text{curvature}} + \frac{\text{power of}}{\text{mirror}} = \frac{\text{final}}{\text{curvature}}$$

$$-\frac{1}{20 \text{ mm}} + \frac{1}{25 \text{ mm}} = \frac{1}{q}$$

$$q = \boxed{-100 \text{ mm}}$$

$$m = -\frac{q}{p} = -\frac{-100 \text{ mm}}{20 \text{ mm}} = \boxed{+5}$$

The image is virtual, erect, 100 mm in back of the mirror, and 5 times as large as the cavity.

Convex mirrors are diverging, or negative, and by a method similar to that which we used for concave mirrors, it can be shown that the focal length of a convex mirror is also equal to half its radius of curvature. Such a mirror always forms a reduced upright virtual image of a real object; convex mirrors are used as safety mirrors, placed where aisles meet in factories (to avoid forklift truck collisions), or placed above strategic spots in stores (to reveal shoplifters).

24-11 Lens combinations

When light passes through several lenses, the image formed by one lens becomes the object for a second lens, and so on for as many lenses as there may be. The over-all magnification is the *product* of the magnifications of the separate lenses. One example will suffice to illustrate the method. We use the method of curvatures in order to visualize how light passes through the system of lenses. The problem can also be solved using $1/p + 1/q = 1/f$, with proper attention to signs.

EXAMPLE 24-18

Compute the position, size, and nature of the final image formed by the system shown in Fig. 24-35.

Lens A:

$$-\frac{1}{50} + \frac{1}{40} = \frac{1}{q}$$

$$q = +200 \text{ cm}$$

$$m_1 = -\frac{+200 \text{ cm}}{50 \text{ cm}} = -4$$

The image is real, inverted, and 4 times as large as the object.

Lens B: The lenses are 220 cm apart, so the converging beam leaving lens A has a chance to reach a focus and diverge again. As the beam strikes B, it has a radius of curvature of 20 cm ($= 220 \text{ cm} - 200 \text{ cm}$) and is diverging.

$$-\frac{1}{20} + \frac{1}{10} = \frac{1}{q}$$

$$q = +20 \text{ cm}$$

$$m_2 = -\frac{+20 \text{ cm}}{20 \text{ cm}} = -1$$

The image formed by lens B is real, inverted (relative to its object), and the same size as its object. The beam leaves lens B headed for a point Q which is 20 cm beyond the lens, i.e., 12 cm beyond lens C.

Lens C: This is a tricky situation. The converging beam leaving lens B doesn't have a chance to reach its focus before lens C intervenes. The radius of curvature is shrinking and is equal to 12 cm when the surface of lens C is reached. Hence the initial curvature for lens C is positive.[*] The lens is negative because of its shape; such a lens tends to remove convergence and/or increase divergence. In this case, the lens actually succeeds in removing all the convergence of the wave front and causes some divergence, as shown by the equation:

$$+\frac{1}{12 \text{ cm}} - \frac{1}{3 \text{ cm}} = \frac{1}{q}$$

$$q = -4 \text{ cm}$$

$$m_3 = -\frac{-4 \text{ cm}}{-12 \text{ cm}} = -\frac{1}{3}$$

---

[*] This is rare; the initial beam can be converging only if the beam has a past history and has already been acted on by one or more lenses or mirrors. The "typical" object sends out diverging rays, and so the initial curvature is usually negative. This is an example of a virtual object.

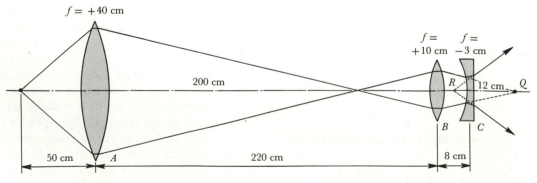

Figure 24-35

The image formed by lens C is virtual, 4 cm to the left of C, inverted relative to its object, and $\frac{1}{3}$ as large as its object.

The final image is at R, halfway between lens B and lens C. It is a virtual image. The over-all magnification is

$$m_1 m_2 m_3 = (-4)(-1)(-\tfrac{1}{3}) = -\tfrac{4}{3}.$$

The three minus signs mean that three inversions have taken place, and hence the final image is *virtual, inverted, and $\frac{4}{3}$ the size of the original object.*

■ SUMMARY

Huygens' principle states that as a wave propagates the new wave front is the envelope of wavelets which spread out from all points of the old wave front. If no surface such as a mirror, prism, or lens intervenes, a plane wave remains plane and a spherical wave spreads out spherically. These facts imply straight-line propagation of light waves; such is actually the case for waves whose wavelength is small compared with the sizes of obstacles or openings.

When light rays strike a smooth surface, both reflection and refraction may take place, in a plane determined by the incident ray and the normal to the surface. The angle of reflection equals the angle of incidence; the angle of refraction is determined by Snell's law, which states that $n_1 \sin \theta_1 = n_2 \sin \theta_2$. The index of refraction n is the ratio of the speed of light in vacuum to the speed in the medium. The variation of the speed of light (and index of refraction) in a medium with frequency (i.e., with color) is called dispersion. In most transparent solids and liquids dispersion is relatively slight.

The critical angle is the angle of incidence, in the more optically dense medium, for which the light emerges into the less dense medium just tangent to the surface. If the critical angle is exceeded, total reflection takes place within the denser medium.

Images formed by converging rays are real, and images formed by diverging rays are virtual. Images can be located by ray-tracing; image size is to object size as image distance is to object distance. Any lens which is thicker at its center than at its edges is a converging, or positive, lens; a lens which is thinner at its center than at its edges is a diverging, or negative, lens. The power of a lens is the reciprocal of its focal length.

The curvature of a wave front is defined as the reciprocal of its radius. A diverging wave front has negative curvature, and a converging wave front has positive curvature. Images can be located by considering changes of curvature caused by the lens; initial curvature plus power of lens equals final curvature.

A plane mirror gives an erect virtual image having the same size as the object, and image distance equals object distance. Concave mirrors are positive, and convex mirrors are negative. The focal length of a mirror equals half its radius of curvature.

■ CHECK LIST

angstrom unit	wave front	anisotropic medium
Huygens' principle	isotropic medium	angle of incidence

angle of reflection
angle of refraction
law of reflection
index of refraction
Snell's law of refraction
critical angle
principal focus
focal length

real image
virtual image
magnification
lens equation
typical case
curvature
diverging wave front
converging wave front

positive lens or mirror
diverging lens or mirror
power of a lens or mirror
diopter
lens-maker's equation
mirror-maker's equation

■ QUESTIONS

24-1 Would you expect Huygens' principle to apply to radio waves?

24-2 Do any radio waves travel in straight lines?

24-3 Give two examples of each of the following processes: (*a*) specular reflection; (*b*) diffuse reflection; (*c*) scattering.

24-4 A beam of white light containing all colors strikes a prism as shown in Fig. 24-36, and the rays of different colors are separated because their speeds in glass are different. From a general knowledge of the relative values of index of refraction, state whether the red rays or the violet rays are deviated the most from their original direction.

24-5 Can a ray in air be totally reflected at a glass surface if the angle of incidence (in air) and the index of refraction are just right?

24-6 Which of the lenses in Fig. 24-37 are positive, and which are negative?

Figure 24-36   Figure 24-37

24-7 Is it possible for a converging wave front to change into a diverging wave front without striking a surface of any kind?

24-8 Is it possible for a diverging wave front to change into a converging wave front without striking a surface of any kind?

24-9 A real image is to be formed by a plane mirror. What must be the nature of the initial wave front which strikes the mirror—converging or diverging?

24-10 Diamond has a higher index of refraction than glass. Is the critical angle for a diamond-air interface larger or smaller than for a glass-air interface? Does this have anything to do with the brilliance of a diamond gem stone?

24-11 Sound waves of wavelength 6 in. are obstructed by a building which is 60 ft wide and 40 ft tall. Is the shadow "sharp"?

24-12 When hot air rises above a radiator, the wall behind the air stream seems to shimmer. Explain. Also explain why stars twinkle.

■ PROBLEMS

24-A1 What is the speed of light in a clear plastic material for which the index of refracton is 2.00?

24-A2 What is the index of refraction of a liquid in which the speed of light is 2.25×10^8 m/s?

24-A3 The speed of light in a certain kind of glass is three-fifths as much as in a vacuum. What is the index of refraction of the glass?

24-A4 Light of frequency 5×10^{14} Hz enters a clear plastic material whose index of refraction is 2.00. What is the wavelength of the light while it is in the plastic?

24-A5 What is the wavelength, in water, of orange light whose wavelength in air is 6000 Å?

24-A6 What is the ratio of the speed of light in benzene to the speed of light in diamond?

24-A7 In Fig. 24-38 a ray of light enters a liquid and is bent toward the normal as shown. Calculate the index of refraction of the liquid.

24-A8 What is the focal length of a spectacle lens of power -12.5 diopters? Describe the lens; is it converging or diverging?

24-A9 A converging lens has focal length 50 cm. Calculate the power of the lens, in diopters. Is it a positive or a negative lens?

24-A10 What is the least height of a plane mirror on a wall in which a man 2.00 m tall can just see his entire image from the top of his head to the soles of his shoes? (*Note:* Do not assume that the man's eyes are at the top of his head.)

24-B1 The rays of the afternoon sun strike the surface of a lake at an angle of 45° with the vertical. At what angle, measured from the vertical, is the refracted ray in the water?

24-B2 A ray of light strikes a piece of glass ($n = 1.50$), making an angle of 30° with the surface. What angle does the refracted ray make with the surface?

24-B3 A ray of light starts from a point on the bottom of a dish of liquid 3 cm deep and strikes a point on the surface which is 4 cm from the point directly above the source. What is the least index of refraction for the liquid which will allow the ray to be totally reflected?

24-B4 A ray of light directed horizontally 60° E of N enters a vertical window pane which is in a N-S plane. If $n = 1.50$ for the glass, what is the direction of the ray in the glass?

24-B5 A flashlight on the bottom of a swimming pool 12 ft deep sends a ray upward and at an angle such that the ray strikes the surface of the water 5 ft from the point directly above the flashlight. What angle does the emerging ray (in air) make with the vertical?

24-B6 A ray of light from a boat's searchlight reaches a skindiver's eye at an angle of 30° from the vertical. What angle in air does the ray make with the *surface* of the water?

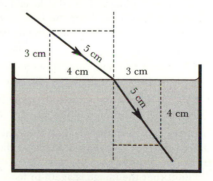

Figure 24-38

24-B7 Calculate the critical angle for a diamond-air interface.

24-B8 What is the index of refraction of a type of glass for which the critical angle at a glass-air interface is 45°?

24-B9 A lamp bulb is placed 400 cm from a spectacle lens of power +0.2 diopter. (*a*) What is the radius of curvature of the wave front just as the light strikes the lens? (*b*) What is the radius of curvature of the wave front just as the light leaves the lens?

24-B10 A point source is 6 m from a converging lens of focal length 2 m. (*a*) Where is the image? (*b*) What is the radius of curvature of the wave front just as the light leaves the lens?

24-B11 An object 6 mm high is 24 cm to the left of a converging lens of focal length 8 cm. Where is the image? Is it real or virtual? Erect or inverted? How large is it?

24-B12 Repeat Prob. 24-B11 for the same object placed 24 cm to the left of a diverging lens of focal length 8 cm.

24-B13 A lens of power +2.5 diopters is held 30 cm above an object 1 mm high. Describe the location, nature, and size of the image.

24-B14 A converging lens is 30 cm from a diverging lens of focal length 20 cm. If a beam of parallel light enters the system from the left, as shown in Fig. 24-39, and the beam is again parallel when it emerges from the second lens, calculate the focal length of the converging lens.

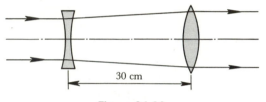

Figure 24-39

24-B15 A projector lens is placed 2 in. from a slide, and the real image is formed on a screen 20 ft from the lens. (*a*) What is the magnification? (*b*) What is the focal length of the lens?

24-B16 An astronaut on the moon focuses a camera lens (of focal length 50 mm) to form an image of the earth. How far must he move the lens toward or away from the film (which?) in order to focus the camera on a fellow astronaut 3.0 m away?

24-B17 The moon's diameter is 3500 km and it is 3.84×10^5 km from the earth. What is the size of the moon's image on the film in a camera whose lens has focal length 50 mm?

24-B18 A lens of power +25 diopters is used to form a real image of a postage stamp on a photographic film which is 5 cm from the lens. (*a*) How far from the lens is the stamp? (*b*) If the stamp is 1 in. tall, how tall is the image?

24-B19 How far from a converging lens of focal length 60 cm must an object be placed if the image is to be a real one, three times as large as the object? (*Hint:* Let x = the unknown object distance; then the image distance is $3x$.)

24-B20 Repeat Prob. 24-B19, for the same converging lens of focal length 60 cm, if the image is to be a virtual one, three times as large as the object.

24-B21 In a laboratory experiment to measure the focal length of a diverging lens, first an image of an object is formed on a screen 50 cm to the right of a converging lens of unknown focal length. The diverging lens is then placed 10 cm to the right of the converging lens, and it is found that the image is again in focus if the screen is moved 20 cm (in which direction?). Calculate the focal length of the diverging lens.

24-B22   Prove that for a mirror, object distance p, image distance q, and focal length f are related by $1/p + 1/q = 1/f$. (*Hint:* Use a diagram similar to Fig. 24-17, but drawn for a mirror instead of a lens. Assume the position of F to be known.)

24-B23   A bird views himself in a convex garden mirror of diameter 1 ft. If the bird is 2 ft from the surface of the mirror, (*a*) where is the image? (*b*) what is the magnification?

24-B24   An object 2 cm high is 30 cm from a convex mirror of focal length 10 cm. Calculate (*a*) the location of the image; (*b*) the nature of the image (real or virtual; erect or inverted; size of image).

24-B25   An object 2 cm tall placed 10 cm from a spherical mirror gives a virtual image 5 cm tall. (*a*) Where is the image? (*b*) Is the mirror concave or convex? (*c*) What is the focal length of the mirror?

24-B26   An object 3 cm high is 50 cm from a concave mirror whose radius of curvature is 20 cm. Calculate (*a*) the focal length of the mirror; (*b*) the location of the image; (*c*) the nature of the image (real or virtual; erect or inverted; size of image).

24-C1   What is the critical angle for a glass-water interface? In which medium must the incident ray be for total reflection? (Assume $n = 1.500$ for the glass used.)

24-C2   A certain thin lens made of flint glass has a power of $+6$ diopters when in air. What is the power of the lens when it is immersed in water?

24-C3   Three adjacent faces of a plastic cube of index of refraction n are painted black, with a clear spot at the painted corner serving as a source of diverging rays. Show that a ray from this corner to the center of a clear face is totally reflected if $n \geq \sqrt{3}$.

24-C4   A positive lens of power 5 diopters forms a real image 5 times as large as the object. How far apart are the object and the image?

24-C5   How far apart are the object and image if a lens of strength $+20$ diopters forms a real image which is 3 times as tall as the object?

24-C6   When an object is moved along the axis of a thin lens, the size of the image is 5 times the size of the object when the object is at point A to the left of the lens, and also if the object is at a point B that is 20 cm farther from the lens. (*a*) Is the lens converging or diverging? (*b*) What is the focal length of the lens? (*c*) Where are the points A and B?

24-C7   An object 5 mm high is placed 15 cm from a lens of focal length $+10$ cm. A second lens, of focal length $+20$ cm, is placed 25 cm beyond the first lens. Compute the position, nature, and size of the final image.

24-C8   An object is placed 20 cm to the left of a diverging lens of focal length 20 cm, and a converging lens of focal length 8 cm is placed 30 cm to the right of the first lens. (*a*) How far is the final image from the object? (*b*) What must be the size of the object, if the final image is 2 mm tall?

24-C9   Two lenses of focal length $+15$ cm and -10 cm are 18 cm apart. (*a*) Where is the final image of an object 60 cm from the converging lens? (*b*) What is the magnification of the combination?

24-C10   What is the radius of curvature of a shaving mirror which gives a twofold magnification of a face 1 ft in front of the mirror?

24-C11   A dedicated sports car buff polishes the inside as well as the outside surface of a hub cap which is a section of a sphere. He sees an image of his face 30 cm in back of the hub cap. He then turns the hub cap over and sees another image of his face 10 cm in back of the cap. (*a*) How far is his face from the hub cap? (*b*) What is the radius of curvature of the hub cap?

24-C12 Calculate the position and nature of the final image in Fig. 24-40. (*Hint:* The light passes through the lens twice.)

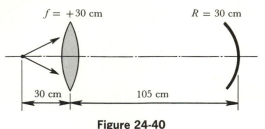

$f = +30$ cm $R = 30$ cm

30 cm 105 cm

Figure 24-40

24-C13 A thin plano-convex lens made of glass of index of refraction n has a curved surface of radius of curvature R. (*a*) What is the power of the lens? (*b*) What is the power of a lens-mirror combination formed by giving a reflecting coating to the plane surface of the lens? (*c*) Repeat, if the reflecting coating is applied to the curved surface of the lens.

24-C14 Use the method of curvatures to prove that the power of two thin lenses which are touching each other is the sum of the powers of the two lenses.

For Further Study

24-12   Derivation of the lens-maker's equation

The method of curvatures for thin lenses is proved by use of a theorem from plane geometry. According to the sagittal theorem[*] the segment x in Fig. 24-41 is given by $x(2R - x) = h^2$. In cases of interest to us x is negligible compared to $2R$, hence $x(2R) \approx h^2$, or

$$x \approx \frac{h^2}{2R} \qquad \text{(sagittal theorem)}$$

Consider now the formation of an image by a lens. In Fig. 24-42 let the radii of curvature of the two lens surfaces be R_1 and R_2; the object distance is p and the image distance is q. While the center of the wave front travels from P to Q in glass, the top of the same wave front travels from A to B in air. From Fig. 24-42 we have

$$AB - PQ \approx x_1 + x_4$$

---
[*] So called because the diagram resembles a bow and arrow (Latin: *sagitta* = arrow). See Appendix, p. 703, for a proof of the sagittal theorem.

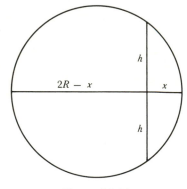

h

$2R - x$ x

h

Figure 24-41

Since the times for paths AB and PQ are equal,

$$\frac{AB}{c} = \frac{PQ}{c/n}$$

$$AB = n\,PQ$$

$$AB - PQ = n\,PQ - PQ$$
$$= (n - 1)\,PQ$$

That is,

$$x_1 + x_4 = (n - 1)(x_2 + x_3)$$

Now we use the saggital theorem. Substituting $x_1 = h^2/2p$, $x_2 = h^2/2R_1$, $x_3 = h^2/2R_2$,

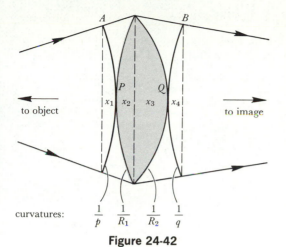

Figure 24-42

$x_4 = h^2/2q$, and canceling $h^2/2$, we obtain

$$\frac{1}{p} + \frac{1}{q} = (n - 1)\left(\frac{1}{R_1} + \frac{1}{R_2}\right) \quad (24\text{-}9)$$

The term on the right involves n, R_1, and R_2, which depend only on the shape of the lens and the index of refraction of the glass. To interpret this term, which is constant for a given lens, let the object be infinitely far from the lens ($p = \infty$); then the image is at the focal point ($q = f$). For this situation $1/p = 0$, $1/q = 1/f$, and from Eq. 24-9 we obtain

$$\frac{1}{f} = (n - 1)\left(\frac{1}{R_1} + \frac{1}{R_2}\right) \quad (24\text{-}6)$$

This is the *lens-maker's equation* by which we can calculate the power ($= 1/f$) of a thin lens when we know the curvatures of its surfaces and the index of refraction of the glass. Note that R_1 and R_2 are considered to be $+$ for convex surfaces, as is true for both surfaces in Fig. 24-42.

■ PROBLEMS

24-C15 The index of refraction of the glass in a lens 50.0 mm in diameter is to be found in the laboratory. The focal length of the lens is $+17.4$ cm, and the lens is 3.22 mm thick at its center. Calculate the index of refraction. [*Hint:* Use the saggital theorem to find $(1/R_1 + 1/R_2)$.]

24-C16 To start a fire, an Arctic explorer makes a "burning glass" out of ice. The lens is plano-convex, 20 cm in diameter, 2 cm thick at its center. Calculate the focal length and power of the lens. (*Hint:* Use the saggital theorem to find the radius of curvature.)

■ REFERENCES

1 Magie, W. F., *A Source Book in Physics* (New York: McGraw-Hill, 1935), pp. 283–94, Huygens' principle; pp. 335–44, speed-of-light determinations by Römer, Bradley, Fizeau, and Foucault.

2 Mulligan, J. F., and D. F. McDonald, "Recent Determinations of the Speed of Light," *Am. J. Phys.,* **20**, 165 (1952); **25**, 180 (1957).

3 Rush, J. H., "The Speed of Light," *Sci. American,* Aug., 1955, p. 62. Experimental methods; the curious trend toward increasing values, after World War II.

4 Satterly, J. W., "An Early Determination of Snell's Law," *Am. J. Phys.,* **19**, 507 (1951). The law of refraction discovered (but not published) by Thomas Heriot and his friends, three years before Snell's work.

5 Smith, F. D., "How Images Are Formed," *Sci. American,* Sept., 1968, p. 97.

6 Herrick, N. J., "Use of Frustrated Total Reflection to Measure Film Thicknesses and Surface Reliefs," *J. Applied Phys.,* **33**, 2774 (1962). Includes an application of total reflection for fingerprinting.

7 Strickland, J., *Straight Wave Reflection from Straight Barriers; Circular Wave Reflection from Various Barriers; Reflection of Waves from Concave Barriers; Refraction of Waves* (films).

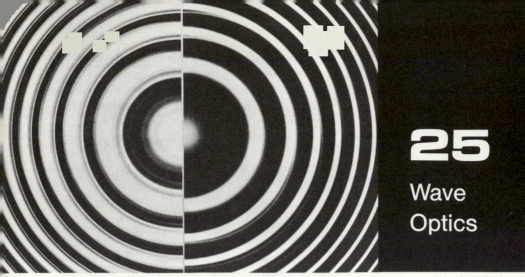

25

Wave Optics

According to the dual theory of light, which we are now approaching, light has wave properties, and also particle properties. More precisely, we can say that sometimes a wave model for light is useful, and at other times a corpuscular (particle) model is useful. Some phenomena can be interpreted by either model. For example, energy is transmitted by a beam of light. We know that a stream of particles can carry kinetic energy from one place to another; and we saw in Chap. 10 that energy can also be transmitted by wave motion. Without specifying the exact nature of light, we therefore define the *intensity* of a beam of light as the rate of flow of energy per unit cross-sectional area. This is the same definition we used for the intensity of a sound wave in Chap. 11. In general, however, neither the wave model nor the corpuscular model is, by itself, successful in describing the entire range of optical phenomena. The new "nonclassical" model is a dual model, combining portions of the older, or "classical," models with some new ideas. In this chapter we shall study the compelling evidence for a wave model, bearing in mind that the evidence for a particle model is yet to come (in Chap. 27).

25-1 Newton's corpuscular theory of light

It was not until about 1800 that clear experimental evidence for the wave nature of light was discovered by Thomas Young (1773–1829), a British physician and experimenter. Let us review the situation at that time. The *possibility* that light is a wave motion had been recognized by Huygens, Hooke, Newton, and others, and indeed the laws of reflection and refraction (geometrical optics) had been derived from Huygens' principle (Chap. 24). It is possible, however, to derive these same laws from a corpuscular theory. Anyone who has played billiards or tennis knows that the angle of incidence equals the angle of reflection, assuming no "English" or top spin on the ball. The law of reflection for perfectly elastic corpuscles was shown by Newton to be a direct consequence of the laws of motion. The corpuscle has components of velocity $\mathbf{v}_x$ and $\mathbf{v}_y$ as it approaches the surface (Fig. 25-1). As it leaves the surface, we find that $\mathbf{v}_x$ remains unaltered in magnitude and direction, since there is no horizontal force and the horizontal momentum of the system is conserved. The vertical component $\mathbf{v}_y$ re-

---

Circular interference fringes formed by the mercury green line of wavelength 5461 Å. The fringes at the left are from natural mercury, a mixture of isotopes; each isotope produces light of a slightly different wavelength, and so the fringes are complex. The fringes at the right are from the single mercury isotope of mass number 198; the sharpness of the fringes indicates that the line is now strictly monochromatic. (U. S. Bureau of Standards)

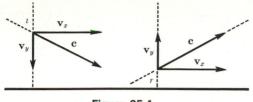

Figure 25-1

mains unaltered in magnitude but reverses in direction; in this way the conditions for perfectly elastic collision are fulfilled. (Velocity of approach equals velocity of separation.) Applying these ideas to the reflection of light, the vector sum of $\mathbf{v}_x$ and $\mathbf{v}_y$ is the velocity of light $\mathbf{c}$; the magnitude of $\mathbf{c}$, and hence the kinetic energy of the bouncing corpuscle, is constant. From the diagram, it is evident that $i = r$, and hence the law of reflection is proved.

Newton also had a corpuscular theory for refraction. He postulated that a corpuscle of light is attracted toward a surface by some sort of force. Thus in Fig. 25-2, the force of the glass on the corpuscle would be directed along the normal. The normal component $\mathbf{v}_y$ increases, but the horizontal component $\mathbf{v}_x$ remains unaltered. The resultant velocity in glass is a vector which makes an angle θ_2 with the normal, and hence the ray of light bends

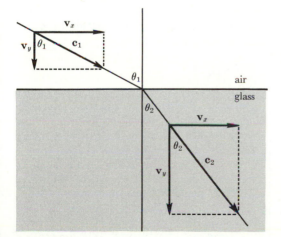

Figure 25-2   Newton's proof of Snell's law, using a corpuscular model for light.

toward the normal. If the supposed force of attraction is a very short-range force, the change in velocity will take place only during the time that the corpuscle is close to the surface, and the bending of the light ray will be abrupt. It is even possible to derive Snell's law from this model:

$$\frac{\sin \theta_1}{\sin \theta_2} = \frac{v_x/c_1}{v_x/c_2} = \frac{c_2}{c_1} = \text{constant}$$

The constant is the *reciprocal* of the constant c_1/c_2 derived in Sec. 24-5 by the wave model and Huygens' principle. However, the essence of Snell's law is that the ratio $\sin \theta_1/\sin \theta_2$ is a constant (any constant) for a given pair of mediums, and so we see that the corpuscular and wave models are equally useful in this case.

We should realize that Newton's corpuscular theory was, in its day, superior to the competing wave theory in certain respects, since his theory used the laws of mechanics which had already proved to be valid in other areas of physics. Newton emphasized the corpuscular aspects of his model on the basis of the then-known facts, especially the straight-line propagation of light. An important experiment to clarify the issue is obvious: one need only measure the speed of light in air and in glass. According to the wave theory, light should slow down as it enters glass (see Fig. 24-8); according to the corpuscular theory, it is attracted toward the surface and should speed up as it enters glass. In Newton's time the speed of light could not be measured in the laboratory, and so either assumption was possible. Not until 1850, over a century after the death of Newton, did Jean Foucault succeed in showing experimentally that the speed of light in water is less than in air; but this verification of the wave theory of light came as an anticlimax. In the early years of the nineteenth century the ingenious experiments of Young, Fresnel, and others had already provided direct evidence for the existence of light waves.

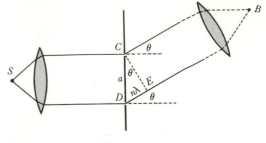

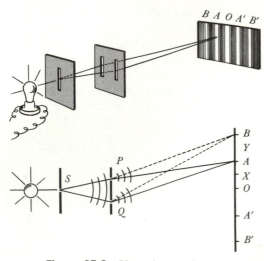

Figure 25-3 Young's experiment.

25-2 Interference

The first experiment in which the interference of light waves was consciously obtained was that of Young (see Fig. 25-3). A narrow vertical slit S is illuminated by a source, and a wave front spreads out from S, striking two narrow slits P and Q, which are close together.[*] Wavelets spread out from P and Q, and any given point on the screen receives light from both slits. It is found that the screen is not uniformly illuminated; some points, such as O, A, B are bright, and others, such as X, Y are dark. Young interpreted this pattern as the result of constructive and destructive interference of light waves. Consider a point such as A. We wish to find the resultant illumination at A due to the two rays PA and QA. Since P and Q are on a wave front whose center is at S, the rays leaving P and Q are in phase with each other. Whether they are in phase when combined at A depends on the path difference. If QA is exactly one wavelength longer than PA, the rays will be in phase at the point A, and the resultant amplitude will be large.[■] In general, *the condition for*

---

[*] The experiment can also be performed with pinholes instead of slits, with some sacrifice in intensity.
[■] Note the implicit use of the superposition theorem here; see Sec. 10-7.

Figure 25-4

a maximum is that the path difference is an integral number of wavelengths. Thus $QO - PO = 0\lambda$; $QA - PA = 1\lambda$; $QB - PB = 2\lambda$, etc., and in general, $QN - PN = n\lambda$ for the nth bright band. (The integer n is called the *order* of interference.) Between the bright bands are dark bands, where the waves are completely out of phase; for these the path difference is $\frac{1}{2}\lambda$, $\frac{3}{2}\lambda$, $\frac{5}{2}\lambda$, etc.

In order to calculate the positions of the bright bands, we simplify the problem by using parallel rays (i.e., plane wave fronts). According to Fig. 25-4, the path difference is DE, and therefore DE must equal $n\lambda$ if constructive interference is to take place. That is, $n\lambda/a = \sin\theta$, where a is the separation of the slits. Thus the general condition for a maximum of a double slit pattern is

$$n\lambda = a\sin\theta \qquad (25\text{-}1)$$

The same equation serves to locate the maxima for a grating consisting of many equally spaced slits.

EXAMPLE 25-1

Parallel rays of wavelength 5000 Å fall upon a pair of slits which are 0.01 mm apart. The interference pattern is focused on a screen 2.30 m from the slits. Calculate the separation of the two 3rd-order bright bands, or images.

First we calculate the angular deviation for the 3rd order. The wavelength is 5000 Å $= 5 \times 10^{-7}$ m, and a is 10^{-5} m.

$$n\lambda = a\sin\theta$$

$$\sin\theta = \frac{n\lambda}{a} = \frac{3(5 \times 10^{-7}\text{ m})}{10^{-5}\text{ m}} = 0.150$$

$$\theta = 8.6°$$

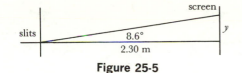

Figure 25-5

The deviation y is found from the tangent of the angle (Fig. 25-5).

$$\tan 8.6° = \frac{y}{2.30 \text{ m}}$$

$$y = (2.30 \text{ m})(\tan 8.6°)$$

$$= (2.30 \text{ m})(0.152) = 0.350 \text{ m}$$

The 3rd-order images are each 0.350 m from the central bright image; hence the distance between them is

$$2y = \boxed{0.70 \text{ m}}$$

When the angular deviation is small, as in Example 25-1, the sine and the tangent are practically equal. In our next example we make use of this approximation.

EXAMPLE **25-2**

Two slits 0.03 mm apart are illuminated by parallel rays, and the 5th-order image is 14 cm from the central image, on a screen 2 m from the slits. Calculate the wavelength of the light used. From Fig. 25-6,

$$\tan \theta = \frac{14 \text{ cm}}{200 \text{ cm}} = 0.07$$

The angle is small, so $\sin \theta$ is also approximately 0.07.

$$n\lambda = a \sin \theta$$

$$\lambda = \frac{a \sin \theta}{n} = \frac{(0.03 \times 10^{-3} \text{ m})(0.07)}{5}$$

$$= 4.2 \times 10^{-7} \text{ m} = \boxed{4200 \text{ Å}}$$

This is violet light (Table 24-1).

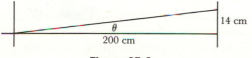

Figure 25-6

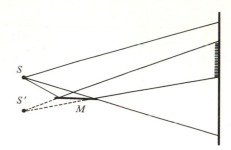

Figure 25-7

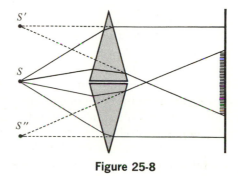

Figure 25-8

Young's experiment does not produce a very bright interference pattern, since most of the light from the source is blocked out and only a small fraction passes through the slits. Two ways of getting a much brighter pattern make use of only one slit and a virtual image of that slit. In Fig. 25-7, a plane mirror M forms a virtual image S', and if the tilt of the mirror is small enough, S and S' are very close together and act as a double slit of small separation. The interference bands are visible in the region where the beams overlap. In the arrangement of Fig. 25-8, two virtual images of a slit are formed by two prisms placed back to back; if the angles of the prisms are small, the separation between S' and S'' is small, and they act as a double slit. Here, too, the interference pattern is visible in the region where the beams overlap.

25-3 Interference of radio, water, and sound waves

The essential feature of interference is that *path difference is equivalent to phase difference.*

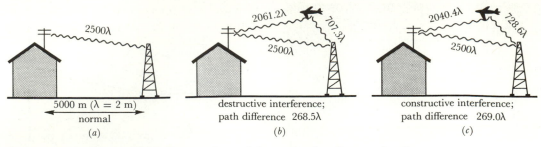

Figure 25-9 Constructive and destructive interference of radio waves.

Suppose that two beams reach a certain point in phase with each other—crest coinciding with crest and trough with trough. Stretching out the path of one beam by an extra half-wavelength is exactly equivalent to retarding its phase at the end-point of the path by half a cycle of vibration. Thus any odd number of half-wavelengths of path difference will produce destructive interference, while an even number of half-wavelengths of path difference will produce constructive interference. These ideas apply to any wave motion, not merely to visible light. We are familiar with interference of radio waves, although perhaps unknowingly so. Normally, a television signal (of wavelength several meters) travels to the receiver by a direct line-of-sight path (Fig. 25-9a). If a plane flies past, a reflected beam may also reach the receiving antenna. If the situation is as in Fig. 25-9b, the two waves interfere destructively, but a few seconds later (c) as the plane moves, the path difference becomes an integral number of wavelengths and constructive interference takes place. The resultant signal

fluctuates in intensity while the plane is flying by, and the picture flutters. A similar effect can be caused at a slower rate by reflection from a drifting cloud or even by reflection from a trail of ions left by a meteor falling through the atmosphere.

Broadcasters make ingenious use of interference to beam their signals into a desired area. For instance, it would be uneconomical for a station near the seacoast to broadcast equally in all horizontal directions, as would be the case if a single vertical antenna were used. Instead, a pair of antennas spaced $\frac{1}{4}\lambda$ apart are fed from the same oscillator, but by electrical means the current in one antenna is caused to reach its peak just $\frac{1}{4}$ cycle later than that in the other antenna (Fig. 25-10). Point Y is $\frac{1}{4}\lambda$ farther from the antenna at A than from that at B. Also, A is caused to radiate $\frac{1}{4}$ cycle later than B, which is equivalent to another $\frac{1}{4}$ of path difference. These two quarter-wavelengths add up to $\frac{1}{2}\lambda$, and hence at Y the two beams interfere destructively, and no energy is radiated out to sea at Y. Just the opposite is true at X. The tower at

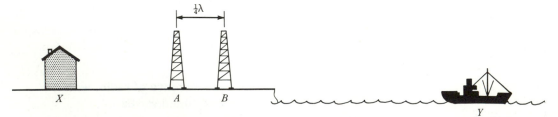

Figure 25-10 Tower A radiates $\frac{1}{4}$ cycle later than tower B; radiation pattern is stronger at X than at Y.

Figure 25-11 Constructive and destructive interference between water waves spreading out from two sources.

E. Leybold's Nachfolger, courtesy J. Klinger

of a wave is determined by the maximum magnitude E of the electric field vector (Fig. 23-4), and the intensity is proportional to E^2.

Interference is by no means confined to light, radio, and other e-m waves. In the ripple tank, a source of light shines through a shallow glass-bottomed tank filled with water, and the crests of ripples act as converging lenses to focus the light on a screen. Another arrangement uses mercury as a liquid, and the images are formed by reflection instead of refraction. In either case, the progress of the wave fronts can be seen. The interference of two wavelets spreading out from adjacent sources is shown in Fig. 25-11; this is Young's experiment for water waves.

The constructive and destructive interference of sound waves is, of course, the basis of the stationary waves in pipes which were discussed in Sec. 11-3. One application of interference to direction finding is shown in Fig. 25-12. It is an interesting physiological fact that the brain can perform the function of mixing the output of the two ears, with the sensation depending upon the resultant amplitude, which in turn depends on the phase difference between the sounds heard in the separate ears. If the path difference x is a half-wavelength, the outputs of the two microphones are out of phase and the observer hears nothing. In use, the microphone

A is $\frac{1}{4}\lambda$ closer to X than is B, but it is fed $\frac{1}{4}$ cycle later than B, so the waves reach X in phase, and the resultant amplitude is twice as much as from either antenna separately.

Considering interference from an energy viewpoint, we expect that the total energy radiated from the antennas must appear somewhere in the area surrounding the towers. Detailed calculation of such "radiation patterns" show that this is true. All the energy is accounted for, but the *distribution* of the energy depends on the number of antennas, their spacing, and their phasing. In a similar fashion, the total light energy falling on the two slits in Young's experiment is fully accounted for, even though certain places on the screen receive no light and other places are brighter than they would have been in the absence of the slit system. For visible light, as for radio waves, the amplitude

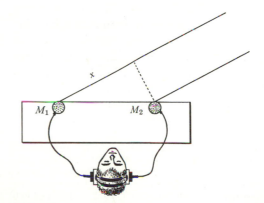

Figure 25-12 Sonic direction finder makes use of interference of sound waves.

assembly is rotated until the sound is a maximum; the path difference is then zero, and the source is broadside to the line M_1M_2.[*]

In dealing with mechanical waves, we defined a node as a place where two vibrations combine to give zero net amplitude at all times. The nodes and antinodes in a pipe or on a string are caused by the interference of one wave with an equal reflected wave traveling in the opposite direction. In this chapter we have been studying interference of waves which are traveling in more or less parallel directions. When two such waves are brought together on a screen or on the retina of the eye, the places where they destructively or constructively interfere are also called "nodes" and "antinodes." Thus we consider an interference pattern on the screen or retina to be a pattern of nodes and antinodes. For light, the resultant electric vector of two interfering e-m waves is analogous to the resultant displacement vector of two interfering mechanical waves on a string or on water. In each case, nodes occur wherever the path difference between the two waves causes them to be permanently out of phase with each other.

25-4   The grating

A series of many slits, equally spaced, is called a *grating*. Parallel rays falling on a grating form sharper and brighter maxima than they would with a pair of slits, but the basic principle is the same. In Fig. 25-13 we illustrate the formation of the 3rd-order maximum. The path length of each ray differs by $\pm 3\lambda$ from that of its neighboring ray. Thus ray DD' is 3λ longer than ray CC', and it is 3λ

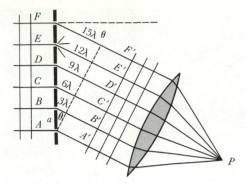

Figure 25-13 Grating with six slits; the lens might represent the eye lens, and P might be a point on the retina.

shorter than ray EE'. Since the rays are all in phase on the wave front $ABCDEF$, they will again be in phase on the wave front $A'B'C'D'E'F'$, if the angle θ is chosen just right. The condition is

$$3\lambda/a = \sin\theta, \quad \text{or} \quad 3\lambda = a\sin\theta$$

where a is the space between adjacent slits, called the *grating space* or *grating constant*. This is the same equation as for a pair of slits. In fact, a pair of slits can be considered to be a rudimentary grating.

It should be noted that Fig. 25-13, and many others in this chapter, are simplified to show only those rays that are of interest. To be sure, wavelets *do* spread out from each slit. This means that rays diverge in all directions from each slit, as shown for the next-to-the-top slit. Our procedure is to pick a direction of observation, indicated by θ; this is done when we orient the lens and select a point P at which we observe the resultant amplitude (if any). This point might be on the retina of the eye or on a photographic plate. If it is objected "How do the rays know enough to bend at an angle θ?" we answer that we can later investigate the resultant amplitude in any other direction by the same method, but for the present we are concerned only with those rays that are eventually brought together at P. In thus drawing selected rays rather than wave fronts, we are

[*] While it is true that a maximum would also be heard if the path difference x were 1λ, 2λ, 3λ, etc., in practical use this is no source of confusion. Most sounds contain components of many frequencies and have many wavelengths. The path difference x could not be equal to $n\lambda$ for all wavelengths at the same time, and no clear maximum would be heard except at the broadside position for which the path difference is 0λ. The condition $x = 0\lambda$ *can* be satisfied for all λ's simultaneously.

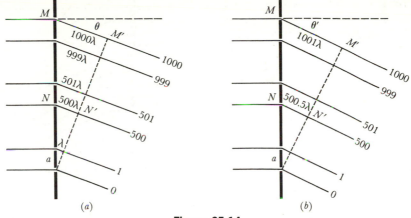

(a)

(b)

Figure 25-14

making full use of the wave theory of light.

The sharpness of the "image" formed by a grating can be seen from the following argument: If there are 1000 slits and we are considering the 1st-order maximum, all the rays cooperate if θ is such that the top ray's extra path (MM') is exactly 1000λ (Fig. 25-14a). If we now increase the angle very slightly to θ' (Fig. 25-14b), just enough to make $MM' = 1001\lambda$, the middle ray's extra path NN' will become 500.5λ. Hence ray 0 and ray 500 are 180° out of phase and cancel each other; likewise ray 1 cancels ray 501, ray 2 cancels ray 502, etc. Every ray from a slit in the bottom half of the grating cancels one in the top half. The resultant intensity is zero at the new angle, θ', which is very close to the former angle θ for which the intensity is a maximum. This means that the maximum is a sharp one. If we had only two slits, a similar small change of angle would cause the top ray to be only 0.001 cycle out of phase (Fig. 25-15), and it would still cooperate with the bottom

ray to give a resultant intensity almost as large as at the peak. The qualitative difference between the interference pattern of a grating and that of a pair of very narrow slits is shown in Fig. 25-16.

We have so far tacitly assumed that the light striking a grating is *monochromatic*, that is, that it consists of a single color and has a single wavelength. If a mixture of colors falls on a grating, the equation $n\lambda = a \sin \theta$ tells us that to each different λ there corresponds a different θ. Since the colors are separated, a grating can serve the same function as a prism, though in an entirely different way. The angular separation between two colors can be quite large, if the grating space a is small enough. We shall consider the grating spectroscope in the next chapter, as an example of an optical instrument.

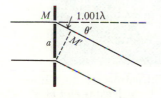

Figure 25-15

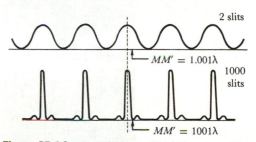

Figure 25-16 Comparison of interference pattern of two slits and that of a grating. Between the sharp maxima formed by the grating are many subsidiary maxima, too weak to be drawn to scale.

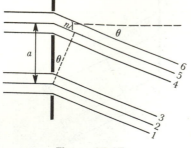

Figure 25-17

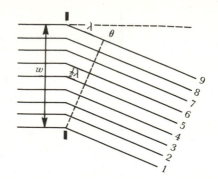

Figure 25-18 Single slit, 1st-order minimum.

25-5 The single-slit diffraction pattern

In our discussions of Young's experiment and the grating, we have always assumed *narrow* slits. In this way we could consider all the rays passing through a given slit to be in phase with each other; this is true enough if the slit width is small compared with the wavelength. In other words, rays 1, 2, 3 in Fig. 25-17 are sufficiently in phase with each other to act as a single ray; and this "super ray" is in phase with the "super ray" composed of rays 4, 5, 6 if $n\lambda = a \sin \theta$. Of course, rays 1 and 3 do differ by a slight (but negligible) amount for any real (but narrow) slit.

Turning now to a wider slit, we must take account of phase differences between rays passing through various parts of the slit. We have to deal with infinitely many rays, of which only nine are shown in Fig. 25-18. The term *diffraction* refers to the interference

between the component rays of a single broad wave front, as in this figure. If we consider the diffraction pattern formed by rays leaving the slit at an angle θ, and choose θ so that $\lambda = w \sin \theta$, where w is the width of the slit, then we can show that the resultant amplitude of all the rays is zero. Ray 1 is 180° out of phase with ray 5, ray 2 is 180° out of phase with ray 6, and so on. Thus each ray in the bottom half of the slit cancels out a corresponding ray in the upper half, just as in Fig. 25-14*b* for a grating at an angle slightly different from the angle giving an interference maximum. The only difference is that now we have a continuous group of infinitely many rays, instead of just 1000 or so.

The resultant amplitude from a single slit falls to zero also at larger angles. For instance, if the extreme path difference is 3λ, as in Fig. 25-21, we can divide the slit into three

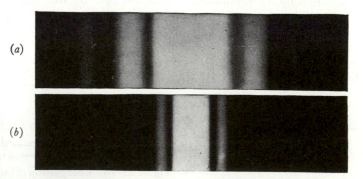

(*a*)

(*b*)

Figure 25-19 Diffraction pattern of a single slit. (*a*) Pattern of narrow slit; (*b*) pattern of wide slit. (By permission from *Fundamentals of Physical Optics*, 3rd ed., by F. A. Jenkins and H. E. White. Copyright 1957. McGraw-Hill Book Co.)

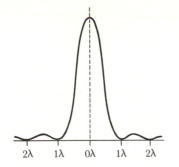

Figure 25-20 Intensity distribution for a single-slit diffraction pattern.

equal segments, each giving zero intensity as before. Thus the general condition for a minimum of a single-slit diffraction pattern is

$$n\lambda = w \sin \theta \qquad (25\text{-}2)$$

Between these minima, the pattern has bright maxima (Figs. 25-19 and 25-20). A special case occurs if $n = 0$ and $\theta = 0$. No path differences exist, all of the many rays are in phase, and in this case they add up to give an intense central maximum. The photograph shows clearly that the central maximum of the diffraction pattern is twice as wide as the others.

An interesting paradox is illustrated in Fig. 25-22. By blocking out most of the single slit to leave only two narrow slits, the intensity in the direction shown is *increased!* Do not imagine that we get something for nothing. The total intensity from the single slit

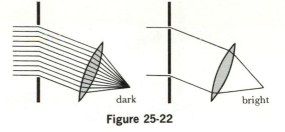

Figure 25-22

(area under the curve) is far greater than the total for two slits, as seen from Fig. 25-23, in which the two patterns are superposed.

Let us restate Eq. 25-2 in another way. Most of the energy of the wave lies in the central bright band, for which the 1st-order minimum is given by $\sin \theta = \lambda/w$. This leads to an important generalization:

The ratio of wavelength to size of opening determines to what extent light or any other wave fails to travel in a straight line.

Similarly, the ratio λ/w determines the amount of diffraction when a wave front is limited by an opaque obstacle.

The diffraction patterns of some other apertures and obstacles are reproduced in Fig. 25-24. In each case, the pattern can be calculated by the methods of wave optics which we have illustrated for the single slit. The shadow of a straight edge is surrounded by a series of bands of uneven spacing; this is the "bending around corners" that is expected for wave motion. Such effects had actually been observed in the seventeenth century by

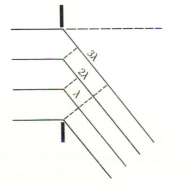

Figure 25-21 Single slit, 3rd-order minimum.

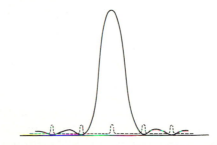

Figure 25-23 Diffraction pattern for a single slit of width w (solid curve) and pattern from a pair of narrow slits of separation w (dashed line).

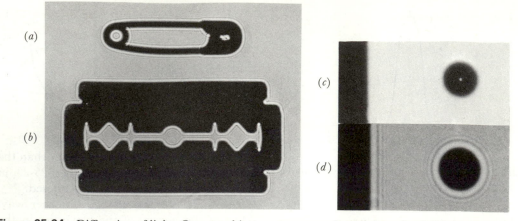

Figure 25-24 Diffraction of light. Opaque objects were mounted on a glass plate with wax and placed 4 m from a point source. A photographic plate was exposed at a position 2 m beyond the obstacles. No lenses were used. Note the diffraction bands at all edges; no shadow is sharp. (*a*) Safety pin. (*b*) Razor blade. (*c*) Ball bearing and edge of razor blade; printed with short exposure to show only densest parts of shadows. (*d*) Same as *c*, printed with longer exposure to show fainter parts of shadows. The Arago bright spot, seen in *c*, is overexposed in *d* and cannot be seen.

Francesco Grimaldi, who failed to make the wave interpretation of his results. One difficulty in all such experiments is a geometrical one. Shadows are usually far from sharp because the source is broad. For instance, the sun subtends an angle of about $\frac{1}{2}°$ in the sky; different parts of the sun cast different shadows which overlap to give a "fuzzy" edge. The shadow of a straight edge may well be diffuse enough to mask the diffraction effects unless special precautions are taken to use a small ("point") source. Figure 25-24*c* is especially interesting. The French physicist Siméon Poisson is said to have objected to the wave theory, arguing that the theory would predict a bright spot at the center of the shadow of a perfectly circular obstacle— a seemingly absurd result.[*] When the experiment was performed by Dominique Arago in 1818, the bright spot was promptly found, and to this day it is called the Arago spot. No more dramatic proof of the wave nature of light can be found. A hypothesis (the wave theory) led to a new and hitherto meaningless experiment, with results contrary to

[*] The proof of this prediction is too advanced to be included in this book.

"intuition" and yet in perfect agreement with the prediction of the theory.

25-6 Applications of interference

Some of the evidence for interference of light waves is familiar in everyday life and indeed was known to but inadequately interpreted by Newton and his contemporaries. Interference fringes can be observed if light strikes an air wedge, as shown in Fig. 25-25. A fine wire or a sheet of tissue paper at *X* forms a thin wedge of air between two flat pieces of glass. At any point such as *A* or *B*

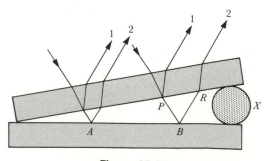

Figure 25-25

rays are reflected both at the upper surface of the air wedge (ray 1) and at the lower surface of the wedge (ray 2). When these rays are brought together on the retina of the eye, the resultant amplitude is large or small depending on the path difference PBR. Since the path difference gradually increases as B moves out along the wedge, the observer sees a series of alternately bright and dark bands. For any two adjacent bands, the total (round-trip) path difference at B is just 1λ longer than the path difference at A. That is, the air wedge is $\frac{1}{2}\lambda$ thicker at B than at A. The wavelength of light can be found by a simple experiment of this sort. In the optical shop, a glass surface can be tested for flatness by laying it on top of a standard surface of known flatness known as an optical flat. If the fringes are wavy (Fig. 25-26a), the air film is of uneven thickness and the high spots of the surface under test can be polished until a parallel set of fringes is obtained (Fig. 25-26b).

A seeming paradox arises in air-wedge experiments. The resultant is dark where the pieces of glass are in contact, and yet according to the wave theory we would expect a bright band here since the path difference is zero. This illustrates a change of phase which takes place during reflection. Recall that, when a sound wave is reflected at the end of a pipe, the result differs according to whether the reflection takes place at a closed end or an open end. At a closed end, a compression is reflected as a compression, which requires an abrupt reversal of the direction of motion of the air molecules. That is, there is a change of phase of half a cycle (180°) when reflection takes place at the closed end of a pipe. On the other hand, at an open end a compression is reflected as a rarefaction, and the air molecules move right on out into the open, with no change of phase. A similar situation exists when light waves are reflected at an optically "more dense" medium (Fig. 25-27a). Ray B is 180° out of phase with ray A, simply because it is reflected in air against a glass surface. How-

(a)

(b)   Bausch & Lomb

Figure 25-26 Testing an optical surface by interference. (a) Initial condition of surfaces, showing three high spots; (b) surfaces polished flat to within a fraction of a wavelength.

ever, in Fig. 25-27b, ray C is in phase with ray D, since this reflection is in glass against air—similar to the reflection of a sound wave at the open end of a pipe. We can now understand why the interference pattern is dark at the vertex of the air wedge. The path difference is zero, but one ray suffers the 180° phase change while the other does not. This

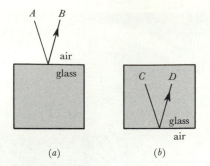

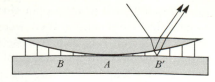

Figure 25-29

Figure 25-27 Light waves undergo a 180° phase change when reflected as in (a), but not when reflected as in (b).

is equivalent to an extra $\frac{1}{2}\lambda$ in the path of one ray, and destructive interference results.

EXAMPLE 25-3

A plane glass plate 10 cm long is separated from another plane glass plate at one end by a piece of aluminum foil 0.004 cm thick. How far apart are the successive interference bands, viewed in light of wavelength 6000 Å?

The foil creates an air wedge, and there is a dark band at the vertex A (Fig. 25-28), where the path difference is 0λ. The next dark band is at B, where the total path difference is 1λ. Hence

$$BC = \tfrac{1}{2}\lambda = \tfrac{1}{2}(6 \times 10^{-7}\,\text{m}) = 3 \times 10^{-7}\,\text{m}$$

We find the distance x by a proportion:

$$\frac{x}{10\,\text{cm}} = \frac{3 \times 10^{-7}\,\text{m}}{4 \times 10^{-5}\,\text{m}}$$

$$x = 0.075\,\text{cm} = \boxed{0.75\,\text{mm}}$$

The bands are spaced 0.75 mm apart and can easily be seen with a magnifying glass. The total number of dark bands in the 10 cm length of glass is

$$\frac{10\,\text{cm}}{0.075\,\text{cm/band}} = 133\,\text{bands}$$

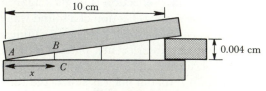
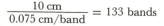

Figure 25-28

One way to obtain a thin air film of variable thickness is to place a long-focus lens on a plane glass surface. Figure 25-29 shows the arrangement, with the curvature much exaggerated for clarity. If the curved surface of the lens is almost flat, the air film varies from zero thickness (at A) to a dozen or so wavelengths (at the edge). When viewed by reflected light, a pattern of dark rings is observed; for any given circular ring, such as at B, the air film is of constant thickness. Ironically Newton himself discovered these interference rings—they are called Newton's rings—but entirely missed their significance. He postulated that corpuscles of light had "fits of easy reflection and easy transmission" to explain the fact that light is partially reflected and partially transmitted at a glass-air surface. He had no consistent theory for the rings. After two and a half centuries it is easy for us to see that Newton's model was inadequate because it led nowhere and predicted no new results (compare the fruitfulness of the wave theory as exemplified by the Arago spot). The intellectual climate of Newton's day was not yet ready for a model in which invisibly small waves, 2 million of them per meter, behave like the water waves seen by the naked eye.

Other reflections from thin films are often observed. A soap bubble consists of a film of soap solution. If viewed in white light, a very thin bubble appears brilliantly colored. Suppose that for some particular color the thickness is exactly $\frac{1}{2}\lambda$. An extra 180° phase shift is introduced at the front surface but not at the back surface. The *effective* (round-trip) path difference is $1\lambda + \frac{1}{2}\lambda = 1\frac{1}{2}\lambda$, and light of this color is not reflected, since the two rays interfere destructively. (The light goes on through the film.) However, neighboring colors have longer or shorter wavelengths, and the film cannot be $\frac{1}{2}\lambda$ thick for all colors at once. Hence the film reflects some colors more than others. As a result a bubble is colored, and the colors change as the thickness of the film changes during evaporation. Eventually, the bubble's thickness is much smaller than any visible wavelength, the path difference is 0λ for all colors, and the bubble appears black (why not white?) just before it breaks.

EXAMPLE 25-4

A soap bubble 5000 Å thick is illuminated with white light. The index of refraction of the film is 1.35. What colors are *not* reflected?

The first wavelength that is not reflected is one for which 5000 Å $= \frac{1}{2}\lambda$. Then the round trip of 10 000 Å would be 1λ, and the phase change at the front surface would account for another $\frac{1}{2}\lambda$, making a total of $1\frac{1}{2}\lambda$. If $\lambda = 10\,000$ Å in film, the wavelength in air would be

$$(10\,000\text{ Å})(1.35) = \boxed{13\,500\text{ Å}}$$

This is in the infrared region.

Next, we consider a wavelength for which the round trip is 2λ. Then 10 000 Å $= 2\lambda$, and $\lambda = 5000$ Å in film, or, in air,

$$(5000\text{ Å})(1.35) = \boxed{6750\text{ Å}}$$

This is red light.

The next color that is completely removed by interference is one for which 10 000 Å $= 3\lambda$; $\lambda = 3333$ Å in film, or

$$\boxed{4500\text{ Å}}$$

in air. This is blue light. Going to still shorter wavelengths, we find that the next destructive interference is for $\lambda = 2500$ Å in film, or

$$\boxed{3375\text{ Å}}$$

in air. This is in the ultraviolet region.

We conclude that only two visible colors are totally canceled: red light of wavelength 6750 Å and blue light of wavelength 4500 Å.

Other colors are reflected to a greater or lesser degree. Since red and blue are missing, the middle of the spectrum predominates, and the bubble appears greenish.

The colors of oil films on pavements are caused in similar fashion by interference.

25-7 The Michelson interferometer

We conclude our study of interference by describing the interferometer invented by a brilliant American physicist, A. A. Michelson (1852–1931). The instrument uses the air-wedge principle, but still allows the working area to be many centimeters in length instead of a fraction of a millimeter. In Figs. 25-30 and 31 light from a broad source S strikes a glass plate M which is "half-silvered" to be partially reflecting so that half the light is reflected to mirror X and half goes on through to mirror Y. The beams reflected from X and Y emerge together and enter the eye E. Interference takes place between the beams, depending on the path difference.*

---

*An extra glass plate is inserted in the path of beam 2 so that each beam traverses the same thickness of glass. This plate is not silvered.

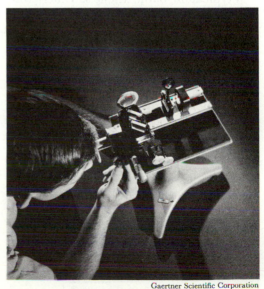

Gaertner Scientific Corporation

Figure 25-30

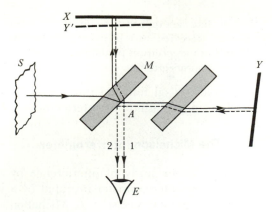

Figure 25-31 Michelson interferometer.

Since M is a plane mirror, it forms a virtual image of Y at Y'. In effect, the two beams 1 and 2 are coming from two mirrors at X and Y'. The great advantage of the method is that the thickness of the air wedge between X and Y' can be adjusted, even to zero if desired, by moving X back and forth by means of a screw. The angle of the wedge can be adjusted by the screws which tilt mirror Y; thus a set of fringes of any desired spacing can be obtained. The fringes move across the field of view as X moves toward coincidence with the image Y'.

A typical use of the Michelson interferometer is to measure the wavelength of light, as described in the following example.

EXAMPLE 25-5

Monochromatic green light is obtained by passing light from a mercury arc through a suitable filter. It is found that moving the mirror X through a distance of 0.273 mm causes a shift of 1000 fringes. Calculate the wavelength of the light.

The passage of 1000 fringes means that the total (round trip) path has been changed by 1000 wavelengths. Thus

$$1000\lambda = 2(0.273 \text{ mm})$$

$$\lambda = \frac{2(0.273 \text{ mm})}{1000} = 5.46 \times 10^{-4} \text{ mm}$$

$$= 5.46 \times 10^{-7} \text{ m}$$

$$= \boxed{5460 \text{ Å}}$$

Michelson used his interferometer to determine the length of the standard meter in terms of the wavelength of a certain red line in the cadmium spectrum. He chose this line because it was the "purest" known at that time. By way of contrast, the sodium yellow "line" is really a close pair of lines, of wavelengths $\lambda_1 = 5890$ Å and $\lambda_2 = 5896$ Å. This is only a 0.1% difference in wavelength; but as a result, two separate interference patterns are obtained simultaneously. Simple arithmetic shows that 500 wavelengths of λ_1 equal $499\frac{1}{2}$ wavelengths of λ_2. Thus an air film of thickness $250\lambda_1$ is of proper thickness to cancel λ_1 [the round trip is $500\lambda_1$, and an extra $\frac{1}{2}\lambda_1$ arises from the phase shift (Fig. 25-27)]. This same air film is just right to reinforce λ_2, and due to this overlapping, no fringes would be seen for a film of this thickness. By pursuing this idea of visibility of fringes, Michelson was able to show the extreme purity of the cadmium red line he used. The mercury green line (5461 Å) is a complex one, as shown by the interference pattern on p. 536, which was obtained using an interferometer which makes sharp circular fringes. Each isotope of mercury produces a line of slightly different wavelength. In recent years it has proved possible to separate the isotopes of mercury, and it has been found that the interference fringes from the ^{198}Hg isotope are even "cleaner" than the cadmium fringes used by Michelson. Precise counting of fringes in an interferometer makes possible the definition of the meter in terms of the wavelength of a certain orange line of one of the krypton isotopes (see Sec. 1-4).

The Michelson interferometer was used in a famous experiment to study the effect of a mechanical "ether" in which light waves were supposed to vibrate. The speed of light *relative to the ether* was supposed to be a constant c, whose value would depend on the physical properties of the ether in much the same way that the speed of sound waves in air depends on the bulk modulus and density of air. Suppose, in Fig. 25-31, that

$AY = AX$, so that the time for light to travel from A to Y and back again exactly equals the time required for the round trip from A to X and back again. Rays 1 and 2 would be out of phase because of the extra 180° phase shift of one of the rays (which one?) and a dark fringe would be seen. However, the earth revolves about the sun, and so is moving through the ether at about 30 km/s; other motions through the ether also doubtless are to be expected. If the whole apparatus moves (with the earth) toward the right with velocity **v**, an "ether wind" blows toward the left, and the resultant speed of light (found by vector addition) is $c - v$ (from A to Y); $c + v$ (from Y to A); and $\sqrt{c^2 - v^2}$ (from A to X or X to A). A similar situation was studied in Prob. 3-B21, where it was found that it takes less time for a swimmer to cross a river and return than for him to swim an equal distance up the river and return. Applied to the optical experiment, ray 1 would get ahead of ray 2, and the resulting phase difference would become apparent as a shift of the interference fringes.

In 1887, Albert A. Michelson and his co-worker Edward W. Morley completed an historic series of experiments designed to determine the speed of the earth through the ether. The interferometer was horizontal and could be rotated about a vertical axis so that the arm AY could be adjusted to be either parallel to, or broadside to, the supposed ether wind. Experiments were made at different seasons of the year and at different times of day. The startling result of the experiment was that no shift of fringes was ever observed. In other words, the speed of light is the same in the directions AX and AY, hence *the speed of light is independent of the motion of the observer.* This negative result of the Michelson-Morley experiment puzzled physicists for almost 20 years. In 1905, Einstein adopted the constancy of the speed of light as one of the postulates of the relativity theory (Sec. 7-4), giving up entirely the idea of an ether. The

negative Michelson-Morley result is consistent with Einstein's postulate, which was formulated for entirely unrelated reasons.

25-8 Polarization of light

The electromagnetic waves described in Sec. 23-5 are *plane-polarized waves,* since the variable quantity is an electric field, or a magnetic field, represented by vectors which lie in a plane. For the electric vector in Fig. 23-4, this is the vertical plane. We have also seen that the Maxwell theory predicts *transverse waves,* since the vectors are perpendicular to the direction of propagation. The fact that the portion of the electromagnetic spectrum which we call light can be polarized was known as long ago as 1669, and the full interpretation came two centuries later, based on the Maxwell theory of electromagnetic waves. In fact, one of the more subtle problems we have to discuss is the fact that most light in everyday life is *not* polarized. The question is not, how can light be polarized, but rather, why shouldn't it always be polarized? How can *un*polarized light exist, when electromagnetic waves by their very nature are plane-polarized?

Light waves are emitted by molecules or atoms which are *excited* by thermal or electrical means. In the classical model which we are using, any given atom vibrates a few thousand or a few million times, sending out a relatively short burst of plane-polarized light called a *wave train.* In the usual source, such as the filament of a lamp bulb or the gas in a discharge tube, the vibrating atom makes many collisions per second with other atoms. If the atom vibrates freely for 10^6 cycles, it emits a wave train containing 10^6 wavelengths. For visible light of wavelength 5×10^{-7} m, this wave train would be emitted during less than 10^{-8} s, and it would be about half a meter long. The wave train is plane-polarized, with the electric vector in a plane determined by the chance orientation of the vibrating atom. If a beam of light from a

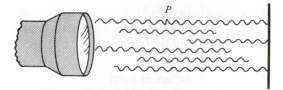

Figure 25-32

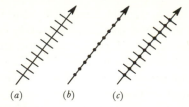

Figure 25-34 (*a*) and (*b*) Rays of polarized light; (*c*) ray of unpolarized light.

flashlight bulb shines on a wall 10 m away, our model leads us to a picture somewhat like Fig. 25-32. Many wave trains fill the space between the flashlight and the wall; they start and stop at random, and the intensity at the wall averages out to a steady illumination. During an atomic collision, the radiation from a single atom might suffer an abrupt change in phase; this would cause a kink in the wave train, as at *P*. Looking head-on at an actual "point" source (actually composed of very many vibrating atoms), we see a superposition of many plane waves with all possible orientations of the electric vector. Each single wave train (such as *XX* in Fig. 25-33) originates from one atom and is plane-polarized with its electric vector perpendicular to the direction of propagation. The net effect averages out, and there is no preferred orientation of the resultant of all these randomly oriented vibrations. Thus we say that ordinary light is unpolarized, but is composed of components polarized in all possible transverse directions. We represent polarized beams as in Fig. 25-34, the bars or dots indicating the direction of the electric vector. Unpolarized light is represented as containing equal proportions of two beams polarized at right angles to each other. For purposes of calculation, any plane-polarized wave can be replaced by components vibrating along two arbitrary axes, as in Fig.

25-35. Thus the unpolarized beam of Fig. 25-33 is equivalent to two equally intense beams polarized along any two arbitrary mutually perpendicular axes. It might seem that we could add up all the vectors of Fig. 25-33 to zero; however, remember that each arrow represents light from a different atom, and there is no *coherence* between the phases of their vibrations. The atoms in an ordinary light source vibrate independently, start and stop at random, and do not pass through zero at the same time. This difficulty has been overcome in the laser (Sec. 28-4) which gives a coherent beam from many cooperating atoms.

One way to determine experimentally the plane of vibration of a beam of light is to look at the source through a sheet of Polaroid film. This material consists of many tiny crystals of a quinine compound, oriented with their axes parallel to each other and spread out on a transparent base. The crystals act like one big crystal and have the unusual property of transmitting only light whose plane of vibration is properly oriented relative to the crystals. Light vibrating in the "wrong" direction is absorbed by the crystals. Thus a sheet of Polaroid can be used as a *polarizer,* to make plane-polarized light out

Figure 25-33

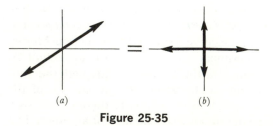

Figure 25-35

Figure 25-36

of ordinary light, and also as an *analyzer,* to determine whether and in what plane a beam of light is polarized.

The polarization of skylight offers a good example of the principles involved in the scattering of light. Unpolarized light from the sun is scattered by air molecules and accounts for the blue sky; simple tests with Polaroid show that skylight is often partially polarized. The effect is most pronounced if the observer looks through Polaroid at right angles to the sun's rays—for instance, if he looks straight up at sunset or sunrise (Fig. 25-36). We consider the two components of polarization separately. Half the light (*A*) can be considered to be polarized with the electric vector horizontal; the electric field exerts a horizontal driving force on electrons in the oxygen or nitrogen molecules, and,

as a result, e-m waves are radiated by these accelerated electrons. The observer is broadside to the miniature antennas represented by the motion of the electrons, and hence he receives scattered light, polarized horizontally of course, since the vibrating charges are moving horizontally. The other half of the original beam (*B*), polarized vertically, also sets electrons into motion, but the observer is looking at these motions head-on, and therefore receives no scattered ray (Sec. 23-6). The over-all result is that the scattered light reaching the observer is polarized.•

Reflection by an electrical insulator such as glass or water also is stronger if the electric vector of the rays is parallel to the surface. Detailed study by means of Maxwell's equations for e-m waves shows that an initially unpolarized beam becomes partially polarized by reflection, with the electric vector parallel to the surface (Fig. 25-37). The analysis is somewhat similar to the discussion of the polarization of scattered light in the previous paragraph. At *Brewster's angle,* given by $\tan \theta = n$, the polarization is strongest. We see here an indication that light is electrical in nature; light cannot be made plane-polarized by reflection from a metal surface. The presence of free electrons in a conductor makes the process of reflection more complicated than for an insulator. Only with the advent of the e-m wave theory of light could a satisfactory theory of metallic reflection be worked out. Much "glare" experienced during ordinary activities can be removed by wearing Polaroid glasses properly oriented. Light reflected from a lake or the ocean is

•Skylight is blue because the efficiency of scattering is proportional to the 4th power of the frequency. Violet light has twice the frequency of red light and is scattered 2^4, or 16, times as readily. This dependence on frequency is true only if the scattering particle is small compared with λ. The condition is fulfilled by molecules of O_2 or N_2, which are about 2 Å in diameter; even local condensations of a dozen or so air molecules which occur at random in the air are small compared with the wavelength and can act as scattering centers. Water drops, which are much larger than λ, scatter all wavelengths of light equally well; hence a cumulus cloud of water drops is white.

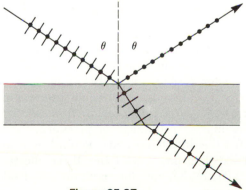

Figure 25-37

partially or completely polarized, with the electric vector horizontal; the Polaroid glasses must be oriented so as to absorb light polarized in this way and to transmit the vertically polarized half of the diffusely scattered (unpolarized) light coming from solid bodies such as boats or fish. The discovery of polarization by reflection was made in 1808 by the French scientist Étienne Louis Malus (1775–1812).

The speed of light in certain transparent crystals—the anisotropic ones—depends on the plane of the electric vector. This is a consequence of the electrical nature of matter. For the crystal shown schematically in Fig. 25-38, it is natural to expect the electric forces along the axis AA' to be different from the forces along the axis BB'. Therefore, as an e-m wave passes through the crystal and secondary wavelets are set up at each atom, we would expect that the direction of the electric vector would be of importance. See Sec. 25-9 for further material on crystal optics.

It is worth noting that the polarization of light waves proves them to be *transverse*. Longitudinal disturbances such as sound waves show all the wave phenomena of interference and diffraction, but they cannot be polarized. It is not difficult to imagine a polarized beam of corpuscles—for instance, a stream of flat stones being "skipped" on the surface of a lake. The orientation of the stone is all-important for the success of the operation. We

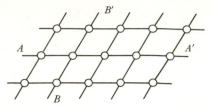

Figure 25-38 Anisotropic crystal (schematic).

could in this way separate a beam of polarized stones from among the unpolarized (randomly oriented) stones striking the water. Some nuclear particles can be polarized by reflection in this way. We should not make the mistake of thinking that polarization, as such, proves the wave nature of light. Light is shown to be (1) transverse and (2) a wave motion by the *combined* evidence of polarization and interference.

In this chapter we have given evidence supporting the wave model for light. Much additional evidence supports a corpuscular model, chiefly in connection with the emission and absorption of light (spectrum analysis, photoelectric effect, etc., to be discussed in Chaps. 27 and 28). Do not imagine, however, that Newton's original corpuscular model has thus been vindicated or reinstated. The simple mechanical "bullet" model of Newton is certainly false, and the newer corpuscular part of the dual model is quite different and more subtle, based on experimental facts unknown in Newton's day.

■ SUMMARY

The early corpuscular theory of light, adopted by Newton, could explain reflection and refraction, on the assumption that light travels faster in a medium than in vacuum. This assumption later turned out to be wrong, but in the meanwhile the corpuscular theory was discredited by the discovery of interference. To obtain interference it is necessary to divide a wave front into two or more parts which come together again after traveling along paths of different lengths. Both constructive and destructive interference are possible, depending on the relative phases of the rays when they recombine. If the effective path difference between two rays is 0λ, 1λ, 2λ, etc., the resultant amplitude is a maximum, and the rays interfere constructively; if the path difference is

$\frac{1}{2}\lambda$, $\frac{3}{2}\lambda$, $\frac{5}{2}\lambda$, etc., the resultant amplitude is zero and the rays interfere destructively.

If two rays strike perpendicularly a pair of narrow slits or a grating, the equation $n\lambda = a \sin \theta$ can be used to calculate the angles θ at which the maxima of intensity are found. For a single slit of width w, the equation $n\lambda = w \sin \theta$ gives the angles for zero intensity in the diffraction pattern. The basic concepts of interference and diffraction apply to all wave motions, including e-m waves, water waves, and sound waves.

Shadows are usually not perfectly sharp because the sources are usually of finite size. Even with a point source, the edge of a shadow shows diffraction bands caused by the wave nature of light. Wave effects are in general more pronounced for larger wavelengths, or for smaller obstacles, since then $\sin \theta \ (= n\lambda/w)$ is larger.

Interference from thin films of air, oil, water, etc., can be applied to many laboratory measurements. Change of phase occurs when a ray is reflected by an optically more dense medium, but no change of phase takes place if a ray is reflected by an optically less dense medium. Certain reflections, therefore, in effect introduce an extra $\frac{1}{2}\lambda$ of path difference.

Interference indicates the wave nature of light, and polarization indicates the transverse nature of light. The electromagnetic wave theory of light predicts light to be transverse waves, and many phenomena such as scattering and reflection show the expected dependence on the direction of the electric vector of the e-m wave. Unpolarized light consists of many short wave trains vibrating in random orientation and phase. Such light is equivalent to equal proportions of two plane-polarized beams whose vibrations are parallel to any two arbitrary mutually perpendicular directions. The passage of light through anisotropic crystals shows a marked dependence upon polarization.

■ CHECK LIST

interference	$n\lambda = w \sin \theta$	Michelson-Morley
$n\lambda = a \sin \theta$	phase change upon reflection	experiment
double-slit pattern	phase difference	coherence
monochromatic light	nth order	polarized light
single-slit pattern	colors of thin films	Brewster's angle
diffraction	Michelson interferometer	

■ QUESTIONS

25-1   What additional experimental fact must be known before the refraction of light can be considered an adequate proof of the wave nature of light?

25-2   What is the path-length condition for the constructive interference of two rays of light?

25-3   Two radio antennas are emitting waves which are in phase at the towers. An observer is 20 000 m from one tower and 22 000 m from the other tower. The wavelength is 500 m. Is reception "good" at the observer's location?

25-4 What change, if any, takes place in the pattern on the screen of Fig. 25-3 if the entire apparatus is immersed in a transparent liquid of index of refraction n?

25-5 If white light shines on two slits as in the upper part of Fig. 25-3, what color (red or violet) appears closest to O in the 1st-order band of color?

25-6 Locate a point in Fig. 25-11 which is 7λ from the left-hand source and 10λ from the right-hand source. Locate a point which is $7\frac{1}{2}\lambda$ from the left-hand source and 10λ from the right-hand source.

25-7 What is the difference, if any, between the interference pattern of a pair of slits separated by 0.001 cm and the pattern of a grating having 1000 slits per cm?

25-8 If you look at a distant line source through a narrow slit, the source appears to become broader as the slit is narrowed. Explain.

25-9 Hold a straight edge, such as a slide rule or a sheet of paper, about 2 ft from the lamp bulb of a desk lamp, and look closely at the edge of the shadow cast on a surface held several inches from the obstacle. Can you explain why the shadow is not perfectly sharp?

25-10 Instead of using the arrangement of Fig. 25-3, in which a pair of slits is illuminated by light from a single slit, an experimenter attempts to get an interference pattern by looking at a pair of sources such as the headlights of a distant automobile. The path length for light from one headlight bulb is exactly 3 wavelengths longer than the path length from the other bulb. The experimenter expects to get an interference maximum, but is disappointed. Explain why no interference takes place under these circumstances.

25-11 In Fig. 25-31, which surface of the diagonal mirror M is the half-silvered one?

25-12 As a soap bubble evaporates, its surface appears black just before it breaks. Explain in detail why it is dark and not bright.

25-13 A film of oil spreads on the surface of a wet pavement (Fig. 25-39), and if it is thin enough the thickness is much less than the wavelength of light. The oil slick appears bright at its thinnest portion. Are the two reflected rays in phase or out of phase? Is the index of refraction of the oil greater or less than 1.33?

25-14 What fact about the nature of light is proved by the phenomenon of polarization?

25-15 A vertical beam of light polarized with the electric vector in an east-west plane travels through a dilute milky liquid and is scattered by small globules of fat (Fig. 25-40). Which observer, A or B, sees a more intense beam of scattered light? Is the observed beam polarized, and, if so, what is the direction of its electric vector?

25-16 Describe the use of Polaroid (*a*) as an analyzer, and (*b*) as a polarizer.

25-17 Refer to Fig. 10-3, in which polarization of waves on a string is illustrated. Which slit is the analyzer and which is the polarizer?

25-18 What is the difference between interference and diffraction?

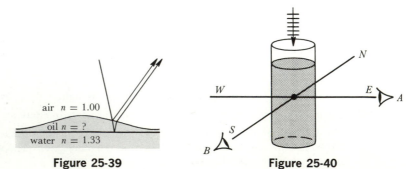

Figure 25-39   **Figure 25-40**

■ PROBLEMS

25-A1 Two slits 4.8×10^{-6} m apart, illuminated by parallel rays, form the 4th-order bright fringe at an angle of 30° from the direct beam. What is the wavelength of the light used?

25-A2 A coarse grating designed for use in the infrared region of the e-m spectrum has slits 0.5 mm apart. What is the sine of the angle of deviation of the 3rd-order image, for radiation whose wavelength is 50 000 Å?

25-A3 Light of wavelength 6×10^{-7} m falls on a pair of slits and the 3rd-order bright fringe occurs at an angle of 30° from the direct beam. How far apart are the slits?

25-A4 (a) A slit 2×10^{-4} cm wide is illuminated perpendicularly with light of wavelength 5000 Å. What is the angular deviation of the first *dark* band on either side of the central maximum? (b) What is the angular deviation of the first *bright* band on either side of the central maximum? (*Hint:* The bright band is approximately halfway between the 1st and the 2nd dark bands.)

25-A5 When light of wavelength 6000 Å falls perpendicularly on a single slit, the first two dark fringes (one on each side of the central bright fringe) are separated by 60°. How wide is the slit?

25-B1 A radio transmitter A operating at 60 MHz is 20 m north of another similar transmitter B which is 180° out of phase with A. How far east of A is the first position where the two beams are out of phase?

25-B2 A small ultrasonic transducer (loudspeaker) and a receiving transducer are 20 cm apart on a table top. A horizontal sheet of plywood 3 cm above the table serves as a reflector. The speed of sound is 340 m/s. What are the three lowest ultrasonic frequencies for which destructive interference takes place?

25-B3 Radio waves of frequency 600 kHz are received at a location 10.0 km from the transmitter. The radio reception temporarily "fades" due to destructive interference between the direct beam and that reflected from a horizontal layer of charged particles (ions) formed in the atmosphere by a passing aircraft. Calculate the minimum height of the aircraft.

25-B4 An observer looks at a source through a grating having 1000 slits per cm. If λ is 600 nm, how many images of the source can be seen? (*Hint:* For what order is $\theta = 90°$?)

25-B5 A grating has 20 000 slits spread out over 10.0 cm.   (a) For what wavelength of light, incident normally as in Fig. 25-14, will the angle between the two 5th-order maxima be 90°? What color is this light?   (b) What is the maximum wavelength observable in the 5th order using this grating? Is this radiation visible?

25-B6 (a) Sunlight falls perpendicularly on a grating having 1000 slits per cm. The rays leaving the grating are focused on a screen 4.00 m from the grating. How far from the central image on the screen is the blue region of the 2nd-order spectrum? (Use an average blue wavelength from Table 24-1.)   (b) Calculate the width of this blue region on the screen. (Find the deviation for each of the two extreme blue wavelengths in Table 24-1.)

25-B7 Light from a flame in which sodium chloride has been placed has an average wavelength 5893 Å. What is the 2nd-order angular deviation of a ray of this light which strikes perpendicularly a grating having 3000 slits per cm?

25-B8 Parallel rays of green light of wavelength 5000 Å strike perpendicularly a grating having 1000 slits per cm. The rays leaving the grating are focused by the eye lens on the retina, which is 2 cm from the eye lens. How far apart, measured on the retina, are the central image and one of the 3rd-order images?

25-B9   In the two parts of Fig. 25-14, $a = 1.00$ μm and $\lambda = 500$ nm. Show that the difference $\theta' - \theta$ is only about $2'$. (*Hint:* Find θ and θ' separately; interpolate in the table of sines in the Appendix as necessary.)

25-B10   Prove that if the wavelength is equal to or greater than the width of a slit, light striking the slit perpendicularly passes through without forming any dark interference bands.

25-B11   Light of wavelength 4.8×10^{-7} m falls perpendicularly on a slit of width 8.0×10^{-7} m. What is the width of the most intense central bright band, on a screen 2 m from the slit? (*Hint:* Most of the energy lies between the two 1st-order minima.)

25-B12   Sound waves of frequency 2000 Hz from a distant outside source enter a tall, narrow open window perpendicularly. The window is 2.2 ft wide and is in the center of a wall of a room 20 ft on an edge. The room is acoustically padded to eliminate reflections from the walls, and a number of observers are seated in a row along the wall opposite the window. How far from the corners of the room are the two observers who hear nothing? (Velocity of sound is 1100 ft/s.)

25-B13   Light of a certain color strikes perpendicularly a slit of width 2.7×10^{-6} m, and on a screen 3 m from the slit the central maximum is 1 m wide. What are the wavelength and color of the light? (*Hint:* The central image is bounded by the two 1st-order dark fringes.)

25-B14   An air wedge is formed by placing a thin blond hair between the edges of two plane glass plates which are in contact at the other end. When viewed by reflected light of wavelength 5461 Å, 78 dark bands are observed. Calculate the diameter of the hair.

25-B15   Two glass plates 15 cm long are in contact at one edge and are separated at the other edge by a piece of aluminum foil 0.0025 cm thick. When the plates are illuminated perpendicularly by light of wavelength 5000 Å,   (*a*) what is the separation of adjacent fringes? (*b*) How many dark fringes are seen?

25-B16   Two glass surfaces are separated by an air film of uniform thickness 9000 Å. If the film is illuminated perpendicularly by parallel light having wavelength 6000 Å, explain why no light of this color is reflected by the air film.

25-B17   Repeat Prob. 25-B16, for the same light illuminating a thin *glass* film of the same thickness as before, namely 9000 Å. The index of refraction of the glass film is 1.50. Do the reflected beams interfere constructively or destructively?

25-B18   A "nonreflecting" coated lens has a thin film of magnesium fluoride ($n = 1.25$) which is deposited on the surface of the glass ($n = 1.56$). What should be the thickness of the film in order that the reflected rays (Fig. 25-41) will be exactly out of phase for green light of wavelength (in air) 5000 Å? Express your answer in μm.

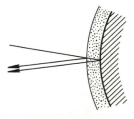

Figure 25-41

25-B19   (*a*) Study the lower part of Fig. 25-26 carefully, and decide at what point on the circumference the two pieces of glass are in contact.   (*b*) If the fringes are viewed using light of wavelength 4358 Å, estimate the thickness of the air film at the point diametrically opposite the point of contact. Express your answer in μm.

25-B20   A soap bubble of thickness 2300 Å has index of refraction $\frac{4}{3}$. Make an analysis similar to that of Example 25-4 to show that the film appears blue when viewed perpendicularly in white light.

25-B21   Light of wavelength 5461 Å is used to illuminate a Michelson interferometer. How many fringes pass the reference mark in the field of view when the movable mirror is shifted forward through a distance of 0.200 mm?

25-B22   The index of refraction of a glass plate is 1.52. What is Brewster's angle when the plate is   (a) in air?   (b) in water of index 1.33?

25-B23   What is the angle of the sun above the horizon when light reflected from a calm lake is most strongly plane-polarized?

25-C1   Light strikes a pair of slits at an angle θ_1 as in Fig. 25-42, and an interference maximum is formed at an angle θ_2. Derive a formula connecting θ_1, θ_2, a, and n for the nth-order maximum of constructive interference.

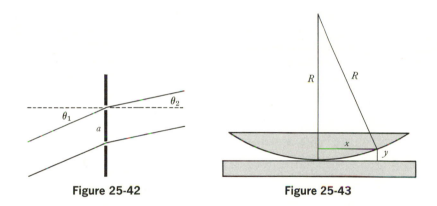

Figure 25-42   **Figure 25-43**

25-C2   Derive a formula for the radius x of the kth dark ring in a Newton's rings experiment. The symbols are defined in Fig. 25-43. [*Hints:*   (a) Obtain a formula for y in terms of R and x.   (b) Solve for x in terms of y (here you can assume that y^2 is negligible compared with Ry).   (c) For destructive interference, y must equal $k(\frac{1}{2}\lambda)$. Substitute this value of y into your formula for x.   (d) Draw the first six rings, if $R = 250$ m, $\lambda = 4 \times 10^{-7}$ m.]

25-C3   Using the formula derived in Prob. 25-C2, discuss the effect of replacing the air film by a film of liquid of index of refraction n, using the same glass surfaces and the same wavelength light. Could you find an unknown index of refraction for a liquid in this way?

25-C4   In a Newton's rings experiment with light of wavelength 500 nm (Fig. 25-43), the 10th dark circular fringe has a radius 2.0 cm when viewed from above.   (a) What is the radius of curvature of the lens? (*Hint:* Use the saggital theorem; see Sec. 24-12.)   (b) If the glass has index of refraction 1.50, what is the focal length of the lens?

25-C5   An experimenter wishes to determine the index of refraction of a piece of cellophane 0.0082 mm thick. He places the cellophane broadside to the beam in one arm of a Michelson interferometer and observes a shift of 18 fringes. He is using the mercury green line, $\lambda = 5461$ Å, as a source. Calculate the index of refraction of cellophane. (*Hint:* There are more wavelengths in 0.0082 mm of cellophane than in the 0.0082 mm of air which it has replaced. For the round trip, the difference is 18λ.)

25-C6   Prove that when light strikes a glass plate at Brewster's angle, the refracted ray is perpendicular to the reflected ray.

25-9 Double refraction

The transparent crystal calcite (Iceland spar) is composed of molecules of calcium carbonate, $CaCO_3$, with the calcium atoms arranged in planes in a crystal lattice. Such crystals show a natural cleavage (Fig. 25-44), with the surfaces making definite angles with each other. By successive cleavages, a calcite crystal can be split into smaller crystals, not necessarily with sides of equal length, but always with the same angles as the larger crystal. This cleavage shows that the cohesive forces (short-range electric forces) between the atoms are not equal in all directions. In other words, the crystals are anisotropic. This same structure leads to the optical phenomenon of double refraction. A single ray may (if properly oriented) split into two rays, as shown in Fig. 25-45. The *ordinary ray*, labeled *O*, behaves as we would expect it to; in the figure it is incident along the normal, and travels straight through the crystal. The *extraordinary ray*, labeled *E*, behaves very peculiarly; it bends *away* from the normal as it enters the crystal (a denser medium), and bends *toward* the normal as it goes from crystal back into the air—exactly contrary to Snell's law. Experiment also shows that the two beams are plane-polarized, with their electric vectors at right angles to each other.

These facts can be explained by Huygens' principle, with an additional assumption about the speed of propagation of light in the crystal. In calcite, the speed of light depends on the plane of vibration of the

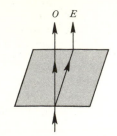

Figure 25-45 Passage of ordinary and extraordinary rays through a calcite crystal.

electric vector of the wave. A certain direction exists in the crystal, called the *optic axis*, such that *the speed is a maximum if the electric vector is parallel to the optic axis.* For calcite, the optic axis (O.A.) is the direction parallel to the line which makes equal angles with the three sides of the obtuse angles at *O* (Fig. 25-46). The O.A. is not just one line through a crystal; an optic axis can be drawn through *any* point in the crystal parallel to the axis through the corner. (Remember that any point inside the crystal could be made into a corner by proper cleavage.) A few of the optic axes of a certain calcite crystal are shown as dotted lines in Fig. 25-46.

The basic assumption is, now, that the speed of light depends on the orientation of the electric vector relative to the optic axis. Let us see how this assumption (which follows from the Maxwell theory of e-m waves) leads to double refraction. We resolve the incident (unpolarized) beam into two components polarized at right angles to each other and choose one component to have its electric vector in the plane determined by the optic axis and the incident ray (Fig. 25-47*a*). While the wave is traveling in air,

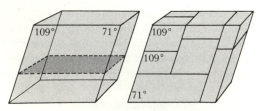

Figure 25-44

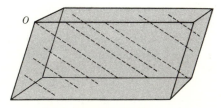

Figure 25-46

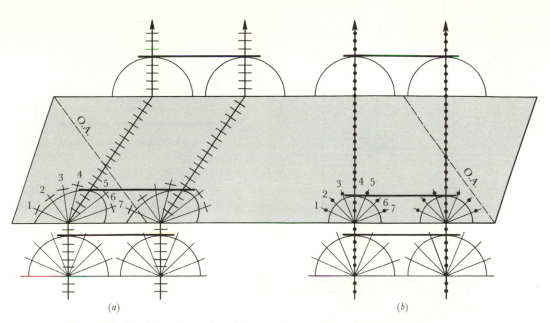

Figure 25-47 Wave fronts in calcite. (*a*) Extraordinary beam; (*b*) ordinary beam.

Huygens' principle predicts straight-line propagation, assuming that the pinhole which limits the size of the beam is large enough compared with λ so that diffraction can be ignored. When the beam reaches the calcite, each Huygens wavelet spreads out in an ellipsoid rather than a sphere. Ray 6 travels fastest, since its electric vector is parallel to the O.A.; ray 2 travels slowest, since its electric vector is perpendicular to the O.A. A similar construction is made for another ray in the incident beam, and the new wave front is drawn tangent to the wavelets. By continued application of Huygens' principle, we see that the beam is propagated through the crystal as shown, still a plane wave, but with the ray not perpendicular to the wave front. Eventually the rays reach the top surface, and wavelets spread out into air. Now, in air, the rays travel equally fast in all directions, and the wave front moves upward, parallel to the original direction. We have described how the extraordinary beam passes through the crystal; as a by-product of the analysis, we are led to expect that the *E* beam in Fig. 25-47*a* is polarized parallel to the plane of the paper.

The ordinary beam is represented by the other half of the incident beam, polarized as in Fig. 25-47*b*. In this case, each of the rays 1, 2, 3, 4, 5, 6, and 7 is vibrating perpendicular to the O.A., since each electric vector is perpendicular to the plane of the paper. Therefore the wavelets in the crystal are spherical, and the wave front moves through in the ordinary fashion. To obtain the complete phenomenon, we must imagine (*a*) and (*b*) superposed. The original unpolarized beam splits into two beams, polarized at right angles to each other.

Two special cases are of interest. If the crystal is cut and polished so that light can enter along the O.A. (Fig. 25-48), then no separation of the beams takes place, and they travel through the crystal with the same speed. If the O.A. is parallel to the surface, we again find that no separation takes place, but in this case the *E* beam travels faster than the *O* beam. Proof of this statement is left as an exercise (Prob. 25-C8).

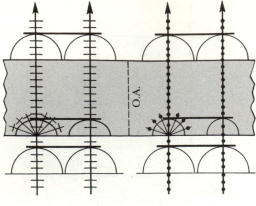

Figure 25-48

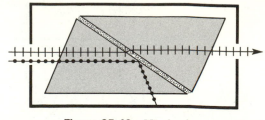

Figure 25-49   Nicol prism.

We have described double refraction in calcite, a typical "negative" crystal. In "positive" crystals, the O.A. is the direction of the electric vector for which the speed of a ray is *least* instead of *greatest*.[*] Still other crystals have two optic axes. Many crystals, such as NaCl, are isotropic, have no O.A., and are not doubly refracting. All the experimental results are rigorously predictable on the basis of the e-m wave model of light.

Since the O beam travels more slowly through calcite, the index of refraction is larger for it than for the E beam. This fact is used in the *Nicol prism,* a device for completely removing half of an initially unpolarized beam. The prism is constructed by cutting a piece of calcite diagonally and then cementing the pieces back together with Canada balsam (Fig. 25-49).[*] The indices of refraction are as follows: for calcite, $n_O = 1.658$, $n_E = 1.486$, and for Canada balsam, $n_O = n_E = 1.530$ (the balsam is an isotropic resin and hence is not doubly refracting). Note that the balsam's index lies between the two indices for calcite. Because of the construction, the rays strike the surface of the cut almost at a grazing angle. Total reflection takes place for the O beam (reflec-

tion at a less dense medium, at an angle greater than the critical angle), but the E beam enters the balsam (it bends toward the normal in a more dense medium) and then re-enters the calcite and eventually emerges into air. The Nicol prism can be used as a polarizer or as an analyzer; it serves the same purpose as a piece of Polaroid film.

25-10   Optical activity

Certain substances have the property of rotating the plane of polarization of a beam of plane-polarized light. This property is called *optical activity*. Quartz, some sugars, and many other organic and inorganic solids and solutions are optically active, but, on the other hand, calcite is not active, in spite of its strong double refraction. In a typical experiment (Fig. 25-50), a beam is rendered plane-polarized by a piece of Polaroid or a Nicol prism P_1, and a second Polaroid or Nicol prism P_2 is adjusted so that no light is transmitted. When the sample is inserted, light reappears, and the analyzer P_2 must be rotated through an angle θ in its own plane to make the intensity zero again. The angle of rotation depends on the material and the path length and is also strongly dependent upon wavelength. If no wavelength is specified, the yellow sodium "D-lines" of average wavelength 589 nm are understood. The specific rotation is measured in deg/mm for a solid, or deg/decimeter for a solution of concentration 1 g/cm^3. Typical values are 22 deg/mm for quartz and 77 deg/dm for a solution of nicotine in water.

---

[*] Ice is such a crystal.

[*] For the exact details of the construction of a Nicol prism, including a slight beveling of the front and back surfaces which we have ignored, see any advanced physics text.

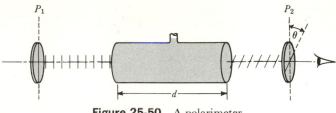

Figure 25-50 A polarimeter.

Optical activity apparently is related to a spiral structure of the individual molecules or of the crystal. Many substances, including quartz, can exist in either a right-handed or a left-handed form. The two sugars *d*-glucose (dextrose) and *l*-glucose are identical except for the direction of rotation. Since the rotation is proportional to the total number of molecules per unit length of path, the concentration of sugar in a commercial syrup can be tested by this means.

Glass and some transparent plastics become optically active when subjected to mechanical stress. Figure 25-51 is a photograph of a plastic model of a bone structure, loaded at one point. The model is placed between Polaroids, and the stress pattern is made visible by the corresponding rotation of the planes of polarization in the stressed regions of plastic. Some substances also become doubly refracting when stressed.

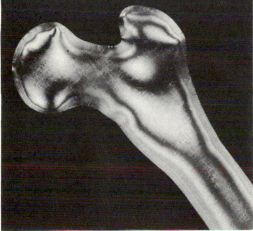

Dr. Henry Milch

Figure 25-51 Photoelastic stress analysis of a model of human bone, using polarized light.

25-11 Holography

Images can be formed without the use of lenses by a process known as *holography*. Unlike images that are formed by lenses or mirrors, a holographic image is truly three-dimensional; it is the result of the reconstruction of the wave fronts which would have reached the eye after an incident beam is scattered by points on the object. To form a hologram the object (traditionally a chess piece!) is illuminated by a broad beam of monochromatic coherent light from a laser. Light scattered from the object strikes a photographic film (Fig. 25-52); every point on the film receives light from every point on the object. Simultaneously, a portion of the illuminating beam is removed by a beam splitter (such as a half-silvered mirror similar to that used in the Michelson interferometer) and passes directly to the film. The film, when developed, is called a hologram; the resultant density at any point of the hologram depends on both the amplitudes and phases of all interfering wave fronts that reach that point.

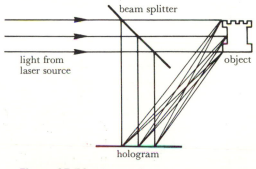

Figure 25-52 Production of a hologram.

To view the picture later, the image is reconstructed by placing the hologram in a laser beam; both a real and a virtual image are formed, in depth (Figs. 25-53 and 25-55). When the viewer's eye is moved from side to side, the "near" parts of the three-dimensional scene are seen to move[*] relative to the more distant parts. It is even possible to make holograms which allow the viewer to move in a 360° path around the image, seeing all sides of the image in a true 3-D representation.

The theory of holography is mathematically complicated. However, we can use our knowledge of interference to gain some insight into what takes place. In Fig. 25-54, XP and $X'P'$ represent parallel rays of a reference beam, and OP and OP' represent rays scattered by some point O on the object. The wave fronts interfere at P to produce a set of fringes; the exposure of the plate varies sinusoidally as in the upper part of Fig. 25-16, with the spacing between maxima dependent on the angle ϕ between the arriving rays. At another point P' closer to the object, the developed plate will also have fringes that are locally sinusoidal, but with a larger spacing between maxima. The hologram of this point source would be as shown at the right of Fig. 25-54.

Now what happens when a single broad reference beam falls on the hologram? Any small portion of the hologram is a grating, but only the 0th-order (straight-through

---

[*] This relative motion is called parallax.

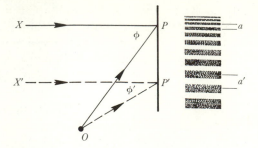

Figure 25-54 Production of hologram of a point object O which is not infinitely far from the photographic emulsion.

beam) and the two 1st-order interference maxima are observed. This is because of the *gradual* (sinusoidal) variations of opacity. (Gratings with sharply bounded openings, as in Fig. 25-13, give *all* orders of interference, as in Fig. 25-16.) Parallel rays (a plane wave front) strike different parts of our hologram "grating," with results shown in Fig. 25-55. B and B' are the 0th-order maxima. A and A' are 1st-order interference maxima, but since the grating space a' at P' is greater than the spacing a at P, the interference angle given by $\sin \theta' = \lambda/a'$ is smaller (see Eq. 25-1; here the order n is 1). Thus $\theta' < \theta$. Rays A and A' form a virtual image which can be seen by an observer who

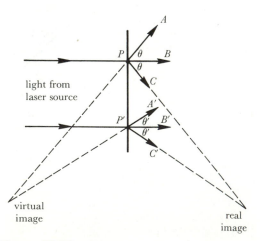

Figure 25-55 Formation of real and virtual images by 1st-order diffracted waves from a hologram.

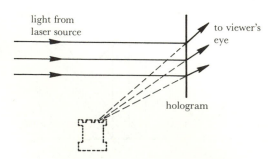

Figure 25-53 Viewing a holographic image.

simply looks through the hologram at a laser beam. The other 1st-order interference maxima C and C' form a real image which can be captured on a photographic plate or viewed with an eyepiece. If the hologram is cut into several pieces, each piece gives the same images as the entire hologram (although with less sharpness), for it is only the *relative* changes in grating space along the hologram that give rise to the images.

We have considered an object which is a single point source. Actual two- or three-dimensional objects consist of many such points; the resulting hologram is then a jumbled-up superposition of "gratings" (Fig. 25-56). Nevertheless, such a hologram, or part of one, can reconstruct the amplitude and phase of the light which originally fell on it from the object (Fig. 25-57). Thus virtual and real images are formed of every point on an extended object.

In our example, a coherent source of light is needed so that the phase difference be-

![image]

William Vandivert, *Scientific American*

Figure 25-57 Reconstructed image from hologram similar to Fig. 25-56c.

tween two rays arriving at P does not fluctuate randomly with time, thereby "washing out" the pattern on the film. The image can be enlarged simply by viewing the hologram in divergent light of wavelength longer than that with which it was made. If coherent x-ray beams can ever be achieved, high resolution can be obtained, and it will be possible to "see" (visually) biological or other structures much smaller than those it is now possible to see using visible light. Holography was first described in 1947 by Dennis Gabor, who received the 1971 Nobel Prize in physics for his pioneering work. With the advent of strong coherent light sources (lasers; see Sec. 28-4) the method has been applied in many areas of physics, chemistry, biology, and engineering. Many of these applications of holography are described in the references at the end of the chapter.

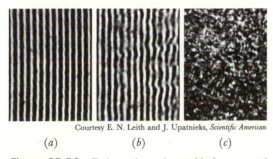

Courtesy E. N. Leith and J. Upatnieks, *Scientific American*

    (a)   (b)   (c)

Figure 25-56 Enlarged portions of holograms of various objects. (a) Very distant point object. (b) Simple transparency as object. (c) Complex diffusely scattering object.

■ PROBLEMS

25-C7 What is the ratio of the speed of the E ray to that of the O ray in calcite? Which travels faster?

25-C8 Prove that, if an unpolarized beam strikes the surface of a crystal whose O.A. is parallel to the surface, the O and the E beams do not separate, but one of them travels faster than the other. (*Hint:* Make a diagram similar to Fig. 25-48, but with a different direction of the O.A.)

25-C9 A polarimeter tube 10 cm long (Fig. 25-50) contains dextrose at a concentration of 2.4 g/cm³. The rotation θ is 126°.   (*a*) Calculate the specific rotation for dextrose. (*b*) What is the concentration of a dextrose syrup which gives a rotation of 72° in the same apparatus?

25-C10 Can a micromole of strychnine ($C_{21}H_{22}N_2O_2$) be detected in a polarimeter tube 2 mm in diameter and 20 cm long, if a rotation of 0.1° can be measured? The specific rotation for strychnine is 128 deg/dm for 1 g/cm³.

■ REFERENCES

1   Magie, W. F., *A Source Book in Physics* (New York: McGraw-Hill, 1935), pp. 294–318. Selections from the original writings of Grimaldi, Newton, Young, and Malus.

2   Cajori, F., *A History of Physics,* rev. ed. (New York: Macmillan, 1929), pp. 89–97. Chiefly Newton's experiments with prisms; his rejection of the wave theory of light. See also pp. 148–71, where the development of the wave theory of light is discussed.

3   Cohen, I. Bernard, "The First Explanation of Interference," *Am. J. Phys.,* **8,** 99 (1940).

4   Pohl, R. W., "Discovery of Interference by Thomas Young," *Am. J. Phys.* **28,** 530 (1960).

5   Sparberg, E. B., "Misinterpretation of Theories of Light," *Am. J. Phys.,* **34,** 377 (1966).

6   Shankland, R. S., "The Michelson-Morley Experiment," *Sci. American,* Nov., 1964, p. 107.

7   Lemon, Harvey B., "Albert Abraham Michelson: the Man and the Man of Science," *Am. Phys. Teacher,* **4,** 1 (1936).

8   Meggers, William F., "Measurement by Mercury," *Sci. American,* Aug., 1948, p. 49.

9   Ingalls, Albert G., "Ruling Engines," *Sci. American,* June, 1952, p. 45. Construction of gratings of high quality.

10   Baumeister, P., and G. Pincus, "Optical Interference Coatings," *Sci. American,* Dec., 1970, p. 58.

11   Waterman, Talbot H., "Polarized Light and Animal Navigation," *Sci. American,* July, 1955, p. 88.

12   Leith, E. N., and J. Upatnieks, "Photography by Laser," *Sci. American,* June, 1965, p. 24.

13   Heumann, S., "How to Make Holograms," in C. L. Stong's "Amateur Scientist" department of *Sci. American,* Feb., 1967, p. 122.

14   Pennington, K. S., "Advances in Holography," *Sci. American,* Feb., 1968, p. 40.

15   Strickland, James, *Superposition of Pulses; Interference of Waves; Effect of Phase Differences Between Sources; Single Slit Diffraction; Multiple Slit Diffraction; Diffraction and Scattering Around Obstacles* (films).

16   Miller, Franklin, Jr., *Diffraction—Single Slit; Diffraction—Double Slit; Michelson Interferometer* (films).

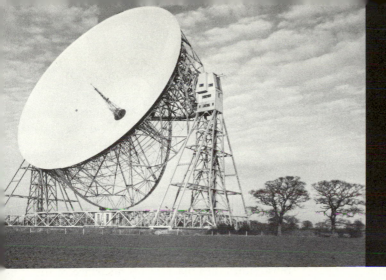

26
Applied Optics

In this chapter we study the applications of the theory of light to a number of common optical instruments. All these instruments use lenses, in one way or another, and so we shall continually rely on our work in geometrical optics (Chap. 24) to find the positions of images (by either of two methods) and their sizes (by the ratio of image distance to object distance). We shall not ignore the disturbing effects of dispersion, nor shall we overlook the inevitable diffraction which follows from the wave nature of light.

26-1   The camera: lens aberrations

The camera is simple in theory but is likely to be complicated in practice. In essence, a converging lens forms an inverted real image on a sensitized film or photographic plate; wherever light strikes, chemical PE is stored which can be released by proper chemical action during the process of developing. We do not wish to inquire further into the chemical and physical processes by which the small grains of metallic silver are caused to appear wherever enough PE has been absorbed by the molecules of the emulsion.

Our problem is to understand why the optical image may not be perfectly sharp, and how to improve it.

In our earlier work, we assumed (without proof) that a point source forms a point image. That is, all the rays which leave a source and pass through the lens come together at a single point. This is only an approximation to the truth, but one good enough for many purposes.[*] However, the precise way to study image formation is to trace the path of each ray, using Snell's law at each of the refracting surfaces. When this is done, a number of defects of a simple camera lens, called *aberrations,* appear. We shall discuss only three of the aberrations of a lens: spherical aberration, off-axis astigmatism, and chromatic aberration.

The surfaces of a simple lens are ordinarily sections of spheres; such surfaces are easy to construct by grinding and polishing. *Spherical aberration* is caused by the fact that such a lens has a longer focal length for rays near the center than for rays passing through an outer

---

[*] If you have studied Sec. 24-12, you will realize that the method of curvatures and the lens equation are based on the sagittal theorem, which is an approximation having greatest validity for thin lenses.

Radio telescope at the Jodrell Bank Experimental Station of the University of Manchester, in England. A tower supports a receiving antenna at the focal point of a metallic "dish" 250 ft in diameter. This telescope responds to electromagnetic radiations of wavelength much longer than that of visible light. (British Information Service)

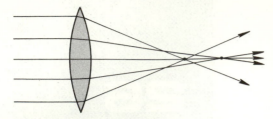

Figure 26-1 Spherical aberration.

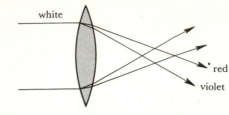

Figure 26-3 Chromatic aberration of a converging lens.

zone (Fig. 26-1). There is no one focal length, but rather a focal region. Most cameras are equipped with an adjustable stop, or diaphragm, which is primarily used to control the amount of light reaching the film. When the sharpest possible image is desired, the photographer reduces the effective diameter of the lens, and compensates for the loss of light by a longer exposure time.

Off-axis astigmatism is a troublesome defect which is noticeable for objects not on the axis of the lens, as in Fig. 26-2. The image of point *P* is *two* short lines: a vertical line *AB* and a horizontal line *CD* closer to the lens. The "focus" is a "circle of least confusion" somewhere between the two line images. To reduce astigmatism, a photographer sees to it that the important part of the picture is centered in the field of view, so that the object points lie near the axis of the lens. For more expensive cameras, systems of lenses have been designed to reduce off-axis astigmatism. The lens surfaces are symmetrical but not necessarily spherical.

A third major lens defect is *chromatic aberration,* which causes color-fringed images even for a lens of small opening and for object and image on the axis. This is one of the effects of dispersion. The index of refraction of glass varies with wavelength, so that a point source

of white light is spread out into a colored line. For simplicity, we consider in Fig. 26-3 rays from a distant source. Bearing in mind that refraction takes place at each surface of the lens, we see that the violet light, for which the index of refraction is greatest, is refracted the most. Thus the focal length is least for violet light and greatest for red light. Study of a diverging lens (Fig. 26-4) shows that the focal point for violet light is closest to the lens in this case too. Chromatic aberration can be corrected for two colors by using a combination of lenses as in Fig. 26-5, called an *achromatic doublet.* The converging lens tends to place the red image to the right of the violet one, while the diverging lens has the opposite effect. With proper choice of the radii of curvature of the surfaces, and with the use of two different kinds of glass for the two lenses, a perfect match can be obtained for any two colors. For visual work, in which the eye rather than a photographic plate is used to receive the image, it is customary to correct the lens for yellow and blue; this will give nearly perfect correction also for green

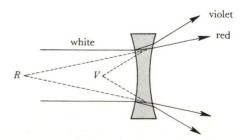

Figure 26-4 Chromatic aberration of a diverging lens.

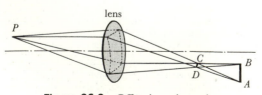

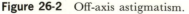

Figure 26-2 Off-axis astigmatism.

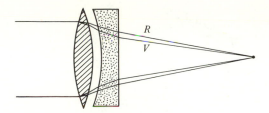

Figure 26-5   Achromatic doublet.

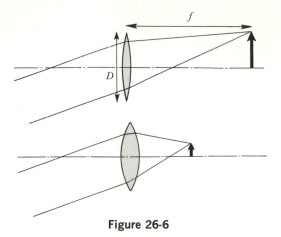

Figure 26-6

light, which lies between yellow and blue, but there may be some residual color at the extreme violet or red ends of the spectrum. An achromatic doublet is essential for a camera of high quality; in many cameras three colors are corrected by a triplet of three lenses.

Another important characteristic of a camera lens is its ability to form a bright image. The brightness or intensity* of an image depends upon the rate at which energy reaches the film. Other things being equal, the intensity of an image of given size is proportional to the area of the lens opening. That is, intensity $\propto D^2$, where D is the diameter of the lens. The focal length of the lens also affects the intensity. Consider two lenses, each forming an image of the same distant object (Fig. 26-6). We have

$$\frac{\text{Image height}}{\text{Object height}} = \frac{\text{image distance}}{\text{object distance}}$$

$$= \frac{f}{\text{object distance}}$$

The image distance in each case is the corresponding focal length, and so the image height is proportional to the focal length f, and the image *area* is proportional to f^2. The smaller image will be brighter, since the same *total* energy per second is shared by fewer molecules of the film. Putting these results together, we see that the intensity I is proportional to D^2/f^2. The important factor is the ratio of focal length to diameter (f/D), which

is called the *f*-number of the lens. A "fast" lens has a large diameter and a short focal length. A simple fixed-focus camera's lens is about $f/16$ or $f/22$, where $f/16$ means that $f/D = 16$. The smaller the *f*-number, the faster is the lens, and the more expensive because of the necessity of more elaborate corrections for various aberrations. A modern camera lens system may have an effective *f*-number of $f/1.6$ or less.

26-2   The human eye

The human eye is not much different in principle from a camera. A converging lens forms a real image on the retina (Fig. 26-7), and by a complex neural mechanism sensations are transmitted to the brain. The retina takes the place of the sensitized film. The camera can focus on objects at different distances from the film; this is accomplished by moving the lens back and forth to change the image distance as necessary. In the eye, however, this focusing adjustment, called *accommodation,* is managed somewhat differently. The lens is fixed in position, about 1.7 cm from the retina. Since it is made of a transparent, flexible tissue, its focal length can be changed by a set of small muscles which change the curvatures of the lens surfaces. In this way the equation $-1/p + 1/f = 1/q$ can be satisfied for any object (within limits), with q about

*Just as for sound waves, the intensity of light is defined as rate of flow of energy per unit cross-sectional area (Sec. 11-5).

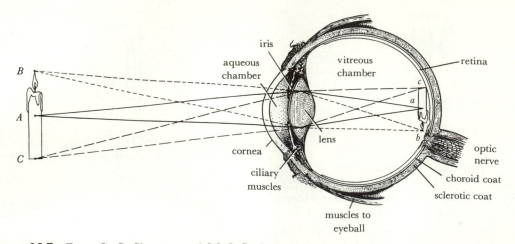

iris

aqueous
chamber

vitreous
chamber

retina

cornea

lens

ciliary
muscles

optic
nerve

choroid coat

sclerotic coat

muscles to
eyeball

Figure 26-7 From G. G. Simpson and W. S. Beck, *Life: An Introduction to Biology,* 2nd ed., Harcourt Brace Jovanovich, 1965.

1.7 cm for a "normal" eye. The accommodation of the lens for various object distances is shown *schematically* in Fig. 26-8. The power of the lens must be greater in (*b*), since the incoming wave front has greater divergence which must be changed into convergence.

Closer study of the eye, considered as an optical instrument, reveals a number of serious defects. Our diagram (Fig. 26-8) was oversimplified by omission of the cornea (Fig. 26-7), which is the hard, transparent protective layer which the light strikes first. Actually, most of the ability of the eye to form an image is due to the cornea and the fluid in the aqueous chamber. In a typical eye, these structures may have a power of +40 diopters, and the lens itself may be able to vary from +20 diopters to +24 diopters. The over-all power of this eye is thus +60 to +64 diopters. This variation is sufficient for most purposes, provided the object is not too close to the eye. If the lens is completely re-

moved, as in an operation for cataract, the patient can still see, although an auxiliary spectacle lens of power +20 diopters or so must be used, to compensate for the loss of the eye lens, and no accommodation is possible.

A common defect of vision is *astigmatism,* in which a point source forms a line image on the retina. The result is the same as for astigmatism of a camera lens, but the cause is different. A cylindrical lens has positive convergence for rays in only one plane, as shown in Fig. 26-9, since it is curved in only one dimension. A cylindrical lens is completely astigmatic;[*] in certain applications this is useful, but not in the camera or eye. Astigmatism of the eye may be caused by

---
[*] Greek: *a* = no, *stigma* = point.

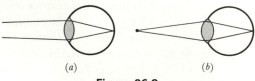

(*a*)

(*b*)

Figure 26-8

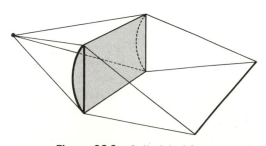

Figure 26-9 Cylindrical lens.

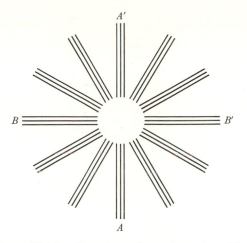

Figure 26-10 Chart for testing astigmatism of the eye.

imperfectly spherical surfaces of the cornea or lens. A quick test for astigmatism can be made with the help of the chart of Fig. 26-10. If, for instance, each point source makes a vertical line image on the retina, the figure will remain sharp in direction AA', but will be blurred in the direction of BB'. To correct for astigmatism, spectacle lenses are used which have a compensating cylindrical curvature, properly oriented. Off-axis astigmatism in the eye is not a problem because mammals and some other vertebrates have adapted to it by scanning the field of view to make the fovea (intersection of the optic axis of the eye and the retina) coincide with the image of any object to be examined in detail.

The other major defects of the eye are an inability to focus clearly on a very distant object (*myopia,* or nearsightedness) or on a very close object (*hypermetropia,* or farsightedness). To see how these defects can be corrected with the aid of spectacles, let us use the method of curvatures. We have seen that the power of the lens-cornea combination can adjust from +60 to +64 diopters in the "normal" eye. When accommodated for distant vision (incoming wave front having curvature of 0 diopters), the lens is completely relaxed, and the power of the eye is the least, namely, +60 diopters, or +60 m^{-1}.

The curvature equation reads

initial curvature	+	power of eye	=	final curvature
0	+	60 m^{-1}	=	$1/q$

This says that the final curvature $1/q$ is +60 m^{-1}; the radius of curvature is $\frac{1}{60}$ m, or about 1.67 cm, just right to converge to the retina. How close can an object approach the normal eye and still be focused on the retina? As the object approaches, the muscles operate to increase the power of the lens-cornea combination from +60 to +64 diopters. While this happens, the curvature of the incident wave front must change from 0 to −4 diopters (Fig. 26-11). At the *near point,* the equation reads

$$-4 + 64 = +60$$

and the final curvature is the same as before (as it must be, if the wave front is to converge on the retina, which is still $\frac{1}{60}$ m from the lens). We see that the near point is about $\frac{1}{4}$ m (25 cm) from the normal eye. (Try it.)

Consider now the myopic eye, for which the eyeball is longer than usual. Even with the lens completely relaxed (Fig. 26-12a), the image is in front of the retina. To obtain a sharp image on the retina, the final curvature must be *less* than +60 m^{-1} (longer radius of curvature)—say, +55 m^{-1}. The required spectacle lens is negative, and the curvature equations in Fig. 26-12b show that a lens of power −5 diopters will allow sharp vision of

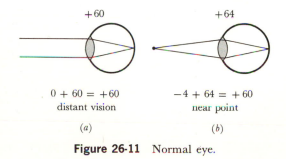

+60   +64

0 + 60 = +60   −4 + 64 = +60
distant vision   near point

(a)   (b)

Figure 26-11 Normal eye.

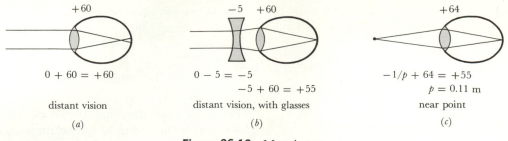

$$+60$$
$$0 + 60 = +60$$
distant vision
(a)

$$-5 \quad +60$$
$$0 - 5 = -5$$
$$-5 + 60 = +55$$
distant vision, with glasses
(b)

$$+64$$
$$-1/p + 64 = +55$$
$$p = 0.11 \text{ m}$$
near point
(c)

Figure 26-12 Myopic eye.

a distant object.* Without glasses, this particular myopic eye has a near point that is calculated in Fig. 26-12c; $-1/p = -9 \text{ m}^{-1}$, and $p = 0.11$ m instead of the normal 0.25 m. This nearsighted person can see distinctly an object that is only 11 cm from his eye.

Hypermetropia can be analyzed in a similar fashion. The eyeball is too short, and a positive lens is used to help the eye lens. In Fig. 26-13 it is assumed that the lens is 1.59 cm from the retina ($\frac{1}{63}$ m), and so the final curvature must be $+63$ diopters for a sharp image. When the near point is calculated as in Fig. 26-13c, the eye lens is made as bulgy as possible, but p works out to be 1 m instead of the normal 0.25 m. This hypermetropic person cannot see an object clearly without glasses unless it is at least 1 m from the eye.

In middle age, most persons gradually lose some of their power of accommodation, as the

eye muscles become weaker. We have stated that the "normal" eye can add 4 diopters to its power by strong muscular effort. However, this ability depends on age; it is not unusual for a child of 10 to be able to add 15 or 20 diopters, while at age 75 the muscles may be able to add only 1 diopter, or even less. Thus people tend to become farsighted as they grow older—a condition known as *presbyopia*. The remedy is the same as for hypermetropia, which is to help out the eye's own lens by adding a positive spectacle lens. When the ability to accommodate is lost, bifocals are worn so that the total power of the system can be altered as required for near and distant vision.* In an inexpensive camera, the lens is of fixed power, and a positive "portrait lens" can be added if it is desired to photograph a nearby subject; this serves the same function as do spectacles for a human who has lost his power of accommodation. Television and movie cameras use "zoom" lenses of variable focal length, in

* Throughout this section we assume that the spectacle lens is close to the eye, so that there is no appreciable change of curvature as the wave front travels between the two lenses.

* Trifocal glasses are by no means uncommon; the upper part of the lens is used for distant vision, the middle part for middle distances, and the lower part for reading.

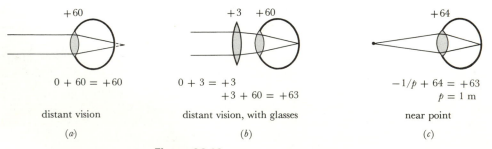

$$+60$$
$$0 + 60 = +60$$
distant vision
(a)

$$+3 \quad +60$$
$$0 + 3 = +3$$
$$+3 + 60 = +63$$
distant vision, with glasses
(b)

$$+64$$
$$-1/p + 64 = +63$$
$$p = 1 \text{ m}$$
near point
(c)

Figure 26-13 Hypermetropic eye.

which the power is changed continuously by changing the spacing between the elements of the lens system.

Pursuing the analogy of the eye and the camera, we note the presence in the eye of a diaphragm, the iris, which is controlled by a set of involuntary muscles as the intensity of the light changes. Chromatic aberration is, fortunately, a minor problem in the eye, although it can be observed. The image on the retina is inverted, as in a camera, but the brain has learned to present a "right-side-up" picture to the consciousness.[*] A simple experiment to prove this is described as follows: Use the point of a pencil to punch a small hole through a piece of thin cardboard or stiff paper. Holding the paper about 4 in. from the eye, look through the hole at a distant lamp or window. Place the pencil point as close to the eye as possible (within an inch) and move it up and down. No lens action is possible, since the object is too close to the eye. When you move the pencil upward, the shadow on the retina also moves upward, but the brain interprets this as a downward motion. Looking through the hole, you will see the shadow appear to move downward.

26-3 The magnifier

The actual size of an object is of small consequence in determining how big it appears to be. For instance, a $\frac{3}{4}$ in. penny held at arm's length (2 ft) looms as large as a 9 in. pie plate at 24 ft, and much larger than a 2000 mi full moon at 240 000 mi. The *angular size* of the object is what counts, and we can use the ratio h/d as a measure of angular size. (This is really the tangent of the angle θ in Fig. 26-14.) Let us compute the angular size of the three objects mentioned above.

[*] A relatively short time is required for the brain to reorganize its "switchboard"; if a subject wears glasses which present everything upside down, he begins to see his world right side up in a week or so. If, after several weeks, the inverting glasses are removed, everything looks upside down again for another week or more.

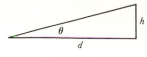

Figure 26-14

Penny: $\dfrac{h}{d} = \dfrac{\frac{3}{4} \text{ in.}}{2 \times 12 \text{ in.}} = \dfrac{1}{32}$

Pie plate: $\dfrac{h}{d} = \dfrac{9 \text{ in.}}{24 \times 12 \text{ in.}} = \dfrac{1}{32}$

Full moon: $\dfrac{h}{d} = \dfrac{2000 \text{ mi}}{240\,000 \text{ mi}} = \dfrac{1}{120}$

The penny and the pie plate have the same angular size and are almost 4 times as "large" as the moon.

The usefulness of a magnifier, a microscope, or a telescope lies in its ability to magnify the *angular* size of an object which is viewed. In Chap. 24 we considered linear magnification m, defined as the ratio (image height)/(object height). Now we define *angular magnification* and denote it by M:

$$M = \text{angular magnification}$$
$$= \frac{\text{angular size of image}}{\text{angular size of object}}$$

The angular magnification produced by an optical instrument is often called its *magnifying power*.

The obvious way to make a small object appear larger is to increase its angular size by bringing it close to the eye. If a stamp collector places a stamp 2 cm from his eye, it seems large enough, to be sure, but the image is hopelessly blurred (try it). This is because the near point of a normal eye is about 25 cm. To see the stamp better, the collector uses a magnifying glass, which is a single converging lens. When he places the stamp just inside the focal point of the lens, a virtual image is formed at least 25 cm from the eye (Fig. 26-15). As seen from the ray diagram, the object and image have the same angular size as seen by the eye. The lens, in effect, moves the stamp back to a distance of

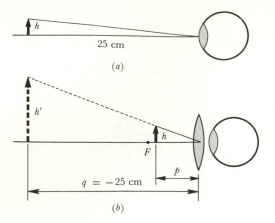

Figure 26-15 (*a*) Without aid, object is of small angular size. (*b*) With aid of magnifier, same object has much larger angular size.

25 cm or more, without at the same time decreasing its angular size.

To compute the magnifying power of the lens, we compare the angular sizes with and without the lens. We first use the method of curvatures to find the object distance; we know that the image is a virtual one, and so the image distance is $q = -25$ cm:[*]

$$-\frac{1}{p} + \frac{1}{f} = \frac{1}{-25}$$

$$\frac{1}{p} = \frac{1}{25} + \frac{1}{f} = \frac{f + 25}{25f}$$

whence

$$p = \frac{25f}{f + 25} = \frac{f}{1 + f/25}$$

As expected, the magnitude of p is just a bit less than f, since the denominator is slightly greater than 1. To find the image size we multiply the object size h by the linear magnification m. According to Eq. 24-4, $m = -q(1/p)$. Therefore

---

[*] In this and similar problems, distances are in cm unless otherwise specified, and the near point is assumed to be 25 cm from the eye.

$$m = -q\left(\frac{1}{p}\right) = -(-25)\left(\frac{f + 25}{25f}\right)$$

$$= \frac{f + 25}{f} = 1 + \frac{25}{f}$$

This gives

$$h' = mh$$

or

$$h' = \left(1 + \frac{25}{f}\right)h \qquad (26\text{-}1)$$

The magnifying power of a magnifier is given by

$$M = \frac{\text{greatest possible angular size (aided)}}{\text{greatest possible angular size (unaided)}}$$

Now we *could* have moved the stamp to within 25 cm of our unaided eye, but no closer (for then we couldn't have seen it clearly). The unaided maximum angular size is therefore $h/25$. Using the magnifier, we have a larger stamp of size h', also at 25 cm, and so the new angular size is $h'/25$.

$$M = \frac{h'/25}{h/25} = \frac{h'}{h}$$

and so we see that in this case the angular magnification is the same as the linear magnification. Using Eq. 26-1 gives

$$M = \frac{h'}{h} = \frac{\left(1 + \frac{25}{f}\right)h}{h} = 1 + \frac{25}{f}$$

Thus, finally, we have found a formula for the greatest angular magnification, or magnifying power, of the lens. If we choose to relax the eye and move the virtual image back to infinity (by placing the object at the focal point of the lens), the image becomes larger, but the angular magnification decreases slightly to $25/f$ in the limit. A numerical example will serve to illustrate these principles, using the lens equation in the "curvature" form $-1/p + 1/f = 1/q$. The "ray" form of the lens equation, $1/p + 1/q = 1/f$, could also be used.

EXAMPLE 26-1

A bug 3 mm tall is viewed by a person having normal vision, using a magnifier of focal length $+2$ cm. Calculate the angular magnification (a) when the lens is adjusted for maximum magnification, (b) when the virtual image is 50 cm from the lens, and (c) when the virtual image is at infinity.

(a) At maximum magnification, the virtual image is at the near point, -25 cm.

$$-\frac{1}{p} + \frac{1}{2} = -\frac{1}{25}$$

$$p = \frac{50}{27} = 1.85 \text{ cm}$$

The bug must be 1.85 cm from the lens, which is just inside the focal point. We next compute the linear size of the image from $h' = h(-q/p)$:

$$h' = h\frac{-(-25)}{\left(\frac{50}{27}\right)} = h(25)(\tfrac{27}{50}) = h(13.5)$$

The image is $(3 \text{ mm})(13.5) = 40.5$ mm high. Without using the lens, we could have placed a 3 mm bug 25 cm from the eye; with the lens, we have a 40.5 mm bug, also 25 cm from the eye. The angular magnification is

$$\frac{40.5 \text{ mm}/25 \text{ cm}}{3 \text{ mm}/25 \text{ cm}} = \boxed{13.5}$$

As a check, we use the formula for M:

$$M = 1 + \frac{25}{f} = 1 + \frac{25}{2}$$
$$= 1 + 12.5 = 13.5$$

(b) If we relax the eye muscles somewhat, to focus on a virtual image 50 cm from the eye, the object distance must be recalculated.

$$-\frac{1}{p} + \frac{1}{2} = -\frac{1}{50}$$

$$p = \frac{100}{52} = 1.92 \text{ cm}$$

Under these circumstances,

$$h' = h\frac{-(-50)}{\left(\frac{100}{52}\right)} = h(50)(\tfrac{52}{100}) = h(26)$$

The height of the image is now $(3 \text{ mm})(26) = 78$ mm. We have not gained much, however, for this larger image is farther away.

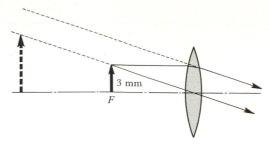

Figure 26-16

$$M = \frac{\left(\dfrac{78 \text{ mm}}{50 \text{ cm}}\right)}{\left(\dfrac{3 \text{ mm}}{25 \text{ cm}}\right)} = \boxed{13.0}$$

(c) If the object is *at* the focal point, the image is at infinity and is infinitely large. The angular size is not infinite (Fig. 26-16), but exactly equals that of a 3 mm object at F.

$$M = \frac{\left(\dfrac{3 \text{ mm}}{2 \text{ cm}}\right)}{\left(\dfrac{3 \text{ mm}}{25 \text{ cm}}\right)} = \frac{25}{2} = \boxed{12.5}$$

This illustrates how the angular magnification approaches $25/f$ as the virtual image recedes to infinity, for a person with "normal" vision.

Minimum eyestrain is achieved if a magnifier is used in such a way that the emerging rays are parallel, as in Fig. 26-16, for then the rays seem to come from infinity and the eye lens is most relaxed.

26-4 The compound microscope

To obtain higher magnification than is possible with one lens, we use a compound microscope. The object is placed near an *objective lens* L_1 of short focal length, and a real image A is viewed through an *eyepiece*, or *ocular*, L_2, which is a simple magnifier (Fig. 26-17). To form the real image at A, the object must be just beyond the focal point of L_1, and since the ratio (image distance)/(object distance) is large, the linear size of the real image A is considerably larger than that of the object.

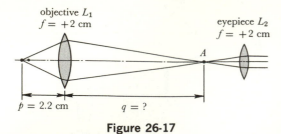

objective L_1
$f = +2$ cm

eyepiece L_2
$f = +2$ cm

A

$p = 2.2$ cm $q = ?$

Figure 26-17

Further angular magnification is obtained when the real image at A serves as an object and is viewed through the eyepiece.

EXAMPLE **26-2**

The objective and the eyepiece of a microscope each have focal length 2 cm. If an object is placed 2.2 cm from the objective, (a) calculate the distance between the lenses when the microscope is adjusted for minimum eyestrain; (b) calculate the magnification.

(a) First we calculate the position of the real image A (Fig. 26-17).

$$-\frac{1}{2.2} + \frac{1}{2.0} = \frac{1}{q}$$

whence $q = 22$ cm. For minimum eyestrain, the eyepiece is adjusted so that parallel rays emerge, and hence A is at the focal point of the eyepiece. The total separation of the lenses is

$$22 \text{ cm} + 2 \text{ cm} = \boxed{24 \text{ cm}}$$

(b) If the object size is h, the real image's size is

$$h' = h \frac{22 \text{ cm}}{2.2 \text{ cm}} = 10h$$

Since the eyepiece is adjusted for minimum eyestrain, its magnifying power is given by $25/f = 25/2 = 12.5$. The over-all magnifying power of the instrument is the product of the two magnifications:

$$M = (10)(12.5) = \boxed{125}$$

In theory, it would seem possible to use a series of lenses to make a real image as large as desired. For instance, the real image A of Fig. 26-17 could have been viewed through a compound microscope instead of a simple magnifying glass. However, the wave nature of light sets an upper limit to the usefulness of any image. When a lens forms an image of a point source, the lens serves somewhat as a single slit, and so the image is surrounded by a diffraction pattern whose angular size is *roughly* $\theta = \lambda/D$, where D is the diameter of the lens. This follows from Eq. 25-2 (Sec. 25-5), which states that the 1st-order minimum comes at an angle given by $\sin \theta = \lambda/w$; remember that for small angles $\sin \theta$ is approximately equal to θ. A circular opening such as a lens does not have a uniform width, however, and detailed analysis shows that a circular opening of diameter D is equivalent to a slit of width $0.82D$. The angular radius of the central part of the circular diffraction pattern is, therefore, given by $\theta = \lambda/0.82D$. Of course, for visible light λ is very small compared with the diameter of a microscope lens, but nevertheless there is an irreducible diffraction pattern surrounding each point of the image. For an objective of a given diameter, each point source makes a blurred image (Fig. 26-18), and two neighboring points of an object would be indistinguishable if their diffraction patterns overlapped. The ability of a lens to form sharp images is called its *resolving power;* we see that the resolving power of a microscope is limited

Figure 26-18 (a) Telescopic image of point source, showing diffraction disk and circular fringes. (b) Two point sources, barely resolved. (Reproduced by permission of Blackie and Son, Ltd., from *Light* by R. W. Ditchburn)

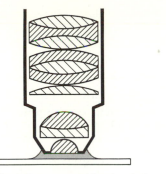

Figure 26-19 Microscope objective (schematic).

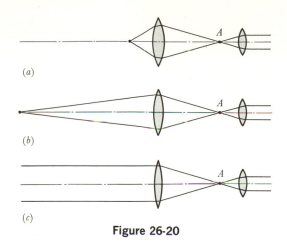

(a)

(b)

(c)

Figure 26-20

by the wave properties of light. Mere magnification is worthless beyond a certain point, for the user would only be magnifying an inherently blurred image. There are two ways to improve the resolving power of a microscope or telescope, and both methods make λ/D smaller: (1) The wavelength can be decreased (some microscopes use ultraviolet light partly for this reason), or (2) the diameter can be increased (this is one reason for large-diameter telescopes).

Our discussion of resolving power has implied that the various lens aberrations have been made so small that diffraction effects are the only remaining source of fuzziness. This is actually true for microscopes, as a result of careful design of the objective lenses. Figure 26-19 shows the objective lens of a modern microscope; it consists of no less than 10 components, highly corrected for many aberrations, and has an effective focal length of a few millimeters. To seek further improvement in a microscope using visible light would be comparable to trying to increase the efficiency of a heat engine beyond the Carnot value. It cannot be done; the nature of light limits the sharpness of the image.

26-5   The astronomical telescope

The difference between a telescope and a microscope is one of degree only. In each case a real image is formed which, in turn, is viewed through an eyepiece. The microscope of Fig. 26-20a forms at A a real image which is larger than the object, whereas the real image in a telescope is smaller than the object (Fig. 26-20b). The moon is several thousand miles in diameter, but its real image may be only an inch or so in diameter. However, the (smaller) real image has the great virtue that it is where we can get close to it; hence its *angular* size can be enormous. Usually a telescope is used to view a distant object (Fig. 26-20c), so the incident rays are practically parallel, and the real image A is at the focal point of the objective lens and also at the focal point of the eyepiece (when adjusted for minimum eyestrain).

To derive a formula for the magnifying power of a telescope, let us consider rays which come from one end P of the object, pass through the center of the lens, and form a real image P' in the focal plane of the objective lens whose focal length is F (Fig. 26-21a). The object is of height h, at some large distance p, and so its angular size as seen with the unaided eye is h/p. When the image h' is viewed with the eyepiece, of focal length f (Fig. 26-21b), the angular size of h'' remains the same as that of h' and is given by h'/f. Thus for the telescope as a whole,

$$M = \frac{\text{angular size of image}}{\text{angular size of object without aid}} = \frac{\dfrac{h'}{f}}{\dfrac{h}{p}}$$

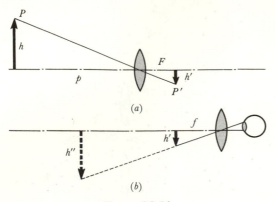

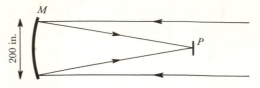

Figure 26-22 Reflecting telescope, curvature of mirror exaggerated for clearness. P represents a photographic plate.

(a)

(b)

Figure 26-21

But, by similar triangles, $\dfrac{h}{p} = \dfrac{h'}{F}$, and hence

$$M = \frac{\dfrac{h'}{f}}{\dfrac{h'}{F}} = \left(\frac{h'}{f}\right)\left(\frac{F}{h'}\right) = \frac{F}{f} \qquad (26\text{-}2)$$

When adjusted for minimum eyestrain for viewing a distant object, the magnifying power of a telescope is the ratio of the focal lengths of the two lenses.

EXAMPLE **26-3**

A comet-seeker's telescope has an objective lens of focal length 40 cm and an eyepiece of power 50 diopters. What is the magnifying power of the telescope?

First we find the two focal lengths, in the same units:

$$F = 40\ \text{cm} = 0.40\ \text{m}$$

$$1/f = 50\ \text{diopters} = 50\ \text{m}^{-1}$$

$$f = \tfrac{1}{50}\ \text{m} = 0.02\ \text{m}$$

$$M = \frac{F}{f} = \frac{0.40\ \text{m}}{0.02\ \text{m}} = \boxed{20}$$

This is a low-power telescope, which is desirable when a large area of the sky is to be scanned.

A concave mirror can be used instead of an objective lens to form the real image, as in Fig. 26-22. All the largest telescopes, such as the 200 in. Hale reflector at Mt. Palomar, are constructed in this way. For one thing,

it would be impossible to obtain an optically uniform piece of glass to make a lens of this size—almost 17 ft in diameter! If supported at the edges, such a lens would sag under its own weight. A mirror need not be transparent, for only the front surface counts, and it is made reflecting by coating with a thin film of silver or aluminum. A large mirror is usually cast in Pyrex or quartz, to take advantage of the low expansion coefficients of these substances; the back is ribbed to provide rigidity. Figure 26-23 shows some of the ways in which the real image can be brought out to an eyepiece. For most astronomical work, a photographic plate or photocell receives the real image, and no eyepiece is used. In the very largest telescopes, the observer can be stationed at the focal point without obstructing an appreciable fraction of the incoming light (Fig. 26-23c).

The reflecting telescope has the great advantage that its objective is entirely free of chromatic aberration, since no light travels through glass. Spherical aberration is eliminated by using a parabolic rather than a spherical mirror; this works well for rays coming in along the axis, but introduces other aberrations for off-axis rays.

The objective diameters are made large for two reasons. First, a large diameter makes λ/D smaller and hence reduces the size of the central "disk" of the diffraction pattern. However, the "geometrical" image of even the closest star (except the sun) is much smaller than the diffraction disk, and hence unobservable. Just as for a microscope, the sharpness of an image is limited by diffraction. This sets a practical limit to the useful

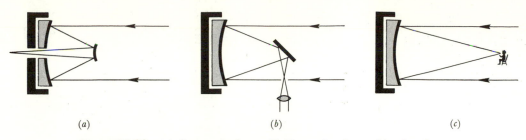

Figure 26-23 (*a*) Cassegrain focus; (*b*) Newtonian focus; (*c*) prime focus.

magnifying power of any telescope. Through careful design and workmanship, the theoretical limit can actually be reached for small telescopes, but the uneven index of refraction of the air, caused by turbulence in the atmosphere, is a serious limitation for large instruments. (The 200 in. telescope at Mt. Palomar has never been used at greater than $\frac{1}{10}$ the "theoretical" resolving power, through no fault of the instrument itself.) Telescopes of large diameter are used for a second reason: they gather more light and make a brighter image. The light-gathering power of a telescope is proportional to the surface area of the objective lens, given by the formula $A = (\pi/4)D^2$. The 200 in. Mt. Palomar telescope of 1951 thus gathers 4 times as much light from a given source as does the 100 in. Mt. Wilson telescope of 1921. Since the brightness of a distant object is inversely proportional to the square of the distance, this means that the larger telescope can "see" twice as far—the fourfold increase due to the increased area of the mirror compensates for the fourfold decrease in brightness due to distance. In this way, the volume of space explored by telescopes using visible light, proportional to the cube of the distance, was increased eightfold in three decades.

Radio telescopes use e-m radiation of wavelength 0.1 m or more, instead of the e-m radiation of wavelength 5×10^{-7} m which we call visible light. The British radio telescope at Jodrell Bank is shown on p. 567; note the parabolic "dish" which reflects radiation to the small receiving antenna at the focal point. Much of the complexity of the structure is due to the necessity of turning the telescope to compensate for the earth's rotation, if a fixed "radio star" is to be tracked. It is found that strong radiations are received from certain "points" in the sky, not necessarily associated with bright visible stars. For radio telescopes, as for optical telescopes, resolving power depends on the ratio (λ/D) of the wavelength to the diameter of the mirror. In spite of the fact that the huge dish of the Jodrell Bank telescope measures some 76 m across, its resolving power is poor. For e-m radiation of wavelength 0.21 m, λ/D is about 0.003, which is far larger than the corresponding ratio for a small optical telescope, or even the human eye. In fact, the diffraction pattern of a point source of 0.21 m wavelength would appear as large as the full moon, and two such sources 0.5° apart would barely be resolved. In spite of such limitations, radio astronomy gives information that optical astronomy cannot give, because e-m radiation having wavelength 0.5 m (for example) easily penetrates distant dust clouds which would completely absorb visible e-m radiation of wavelength 5×10^{-7} m. The energy-gathering power of large radio telescope installations gives astronomers their farthest view of distant space.

26-6 The Galilean telescope

The astronomical telescope suffers from two limitations: it gives an inverted image, and it is rather long and unwieldy. These are

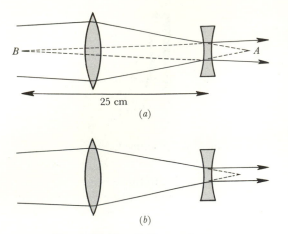

$B \cdots = = = =$ A

25 cm

(a)

(b)

Figure 26-24 Galilean telescope.

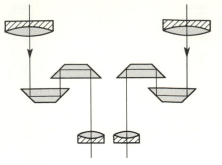

Figure 26-25 Prism binoculars.

not serious faults for astronomical work, but for other uses the original form used by Galileo is often preferable. An objective lens gathers light (Fig. 26-24*a*), but the converging rays are not allowed to reach their focus. Instead, a strong diverging lens changes the curvature to $-4 \ m^{-1}$, and a virtual image is formed at B, 25 cm from the eye. Such a telescope is compact and gives an erect image, but has a rather small field of view. When adjusted for minimum eyestrain, the rays emerge parallel to each other, as in Fig. 26-24*b*, and the virtual image is at infinity. The Galilean telescope is used for low-power instruments such as the opera glass.

In all telescopes, the magnifying power is greater for an objective of long focal length. This leads to the design of prism binoculars, where totally reflecting prisms are used instead of mirrors to give a longer path and allow the objective lens to have longer focal length (Fig. 26-25). The main function of the prisms is, however, to reinvert the image by multiple reflection. Even an inexpensive pair of binoculars makes use of achromatic lens combinations for objective and eyepiece. Numerous other forms of telescopes and binoculars exist. In the "spyglass" the inverted image is made erect by means of an additional lens, which requires a tube of unwieldy length.

26-7   The prism spectroscope

A spectroscope is used to analyze a beam of light by breaking it up into its component colors. Newton experimented with prisms while a student at Cambridge University. He found that white light can be broken up into the colors of the rainbow, and furthermore that any given spectrum color, red for instance, is "pure" and cannot be further modified by a second prism. Light of the various spectral colors, when recombined, gives white light. Newton did not follow up his important discovery, which might well have led to the spectroscope decades in advance of the work of others.

In Chap. 24 we gave Snell's law and stated that the index of refraction of glass is greater for violet light than for red. Thus we can "explain" the action of a prism in separating the rays of different colors, as in Fig. 24-36. This is no explanation at all, of course, but the full theory of dispersion is too advanced for inclusion in this text. Accepting the *fact* of dispersion (i.e., the dependence of n on λ), we now proceed to describe the prism spectroscope, which is illustrated in Fig. 26-26.

A slit S is illuminated by the source, and a *collimator* lens C makes the light parallel. This means that S is at the focal point of C; in the better instruments C is an achromatic combination, so that the focal length does not depend upon color. The parallel rays strike the prism, are bent variously, and

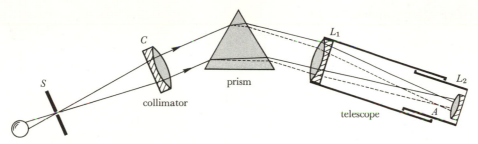

Figure 26-26 Prism spectroscope. Solid rays, red light; dashed rays, violet light.

emerge, still in parallel bundles. An objective lens L_1 forms real images at A, and the eye views these real images of the slit, one in each color, through an eyepiece L_2. The combination L_1 and L_2 is, of course, a telescope, and the observer uses the telescope to view the slit S, which has, in effect, been removed to infinity by the action of the collimator lens. For photographic work, the eyepiece is not used, and the film or plate is placed at A.

We often speak of spectrum "lines." Thus the strong "green line" of the mercury spectrum has a wavelength of 5461 Å. The images are lines only because the slit is tall and narrow. A curved slit would give curved real images at A, one in each color (Fig. 26-27). The thin straight slit is preferable, to avoid overlapping of the images. If the slit is made wider, *each* of the lines in Fig. 26-27a becomes wider, since they are all images of the same slit.

One criterion of a good spectroscope is that it be able to show as separate two colors which have almost the same wavelength. For a spectroscope, *resolving power* is measured by the thinness of the line image of a very narrow slit. Just as for a microscope or telescope, the diffraction of light sets the limit. The size of the prism is usually the controlling factor (Fig. 26-28), since the prism limits the width of the beam in the same way that a single slit would. Two adjacent lines are resolved if the width of each line due to diffraction is less than their separation due to the difference between the two indices of refraction for the two neighboring colors.

In use, a spectroscope is calibrated by light of known wavelengths. The photo on p. 609 shows the sun's spectrum (center), flanked on each side by lines of the iron spectrum which are emitted by an incandescent electric arc between two iron electrodes. In this photograph, the wavelengths are in the visible region. To study the ultraviolet region, for which glass is opaque, quartz prisms can be used down to about 1000 Å. At the other extreme, glass becomes opaque to infrared radiation at about 22 000 Å. Workers studying the far infrared region of the spectrum use prisms of rock salt, sylvite (KCl), and other substances which are transparent at the wavelength used.

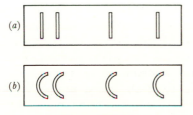

Figure 26-27

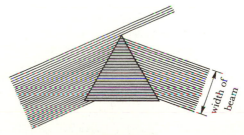

Figure 26-28

26-8　The grating spectroscope

We saw in Sec. 25-4 how a grating forms a series of bright images for any given wavelength. The angle of deviation for the nth-order interference maximum depends on the wavelength, according to the equation $n\lambda = a \sin \theta$. This means, of course, that a grating can replace a prism in a spectroscope. A simple grating spectroscope is shown in Fig. 26-29; the slit S, collimator C, and telescope

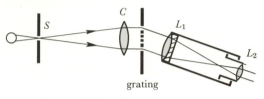

Figure 26-29　Grating spectroscope.

L_1L_2 are the same as for a prism spectroscope. Since $\sin \theta = n\lambda/a$, in any given order violet light (having the shortest λ) is deviated least, and red light is deviated most. Thus the grating spectroscope presents the colors in a sequence opposite to that of a prism spectroscope. The grating spectroscope furnishes several complete spectra, each line occurring in several orders.* The separation between two colors can be considerable, as shown in our next example.

EXAMPLE **26-4**

A grating of 4000 lines per centimeter is used to form a spectrum of a source which consists of two lines, a violet one of wavelength $\lambda_1 = 4000$ Å, and a red one of wavelength $\lambda_2 = 7500$ Å. Calculate $\sin \theta$ for the 1st- and 3rd-order images of each line. What is the angular separation between red and violet in the 3rd order?

The grating space is $\frac{1}{4000}$ cm = 25×10^{-5} cm = 25 000 Å. We compute $\sin \theta$ for each line, using $\sin \theta = n\lambda/a$; for instance, for the red line in the 3rd order:

---

*An interesting phenomenon known as overlapping of orders is considered in Prob. 26-B24.

$$\sin \theta = \frac{3(7500 \text{ Å})}{25\,000 \text{ Å}} = \frac{22\,500 \text{ Å}}{25\,000 \text{ Å}} = 0.900$$

$$\theta = 64°$$

We can tabulate the results as follows:

LINE	$\sin \theta$	θ
violet, 1st order	0.16	9°
red, 1st order	0.30	17°
violet, 3rd order	0.48	29°
red, 3rd order	0.90	64°

The angular separation between red and violet in the 3rd order is $64° - 29° = 35°$.

If we compare the grating spectroscope and the prism spectroscope, we find that each has its advantages and disadvantages. As seen from the preceding example, the dispersion (ability to spread out the colors) of a grating spectroscope can be large, especially in the higher orders. The dispersion of a prism spectroscope depends on the rather small variation of index of refraction, and a spread of 5° between violet and red lines is considered excellent. The dispersion of a grating depends on λ and the grating space a, which can be made as small as desired. The resolving power of a grating can also be made much higher than that of a prism; as shown in Sec. 25-4, the sharpness of the line depends on the total number of rulings, and on the order. In the 2nd order, a well-constructed grating 10 cm long, having 5000 lines per centimeter (a total of 50 000 lines), can separate two colors which differ in wavelength by only one part in 100 000! Still another advantage of the grating is that the lens can be dispensed with entirely, in order to study the extreme infrared or ultraviolet spectra, which would be absorbed by a glass or quartz lens. The spectrograph of Fig. 26-30 has a grating ruled on a concave mirror. Without the rulings the mirror would form a real image of the source at I. The 1st- or 2nd-order interference maximum I' is focused on a blackened thermocouple T, which serves to indicate the presence of infrared e-m radiation.

On the other hand, the grating spectro-

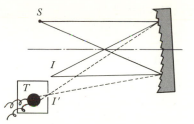

Figure 26-30 Infrared spectrograph.

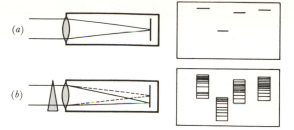

Figure 26-31 The spectra of many stars can be obtained with one exposure by placing a prism in front of the objective lens of a telescope.

scope is wasteful of light, since usually much of the available energy goes into the central (undeviated) image.* For this reason, prisms are used when weak sources must be examined, or when the very highest resolving power or dispersion is not needed. Figure 26-31 shows the spectra of several stars, formed in one exposure on a plate placed at the focal point of an astronomical telescope. In (a) each line is the image of a star; the telescope was purposely driven slightly too slowly, so that the star images were drawn out into lines by the rotation of the earth. For (b), a prism was placed in front of the objective lens, and properly oriented, with the result that each line image was spread out into a tiny spectrum. In astronomy, grat-

---

*The grooves of modern reflection gratings are often shaped to reflect a large fraction of the light in the approximate direction of the desired interference maximum.

ing spectra are used for studying strong sources: the sun and the brighter stars.

In this chapter we have not attempted to describe all of the many optical instruments in use. The six or seven instruments we have studied are interesting and useful. More important for our study of physics, we have been able to use wave optics (and the wave-front interpretation of geometrical optics) as adequate models for predicting the performance of these common optical instruments. On the basis of our studies so far, the wave theory of light has carried the day. We shall see in the next chapter, however, that for a complete interpretation of optical phenomena the photon (corpuscular) model of light must also be used.

■ SUMMARY

The camera and the human eye each form a real image on a sensitive surface. They differ in their means of achieving accommodation; in the camera the lens is moved, whereas in the eye the power of the lens is changed. An ideal lens gives a sharp undistorted image of the object, and any departure from this condition is called an aberration. Spherical aberration can be corrected by using a small lens opening. Chromatic aberration can be corrected by using combinations of lenses made of different kinds of glass. Astigmatism causes a line image of a point source. In the eye, such a defect is usually due to a cylindrical component of the curvature of the cornea and lens. A spherical camera lens suffers from astigmatism of a different sort, for rays which are not along the axis of the lens. Approximate correction of astigmatism is possible, using combinations of lenses which are not necessarily spherical.

The normal eye has a power which adjusts from about $+60$ diopters to $+64$ diopters, and the near point is about $\frac{1}{4}$ meter, or 25 cm, from the eye. Negative spectacle lenses are used to correct for myopia, and positive lenses to correct for hypermetropia and presbyopia. Cylindrical lenses are used to correct for astigmatism.

The simple magnifier forms a virtual image whose distance from the eye is at least 25 cm and may be as much as infinity. The angular magnification, or magnifying power, lies between $1 + 25/f$ and $25/f$, the latter value being for minimum eyestrain when the virtual image is at infinity.

The compound microscope and the telescope are similar in construction; the objective forms a real image of the object, and this image is received on a photographic film or is viewed with a magnifier. In the microscope the real image is larger than the object, while in the telescope the real image is smaller than the object. When adjusted for viewing a distant object with minimum eyestrain, the magnifying power of the telescope is the ratio of the focal lengths of the two lenses.

The resolving power of a microscope or telescope is its ability to show sharp detail; this is not the same as magnifying power, which is the ability to form a large image on the retina of the eye. Resolving power is limited by the diameter of the objective lens, which acts in much the same manner as does a single slit to form a blurred image of a point object. The essential factor which determines the resolving power is λ/D, the ratio of wavelength to lens diameter. In order to obtain the full theoretical resolving power, the lens aberrations must, of course, be negligibly small.

Most large astronomical telescopes use a concave mirror to form the real image directly on a photographic plate, photocell, thermocouple, radio antenna, or other detector of e-m radiation. The intensity of the image is proportional to the area of the objective mirror or lens.

A spectroscope usually consists of a slit, collimator, prism or grating, and telescope. Except for a lack of intensity, the grating spectroscope is superior to the prism spectroscope, for it has greater resolving power and dispersion, and it is adaptable to radiations such as infrared or ultraviolet for which absorption in prisms or lenses would be excessive.

■ CHECK LIST

aberration
spherical aberration
off-axis astigmatism
chromatic aberration
f-number of a lens
accommodation
myopia
hypermetropia
near point

presbyopia
angular size
angular magnification
$M = 1 + 25/f$
$M = 25/f$
compound microscope
objective lens
eyepiece
resolving power of a lens

astronomical telescope
$M = F/f$
minimum eyestrain
Galilean telescope
collimator
dispersion of
    a spectroscope
resolving power of
    a spectroscope

■ QUESTIONS

26-1 Describe three aberrations of a simple spherical lens. Does the lens of the human eye have these aberrations?

26-2 Why does a camera give a sharper image on the film if the lens is "stopped down" to have a smaller diameter?

26-3 Hypermetropia and presbyopia are different conditions, each corrected by a positive spectacle lens. Explain.

26-4 If you shut your eyes for a while in a dark room and then press gently (through the eyelid) on the extreme left part of the left eyeball, you "see" flashes of light at the *right* side of the field of view. Explain.

26-5 Why does a young person who wears glasses usually not require bifocals?

26-6 For any given person, does the magnifying power of a simple magnifier depend very much upon where it is held?

26-7 Compare and contrast the compound microscope and the astronomical telescope.

26-8 Why cannot a microscope be constructed, using visible light, which will give a clear view of a sugar molecule?

26-9 Why is a radio telescope useful for research, in spite of the fact that the resolving power of even a large one is less than that of an optical telescope?

26-10 In what ways is a telescope using an objective mirror superior to one using an objective lens? What are the disadvantages, if any, of the reflecting telescope?

26-11 What are two advantages of the Galilean form of telescope?

26-12 Why are spectrum colors usually referred to as "lines"?

26-13 Explain the difference between the dispersion and the resolving power of a spectroscope.

26-14 What are some of the advantages and disadvantages of a grating spectroscope as compared with a prism spectroscope?

■ PROBLEMS

26-A1 Which has the greater angular size: a football player 6 ft tall, viewed from a stadium seat 80 yd away, or the 2 in. image of the player, viewed by a spectator 10 ft from his TV screen?

26-A2 Which has larger angular size for a viewer on earth: a satellite 52.8 ft in diameter 1000 mi above the earth, or the planet Saturn, which is 70 000 mi in diameter and averages 886×10^6 mi from the earth?

26-A3 A camera lens in a television studio has a diameter of 5 cm and focal length 20 cm. What is the *f*-number of this lens? Is it a "fast" lens?

26-A4 What is the focal length of an $f/4.6$ lens whose diameter is 3 cm?

26-A5 In 1675 the Dutch biologist Antony van Leeuwenhoeck, using a single lens probably of focal length 1.25 mm, discovered bacteria. What was the magnifying power of this early simple microscope?

26-A6 What is the magnifying power of a lens of focal length $+2.5$ cm, when used by a person with normal vision?

26-A7 A microscope's objective lens forms a real image that is magnified 15 times; the eyepiece has an angular magnification of 20. What is the over-all magnification of the microscope?

26-A8 What is the angular magnification of an astronomical telescope which has an objective of focal length 2.50 m and an eyepiece of focal length 1 cm?

26-A9 An astronomical telescope has an objective lens of focal length 60 cm. What magnifying powers can be obtained if two eye lenses are available having focal lengths 2 cm and 0.75 cm?

26-A10 In James Thurber's *Many Moons* (Harcourt Brace Jovanovich, 1943), a little princess who wanted the moon settled for a gold moon the size of her thumbnail which covered the real one. Determine the angular size of your thumbnail, held at arm's length. Is it larger or smaller than that of the moon? (See Sec. 26-3.)

26-B1 A successful picture is taken using an exposure of $\frac{1}{100}$ s and a camera lens of speed $f/11$. The photographer wishes to repeat the shot, with his lens stopped down to $f/22$. What should the exposure time be?

26-B2 What is the approximate f-number of a human eye if the open diameter of the iris is 5 mm and the eye, focused on a distant object, has a power of $+60$ diopters?

26-B3 What is the distance from lens to retina in the myopic eye illustrated in Fig. 26-12a?

26-B4 A certain young girl can adjust the power of her eye's lens-cornea combination between limits of $+57$ diopters and $+65$ diopters. With the lens relaxed, she can see a distant star clearly. (a) How far is this girl's near point from her eye? (b) How far is her retina from her eye lens?

26-B5 A man can see clearly only objects which lie between 50 cm and 250 cm from his eye. Calculate the prescription of bifocal glasses which will enable him to see distant objects (through the top half) and also read a book at a distance of 20 cm (through the lower half).

26-B6 A myopic student wears glasses of power -10 diopters to see distant objects, and the distance from eye lens to retina is 2.0 cm. (a) What is the power of the student's eye, when relaxed? (b) If the student can give an extra $+5$ diopters to his eye lens by muscular effort, where is his near point without glasses?

26-B7 A retired bank president can easily read the fine print of the financial page when the newspaper is held at arm's length (60 cm) from his eye. What should be the focal length of a spectacle lens that will allow him to read at a comfortable distance of 24 cm?

26-B8 The near point for a certain child is 8 cm; the far point (with eye lens relaxed) is 125 cm. The eye lens is 2.00 cm from the retina. (a) Between what limits, measured in diopters, does the power of this child's lens-cornea combination vary? (b) Calculate the power of a spectacle lens this child should use for relaxed distant vision. Is the lens converging or diverging?

26-B9 What is the magnifying power of a magnifying glass of strength 50 diopters, when used by a person having normal vision? The lens is adjusted for minimum eyestrain.

26-B10 What is the magnifying power of a concave shaving mirror of radius of curvature 50 cm, when used by a man whose face is 20 cm from the mirror? (*Hint:* For the angular size without aid, imagine the man views his image in a plane mirror.)

26-B11 Carry through an analysis, similar to that in Example 26-1, for a person whose near point is at 50 cm instead of 25 cm, using the same magnifier of focal length $+2$ cm. Calculate the angular magnification when the lens is adjusted for maximum magnification.

26-B12 What is the power, in diopters, of a magnifying glass which has a magnifying power of 6 when used by a person having normal vision? The lens position is adjusted to give maximum magnifying power.

26-B13 A motion picture projector's lens is $f/1.6$, and is 5 cm in diameter. What would be the maximum magnifying power of this lens if used as a magnifying glass by a person having normal vision?

26-B14 The objective lens of a device has focal length $+2$ cm, and the eyepiece has focal length $+2$ cm. When adjusted for minimum eyestrain, the lenses are 22 cm apart. Is the device a microscope or a telescope?

26-B15 The objective lens of a microscope of focal length 3.9 mm is placed 4.0 mm from a specimen. The image is viewed with an eyepiece of focal length 10 mm, adjusted for minimum eyestrain. (a) How far apart are the lenses? (b) What is the over-all magnification of the instrument?

26-B16 A pocket microscope used for viewing a gem has an objective of focal length $+0.5$ cm and eyepiece of focal length $+2$ cm. When adjusted for minimum eyestrain, the lenses are 5 cm apart. Calculate (a) the object distance for the objective lens, and (b) the over-all magnification of the microscope.

26-B17 An astronomical telescope has an objective lens of power 1.25 diopters, and the eyepiece is 82 cm from the objective when adjusted for minimum eyestrain. (a) Calculate the angular magnification of this telescope. (b) How close would a man a kilometer away seem to be when viewed through the telescope?

26-B18 An amateur astronomer's telescope has a mirror which is 6 in. in diameter, with a radius of curvature of 120 cm. When he uses an eyepiece of focal length 8 mm, what is the angular magnification of his instrument?

26-B19 An elderly sailor is shipwrecked on a desert island but manages to save his spectacles. The lens for one eye is of power $+2$ diopters and the other lens has a power of $+20$ diopters. (a) What is the magnifying power of the telescope he can construct with these lenses? (b) How far apart are the lenses when the telescope is adjusted for minimum eyestrain?

26-B20 In the prism binoculars shown in Fig. 26-25 the optical path from objective to eyepiece is 45 cm, but the length of the instrument is much less than this because of the use of prisms to fold the path while also reinverting the image. (a) The eyepiece has focal length 5 cm; what is the magnifying power when adjusted for minimum eyestrain? (b) Design a terrestrial (non-inverting) straight-tube telescope using the same lenses as the binoculars, with an additional "erecting" lens of focal length $+9$ cm inserted in the tube beyond the focal point of the objective at such a place that it forms a real image (with $m = -1$) at the focal point of the eyepiece. What is the overall length of such a telescope?

26-B21 An amateur astronomer makes a reflecting telescope whose mirror has a radius of curvature 3.00 m and whose eyepiece lens has focal length 1.2 cm. (a) What is the approximate over-all length of the telescope when adjusted for minimum eyestrain? (b) What is the magnifying power? (c) How close would the moon seem when viewed through the telescope if the actual distance is 386 000 km?

26-B22 A spotting scope in the form of a Galilean telescope has two lenses 25 cm apart when adjusted for viewing a distant object with minimum eyestrain. If the eyepiece has a power of -25 diopters, what is the focal length of the objective?

26-B23 A grating having 5000 lines per centimeter forms a spectrum of a source of white light. Calculate the angular separation of the yellow and blue regions of the spectrum, in the 2nd order. (Use $\lambda = 5800$ Å for yellow light and $\lambda = 4700$ Å for blue light.)

26-B24 A source emits three lines: a violet line of wavelength 4000 Å, a green line of wavelength 5500 Å, and a red line of wavelength 7000 Å. Images of these lines are formed by a grating whose grating space is 25 000 Å. Calculate $\sin \theta$ for all observable orders of each line, and arrange the images (13 in all) in sequence according to their $\sin \theta$ values, thus: V_1, G_1, R_1, V_2, ... The resulting sequence illustrates a phenomenon known as the "overlapping of orders" in a grating spectrum of visible light.

26-C1 A compound microscope has an objective lens of focal length 1 cm, and an eyepiece of focal length 2 cm. The object is a bug 0.2 mm in diameter on a slide which is 1.10 cm from the objective lens. (a) How far apart are the lenses when the object is viewed with the lenses adjusted for minimum eyestrain? (b) What is the angular size of the magnified bug? (c) For classroom demonstration, a real image of the bug is formed on a screen 100 cm from the eyepiece. How far from its original position must the eyepiece be moved, and in which direction? (d) What is the linear size of the image of the bug on the screen?

26-C2 A laboratory (astronomical) telescope is used to view a scale that is 400 cm from the objective lens. The objective has focal length 20 cm, and the eyepiece has focal length 2 cm. Calculate the angular magnification when the telescope is adjusted for minimum eyestrain. (*Hint:* The object is not at infinity, and so the simple expression F/f is not quite accurate for this problem.)

26-C3 The lenses of an astronomical telescope are 128 cm apart, when adjusted for viewing a distant object with minimum eyestrain. The angular magnification is 63. Compute the focal length of each lens.

26-C4 An opera glass (Galilean telescope) is used by a person whose near point is at 25 cm to view a performer 2 m tall, 40 m from the objective lens. The focal length of the objective is $+30$ cm and that of the eyepiece is -2.5 cm. When the telescope is adjusted for maximum magnification, compute (a) the size of the real image that would have been formed by the objective; (b) the virtual object distance for the negative lens; (c) the distance between the lenses; (d) the over-all angular magnification.

26-C5 An achromatic lens (Fig. 26-32) has a plano-convex crown glass lens of radius of curvature R_C in contact with a plano-concave flint glass lens of radius of curvature R_F. The indices of refraction for red light (656 nm) and blue light (486 nm) are: crown glass, $n_R = 1.5145$, $n_B = 1.5240$; flint glass, $n_R = 1.6221$, $n_B = 1.6391$. (a) Use the lens-maker's formula and Prob. 24-C14 to show that the power $1/f$ of the combination is $(n_C - 1)/R_C - (n_F - 1)/R_F$. (b) Show that the power is independent of color, for the two colors involved, if $\Delta n_C/R_C = \Delta n_F/R_F$, where $\Delta n_C = n_B - n_R$ for crown glass and $\Delta n_F = n_B - n_R$ for flint glass. (c) If the crown lens has $R_C = 20.00$ cm, calculate the radius of curvature R_F. (d) Calculate $1/f$ for the combination for each of the two colors, and verify that the lens is achromatic.

Figure 26-32

26-C6 A grating 15 cm long has 6000 lines per centimeter. When used in the second order, could two lines of wavelengths 6000.00 Å and 6000.03 Å be separated?

26-C7 Could a person see as separate the two headlights of an automobile which is 8 km away, if the lamps are 1.5 m apart? Assume yellow light for which $\lambda = 5800$ Å, and assume the pupil of the eye to be 4 mm in diameter. (*Hint:* Calculate the angular separation of the lamps, and compare with the angular radius of the diffraction circle given by $\lambda/0.82D$.)

26-C8 What must be the diameter of a radar "dish" antenna in order for it to distinguish between two targets 400 m apart at a distance of 40 km   (*a*) if the wavelengths is 10.2 cm? (*b*) if the wavelength is 3.2 cm?

26-C9 Estimate the resolving power of the human eye, insofar as it is limited by the diameter of the pupil of the eye. Assume light of wavelength 5500 Å and an opening 3 mm in diameter.   (*a*) Calculate the angular diameter of the central disk of the diffraction pattern of a point source (Fig. 26-18). Express your answer in radians and in minutes of arc. (*b*) If the image is formed on a retina 17 mm from the lens, calculate the approximate linear diameter on the retina of the diffraction disk; express your answer in microns. (*c*) Compare with the human eye as it has actually evolved: near the fovea centralis (region of sharpest vision) the retina has about 250 000 cones per mm²; each cone is about 2 μ in diameter. Show that the diffraction disk covers about 10 cones.   (*d*) Compute the angular size of a single cone, and show that this is somewhat smaller than the actually found best resolving power of about 0.7′.

■ REFERENCES

1 Wald, George, "Eye and Camera," *Sci. American,* Aug., 1950, p. 32.

2 Hirsh, F. R., Jr., and E. M. Thorndike, "On the Pinhead Shadow Inversion Phenomenon," *Am. J. Phys.,* **12,** 164 (1944).

3 Boring, Edwin G., *Sensation and Perception in the History of Experimental Psychology* (Appleton Century, 1942), p. 237. A description of the experiment of G. M. Stratton on the psychological adjustment of a person wearing inverting glasses.

4 Erismann, T., and I. Kohler, *Upright Vision Through Inverting Spectacles* (film). The Pennsylvania State University, Audio Visual Aids Library, University Park, Pa.

5 Jacobs, S. F., and A. B. Stewart, "Chromatic Aberration in the Eye," *Am. J. Phys.,* **20,** 247 (1952).

6 Andrade, E.N. da C., "Robert Hooke," *Sci. American,* Dec., 1954, p. 94. Contains a description of Hooke's early microscopic investigations.

7 Burns, D. M., and S. G. G. Macdonald, *Physics for Biology and Pre-Medical Students* (Reading, Mass.: Addison-Wesley, 1970), Chap. 22, "The Optical Microscope," Chap. 23, "Specialist Microscopy."

8 Waaland, J. R., "Fraunhofer and the Great Dorpat Refractor," *Am. J. Phys.,* **35,** 344 (1967). The first modern, achromatic, refracting telescope, of aperture 9½ in.

9 Kraus, John D., "Radio Telescopes," *Sci. American,* March, 1955, p. 36.

10 Lovell, Donald J., "Principles of Colorimetry," *Am. J. Phys.,* **18,** 104 (1950).

11 Miller, Franklin, Jr., *Resolving Power* (film).

You are familiar with the general idea of atomic structure—the massive positively charged nucleus, surrounded by a cloud of negatively charged electrons which make up the outer atom. In the next four chapters we consider the finer details of the picture, study the ways in which knowledge of atomic structure has come to us, and see how that knowledge can be put to use. We shall find that the nucleus is complex and often unstable. We shall find that the emission and absorption of light and other e-m radiation are closely related to the electrical structure of the outer atom, but in a strange and "nonclassical" way. To begin our study of the atom we consider the nature of the electron and of the photon.

27-1 The charge of the electron

The facts of electrolysis (Sec. 17-6) strongly suggest that the charges of ions such as Cu^{++} are multiples of some fundamental unit of charge, which we call the electron. Strictly speaking, electrolysis experiments prove only that the *average* charge per uni-

valent ion is 1.6×10^{-19} C. The individual charges might differ greatly among themselves and still average out to 1.6×10^{-19} C when many ions are considered. (Even in the most delicate weighing experiments, we are dealing with some 10^{16} ions.) We have already met with such a statistical averaging-out effect in our study of the kinetic theory of gases. At any given temperature, the molecules of a gas have many different speeds, ranging from very slow to very fast, and yet the average speed has a definite value.

In his oil-drop experiment, Robert A. Millikan (1868–1953) succeeded in measuring the effect of single electrons. Tiny drops of mineral oil from an atomizer are allowed to fall into the space between two charged plates; many of the drops are charged by friction during the atomization process. In Fig. 27-1 the observer has selected a negatively charged drop which he holds stationary by applying a positive potential to the upper plate. The drop is subject to two forces: its weight W, due to the downward gravitational field, and an upward force $F = qE$ due to the electric field E. The electric field's intensity, in new-

Cathode rays (electrons) are deflected downward as they move toward the right through a region of magnetic field. Can you tell in which horizontal direction is the magnetic field that is supplied by the horseshoe magnet? (Fundamental Photographs)

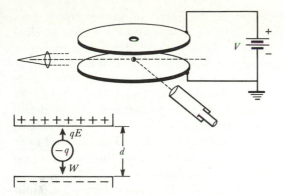

Figure 27-1 Millikan's oil-drop experiment to measure the charge of the electron.

tons per coulomb, equals the potential gradient V/d, in volts per meter (Sec. 18-5). When the electric field is adjusted until the drop is motionless,

$$\text{net } F = 0$$

$$q\frac{V}{d} - W = 0$$

$$q = \frac{Wd}{V} \qquad (27\text{-}1)$$

Since d and V are easily measured, the problem of finding q reduces to that of finding W. In the Millikan experiment the weight of each drop is measured indirectly by turning off the field and allowing the drop to fall by the action of gravity alone. The terminal velocity (Sec. 12-9) depends on the size of the drop and the known coefficient of viscosity of air (see Example 1-2). By judicious use of a switch to turn the electric field on or off, a single drop can be kept moving up and down in the field of view for hours at a time, and its radius can be computed.[*] Having determined the drop's size, we can then calculate its weight from the density of the oil. Equation 27-1 can then be used to compute the charge q on any drop.

Millikan found the charge on each drop to be an integral multiple of 1.60×10^{-19} C. For instance, five successive experiments on

five drops might give charges as follows (all in units of 10^{-19} C): -3.20; -8.00; $+4.80$; -1.60; $+3.20$. No drop having charge less than 1.60×10^{-19} C was ever observed. In addition, any given drop will occasionally be seen to change its charge abruptly, always by a multiple of 1.60×10^{-19} C. We interpret these facts as showing the existence of an elementary charge which we call the electron. A drop gains or loses an integral number of electrons, either by friction or by some other means. If an ion bumps into a drop that is just balanced between the plates, one or more electrons may be transferred to or from the drop, which then takes off suddenly either upward or downward, depending on the sign of its charge and the direction of the electric field. The PD between the plates must then be adjusted to a new value to balance the drop with its new charge, and a new calculation can be made.

The oil-drop experiment shows the atomic[*] nature of charge. At least in this experiment and, by inference, in general, we conclude that electric charge is *quantized,* rather than subdivisible into infinitely small fragments. The experiments of electrolysis are consistent with the quantization of charge, but do not require it. The oil-drop experiment sharpens our concept of electric charge by showing it to come in small units or packages. This amount of charge is the electronic charge, of magnitude $e = 1.60 \times 10^{-19}$ C.

27-2 The mass of the electron

Nothing in the preceding section implies that electrons are themselves particles having mass. Evidence for small negatively charged particles comes from studies of the discharge of electricity through gases. A typical experiment is shown in Fig. 27-2. The tube contains air at low pressure (about 10^{-4} torr). When the connections are as shown, charge flows through the tube, and the gas glows.

[*] The drop is so small that diffraction effects prevent direct measurement of its radius. Only a point of light is seen in the microscope.

[*] We use the word in its general sense, meaning indivisible.

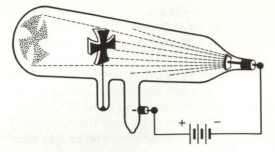

Figure 27-2 Cathode rays.

Using the modern interpretation, we identify the carriers of charge as positive and negative ions of oxygen and nitrogen, as well as electrons. In addition, a stream of particles leaves the negative terminal (the cathode) and proceeds in straight lines, perpendicular to the surface of the cathode. The passage of the particles is made evident by a bluish glow in the gas, and their straight-line propagation is shown by the sharp shadow cast by an obstacle. The particles evidently have inertia, for they exert a measurable force on the obstacle, and they carry energy, since a thin metallic target can be made red-hot by the impact of the particles. The moving particles can be deflected by a magnetic field, and so they must carry electric charge (Sec. 21-1). The direction of the magnetic force on the moving particles shows them to be negative. We now know that these *cathode rays* are electrons which have been ejected from the cathode by the impact of the heavy ions. Exactly similar mechanical and heating effects are observed for a stream of electrons emitted by a heated metal. Cathode rays were studied extensively by the British physicist J. J. Thomson (1856–1940), who is commonly regarded as the discoverer of the electron.

To measure the mass of the electron, we must somehow apply Newton's second law. The essential features of the deflection method are shown in Fig. 27-3. Electrons, emitted from an indirectly heated cathode K in an evacuated tube, are accelerated to some velocity $\mathbf{v}$ by an applied potential

difference V. Some of the electrons pass through a small hole in the accelerating electrode and enter a region where a magnetic induction $\mathbf{B}$ is applied at right angles to their motion. The magnetic force $\mathbf{F}$ serves as a centripetal force to cause the particles to move in a circle whose radius R is determined by the slits. To make a measurement, the magnetic induction $\mathbf{B}$ is adjusted until the electron current into the collector cup shows a sharp maximum.

The magnitude of the velocity v is found by the energy principle. Each electron (of charge e) falls through a PD equal to V, and hence loses an amount of PE equal to Ve.

$$\text{Increase of KE} = \text{decrease of PE}$$

$$\tfrac{1}{2}mv^2 = Ve$$

$$v^2 = \frac{2Ve}{m} \qquad (27\text{-}2)$$

The centripetal acceleration is therefore

$$a = \frac{v^2}{R} = \frac{2Ve}{mR}$$

Now we apply Newton's second law to the circular motion:

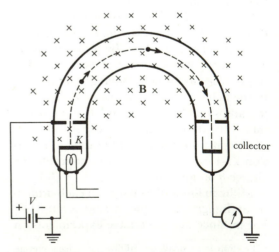

Figure 27-3 Magnetic deflection method for measurement of the electron's e/m ratio.

$$\text{net } F = ma$$

$$Bev = ma$$

$$Be\sqrt{\frac{2Ve}{m}} = m\left(\frac{2Ve}{mR}\right)$$

$$\frac{e}{m} = \frac{2V}{B^2R^2} \qquad (27\text{-}3)$$

Using Eq. 27-3, we can determine the *ratio* of charge to mass for an electron by measuring V, B, and R. Experiments of this sort show that all electrons have the same e/m ratio for nonrelativistic speeds; and at higher speeds the ratio e/m decreases by just the right amount as demanded by relativity, since the mass increases according to the formula $m = m_0/\sqrt{1 - v^2/c^2}$. The experimental value of e/m_0 is 1.759×10^{11} C/kg. Combining this with the value of e gives for the rest mass of the electron

$$m_0 = \frac{e}{e/m_0} = \frac{1.602 \times 10^{-19} \text{ C}}{1.759 \times 10^{11} \text{ C/kg}}$$

$$= 9.11 \times 10^{-31} \text{ kg}$$

The combined result of the experiments described so far in this chapter show that (1) charge is quantized, and (2) free electrons exist, all of which are identical, each carrying one unit of charge always associated with the same amount of mechanical inertia.

27-3 The photoelectric effect

With the development of high-vacuum techniques in the late nineteenth century it became possible to study the emission of electrons from a surface under the influence of light. In the photocell circuit of Fig. 27-4, light shining on a cathode K may cause electrons to leave the surface and travel through the vacuum to the positive electrode A. The electrons then flow through the battery, the resistor R, and the galvanometer G back to K again. As might be expected, the current is proportional to the intensity (power per unit area) of the light, and so the photocell current can actuate a relay to open doors,

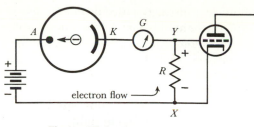

Figure 27-4 Photocell circuit.

ring alarms, and so on (Fig. 27-5). One way of amplifying the effect is to apply the PD between X and Y to the grid and the cathode of a triode; this PD equals the IR drop in the resistor R. If R is large enough, even a weak current may cause the grid potential to change by several volts. In a gas-filled photocell, the electrons create ions by collision as they move between the electrodes, and the total current is much greater than for a vacuum photocell.

Quantitative study of the photocell current reveals some surprising facts. For any surface, there is a *threshold frequency* for the light; if the frequency is lower than the threshold, no

Figure 27-5 Photoelectric cell.

photoelectrons are emitted no matter how intense the light. Also, if the frequency is above the threshold, all the photoelectrons which are ejected from the top layers of a surface have the same KE, regardless of the intensity of the light.• One might have expected a bright light to eject electrons with a greater velocity than a less bright light, but experiment shows that a bright light merely ejects *more* electrons. To illustrate this strange situation, let us imagine a photocell surface whose threshold frequency is 6×10^{14} Hz. This corresponds to green light of wavelength 5×10^{-7} m. If the surface is illuminated by strong *red* light, not a single photoelectron is ejected, since the frequency of red light is less than the threshold frequency. On the other hand, the faint light from a distant *blue* star can eject electrons, since the frequency of blue light is greater than the threshold frequency; the KE of these ejected electrons does not depend on the star's brightness. Classical physics is unable to cope with these experimental facts.

We discussed one of the early symptoms of the inadequacy of classical physics in Sec. 9-6, where we stated that angular motions and vibrations are quantized. That is, a rotating molecule can have 0, 1, 2, 3, . . . quanta of kinetic energy, but the in-between values are forbidden. In Sec. 16-8 we saw how this quantum idea explains the decreased specific heat of hydrogen gas at very low temperatures, for the energy of the collisions is then insufficient to give a molecule even one quantum of rotational energy. In 1900 the German physicist Max Planck (1858–1947) was able to explain the shape of the black-body radiation curves (Fig. 14-4) by making a quantum assumption. An incandescent solid body emits e-m radiation because it contains many oscillators (of atomic dimensions); according to Sec. 23-6 each

oscillator serves as an antenna to radiate an e-m wave. Planck's new assumption was that any given tiny oscillator cannot vibrate with an arbitrary amplitude; its energy E is quantized according to the equation

$$E = nh\nu \qquad (27\text{-}4)$$

In this equation n is an integer, ν is the frequency of the oscillator (determined by its mass and its force constant), and h is a universal constant which has since been known as *Planck's constant*. The value of h is 6.63×10^{-34} J·s.

Einstein boldly applied Planck's quantum hypothesis to the radiation itself: he assumed (in 1905) that the energy of e-m radiation is absorbed one quantum at a time. Such a quantum of radiation energy is now called a *photon*. For light of any frequency ν, the energy of a photon is given by

$$E = h\nu \qquad (27\text{-}5)$$

Thus light comes in packages; and the size of the package depends on the color of the light.

EXAMPLE **27-1**

What is the energy of a photon of green light?
If we take 5000 Å as the average wavelength of green light, then

$$\nu = \frac{c}{\lambda} = \frac{3 \times 10^8 \text{ m/s}}{5 \times 10^{-7} \text{ m}} = 6 \times 10^{14} \text{ s}^{-1}$$

The energy of a photon for this light is

$$E = h\nu$$
$$= (6.63 \times 10^{-34} \text{ J·s})(6 \times 10^{14} \text{ s}^{-1})$$
$$= \boxed{3.98 \times 10^{-19} \text{ J}}$$

Einstein's photoelectric equation states that the KE of the ejected electron equals the energy of the incoming photon, minus the work w, called the *work function,* needed to remove the electron from the surface. In symbols,

$$\tfrac{1}{2}mv^2 = h\nu - w \qquad (27\text{-}6)$$

• This is an idealization, true for thin metallic films. In actual experiments, an electron may lose some of its energy while traveling through the metal to reach the surface.

We see from this equation that no electrons can be emitted unless $h\nu$ is greater than w; hence we infer the existence of some threshold frequency. The equation also tells us that the KE of an ejected electron depends only on the frequency of the light and the work function of the surface. We interpret a bright red light as a stream of many relatively weak photons; a faint violet light as a stream of a few, more energetic, photons.

At this point we introduce a new unit of energy called the *electron volt*, which we shall use repeatedly in our study of atomic physics. *An electron volt* (eV) *is the energy acquired by an electron in moving through a PD of one volt.* To calculate its value, we use Eq. 18-3:

$$W_{B \to A} = QV_{AB}$$
$$1 \text{ eV} = (1.60 \times 10^{-19} \text{ C})(1 \text{ J/C})$$
$$= 1.60 \times 10^{-19} \text{ J}$$

This is a handy unit for photoelectricity, since the work functions of most metals lie in the range of 1 to 10 eV. For the sake of convenience, we also calculate the energy of a photon in electron volts, as follows:

$$E = h\nu = h\frac{c}{\lambda}$$

$$= (6.63 \times 10^{-34} \text{ J} \cdot \text{s})\left(\frac{3 \times 10^8 \text{ m/s}}{\lambda}\right)$$

$$\times \left(\frac{1 \text{ Å}}{10^{-10} \text{ m}}\right)\left(\frac{1 \text{ eV}}{1.60 \times 10^{-19} \text{ J}}\right)$$

where λ is the wavelength *in angstroms*. Simplifying, we obtain

$$E = \frac{12\,400 \text{ eV} \cdot \text{Å}}{\lambda} \qquad (27\text{-}7)$$

If λ is in angstroms, E is in electron volts. The usefulness of the electron volt as an energy unit is seen in the following example.

EXAMPLE **27-2**

The work function for sodium metal is 2.46 eV. With what KE are photoelectrons ejected from a thin sodium surface illuminated by light of wavelength 4000 Å?

First we calculate the photon energy in electron volts.

$$E = \frac{12\,400 \text{ eV} \cdot \text{Å}}{4000 \text{ Å}} = 3.10 \text{ eV}$$

This is the energy of each photon of the light of wavelength 4000 Å. Now, using Einstein's photoelectric equation (Eq. 27-6), we find

$$\text{KE} = h\nu - w$$

$$= 3.10 \text{ eV} - 2.46 \text{ eV} = \boxed{0.64 \text{ eV}}$$

In practice, this is an upper limit; some photoelectrons may have less than this energy if they lose energy by collisions as they make their way through the metal to the surface.

EXAMPLE **27-3**

What is the threshold wavelength for a sodium surface? At the threshold, no energy is available for KE. Hence

$$0 = h\nu - w$$

and the photon energy at the threshold is

$$E = h\nu = w = 2.46 \text{ eV}$$

$$\lambda = \frac{12\,400 \text{ eV} \cdot \text{Å}}{E} = \frac{12\,400 \text{ eV} \cdot \text{Å}}{2.46 \text{ eV}} = \boxed{5040 \text{ Å}}$$

The threshold is in the green region of the spectrum (see Table 24-1).

EXAMPLE **27-4**

Through what PD would an electron have to fall in order to acquire as much energy as a dust particle has, if the dust particle has a mass of 0.0004 g and is falling at a speed of 1 cm/s?

The KE of the dust particle is first calculated in joules, and then converted to electron volts.

$$\text{KE} = \tfrac{1}{2}mv^2 = \tfrac{1}{2}(4 \times 10^{-7} \text{ kg})(10^{-2} \text{ m/s})^2$$

$$= 2 \times 10^{-11} \text{ J}$$

$$= (2 \times 10^{-11} \text{ J}) \times \left(\frac{1 \text{ eV}}{1.6 \times 10^{-19} \text{ J}}\right)$$

$$= 1.25 \times 10^8 \text{ eV} = 125 \times 10^6 \text{ eV}$$

$$= \boxed{125 \text{ MeV}}$$

The dust particle's KE is 125 million electron

volts (125 MeV), and an electron would have to fall through a PD of 125 million volts to have as much energy.

27-4 The photon as a corpuscle

The first subatomic particle to be discovered was the electron. We have seen two rather different attributes of the electron: it is a quantum of charge (1.60×10^{-19} C); it is also a corpuscle of matter (9.11×10^{-31} kg), and hence, because of its inertial mass, an electron can have momentum and kinetic energy. Presumably, an electron also has gravitational mass (see the last sentence of Sec. 4-8) and therefore has weight and can have gravitational potential energy. Let's now consider the corpuscular nature of the photon—it is a quantum of *energy*. To be sure, some photons carry more energy than others (according to $E = h\nu$), but the photoelectric effect shows clearly that the energy is absorbed a quantum at a time. By itself, the photoelectric effect shows nothing one way or the other about the possibility that a photon is a corpuscle with localized *mechanical* properties such as momentum and mass. Such evidence came from x-ray photons.

We will study the production of x rays in Sec. 28-5. At this time we need only know that x rays are electromagnetic radiation whose photons have high energy (typically 5000 to 100 000 eV) and correspondingly short wavelength (typically 0.1 to 2 Å). In his study of the scattering of x rays, the American physicist Arthur Compton in 1922 observed a small change in wavelength. In a typical case, the molybdenum $K\alpha$ line of wavelength 0.710 Å was changed to 0.734 Å when scattered by the electrons in a piece of carbon. Although this is a small difference (about 3%), Compton's analysis of his experiment was destined to play a major role in the development of the dual theory of light.

The experimental setup is shown in Fig. 27-6. X rays from a molybdenum target

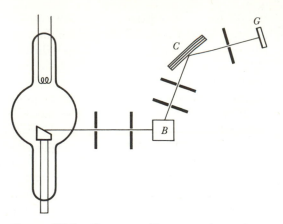

Figure 27-6 Compton effect: experimental arrangement for studying change of wavelength of x-ray photons scattered by carbon block B.

strike a carbon block B, and the scattered rays are analyzed by a crystal C, which serves as a grating. The x-ray beam is defined by a series of slits, and the scattered x rays are detected by a Geiger counter (Sec. 29-4), photographic plate, or ionization chamber G. A man-made grating would be too coarse to give a diffraction pattern, for to be effective the grating space must be comparable with the wavelength. Fortunately, the atoms in a crystal are about an angstrom apart, and so the layers of atoms—so-called crystal planes—can serve as a natural grating for x-ray diffraction. In England, before World War I, the Braggs, father and son,[*] had developed the crystal spectrometer used by Compton; as so often happens, improved instrumentation was a necessary prelude to a critical experiment.

Accurate data are not enough; physical insight and intuition are also needed to build a theory. Compton successfully applied the laws of mechanics to the elastic collision between an x-ray photon and a "free" electron in the carbon.[*] The photon is deflected by

---

[*] Sir William Henry Bragg and Sir William Lawrence Bragg.

[*] Since carbon has an atomic number of only 6, even an electron in an inner (K) shell is loosely bound, and the few hundred eV required to remove it is negligible compared with the energy of the x-ray photon.

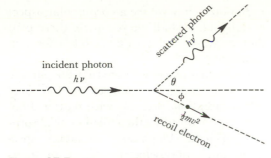

incident photon
$h\nu$

scattered photon
$h\nu'$

θ

ϕ

$\frac{1}{2}mv^2$

recoil electron

Figure 27-7 Conservation of energy during scattering of a photon by an electron.

the electron, and the electron acquires a recoil velocity in some direction, as shown in Fig. 27-7. To apply the law of conservation of momentum to this collision, we must have a formula for the momentum of a photon. Such a formula comes from the relativistic mass equation. The rest mass m_0 of a photon is 0, since a photon cannot be at rest in any frame of reference (Einstein's 2nd postulate). However, the total *relativistic* mass m is not zero. In fact, from $E = mc^2$, we see that a photon's mass is proportional to its energy: $m = E/c^2$. The momentum is, as always, the product of mass and velocity.

$$p = mc = \frac{E}{c^2} \cdot c = \frac{h\nu}{c} = \frac{h}{\lambda}$$

where we use the fact that the wavelength λ is c/ν. Thus the magnitude of a photon's linear momentum is

$$p = \frac{h}{\lambda}$$

or

$$\text{Momentum} = \frac{h}{\text{wavelength}} \qquad (27\text{-}8)$$

Now we are ready to apply the conservation laws to the collision of a photon with an electron which is originally at rest. By the law of conservation of energy,

Energy of original photon	=	energy of scattered photon	+	KE of recoil electron	
$h\nu$	=	$h\nu'$	+	$\frac{1}{2}mv^2$	(27-9a)

This equation has two unknowns, ν' and v. The law of conservation of momentum gives two additional equations, one for the x-components and one for the y-components of momentum (Fig. 27-8).

$$\frac{h\nu}{c} = \frac{h\nu'}{c} \cos\theta + mv\cos\phi \qquad (27\text{-}9b)$$

$$0 = \frac{h\nu'}{c}\sin\theta - mv\sin\phi \qquad (27\text{-}9c)$$

These three equations can be solved simultaneously, giving ν' and v as well as the angle ϕ at which the electron recoils. Experimentally, the slits are adjusted to select a given angle θ for the scattered photon; the equations then predict that electrons will recoil only in the calculated direction and will have a definite energy $\frac{1}{2}mv^2$. The theory was confirmed by Compton in two ways. According to Eq. 27-9a, the scattered photon always has less energy ($h\nu'$) than the original photon ($h\nu$); therefore, since ν decreases, λ increases, as is actually observed. In later experiments, Compton also measured the recoil electron's energy and direction, using a Geiger counter as a detector, and found them to be in precise agreement with the theory. Compton and his co-workers were also able to show that the scattered photon and the recoil electron entered their respective Geiger counters at the same instant (the time of flight was negligibly short), and so they must be the end results of a single collision process.

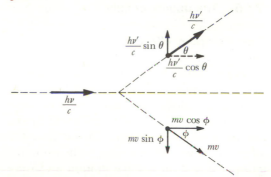

Figure 27-8 Conservation of momentum during scattering of a photon by an electron.

The Compton effect is important because it shows the photon to have corpuscular properties, in the same sense that a bullet or billiard ball has. This goes beyond the concept of a photon that was built up to explain the photoelectric effect. In addition to considering the photon to be a quantum of energy, now we must consider it also to have mechanical properties such as inertia and momentum. The model of light as a stream of bullets seems inescapable.

Let us stop for a moment and take stock of the properties of the electron and the photon, as they have been revealed by experiments described so far in this chapter. The electron has mass, and a negative charge. The photon has no rest mass, but it has energy and momentum. These corpuscles can interact in collisions. It is perhaps useful to recall that the theory of relativity doesn't make a sharp distinction between mass and energy. Thus the corpuscular nature of mass and energy are no doubt intimately related. The quantum nature of charge is still not well understood, nor is it clear whether charge and energy (i.e., mass) are fundamentally related to each other. These questions, like those about the basis of gravitation, are challenges to the best physicists of today.

27-5   The duality of light

The Compton effect (Fig. 27-6) illustrates the wave-particle duality of light in an extreme form. In the very apparatus by which the bulletlike character of x-ray photons is revealed, these same x rays show their wave character when reflected by the crystal grating. In Chap. 25 we took the phenomenon of interference to be an indisputable proof for a wave motion and were at great pains to build up a wave model for light (including x rays and all other e-m radiation). Now we see a corpuscular aspect of light. How shall we reconcile these conflicting wave and particle models for e-m radiation?

In the first place, we note that the particle aspects of radiation come to the fore only when radiation interacts with matter. Photons are involved in the emission or absorption of light (black-body radiation, spectrum lines, photoelectric effect), or in its scattering by an electron (Compton effect). On the other hand, the wave aspects are evident when light travels from one place to another without losing energy (Huygens' principle, interference, diffraction). The two aspects of light seem to be important at different times. One way out of the difficulty is to cry "modelitis." The function of a model is to make a new phenomenon intelligible in terms of familiar laws and events of our everyday, large-scale world. Note how we have used large-scale models in our description of e-m waves (comparing them with water waves) and also in our description of photons (comparing them with billiard balls). It may well be, however, that *no* intuitively satisfying model for the small-scale phenomenon of light can exist. There is no reason why the behavior of a single photon should be the same as that of *any* large-scale object with which we are familiar. Many physicists prefer to accept light for what it is and attempt no models. In fact, half a century of living with the strange duality of photons has bred a certain familiarity with light; advanced workers now use photons as "understandable" models for still more obscure phenomena!

Insofar as a model is possible, we can visualize the e-m waves as being "guide waves" which tell the photons where to go. According to this model, the e-m waves of Maxwell are *probability waves*. There is a large probability of finding a photon at any point where the amplitude of the wave is large. If two probability waves interfere destructively to form a node, the probability

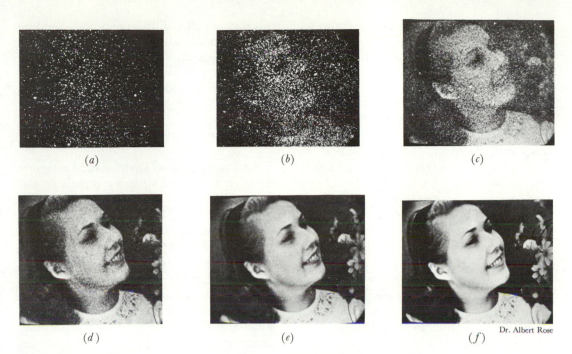

Figure 27-9 Photons of light needed to form a complete picture. The numbers given below tell how many photons were used to form each of these reproductions of the same image. (*a*) 3×10^3 photons. (*b*) 1.2×10^4 photons. (*c*) 9.3×10^4 photons. (*d*) 7.6×10^5 photons. (*e*) 3.6×10^6 photons. (*f*) 2.8×10^7 photons.

Dr. Albert Rose

of finding a photon at the node is zero. One example will illustrate how photons are guided by probability waves. Suppose a pair of slits and a screen are placed near a source, as in Young's experiment (Fig. 25-3). The mere presence of the slits and screen sets up a pattern of probability waves, which interfere constructively to form an antinode at A and a node at Y. The spacing of these nodes of probability depends on the spacing of the slits, etc. If now the source is turned on, photons are emitted, one at a time, and reach the screen. We can't trace the path of any photon from S through either slit to A, but we know that it is highly probable that a photon will appear at A. Let us assume that the screen is a photographic plate with many silver bromide molecules; each photon that is absorbed gives up its energy $h\nu$ to a single grain of the plate. As soon as the source is turned on, silver bromide grains become

activated,[*] more or less at random; after a while, the pattern becomes evident, and we realize that A is a probability antinode, and X a probability node. Experiments of this sort have actually been performed, using very weak sources of light, and a row of Geiger counters in place of the photographic plate. Figure 27-9 shows how an image formed by a lens is actually a probability distribution for the arrival of photons.

In this way we can compare a slit system, a grating, a lens, or a prism with a traffic light which directs the flow of cars at an intersection on the highway. The light flashes red and green even in the dead of night, when no cars are on the road. The probability waves are there, ready to guide whatever car

[*] The photon's energy $h\nu$ is stored in the grain as chemical PE and is used in the development process to help along the chemical reaction which makes metallic silver.

may come along. Guide waves are also involved in the photoelectric effect. Even before the Compton effect showed that photons have mass and are concentrated bundles of energy, it was known that very weak light can eject photoelectrons, if the frequency is above the threshold frequency as determined by the work function. A serious difficulty arises when the process is considered from the wave point of view. For instance, let a very weak source shine on a metal surface of area 1 m². About 10^{20} atoms lie in a surface of this area. To eject a photoelectron might require about 10^{-18} J (≈ 6 eV), and, if the intensity of the light is 10^{-6} J/s·m², each of the 10^{20} atoms in the surface could absorb only 10^{-26} J/s. It would require 10^8 s (over 3 years!) for an atom to absorb enough energy to eject a photoelectron. Yet experiment shows that *some* photoelectrons are emitted within a fraction of a second. Once again, we are forced to assume that the energy of the incoming light is carried by small bullet-like photons, which, in the case of a very weak source, are powerful enough but few and far between. We visualize the e-m wave as guiding the photons in such a way that it is equally likely for a photon to be absorbed anywhere on the surface, and hence photoelectrons are emitted at random. This is the way in which a uniform beam of light is interpreted according to the guide wave model.

If a similar duality were true for water waves, we would have the following strange phenomenon: A Coke bottle dropped off a New York pier which is 20 ft above water level sends out a wave front which spreads out in ever-widening circles until it reaches Lisbon, Bordeaux, and Dublin. At some unpredictable place, say Lisbon, a similar Coke bottle floating near a pier suddenly pops up 20 ft out of the water, and simultaneously the very weak and undetectable water wave which was in the form of a large circular wave front disappears, all of its energy being given to the Lisbon bottle. If

this sort of thing did happen, we would be tempted to talk about water waves as guide waves. The photoelectric effect for weak light is no stranger than this, and equally inexplicable by a wave model.

27-6 The duality of matter

According to the dual model of light, a "guide wave" is associated with every photon, and the wavelength of the guide wave can be measured by interference experiments with gratings and the like. In 1924 the French student Louis de Broglie proposed that guide waves are also associated with "ordinary" particles of matter, such as electrons and baseballs. To pursue the analogy between photon waves and matter waves, let us first recall the relativistic result for a photon's momentum. Equation 27-8 can be written as

$$\lambda = \frac{h}{\text{momentum}}$$

This says that the wavelength of the photon's guide wave equals Planck's constant divided by the photon's momentum.

De Broglie assumed that the same equation holds for matter waves; that is, a guide wave is associated with *any* moving body which has momentum. For matter waves, the wavelength of the guide wave is given by

$$\lambda = \frac{h}{mv} \tag{27-10}$$

As our next example shows, the wavelengths associated with ordinary bodies are much too small to be observed.

EXAMPLE 27-5

Calculate the de Broglie wavelength associated with a 20 g ping-pong ball which is moving at 400 cm/s.

$$\lambda = \frac{h}{mv} = \frac{6.63 \times 10^{-34} \text{ J·s}}{(20 \times 10^{-3} \text{ kg})(4.00 \text{ m/s})}$$

$$= \boxed{8.29 \times 10^{-33} \text{ m}}$$

This fantastically short wavelength is unobservable. A ping-pong ball headed in the general direction of an open window proceeds in a straight line with no fuzzy "shadow." It either goes out the window or is reflected back, depending on how it is aimed. Relative sizes are important. The wave nature of visible light escaped notice for so long because λ is so much smaller than the size of an ordinary opening such as a window. Therefore, according to Eq. 25-2, the angle of diffraction (1st-order minimum given by $\sin \theta = \lambda/w$) is so small that the light travels in straight lines as far as one can tell in this case. The diffraction is there, to be sure, but it is too small to be observed; λ/w is too small. Only when λ approaches the size of the slit or obstacle does the diffraction of light become evident. The same argument applies with even greater force to the ball of Example 27-5. The ball's guide wave has a wavelength which is only about 10^{-26} as large as a light wave. We are quite justified in using Newton's laws of motion for the motion of a ping-pong ball.

Passing now to particles of smaller mass, we see from Eq. 27-10 that the wavelengths of the guide waves associated with moving electrons may well be observable. In 1927, C. J. Davisson and L. H. Germer in the United States interpreted their experiments on the scattering of electrons by a metal surface as indicating the wave nature of the electron. Independently, in England, wave effects were sought for and found in 1927 by G. P. Thomson, son of J. J. Thomson (Sec. 27-2). The 1937 Nobel prize in physics was shared by Davisson and G. P. Thomson for their discovery of the *wave* nature of the electron; it is interesting to note that in 1907 J. J. Thomson had received the Nobel prize for his experimental proof of the *corpuscular* nature of the electron. The Thomsons, father and son, brilliantly demonstrated two opposite, but complementary, aspects of the nature of matter.

The wave nature of matter has also been shown experimentally for nuclear particles, such as the neutron.

EXAMPLE **27-6**

Calculate the KE and the wavelength of a neutron whose speed is 10^3 m/s.

The neutron mass is 1.67×10^{-27} kg; hence

$$\text{KE} = \tfrac{1}{2}mv^2$$
$$= \tfrac{1}{2}(1.67 \times 10^{-27} \text{ kg})(10^3 \text{ m/s})^2$$
$$= 8.35 \, 10^{-22} \text{ J}$$
$$= (8.35 \times 10^{-22} \text{ J})\left(\frac{1 \text{ eV}}{1.60 \times 10^{-19} \text{ J}}\right)$$
$$= \boxed{5.22 \times 10^{-3} \text{ eV}}$$

The wavelength of the guide wave is calculated by the de Broglie equation:

$$\lambda = \frac{h}{mv} = \frac{6.63 \times 10^{-34} \text{ J·s}}{(1.67 \times 10^{-27} \text{ kg})(10^3 \text{ m/s})}$$
$$= 3.97 \times 10^{-10} \text{ m} = \boxed{3.97 \text{ Å}}$$

This wavelength is comparable with the distances between atoms in crystals, and hence a beam of such neutrons passing through a crystal can give rise to interference and diffraction just as can a beam of x rays. Figure 27-10*a* shows an x-ray diffraction pattern produced by a NaCl crystal, and Fig. 27-10*b* shows a neutron diffraction pattern of a NaCl crystal. The arrangement of spots is determined by the crystal structure of NaCl; the patterns show many similarities. Diffraction of x rays and neutrons is a powerful tool for crystal structure analysis.

An ordinary microscope such as we discussed in Sec. 26-4 uses visible light of wavelength several thousand angstroms. We can call such an instrument a *photon microscope*, in recognition of the fact that a ray of light is a beam of photons. The *electron microscope* uses a beam of electrons instead of a beam of photons. Electrons are reflected from a specimen and focused by electric and magnetic fields in such a way that an enlarged image of the object is formed. The image is made visible when the focused

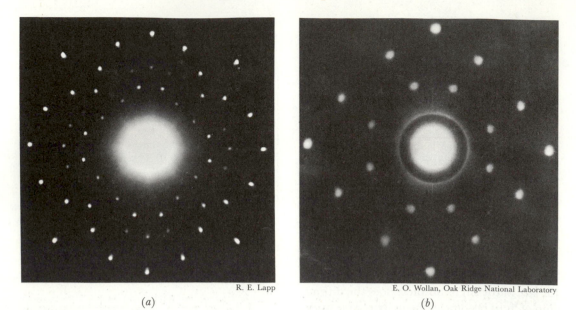

R. E. Lapp

(a)

E. O. Wollan, Oak Ridge National Laboratory

(b)

Figure 27-10 Wave nature of photons and neutrons. (*a*) X-ray diffraction by a single crystal of NaCl. (*b*) Diffraction of neutrons by a single crystal of NaCl.

beam of electrons strikes a fluorescent screen similar to the face of a television picture tube (Fig. 27-11). The great advantage of an electron microscope is its high resolving power. Objects as small at 2.3 Å have been resolved, a feat forever beyond the capability of a microscope using ordinary light. We saw in Sec. 26-4 that high resolving power requires a small ratio of wavelength to diameter of lens. In the electron microscope λ/D can be made negligibly small, and the quality of the image is limited only by the various aberrations. We have been comparing two kinds of microscopes: the photon microscope, which uses guide waves several thousand angstroms long, and the electron microscope, which uses guide waves less than one angstrom long.

27-7   The uncertainty principle

The dual nature of light implies an uncertainty principle which would have no meaning in a single theory, whether a wave theory

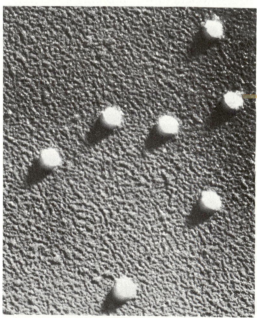

Samuel C. Silverstein, The Rockefeller University

Figure 27-11 Electron microscope photograph of reovirus particles; total magnification about 60 000. Extremely high resolving power is possible because of the small wavelengths associated with electrons. Detail down to about 50 Å is clearly shown in this photograph.

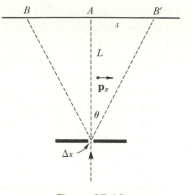

Figure 27-12

or a particle theory. As applied to photons, the uncertainty principle can be described with the help of Fig. 27-12, in which we illustrate the diffraction of light waves by a single slit. According to Sec. 25-5, almost all the light energy goes into the central band BB' of the diffraction pattern. The narrower the slit, the wider the central broad diffraction band on the screen. In symbols, $\sin \theta = \lambda/\Delta x$, where Δx is the width of the slit. Let us look at this experiment from the photon point of view. The fact that the photons spread out means that while passing through the slit they have some sideways momentum (p_x) as well as forward momentum.[*] Since photons reach all parts of the central diffraction band BB', they must have different values of p_x. In fact, the sideways momentum of a photon may lie anywhere between 0 (for a photon which reaches A) and Δp (for a photon which reaches B or B'). That is, there is an uncertainty of Δp in our knowledge of the momentum of the photons which pass through the slit. The position of the photons is also uncertain; they may pass through any portion of a slit whose width is Δx. If we try to locate the position of any given photon more precisely, by narrowing the slit, we decrease the uncertainty in x, but, according to $\sin \theta = \lambda/\Delta x$, the beam spreads

[*] We use the symbol p for linear momentum mv.

out, and the uncertainty in the sideways momentum becomes greater. Thus when Δx is small, Δp is large, and vice versa. The relationship between Δp and Δx is a simple one, easily derived from Fig. 27-12. We treat the photon as a projectile, just as in Sec. 3-11. The x-motion and the y-motion are separate, both taking place during the same time. If t is the time of flight from slit to screen, we have $L = ct$, and $s = (\Delta v_x)t = (\Delta p/m)t$. We start from the experimentally verified formula for the sine of the angle of diffraction (see Eq. 25-2):

$$\sin \theta = \frac{\lambda}{\Delta x}$$

Hence, approximately,

$$\frac{s}{L} \approx \frac{\lambda}{\Delta x}$$

Substituting for s and L, we get

$$\frac{(\Delta p/m)t}{ct} \approx \frac{\lambda}{\Delta x}$$

whence

$$\Delta p \, \Delta x \approx mc\lambda$$

Since $\lambda = h/mc$ (see Sec. 27-4), we get

$$\Delta p \, \Delta x \approx mc \left(\frac{h}{mc}\right)$$

or, for this experiment,

$$\Delta p \, \Delta x \approx h \qquad (27\text{-}11)$$

The product $\Delta p \, \Delta x$ might be greater than h, but no experiment has ever been imagined which could give a value of $\Delta p \, \Delta x$ much less than Planck's constant.[*]

Werner Heisenberg in about 1925 proposed that Eq. 27-11 represents a natural limitation to our knowledge. We have met with such natural limitations before: the maximum efficiency of a Carnot heat engine is $(T_2 - T_1)/T_2$; the maximum

[*] The precise formulation of the principle depends upon how "uncertainty" is defined. If the uncertainties are defined by usual statistical procedures, the inequality reads $\Delta p \, \Delta x \geqslant h/4\pi$.

speed of a particle is 3×10^8 m/s; the sharpness of the image of a distant star is limited by the ratio λ/D. These limitations are inherent in nature and are not due to heat losses, friction, lens imperfections, or the like. Heisenberg's uncertainty principle applies to matter as well as to photons; it is impossible to know *precisely* both the position and the momentum of a particle. As an armchair experiment, let us try to determine the exact position of an electron by looking at it through a microscope. To achieve high precision, we would try to avoid diffraction effects by using light of very short wavelength, perhaps x rays in an (imaginary) x-ray microscope. Photons of such light have high energy, according to $E = h\nu$, and would give the object a recoil velocity, thus rendering its position uncertain (Compton effect). If we use long-wavelength light to avoid recoil, then we lose resolving power. It turns out that here, too, the Heisenberg uncertainty principle gives the relationship between the uncertainty in position and the uncertainty in momentum for a particle of matter.

We have come a long way from the simple cause-and-effect mechanics of Newton's day. It was thought then that one could exactly predict the future position and momentum of any body, if one knew its original position and momentum precisely enough. It should be sufficient to apply Newton's laws of motion, and the only reason that one could not predict the future course of the universe was thought to be that the mathematical solution of so many simultaneous equations (one for each particle in the universe) would be too difficult. Now we see that the future path of a particle cannot be exactly predicted. The unavoidable uncertainty in our knowledge of the initial position or momentum, or both, affects our knowledge of the future. It has often been said that the law of causality has been abandoned; this is a true statement only when applied to a single particle's mo-

tion. The predictions of modern physics are probabilistic, as discussed at the end of Sec. 27-5. We cannot say where any given photon will go, but the *observable* photograph of a double-slit interference pattern is quite definite, being an average effect for many billions of photons. An exactly similar interpretation can be made when a stream of particles passes through a pair of slits; the de Broglie waves are waves of probability, and, although the path of any single particle is subject to uncertainty, the average effect for many particles is exactly predictable (Fig. 27-13). We have seen that the very act of measuring position introduces an unavoidable error in our knowledge of momentum. The uncertainty principle is a limitation upon our *knowledge* of nature. Whether nature is "actually" determinate is an unanswerable question, since only measurable quantities (probabilities) can enter into a physical theory.

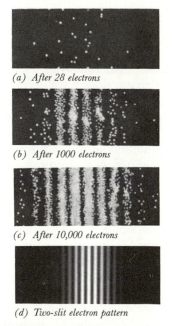

(a) *After 28 electrons*

(b) *After 1000 electrons*

(c) *After 10,000 electrons*

(d) *Two-slit electron pattern*

Figure 27-13 Double-slit interference pattern for electrons. (*a*), (*b*), (*c*) Computer simulation of growth of pattern. (*d*) Actual photograph of two-slit electron diffraction pattern. (*a, b, c* by Dr. Elisha Huggins; *d* by Dr. C. Jönsson)

All electrons are identical, in both charge (Millikan) and rest mass (J. J. Thomson), regardless of their origin.

Electromagnetic radiation is emitted or absorbed a photon at a time, with the photon energy given by $E = h\nu$ (Planck). The work function is the energy required to remove an electron from a metal surface. The KE of a photoelectron equals the energy of the incoming photon minus the work function (Einstein). At the threshold, the photon energy just equals the work function, and the electrons are emitted with zero KE. Collision experiments show that x-ray photons are like concentrated bullets, having momentum as well as energy (Compton). The dual nature of light is shown by such experiments. It is not necessary to make a model for light, but if a model is desired, the interpretation of e-m waves as guide waves, or probability waves, for the photons is as good as any.

Matter as well as light has both wave and particle properties. For each, the wavelength of the guide wave is given by $\lambda = h/\text{momentum}$ (de Broglie, Davisson, G. P. Thomson). Matter waves have such small wavelengths that they can be ignored except when dealing with single elementary particles such as electrons, protons, or neutrons.

According to the uncertainty principle (Heisenberg), the product of the uncertainty in position and the uncertainty in linear momentum is never much less than Planck's constant, and may be much more. The principle applies both to photons and to particles of matter. The law of causality in physics is a statistical (probabilistic) one, and the motions of individual particles or photons cannot be predicted with certainty.

■ CHECK LIST

Planck's constant	$E = 12\ 400\ \text{eV} \cdot \text{Å}/\lambda$	$\lambda = h/mv$
photon	$\frac{1}{2}mv^2 = h\nu - w$	$\Delta p\ \Delta x \approx h$
$E = h\nu$	Compton effect	Heisenberg uncertainty
work function of a surface	guide waves	principle
electron volt		

■ QUESTIONS

27-1 Suppose there were two kinds of electrons: type A of charge 1.00×10^{-19} C, and type B of charge 2.20×10^{-19} C. Suppose also that the two kinds of electrons occurred at random in nature, equally plentifully in all kinds of atoms. Could this hypothesis about two kinds of electrons be tested by the oil-drop experiment? Could it be tested by electrolysis experiments?

27-2 In Fig. 27-2, in what direction (up or down) will the shadow move if a magnetic field directed toward the reader is turned on?

27-3 Do all electrons have the same mass? Do all electrons have the same rest mass?

27-4 Ordinary photographic film is least sensitive to red light, hence the use of red safelights in darkrooms. Explain, using the photon model of light.

27-5 It is harder to remove a free electron from copper than from sodium. Which metal has the greater work function? Which has the higher photoelectric threshold frequency?

27-6 If a proton is accelerated through a PD of 10^5 V, what is its KE, in electron volts?

27-7 If traveling at the same speed, which particle has a guide wave of longer wavelength: a proton or an electron?

27-8 Has the theoretically possible high resolving power of the electron microscope been achieved as yet?

27-9 Is the future of the physical world predictable? Discuss this question, both pro and con.

■ PROBLEMS

27-A1 Calculate the energy of a photon of ultraviolet light whose wavelength is 1220 Å. Express your answer both in joules and in electron volts.

27-A2 Calculate the wavelength of a photon whose energy is 2.30 eV. Express your answer both in meters and in angstrom units.

27-A3 A particle has an energy of 1 μerg. Express this energy in million electron volts (MeV).

27-A4 A speck of dust of mass 10 μg is lifted 1 cm. Calculate the increase in gravitational PE (a) in joules; (b) in million electron volts (MeV).

27-A5 Calculate the wavelength associated with a particle of mass 10^{-19} kg moving at a speed of 10^{-6} m/s.

27-A6 Calculate the wavelength associated with an electron of mass 9.11×10^{-31} kg moving through a vacuum tube with speed 2×10^7 m/s.

27-B1 An oil drop for which the density is 800 kg/m³ is balanced between two plates 6 mm apart. The PD between the plates is 1200 V. If the drop has a charge of seven electrons, calculate its radius. (*Hint:* First find the drop's mass, then its volume.)

27-B2 Calculate the mass of an oil drop carrying a charge of five electrons, which is just balanced between two plates which are 2 cm apart. The PD between the plates is 3000 V.

27-B3 Using the known value of e/m_0, calculate the radius of circular path in Fig. 27-3 for electrons which have been accelerated through a PD of 300 V. The magnetic field is 3×10^{-4} T. (Relativity effects can be ignored.)

27-B4 If the magnetic field in Fig. 27-3 is that of the earth ($B = 6 \times 10^{-5}$ T), what accelerating voltage V would give the electron beam a radius of curvature 0.25 m?

27-B5 What is the KE in pJ of a helium ion He^{++} which is accelerated through a PD of 1 GeV?

27-B6 Use Table 14-4 to find the approximate ratio of the energy of an x-ray photon to that of a photon emitted by an FM broadcasting station. Why is the photon nature of the radio radiation not easily observed?

27-B7 What is the maximum KE of the photoelectrons ejected from a piece of tungsten, for which the work function is 4.52 eV, by ultraviolet radiation of wavelength 2500 Å?

27-B8 The work function of a surface is 2.10 eV. Calculate the maximum KE of electrons ejected from this surface by radiation of wavelength 2100 Å.

27-B9   Light of wavelength 4000 Å shines on a thin metal surface, and the ejected photoelectrons have KE equal to 0.52 eV. Calculate the work function of the surface.

27-B10   Calculate the work function of a surface from which the yellow light (of wavelength 5893 Å) from a flame containing sodium ejects photoelectrons of maximum KE 0.56 eV.

27-B11   What is the threshold wavelength for a tungsten surface whose work function is 4.52 eV?

27-B12   What is the work function of a surface whose threshold wavelength is 5500 Å?

27-B13   Monochromatic visible light shines on the cathode of a photocell for which the work function is 1.14 eV. The negative "stopping potential" needed to prevent photoelectrons from reaching the anode is 1.80 V. What color is the light?

27-B14   Can the yellow "sodium light" ($\lambda = 589$ nm) eject photoelectrons from a clean metallic sodium surface for which the work function is 2.46 eV?

27-B15   As the wavelength of light striking a surface is gradually decreased, emission of photoelectrons starts when λ reaches 400 nm. What is the maximum KE of the emitted photoelectrons when λ reaches 300 nm?

27-B16   (*a*) What is the momentum of an x-ray photon whose wavelength is 1.54 Å?   (*b*) What is the (relativistic) mass of this photon?

27-B17   Revise the photoelectric equation (27-6) to be applicable to relativistic electrons with speeds not negligible compared to *c*.

27-B18   Calculate the momentum of a photon of the yellow sodium light whose wavelength is 5890 Å.

27-B19   An x-ray photon of wavelength 1.00 Å strikes a free electron, and the wavelength of the scattered photon is found to be 1.01 Å. What is the KE of the recoiling electron?

27-B20   Calculate the wavelength of the guide waves associated with a rifle bullet of mass 25 g moving at 800 m/s. Does the bullet show any observable wave behavior?

27-B21   (*a*) Suppose that the position of the bullet in Prob. 27-B20 is known to within 2 cm (the diameter of the rifle barrel); what is the least uncertainty of its momentum?   (*b*) If the gun is aimed at a target 1500 m away (time of flight 3 s), by how much might it miss the target due to the Heisenberg uncertainty principle?   (*c*) Why do some bullets miss the target by much more than the amount calculated in (*b*)?

27-B22   Derive a formula for the de Broglie wavelength of a particle of charge *Q* moving in a circular path of radius *R* in a magnetic field *B*.

27-B23   What is the ratio of the KE of an electron to that of a proton which has the same de Broglie wavelength? Assume nonrelativistic speeds. (*Hint:* Solve first in symbols, then use data from the Appendix.)

27-B24   Calculate the de Broglie wavelength of an average O_2 molecule at 0°C (see Example 16-1). Compare your answer with the size of the molecule (see Sec. 9-6).

27-C1   If the wavelength of light striking a certain surface is 1000 Å less than the threshold wavelength, the maximum KE of the ejected photoelectrons is 1 eV. Calculate the threshold wavelength for this surface.

27-C2   An explorer on the moon turns on a 100 W lamp, and the light from the bulb spreads out equally in all directions. Assuming that 1% of the energy is radiated as visible light (of average wavelength 5550 Å), about how many photons per second of visible light from the bulb would strike the mirror of a large telescope on the earth? (Diameter of the mirror = 5 m; moon's distance = 3.8×10^8 m; neglect absorption in the earth's atmosphere.)

27-C3   What is the wavelength of a photon whose momentum is the same as that of a 5 eV electron? In what region of the e-m spectrum (Table 14-4) is this photon?

27-C4 Example 27-2 surely illustrates conservation of energy, which is the basis of the photoelectric equation. Study this same example with respect to *momentum* conservation. Calculate, in mks units: (*a*) the momentum of the incoming photon; (*b*) the momentum of the ejected electron. (*c*) Explain how the electron's momentum can be greater than the photon's momentum, as in this example.

27-C5 Ideally, in Fig. 27-3 the electron's horizontal velocity v_x is 0 as it passes vertically through the first slit. The width Δx of the first slit introduces an uncertainty Δv_x which in turn introduces an uncertainty $\Delta x'$ in the horizontal position of the beam at the second slit. Using data from Prob. 27-B3, find the time of flight and hence $\Delta x'$, if Δx is 0.001 mm. From these data, decide whether the uncertainty principle is significant for this experiment.

■ REFERENCES

1 Magie, W. F., *A Source Book in Physics* (New York: McGraw-Hill, 1935), pp. 578–79, discovery of photoelectric effect by Hallwachs in 1888; pp. 583–97, discovery of the electron by J. J. Thomson, a method different from that described in the text—his conclusions are on pp. 596–97; pp. 600–10, Roentgen's original papers relating the discovery of x rays; pp. 354–56, Kirchhoff's analysis of the origin of Fraunhofer lines; pp. 360–65, Balmer's empirical formula for the wavelengths of the visible hydrogen lines.

2 Millikan, R. A., *Electrons* (+ *and* −)···, (Univ. of Chicago Press, rev. ed., 1947).

3 Thomson, G. P., "J. J. Thomson and the Discovery of the Electron," *Physics Today*, Aug., 1956, p. 19. Reminiscences by his son.

4 Dunnington, F. G., *Phys. Rev.*, **43**, 404 (1933); **52**, 475 (1937). Two research papers giving the determination of e/m by the method described in the text.

5 Darrow, K. K., "The Quantum Theory," *Sci. American*, Mar., 1952, p. 47. The work of Planck, Bohr, and Compton.

6 Furth, R., "The Limits of Measurement," *Sci. American*, July, 1950, p. 48.

7 Gamow, G., "The Principle of Uncertainty," *Sci. American*, Jan. 1958, p. 51. See also a book review (p. 111–116 of the same issue) by J. R. Newman of D. Bohm's *Causality and Chance in Modern Physics*—the argument of a physicist who does not accept the uncertainty principle.

8 Darrow, K. K., "Davisson and Germer," *Sci. American*, May, 1948, p. 51. The experimental discovery of the wave nature of electrons, independently of the work of G. P. Thomson in England.

9 Walters, A. E., *The Photoelectric Effect* (film).

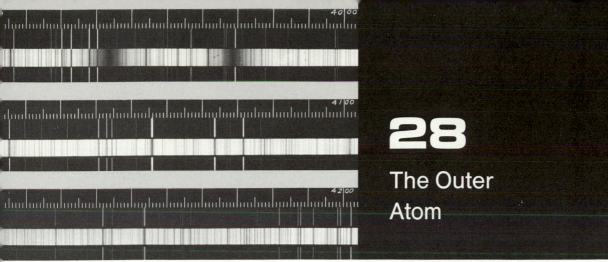

28

The Outer Atom

Almost all of our knowledge of the outer atom—the cloud of electrons surrounding the nucleus—has come to us through analysis of the light emitted or absorbed by an atom. From this analysis have come the concept of energy levels and an orbital model which considered the atom to be a miniature solar system. This model is no longer capable of giving a full account of atomic behavior, but it is a valuable first stage which gives much useful insight. We start in the year 1911, at a time when it was known that electrons and photons somehow interact on the atomic scale of things, but it was not surely known just how the positive and negative charges are distributed in an atom. Through the Rutherford scattering experiment, the atomic nucleus was discovered.

28-1 Distribution of charge in the atom

One of the great experimental physicists of the past century was Ernest Rutherford (1871–1937), a New Zealander who worked in Canada at McGill University and in England at Manchester, and later was head of the Cavendish Laboratory at Cambridge University. It was Rutherford and his co-workers who in 1908 identified the α particles shot out by radioactive atoms as helium ions. At Manchester, he directed the work of two of his students in a crucial experiment on the scattering of α particles by thin metallic foils.

It is hard for us to realize that only 65 years ago the accepted picture of an atom was quite different from the orbital model we know today. There was no doubt that the electrically neutral atom consisted of equal amounts of positive and negative charge, as shown by the existence of ions. It was also realized that electrons could be separated from an atom, as shown by the photoelectric effect and by J. J. Thomson's measurements of e/m_0 for "the corpuscle," as the electron was then called. The Bohr theory had not yet been worked out, and most physicists believed that the atom's structure must somehow allow the electrons to vibrate back and forth as they emitted e-m radiation. One model, due to Kelvin and Thomson, pictured a hydrogen atom as a sort of spherical ball of jellylike positive charge in which were embedded one or more electrons. A

Absorption spectrum of the sun (dark lines) in the violet region, 3900 Å to 4200 Å. The bright lines on either side of the sun's spectrum are from a laboratory iron arc, used for calibration purposes. Many lines coincide, showing the presence of iron vapor in the sun's atmosphere. The two wide dark lines in the upper portion are due to absorption by calcium in the sun's atmosphere. (Mt. Wilson and Palomar Observatories)

single electron's normal position would be at the center of the ball, where the net force on it would be zero; if pulled aside, the electron would oscillate back and forth in SHM and give out an e-m wave. The main problem was to explain the existence of so many spectrum lines, each of a different frequency (Fig. 28-4), for most elements emit infinitely many lines converging to various series limits.

Rutherford in 1911 explored the structure of the atom by shooting α particles through a thin metallic foil. According to the Thomson model, most of the particles would be scattered somewhat, due to the Coulomb force of repulsion between the positively charged α particles and the positive balls of atomic charge. An α particle could hardly miss coming close to some like (positive) charge as it passed through the metal, thus being deflected. According to Coulomb's law (Sec. 17-5) any repulsive force is inversely proportional to the square of the distance; in the Thomson model, the α particle is at no time very close to any concentrated bit of charge, and so it could experience no large repulsive force. Hence, according to this model, almost every α particle would suffer a moderate deflection, not exceeding a few degrees (Fig. 28-1a).

The experiment showed an entirely different result. Most of the α particles went right on through the foil with no change of direction, but a few were deflected through large angles (Fig. 28-1b). Some even bounced straight back, a deflection of 180°. To explain the large forces that cause such large deflections, Rutherford had to assume that an α particle is itself small, and that it can, if aimed correctly, approach to within about 10^{-14} m of a concentrated "center of charge" which he called the *nucleus*. From the data of crystal structure, the atoms themselves were known to be about 10^{-10} m (1 Å) apart. Thus Rutherford was led to the nuclear model of an atom: a small, heavy, positively charged central nucleus, with negative electrons separated from the nucleus by

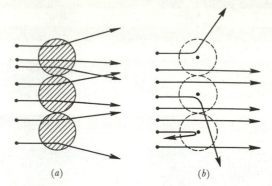

Figure 28-1 Scattering of α-particles according to two models of atomic structure. (*a*) Thomson's model; (*b*) Rutherford's nuclear model.

much empty space. To visualize these open spaces, let us imagine an atom blown up to the size of a football field. The nucleus would then be a small heavy glob of matter the size of a lead-pencil eraser; the rapidly moving electrons would be still smaller, but (averaged over a period of time) there would be a reasonably uniform probability of finding an electron at any point within the football field. To complete the picture, imagine firing another lead-pencil eraser at the football field, and add a strong repulsive force which is effective only when the two erasers are within a foot or so of each other. Small wonder that so few α particles are scattered in the Rutherford experiment!

By mathematical analysis of his scattering experiments with gold, silver, and copper foils, Rutherford was able to determine the magnitude of the central charge in each case. Expressed in units of the electron charge e, the gold nucleus turned out to have a measured charge of $+77.4e$, silver $+46.3e$, and copper $+29.3e$. Within experimental error, these numbers are the atomic numbers (Z) of the elements; for instance, silver is the 47th element in the Periodic Table. Hence there must be Z electrons outside the nucleus, to make up the volume of the atom, and they must be in planetary orbits to keep from falling into the nucleus. The scattering of x-ray photons confirms this; such photons are scattered by the electron cloud, and Barkla (also in 1911) showed experimentally that the scattering of x rays is proportional to the atomic number Z.

The Rutherford experiment is an outstanding example of how important results could (in 1911) be obtained with simple apparatus. The scattered α particles struck a fluorescent screen, which was simply a layer of small zinc sulfide crystals, and the tiny flashes of light* produced by impacts were observed with a magnifying glass. A speck of radioactive material, some slits, and the foil itself completed the apparatus by which Rutherford established the nuclear model for the atom. As so often happens, the experiment raised new problems. If the electrons move in orbits outside the nucleus, a serious question of stability arises: why should not such an electron continuously emit radiation on account of its centripetal acceleration (Sec. 23-8)? Physicists at least were now asking the right questions; and among those physicists was a young Danish student in Rutherford's laboratory at Manchester named Niels Bohr (1885–1962). We shall see in the next section how Bohr's model of the hydrogen atom followed in 1913, just two years after Rutherford's historic experiment.

28-2 Emission spectra

The quantum hypothesis which had proved so successful in explaining black-body radiation (Planck, 1900) and the photoelectric effect (Einstein, 1905) was applied in 1913 by Bohr to achieve a breakthrough in still another perplexing problem for which classical physics was inadequate. Bohr succeeded

in calculating the wavelengths of the lines in the spectrum emitted by hydrogen atoms.

In a solid, a liquid, or a dense gas, the atoms are close to each other, and they interact to give a *continuous spectrum* with all frequencies (and wavelengths) represented. However, the spectrum emitted by atoms of an incandescent gas is *discrete;* that is, there are sharp lines of definite wavelength, with no light emitted at the intermediate wavelengths. Some representative spectra are shown in Fig. 28-2. Presumably the atoms in a gas are free enough to be able to emit their own characteristic frequencies, much as a piano string vibrates with certain natural frequencies. Some energy must be supplied to start a string vibrating; likewise, an atom must absorb energy in order to emit radiation. There are many ways in which an atom can be given the necessary energy. For instance, the sodium spectrum is obtained if solid NaCl is placed in a flame; the substance breaks down into Na and Cl atoms, which absorb energy by thermal bombardment. The copper spectrum can be formed by passing an electric arc or a spark between metallic copper electrodes; the radiation comes from copper vapor between the electrodes. Gaseous substances such as helium or hydrogen can be excited electrically when enclosed in a glass tube at low pressure.

The spectrum of a *solid* or a *liquid* depends only on its temperature; molten iron in a furnace and the ceramic bricks lining the furnace look about the same—"red-hot" in each case—and they emit the same black-body spectrum, which is continuous (Sec. 14-6). On the other hand, the frequencies emitted by any incandescent *gas* are characteristic of the substance, and temperature affects only the relative intensities of the lines. The line spectrum of a gas therefore can be used as a means of identification. The spectrum of hydrogen molecules (H_2) is considerably more complex than that of hydrogen atoms (H); in addition to many

---

*You can see the random flashes caused by individual α particles from radium if you hold a luminous wrist watch very close to the eye in a dark room; the flickering can be seen without a magnifying glass, even though the watch dial is badly out of focus. It is essential that the eye become dark-adapted, which may take several minutes.

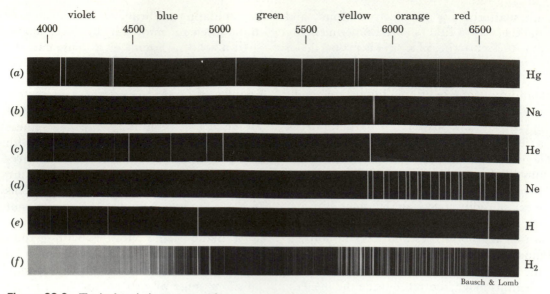

Figure 28-2 Typical emission spectra of gases; wavelengths in angstroms. The spectrum of the hydrogen molecule (f) shows two of the lines of atomic hydrogen (e), since some of the molecules in the discharge tube are dissociated into atoms; H_2 also shows a continuous spectrum in the blue-violet region.

closely spaced lines there is a continuous emission in the blue, violet, and ultraviolet region (Fig. 28-2f). In general, the term *band spectrum* is used to describe the very closely spaced line spectrum emitted by molecules that contain more than one atom.

In most spectra, the lines seem to be distributed almost at random, but hydrogen, the simplest atom, gives a spectrum which shows some regularity. For this reason, the hydrogen spectrum has long been the testing ground for new theories, and it was this spectrum which Bohr attacked. Pursuing the acoustical analogy, we might call the lowest frequency the fundamental, and the higher frequencies (shorter wavelengths) the overtones. However, the hydrogen frequencies tend toward a limit and cluster together near 8.2×10^{14} Hz; this cannot be explained on any classical model, according to which the frequencies of overtones increase without limit. Another difficulty with the classical model is that the electron in hydrogen could not be stationary, for it would then fall toward the positive nucleus.

This leads to the expectation that the electron would have to move in some sort of orbit; but then its centripetal acceleration would cause e-m radiation (Sec. 23-8), and the electron would spiral into the nucleus as it lost energy. We know, however, that hydrogen atoms can exist for long times without radiating energy.

Bohr made a virtue of necessity by *postulating* that an electron can revolve around a nucleus without radiating. Since the atom can absorb energy and re-emit it as radiation, Bohr assumed that an atom can exist in any one of a number of radiationless *energy states*, E_1, E_2, E_3, ... Next, Bohr postulated that Einstein's photoelectric equation for the energy of a photon applies within the atom. When an atom changes from one energy state E_2 to another lower energy state E_1, the difference in energy is $E_2 - E_1$. Bohr's assumption was that all this energy is used in forming a single photon. Thus the energy of the emitted photon is given by

$$h\nu = E_2 - E_1 \qquad (28\text{-}1)$$

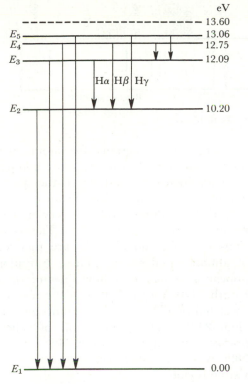

eV
- - - - - - - - - - - - - - - - - - 13.60
E_5 ———————————————— 13.06
E_4 ———————————————— 12.75
E_3 ———————————————— 12.09

Hα |Hβ| Hγ

E_2 ———————————————— 10.20

E_1 ———————————————— 0.00

Figure 28-3   Energy levels for hydrogen.

An *energy-level diagram* such as Fig. 28-3 helps us visualize these postulates. Each allowed energy state is represented by a horizontal line; we choose our reference level of zero energy to be the normal (unexcited) state of the atom. The hydrogen atom can exist in radiationless states having energies 0 eV, 10.20 eV, 12.09 eV, etc., but an energy of 5.00 eV, for instance, is impossible. The arrows represent *transitions,* or *jumps,* between energy states, and the length of any arrow is proportional to the frequency of the photon which is emitted during the transition, according to the equation $h\nu = E_2 - E_1$. Any given atom emits only one photon at a time, and the complete spectrum containing many lines is an average effect formed by many atoms emitting many photons. Imagine a series of photon counters replacing the photographic plate in a spectroscope, with the counters arranged so that each counter receives light of one of the characteristic wavelengths. The counters click at random, each click corresponding to an atomic process during which some one atom's electronic energy changes abruptly by some amount ΔE. If, say, the mercury green line is twice as bright as one of the yellow lines, this merely means that one transition is twice as probable as the other, and more atoms per second emit the green photons. On the average, the light of either color appears to be steady. In a similar fashion, heavy rain on a tin roof makes a steady sound, although the noise is actually due to a series of discrete raindrops, falling at random.

These two postulates of Bohr are as good today as in 1913. His third postulate gave a rule for finding the values of the "allowed" energy levels. Bohr was led to assume that the allowed energies correspond to certain allowed orbits for the electrons; those in orbits of larger radius have larger energies. The third Bohr postulate states that the allowed orbits are those for which the angular momentum ($I\omega$; see Sec. 8-7) of the electron is an integral multiple of Planck's constant divided by 2π. That is,

$$\text{Angular momentum} = n\frac{h}{2\pi} \qquad (28\text{-}2)$$

where n is an integer. Using this postulate, Bohr was able to calculate the energies of the hydrogen atom's energy levels, and hence, using the differences between these energies, he could calculate the frequencies of the emitted photons. He obtained exact agreement, within experimental error, with the observed frequencies. The Bohr theory for the hydrogen atom is discussed in greater detail, using elementary calculus, in Sec. 28-7.

As so often happens, Bohr's third postulate (Eq. 27-9) turned out to be an approximation. Modern quantum mechanics uses a different, more general postulate which (of course) contains Bohr's as a special case.

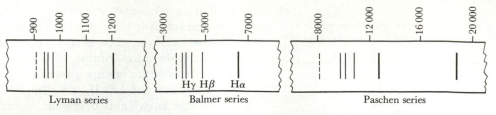

Figure 28-4 Spectral series in hydrogen.

This is the usual course of events, and should cause us no concern. Bohr's bold and imaginative theory of the hydrogen atom was a necessary first step toward a complete theory valid for all atoms.

The lines in the hydrogen spectrum occur in several groups known as "series" (Fig. 28-4), which are named after the men who first investigated them, before the Bohr theory. The energy level diagram gives a ready explanation for these series. For instance, the Balmer series contains lines emitted during quantum jumps $E_3 \rightarrow E_2$, $E_4 \rightarrow E_2$, $E_5 \rightarrow E_2$, . . . , always with the second energy level E_2 as the final state. For the lines of the Lyman series, it is the level E_1 that is the final state. The Lyman series lies entirely in the ultraviolet region, since the quantum jumps are large and the photons have high frequency and short wavelength. The Paschen series, for which E_3 is the final state, and other higher series lie entirely in the infrared region. It so happens that the entire *visible* spectrum of hydrogen contains only members of one series, the Balmer series. This is why the hydrogen spectrum appears so simple.

EXAMPLE **28-1**

From Fig. 28-3 estimate the wavelength of the second Balmer line, Hβ in Fig. 28-4.

Since $h\nu = E_4 - E_2$ for the second line in the series, the photon's energy is 12.75 eV − 10.20 eV = 2.55 eV.

$$\lambda = \frac{12\,400}{E} = \frac{12\,400}{2.55} = \boxed{4860 \text{ Å}}$$

This line is in the blue-green region of the spectrum.

Energy-level diagrams for atoms with more than one electron are more complex than for hydrogen, and hence the spectra are also more complex. For instance, two of the mercury energy levels are very close together (E_8 and E_9 in Fig. 28-5). The transitions $E_8 \rightarrow E_5$ and $E_9 \rightarrow E_5$ are therefore of almost equal energy. These transitions appear as a pair of yellow lines of wavelength 5770 Å and 5790 Å in the mercury spectrum (Fig. 28-2a). The sodium spectrum (Fig. 28-2b) has a strong pair of yellow-orange lines of wavelength 5890 Å and 5896 Å, and 13 other very weak lines in the visible region.

28-3 Absorption spectra

An atom must, somehow, absorb energy if it is to be able to emit radiation at a later time. When light is absorbed by a gas atom, a photon disappears and the atom's internal energy increases. Since the atom's energy can change only by discrete amounts, such as ΔE in Fig. 28-5, only certain photons can be absorbed. In fact, it should be apparent that *an atom can absorb only those photons it is capable of emitting*, since the equation $\Delta E = h\nu$ applies equally well for a transition in either direction. It was discovered by Joseph von Fraunhofer in 1814 that the sun's spectrum consists of dark lines across a continuous background (p. 609); these lines are caused by absorption by the sun's atmosphere, which contains Na, Fe, H, and over 60 other elements in gaseous form. The frequency of any absorption line is exactly equal to that

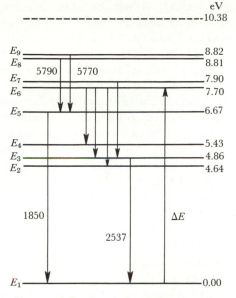

Figure 28-5 Energy levels for mercury.

of a line emitted by the same atom when excited. This reminds one of the phenomenon of acoustic resonance (Sec. 11-4), but the quantum explanation avoids using a mechanical model such as a resonating string or air column. The absorption spectrum is a powerful tool for chemical analysis, since the characteristic dark lines are caused by a gas which is not necessarily hot enough to emit light.

To understand the relation between absorption and emission we must pay close attention to probabilities and lifetimes. At ordinary temperatures, an atom spends almost all its life in its lowest energy state (E_1), where its total energy is least. If by chance an incoming photon gives up its energy to the atom, or if energy is available during a collision, the atom can be *momentarily* raised to one of its allowed excited states. An atom usually stays in an excited state for only about 10^{-8} s, and then if left to itself emits a photon and spontaneously "jumps" to a lower state. After a very short time, the atom has been "shaken down" to its normal state, and one or more photons

have been emitted in the process. For example, after mercury vapor has been illuminated with photons of energy $E_6 - E_1$ (Fig. 28-5), an atom which has been excited to the state E_6 can radiate any one of several photons in returning (by stages) to its lowest level. The emission of light which is thus dependent on the absorption of other light is called *fluorescence*. We see from the energy-level diagram that fluorescent radiation always has a longer wavelength than the radiation that caused the excitation, since the fluorescent photons are emitted during smaller energy jumps. Fluorescence in solids is more complex than in gases, and if the fluorescent photons are emitted after a measurable time interval, we speak of the phenomenon as *phosphorescence*. There is no sharp dividing line between fluorescence and phosphorescence. The fluorescent lamps used in homes and factories are coated with certain "phosphors" which absorb ultraviolet photons (having large hv values) from the mercury arc in the gas within the tube, and later on emit photons of visible light (low hv values).

Certain energy states have "long" lifetimes of 10^{-3} s or even, in some cases, 1 s or more. These energy levels are called *metastable states* (they are "almost stable"). In practice an atom in a metastable state is more likely to return to a lower state by giving up its excitation energy in a collision with a gas atom, ion, or free electron than by radiating a photon. In mercury the energy level E_2 at 4.64 eV is a metastable state; note that no photon transition $E_2 \rightarrow E_1$ is drawn in Fig. 28-5. The mercury atom can, of course, absorb 4.64 eV in a collision process, or if it is already in state E_2 it can give up 4.64 eV of energy in a collision and thus return to the lowest state. A famous example of metastable states occurs in ionized oxygen atoms: high in the earth's atmosphere the density of air is so small that collisions are very infrequent, and the excited metastable state can endure long enough (1 s, on the average) for photons to be emitted. This accounts for spectrum lines first observed in the aurora borealis which cannot easily be observed in sources in the laboratory where collisions are more frequent.

If enough energy is available either from collision or from an incoming photon, an electron can be removed from an atom, leaving behind a positive ion. The dotted lines at the top of the energy-level diagrams of Figs. 28-3 and 28-5 represent ionization energies. For instance, 12.75 eV of energy is insufficient to ionize hydrogen; but this amount of energy can be absorbed and retained for a short time. On the planetary model, we say that the electron has been placed in a larger orbit. If exactly 13.60 eV is absorbed, the electron is just removed from the atom, and if 13.80 eV is absorbed, the electron is removed to infinity (which in this case means 10^{-8} m or so) and given an extra 0.20 eV of KE to boot, as it flies away. We speak of 13.60 V as the *ionization potential* of hydrogen; according to the diagram, it is just equal to the photon energy (in eV) of the most energetic line of the Lyman series (right at the series limit). Experiment confirms the spectroscopic value of the ionization potential. Conduction of electricity through a gas depends on the presence of ions, and the current through a hydrogen-filled tube suddenly increases when the applied PD reaches 13.60 V.

28-4 Coherent light—the laser

Let us look a little more closely at the way in which a photon interacts with an atom. If a photon's energy $h\nu$ is not equal to the energy difference ΔE between two allowed energy states of an atom, either the photon goes right on by the atom and there is no interaction, or, if it collides with the atom, the collision is perfectly elastic and no energy is transferred. If, however, the photon's energy $h\nu$ exactly matches the energy difference ΔE, it will be able to cause a transition in *either* direction. Such a transition caused by a photon is called a *stimulated transition*. This is a form of resonance; the incoming photon is "tuned" (in energy) to a pair of energy levels in the atom. In discussing absorption, we assumed that the incoming photon of the proper energy induces or stimulates an upward transition, for instance $E_1 \rightarrow E_6$ in Fig. 28-5. But the same photon would be equally effective in causing a downward transition $E_6 \rightarrow E_1$, because in this case the photon's energy also matches the energy-level difference. In a downward stimulated transition, after the encounter there are two identical photons (the incoming one and the emitted one) each of energy $h\nu$ $(= E_6 - E_1)$; the energy of the extra photon has come from the stored PE of the atom as it changes from state E_6 to the lower energy state E_1.

Suppose now a beam of photons of the proper energy passes through a container in which atoms are enclosed. Some photons will stimulate certain atoms to absorb energy $(E_1 \rightarrow E_6)$; other photons will stimulate other atoms to emit radiation $(E_6 \rightarrow E_1)$. Even though the probability of a stimulated emission exactly equals the probability for stimulated absorption, there is a *net* absorption because normally the population of a higher energy level is much less than that of a lower energy level. In a *normal population* there are more atoms in state E_1 (ready to absorb) than there are in state E_6 (ready to emit); the relative numbers depend strongly on the temperature. If, somehow, an *inverted population* can be achieved so that there are more atoms in the upper state than in the lower state, then instead of absorption there can be a net emission of photons; this build-up of the number of photons in the container amounts to an amplification process.

Our understanding of this light amplification process has led to the development of a new source of extremely intense coherent light—the *laser*. To obtain laser action (the word is an acronym for *l*ight *a*mplification by *s*timulated *e*mission of *r*adiation) three conditions must be met. (1) The atoms must have an upper energy state in which electrons linger longer than usual—a metastable state—so that there is time for the stimulating impact before the normal spontaneous emission takes place. (2) There must be a means of raising far more electrons to this excited metastable state than would be there normally as a result of ordinary temperature-limited collision processes, so that this upper state *alone* becomes as highly populated as it would be at an extremely high temperature, but without actual heating of the material. This is an inverted population. Finally, (3) the emitted photons must be reflected back and forth through the excitable atoms very many times so that stimulated emission can make up for the escape of photons and keep the process going.

The third condition is met in practice by enclosing the excitable atoms in a box—a resonant cavity—with both ends consisting of perfectly parallel mirrors, but with one or both mirrors slightly transparent to "leak out" enough photons to be used as an external beam. The resulting laser radiation has several valuable features. Because of the reflecting box system, the beam has a very small angular divergence (limited only by diffraction effects related to the diameter of the mirror considered as an aperture; and here λ/D is extremely small). Other sources of radiation emit from exposed surfaces in all directions with considerable divergence. This parallelism of the laser beam allows greatly increased intensity, that is, more energy per second from a laser beam can be focused on a smaller spot. In addition, the phase of the stimulated emission is identical with that of the incident radiation, so that the entire emerging beam consists of coherent radiation (Sec. 25-8). The laser wave fronts are at all times surfaces of constant phase,

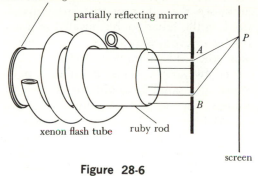

Figure 28-6

in contrast to the wave fronts of ordinary, "incoherent" light.

A simple experiment demonstrates the coherence of a broad laser beam: an interference pattern can be obtained merely by putting two slits in front of a laser (Fig. 28-6). A maximum occurs at P if the path difference $BP - AP$ equals an integral number of wavelengths. Even though rays passing through A and B come from different parts of the laser, their phase relationship remains constant. In exactly similar fashion, the phases of two radio transmitters can be maintained constant (by electrical means) to give an interference pattern such as was discussed in connection with Fig. 25-10 on p. 540. If, instead of a laser, we used a broad source of ordinary light (made monochromatic by use of a filter), no interference would be observed (Fig. 28-7). To be sure, a momentary pattern on the screen might

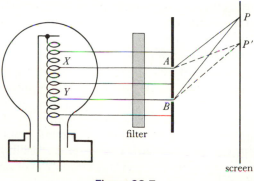

Figure 28-7

endure for a nanosecond or so (10^{-9} s) if we could see the effect of just two separate atoms in the broad source, located at X and Y, which were at that moment in phase with each other. Due to random charges in phase caused by atomic collisions in the source, atoms at X and Y would surely have an entirely different phase relationship a microsecond later (10^{-6} s), so a new point P' would be a maximum, with a change in path difference compensating for a change in phase. The combined effect of the superposed interference patterns of many atoms in the source emitting waves having random phases would average out to give a uniformly illuminated screen, and no pattern would be seen. Interference *can* be observed with ordinary light, but only if a single wave front is split into several portions which travel different path lengths and are recombined; study again the various arrangements illustrated in Chap. 25 (Figs. 25-3, 25-4, 25-7, 25-8, 25-25, 25-29, 25-31) with this point in mind.

Practical lasers (Fig. 28-8) fall into two classes. Systems with atoms in the gaseous phase, such as the helium-neon laser, may be operated so as to emit continuously. These have rather low total power output (although they still have beam intensities exceeding those of the best searchlights, per unit power input, owing to their much smaller angular divergence). Such lasers are stable, convenient to use, and emit very sharp spectrum lines. Systems with atoms in the solid state, such as the ruby laser, usually have to be operated in discrete flashes. In these lasers the electrons are often raised to metastable states by bathing them in an intense flash of (incoherent) white light from a xenon flash tube similar to a photographic "strobe" light. Solid-state lasers have been employed to produce phenomenal intensities (when focused down to small areas) capable of vaporizing diamonds, drilling fine holes in steel or tungsten, repairing detached retinas, and similar feats. On a

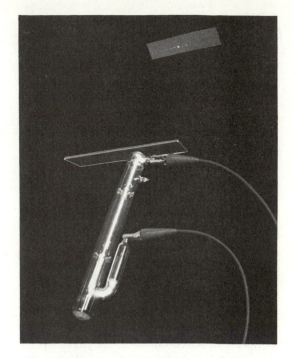

Figure 28-8 Coherent light from helium-neon laser tube passes through grating ruled on glass. Several orders of interference are seen on each side of the central image.

less spectacular scale, the large electric field intensities in laser beams have opened up entirely new fields of fundamental physics. For instance, it has been found that the dielectric constant of any given substance is not constant, but varies with the electric field strength. An industrial use of lasers is shown in Fig. 28-9.

Some other applications of laser radiation are in telecommunications and in photography. The phase coherence of laser light makes it exactly similar to e-m radiation generated by electrical circuits, except for a tremendously higher frequency. Many new channels of communication can be utilized for voice, pictures, or data transmission by modulating the electric field of a coherent laser beam. The coherence of laser light makes possible photographic images without lenses. These images, formed with the help of holograms, are discussed in Sec. 25-11.

Figure 28-9 A sharply focused laser beam produces high temperatures when it passes through matter.

For further applications of lasers and masers to a variety of problems, see the references at the end of this chapter.

28-5   X rays

The electrons in any atom are arranged around the nucleus in such a way that some are much harder to remove than others. If we use the Bohr planetary model, we group the electrons into "shells" called the K-, L-, M-, N-, etc., shells. Electrons in the K-shell are closest to the nucleus and are hardest to remove. An L-electron has more energy than a K-electron in the same atom; it can be compared to a satellite in orbit, partly removed from the earth. Chemical differences between atoms are due to their electronic structures. For instance, sodium has 2 K-electrons, 8 L-electrons, and 1 M-electron. The M-electron, being in an outer shell, is easily removed; this is why sodium is so active chemically and readily forms a Na$^+$

ion. The next atom in the Periodic Table (p. 721) is magnesium, which has 2 K-electrons, 8 L-electrons, and 2 M-electrons. With 2 electrons in its outer shell, Mg is chemically quite different from Na.

It might seem that all 11 electrons of a sodium atom, for instance, should be K-electrons, since an electron tends to fall to the lowest available energy level. All atoms would then be chemically similar. To explain why this is not so, a new postulate was introduced by the Austrian physicist Wolfgang Pauli. The Pauli exclusion principle limits the number of electrons in each shell —2 in the K-shell, 8 in the L-shell, 18 in the M-shell, etc. The 11th electron in sodium is an M-electron, whose energy is considerably greater than that of the "lower" K- and L-electrons, and yet it cannot fall in because the inner shells are "full." The Pauli exclusion principle is a powerful description of one aspect of nature; from it can be deduced the Periodic Table and, in principle, much of chemistry.

With this background, we are ready to consider the origin of x rays. The lines of the visible spectrum of an atom arise when the *outer* electrons gain and then lose a few electron volts of energy. Far greater photon energies can be obtained by first removing one of the inner electrons. This can be accomplished in an x-ray tube (Fig. 28-10) in which high-speed electrons strike a target. The energy required to remove an electron from an atom is called the *binding energy* of that electron. To create an ion by knocking out an inner electron of an atom of the target material requires much work; for instance, the binding energy of a K-electron in molybdenum is some 20 000 eV (hence the need for a high voltage to operate an x-ray tube with a molybdenum target). Once a K-electron has been removed, an energy level is available, and an L-, M-, or N-electron can fall down. During this rearrangement of the electrons, a photon is emitted whose energy is given by $\Delta E = h\nu$.

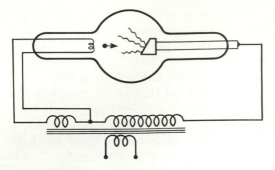

Figure 28-10 X-ray tube. Electrons, emitted from the heated filament, are accelerated during the half-cycles in which the target is positive relative to the filament.

In molybdenum, some of the many possible transitions have energies which are shown in Table 28-1. The photons which are emitted are called *x rays,* and because of their high energies they penetrate solid matter much more easily than do the photons of visible light. The wavelengths, calculated from Eq. 27-7, turn out to be about 1 Å— less than $\frac{1}{1000}$ the wavelength of visible light.

Different x-ray photons are obtained if an *L*-electron is knocked out; this requires less work than to knock out a *K*-electron. When an *M*- or *N*-electron falls to an empty place in the *L*-shell, ΔE is less than for the jump to a *K*-shell; hence these "*L*-series" photons have considerably less energy (and longer wavelength) than those of the "*K* series" listed in Table 28-1.

X-ray spectra give valuable information about atomic structure, since the wavelengths of x rays are characteristic of the atom, just as are the wavelengths of optical

TABLE **28-1**

Approximate wavelengths of *K*-series lines in the x-ray spectrum of molybdenum

Transition	Energy change, eV	Wavelength, Å	Designation of line
$L \to K$	17 500	0.71	$K\alpha$
$M \to K$	19 800	0.63	$K\beta$
$N \to K$	20 000	0.62	$K\gamma$

spectra. X-ray spectra depend on the inner parts of the electron cloud, while visible, ultraviolet, and infrared spectra depend on the outer parts of the electron cloud where the binding energies are small. X rays can be produced by fluorescence, if a photon of the right energy happens to be absorbed by an inner electron. The absorbed photon's energy must be sufficient to take the inner electron all the way out, leaving the atom ionized with an inner vacancy; thus x-ray photons are needed to give x-ray fluorescence.

In an x-ray tube, an electron which strikes and enters the target undergoes many deflections as it passes near the positively charged nuclei. During each deflection the electron's path is curved, giving rise to a centripetal acceleration. We saw in Sec. 23-8 that acceleration of charge is associated with the production of e-m radiation. Thus one or more photons are radiated during each encounter; this radiation is called *bremsstrahlung.*• The electron gradually loses its KE in this series of collision events, eventually becoming slow enough to be captured by an ion or to become a free electron in the metallic target material.

The bremsstrahlung (Fig. 28-11) is a base

• German: braking radiation.

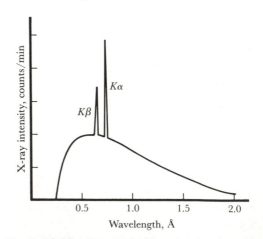

Figure 28-11 X-ray spectrum of molybdenum in a tube operated at 50 kV. The *K*α and *K*β lines are shown, superposed on the bremsstrahlung.

upon which the sharp x-ray lines of the K, L, . . . , series are superposed. Bremsstrahlung is a continuous function of wavelength, independent of the target material, analogous to the continuous optical spectrum of a black body (Fig. 14-4), which is also independent of the source material. Note however, that for any given PD across the x-ray tube, the bremsstrahlung has a short-wavelength limit, corresponding to a highest-energy photon. This limit represents the conversion of the electron's initial KE entirely into the energy of a single photon.

28-6   Quantum mechanics

The Rutherford-Bohr model of the atom gave, as we have seen, a dramatically successful initial interpretation of the hydrogen spectrum and the general processes of emission and absorption of photons by atoms. However, in the dozen years following 1913, it became apparent that the details of the spectra of atoms other than hydrogen—even of such a simple one as $_2$He—could not be explained. The Bohr theory gave no explanation of the probabilities of transition between energy levels, which are related to the intensities of spectral lines. Other basic phenomena were outside the framework of the Bohr theory; these include the chemical bond and solid-state phenomena, superconductivity, electron spin and the related magnetic properties of matter, and radioactive decay. The Bohr theory was nonrelativistic. In 1924 de Broglie's proposal of the wave-particle duality of matter set the stage for a development in physics comparable to the Newtonian synthesis of 1666 in its generality and power.

About 1925, Heisenberg, Schroedinger, Born, Jordan, Dirac, and others applied the idea of duality to develop a new science of mechanics called wave mechanics, or quantum mechanics. Newton's laws are a special case of quantum mechanics, valid for collections of many particles, such as a baseball. Bohr's third postulate is also a special case, valid for circular orbits in the hydrogen atom only. Quantum mechanics, which is the mechanics of matter waves, is highly mathematical. It is a successful theory, and it is fair to say that the physicist's knowledge of the outer atom is, in principle, complete. The only remaining difficulties seem to be mathematical. To illustrate the contrast between the Bohr quantum theory and modern quantum mechanics, consider the description of the first allowed orbit of the electron in the hydrogen atom. According to Bohr, the electron moves in a circular orbit with a definite momentum and a definitely assigned position at all times. According to quantum mechanics, since the momentum is known precisely, the position must be highly uncertain. There is *some* probability of finding the electron at any point in space, even as far as a mile or so from the nucleus; but, upon working out the probabilities, it turns out that the electron is *most likely* to be at a distance of 0.53 Å from the nucleus, which is exactly the radius of the definite orbit that Bohr calculated. The uncertainty principle leads therefore to the concept of a "smeared-out" atom. The electron is still supposed to be a point charge, but when joined with a proton to form a H atom, the probability of finding it in any given small volume is less than 100%. Thus the electron in a H atom can never be assigned a precise position. According to our probabilistic model, it is the guide waves, not the electron, that are smeared out.

■ SUMMARY

The nucleus of an atom is of the order of 10^{-14} m in size, which is about 10^{-4} the size of the electron cloud. When α particles bombard a thin metallic foil, only a few are scattered, but they are scattered through large angles. This shows the concentration of charge in a central nucleus (Rutherford).

A hot solid, liquid, or dense gas emits a continuous spectrum which depends only on temperature. A hot gas emits a line spectrum which depends on the nature of the gas. The Bohr theory of the line spectrum of hydrogen is based on three postulates: (1) An atom can exist without radiating; (2) when the energy state of an atom changes by ΔE, the frequency of the photon which is emitted or absorbed is given by $\Delta E = h\nu$; (3) the allowed energies are those for which the electron's angular momentum is an integral multiple of $h/2\pi$.

An atom can remain in an excited state for less than a microsecond, except in rare cases when it is in a metastable state having a considerably longer lifetime. Hence at ordinary temperatures, most of the atoms in a gas are in their lowest energy state (normal population). Absorption of a photon can raise an atom to a higher state, and one or more fluorescent photons can be emitted as the atom returns (in steps) to its normal state. An atom can absorb only those wavelengths it is capable of emitting. Stimulated emission of radiation leads to light amplification if photons interact with atoms which are in an inverted population condition. The laser uses stimulated emission to produce an intense beam of coherent light.

In any atom, the orbital electrons are grouped into shells, all electrons in any shell having roughly the same energy. According to the exclusion principle (Pauli), the maximum number of electrons in any shell is limited: 2 in the K-shell, 8 in the L-shell, etc. X rays are produced when an inner electron is removed. This requires a high-speed impact, or absorption of an energetic (high-frequency) photon. X-ray photons ordinarily have energies of tens of thousands of electron volts and wavelengths of the order of magnitude of an angstrom. Since atoms in a crystal are about an angstrom or so apart, a crystal can be used as a grating to form an x-ray spectrum.

Quantum mechanics, or wave mechanics (Heisenberg, Schroedinger, Born, Jordan, Dirac), is a logical development based on the duality of matter. Newtonian mechanics is a special case of quantum mechanics.

■ CHECK LIST

continuous spectrum
discrete spectrum
Bohr postulates
energy state or energy level
Lyman series
Balmer series
Paschen series
metastable state
fluorescence

phosphorescence
ionization potential
normal population
inverted population
stimulated transition
laser
K-, L-, M-, etc., shells
bremsstrahlung

28-1 The Rutherford scattering experiment may be described with the phrase "too little, too much." Explain, in terms of the angle of scattering and the numbers of scattered α particles.

28-2 What can you say about the nature (solid or gaseous) of a source which emits a line spectrum?

28-3 What sort of spectrum, line or continuous, gives information about the chemical composition of a source? How can the temperature of a source be found from a study of its spectrum? It this always possible for all sources?

28-4 Using Fig. 28-2, explain why a neon sign is red. What color do you except a helium sign to have?

28-5 Why do so many dark lines in the sun's spectrum on p. 609 coincide in position with adjacent bright lines? How do you account for the dark lines which do *not* match up?

28-6 A dancer's costume, coated with fluorescent material, is illuminated with ultraviolet radiation, so-called "black light," and visible light is emitted. Could visible fluorescence also be produced by infrared radiation shining on a suitable substance?

28-7 What is the difference between a downward spontaneous transition and a downward stimulated transition?

28-8 By which, if any, of the following methods can a mercury atom reach the excited metastable state E_2 shown in Fig. 28-5? (*a*) Absorption of a photon of energy $E_2 - E_1$; (*b*) spontaneous emission of a photon of energy $E_6 - E_2$; (*c*) absorption of energy in a collision process; (*d*) emission of a photon of energy $E_6 - E_2$ in a stimulated transition.

28-9 Explain the operation of a laser, and describe the characteristics of the light emitted by a laser.

28-10 What feature of atomic structure makes iron chemically different from aluminum?

28-11 Describe a sequence of events, on an atomic scale, which might result in the production of an x-ray photon.

28-12 Is any bremsstrahlung radiation emitted during the α-particle scattering events shown in Fig. 28-1*b*?

28-13 Are any of Bohr's postulates of 1913 retained in the quantum mechanics of 1925?

28-14 The Balmer series lines appear in the absorption spectrum found in hot stars, but not in absorption spectra obtained in room-temperature experiments on the earth. Explain.

■ PROBLEMS

28-A1 Using the photograph on p. 609, determine to five significant figures the wavelengths of the three most intense iron spectrum lines in the region 4000 Å to 4100 Å.

28-A2 In state A, the electronic energy of an atom is 5.87 eV, and in state B its electronic energy is 3.24 eV. What is the energy of the photon emitted in the transition from A to B?

28-A3 Calculate the energy and wavelength of the photon emitted when a mercury atom's energy changes from E_6 to E_2 (Fig. 28-5).

28-A4 Calculate the maximum KE (in J) of an electron for which a collision with a hydrogen atom in its normal state is surely an elastic one.

28-A5 How much energy is needed to ionize a mercury atom which is already in the metastable state E_2 (Fig. 28-5)?

28-B1 Use the energy-level diagram of Fig. 28-3 to calculate the wavelength of the Lyman α line (the line in the Lyman series having the longest wavelength). Is this line visible?

28-B2 Using Fig. 28-3, calculate the wavelength of the limit of the Lyman series (Fig. 28-4) in hydrogen. In what region of the e-m spectrum is this limit?

28-B3 Using the energy-level diagram of Fig. 28-3, estimate the energy and wavelength of the line of next-to-longest wavelength in the Paschen series of hydrogen (all transitions in this series terminate on E_3). Check your answer by referring to Fig. 28-4. Is this line visible?

28-B4 The fourth line in the hydrogen Balmer series (Hδ) has a wavelength of 4101 Å. (*a*) Identify the line in Fig. 28-4. (*b*) Identify the line in the hydrogen emission spectrum in Fig. 28-2e. (*c*) Identify the line in the sun's absorption spectrum on p. 609. (*d*) One of the energy levels involved in the emission or absorption of this line is shown in Fig. 28-3; use the wavelength of the line to compute the value of the other energy level involved.

28-B5 Determine which two energy levels in the mercury atom (Fig. 28-5) are involved in the emission of the green line whose wavelength is 5461 Å. (*Hint:* First find the energy of the photons, in electron volts.) Identify the line in Fig. 28-2a.

28-B6 A mercury line has wavelength 4358 Å. Identify the two levels in Fig. 28-5 involved in the production of this line.

28-B7 Refer to the rotating oxygen molecule discussed in Sec. 9-6. Express in electron volts the rotational energy of the oxygen molecule, which has three quanta of rotational energy. If an oxygen molecule absorbs a photon sufficient to give it this much energy, what is the wavelength of the incident radiation? Is this radiation visible?

28-B8 What is the energy of the least energetic photon that can be absorbed by a hydrogen atom at room temperature? (Use Fig. 28-3.) Identify the line in Fig. 28-4.

28-B9 What is the fractional decrease in mass of a mercury atom when a photon of the ultra-violet line of wavelength 2537 Å is emitted (Fig. 28-5)?

28-B10 Suppose the orbital angular momentum of the earth were quantized according to the Bohr relation (Eq. 28-2). What would the quantum number n be? Could such a quantization be detected experimentally? (See Appendix for data.)

28-B11 Use the information in Table 28-1 to calculate the energy and wavelength of the molybdenum Lα x-ray line (which results from the transition of an electron from the M-shell to a vacancy in the L-shell). (*Hint:* First make an energy-level diagram for molybdenum.)

28-B12 What is the short-wavelength limit, in Å, of the bremsstrahlung radiation used in cancer therapy, if the x-ray tube is operated at 200 kV?

28-B13 What is the wavelength, in Å, of the highest-energy x rays emitted when an electron beam is accelerated through a PD of 30 kV and strikes the face of a color TV tube?

28-B14 Show that an electron beam of energy 18 000 eV has insufficient energy to cause a molybdenum target to emit its Kα line. (*Hint:* Estimate from Table 28-1 the energy required to ionize molybdenum by removing a K-electron.)

28-B15 Show that the Bohr postulate, Eq. 28-2, can be obtained by assuming that there are stationary de Broglie waves in the orbit, such that n wavelengths are contained in the circumference of the orbit. *Note:* This result, while interesting, is not as powerful as it seems. According to wave mechanics, the electron cannot be in a plane orbit for then both its angular momentum and the corresponding position coordinate (orientation of the axis of revolution) would be known precisely, in violation of the uncertainty principle. The extension of the idea of this problem to three dimensions involves complicated mathematics and leads to three quantum numbers instead of one.

28-C1 A large ruby laser with end mirror of diameter 10 cm sends out an almost parallel beam of wavelength 6943 Å. (*a*) Calculate the angular divergence of the beam, which is limited by the end mirror serving as a circular aperture (see the last two paragraphs of Sec. 26-4). (*b*) If pointed at the moon, 3.8×10^8 m away, how large would be the diameter of the central part of the circular diffraction pattern on the moon's surface?

For Further Study

28-7   The Bohr theory for hydrogen

In this section, we use the Bohr quantum postulate to calculate the numerical values of the hydrogen energy levels given in Fig. 28-3. We will see that the theory leads to exact agreement between the calculated ionization energy and the experimental value of 13.60 eV.

For generality, we will consider a *hydrogen-like* atom (Fig. 28-12) consisting of a single electron moving in a circular orbit around a nucleus of charge $+Ze$ (here Z is the atomic number). Such an atom might be ionized helium, $_2\text{He}^+$. Helium is the second element in the periodic table and normally has two orbital electrons; the removal of one electron leaves an ion which, like hydrogen, has a single electron but a nuclear charge $+2e$ instead of $+1e$ which is the charge of a proton in the hydrogen atom. Similarly, the hydrogenlike ionized atoms $_3\text{Li}^{2+}$, $_4\text{Be}^{3+}$, etc., contain only one orbital electron.

Bohr's assumption (Eq. 28-2) was that the angular momentum of the electron can only

be an integral multiple of $h/2\pi$. In Chap. 8 we saw that angular momentum equals (moment of inertia) $\times$ (angular velocity); here the moment of inertia is simply mr^2, and the angular velocity is v/r. Hence the angular momentum of the electron is $I\omega = (mr^2)(v/r) = mvr$. Bohr's postulate is, therefore,

$$mvr = \frac{nh}{2\pi} \qquad (28\text{-}3)$$

where n is an integer.

The centripetal acceleration of the electron is v^2/r, and the centripetal force is supplied by the Coulomb attraction between the nucleus and the electron (Eq. 17-1). Therefore, applying Newton's second law, $F = ma$, we find

$$\frac{k(Ze)(e)}{r^2} = \frac{mv^2}{r} \qquad (28\text{-}4)$$

We now have two equations with v and r as the two unknowns. Solving simultaneously[*] for v and r, we obtain

$$v = \frac{2\pi kZe^2}{nh} \qquad (28\text{-}5)$$

$$r = \frac{n^2h^2}{4\pi^2mkZe^2} \qquad (28\text{-}6)$$

We can immediately calculate the radii of the allowed orbits. From Eq. 28-6 we see that the radii are proportional to the square of the quantum number n. Using the numerical values of the various constants

---

[*] A good way to start is to solve (28-4) for mv^2r and then divide by (28-3) to get v.

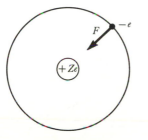

Figure 28-12

(and setting $Z = 1$ for hydrogen), we obtain for the radius of the smallest orbit

$$r_1 = \frac{(1)^2(6.63 \times 10^{-34} \text{ J} \cdot \text{s})^2}{(4\pi^2)(9.11 \times 10^{-31} \text{ kg})(9.00 \times 10^9 \text{ N} \cdot \text{m}^2/\text{C}^2)(1)(1.60 \times 10^{-19} \text{ C})^2}$$
$$= 0.529 \times 10^{-10} \text{ m} = 0.529 \text{ Å}$$

This is a remarkable achievement: the radius of the first Bohr orbit is calculated using only the fundamental constants h, m, k, and e, and the value so obtained is in agreement with various indirect experimental estimates of the size of the H atom in its normal state.

To obtain the energy levels, we must calculate both KE and PE. The kinetic energy is $\frac{1}{2}mv^2$, and from (28-4) we find that

$$\text{KE} = \frac{kZe^2}{2r} \qquad (28\text{-}7)$$

The PE is electrical in nature. The system gains energy if the electron is moved from its orbit to infinity against an opposing (attractive) force kZe^2/r^2 of the positive nucleus. Similarly, the system loses PE if an electron falls down from ∞ into some orbit. To calculate this energy difference, we note that the work required to move the charge e from r to ∞ is the area under the force graph (Fig. 28-13). Thus

$$W = \lim_{\Delta r \to 0} \sum \Delta W$$

$$= \lim_{\Delta r \to 0} \sum_{r=r}^{r=\infty} \frac{kZe^2}{r^2} \, \Delta r$$

$$= kZe^2 \lim_{\Delta r \to 0} \sum r^{-2} \, \Delta r$$

The limit of the sum is a definite integral:

$$W = kZe^2 \int_r^\infty r^{-2} \, dr$$

The table of integrals in the Appendix tells us that $\int x^n \, dx = (x^{n+1})/(n + 1)$; here x is replaced by r and $n = -2$. Thus

$$W = kZe^2 \left[\frac{r^{-1}}{-1} \right]_r^\infty = kZe^2 \left[\frac{\infty^{-1}}{-1} - \frac{r^{-1}}{-1} \right]$$

or

$$W = \frac{kZe^2}{r}$$

This is the magnitude of the change in PE

of the system as the charge is moved from r to ∞.

Now we must decide on a reference level from which to measure the PE. We choose $\text{PE} = 0$ when the electron is at $r = \infty$ (removed from the atom). (Recall that only *changes* of PE are observable; we are free to select any convenient configuration of the system as the reference level). The system has $\text{PE} = 0$ when the electron is at infinity, it loses energy when the electron falls toward the nucleus, so the PE when the electron is in a circular orbit of radius r is negative. Thus

$$\text{PE} = -\frac{kZe^2}{r} \qquad (28\text{-}8)$$

The PE is twice as large as the KE and is negative. The total energy is the sum of Eqs. 28-7 and 28-8.

$$E = \text{KE} + \text{PE}$$
$$= \frac{kZe^2}{2r} - \frac{kZe^2}{r}$$

or

$$E = -\frac{kZe^2}{2r}$$

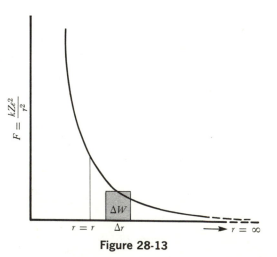

Figure 28-13

Finally, we substitute the value of r for the nth orbit (Eq. 28-6) to obtain the Bohr energy formula:

$$E_n = -\frac{2\pi^2 m k^2 Z^2 e^4}{n^2 h^2} = -\frac{\text{constant}}{n^2} \tag{28-9}$$

To illustrate, for $n = 1$,

$$E_1 = -\frac{2\pi^2(9.11 \times 10^{-31}\ \text{kg})(9.00 \times 10^9\ \text{N} \cdot \text{m}^2/\text{C}^2)^2(1)^2(1.60 \times 10^{-19}\ \text{C})^4}{(1)^2(6.63 \times 10^{-34}\ \text{J} \cdot \text{s})^2}$$

$$= -2.17 \times 10^{-18}\ \text{J}\left(\frac{1\ \text{eV}}{1.60 \times 10^{-19}\ \text{J}}\right)$$

$$= -13.6\ \text{eV}$$

When values of m, k, e, and h to four significant figures precision are used, the calculation gives $E_1 = -13.60$ eV. This is exactly equal to the experimentally determined ionization energy for hydrogen as determined from the analysis of its spectrum.

From the point of view adopted in this section, the total energy of the atom is negative because we have chosen $E = 0$ for the ionized state, formerly called $E = +13.60$ eV. The two notations in Fig. 28-14 are equivalent, and the energy *differences* are identical. Thus the first Balmer series line (shown as a transition between E_3 and E_2) has a photon energy corresponding to 1.89 eV in either scheme.

The energy of any transition is found from the difference in energy levels. When expressed in electron volts, the formula for the emitted photon's energy is

$$h\nu = 13.60\left(\frac{1}{n^2} - \frac{1}{m^2}\right) \tag{28-10}$$

where n and m are integers. The Lyman series has $n = 1$, with $m = 2, 3, 4, \ldots$; the Balmer series has $n = 2$, with $m = 3, 4, 5, \ldots$; and similarly for the other series.

A nice verification of the Bohr theory comes from a study of the spectrum of He$^+$ for which $Z = 2$ and the constant in Eqs. 28-9 and 28-10 is $2^2(13.60)$ eV. Some of the spectrum lines of He$^+$ should coincide with certain lines in the H spectrum. For example, we can use suitable values of n and m to calculate the following photon energies:

H spectrum:

$$h\nu = (13.60)(1)^2\left(\frac{1}{2^2} - \frac{1}{4^2}\right) = (13.60)\left(\frac{3}{16}\right)$$

He$^+$ spectrum:

$$h\nu = (13.60)(2)^2\left(\frac{1}{4^2} - \frac{1}{8^2}\right) = (13.60)\left(\frac{3}{16}\right)$$

The lines should therefore have identical energy (and wavelength), even though they are emitted by different substances. Actually there is a very slight difference (less than 0.1%), because the motion of the nucleus (which we have neglected) is less for the He$^+$ nucleus which is more massive. Historically, helium was first discovered through

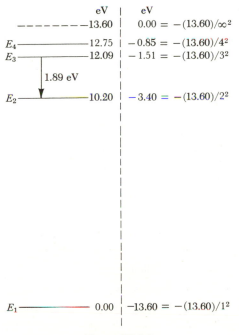

eV		eV
$------$ -13.60		$0.00 = -(13.60)/\infty^2$
E_4 ———— 12.75		$-0.85 = -(13.60)/4^2$
E_3 ———— 12.09		$-1.51 = -(13.60)/3^2$
1.89 eV		
E_2 ———— 10.20		$-3.40 = -(13.60)/2^2$
E_1 ———— 0.00		$-13.60 = -(13.60)/1^2$

Figure 28-14

its dark lines in the absorption spectrum of the sun*; some workers thought it to be a

form of hydrogen because of this near-coincidence of lines which was explained later by the Bohr theory.

* Greek: *helios* = sun.

■ PROBLEMS

28-C2 Solve Eqs. 28-3 and 28-4 to obtain the equations for v and r given in Eqs. 28-5 and 28-6.

28-C3 There is a gravitational attraction between the nucleus and the electron. Explain quantitatively why we can ignore the contribution of gravitational PE to the total energy of the atom.

28-C4 Show that the ratio of the wavelengths (which is the same as the ratio of the frequencies) of the first two lines in the Balmer series is $27:20$.

28-C5 What is the approximate wavelength of the H and He$^+$ lines which were discussed in the last paragraph of Sec. 28-7? Identify the line in Fig. 28-2e. Why is this line not found in Fig. 28-2c?

28-C6 What is the electric potential in V, at a point in the first Bohr orbit of a hydrogen atom?

28-C7 What would be the quantum number for a circular orbit in a hydrogen atom for which the radius of the atom is large enough to be seen in a visible-light microscope which can resolve details down to 4000 Å? (*Hint:* Use a proportion; how is r_n related to r_1?)

28-C8 Show that for an electron in the lowest-energy circular orbit in hydrogen, v/c is approximately $1/137$, thus verifying the approximate validity of using nonrelativistic mechanics in the Bohr theory.

■ REFERENCES

1   Andrade, E. N. daC., "The Birth of the Nuclear Atom," *Sci. American,* Nov., 1956, p. 93.

2   Osgood, T. H., and H. S. Hirst, "Rutherford and his Alpha Particles," *Am. J. Phys.,* **32,** 681 (1964).

3   Struve, O., "The Fraunhofer Lines," *Sky and Telescope,* **10,** 117 (1951); see also the back cover of this issue, which has a photograph of the solar spectrum 4 ft long.

4   Ufford, C. W., "Spectroscopy," *Am. Phys. Teacher,* **6,** 299 (1938).

5   Behrens, C. E., "Atomic Theory from 1904–1913," *Am. J. Phys.,* **11,** 60 (1943); "The Early Development of the Bohr Atom," *Am. J. Phys.,* **11,** 135 (1943); "Further Development of Bohr's Early Atomic Theory," *Am. J. Phys.,* **11,** 272 (1943).

6   Crawford, B., Jr., "Chemical Analysis by Infra Red," *Sci. American,* Oct., 1953, p. 42.

7   Watson, E. C., "The Discovery of X Rays," *Am. J. Phys.,* **13,** 281 (1945). A full translation of Roentgen's original papers; also found in Magie (see above).

8   Jauncey, G. E. M., "Birth and Infancy of X Rays," *Am. J. Phys.,* **13,** 362 (1945).

9   Schawlow, A. L., "Laser Light," *Sci. American,* Sept., 1968, p. 120.

10 Herriott, D. R., "Applications of Laser Light," *Sci. American,* Sept. 1968, p. 141.

11 Miller, S. E., "Communication by Laser," *Sci. American,* Jan., 1966, p. 19.

12 National Film Board of Canada, *Thomson Model of the Atom; Rutherford Scattering* (films).

13 Miller, F., Jr., *Absorption Spectra* (film).

29

The Nucleus

The closing years of the nineteenth century were extraordinarily fruitful in several ways. In 1890, atoms were thought to be indivisible, and light was thought to be an e-m wave with no corpuscular properties. By the year 1900, J. J. Thomson had discovered the free electron, Roentgen had produced x rays, and Planck had postulated a quantum mechanism for the emission of light by a black body. This busy decade marked the beginning of a new era in physics. No longer was it possible to describe nature entirely with models based upon Newton's laws and with appeal to everyday experience.

29-1   Early discoveries about radioactivity

In 1896 the French physicist Henri Becquerel was studying the fluorescence and phosphorescence of various minerals which had been exposed to the newly discovered x rays. When, in the course of his work, he came to the uranium compounds, he discovered, by chance, that even unexposed uranium compounds emit penetrating radiations which can blacken a photographic plate or ionize

a gas such as air. These properties are shared by x rays, but Becquerel's "rays" were continuously emitted even in the absence of a source of energy. At first, Becquerel suspected that the radiations were a form of invisible fluorescence. The uranium might merely be releasing energy which it had stored up in some way from sunlight. To test this hypothesis he examined a piece of uranium ore which had been kept in a lead-shielded, light-tight box for several months. The sample was able to produce continuous ionization of a gas and to affect a photographic plate wrapped in black paper, and yet there was no apparent source for this energy. The phenomenon was called radioactivity, and it posed a serious problem, for it apparently violated the law of conservation of energy.[*]

---

[*] The discovery of radioactivity did clear up one mystery. Although geological evidence showed the age of the earth to be of the order of a billion years, physicists had protested that the earth could not be this old; according to various estimates, it would have cooled off from a molten state to its present temperature in 20 to 200 million years, unless there were a source of heat energy inside the earth. The radioactive minerals supply such a source, and so the great debate of 1862–1900 was resolved in favor of the geologists.

High-energy particles traversing a portion of a cloud chamber located 195 ft from a brass target which is inside a billion-volt atom-smasher. Tracks of widely different density are visible. (Brookhaven National Laboratory)

Many workers pushed forward the study of radioactivity in the next decade. Pierre and Marie Curie processed many tons of impure uranium ore and isolated a fraction of a gram of two new elements, polonium and radium, both of which are considerably more active (and less abundant) than uranium. Several dozen radioactive substances, all of high atomic weight, were discovered, and it was found that just three kinds of radiation are emitted. These radiations were labeled α, β, and γ rays, in order of increasing penetrating power. We can describe these "rays," using modern terminology.

Alpha rays are a stream of positively charged particles. The alpha particle (α) is a doubly charged helium ion $_2^4\text{He}^{++}$, i.e., a bare helium nucleus. Alpha particles are easily stopped by a sheet of paper or by a few centimeters of air.

Beta rays are a stream of high-speed electrons. The β particles have the same e/m_0 ratio as do the free electrons which evaporate from a heated filament. Beta rays are much more penetrating than alpha rays; they can usually penetrate thin metal sheets such as the wall of a Geiger counter, but cannot pass through a millimeter of lead or a few centimeters of flesh. We attribute the greater penetrating power of β particles to their high speeds; the slower α particles are more likely to lose energy by creating ions during their travel through matter, since they take a longer time to pass through an atom. Both α and β particles penetrate farther if they have greater KE to start with. Ultimately, an α particle picks up two electrons and becomes a neutral helium atom,[*] and a β particle attaches itself to the electron cloud of some atom or ion, or enters a metal to become a free electron.

Gamma rays are high-energy photons. The guide waves of γ-ray photons have such short wavelengths that diffraction and inter-

ference can be observed only with difficulty; hence the corpuscular aspects of γ rays dominate the picture. For instance, the γ rays from radium are photons of energy about 2×10^5 eV, or 0.2 MeV. Like all other photons, γ rays are not deflected by magnetic or electric fields.

One outstanding early observation about radioactivity was its independence of external conditions. The nature of the radiations from uranium, for instance, and the strength (activity) of a given amount of uranium were found to be completely unaffected by temperature, pressure, and other physical changes; neither could any effect of chemical combination be found. The activity of a mole of UCl_3 exactly equals that of a mole of UO_2 or a mole of UF_6 gas.

29-2 Radioactive decay

The nuclear or planetary model of an atom developed by Rutherford in 1911 explains why radioactivity is so unaffected by external conditions. Chemical reactions take place when the outer parts of electron shells come close together. During collisions between atoms in a gas, the outer shells may become momentarily distorted. However, during such physical or chemical changes the nucleus is protected by the outer electron cloud; the nuclei never get close enough to each other to react, except at high speeds corresponding to temperatures of over 10^6 °K. Since radioactivity is practically unaffected by chemical state, temperature, pressure, and other conditions, we conclude that it is a nuclear property. The α, β, and γ rays are, therefore, emitted from the nucleus. Let us now consider the structure of such a nucleus.

According to our present understanding, the nucleus contains heavy particles, protons and neutrons, which are collectively called *nucleons* (Sec. 2-3). Any given combination of protons and neutrons is called a

---
[*] The spectrum of helium is obtained from a sealed glass tube which has contained an α emitter for some time.

nuclide; several nuclides all of the same atomic number are called *isotopes* of the element in question. The symbol Z is used for the *atomic number,* or the number of protons in the nucleus, and A represents the *mass number,* or total number of nucleons. Hence the number of neutrons in any nucleus is given by $A - Z$. A also represents the atomic mass (to the nearest integer) of the nucleus. Atomic numbers Z are given in the Periodic Table (p. 721). The symbol for a nuclide has a subscript for Z and a superscript for A: $_{26}^{56}\text{Fe}$ has $Z = 26$ and $A = 56$. Information about Z is, of course, contained in the chemical symbol; in advanced work the subscript for Z is usually omitted.

Not all combinations of protons and neutrons can stick together to form a stable nuclide. Roughly speaking, stable isotopes of light elements have about 1 neutron for every proton (example: $_8^{16}\text{O}$ has 8 protons and 8 neutrons). In the heavier elements, the trend is toward somewhat more neutrons than protons (example: $_{80}^{200}\text{Hg}$ has 80 protons and 120 neutrons). If a nuclide deviates too much from the norm, it is certain to be unstable and radioactive.

We shall use the isotopes of chromium to illustrate nuclear stability and several types of instability. Eleven isotopes of chromium are known to exist; the heaviest is $_{24}^{56}\text{Cr}$, which has 24 protons and $56 - 24 = 32$ neutrons, and the lightest is $_{24}^{46}\text{Cr}$, which has 24 protons and 22 neutrons.

Stable isotopes. Ordinary chromium is a mixture of four stable isotopes. The masses of the individual isotopes can be found by use of the *mass spectrometer,* which is similar in principle to the e/m apparatus pictured in Fig. 27-3 for electrons. Many types of mass spectrometer have been developed, in which combinations of magnetic and electric fields focus ions on a collector. Figure 29-1a is a "photograph" of the chromium mass spectrum; each Cr^+ ion sticks to a glass plate at a place determined by its

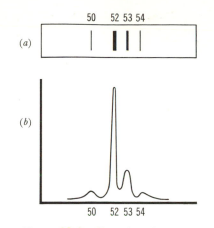

Figure 29-1 Chromium isotopes.

e/m ratio. Thus isotopes of different masses are brought to different places.[*] In another type of mass spectrometer the Cr^+ ions strike a metallic collector, they are neutralized by free electrons, and the electron current is a measure of the abundance of the isotope (Fig. 29-1b). Such experiments show four stable isotopes of chromium, with relative abundances as follows: $_{24}^{50}\text{Cr}$ (4.3%), $_{24}^{52}\text{Cr}$ (83.8%), $_{24}^{53}\text{Cr}$ (9.5%), and $_{24}^{54}\text{Cr}$ (2.4%). The chemical atomic mass of chromium (51.996) is the weighted average of these values; it turns out to be so close to 52 because the isotope of mass number 52 is by far the most abundant in nature. Careful use of a precise mass spectrometer reveals that the atomic masses of the isotopes are not exactly integers. Thus a neutral atom of ^{50}Cr has mass 49.94606, in terms of 12.00000 for the mass of $_6^{12}\text{C}$, a neutral carbon atom of mass number 12. We shall see in our later work that precise knowledge of nuclear masses is essential to our understanding of radioactive decay and nuclear transmutations.

Beta decay; half-life. Turning now to the other isotopes of Cr, let us first consider $_{24}^{55}\text{Cr}$. This nuclide has more than the normal number of neutrons; there is a high proba-

---

[*] See Fig. 30-7 on page 663 for a mass spectrogram of cadmium isotopes.

bility that one of its neutrons will spontaneously change into a proton, increasing the atomic number of the nuclide by 1 unit. According to the law of conservation of charge, a negative charge is created to balance the newly created extra positive charge in the nucleus. This negative charge is emitted as a β particle, or negative electron, also called a *negatron*• ($_{-1}^{0}e$, or β^-). Since the nucleus now has an atomic number of 25, we know that it has become an isotope of manganese. The total number of nucleons has remained unaltered during the changeover of a neutron into a proton, so the mass number is still 55. Symbolically,

$$_{24}^{55}\text{Cr} \rightarrow {}_{25}^{55}\text{Mn} + {}_{-1}^{0}e \quad (3.52 \text{ min})$$

Note how this equation balances in two ways: the charge balances: $24 = 25 + (-1)$; and the total number of nucleons balances: $55 = 55 + 0$. This sort of spontaneous change is called *radioactive decay*.

The *half-life* $T_{1/2}$ of a radioactive decay process equals the time required for half the nuclei in a sample of a nuclide to make the transition. For the decay of $_{24}^{55}\text{Cr}$, experiment shows the half-life to be 3.52 min; this value is written at the right of the decay equation. This means that if, say, 8 million $_{24}^{55}\text{Cr}$ nuclei are present at a certain time, 3.52 min later only 4 million will still be $_{24}^{55}\text{Cr}$; and after another 3.52 min, the number will be down to 2 million, etc. The nuclei are much like birds sitting on a fence. Assuming that the birds take off at random, we can estimate quite closely how many will be left after, say, 3.52 min or 7.04 min, but we cannot say *which* bird will take off first, or which will be the last to go. We see that radioactive decay is intimately related to probability, and only a statistical model is possible. Half-lives for β decay are 10^{-2} s or longer. Considering all types of decay, the measured half-lives of

————
•From now on, we speak of the "ordinary" electron, with which you are familiar, as a negatron, paving the way for introduction of the positron, or positive electron, later on.

known nuclides vary from less than a microsecond (high probability of decay) to more than 10^{17} years (very low probability of decay).

EXAMPLE **29-1**

A microgram of the nuclide $_{11}^{25}\text{Na}$ (half-life 60 s) is in a test tube at noon. The material decays with emission of a negatron. (a) Identify the products of the decay. (b) How many ^{25}Na atoms are present at the start? (c) How many are present at 12:10 P.M.? (d) Estimate how long it will be before all the atoms are transformed.

(a) The reaction is

$$_{11}^{25}\text{Na} \rightarrow {}_{Z}^{25}\text{X} + {}_{-1}^{0}e$$

To balance the charges, we see that Z must be 12, and hence (using the Periodic Table) we determine that the products are the magnesium isotope $_{12}^{25}\text{Mg}$ and a negatron.

(b) The atomic weight is 25, and we use Avogadro's number to find how many atoms are present.

$$\frac{6.02 \times 10^{23}}{25 \text{ g}} = \frac{N}{10^{-6} \text{ g}}$$

$$N = \boxed{2.408 \times 10^{16} \text{ atoms}}$$

(c) At the end of one half-life, which in this case is 1 min, only half the original number of atoms is present, and so on. We can make a table of the number present at any time:

12:00	2408×10^{13}
12:01	1204×10^{13}
12:02	602×10^{13}
12:03	301×10^{13}
etc.	etc.

After 10 half-lives (10 min) the number is $1/2^{10}$ or $1/1024$ of the original number. That is,

$$N' = \left(\frac{1}{1024}\right)(2408 \times 10^{13})$$

$$= \boxed{2.35 \times 10^{13} \text{ atoms}}$$

(d) To within a few per cent, each 10 half-lives reduce the number by a factor of 10^3.

Thus, approximately, the amount of ^{25}Na decays as follows:

12:00	2.4×10^{16}
12:10	2.4×10^{13}
12:20	2.4×10^{10}
12:30	2.4×10^{7}
12:40	2.4×10^{4}
12:50	$2.4 \times 10 = 24$ atoms
12:51	12
12:52	6
12:53	3
12:54	?

By 12:54, only one or two ^{25}Na atoms remain, and the chances are 50-50 that if *one* is left at 12:55, it will decay during the next 60 s. We are running into a statistical situation here, and exact predictions cannot be made. (It would be meaningless to say that $1\frac{1}{2}$ atoms remain at 12:54!) However, we can be reasonably sure that by 1:15 P.M. there are *no* remaining ^{25}Na atoms unless by a wild million-to-one chance.

The mathematics of radioactive decay are considered in more detail in Sec. 29-9, where an equation is derived by which the number of nuclei present at any time can be calculated.

Positron emission; electron capture. Another of the unstable isotopes of chromium is $^{51}_{24}$Cr, which disintegrates according to the equation

$$^{51}_{24}\text{Cr} + {}^{0}_{-1}\text{e} \rightarrow {}^{51}_{23}\text{V} \quad (27.8 \text{ days})$$

This is an example of electron capture, represented by the symbol ϵ, in which the ^{51}Cr nucleus "sucks in" one of the electrons from its external electron cloud.[*] The electron disappears, and at the same time (thus conserving charge) a proton changes into a neutron inside the nucleus, making an isotope of element number 23, vanadium.

Still another chromium isotope decays thus:

$$^{49}_{24}\text{Cr} \rightarrow {}^{49}_{23}\text{V} + {}^{0}_{+1}\text{e} \quad (41.9 \text{ min})$$

[*] Capture from the *K*-shell is usually much more likely than capture from the *L*-shell.

Here the emitted particle is a *positron,* which is a positive electron $({}^{0}_{+1}\text{e}$, or $\beta^{+})$. Although positrons were first discovered as a component of cosmic rays, many unstable nuclides such as $^{49}_{24}$Cr emit positrons spontaneously. All such nuclides have rather short half-lives and have to be made artificially. In a sense, positron emission and electron capture are alternative processes, since in each case the daughter nucleus would be the same. Some nuclides decay in both ways at once; that is, each process has a certain probability of occurring. Positrons $({}^{0}_{+1}\text{e})$ and negatrons $({}^{0}_{-1}\text{e})$ are identical except for the sign of their charge. We consider them to be the same particle, the electron, in two different states of charge; more precisely, the positron is the "antiparticle" of the negatron.[*] To complete the picture, we list in Table 29-1 the eleven isotopes of chromium. A similar table for the light elements, $_1$H to $_{12}$Mg, is in the Appendix.

Alpha decay; fission. The unstable isotopes of chromium illustrate the three major ways in which a light nucleus decays: negatron emission, positron emission, and elec-

[*] The emission of either a negatron or a positron, or electron capture, is called beta decay.

TABLE 29-1 **Isotopes of $_{24}$Cr**

Nuclide	Abundance or mode of decay	Half-life
$^{46}_{24}$Cr	β^{+}	1.1 s
$^{47}_{24}$Cr	β^{+}	0.4 s
$^{48}_{24}$Cr	ϵ	23 h
$^{49}_{24}$Cr	β^{+}	42 m
$^{50}_{24}$Cr	4.3%	
$^{51}_{24}$Cr	ϵ	27.8 d
$^{52}_{24}$Cr	83.8%	
$^{53}_{24}$Cr	9.5%	
$^{54}_{24}$Cr	2.4%	
$^{55}_{24}$Cr	β^{-}	3.5 m
$^{56}_{24}$Cr	β^{-}	5.9 m

tron capture. Most unstable nuclides decay in one or more of these three ways. Other modes of decay are possible, however, especially among the heavy elements. *Alpha-particle emission* was among the first to be observed by Becquerel and the Curies; a typical example is the decay of one of the radium isotopes into the noble gas radon[*] (Rn):

$$^{226}_{88}\text{Ra} \rightarrow {}^{222}_{86}\text{Rn} + {}^{4}_{2}\alpha \quad (1622 \text{ years})$$

The α particle is the same as a helium nucleus ${}^{4}_{2}\text{He}$, and the equation balances both "above and below." Apparently the combination of two protons and two neutrons is so stable that such a group can stick together inside the nucleus and be emitted as a single superparticle, the α particle. Less frequent is decay by *neutron emission* from the nucleus. Some heavy nuclei (rarely) break into two large fragments; this is called *spontaneous fission*. A typical example is ${}^{238}_{92}\text{U}$, whose half-life for α emission is 4.5×10^9 years, and whose half-life for spontaneous fission is 8.3×10^{15} years.

Gamma ray photons from nuclei. We have seen in Chap. 28 that the energy of the outer electron cloud of an atom is quantized; energy-level diagrams such as Fig. 28-5 show the allowed energies of the entire system of electrons, and, when the energy of the electron cloud changes by ΔE, a photon is radiated whose frequency is given by $\Delta E = h\nu$. These changes take place in the electron cloud. In a similar fashion, the nucleus, too, can exist in various energy states, and a transition from one state to another results in emission of a photon. The *nuclear energy levels* are separated by thousands or millions of electron volts, and so the photons that are emitted have high energy. These high-energy photons that result from nuclear "de-excitation" or "shake-

down" are called gamma rays.[*] Most γ rays accompany α or β decay or electron capture. For instance, the ${}^{238}_{92}\text{U}$ nucleus emits γ-ray photons of energy 0.045 MeV as well as several groups of α particles. (In some cases it happens that the energy of the excited nucleus is given directly to one of the K-, L-, M-, . . . orbital electrons which is then ejected from the atom with a definite KE. This de-excitation process is called *internal conversion.*) Among the chromium isotopes, the nuclide ${}^{48}_{24}\text{Cr}$ emits two γ rays when it captures a K-electron; ${}^{49}_{24}\text{Cr}$ emits three γ rays during positron decay; ${}^{51}_{24}\text{Cr}$ emits one γ ray during K-capture; but ${}^{55}_{24}\text{Cr}$ emits no γ ray during its negatron emission. Sometimes a nucleus can remain in an excited state for a measurable time; such an excited nucleus (in a metastable state) is called an *isomer.* An isomer can decay to a stable state (with emission of a γ-ray photon), or, if its lifetime is long enough, the isomer may decay in any of the standard ways without first going to its stable state. Nuclear γ rays offer one of the best means of mapping the structure of nuclear energy levels. Nuclear spectroscopy is still in the data-gathering stage, similar to optical spectroscopy prior to the Bohr theory. Energy levels exist and can be measured, and physicists hope that a breakthrough in nuclear theory is near at hand.

A portion of a chart of nuclides is shown in Fig. 29-2; both negatron (β^-) and positron (β^+) emission are represented, as well as electron capture (ϵ). Two of the nuclides, ${}^{58}\text{Co}$ and ${}^{60}\text{Co}$, have isomeric states. The relative abundances of the stable nuclides are shown, in per cent.

Let us summarize the various modes of disintegration of a nucleus: It may emit an α-particle (${}^{4}_{2}\text{He}$ nucleus), a β^- particle (negatron, or negative electron), a β^+ particle (positron), or a neutron. It may capture an orbital electron, usually from the K-shell. It

---

[*]This element, of atomic number 86, is a chemically inert gas sometimes called emanation (Em). All the isotopes of radon are unstable; among them are ${}^{219}_{86}\text{Rn}$ (actinon), ${}^{220}_{86}\text{Rn}$ (thoron), and ${}^{222}_{86}\text{Rn}$.

[*]There is no difference between a 0.100 MeV γ-ray photon and a 0.100 MeV x-ray photon; the distinction, if any, is usually based upon origin.

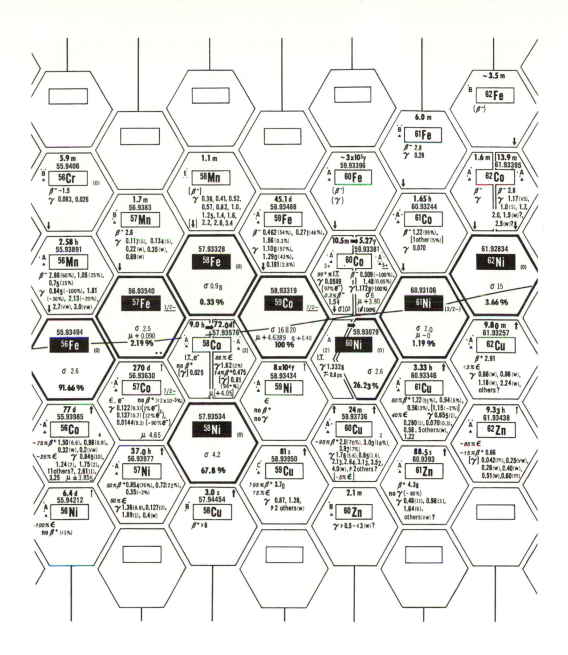

Figure 29-2 Stable and unstable nuclides with mass numbers 56 through 62. [Adapted from *Trilinear Chart of Nuclides,* by W. H. Sullivan, Oak Ridge National Laboratory (U.S. Government Printing Office, 1957).] Masses, in u, are relative to $^{12}_{6}C = 12.00000$. For each stable nuclide, relative abundance is shown in per cent. For each unstable nuclide, modes of decay are given and half-lives are shown in years (y), days (d), hours (h), minutes (m), or seconds (s). Numbers following β or γ are energies in MeV of emitted β particles or γ photons. Other symbols: ϵ, electron capture; I.T., isomeric transition; σ, cross section in barns for slow neutron capture; e^-, internal conversion electrons; μ and q refer to magnetic properties of nuclide; A, B, C, etc., at left of nuclide symbol refer to method of experimental identification.

may undergo spontaneous fission. During any of these changes, a γ ray (photon) may be emitted from the nucleus. A γ ray is also emitted during an isomeric transition when a nucleus falls from an excited state to a stable state. About the only imaginable mode of decay which has not been observed is the direct emission of a proton or neutron from the nucleus, with a measurably long half-life, although these particles are emitted by some nuclides promptly after β decay to excited states of the nuclides.

29-3 Stable and unstable nuclides in nature

You may well wonder just why some isotopes of any given element are stable and others are unstable. This is indeed a puzzling question upon which much work is being done. Using chromium again as an illustration, we find that isotopes of mass number 50, 52, 53, and 54 are stable, and 46, 47, 48, 49, 51, 55, and 56 are unstable. However, the terms "stable" and "unstable" are relative. When we say that ^{50}Cr is "stable," we mean that the probability of a disintegration is zero, or else is so small that no decay is observable. (See Prob. 29-B6.) A nuclide having a half-life of more than 10^{18} years would, for all practical purposes, be stable. At the other extreme, when we say that ^{57}Cr does not exist, we mean that if it does exist its half-life is so short that it decays before it can be observed. From this point of view all nuclides are *possible*, although most of them are so unstable as to be unobservable.

In the beginning, i.e., at a time estimated to be some 10^9 to 10^{10} years ago when the universe acquired its present form, probably all nuclides were formed in equal abundance and the highly unstable ones immediately decayed, becoming eventually long-lived "stable" or "nearly stable" nuclides. The sun and earth were apparently formed about 5×10^9 years ago. Thus any ^{55}Cr

(half-life 3.52 min) that was in the solar system at that time has long since decayed (see Example 29-1). At the other extreme, ^{238}U has a half-life of 4.5×10^9 years; therefore, about half the ^{238}U originally present is still in existence. With such a long half-life, the probability of its decay is small, and so ^{238}U is not particularly "active." The extremely active radium isotope $^{226}_{88}$Ra, discovered and isolated with so much effort by Madame Curie, has a half-life of 1622 years; any "primeval" radium would have long since decayed. To understand why ^{226}Ra and some nuclides of even shorter half-life are still found in nature, we consider the *radioactive series* which starts with the long-lived nuclide $^{238}_{92}$U (Fig. 29-3). This nuclide emits an α particle and becomes $^{234}_{90}$Th; half the original atoms undergo this decay in 4.5×10^9 years. We indicate this process by a diagonal line on the diagram. The $^{234}_{90}$Th in turn decays by β^- emission to $^{234}_{91}$Pa, an isotope of protactinium. This negatron emission is represented by a short horizontal line on the diagram. The half-life for this decay is 24 days. Next, $^{234}_{91}$Pa decays into $^{234}_{92}$U by another negatron emission, this time with a half-life of 1.2 min. Now a long series of α emissions takes place, with $^{226}_{88}$Ra as one stage in the life history of the nucleus. The end result is the stable lead isotope $^{206}_{82}$Pb (some of the alternative β decays of small probability are omitted). The Curies found small amounts of radium in the uranium ore (pitchblende) which they purified. This radium, of relatively short half-life, was continually being replenished by radioactive decay of nuclei higher in the series. The whole series is dependent on one nuclide ($^{238}_{92}$U) having a half-life comparable with the age of the earth.

Other radioactive series start with $^{235}_{92}$U (half-life 7.1×10^8 years) and $^{232}_{90}$Th (half-life 1.4×10^{10} years). Many radioactive series exist whose longest-lived member's half-life is relatively short; such series can now be studied by making the necessary parent nuclide artificially. Before the nature of radioactivity was well understood, special names were given to certain nuclides occurring in the various natural series. For instance, in the uranium series of Fig. 29-3, $^{234}_{90}$Th was called "uranium X1" (UX$_1$); $^{210}_{82}$Pb was called "radium D" (RaD); $^{210}_{83}$Bi was called "radium E" (RaE); and so on.

These names are now of only historical interest.

Every element has at least one unstable isotope. The hydrogen isotope $^3_1$H, of mass number 3, is known as tritium; it is a negatron emitter with a half-life of 12.3 years. Even the neutron itself is unstable when it is not bound together in a nucleus with

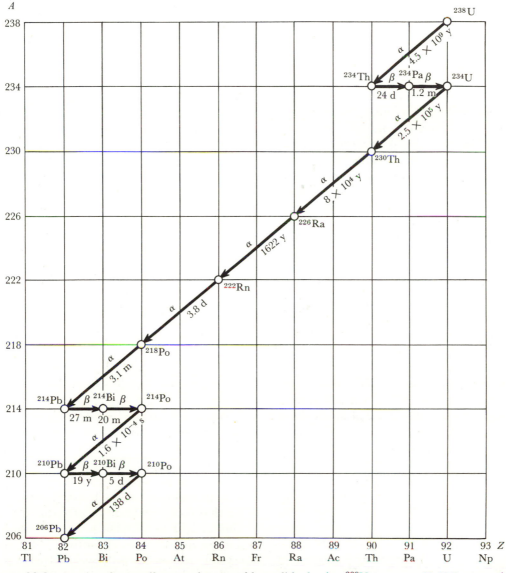

Figure 29-3 A series of naturally occurring unstable nuclides having ^{238}U as parent. Half-lives are given in years (y), days (d), hours (h), minutes (m), or seconds (s).

other particles; the free neutron is a negatron emitter with a half-life of 13 min. On
the other hand, several elements of medium
mass and all the heavy elements have *no*
stable isotopes. These include technetium
(whose isotopes of longest half-life are $^{98}_{43}Tc$
and $^{99}_{43}Tc$, each of half-life about 10^5 years),
and promethium (whose stablest isotope is
$^{145}_{61}Pm$, about 30 years), and all elements of
atomic number 84 and higher.

29-4   Experimental techniques

Over the years, physicists have developed
many devices for detecting the particles and
photons emitted by radioactive atoms. Most
of these methods depend for their success
upon the ions produced by a moving charged
particle. As a charged particle, such as an
α particle or an electron, passes through air,
its electric field distorts the air molecules,
and some molecules become ionized. As a
consequence, the air becomes a better conductor of electricity. In the early days, a
radioactive sample was placed directly inside
a charged gold-leaf electroscope, and the
leaves collapsed at a rate proportional to the
strength of the sample. Modern electronic
amplifiers can be used to detect the ionization pulse due to a single, heavy, charged
particle such as an α particle or a proton
passing through an *ionization chamber.*

The Geiger-Müller tube counter (Fig.
29-4) (invented in 1928, now called simply
a *Geiger counter*) makes use of an avalanche

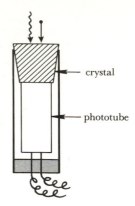

Figure 29-5   Scintillation counter.

effect. The tube contains a gas such as argon
at a pressure of, say 0.1 atm, and the thin
central wire is about 1000 V positive with
respect to the metal wall. If the incoming
particle creates just one ion pair, the negative ions are attracted to the + wire, creating more ions by collision. A current rapidly
builds up, and the *IR* drop in the resistor *R*
may amount to several volts, to be further
increased by the amplifier tube. The Geiger
counter responds only to charged particles.
A photon is uncharged, and therefore produces practically no ions in the gas. Photons may, however, eject photoelectrons
from the metal of the wall, and these photoelectrons may initiate an avalanche. The
efficiency of a Geiger counter may be as
high as 99% for electrons which penetrate
the outer wall, but less than 1% for x rays
or γ rays.

We have mentioned earlier the scintillation
detector of Rutherford. The modern development of this device is the *scintillation
counter* of Fig. 29-5, which uses a single
crystal such as NaI. The crystal, perhaps an
inch in depth, is so much more dense than
air that there is a good chance that a γ-ray
photon will be absorbed and will emit a
fluorescent photon of visible light. The
crystals are usually "sensitized" by controlled addition of an impurity such as
thallium. The crystal is in a light-tight

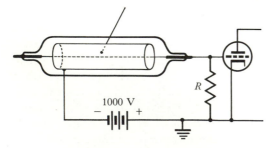

Figure 29-4   Geiger counter.

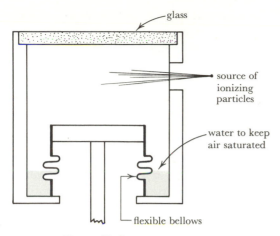

Figure 29-6 Cloud chamber.

enclosure, and the faint splash of light is picked up and amplified by a sensitive phototube. No avalanche process is involved, and the scintillation counter is

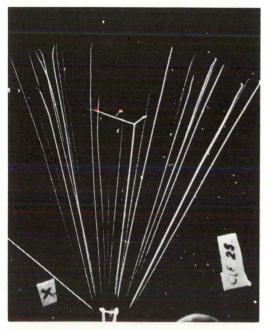

Figure 29-7 Scattering of α particle by a heavier nucleus of fluorine. Note also that all the unscattered α particles have about the same range in the gas of the cloud chamber. (I. K. Bøggild, from *An Atlas of Typical Expansion Chamber Photographs,* by W. Gentner, H. Maier-Leibnitz, and W. Bothe, Pergamon Press.)

Figure 29-8 Transmutation of nitrogen by α particle, according to the reaction ${}^{14}_{7}N + {}^{4}_{2}He \rightarrow {}^{17}_{8}O + {}^{1}_{1}H$. The ejected proton travels downward and to the left, and the oxygen nucleus leaves a short track, upward and slightly to the right. The source of α particles is a mixture of two nuclides: ${}^{212}Bi$ (emitting α particles of shorter range) and ${}^{212}Po$ (emitting α particles of longer range). (P. M. S. Blackett and D. S. Lees, from *An Atlas of Typical Expansion Chamber Photographs,* by W. Gentner, H. Maier-Leibnitz, and W. Bothe, Pergamon Press.)

fast-acting, rugged, and of high efficiency both for charged particles and for γ rays.

The *cloud chamber* (Fig. 29-6) makes visible the path of an ionizing particle. When a charged particle passes through air, it leaves a trail of ions, and, if the region is supersaturated with water vapor, the ions serve as centers upon which visible drops of water may condense (Figs. 29-7 and 29-8). The effect is similar to the formation of "vapor trails" in the sky by a passing jet aircraft. The *bubble chamber* uses a liquid such as propane or liquid hydrogen and the passage of an ionizing particle is made evident by a trail of tiny bubbles (Fig. 30-13).

The *spark chamber* contains a gas and a set of many parallel thin foils separated by a few mm or cm. The ionization path left by a particle serves as a path for sparks which pass from foil to foil and thus show the particle's trajectory. Spark chambers are faster acting than bubble chambers, and unlike bubble chambers they can be triggered to be "flashed" only when an event of interest has just occurred. All three of these chambers give a great deal of information about the particle which makes a track. If the chamber is placed in a magnetic field, moving charged particles are bent in arcs of circles, and so their energies can be found. Particles can be identified by the density of drops or bubbles along the path. The slow, heavy α particles create many more ion pairs per centimeter of path, and give a denser trail, than do the light electrons.

Another detection device is the *photographic emulsion;* upon development, grains of silver are deposited where ionizing particles have passed. The tracks are viewed with a microscope (Fig. 30-14).

29-5 Conservation laws in radioactivity

Our study of classical physics revolved around four conservation laws: the laws of conservation of linear momentum (Sec. 6-3), mass-energy (Sec. 7-7), angular momentum (Sec. 8-8), and charge (Sec. 17-4). All these laws are valid during nuclear processes; indeed, this universality is one reason we have stressed the conservation form for our general principles.

Conservation of charge is illustrated by distintegration equations such as $^{55}_{24}\text{Cr} \rightarrow$ $^{55}_{25}\text{Mn} + ^{0}_{-1}\text{e}$. Here the essential physical process is the "creation" of a negative electron while the neutron changes into a proton. It is easy to see *mathematically* that charge is conserved in the process; but you may feel uneasy that an electron's charge has been created out of nothing, so to speak.

For this reason you may have the impression that a neutron is a "proton plus an electron," and that the electron is simply ejected from the neutron. This is incorrect, however, for there are several reasons for believing that an electron cannot exist inside a nucleus. We can consider the process to be the change of a nucleon from one state to another. When it is neutral we call it a neutron, and when it is charged positively we call it a proton. The situation is somewhat analogous to the emission of a photon of yellow light when a sodium atom's electron cloud changes from a certain energy state to a certain lower energy state. The photon is materialized during the transition. Similarly, the electron does not "exist" inside the nucleus, ready to be ejected like a bullet from a gun; it is created when a nucleon changes its state. Even this picture may suffer too much from modelitis. A nucleus may be just a sort of melting pot in which individual nucleons lose their identity.

Conservation of mass and conservation of energy are intimately related, as we saw in Sec. 7-7. If we use relativistic mass in our calculations, mass and energy are each (separately) conserved. Usually, however, we deal with *rest mass,* and find that rest mass (m_0) can be converted into energy according to the Einstein equation $\Delta E = (\Delta m_0)c^2$. Instead of using joules for energy and kilograms for mass, we introduce units which are much more convenient for nuclear physics. For energy, we use MeV (million electron volts). According to the definition in Sec. 27-3, 1 MeV = $10^6 \times 1.60 \times 10^{-19}$ J = 1.60×10^{-13} J. For mass, we use the *unified atomic mass unit* (u), which is defined so that the mass of the neutral $^{12}_{6}\text{C}$ atom (nucleus plus 6 electrons) is exactly 12.000000 u. The Einstein formula is used to obtain the conversion factor between MeV (an energy unit) and u (a mass unit). One unified atomic mass unit is $\frac{1}{12}$ the mass of a carbon-12 atom, which is found by use of Avogadro's number:

$$1 \text{ u} = \frac{1}{12}\left(\frac{12.000000 \text{ g}}{6.02 \times 10^{23}}\right) = 1.66 \times 10^{-24} \text{ g}$$

$$= 1.66 \times 10^{-27} \text{ kg}$$

Hence

$$\Delta E = (\Delta m_0)c^2 = (1.66 \times 10^{-27} \text{ kg})$$
$$\times (3.00 \times 10^8 \text{ m/sec})^2$$
$$= 1.49 \times 10^{-10} \text{ J}$$
$$= 9.31 \times 10^8 \text{ eV} = 931 \text{ MeV}$$

Thus *the energy equivalent of 1 u is 931 MeV.* The electron's mass is 9.11×10^{-31} kg, which is 0.000549 u, equivalent to 0.51 MeV.

For every nuclear event, a mass-energy balance is possible. To illustrate this, let us once more study the decay $^{55}_{24}\text{Cr} \rightarrow {}^{55}_{25}\text{Mn} + {}^{0}_{-1}\text{e}$, this time considering the rest masses in detail.

BEFORE DECAY	
$^{55}_{24}\text{Cr}$ nucleus	54.9279 u
24 electrons	0.0132
	54.9411

AFTER DECAY	
$^{55}_{25}\text{Mn}$ nucleus	54.9244 u
24 electrons	0.0132
1 negatron (emitted from nucleus)	0.0005
	54.9381

The 24 outer electrons of $^{55}_{24}\text{Cr}$ still surround the $^{55}_{25}\text{Mn}$ nucleus just after the decay process; thus $^{55}_{25}\text{Mn}$ is formed as a positive ion, having a deficiency of one electron. Eventually, it picks up an electron, or it takes part in some chemical reaction. Note that the mass total after decay includes 25 electrons, just enough for a neutral Mn atom. For this reason, masses of *neutral* atoms can be used in calculations of this sort. The mass balance is not exact. As we see, the total rest mass before decay is greater than the total rest mass after decay. The difference is 0.0030 u, which is equivalent to 2.8 MeV. This is the energy available for KE of the ejected particle (or particles) and for the γ-ray photons. Experiment shows that the maximum KE of the emitted negatrons is 2.8 MeV, which agrees with the available energy calculated from the mass data. No γ rays are observed to accompany the decay of $^{55}_{24}\text{Cr}$, but many other nuclides emit γ rays during the decay process.

The decay of ^{55}Cr is typical of the conversion of rest mass into energy.[*] Detailed calculations of this sort for hundreds of nuclear processes strengthen our confidence in the theory of relativity.

One major problem arises in the study of β-ray energies: particles are emitted with a continuous range of energies from zero up to the maximum given by the known mass difference. At first it was supposed that all the particles start out with the same KE (2.8 MeV in the example we are considering) but that some of them give up a portion of their KE to the nucleus on the way out. This would show up as heat energy which should be observable by a calorimeter experiment. Surprisingly, no such production of heat was observed. Furthermore, the missing energy cannot be radiated in the form of γ rays, as shown by experiments in which the calorimeter cup was made of thick lead, which would absorb γ rays and produce heat. For a while it was seriously proposed that the law of conservation of energy be given up for β decay. The present view is that the missing energy is carried away by *neutrinos,* neutral particles of small (probably zero) rest mass.[*] Such particles do not affect a counter or a cloud chamber and are almost impossible to detect. Every β decay is, therefore, the simultaneous emission of an electron-neutrino pair. If 2.8 MeV is the energy released during the decay of a nucleus of $^{55}_{24}\text{Cr}$, and if the observed

---

[*] The *total* (relativistic) mass of the system has been conserved, for the fact that the emitted negatron has 2.8 MeV of KE means that this negatron's mass is greater than if it were at rest.

[*] If you are bothered by the idea of a particle of zero rest mass which has energy and momentum, recall that the photon is just such a particle, as shown by the Compton effect (Sec. 27-4).

electron has 2.5 MeV, then the neutrino carries the other 0.3 MeV. Another $^{55}_{24}$Cr nucleus might eject a negatron of energy 2.1 MeV, along with a neutrino of energy 0.7 MeV. In this way the law of conservation of energy can be satisfied for each individual disintegration event. Only the maximum energy of any given β-particle group is listed in tables such as Fig. 29-2.

Although neutrinos are elusive, they do carry energy and hence have momentum. Neutrinos are also emitted during electron capture, and the recoil of the nucleus can be observed. Neutrinos can be captured (very rarely) by protons or neutrons. The "reality" of the neutrino is no longer in question.

The complete equation representing a β decay includes the neutrino:

$$^{49}_{24}\text{Cr} \rightarrow ^{49}_{23}\text{V} + ^{0}_{+1}\text{e} + \nu_e$$

or

$$^{55}_{24}\text{Cr} \rightarrow ^{55}_{25}\text{Mn} + ^{0}_{-1}\text{e} + \bar{\nu}_e$$

where ν_e and $\bar{\nu}_e$ represent the electron's neutrino and antineutrino (which differ only in the direction of their spins). However, the neutrinos are often omitted from the equations, their presence being tacitly understood.

Conservation of linear momentum is illustrated in Fig. 29-7, which shows the elastic scattering of an α particle by a heavier nucleus of fluorine. The vector sum of the momentums before collision equals the vector sum after collision. Note how similar this event is to the collision of billiard balls; the resemblance is more than accidental, since linear momentum is conserved in each process.

The angular momentum of a particle such as a proton, electron, or neutron is considered to be a spin, analogous to the rotation of the earth on its axis. In the Bohr theory (Sec. 28-2), the *orbital* angular momentum is quantized, in multiples of $h/2\pi$. Similarly, *spin* angular momentum of a particle is

quantized, but in multiples of $\frac{1}{2}(h/2\pi)$. Angular momentum is conserved in nuclear processes, as it is in large-scale physics (see Sec. 29-8).

Still another conservation law is the law of conservation of *parity;* this deals with the symmetry properties of the particles or atoms and has no counterpart in large-scale physics, being essentially quantum-mechanical in nature. Conservation of parity requires that if an experiment takes place, the mirror image of the same experiment can also take place. Thus if a "left-handed" neutrino emitted during β^+ decay spins counterclockwise when viewed from behind (spin vector opposite to linear momentum vector), then we would also expect "right-handed" neutrinos (spin vector parallel to linear momentum vector) to be emitted during β^+ decay. Until recently, it was supposed that conservation of parity applied to all nuclear processes, as it is known to do with regard to the outer electrons of an atom. In 1956 it was shown, as a result of experiments suggested by the theoretical physicists T. D. Lee and C. N. Yang, that this law is not true, during certain "weak interactions"; β decay is such an interaction. Experiment shows that neutrinos emitted during β^+ decay are *all* left-handed, and antineutrinos emitted during β^- decay are *all* right-handed. The non-conservation of parity for weak interactions was sensational news, and Lee and Yang received the 1957 Nobel prize in physics for their imaginative work.

We have been considering the validity of five conservation laws: those dealing with charge, mass-energy, linear momentum, angular momentum, and parity. Of these, the law of conservation of parity has been shown to be less generally valid than had been previously thought. Far from being a blow to progress, the breakdown of parity conservation has opened the way to bold speculation about the validity of other conservation laws. So far, the remaining four conservation laws appear to be valid in all nuclear processes, as they are in large-scale, everyday life. It is entirely possible, of course, that new experiments may some day reveal areas in which the law of conservation of charge, for instance, is not true. As always, a well-established law must be modified or abandoned if even one experimental fact is at variance with it. Physicists welcome the challenge of these new ideas.

29-6 Binding energy

Work must be done in order to "take apart" any assemblage of particles; the amount of energy that must be supplied is called the *binding energy* (BE) of the system. A large-scale example of binding energy is the work required to separate a space ship from the earth. In this case, the BE can be calculated relatively easily, since the gravitational force is a well-known function of distance, Newton's law of gravitation.

On the atomic scale of things, the binding energy of the hydrogen atom is found by considering the energy levels of its electron. According to Fig. 28-3, if the electron is originally in its lowest energy state, 13.60 eV is needed to remove the electron from the electrical influence of the proton. Thus the BE of the H atom is 13.60 eV. This value can be calculated from Coulomb's law, as in Sec. 28-7. The binding energies of organic molecules are more difficult to calculate, but they can be measured in the laboratory and are from 2 eV to 20 eV.

On the nuclear scale of things, binding energies are much larger; they are not easily calculated since the nuclear forces are not so well understood. Fortunately, experimental values of nuclear masses can be used to calculate nuclear binding energies.

EXAMPLE 29-2

What is the binding energy of the least strongly bound neutron in a $^{12}_{6}C$ nucleus?

In effect, we are asked for the difference in

energy between $^{12}_{6}C$ and $(^{11}_{6}C + ^{1}_{0}n)$. We use the nuclear masses from Table 8 in the Appendix.

BEFORE TAKING APART

$^{12}_{6}C$	12.000000

AFTER TAKING APART

$^{11}_{6}C$	11.011432
$^{1}_{0}n$	1.008665
	12.020097

Mass difference 0.020097 u = $\boxed{18.7 \text{ MeV}}$

Since the mass of the fragments is greater than the mass of the original $^{12}_{6}C$ nucleus, energy must be supplied in the amount of 18.7 MeV to remove the neutron from the nucleus.

EXAMPLE 29-3

Calculate the binding energy of an *average* nucleon in $^{12}_{6}C$.

Let us imagine breaking the $^{12}_{6}C$ nucleus up into its 12 constituent nucleons, i.e., into 6 protons and 6 neutrons.

BEFORE

$^{12}_{6}C$	12.000000 u

AFTER

$6 \, ^{1}_{1}H = 6(1.007825)$	6.046950 u
$6 \, ^{1}_{0}n = 6(1.008665)$	6.051990
	12.098940

Mass difference = 0.098940 u = 92.1 MeV

BE per nucleon = 92.1/12

$= \boxed{7.7 \text{ MeV per nucleon}}$

To remove only one neutron from $^{12}_{6}C$ (Example 29-2) would require 18.7 MeV, more than twice the average amount per nucleon to break up $^{12}_{6}C$ into 12 separate nucleons.

Binding energies per nucleon (Fig. 29-9) show a rising trend for the lightest nuclides, and are remarkably constant at about 8 MeV per nucleon for most nuclides above about mass number $A = 20$. The pronounced peak at $A = 4$ (for the alpha particle $^{4}_{2}He$) shows that the combination of 2 protons and 2 neutrons is exceptionally stable (has large BE).

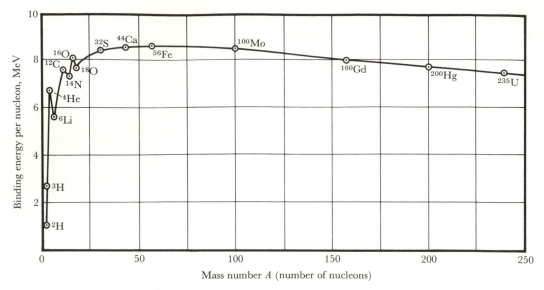

Figure 29-9 Nuclear binding energy curve.

This is correlated with the fact that α particles are ejected as a single stable entity during radioactive decay of heavy elements such as radium.

The nuclear binding energy curve has a broad maximum at about $A = 50$ to 60. Thus $^{56}_{26}\text{Fe}$, a common constituent of metallic meteorites, is more stable than either lighter or heavier nuclides.

These experimental facts about nuclear binding energies are data which must serve as a basis for any theory about nuclear structure. We will return to these matters in Chap. 30 when we discuss interactions between the nucleons (protons and neutrons) which are the constituents of the "nuclear matter" discussed first on page 20.

Example 29-3 shows that energy is needed to separate a light nuclide such as $^{12}_{6}\text{C}$ into its component parts. The reverse is also true: energy is released when $^{12}_{6}\text{C}$ is built up out of its parts. Such a process, called *fusion*, releases energy because the BE curve is rising —has a positive slope—for light elements. The opposite is true for heavy elements. Here the BE curve is decreasing—has a negative slope—and energy is released when

a nuclide such as $^{235}_{92}\text{U}$ splits into two or more fragments. The BE curve is thus of utmost importance for the controlled release of nuclear energy by fission (Sec. 30-4) and by fusion (Sec. 30-5).

29-7 Transmutations

The artificial production of gold from mercury was one of the dreams of the alchemists, but by 1895 it had generally been agreed that chemical elements are stable and unalterable. With the discovery of naturally occurring radioactivity, it was realized that the elements can and do change. The modern search for artificial transmutation of the elements began with the use of high-speed particles as projectiles. Rutherford in 1919 allowed α particles from $^{214}_{84}\text{Po}$ (KE = 7.68 MeV) to pass through air, and observed scintillations which were caused by protons. The reaction is with the nitrogen atoms in the air:

$$^{14}_{7}\text{N} + ^{4}_{2}\text{He} = ^{17}_{8}\text{O} + ^{1}_{1}\text{H}$$

Here, as in radioactive decay, the equation

balances for both charge and mass number.

The α particle (which has a charge of $+2e$) is strongly repelled as it approaches the $^{14}_{7}$N nucleus (which has a charge of $+7e$). This is one reason a charged projectile must have a high energy, for otherwise it will never get close enough to its target to "react."[*] The critical distance is about 10^{-14} to 10^{-15} m, the "radius" of a nucleus.

A detailed calculation using exact masses shows that some energy is available for the KE of the proton and for KE of the ^{17}O nucleus as it recoils (in the calculation, all masses are for the neutral atoms of ^{14}N, ^{4}He, etc.; thus 9 outer electrons in all are included both before and after).

BEFORE COLLISION

$^{14}_{7}$N	14.003074 u
$^{4}_{2}$He	4.002603
Original KE of particle (equivalent to 7.68 MeV)	0.00824
	18.013917

AFTER COLLISION

$^{17}_{8}$O	16.999133 u
$^{1}_{1}$H	1.007825
	18.006958

The difference is 0.00696 u, or 6.48 MeV, and this energy is shared between the proton and the ^{17}O nucleus. The cloud chamber photograph (Fig. 29-8) shows the collision. Even if the mass balance is right for a reaction to take place, we still have to deliver the missile to the target, which requires that the projectile have high energy. In the nitrogen reaction described above, we see that the initial KE of the α particle is also needed for another reason; without some of the 0.00824 u represented by this KE, the total mass before collision would have been insufficient to supply the total mass after collision, as simple arithmetic shows. This is an example of a reaction for which there is a threshold energy.

As an illustration of nuclear transmutation, we have studied one reaction in detail. To save space, nuclear reactions are often written in abbreviated form; the bombardment of ^{14}N by alpha particles can be written as

$$^{14}\text{N}(\alpha,\text{p})^{17}\text{O}$$

The atomic numbers are omitted, since they are really superfluous. The necessary information about atomic number is given by writing the chemical symbol. The first symbol inside the parentheses represents the projectile, an α particle, and the second symbol represents the emitted particle, a proton. Many other types of reaction are possible. Bombardment with neutrons is very effective, since the neutron has no charge and can easily come close to the target nucleus. In fact, simple capture of slow neutrons works best of all, because the projectile spends more time as it passes by the target. In this connection a "slow" neutron means one whose KE is less than 1 eV. (Compare with the millions of electron volts usually required for a charged particle to cause a transmutation!) A few examples will serve to illustrate the variety of nuclear reactions. Our symbols are: α for alpha particle ($^{4}_{2}$He); p for proton ($^{1}_{1}$H); d for deuteron ($^{2}_{1}$H); n for neutron ($^{1}_{0}$n); γ for gamma-ray photon ($h\nu$).

1. $^{9}\text{Be}(\alpha,\text{n})^{12}\text{C}$. This is the reaction by which neutrons were first produced and recognized as such by Chadwick, in England (1932). Convenient, portable neutron sources are made by mixing powdered beryllium with a long-lived α emitter such as radium or plutonium.

2. $^{50}\text{Cr}(\text{n},\gamma)^{51}\text{Cr}$. This is a simple capture reaction and illustrates one way of producing the unstable nuclide ^{51}Cr we discussed in Sec. 29-2. Ordinary chromium metal, containing stable ^{50}Cr, is placed in the vicinity of a fission reactor which produces large numbers of neutrons.

3. $^{2}\text{H}(\text{d},\text{n})^{3}\text{He}$. The bombardment of deu-

---

[*] That is, for the short-range nuclear force of attraction to be greater than the long-range Coulomb repulsion.

terons (in the form of a target of heavy ice) by deuterons. This reaction is remarkable because of its large yield, even for bombarding energies as low as 25 000 eV. The product ^{3}He is stable.

4. ^{7}Li(p,α)^{4}He. This is an abbreviation for ^{7}Li $+$ ^{1}H $=$ 2^4He. Two α particles are produced, with a total energy release of 17 MeV.

5. ^{14}N(n,p)^{14}C. The ^{14}C, a negatron emitter with half-life of 5770 years, is useful as a tracer in studying chemical reactions and in dating by the radiocarbon method (Sec. 30-1).

Over 1000 different nuclear reactions have been studied. Whether or not any proposed reaction is possible depends on the mass balance, so the importance of precise knowledge of nuclear masses is obvious. Conversely, masses of unstable nuclides can be determined by study of the energy balance for reactions that are known to take place.

Our notation for transmutations is really a shorthand for a complex process. The present view is that a *compound nucleus* is formed, which has an unobservably short lifetime, estimated to be about 10^{-15} s. Thus reaction 5 above should be written in two steps:

$$^{14}_{7}\text{N} + {}^{1}_{0}\text{n} = [{}^{15}_{7}\text{N}]$$
$$[{}^{15}_{7}\text{N}] \rightarrow {}^{14}_{6}\text{C} + {}^{1}_{1}\text{H}$$

The brackets around $[{}^{15}_{7}\text{N}]$ indicate that this is a compound nucleus formed in a highly excited state having extremely short half-life. We should realize that any description, or model, for a nuclear transmutation is subject to the limitations of all models, and should not be taken too literally.

For any reaction, the nature of the target, the product, or the particles can be found by applying the laws of conservation of charge and conservation of mass number. The Periodic Table (p. 721) must be used to identify the element when the atomic number is known, and vice versa.

EXAMPLE 29-4

The phosphorus isotope $^{32}_{15}$P has a half-life of 14.3 days and is useful in biological studies. It can be produced by a (n,p) reaction. What target must be used?

The equation must read

$$^{A}_{Z}\text{X} + {}^{1}_{0}\text{n} = {}^{32}_{15}\text{P} + {}^{1}_{1}\text{H}$$

and it is evident that $Z = 16$ and $A = 32$. The target nuclide is, therefore, $^{32}_{16}$S.

EXAMPLE 29-5

When ^{24}Mg is bombarded with deuterons, α particles are emitted. Identify the product nuclide.

We look up the atomic number of magnesium in the Periodic Table and write $^{24}_{12}$Mg $+$ $^{2}_{1}$H $=$ $^{A}_{Z}$X $+$ $^{4}_{2}$He. The equation balances for $Z = 11$ and $A = 22$. Hence the product is $^{22}_{11}$Na.

EXAMPLE 29-6

The sodium isotope produced in the reaction of Example 29-5 is unstable. Make a guess as to the nature of its decay.

If $^{22}_{11}$Na is unstable, it probably has too few neutrons, since the trend is toward slightly more than one neutron for every proton. Hence we expect a proton to turn into a neutron, and the expected decay is $^{22}_{11}$Na $\rightarrow$ $^{22}_{10}$Ne $+$ $^{0}_{+1}$e (positron emission); perhaps the decay is $^{22}_{11}$Na $+$ $^{0}_{-1}$e $\rightarrow$ $^{22}_{10}$Ne (electron capture). Since the nuclide is a light one, compared with uranium, radium, etc., we rule out the possibility of an α decay. Reference to Table 8 in the Appendix shows that ^{22}Na is indeed a positron emitter, with a half-life of 2.6 years.

■ SUMMARY

Radioactive decay takes place when a neutron changes into a proton (negatron emission) or when a proton changes into a neutron (positron emission). A group of two neutrons and two protons is exceptionally stable and can be emitted as an α particle by a heavy nucleus. During electron capture a nucleus captures an orbital electron, usually from its K-shell. Less frequent modes of decay are spontaneous fission and neutron emission. During any nuclear change, γ rays may be emitted. These are high-energy photons whose energy equals the difference between two energy levels in the nucleus. Nuclear disintegration is a statistical process, and the half-life is the time required for half of any given number of nuclei to decay. Natural radioactive series begin with ^{238}U, ^{235}U, and ^{232}Th, whose half-lives range from 0.7 to 14 billion years. Since these lives are comparable with the age of the earth, some of the parent nuclides still remain and are continually producing a chain of daughter products having shorter half-lives.

The laws of conservation of charge, mass-energy, and momentum are obeyed during nuclear events. In nuclear disintegrations and in transmutations, the energy which is released as KE comes from the transformation of rest mass into energy according to $\Delta E = (\Delta m_0)c^2$. Conversely, if the rest mass of the system increases during a reaction, energy must be supplied. One unified atomic mass unit (u) is equivalent to 931 million electron volts (MeV). During β decay, some of the available energy may be given to a neutrino, which has no charge and (probably) no rest mass. Many types of transmutation reaction are possible, if the mass balance is favorable. Capture of slow neutrons is effective because of the absence of Coulomb repulsion between projectile and target.

■ CHECK LIST

alpha ray	positron	bubble chamber
beta ray	negatron	spark chamber
gamma ray	internal conversion	MeV
nucleon	isomer	u
nuclide	radioactive series	neutrino
isotope	ionization chamber	binding energy
atomic number	Geiger counter	compound nucleus
mass number	scintillation counter	
half-life	cloud chamber	

■ QUESTIONS

29-1 Explain why the early observations of radioactivity seemed to contradict one of the conservation laws. Is this objection still valid?

29-2 Compare and contrast α, β, and γ rays as regards (a) penetrating power; (b) charge; (c) mass; (d) deflection by a magnetic field.

29-3 What is the approximate ratio of the rest mass of an alpha particle to that of a beta particle?

29-4 Why are the chemical atomic weights of some elements not very close to integers?

29-5 The noble gas radon has half-life 3.8 days and has been used in cancer therapy for many years. Explain how a nuclide of such short half-life can occur naturally. (See Fig. 29-3.)

29-6 When $^{66}_{29}$Cu decays to $^{66}_{30}$Zn by negatron emission, is the zinc atom (just after the decay) probably formed as a neutral atom, a positive ion, or a negative ion?

29-7 In the portion of the chart of nuclides shown in Fig. 29-2, there is no example of two stable nuclides in vertically adjacent hexagons. Show that this must be true, throughout the entire chart of nuclides.

29-8 Describe three devices for detecting the passage of a charged particle.

29-9 Does a gamma ray have (*a*) any mass? (*b*) any momentum? (*c*) any charge?

29-10 What is the precise meaning of the statement: "$^{18}_{8}$O is a stable isotope of oxygen"?

29-11 In Fig. 29-4, when an ionizing particle passes through the counter, does the potential of the grid of the triode (relative to ground) increase or decrease?

29-12 Does a neutrino have (*a*) mass? (*b*) rest mass? (*c*) linear momentum? (*d*) KE?

29-13 What is the evidence for believing that radioactivity is a nuclear process?

29-14 Explain why bombardment of a nucleus with neutrons is so much more effective than bombardment with protons.

■ PROBLEMS

Note: Refer when needed to Periodic Table on p. 721 and to the table of masses of light nuclides on p. 723.

29-A1 Identify the elements which have the following isotopes: $^{118}_{50}$X; $^{3}_{2}$X; $^{50}_{22}$X; $^{197}_{79}$X.

29-A2 How many neutrons are in each nuclide of Prob. 29-A1?

29-A3 Identify the elements which have the following stable isotopes: $^{12}_{6}$X; $^{22}_{11}$X; $^{200}_{80}$X; $^{107}_{47}$X.

29-A4 Consider the nuclide $^{60}_{28}$Ni. How many protons does it have? How many neutrons? How many nucleons?

29-A5 The half-life of ^{55}Fe is 3 years. What fraction of a sample of this nuclide remains after 12 years?

29-A6 A sample of radioactive material gives 6000 counts per minute at 1:00 P.M., and it gives 1500 counts per minute at 4:00 P.M. What is its half-life?

29-A7 A reactor produces 256 μg of a certain unstable nuclide, and only 16 μg remains after 5 days. What is the half-life of the nuclide?

29-A8 A sample of ^{100}Pd (half-life 4 days) has mass 1.776 g at 3:00 P.M. on July 4th; what mass of ^{100}Pd remains at 3:00 P.M. on July 12th?

29-A9 Extend the table in Example 29-1 to 1:15 P.M. and verify the "million-to-one chance" mentioned in the last line of the example.

29-A10 Identify the daughter nuclides formed, and write the equation for each of these decays: (*a*) ^{59}Ni captures a *K*-electron; (*b*) ^{40}K emits a negatron; (*c*) ^{25}Al emits a positron; (*d*) ^{210}Po emits an α particle; (*e*) ^{113}Cd makes an isomeric transition and emits a γ ray.

29-A11 Write an equation for each of the following decays: (*a*) ^{20}Na emits a positron; (*b*) ^{38}Cl emits a negatron; (*c*) ^{218}Po emits an α particle; (*d*) ^{55}Fe captures a *K*-electron.

29-A12 The nuclide $^{64}_{29}$Cu can decay by β^- emission, by β^+ emission, and by electron capture. What is the final nuclide in each case?

29-A13 During a certain transmutation, 0.030 u of rest mass has disappeared. How much nuclear energy has been released (in MeV)?

29-A14 How many u of rest mass must disappear in a nuclear reaction in which 185 MeV of energy is released?

29-A15 ^{69}Zn decays by an isomeric transition to ^{69}Zn, with a half-life of 13.8 h, emitting a γ ray of energy 0.436 MeV. Which isomer of ^{69}Zn has greater mass? What is the mass difference between these two isomers?

29-B1 In Fig. 29-2, (a) which of the elements have only one stable isotope? (b) Which unstable nuclide has the longest half-life? (c) For which nuclide are positron emission and electron capture equally likely? (d) Are any nuclides shown which decay only by β^- decay? only by β^+ decay? only by electron capture? only by γ-ray emission? (e) Do any nuclides decay by emission of more than one β^- particle?

29-B2 Compute the energy of the negatron emitted during the decay of a free neutron. The reaction is $^1_0\text{n} \rightarrow ^1_1\text{H} + _{-1}^{\ 0}\text{e}$.

29-B3 Tritium (hydrogen of mass number 3) decays by negatron emission. Identify the daughter nuclide, and compute the energy of the emitted negatron.

29-B4 Compute the energy of the negatron emitted during the decay of ^{14}C.

29-B5 Using data from Fig. 29-2, show that the energy of the most energetic β^- particle from ^{59}Fe is consistent with the mass difference between ^{59}Fe and ^{59}Co.

29-B6 Show that ^{50}Cr is stable in the sense that the probability of β^- decay is zero. How about α decay? Electron capture? Neutron emission? Masses of neutral atoms are: ^{50}Cr 49.94606, ^{50}Mn 49.95422, ^{50}V 49.94716, ^{49}Cr 48.95127, ^{46}Ti 45.95263.

29-B7 Using data from Fig. 29-2, (a) calculate the maximum energy available in the negatron decay of ^{56}Mn. (b) For this decay, find a β ray and a γ ray whose combined energy is within 0.03 MeV of the maximum energy found in part (a).

29-B8 What is the KE (in MeV) of an α particle whose speed is 2.00×10^7 m/s? (*Hint:* First convince yourself that the classical formula for KE is valid in this case.)

29-B9 The radium isotope used in therapy is $^{226}_{88}$Ra which decays by α-particle emission to radon, $^{222}_{86}$Rn (see p. 637). Masses of neutral atoms are: $^{226}_{88}$Ra = 226.0254; $^{222}_{86}$Rn = 222.0175. What energy is released during the decay?

29-B10 According to mass spectrometer data, the masses of ^{209}Bi and ^{205}Tl are 208.9804 and 204.9745, respectively. Would these masses allow α decay? If so, with what energy would the α particle be ejected?

29-B11 Show that the ^{2}H(d,n)^{3}He reaction (item 3 of Sec. 29-7) has no energy threshold and can take place for very low bombarding energies. What is the minimum total KE of the fragments in this reaction?

29-B12 Calculate the BE of the least strongly bound proton in a $^{12}_{6}$C nucleus.

29-B13 Calculate the average BE per nucleon in a $^{14}_{7}$N nucleus.

29-B14 Compare the stability of ^{4}He and ^{7}Li as follows: (a) Using nuclear mass data, calculate the work required to remove one neutron from $^4_2$He to form $^3_2$He. (b) Calculate the work required to remove one neutron from $^7_3$Li to form $^6_3$Li. (*Note:* The fact that a neutron is almost three times as strongly bound to a helium nucleus as it is to a lithium nucleus is evidence for a "closed shell" in helium, and explains the stability of the α particle.)

29-B15 Fill in the missing symbols: ^{10}B(d,?)^{11}C; ^{35}Cl(n,p)??; ?Mn(p,n)^{55}?; ??(d,2n)^{65}Zn; ^{59}Co(n,?)^{60}Co.

29-B16 Fill in the missing symbols: $^{10}B(n,\alpha)??$; $^{27}Al(\alpha,?)^{30}P$; $??(n,\gamma)^2H$; $?Mg(\gamma,p)^{24}?$; $^{113}Cd(?,\gamma)^{114}Cd$.

29-B17 Slow neutrons create no ions and thus do not affect an ordinary Geiger counter. If a counter is filled with the gas boron trifluoride, the capture of neutrons by ^{10}B nuclei is made evident by a $^{10}B(n,\alpha)$ reaction; the α particles are counted with high efficiency. Identify the product nuclide in this reaction, and calculate the energy of the particles.

29-B18 Calculate the minimum energy of an α particle that could cause the $^{14}N(\alpha,p)^{17}O$ reaction to take place. (For masses, see Sec. 29-7, or Table 8 in Appendix.)

29-C1 The cloud chamber photograph (Fig. 29-8) shows the 7.68 MeV α particle to be traveling in a well-defined path. To see whether the Heisenberg uncertainty principle would predict a measurable uncertainty in position for such a particle, calculate the de Broglie wavelength of the guide waves associated with the motion of an α particle of energy 7.68 MeV. (*Hint:* The velocity is low enough so that you need not use any relativistic formulas.)

29-C2 Alcohol vapor is placed in a cloud chamber to provide carbon nuclei which can serve as targets for bombardment by α particles. What is the total KE of the fragments in a $^{12}_{6}C(\alpha,p)^{15}_{7}N$ reaction if the energy of the α particles is 7.68 MeV?

29-C3 (*a*) Use mass data from Fig. 29-2 to show that ^{57}Co can decay by electron capture but cannot decay by positron emission. (*Hint:* Make a table similar to that on p. 641. Masses in Fig. 29-2 are for neutral atoms.) (*b*) Make a similar calculation to show that ^{56}Co can decay by positron emission as well as by electron capture.

29-C4 A possible nuclear energy-level diagram for ^{56}Fe has levels at 0 MeV (= normal state of ^{56}Fe), 0.84 MeV, 2.65 MeV, 2.96 MeV, and 3.70 MeV (= normal state of ^{56}Mn). (*a*) Verify that the maximum energy available in the decay of ^{56}Mn is consistent with the mass difference given in Fig. 29-2. (*b*) Draw the energy levels to scale, and show numerically that all eight of the β^- and γ emissions listed in Fig. 29-2 for ^{56}Mn are accounted for by these levels.

29-C5 The nuclide ^{60}Co is commonly used in cancer therapy. (*a*) What total energy is available when the 5.27 y isomer decays, in two stages, to stable ^{60}Ni? (Use masses from Fig. 29-2). (*b*) Show quantitatively how the total available energy is apportioned among the products of the decays.

29-C6 The 9.58 min ^{27}Mg decays with negatrons of maximum energies 1.59 MeV and 1.75 MeV. Gamma rays of energies 1.01 MeV and 0.84 MeV are also observed. The masses of the neutral atoms are: ^{27}Mg 26.98435 and ^{27}Al 26.98154. (*a*) What is the total energy of the decay? (*b*) Draw an energy-level diagram with three excited states.

29-C7 Construct a possible energy-level diagram with three excited states which can account for all six of the β^- and γ emissions from ^{59}Fe, using data from Fig. 29-2. (*Hint:* First use the mass difference to find the maximum available energy.)

For Further Study

29-8 Conservation of angular momentum in β decay

Let us consider the decay of $^{55}_{24}Cr$ once again, this time paying attention to the spins of the particles. The nucleus contains 24 protons and 31 neutrons. In units of $h/2\pi$, the angular momentum (spin) of each proton is $\frac{1}{2}$, and the spin of each neutron is also $\frac{1}{2}$. We make a model in which the axes of rotation of the particles inside the nucleus are parallel to one another but can point in either direction. Thus the total spin of all 55 particles *could* be $\frac{1}{2}, \frac{3}{2}, \frac{5}{2}, \frac{7}{2}, \cdots \frac{55}{2}$, always an odd multiple of $\frac{1}{2}$ since there is an odd number of particles. Of all these possibilities, it appears that the spin of $^{55}_{24}\text{Cr}$ is $\frac{3}{2}$; this corresponds to 29 nucleons spinning in one direction and 26 nucleons spinning in the opposite direction. The resultant spin is $\frac{29}{2} - \frac{26}{2} = \frac{3}{2}$, as observed. Similarly, the spin of $^{55}_{25}\text{Mn}$ must be an odd multiple of $\frac{1}{2}$; experiment shows it to be $\frac{5}{2}$. It is known also that the electron's spin is $\frac{1}{2}$ (this applies to both the negatron and the positron). We now write the equation for the β decay of $^{55}_{24}\text{Cr}$:

$$^{55}_{24}\text{Cr} \rightarrow \, ^{55}_{25}\text{Mn} + \, ^{\,0}_{-1}e + \bar{\nu}_e \qquad (29\text{-}1)$$

$$\tfrac{3}{2} \rightarrow \quad \tfrac{5}{2} \; + \; \tfrac{1}{2} \; + \; ? \qquad (29\text{-}2)$$

We have already postulated neutrinos (ν_e) and antineutrinos ($\bar{\nu}_e$) to help conserve energy in β decay; now we see that these particles are equally important for the conservation of angular momentum. Equation 29-2 is a vector equation, and angular momentum can be conserved if the antineutrino's spin is $\frac{1}{2}$ and if the negatron and the antineutrino are each emitted with their spins opposite to that of the $^{55}_{25}\text{Mn}$ nucleus. Thus, using $+$ and $-$ signs to indicate possible directions of the spins, we can write

$$\tfrac{3}{2} \rightarrow \tfrac{5}{2} + (-\tfrac{1}{2}) + (-\tfrac{1}{2}) \qquad (29\text{-}2')$$

While it is true that a spin of $\frac{3}{2}$ for the antineutrino would also be possible for the reaction of Eq. 29-1 ($\frac{3}{2} = \frac{5}{2} + \frac{1}{2} - \frac{3}{2}$), other evidence indicates the value $\frac{1}{2}$ for the spin of both the neutrino and the antineutrino. It is worthy of note that the neutrino must have *some* half-integral spin; under no circumstances could a spin of 0 or 1 satisfy the vector equation 29-2, which illustrates the law of conservation of angular momentum for a nuclear process.

29-9 Formula for radioactive decay

The decay of a radioactive nucleus is a matter of chance. If N nuclei are present at a time t, we cannot tell when a given nucleus will decay, but there is a definite probability that ΔN of these nuclei will decay during a small time interval Δt which extends from t to $t + \Delta t$. The number decaying is proportional to the number present (N) and to the time interval (Δt). The proportionality constant is the *disintegration constant* λ, defined by the equation

$$\Delta N = -\lambda N \, \Delta t \qquad (29\text{-}3)$$

The $-$ sign is used because N decreases when t increases.

We now use Eq. 29-3 to derive an equation for N as a function of t. First we rearrange Eq. 29-3 to obtain

$$\frac{\Delta N}{N} = -\lambda \, \Delta t$$

$$\lim_{\Delta N \to 0} \sum \frac{\Delta N}{N} = -\lambda \left(\lim_{\Delta t \to 0} \sum \Delta t \right)$$

$$\int_{N_0}^{N} \frac{dN}{N} = -\lambda \int_0^t dt$$

The initial number of nuclei is N_0, and the lower limits are chosen to conform to the fact that $N = N_0$ when $t = 0$.

We carry out the integrations using the table of integrals in the appendix, obtaining

$$[\log_e N]_{N_0}^{N} = -\lambda [t]_0^t$$

$$\log_e N - \log_e N_0 = -\lambda t \qquad (29\text{-}4)$$

$$\log_e \frac{N}{N_0} = -\lambda t$$

$$\frac{N}{N_0} = e^{-\lambda t}$$

$$N = N_0 e^{-\lambda t} \qquad (29\text{-}5)$$

where e is 2.718, the base of natural logarithms.

Thus we see that radioactive decay is governed by an exponential equation (Fig. 29-10a). Also, Eq. 29-4 shows that a graph

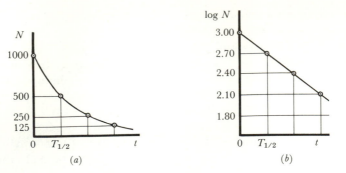

Figure 29-10 Radioactive decay. (*a*) Graph of N as a function of t. The value of N decreases by a factor of 2 in any time interval equal to a half-life $T_{1/2}$. (*b*) Graph of log N versus t. Since log $2 = 0.30$, the value of log N decreases by 0.30 in any time interval equal to a half-life $T_{1/2}$.

of log N versus t is a straight line (Fig. 29-10*b*). Note that log N means $\log_{10} N$, which is proportional to $\log_e N$.

The half-life $T_{1/2}$ is the time required for the number of nuclei to decrease from N_0 to $\frac{1}{2}N_0$. To obtain a relationship between the half-life and the disintegration constant we first multiply Eq. 29-4 by -1 and then substitute the values $t = T_{1/2}$ and $N = \frac{1}{2}N_0$:

$$-\log_e \tfrac{1}{2}N_0 + \log_e N_0 = \lambda T_{1/2}$$

$$\log_e \frac{N_0}{\frac{1}{2}N_0} = \lambda T_{1/2}$$

$$\log_e 2 = \lambda T_{1/2}$$

$$T_{1/2} = \frac{\log_e 2}{\lambda} = \frac{0.693}{\lambda} \qquad (29\text{-}5)$$

The activity A of a sample is the rate of decay measured, say, in disintegrations per second.[*] Thus A is the absolute magnitude of $\Delta N/\Delta t$, which from Eq. 29-3 is

$$A = \lambda N \qquad (29\text{-}6)$$

The activity is the product of the disintegration constant and the number of nuclei present. Since A is proportional to N, the activity of a sample decreases in the same

fashion as does N, and we can write

$$A = A_0 e^{-\lambda t} \qquad (29\text{-}7)$$

EXAMPLE 29-7

Calculate the activity on May 15 of a sample of ^{32}P which gave 2400 counts per minute in a Geiger counter on May 5. The half-life of ^{32}P is 14 days.

First we calculate the disintegration constant:

$$\lambda = \frac{0.693}{T_{1/2}} = \frac{0.693}{14 \text{ d}} = 0.0495 \text{ d}^{-1}$$

The elapsed time during decay has been 10 d.

$$A = A_0 e^{-\lambda t}$$
$$= 2400 \, e^{-(0.0495 \text{ d}^{-1})(10 \text{ d})}$$
$$= 2400 \, e^{-0.495} = 2400(0.610)$$
$$= \boxed{1465 \text{ counts/min}}$$

To find $e^{-0.495}$ a table of exponentials can be used, or one can use a log table and the identity $\log_{10} (e^x) = 0.434 \, x$.

EXAMPLE 29-8

How many disintegrations per second take place in a 2 μg sample of ^{226}Ra?

The number of nuclei in the sample is calculated using Avogadro's number; 1 mole of ^{226}Ra is 226 g.

$$\frac{N}{2 \times 10^{-6} \text{ g}} = \frac{6.02 \times 10^{23} \text{ nuclei/mole}}{226 \text{ g/mole}}$$

$$N = 5.33 \times 10^{16} \text{ nuclei}$$

---

[*] Activities can also be expressed in counts per second or counts per minute; for any given counter system there is a definite proportionality between the number of disintegrations and the number of recorded counts.

The disintegration constant is found from the half-life which is given in Fig. 29-3: $T_{1/2} = 1622$ y $= 5.12 \times 10^{10}$ s. Hence

$$\lambda = \frac{0.693}{5.12 \times 10^{10} \text{ s}} = 1.35 \times 10^{-11} \text{ s}^{-1}$$

From Eq. 29-6 we find the activity:

$$A = \lambda N$$
$$= (1.35 \times 10^{-11} \text{ s}^{-1})(5.33 \times 10^{16} \text{ nuclei})$$
$$= \boxed{7.20 \times 10^5 \text{ nuclei/s}}$$

From this invisibly small sample almost a million radium nuclei disintegrate each second.

■ PROBLEMS

29-C8 Use the law of conservation of angular momentum to show that a neutrino is emitted during electron capture.

29-C9 The nuclide $^{48}_{24}$Cr decays by positron emission to form $^{48}_{23}$V; the $^{48}_{23}$V then decays by another positron emission to form stable $^{48}_{22}$Ti. The spin of $^{48}_{22}$Ti is known to be 0. On the basis of these data, what spins are possible for $^{48}_{23}$V?   for $^{48}_{24}$Cr?

29-C10 The spin of $^{49}_{22}$Ti is known to be $\frac{7}{2}$; how many nucleons spin in one direction, and how many in the other?

29-C11 4.00×10^{20} nuclei are in a sample of ^{60}Cu at 4:00 P.M.; how many ^{60}Cu nuclei are present at 4:12 P.M.? (See Fig. 29-3 for the half-life.)

29-C12 How long is required for a sample of ^{32}P (half-life 14 days) to lose $\frac{1}{3}$ of its activity?

29-C13 What fraction of the ^{40}K atoms in existence at the formation of the earth 4.5×10^9 y ago still survive? The half-life of ^{40}K is 1.3×10^9 y.

29-C14 A quantity of the iodine nuclide ^{123}I, of half-life 13 h, is manufactured in a cyclotron and shipped by air to a hospital. If the shipment requires 4 h, what fraction of the produced ^{123}I reaches the hospital?

29-C15 A cobalt "bomb" containing 3 g of ^{60}Co is used for cancer therapy. At what rate, in watts, is heat produced in a thick-walled storage container which absorbs all radiation? (*Hint:* See Prob. 29-C5 for the energy released per disintegration.)

29-C16 How many disintegrations per second occur in a 90 g sample of pure carbon in the form of charcoal which contains radioactive carbon ^{14}C to the extent of 1 part in 10^{12}? The half-life of ^{14}C is 5730 years.

■ REFERENCES

1 Magie, W. F., *A Source Book in Physics* (New York: McGraw-Hill, 1935), pp. 610–13, Becquerel's experiments with uranium; pp. 613–16, discovery of polonium and radium by the Curies.

2 Behrens, C. E., "Atomic Theory from 1904 to 1913," *Am. J. Phys.*, **11**, 60 (1943).

3 Jauncey, G. E. M., "The Early Years of Radioactivity," *Am. J. Phys.*, **14**, 226 (1946).

4 Korff, S. A., "Counters," *Sci. American*, July, 1950, p. 40.

5 Collins, G. B., "Scintillation Counters," *Sci. American*, Nov., 1953, p. 36.

6 Glaser, D. A., "The Bubble Chamber," *Sci. American*, Feb., 1955, p. 46.

7 O'Neill, G. K., "The Spark Chamber," *Sci. American*, Aug., 1962, p. 37. Also discusses Geiger counter, cloud chamber, and bubble chamber.

8 Morrison, P., "The Overthrow of Parity," *Sci. American*, Apr., 1957, p. 45.

9 Miller, F., Jr., *Radioactive Decay; Scintillation Spectroscopy* (films).

30

Applied Nuclear Physics

30-1 Some uses of radioactivity

We have described the basic physical facts about the nucleus in Chap. 29; now we look at some of the ways in which this knowledge has been put to practical use. Some of the most spectacular engineering of the past few decades has been nuclear engineering, and more is yet to come. In this chapter we shall study devices capable of producing large-scale conversion of rest mass into energy. Our first topic in applied nuclear physics deals, however, with applications of radioactivity to scientific investigation and therapy.

Every element has at least one unstable isotope, and, if its half-life is not too short or too long, such an isotope can be used in many ways. All high-energy radiations to some extent damage biological tissues. These effects depend in a complicated way upon the nature of the tissue and on the kind and energy of the radiation. In general, the damage is related to the production of ions in the tissue. It was early found that rapidly growing cancerous cells are more susceptible

to γ rays than are normal cells. Thus impure radium and its daughter radon were from the beginning of the twentieth century used for cancer therapy (most of the effect was from the γ rays emitted by ^{214}Pb and ^{214}Bi, which are descendants of ^{226}Ra; see Fig. 29-2). The patient may recover if the dose is sufficient to kill all the cancerous cells while killing only a fraction of the intermingled normal cells. Since the advent of nuclear reactors, strong samples of many radioactive nuclides such as ^{60}Co (half-life 5.3 years) can be manufactured and used for therapy. In fact, ^{60}Co can replace radium for almost all purposes; its γ rays have energies of 1.2 and 1.3 MeV (see Fig. 29-3). The iodine isotope ^{131}I (8 days; 0.4 MeV) is used to treat cancer of the thyroid gland; it is selectively absorbed by the thyroid, and so the radiation is concentrated right where it is needed. These are just a few of the nuclides now being used for therapy.

Many unstable nuclides are used as *tracers*. One example is in the study of the wear of piston rings in an automobile engine. Before the piston ring is placed in the engine, it is

Looking down into the tank of a nuclear reactor. Fuel elements comprising the "core" stand on pins on the round grid plate at the bottom of the tank. Five neutron-absorbing rods, used to control chain reaction of assembled core, are in position among the fuel elements, inserted to various depths. When the reactor is operating, the tank is filled with water, which is used as a moderator. (Courtesy Babcock and Wilcox Company)

irradiated with neutrons, and some ^{59}Fe is formed by the ^{58}Fe$(n,\gamma)^{59}$Fe reaction. As the piston rings wear during the test, some of the iron atoms which rub off into the oil are ^{59}Fe, which can be detected when the oil is later tested with a Geiger counter. The method is far more sensitive than would be chemical analysis of the oil, and only a few minutes are required. In addition, the only radioactive iron in the oil must have come from the piston ring, and iron from other parts of the engine does not influence the analysis.

Many compounds can be "tagged" with an unstable atom and traced through a chemical reaction. For instance, the methane-producing bacterium *Methanobacterium omelianski* produces acetic acid and methane from ethyl alcohol and carbon dioxide according to the reaction

$$2C_2H_5OH + CO_2 \rightarrow 2CH_3COOH + CH_4$$

Into which of the product molecules has the carbon of the CO_2 gone? By making the CO_2 with a radioactive isotope of carbon, it has been shown that the "labeled" carbon goes into the methane (CH_4), none of it appearing in the acetic acid (CH_3COOH). The long-lived carbon isotope ^{14}C (half-life 5730 years) is of great importance for biological research, as is tritium, the hydrogen isotope ^{3}H (half-life 12 years).

Tracer experiments can also be performed using stable isotopes; for instance, a protein can be enriched in ^{15}N, which is normally present to an extent of only 0.4% (see table on p. 723. The biological utilization of the protein can be determined by analyzing nitrogen from various tissues, using a mass spectrometer. In general, work with radioactive tracers is much easier than with stable isotopes.

The unchanging disintegration rate of an unstable nuclide can serve as an atomic clock reaching far back into the past. An early example was the determination of the age of the earth (more precisely, of the time

that has elapsed since the solidification of the crust). As shown in Fig. 29-3, ^{238}U (half-life 4.5×10^9 years) ultimately forms the stable lead isotope ^{206}Pb. Thus if a uranium ore contains only a little ^{206}Pb, it must be relatively young. Careful analysis using several long-lived nuclides, including the potassium isotope ^{40}K (half-life 1.3×10^9 years), gives a consistent figure of about 5×10^9 years for the age of the earth's crust. The carbon isotope ^{14}C is of shorter half-life (5730 years) and is useful for dating archeological finds such as a wooden coffin in a Pyramid, or prehistoric charcoal or cloth. Because of its short half-life, no primeval ^{14}C remains, but it is constantly being created in the atmosphere by the action of cosmic rays in a ^{14}N$(n,p)^{14}$C reaction. Thus a certain small fraction of the CO_2 molecules in the air contain ^{14}C; most of them of course contain the ordinary stable isotope ^{12}C. Living plants take up this same small fraction of ^{14}C, and animals feed upon the plants. Thus all *living* organisms contain about 1 atom of ^{14}C for every 10^{12} atoms of ^{12}C. When the organism dies, however, the ^{14}C decays to ^{14}N by negatron emission and is not replaced. The time since the specimen died can be determined by measuring the ratio of ^{14}C to ^{12}C. One more example illustrates the usefulness of radioactive dating methods. Certain meteorites fell in Kansas and Nebraska in 1948. One of the stony fragments was analyzed by the ^{40}K method and found to be about 4.2×10^9 years old, approximately the same as the age of the earth. By studying the amount of ^{3}He formed by cosmic ray action during the stone's flight through space, it was determined that the meteorite had been exposed to cosmic rays for only 0.28×10^9 years. The conclusion was that the rock had been formed a few billion years ago, and lay buried in a planet or asteroid which broke into fragments a few hundred million years ago. Thus nuclear science in this case supports the idea of a catastrophic collision in space, perhaps the breakup of a planet to form the asteroids which lie mostly between the orbits of Mars and Jupiter.

30-2   Particle accelerators

In Sec. 30-1 we assumed that many unstable isotopes are available and that they can be produced in quantity when needed. Broadly speaking, nuclear reactions take place in particle accelerators (or "atom-smashers"), and in chain-reacting nuclear reactors. We shall discuss these devices in the next three sections.

The simplest accelerator is the *Van de Graaff generator,* shown in Fig. 30-1. The pointed electrode A is connected to a rectifier circuit V, which maintains a PD of perhaps 50 000 V, and positive charge is sprayed onto a moving silk belt. The charge is removed at C, and replaced by negative charge at D. In an endless process, electrons are moved from the sphere S to ground, and the sphere becomes charged to a high positive potential. To use the generator as an atom-smasher, a

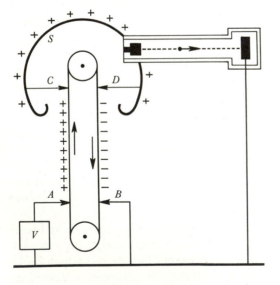

Figure 30-1   Van de Graaff generator (schematic).

source of H^+ or He^+ or He^{++} ions is mounted inside the sphere, and the positive ions are repelled toward a grounded target by the positively charged sphere. Protons of energy up to about 30 MeV can be produced if three Van de Graaff generators are used in tandem.

The *cyclotron,* developed in 1930 by E. O. Lawrence and M. S. Livingston, uses a magnetic field to bend the particles in orbits (Fig. 30-2). In this way a modest PD of, say, 50 000 V can be applied over and over again to the same particle, to build up a large final energy. The proton source injects H^+ ions at low speed, and the particles move inside two semicircular flat cavities, called "dees" because of their shape. The dees are enclosed in an evacuated chamber placed between the pole pieces of a powerful electromagnet, so that the particles cut across a vertical magnetic field. As we saw in Chap. 21, the magnetic force on a moving charge is perpendicular both to the field and to the motion; therefore each proton moves in a circular orbit, for which the centripetal force is supplied by the magnetic force. The dees are connected to a rapidly varying source of alternating potential, say, $\pm 50\,000$ V at a frequency of 10^7 Hz. If a proton reaches A at just the right time, when D_1 is negative relative to D_2, it will be accelerated across the gap and receive an additional 50 000 eV of energy. Since it is now moving faster, the magnetic field bends it in a semicircle of larger radius. By the time the proton reaches C, the polarity of the dees has reversed, and D_2 now is negative relative to D_1 and the proton gains another 50 000 eV. In this way the proton spirals outward in a series of semicircles, and eventually strikes a target T.

The cyclotron works only because of a lucky coincidence: the time for each semicircular journey is the same for all particles, whether they are moving slowly (near the center) or fast (near the outside). To prove

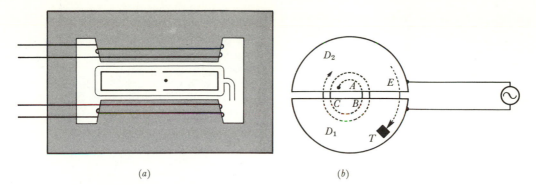

(a)   (b)

Figure 30-2   Cyclotron (schematic): (*a*) side view; (*b*) top view of dees.

this, we use Newton's second law and Eq. 21-4 for the magnetic force on a moving charge.

$$\text{net } F = ma$$

$$Bqv = m\frac{v^2}{r}$$

whence

$$v = \frac{Bqr}{m}$$

This equation is valid for particles of any speed, if the relativistic mass *m* is used. The time for half a revolution is given by

$$T = \frac{\text{distance}}{\text{speed}} = \frac{\frac{1}{2}\,(2\pi r)}{(Bqr)/m}$$

or

$$T = \frac{\pi m}{Bq} \qquad (30\text{-}1)$$

The radius has canceled out; therefore the time for one semicircular trip depends only on *B*, *m*, and the charge *q*. If these quantities are constant, then *T* is constant, and an oscillator of fixed frequency can accelerate *simultaneously* both the slow protons at *A* and the faster ones at *E*, which have already made many trips in their spiral path.

One might think that there is no limit to the energy that can be given to a particle by cyclotron action. To obtain large final energy, a strong magnetic field must be used, and the diameter of the pole pieces

must be large so that each particle can make many trips as it spirals outward. However, aside from cost and other practical limitations on size, there is an unavoidable limitation due to the relativistic increase of mass of the high-speed particles. Equation 30-1 shows that the time for a semicircular trip increases if mass increases, and so, as the particles gain energy—and mass—they fall behind and get out of step with the oscillating voltage applied to the dees. A 20 MeV proton moves at about $\frac{1}{5}$ the speed of light, and its mass is 2% greater than its rest mass. This is enough to spoil the regularity of the cyclotron action. The difficulty can be overcome in the *synchrocyclotron* by letting protons move through the machine in short bursts, and increasing the time between alternations of the dees as the particles spiral out. In this way, Eq. 30-1 remains satisfied throughout the spiral path of any group of particles. Another solution is to increase the magnetic induction *B* while the group of particles is spiraling outward, allowing *T* to remain fixed. By such methods, proton energies up to 700 MeV have been obtained.

Still higher energies are equivalent to 10^9 volts or more. The term "billion" is ambiguous—it means 10^9 in the U.S. and 10^{12} in Great Britain—so the metric prefix giga- (G) is coming into general scientific use to represent 10^9. To reach the giga-electron-volt (GeV) range, physicists took advantage of the facts of relativity which previously were a limitation. No particle can have a speed greater than the speed of light. A 2 MeV electron is already moving at a speed of $0.98c$, and further increase of energy goes mainly into increasing its mass. Thus *all* very high-energy particles move at the same speed, about 3×10^8 m/s, regardless of rest mass or energy (Table 30-1). The *synchrotron* takes advantage of this simplicity. If the particles are somehow brought up to nearly the speed of light, then as they gain energy their speed remains constant and the time for each revolution remains constant. The magnetic induction must be gradually increased as the particles pick up energy and mass (see Eq. 30-1). The great advantage of the synchrotron principle is that the radius of the orbit is practically constant, and so the magnetic induction need be applied only at the circumference, instead of throughout the entire area of the orbit as in the cyclotron. The beam of particles is confined to a doughnut-shaped chamber which lies between the poles of a series of strong electromagnets (Fig. 30-3; see also Fig. 8-11). In each circular trip the beam goes through a number of radiofrequency cavities (operating at, say, 50 MHz) in which energy is imparted to the particles during

Figure 30-3 Synchrotron under construction at Argonne National Laboratory. The gap in the magnet is for the vacuum chamber containing the proton beam. Protons gain energy each time they pass through the radiofrequency cavity at upper left. Before injection into the main ring, protons are accelerated by a linac, whose exit portal is seen at top center. The man on the ladder is using an optical alignment device.

each pass. Before entering the main ring the beam is brought up to the "injection energy" by a series of auxiliary accelerators of various types, including a linear accelerator (see below).

As of 1972, twelve electron synchrotrons have been constructed in seven countries which yield electrons of energy greater than 1 GeV, and thirteen proton accelerators in five countries are in the 1 to 200 GeV range. The large accelerator in Batavia, Illinois (Fig. 30-4), to be named the Fermi National Accelerator Laboratory, has diameter 2.00 km. The original design of the main ring was for acceleration of protons to energies of 200 GeV. High intensity (a beam of many particles per second) is as important as high energy per particle; the Batavia accelerator is designed to give 1.3×10^{13} protons per second.

A major factor is the cost of installing and running the magnets which keep the beam in its circular path. For a given radius, the time for 1 revolution is essentially con-

TABLE **30-1**

Dependence of speed of various particles upon their kinetic energy*

Kinetic energy	Electron speed	Proton speed	Deuteron speed
0.1 MeV	.553	.0146	.0103
1 MeV	.941	.0462	.0326
10 MeV	.9988	.145	.103
100 MeV	.999987	.429	.316
1 GeV	.99999987	.876	.760
10 GeV	.9999999987	.9963	.988
100 GeV	.999999999987	.999957	.99983

*All speeds are expressed as fractions of the speed of light.

Figure 30-4   Aerial view of the National Accelerator Laboratory site showing the main ring, of diameter 2.00 km (1.24 miles). In the foreground are the linac, the booster, and the beam transfer region.

stant (since $v \approx c$). Then Eq. 30-1 shows that the mass of a particle (i.e., its energy) is proportional to the magnetic induction B. Recent advances in the technology of solid-state rectifiers (Sec. 22-10) make possible much larger magnet currents, using existing power transformers. Thus proton energies of 400 to 500 GeV will be available shortly after start-up of the main ring at Batavia, and perhaps 1000 GeV if superconducting magnets are added later. A synchrotron planned at CERN (European Organization for Nuclear Research) at Geneva, Switzerland, can be operated at 150 GeV initially, later at 300 GeV, and up to 800 GeV as funds for superconducting magnets become available.

We shall briefly mention two other types of accelerators. The *betatron* uses transformer action to give energy to a stream of electrons. This was the first of the doughnut-chamber accelerators; it is practical because electrons reach essentially constant speed (nearly equal to c) at moderate energies. In effect, the circular electron beam takes the place of the secondary coil of a transformer in which the primary is a large central magnet whose field changes at 60 Hz. A similar machine using protons would require an impossibly large central magnet. X rays are produced as bremsstrahlung when the electron beam strikes a target. In fact, betatrons now find their greatest use in generating high-energy x rays for industrial and medical uses.

A natural limitation still remaining in all accelerators, even if the relativistic increase in mass is taken care of, is the loss of energy by radiation. A proton or other charged particle moving in a circular orbit in an accelerator has a centripetal acceleration, and so it radiates energy in the same way as discussed in Sec. 28-2 for a pre-Bohr atom. In a circular accelerator, however, no radiationless states exist to save the day, as was the case for an electron's orbit around the nucleus. The maximum energy of a circular accelerator is reached, therefore, when the rate of e-m radiation, called *synchrotron radiation,* equals the power input to the beam of particles. These radiation losses are inevitable as long as the particle has centripetal acceleration (the nonrelativistic formula is v^2/r). They can be reduced by using a path of large radius—another reason for the cult of largeness in the construction of accelerators.

In the *linear accelerator* (linac), particles move in a straight line between many properly spaced electrodes, instead of returning over and over to the same gap; no magnetic field is needed. There is no centripetal acceleration, and radiation losses occur only while the particles are crossing the narrow gaps where they gain energy. The Stanford linear accelerator (2 miles long) became operational in 1966 and produces an intense beam of electrons of energy 21 GeV. The Heavy Ion Linear Accelerator (Hilac) at Berkeley, California, can accelerate ions of all masses, 1 through 238, to an energy which is variable between 2.5 to 8.5 MeV per nucleon. A search is being made for $^{298}_{114}X$, a superheavy nuclide predicted to have exceptional stability,[*] by bombarding uranium with heavy-ion projectiles such as partially stripped uranium ions $^{238}_{92}U^{39+}$.

We have given only the broad outlines of the various particle accelerators. All told, 28 accelerators giving particle energies in the GeV range are now (1972) in use. Their successful operation calls for precise and imaginative design. Not the least problem is that of focusing the particles into a narrow, stable beam of high intensity. Much current research is devoted to such questions of stability and intensity.

30-3 Colliding beams

There is a difficulty with all accelerators that is clearly apparent even in a classical nonrelativistic analysis. In the accelerators so far described we are dealing with an inelastic collision, say between a proton moving at high speed and a target proton which is at rest. We saw in Chapter 6 that in such a collision, momentum is conserved, but some kinetic energy inevitably disappears. In a nuclear experiment this "lost" kinetic energy is available for the rest mass of new particles, and it would be desirable to use *all* of the KE of the incoming particle in this way. Unfortunately, to conserve the linear momentum of the system, the combined mass (or the masses of the fragments) must recoil in the forward direction. Especially at high energies, this recoiling target or the recoiling products of the collision carry away a large fraction of the incoming KE. For example, calculation shows that if a 28 GeV proton collides with and sticks to a stationary target proton, the combined mass moves forward with 20.5 GeV of KE; only 7.5 GeV is available for transformation into the mass of new particles.

The total KE of a collision can be made "useful" by a clever stratagem, if two equal particles moving with equal speeds in opposite directions collide and stick together. In the laboratory frame of reference the momentum of the system is zero both before and after the collision; there is no KE after the collision and 100% of the KE of the colliding particles is available.

---

[*] Nuclear theory suggests that such a nuclide would be an α emitter with a half-life of perhaps 1000 years.

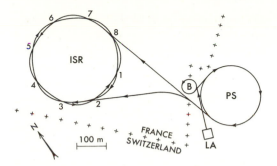

Figure 30-5 Beam paths for Intersecting Storage Rings (ISR) at CERN. Shown are the linear accelerator (LA), which operates at 50 MeV, the booster (B) at 800 MeV, and the proton synchrotron (PS) at 28 GeV.

In 1971 the Intersecting Storage Ring (ISR) went into operation at CERN at the border between Switzerland and France near Geneva. A 28 GeV proton synchrotron is used to inject beams of protons into two almost-circular rings about 300 m in diameter which intersect at 8 points (Figs. 30-5 and 6). The beams can be stored and used for about a day before they must be replenished from the synchrotron. The two

Figure 30-6 Intersection point I-2 at the CERN Intersecting Storage Rings. Here the tunnel is widened and the floor lowered, to accommodate experimental equipment later installed at the intersection. At the right, beneath the number 2 on the wall, is the end of a beam-transfer tunnel which brings protons from the synchrotron to the ISR.

colliding beams, each of energy 28 GeV, release 56 GeV of energy, equivalent to a 1700 GeV beam striking a stationary target. The engineering problems associated with such machines are great, but the reward, in effective energy, is tremendous.

30-4 Fission

As we all know, nuclear fission releases energy, and the process can be made self-sustaining in a chain reaction. Fission was discovered in 1939 by the German chemists and physicists Hahn, Strassmann, Meitner, and Frisch; Hahn's own account[*] makes interesting reading and shows that other workers of the period were on the verge of the discovery. The "liquid drop" model of the nucleus, due to Bohr, pictures the nucleus as a sort of drop of liquid able to become constricted and to oscillate. The addition of one more nucleon might cause certain nuclei to oscillate too wildly, resulting in a breaking-up process called *fission*—the splitting of a nucleus into relatively large fragments. The typical and most familiar fission reaction takes place when slow neutrons strike ^{235}U:

$$^{235}_{92}U + ^{1}_{0}n \rightarrow [^{236}_{92}U] \rightarrow$$
$$A + B + \text{neutrons} + \text{energy}$$

The compound nucleus $[^{236}_{92}U]$ splits into two fragments A and B, with some neutrons also released during the fission. To see how this works out, let us first make the simple assumption that the nucleus splits into two equal fragments and that no neutrons are emitted. Half of 92 is 46, and the element of atomic number 46 is palladium. Thus a conceivable reaction would be

$$^{235}_{92}U + ^{1}_{0}n \rightarrow ^{118}_{46}Pd + ^{118}_{46}Pd$$

Reference to a table of nuclides shows that

[*] "The Discovery of Fission," by O. Hahn, *Sci. American*, Feb., 1958, p. 76.

the stable Pd isotopes range from $^{102}_{46}$Pd to $^{110}_{46}$Pd; $^{118}_{46}$Pd would have at least eight extra neutrons and in fact is so unstable that it has never been observed. A more likely possibility would be

$$^{235}_{92}U + ^{1}_{0}n \rightarrow ^{110}_{46}Pd + ^{110}_{46}Pd + 16\ ^{1}_{0}n$$

This reaction does take place, but experiment shows that the two fragments are most likely to be of somewhat different mass. As a typical and important fission reaction, let us study

$$^{235}_{92}U + ^{1}_{0}n \rightarrow ^{90}_{38}Sr + ^{136}_{54}Xe + 10\ ^{1}_{0}n$$

This is just one of hundreds of possible fission reactions; it is important for human survival because the strontium isotope $^{90}_{38}$Sr is a negatron emitter with half-life of 28 years, and is a major constituent of fallout from nuclear explosions. To see whether the reaction is possible, we check the masses.

BEFORE FISSION

1 neutron	1.0087 u
$^{235}_{92}$U	235.0439
	236.0526

AFTER FISSION

$^{90}_{38}$Sr	89.9073 u
$^{136}_{54}$Xe	135.9072
10 neutrons	10.0867
	235.9012

MASS DIFFERENCE

$$0.1514\ u = 141\ MeV$$

Since the total rest mass of the system *decreases*, the reaction is possible. The mass calculation shows that 0.1514 u of rest mass disappears during this fission. At the rate of 931 MeV/u, this is equivalent to about 141 MeV of energy for each fissioning ^{235}U nucleus. Other fission reactions release more or less energy from ^{235}U; the average is about 200 MeV per atom of ^{235}U.

It is the release of neutrons which makes the chain reaction possible. The number of neutrons per fission is considerably less than

ten, for in many cases the fission products convert their excess neutrons into protons by negatron emission. Nevertheless, more than one neutron is formed, on the average, and so the process builds up. If the multiplication factor were exactly 2, then one incoming neutron would release two neutrons, which in turn would strike two ^{235}U nuclei and release four neutrons, and so on until an explosion resulted. For controlled release of nuclear power, which is what we want from a nuclear reactor, the multiplication factor is adjusted to be only slightly greater than 1. Uranium-235 is most easily fissioned by a neutron that is moving slowly, for in this way the neutron has more time to interact with the nucleus. We saw in Chap. 16 that the average translational KE per molecule of any solid, liquid, or gas is given by $(\frac{3}{2}RT)/N_A$, which works out to be about 0.038 eV at room temperature. Neutrons of about this energy are called slow neutrons, or *thermal neutrons*.

To understand the operation of a nuclear reactor, we first consider what might happen when a neutron strikes a nucleus. Several possibilities exist. (1) The neutron may simply bounce off, transferring some of its KE to the target nucleus. This is called a scattering process. (2) The neutron may be captured, probably with emission of γ rays, in a (n,γ) reaction. (3) The neutron may induce some nuclear reaction such as (n,α) or (n,p). (4) Finally, the neutron may be like the straw that broke the camel's back and induce fission. The probabilities of these events are measured in terms of *cross sections*. The atoms of a solid are about 3×10^{-10} m apart, and so to each atom there belongs an area (for an oncoming neutron) of about $(3 \times 10^{-10}\ m)^2$, or about 10^{-19} m^2. Similarly, the "geometrical cross sections" of the nuclei range from about 0.4×10^{-28} m^2 for $A = 20$ to about 1.8×10^{-28} m^2 for $A = 240$, where A is the mass number. To deal with such small areas we use the *barn,* defined as a unit of area equal to 10^{-28} m². We see that a *nucleus* has a geometrical cross section of about 1 barn, and an *atom* has a geometrical cross section of about 10^9 barns. If only 1 out of 1000 nuclei scatters a slow neutron, we say that the cross section for scattering is $\frac{1}{1000}$ of the geometrical cross section, that is, about $\frac{1}{1000}$ barn. The cross section thus is the "effective area" for any given process, and the same nucleus may have a number of different cross sections for different processes. Table 30-2 gives some cross sections that are important for reactor design. Nuclei differ widely in their capture cross sections; the very large capture cross section for $^{113}_{48}$Cd means that in effect this nucleus "reaches out" and captures slow neutrons that don't really strike within its geometrical area. This property of one cadmium isotope (comprising 12% of naturally occurring cadmium) makes cadmium useful as a shield or absorber of slow neutrons (see Fig. 30-7). The boron nuclide $^{10}_{5}$B is also used for the same purpose; in this case the neutrons are removed by a (n,α) reaction. The high fission cross section of ^{235}U indicates that this isotope is useful in nuclear reactors. Unfortunately, it occurs only to the extent of 1

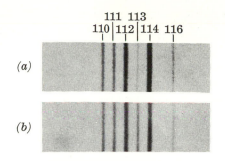

Figure 30-7 Mass spectra of cadmium, showing that the large cross section for absorption of neutrons (Table 30-2) is due to a ^{113}Cd(n,γ)^{114}Cd capture reaction. (*a*) Cadmium scraped from the surface of an irradiated sample of the metal; ^{113}Cd has been changed to ^{114}Cd by neutron capture. (*b*) Cadmium from interior of sample, to which neutrons have not penetrated; normal abundance of isotopes. (A. J. Dempster, from *Nuclear Radiation Physics,* by R. E. Lapp and H. L. Andrews, Prentice-Hall, Inc.)

part in 140; ordinary metallic uranium is almost all ^{238}U, and ^{233}U has such a short half-life that it is not found in nature. Several nuclides, including ^{238}U, undergo fission when bombarded with *fast* neutrons, but such reactions cannot be controlled and are useful only for explosive purposes.

We are ready now to consider briefly some problems to be overcome in constructing a nuclear reactor. The fuel elements are slugs of fissionable material, usually ^{235}U which has been separated from natural uranium. The uranium slugs are surrounded by a *moderator* such as heavy water or carbon (graphite), whose function is to slow down the fast neutrons released during fission. As discussed in Sec. 6-11, the most effective way to slow down a neutron to thermal velocity is to let it strike another particle of mass equal to that of the neutron. The proton would be ideal for the purpose if it did not have such a large capture cross section (Table 30-2). Using ordinary water as a moderator is inefficient, for many of the precious fission neutrons are lost by absorption in the reaction ^{1}H(n,γ)^{2}H, forming deuterium. The best practical moderator is

TABLE 30-2
Nuclear cross sections

Nuclide	Cross section for capture of slow neutrons, barns	Cross section for fission by slow neutrons, barns	Half-life for α-emission
$^{1}_{1}$H	0.33		
$^{2}_{1}$H	0.0006		
$^{10}_{5}$B	4000		
$^{12}_{6}$C	0.0037		
$^{113}_{48}$Cd	27000		
$^{232}_{90}$Th	7.5	0	1.4×10^{10} y
$^{233}_{92}$U	50	530	1.6×10^5
$^{235}_{92}$U	100	580	7.1×10^8
$^{238}_{92}$U	2.7	0	4.5×10^9
$^{239}_{94}$Pu	270	740	2.4×10^4

heavy water containing deuterium, for which the capture cross section is small. No isotopes of He, Li, Be, or B are suitable. The abundant isotope of carbon, ^{12}C, has a satisfactorily low capture cross section, and many reactors use graphite moderators. Carbon of extreme purity is required; traces of boron will use up neutrons and poison the carbon as a moderator.

Before the reactor can operate, it must be built up to some critical size, for otherwise too many neutrons will escape before they have had a chance to cause a second fission. The critical size for a ^{235}U reactor using heavy water as a moderator is surprisingly small; one design is 21 cm in radius and uses only 242 g of ^{235}U. Rods of cadmium or boron are inserted in the reactor to control the multiplication factor and hence control the power level. To turn the reactor on, these rods are pulled out just far enough to allow the multiplication factor to become greater than 1; to shut down, the rods are dropped back in. Neutron-absorbing control rods are visible in the photo on p. 654, which shows the interior of a reactor under construction.

As a research tool, nuclear reactors supply intense beams of neutrons. Thus many useful unstable nuclides can be prepared simply by placing a flask of the target material in the vicinity of an operating reactor for a few days or months. The uranium slugs (which are sealed in cans) must be removed periodically and purified of the fission products; these products can be separated to give many radioactive nuclides of medium atomic weight.

Until 1940, the Periodic Table ended with uranium, element number 92. Thirteen new elements have since been produced and identified, the latest being hahnium, of atomic number 105. The production of two of these new elements, neptunium and plutonium, is of interest. In a reactor using natural uranium, the chain reaction is kept going by the fission of the relatively few ^{235}U nuclei present. The abundant isotope

^{238}U does not undergo slow neutron fission, but these nuclei do capture neutrons, forming unstable ^{239}U, which forms neptunium and plutonium as follows:

$$^{238}_{92}U + ^{1}_{0}n \rightarrow ^{239}_{92}U$$

$$^{239}_{92}U \rightarrow ^{239}_{93}Np + ^{0}_{-1}e \quad (24 \text{ min})$$

$$^{239}_{93}Np \rightarrow ^{239}_{94}Pu + ^{0}_{-1}e \quad (2.33 \text{ days})$$

The plutonium isotope $^{239}_{94}Pu$ is an α emitter of half-life 24 000 years; it is therefore stable enough to be separated chemically from uranium, neptunium, and all the fission products in the reactor. It happens that $^{239}_{94}Pu$ is fissionable by slow neutrons (see Table 30-2), and so it can serve the same purposes as ^{235}U. A plutonium reactor could thus convert all the "inert" ^{238}U into "useful" ^{239}Pu. This process, called *breeding*, effectively extends the world supply of fissionable uranium by a factor of over a hundred. A similar breeding process converts the plentiful $^{232}_{90}Th$ into fissionable $^{233}_{92}U$ of half-life 160 000 years.

The application of fission to nuclear weapons is well known. A firing mechanism brings together several small pieces of ^{235}U or ^{239}Pu metal, and an explosive release of energy takes place when the critical mass is thus formed. The production of nuclear power for peaceful uses is indirect. In one design, water pipes are buried in the reactor, and the heat of the moderator is used to make steam which turns a generator. The technology of nuclear power is being rapidly developed, and electric power generated by nuclear reactions has become economically competitive with conventional sources of power, which are also being developed to higher efficiencies. During the next few years we look forward to a break-through in this field; what is needed is a method of converting heat energy directly into electric energy without the intervening low-efficiency processes. Even more desirable would be a method of converting nuclear energy directly into electric energy without becoming involved at all with heat engines and their inherently low (Carnot-limited) efficiencies.

30-5 Fusion

Release of nuclear energy also takes place when the lighter nuclei are joined together. This process of *fusion* is represented by the simple case of the capture of a neutron by a proton, the process which lowers the efficiency of ordinary water as a moderator in a fission reactor. The reaction is

$$\text{{}_{1}^{1}H} + \text{{}_{0}^{1}n} \rightarrow \text{{}_{1}^{2}H}$$

and, by using masses from the table of light nuclides on p. 723, it is easy to show that 0.00239 u disappears during this capture. Therefore, the energy release is 931 × 0.00239, or 2.23 MeV; this energy is available for KE of the deuteron and for the emission of γ rays. The conversion of rest mass into energy also takes place during other fusion reactions between light nuclei. Some reactions involving deuterium and tritium are

(a) $\text{{}_{1}^{2}H} + \text{{}_{1}^{2}H} \rightarrow \text{{}_{1}^{3}H} + \text{{}_{1}^{1}H}$ (4.0 MeV)

(b) $\text{{}_{1}^{2}H} + \text{{}_{1}^{2}H} \rightarrow \text{{}_{2}^{3}He} + \text{{}_{0}^{1}n}$ (3.3 MeV)

(c) $\text{{}_{1}^{2}H} + \text{{}_{1}^{3}H} \rightarrow \text{{}_{2}^{4}He} + \text{{}_{0}^{1}n}$ (17.6 MeV)

(d) $\text{{}_{1}^{3}H} + \text{{}_{1}^{3}H} \rightarrow \text{{}_{2}^{4}He} + 2\text{{}_{0}^{1}n}$ (11.3 MeV)

A desirable fusion reaction is one in which four protons combine to form helium plus two positrons:

$$\text{{}_{1}^{1}H} + \text{{}_{1}^{1}H} + \text{{}_{1}^{1}H} + \text{{}_{1}^{1}H} \rightarrow \text{{}_{2}^{4}He} + 2\,\text{{}_{+1}^{0}e}$$
$$(26.7 \text{ MeV})$$

The reaction is really between nuclei, not neutral atoms. That is,

$$4p \rightarrow \alpha + 2\,\text{{}_{+1}^{0}e} + \text{energy}$$

To obtain an equation involving the masses of the neutral atoms we add 4 electrons to each side of this equation:

$$(4p + 4e) \rightarrow (\alpha + 2e) + (2\,\text{{}_{+1}^{0}e} + 2\,\text{{}_{-1}^{0}e})$$
$$+ \text{ energy}$$

or

$$4\,\text{{}_{1}^{1}H} \rightarrow \text{{}_{2}^{4}He} + \text{energy} + \text{energy}$$

In writing the last equation, we have made use of the fact that the two positrons and the two negatrons will come together and be annihilated, transforming their masses entirely into energy. Thus the total rest mass converted into energy during this fusion can be computed as the difference between the mass of four neutral hydrogen atoms and one neutral helium atom.

BEFORE FUSION
$$4\,\text{{}_{1}^{1}H} \quad 4.03130 \text{ u}$$

AFTER FUSION
$$\text{{}_{2}^{4}He} \quad 4.00260$$

MASS DIFFERENCE $\overline{0.02870 \text{ u}} = 26.7 \text{ MeV}$

Spectroscopic evidence shows that the sun consists almost entirely of hydrogen, with some helium and traces of heavier elements. The energy radiated by the sun and other stars must come from nuclear fusion reactions, but the simultaneous collision of four protons would be rare indeed, even inside the sun. However, several chain reactions are possible in which the $\text{{}_{2}^{4}He}$ is built up one step at a time. The *proton-proton chain* forms helium out of hydrogen as follows:

(e) $\text{{}_{1}^{1}H} + \text{{}_{1}^{1}H} \rightarrow \text{{}_{1}^{2}H} + \text{{}_{+1}^{0}e}$

(f) $\text{{}_{1}^{2}H} + \text{{}_{1}^{1}H} \rightarrow \text{{}_{2}^{3}He}$

(g) $\text{{}_{2}^{3}He} + \text{{}_{1}^{1}H} \rightarrow \text{{}_{2}^{4}He} + \text{{}_{+1}^{0}e}$

or

(h) $\text{{}_{2}^{3}He} + \text{{}_{2}^{3}He} \rightarrow \text{{}_{2}^{4}He} + 2\,\text{{}_{1}^{1}H}$

Another possible chain starts with $\text{{}_{6}^{12}C}$ and builds up helium out of hydrogen in a series of steps. The $\text{{}_{6}^{12}C}$ is regenerated in the last step, and so it is not used up in the reaction. This *carbon cycle* is discussed in Prob. 30-B9.

The basic reactions *e*, *f*, and *g* have been studied in the laboratory, and the cross sections are known. We must bear in mind that these experiments use projectiles of energies of 1 MeV or so; such energies are needed so that protons can approach within nuclear distances of other protons which are also positively charged. At best, in the laboratory, we use beams of low intensity, perhaps just a few billion projectiles per second, and make up in efficiency (i.e., in probability or cross section) what we lack in numbers. Inside the sun, the situation is reversed. The energies are very low, in comparison with those of protons produced by a cyclotron, and so the cross sections are extremely small. However, the sun is almost all hydrogen, and the collisions occur so often that many reactions take place per second, in spite of the small probability. To fix these ideas, let us compute the average energy of a proton inside the sun.

EXAMPLE 30-1

Calculate the average energy of a proton at the center of the sun where the temperature is 15 000 000°K.

Using Eq. 16-5, we find

$$\text{KE per mole} = \tfrac{3}{2}RT$$
$$= \tfrac{3}{2}(8.31 \text{ J/mole} \cdot \text{K}°)$$
$$\times (1.5 \times 10^7 °\text{K})$$
$$= 18.7 \times 10^7 \text{ J/mole}$$

$$\text{KE per atom} = \frac{18.7 \times 10^7 \text{ J/mole}}{6.02 \times 10^{23} \text{ atoms/mole}}$$
$$\times \left(\frac{1 \text{ eV}}{1.60 \times 10^{-19} \text{ J}}\right)$$
$$= \boxed{1.94 \times 10^3 \text{ eV/atom}}$$

The average KE of a proton inside the sun is only 1940 eV, or 0.00194 MeV; the sun is a puny atom-smasher indeed. Nuclear reactions at such low energies are unobservable on earth. It is a triumph of nuclear theory that the predicted cross sections for such reactions give just the observed amount of radiation from the sun. To supply this energy, the sun loses 4 million tons of mass per second, through fusion reactions such as we have described.

30-6 Nuclear power

In this section we compare and contrast fission and fusion power reactors in the light of both present technology and long-range outlook. We have described the nuclear reactor which uses the fission of ^{235}U, ^{239}Pu, or ^{233}U. Control of the process demands use of slow neutrons which can be absorbed by cadmium or boron rods. The power appears as heat, and hence a series of auxiliary heat exchangers, steam boilers, and turbines are necessary to produce easily managed electrical or mechanical power. We list some of the uses of fission reactors: (1) production of ^{239}Pu and ^{233}U by breeding, starting from inert ^{238}U and ^{232}Th; a breeder-reactor could actually produce more ^{239}Pu than the amount of ^{235}U used to keep the reactor going; (2) production of artificially radioactive nuclides for tracer studies and medical therapy, and possible industrial uses such as preservation of foods; (3) production of neutron beams of high intensity for use in experimental work; (4) production of power for turning electric generators or for propelling ships and submarines; (5) production of vast quantities of heat for desalinization of sea water.

Some of the disadvantages of fission reactors are: (1) the heavy shielding required to protect personnel from radiation; (2) the scarcity of fuel, and the ultimate exhaustion of the world supply of fissionable material; and (3) the difficulty of disposing of highly active fission products. It is estimated that if all the power needs of the United States today were supplied by fission reactors, the nuclear waste would have an activity equal to that of the fallout from some 200 000 atomic bombs* each year. The consumption of power in the United States is expected to increase fortyfold by the year 2000, and it is doubtful if oil, coal, gas, and hydroelectric

---

*The standard of comparison is the bomb exploded at Hiroshima in 1945, which was equivalent in energy to 20 kilotons of TNT (3.8×10^9 J).

installations can supply this much power. Fission power, despite its hazards, may be a necessary stopgap while fusion power is being developed.

Controlled fusion reactors present a different set of problems which have not yet been solved. *Fission* chain reactions are possible at low temperatures only because of the release of neutrons during fission, and because of the fortunate circumstance that slow neutrons can approach close to the nucleus without suffering electrostatic repulsion. In a *fusion* reactor, the particles will have to have high energies to be able to come close enough to react. This calls for high temperatures—in fact, the problem is nothing less than constructing an artificial star, at a temperature of millions of degrees. At such temperatures, all atoms break down into *plasma,* which consists of a mixture of nuclei and free electrons. Fortunately, the density of such a plasma need not be great, and it is estimated that a powerful fusion reactor would use a plasma which is less dense than the air we breathe. The problem is how to contain such a hot plasma without allowing it to touch the wall of the vessel. Contrary to popular belief, such a reactor would not be dangerous; the "fire" would actually go out if the plasma touched the wall (see Prob. 30-C5). Many schemes for using magnetic fields to contain the plasma are being investigated, all aimed at keeping the plasma from touching any material object to which it might lose thermal energy.

The fusion processes are called *thermonuclear reactions* because of their resemblance to the ordinary burning of coal or wood. There is no critical size, except possibly to allow for heat leakage, and no theoretical upper limit to the rate at which energy can be produced from mass. It is only necessary to produce a high temperature in a controlled thermonuclear reactor (CTR), and then (figuratively) to shovel in the fuel, as much or as little as may be desired. The rate of release of nuclear energy depends on the temperature, and the fuel "ignites" if this rate equals or exceeds the rate of energy loss through radiation, conduction through the walls, etc. The ignition temperatures are calculated to be 350×10^6 °K for the deuteron-deuteron reaction, and only 50×10^6 °K for the deuteron-triton reaction. The nuclear fuel of the future will in all likelihood be deuterium; even though this isotope of hydrogen is rare, it is present to the extent of 1 part in 7000 in naturally occurring hydrogen (Table 8 in the Appendix). As shown in Probs. 30-C4 and 30-C5, each gallon of sea water contains enough deuterium to give the energy equivalent of about 200 gallons of high-test gasoline.

A further advantage of fusion power over fission power is its absolute freedom from long-lived radioactive waste. In the equations of Sec. 30-5, only tritium ($^3_1$H) is unstable. It emits no γ rays, and its negatrons are easily shielded; in any case, the half-life is 12 years, and so it is stable enough to be used up as a fuel element yielding stable $^4_2$He (Eqs. *c* and *d*), or it can be stored long enough to decay to a harmless radiation level. Containment of a large amount of radioactive material such as tritium will pose a serious problem, since structural failure may occur due to radiation damage caused by induced radioactivity in the container material; a similar problem also arises in the walls of the CTR itself.

To summarize, the advantages of fusion power are: (1) inexhaustible and cheap fuel supply, and (2) no waste disposal problem, if the container problem can be solved. Among the disadvantages are: (1) extremely high ignition temperature; (2) the difficulty of confining the reaction in a vessel; and (3) the size and complexity of installations, unsuited for small-scale uses such as propulsion.

One thing is certain: power consumption is growing by leaps and bounds in our technological society, and some form of nuclear power will be an economic necessity within the foreseeable future.

30-7   Biological effects of radiation

In a purely physical sense, the activity of a radioactive source is measured in *curies:* 1 curie (Ci) is defined as exactly 3.7×10^{10} disintegrations per second (dis/s). This is approximately the activity of 1 gram of $^{226}_{88}\text{Ra}$. Radiation sources in the laboratory are often measured in microcuries (μCi).

EXAMPLE **30-2**

In a tracer experiment in physiology, a sample of muscle tissue has 0.022 μCi of the radioactive phosphorus isotope $^{32}_{15}\text{P}$, of half-life 14 days. What counting rate would be registered by a Geiger counter which receives 40% of the β particles emitted by the $^{32}_{15}\text{P}$ in the sample? Assume that each β particle is counted with 95% efficiency.

Rate

$$= (0.022 \times 10^{-6} \text{ Ci}) \left(\frac{3.7 \times 10^{10} \text{ dis/s}}{1 \text{ Ci}} \right)(.40)(.95)$$

$$= 310 \text{ counts/s}$$

The *biological* effect of a source, or dose, of radiation is determined in part by the total energy absorbed; this energy, delivered by an incoming particle, causes the breakup of a protein or other complex molecule, or it causes genetic damage through altering a gene in a chromosome. Since it takes less than 100 eV to create an ion pair in a complex organic molecule, it is obvious that a single β particle or γ-ray photon of energy up to several MeV can create many ion pairs. Thus radiation damage depends on the energy of the radiation and also on the ease of absorption of the radiation by the type of tissue (muscle, bone, etc.) considered. Because of its mass, an α particle, for example, of a given energy is more effective that a β particle of the same energy. We see the need for a factor, the *relative biological effec-*

tiveness (RBE), so that different kinds of radiation can be compared. The RBE for x rays, gamma rays, and β rays is assumed to be 1, and that for α particles is about 10 or 20. The *radiation absorbed dose* (rad) is the standard unit of measurement for radiation exposure; it corresponds to the absorption of 100 ergs of energy in 1 gram of tissue. The purely physical unit *roentgen* (R) is also used; it is approximately the same as the rad, for x rays or gamma rays (Prob. 30-B14). The most useful unit is the *radiation equivalent, man* (rem) which is the product of the number of rads and the RBE.

Biological radiation damage falls into two categories. *Somatic damage* (to parts of the body other than the reproductive cells) is in most (not all) cases tolerable in small doses. For example, the body recovers from the somatic damage of a minor sunburn or a cut finger. This means that there is often a *threshold* for somatic damage; presumably organisms have evolved so as to be only slightly affected somatically by the normal, ever-present background of radiation. This background is due to several causes. At sea level, cosmic rays subject the body to about 100 mrad (millirads) per year (but as much as 170 mrad at mile-high Denver, Colorado). In addition, there is a natural radiation in the air due to the decay products of the gas $^{222}_{86}\text{Rn}$ (Sec. 29-3). Equally important is the body burden of β radiation from $^{40}_{19}\text{K}$ (half-life 1.3×10^9 years). A 70 kg adult has about 0.13 kg of potassium, chiefly in the muscle, and the normal $^{40}_{19}\text{K}$ content of the human body is about 0.1 μCi, which delivers about 20 mrad/year to the gonads and other soft tissues and about 10 mrad/year to bone. Other natural sources of radiation include drinking water and foods. We have lived with these and other somatic challenges for millions of years; we should be concerned at any increased radiation levels which might, say, double the body burden of radioactivity.

We have mentioned the use of radioactive nuclides in cancer therapy (Sec. 30-1). The

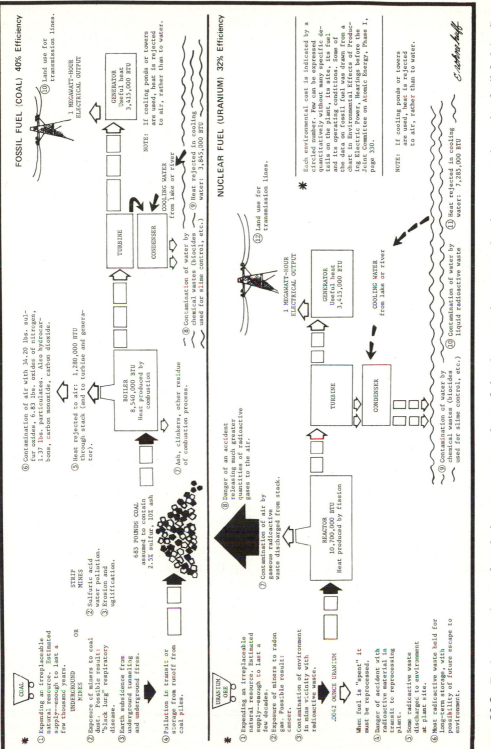

Figure 30-8 Environmental cost of producing one megawatt-hour of electric energy in a steam electric power station, using fossil fuel (coal) or nuclear fuel (^{235}U). Chart by D. E. Abrahamson, Scientists' Institute for Public Information.

relation of radiation to cancer is twofold, for it has become apparent that exposure to low levels of radiation actually causes certain cancers of the blood (leukemia), bone, thyroid, and lung.

Genetic damage seems to have no threshold; *any* dose to a gonadal chromosome, however small, produces *some* genetic damage. Most mutations caused by gene damage are harmful, although these effects, even lethal ones, may not become overt until many generations have gone by. Thus very minor exposure to low levels of radiation can, in the long run, cause large genetic effects in a population. A little mutation is a good thing; it is probable that the course of evolution depends in part on mutations caused by the cosmic ray background. But there is considerable concern about genetic damage to the human race from excessive medical use of diagnostic x rays. Radiation thus far introduced into the atmosphere by the testing of nuclear weapons is of lesser magnitude, but if such testing leads to actual warfare, the somatic and genetic effects will be catastrophic. The isotopes $^{90}_{38}$Sr (a fission product, half-life 28 years) and $^{14}_{6}$C (a fusion by-product, half-life 5730 years) are especially troublesome in this regard. Genetically speaking, for the population as a whole it is the *total* dose to the gonads that counts; it probably makes little difference whether radiation exposure of an individual is at a low level for a long time, or at a high level for a short time (as in nuclear warfare). Contamination of atmosphere, water, or food that affects large segments of the world's population is harmful precisely because of the large number of persons involved. The calculated risk taken by workers in nuclear energy installations is, by contrast of less importance, but no one is quite sure that these imponderables are being correctly assessed.

In spite of biological dangers, man-made radiation sources have so far conferred far more identifiable benefit than harm. Medical use of x rays has saved countless lives. Nuclear power plants hold out the promise of cheap power, with less environmental contamination (Fig. 30-8), and this power can be used for social goods such as massive desalinization of sea water. Industrial x-ray inspection of aircraft wings for hidden flaws reduces the hazard of air travel. There are many similar examples. Nevertheless, society must eliminate excessive or needless exposure to ionizing radiation. Groups of scientists and engineers in several countries are active in presenting these issues to the general public.

30-8 High-energy physics

Contemporary physics seems to be preoccupied with bigger and bigger machines which give higher and higher energies to particles. Let us look at the motivation for such a trend and see what some of the results have been. We can consider an atom-smasher as a tool with three fairly distinct uses: (1) to probe the structure of nuclear matter; (2) to create new particles; and (3) to study the interactions between particles.

To study the *structure of nuclear matter,* high-energy particles are shot at nuclei whose structure is to be determined. In order to "see" clearly, we must use a fine enough probe. More specifically, to obtain high resolving power we need to use a wavelength that is much smaller than the object which we are investigating, for only in this way can the effects of diffraction be made negligible. From this point of view, we can interpret Rutherford's scattering experiment in a new fashion. He used α particles of energy about 9 MeV and was able to deduce something about the nucleus. The de Broglie wavelength of any particle is given by $\lambda = h/mv$, where mv is the momentum of the particle; it is not hard to show that 9 MeV α particles have a velocity of about 2.1×10^7 m/s and a wavelength of about

5 × 10⁻¹⁵ m (= 5 fermis).* We now know that the radius of a gold nucleus is about 7 fermis, and so we see that Rutherford's probing beam which penetrated a gold foil was just fine enough to reveal some detail. In fact, Rutherford was able to deduce the existence of the nucleus, but not much more.

As higher energies have become available, the de Broglie wavelengths of particles used have become shorter, and the probing action has become more delicate. A probe of high-energy electrons is preferable to a probe using high-energy protons, because electrons respond only to electromagnetic forces which are better understood than nuclear forces. The newest accelerators take the projectiles far into the relativistic range, where a particle's mass is much larger than its rest mass. In a sense, the relativistic mechanics is here simpler than Newtonian mechanics, since the velocity $\approx c$ and cannot increase. Under these circumstances, the de Broglie wavelength is

$$\lambda = \frac{h}{mv} \approx \frac{h}{mc} = \frac{hc}{mc^2} = \frac{hc}{E}$$

and so the wavelength of any fast particle is inversely proportional to its energy. Electrons of energy 1.3 GeV produced in a linear accelerator have energy 2.1×10^{-10} J, and their wavelength is given by

$$\lambda = \frac{hc}{E}$$
$$= \frac{(6.63 \times 10^{-34} \text{ J·s})(3.00 \times 10^8 \text{ m/s})}{2.1 \times 10^{-10} \text{ J}}$$
$$= 0.95 \times 10^{-15} \text{ m} = 0.95 \text{ fm}$$

Since the wavelength is $\frac{1}{7}$ the radius of a gold nucleus, the resolving power is sufficient to show some detail in the structure as shown in Fig. 30-9, which reveals a uniform

---

*The metric prefix for 10^{-15} is femto-; thus the abbreviation fm (femtometer) also stands for the fermi, in honor of Enrico Fermi (1901–1958).

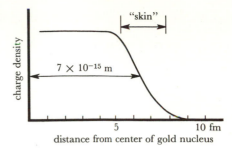

Figure 30-9 Distribution of electric charge in a gold nucleus; graph based on high-energy electron scattering experiments of Hofstadter. The electron energy was about 1.3 GeV, and the de Broglie wavelength about 1 fm.

charge density in the interior and a "skin" at the outer edge. The electric structure of the proton and the neutron have been studied, using electron beams of energy up to 21 GeV. Experiment shows that the proton does not have a sharp boundary (Fig. 30-10); its effective radius is about 0.8 fm. The neutron appears to be electrically neu-

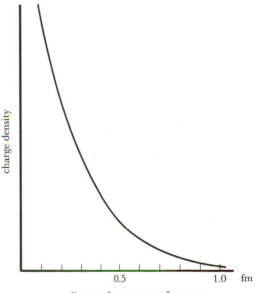

Figure 30-10 Distribution of electric charge in a proton; graph based on scattering of 6 GeV electrons, of de Broglie wavelength about 0.2 fm.

tral throughout its volume. There is some evidence that protons and neutrons have a complex internal structure consisting of scattering centers much smaller than the size of the nucleon. These pointlike entities are tentatively called *partons.* It is all very reminiscent of Rutherford's original study of atomic structure.

A second use of very high energies is in the *production of subatomic particles.* The rest mass of an electron is 0.00055 amu, which is equivalent to 0.51 MeV, so we might expect that a 0.51 MeV photon can be converted into an electron according to $\Delta E = (\Delta m_0)c^2$. This would be an impossible process, however, since charge would not be conserved. A possible process is *pair production,* in which a negatron and positron are formed from a γ ray of energy 1.02 MeV or more. (Any excess photon energy is shared as KE of the two electrons.) Figure 30-11 illustrates this process; the incoming photons leave no tracks, since they are uncharged and produce no ions. When the negatron and positron are materialized, they are both traveling in a forward direction, thus carrying most of the initial momentum of the photon, which has disappeared.[*] The cloud chamber is in a magnetic field, which gives opposite curvatures to the two tracks, showing that they are oppositely charged. In this one photograph we see illustrated three of the conservation laws: the conservation of mass-energy, linear momentum, and charge. Angular momentum (spin) is also conserved, but this fact cannot be determined from the photograph.

The rest mass of a proton is about 1.008 u, equivalent to 938 MeV (almost 1 GeV). A proton can be materialized if sufficient energy is available. Once again, conservation of charge requires that two particles be produced. The proton-antiproton pair was

---

[*] Pair production must take place near some particle such as a nucleus, which takes up a small part of the photon's momentum. The track of this nucleus is not seen in Fig. 30-11.

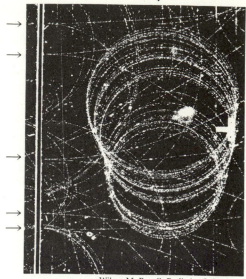

Wilson M. Powell, Radiation Laboratory, University of California

Figure 30-11 Conversion of energy into mass during the production of negatron-positron pairs. The cloud chamber is in a strong magnetic field, perpendicular to the plane of the photograph; negatrons curve clockwise, and positrons curve counterclockwise. Gamma rays enter the chamber from the left. At least five pair-production events can be seen originating in the absorbing plate which runs vertically through the chamber. Another pair is created in the gas at top center of the photo about 6 mm below the upper arrow; the negatron member of the pair spirals about 36 times in the magnetic field before coming to rest.

first produced in 1955 by the 6 GeV synchrotron at Berkeley, California, in a high-energy collision between two protons. A total of 6 GeV is needed: 2 GeV to create the mass of the proton-antiproton pair, and 4 GeV for the kinetic energies of the various particles after collision. The positron and antiproton are "antimatter"—exactly similar to their normal counterparts except for the sign of their charge. These particles are stable, in the sense that they do not undergo radioactive decay; but their lifetimes are very short, at least under laboratory conditions on the earth. A positron which comes near an electron will be annihilated, and the mass equivalent of the two particles becomes *annihilation radiation,* which usually consists of two γ rays each of energy 0.51 MeV. Similarly, an antiproton cannot long endure in an environment which contains matter made up of ordinary protons and neutrons. Atoms made of antimatter actually exist; for instance, a positron can circulate around an antiproton to form "antihydrogen." Nuclear antimatter has been created; thus an antideuteron is a combination of an antiproton and an antineutron. However, there is no evidence for antigravity; antimatter behaves normally in the earth's gravitational field. Such experiments are at the growing edge of contemporary physics. Many other particles have been materialized artificially with the energies now available, and no one yet knows what new particles may come into existence when experiments at an effective energy of 1700 GeV are performed with colliding proton beams.

30-9 Subatomic particles

More than 100 particles, having masses up to several times that of the proton, have been discovered. Most of them are unstable, with lifetimes as short as 10^{-23} s. The key to understanding and classifying this bewildering array of particles is the way in which they interact. In Sec. 18-3, we introduced a model in which the electric field is described by the exchange of "field particles" —photons. Physicists now recognize *four* distinctly different types of field, each with its own field particle. These interactions have radically different strengths (Table 30-3).

First, let us consider the more familiar interactions, the long-range electromagnetic and gravitational interactions. The first field particles to be discovered were photons, which are easy to produce. All you have to do is to accelerate a charge and some photons are "shaken loose" from the swarm that is continually being exchanged

between that charge and some other charge. This is how photons associated with radio waves come into existence as electrons move back and forth in a transmitting antenna; and it is how x-ray photons come into existence when a beam of high-speed electrons strike a stationary target in an x-ray tube. The model can be extended to cover the emission of photons during quantum jumps in an atom; one of the photons being exchanged between the nucleus and the orbital electrons is materialized when an electron changes its orbit.

Gravitational forces are inherently extremely weak compared with e-m forces (see Prob. 17-B1), and the gravitational field particles (gravitons) are probably beyond the reach of experimental techniques. Both the e-m and the gravitational interactions are "long range," decreasing gradually, inversely as the square of the distance between the interacting particles.

The two nuclear interactions S (strong) and W (weak) are of short range, and they decrease abruptly to practically zero when the distance between particles exceeds a few fermis. This distance is approximately the size of a nucleus; thus nuclei attract each other only when they are essentially in contact with each other. For the moment, let us skip the weak interaction and consider the first entry in Table 30-3. Strong interactions are truly nuclear—for instance, an attraction between two protons sufficient to overcome the e-m (Coulomb) repulsion of like charges, thus keeping the $^4_2$He nucleus (α particle) from flying apart. The field particles exchanged by nucleons are *mesons;* the lightest mesons, known as pions, have rest mass almost 300 times the mass of an electron (Table 30-4). We picture a nucleus as surrounded by a cloud of mesons (such as pions) which venture out a little way and are then reabsorbed within about 10^{-23} s. Traveling at nearly the speed of light, a meson could move only about 3×10^{-15} m (3 fm) from its parent nucleus during this short interval

TABLE 30-3   Interactions between particles

Symbol	Type of interaction	Relative strength	Range of force	Field particle	Rest mass of field particle
S	strong	1	short ($\approx$1 fm)	meson	$\geqslant 264\ m_e$
E	electromagnetic	10^{-2}	long ($1/r^2$)	photon	0
W	weak	10^{-15}	short ($\approx$1 fm)	W particle (?)	$>4000\ m_e$
G	gravitational	10^{-40}	long ($1/r^2$)	graviton (?)	0

of time. This explains why the nuclear force is of such short range. In our "boys-playing-catch" analogy, the repulsive force between the two boys would become zero if their separation were greater than the distance to which they could throw the ball. Free pions and other mesons can be materialized if enough energy is concentrated near a nucleus; for example, the pion's rest mass is equivalent to 135 MeV, and, if a fast projectile such as a proton comes within a few fermis of a nucleus, some of the energy of the projectile may be transformed into a pion's rest mass. The pions are shaken loose from the nucleons, just as photons are shaken loose when electrons are accelerated or undergo collision. The existence of pions was predicted by the Japanese physicist H. Yukawa several years before they were first found in the cosmic-ray debris at sea level. When sufficient energy became available from accelerators, pions and other mesons were created artificially in copious quantities.

To understand why mesons are unstable (having a half-life of about 10^{-8} s for $\pi^\pm$, 10^{-16} s for π^0), we introduce the idea that the strength of an interaction determines the stability of a particle. A strongly interacting particle, left to itself, tends to break up (disintegrate) as rapidly as it can. We can estimate this time to be about 10^{-23} s, which is the time for a high-speed particle to travel across the diameter of a nucleus. During this time, the strong interaction will cause a particle to decay *unless something prevents this from happening*. We therefore expect the half-life of any strongly interacting particle to be

about 10^{-23} s unless it is in something like a metastable state and is prevented from breaking up. Such a particle could then live long enough for a less likely process to take place, governed by the e-m interaction which is 10^{-2} as strong as the strong interaction. Such a particle would break up in about 10^{-20} s unless once again it is in a metastable state relative to the e-m interaction. The weak interaction eventually takes over if the first two decays are "forbidden"; typical lifetimes for particles which decay by the weak interaction are about 10^{-10} s. Even then, if the available energy (i.e., mass difference) is small, the lifetime of a weakly interacting particle can be much longer than 10^{-10} s; this is how the measurable half-lives of radioactive nuclei range from a fraction of a second to as long as 10^{18} years. The field particle for weak interactions, the W-particle, has not yet been found; the energy needed for its creation is more than can be supplied even by a 33 GeV accelerator. It is hoped that the effective energy of 1700 GeV available from the CERN colliding beams (Sec. 30-3) will be sufficient. Finally, the stable nuclides such as $^{12}_{6}$C, $^{1}_{1}$H (the proton itself), and the electron are stable against decay by any of the first three interactions. It is a matter of speculation whether the very weak gravitational interaction can give rise to spontaneous decay; at any rate, none has yet been detected.

The subatomic particles listed in Table 30-4 include all known particles which are stable against the strong interaction and hence have lifetimes of at least 10^{-20} s. The

TABLE 30-4   Subatomic particles*

Group	Members	Symbol	Spin	Rest mass	Half-life, s	Typical decay products
Gravitational field particle	graviton	—	2	0	stable	
Electromagnetic field particle	photon	γ	1	0	stable	
Leptons	electron's neutrino	$\nu_e(\bar{\nu}_e)$	$\frac{1}{2}$	0	stable	
	muon's neutrino	$\nu_\mu(\bar{\nu}_\mu)$	$\frac{1}{2}$	0	stable	
	electron	$e^-(e^+)$	$\frac{1}{2}$	1	stable	
	muon	$\mu^-(\mu^+)$	$\frac{1}{2}$	207	10^{-6}	$e^- + \bar{\nu}_e + \nu_\mu$
Mesons	pion	π^0	0	264	10^{-16}	$\gamma + \gamma$
		$\pi^+(\pi^-)$	0	273	10^{-8}	$\mu^+ + \nu_\mu$, $e^+ + \nu_e$ (rarely)
	kaon	$K^+(K^-)$	0	966	10^{-8}	$\mu^+ + \nu_\mu$; $\pi^+ + \pi^0$; others
		K_1^0	0	974	10^{-10}	$\pi^+ + \pi^-$; $\pi^0 + \pi^0$
		K_2^0	0	974	10^{-8}	$e^\pm + \overset{(-)}{\nu}_e + \pi^\mp$; $\mu^\pm + \overset{(-)}{\nu}_\mu + \pi^\mp$; $\pi^+ + \pi^- + \pi^0$; $\pi^0 + \pi^0 + \pi^0$
	eta	η	0	1074	10^{-18}	$\gamma + \gamma$; $\pi^0 + \pi^0 + \pi^0$; $\pi^+ + \pi^- + \pi^0$
Baryons	nucleons: proton	$p(\bar{p})$	$\frac{1}{2}$	1836.1	stable	
	neutron	$n(\bar{n})$	$\frac{1}{2}$	1838.6	10^3	$p + e^- + \bar{\nu}_e$
	hyperons: lambda	$\Lambda^0(\overline{\Lambda^0})$	$\frac{1}{2}$	2183	10^{-10}	$p + \pi^-$; $n + \pi^0$
	sigma	$\Sigma^+(\overline{\Sigma^+})$	$\frac{1}{2}$	2328	10^{-10}	$n + \pi^+$; $p + \pi^0$
		$\Sigma^0(\overline{\Sigma^0})$	$\frac{1}{2}$	2333	$<10^{-11}$	$\Lambda^0 + \pi$
		$\Sigma^-(\overline{\Sigma^-})$	$\frac{1}{2}$	2343	10^{-10}	$n + \pi$
	xi	$\Xi^0(\overline{\Xi^0})$	$\frac{1}{2}$	2572	10^{-10}	$\Lambda^0 + \pi^0$
		$\Xi^-(\overline{\Xi^-})$	$\frac{1}{2}$	2585	10^{-10}	$\Lambda^0 + \pi^-$
	omega	$\Omega^-(\overline{\Omega^-})$	$\frac{3}{2}$	3276	10^{-10}	$\Xi^0 + \pi^-$; $\Lambda^0 + K^-$; $\Xi^- + \pi^0$
Weak interaction field particle	W particle	$W^+(W^-)$	1	>4000	$\sim 10^{-17}$	$\mu^+ + \nu_\mu$; $e^+ + \nu_e$ (?)

* As of 1972.

Notes:
1. Particles (resonances) which decay by the strong interaction, with half-lives less than 10^{-20} s, are not listed.
2. Antiparticles are shown in parentheses. γ, π^0, K_1^0, K_2^0, and η are their own antiparticles.
3. All spins are in multiples of $h/2\pi$. All charges are numerically equal to the electron charge, or zero. All masses are in multiples of the electron's rest mass.
4. Decay products shown for the particles; decays for the antiparticles are analogous. The most prevalent modes of decay are shown.
5. The graviton and the W particle have been predicted but not yet observed.

particles, almost 40 in number, fall into several broad classifications. *Leptons* are lighter than nucleons and have spin $\frac{1}{2}$, measured in units of $h/2\pi$; *mesons* are intermediate between electrons and nucleons and have integral spin; *baryons,* of half-integral spin ($\frac{1}{2}$, $\frac{3}{2}$, $\frac{5}{2}$ . . .), include *nucleons* (proton and neutron) and more massive particles called *hyperons.*• The various groups of particles are also distinguished by their interactions, listed in Table 30-3. Leptons interact via G and W (and, if charged, by E); mesons interact via G and S (and, if charged, by E); baryons interact via G, S, and W (and, if charged, by E). The strongly interacting particles—mesons and baryons—are collectively known as *hadrons*. An electron, which is a charged lepton, has only the G, W, and E interactions; it is unaffected by the strong interaction, and it plows right through a nucleus except for the electric forces—this is why Fig. 30-9 gives the *charge* distribution in a nucleus. In effect, the electron ignores all nuclear forces because the weak interaction which it and the nucleus experience is only 10^{-13} as strong as the e-m interaction which dominates the picture.

In Table 30-4 we have listed (somewhat arbitrarily) only those particles which are stable against decay by the strong interaction. Most of the particles in Table 30-4 endure for at least 10^{-10} s. This is a "long" time—a nucleus moving at almost the speed of light could, in this time, travel about 3 cm, which is 10^{12} times its own diameter. But the table by no means exhausts the list of known particles, if we are willing to accept particles of very short lifetime. As a typical example, pions are fired at protons (in the liquid hydrogen of a bubble chamber). A pion and a proton may stick together as a unit, becoming (momentarily) a particle which breaks up in about 10^{-23} s, because of the strong interaction; in this case no metastable state exists to save the day. We have the rest masses of the pion and the proton plus some definite amount of KE which is transformed into mass to make up the mass of the new entity called a *resonance*• (= an unstable state of strongly interacting matter). This resonance is observed experimentally by measuring the energies and momentums of the particles after break-up; there is a strong preference for the total mass-plus-energy of the fragments to add up to a constant, which is the mass of the super-unstable particle. In the pion-proton collision the resonance, an excited state of the nucleon, is called an N particle or N resonance. Among the nucleon resonances are some "particles" with double charge; other nucleon resonances are neutral or negative; all are extremely short-lived. The list of nucleon resonances so far discovered includes at least 28 particles: in simplified notation these are N_1^{++}, N_1^+, N_1^0, N_1^-, all of mass about 2425 times the electron mass; N_2^+, N_2^0, of mass 2960; N_3^{++}, N_3^+, N_3^0, N_3^-, of mass 3330; N_4^+, N_4^0, also of mass 3330; N_5^{++}, N_5^+, N_5^0, N_5^-, of mass 3760; N_6^+, N_6^0, of mass 4300; N_7^{++}, N_7^+, N_7^0, N_7^-, of mass 4620; N_8^+, N_8^0, of mass 5180; N_9^{++}, N_9^+, N_9^0, N_9^-, of mass 5530. To these we should add the "ground state," which we can denote as N_0^+ and N_0^0, the ordinary proton and neutron of mass about 1837 units. From this point of view it really makes very little sense to consider the neutron and proton as any more elementary than their 28 excited states, except for their stability.

• Greek: *leptos* = small, tiny; the smallest Greek coin is a lepton, equivalent to about 0.013 cent. *Mesos* = in the middle, moderate; as in "mezzo-soprano" and many biological terms such as "mesoderm." *Barys* = heavy; as in "barium," "baritone." *Hyper* = over, above, beyond; as in "hypersensitivity," "hypertrophy." *Hadros* = thick, stout, strong.

• The time 10^{-23} s ($\approx 3 \times 10^{-15}$ m/3×10^8 m/s) is a "natural" time for a nuclear collision, equal to the time for a high-speed particle to travel the diameter of a nucleus at almost the speed of light. The reaction may continue to "ring" or "resonate" for 5, 10, 20, or more times the natural unit—still less than 10^{-20} s—if the exchanged meson makes a number of to-and-fro trips before the particles separate.

The discovery of resonances has proceeded apace; in addition to the N resonances, many different groups are known bearing such esoteric symbols as ρ, ω, Δ, Y^*, Ξ^*, etc. The discovery of resonances has actually simplified high-energy physics, for several regularities have been found among their masses (with a resulting challenge to the theoreticians). We are reminded of the analysis of the hydrogen optical spectrum in terms of energy levels (Fig. 28-3). The analogy would be complete if we were to give different names to the H atom in its various energy states—call it H′ in its lowest state E_1 and H″ in the excited state E_2 (which has a lifetime of about 10^{-8} s relative to emission of a photon of ultraviolet light). There *is* a mass difference between H′ and H″ because 10.20 eV of energy has been added and $\Delta E = (\Delta m)c^2$, but the mass difference is so small as to be unmeasurable. In the case of the subatomic particles, the excited states and the resonances have considerable extra mass, and their lifetimes are $\approx 10^{-23}$ s. The decay processes differ: the H atom decays from its excited state by emitting photons, but resonances decay by emitting charged or neutral particles rather than photons. For example, two of the nucleon resonances decay as follows:

$$N_1^{++} \rightarrow N_0^+ + \pi^+ + KE \qquad (10^{-23} \text{ s})$$

and

$$N_4^- \rightarrow N_0^0 + \pi^- + KE \qquad (10^{-23} \text{ s})$$

The rapid strong-interaction decay of N_0^0 (the ordinary neutron) is forbidden, so it decays by the weak interaction to a proton:

$$N_0^0 \rightarrow N_0^+ + e^- + \bar{\nu}_e \qquad (750 \text{ s})$$

The other low state of the nucleon, N_0^+ (the ordinary proton), cannot decay at all because there is no state of lower energy (i.e., of lower mass) for it to "fall" to.

It remains to discuss *leptons,* which do not interact by the strong interaction. The muon seems to be simply a heavy electron, and an unstable one at that. It decays (with a half-

life of about a microsecond) into an electron and two neutrinos; the ordinary electron cannot decay because there is no less massive lepton available except the neutrino, and $e \rightarrow \nu$ would not conserve electric charge. The antineutrino $\bar{\nu}$ differs from the neutrino only in that its spin is oppositely directed relative to the direction of its motion. It has recently been recognized that there are two distinct types of neutrinos, the electron's neutrino ν_e and the muon's neutrino ν_μ. Neutrinos that appear in β decay are associated with electrons, as in the decay of ^{55}Cr (p. 642), or in the decay of the free neutron:

$$n \rightarrow p + e^- + \bar{\nu}_e$$

The other neutrino appears in the decay of a pion:

$$\pi^+ \rightarrow \mu^+ + \nu_\mu$$

One of each is involved in the decay of a muon:

$$\mu^- \rightarrow e^- + \bar{\nu}_e + \nu_\mu$$

Free neutrinos react (very rarely) with matter, and can be observed to initiate reactions such as

$$\bar{\nu}_e + p \rightarrow n + e^+$$

and

$$\nu_\mu + n \rightarrow p + \mu^-$$

30-10　Cosmic rays

The powerful atom-smashers we have described are expensive and complicated, but fortunately we are living in a perpetual high-energy laboratory, which operates day and night free of charge. We are referring to the cosmic rays, first discovered about 1900 when it was found that ions are continually being produced in the atmosphere, making the air a slight conductor of electricity. Since then, cosmic rays have been studied at high and low altitudes, with electroscopes,

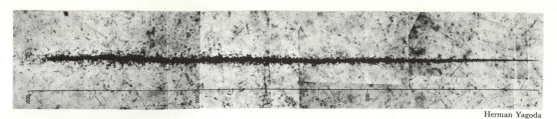

Figure 30-12 Primary cosmic ray particle. An iron nucleus with 26 positive charges enters a photographic emulsion from the left, and the track tapers to a point as the ion captures electrons and finally comes to rest as a neutral iron atom. Total path length, 1.5 mm.

Geiger counters, cloud and bubble chambers, and photographic emulsions. Before the development of large accelerators, the only source of high-energy collisions was in the upper atmosphere, and many of the elementary particles listed in Table 30-4 were first discovered in cosmic rays.

The *primary cosmic ray particles* which strike the earth from outer space are mostly extremely high-energy protons, with a sprinkling of deuterons and other positively charged light nuclei. The track of a primary particle is shown in Fig. 30-12. The photographic plate was exposed during a high-altitude balloon flight, and records the stopping of an iron nucleus in the emulsion.

The whole track is about 1.5 mm long. As the iron nucleus traveled through the emulsion, originally a bare nucleus with 26 + charges, it picked up electrons and finally came to rest at the right as a neutral atom. The track is thicker at the start because the highly charged nucleus caused more ionization, evidenced by the many side tracks of secondary electrons. Very few of the cosmic ray primaries reach sea level; instead, we observe a complex mixture of *secondary particles* including negatrons and positrons, photons, neutrons, protons, and a variety of other particles. Among these are the charged pion (π^- or π^+), and the muon (μ^- or μ^+). Charged pions, with half-life about 2 $\times$

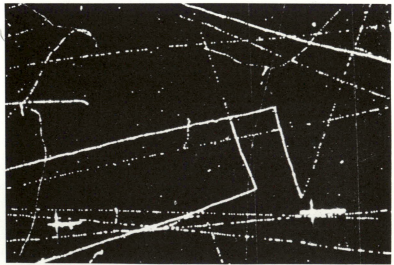

Figure 30-13 Bubble chamber photograph showing two $\pi \to \mu \to e$ decay processes (heavy tracks with right-angle bends). The electron tracks are characterized by a small density of bubbles.

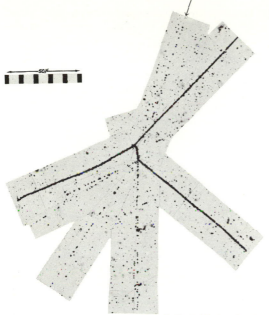

C. F. Powell, Physical Society, London

Figure 30-14 Cosmic ray star in photographic emulsion. A primary cosmic ray proton initiates a star, consisting of about five light particles and two heavier particles. One of these is a pion which, after a short upward journey, strikes another nucleus in the emulsion and causes a second disintegration.

10^{-8} s, decay into muons, and the muons in turn, with half-life about 2×10^{-6} s, decay into electrons (Fig. 30-13).

Cosmic ray *stars* originate when a primary particle strikes a nucleus head-on (Figs. 30-14 and 15). Such collisions are, naturally, more frequent at high altitudes where there are more primary particles. Many primary cosmic ray particles have energies of more than 1000 GeV. At sea level, we see only the secondary effects. Often *showers* of many thousands of electrons are observed, apparently originating from a single point above the observer (Fig. 30-16). Some extensive showers originate from single primary particles of energy as high as 100 000 000 GeV (10^{17} eV). The source of this tremendous energy is unknown; mere conversion of rest mass to energy seems unlikely, since the complete annihilation of a heavy nucleus

such as ^{238}U would give only 238 $\times$ 931 MeV, about 200 000 MeV or 200 GeV. It is possible that a gigantic cyclotron or synchrotron action takes place in the weak magnetic fields which surround some stars, or in the stronger magnetic fields associated with sunspots. There is some suspicion that the gravitational interaction, weak as it is, may be the clue to understanding cosmological processes which operate on an extremely large scale. It is possible that the primary cosmic radiation is associated with large-scale gravitational processes which are unobservable on the small scale represented by a laboratory or indeed by a solar system. The origin of the cosmic ray primaries is one of the most challenging problems of contemporary physics.

Instruments carried by balloons, artificial satellites, and lunar probes have measured

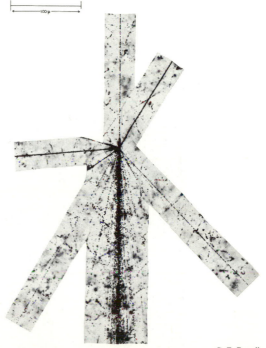

C. F. Powell

Figure 30-15 Cosmic ray star in emulsion, initiated by an α particle of about 3000 GeV energy. In addition to several heavy particles, a jet of about 150 secondary charged mesons is observed.

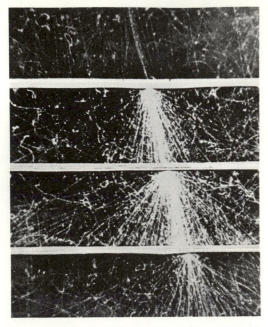

Figure 30-16 Cascade shower of electrons initiated by a high-energy cosmic ray particle. Several lead plates are placed within the cloud chamber to increase the probability of interaction. Total energy of the event, about 30 GeV. (M. Deutschmann, from *An Atlas of Typical Expansion Chamber Photographs,* by W. Gentner, H. Maier-Leibnitz, and W. Bothe, Pergamon Press.)

the intensities of cosmic rays and other radiations at high altitudes. In 1958, intense radiation was discovered at very high altitudes consisting of high-speed particles in the GeV range of energies. The earth is surrounded by two layers of "trapped" particles—electrons, protons, or both—which are kept from reaching the surface by the earth's magnetic field. Two distinct, widely separated doughnut-shaped regions of radiation exist, called the Van Allen belts, centered at about 6000 mi and 14 000 mi above the earth's magnetic equator. The radiation extends to about 40 000 mi above the equator but is much less intense over the magnetic poles. Figure 30-17 shows how radiation data were obtained from an early specially instrumented satellite. Two sources for these trapped particles have been suggested: (1) the decay of cosmic ray neutrons (n → p + e) at low altitude, and (2) bombardment by solar plasma (protons and electrons) shot out from the sun. It is probable that the inner and outer zones have different physical origins, with the inner zone being due to process 1 and the stronger and

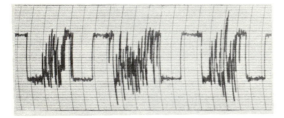

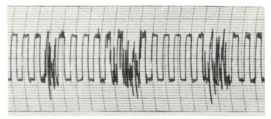

(Recording courtesy of J. A. Van Allen)

(*a*) 08^{h}20 20 Greenwich mean time. Satellite at 35°S. lat., 68°W. long., altitude 940 mi.

(*b*) 08^{h}26^{m}35^s Greenwich mean time. Satellite at 22°S. lat., 53°W. long., altitude 725 mi.

Figure 30-17 Radio signals from satellite 1958 Epsilon (Explorer IV). The signals were received on Aug. 1, 1958, at Santiago, Chile, and show the increased intensity (*b*) as the satellite entered the equatorial radiation belt. *Interpretation:* Time axis is horizontal; 1 division = 0.5 s. Vertical axis, frequency of an audio oscillator on board the satellite. A Geiger counter controls the oscillator's frequency, which changes abruptly from 675 Hz to 785 Hz, or vice versa, whenever 1024 counts have been accumulated. These audible tones modulate a carrier wave of frequency 108.03 MHz. The carrier wave is radiated by a dipole antenna consisting of the two halves of the satellite itself, which are electrically insulated from each other. As it moves along its orbit, the satellite tumbles end over end with a period of about 7 s. When it is pointing head-on or away from the observer, reception is poor (see Sec. 23-6) and "hash" appears every 3.5 s.

larger outer zone being due to process 2. Space probes have established the fact that the planet Venus does not have a trapped radiation belt; this is consistent with the observed absence of a magnetic field for that planet.

■ SUMMARY

Radioactive nuclides find use in therapy, as tracers, in industry, and in dating experiments.

Particle accelerators of several types are used to achieve high energies. The Van de Graaff is an electrostatic generator. The cyclotron, synchrocyclotron, synchrotron, and betatron use magnetic fields to bend the particles into orbits. No magnetic field is used in the linear accelerator. Limitations on maximum energy are size of machine and cost; fundamental theory limits energies in some machines because of the relativistic increase of mass at high speeds, and in all machines radiation losses from accelerated particles constitute a limit. Use of colliding beams makes the full KE of the particles available for causing reactions or creating new particles.

Slow neutrons of thermal energy (about 0.04 eV) cause several heavy nuclides to undergo fission. Other heavy nuclides are fissionable by fast neutrons only. Chain reactions are possible because of the production of excess neutrons during fission; in a reactor, a moderator reduces the speeds of these excess neutrons to thermal values.

The cross section of a nucleus for a given process is the effective area of the nucleus for that process. A large cross section means a high probability for the process to happen. Only three nuclides have high cross sections for fission by slow neutrons and also a usefully long half-life. These nuclides are ^{235}U, which can be separated from natural uranium, and ^{239}Pu and ^{233}U, which can be manufactured in breeder-reactors.

Fusion of light nuclides releases energy, since rest mass disappears. In the sun, four protons are built up into an α particle; the proton-proton chain and the carbon cycle are possible mechanisms for the process. Fusion reactors on earth will require high temperatures, but have the advantages of plentiful fuel and freedom from radioactive waste products.

Somatic and genetic radiation damage to tissue depends on the nature and energy of the absorbed radiation, as well as on the amount of dosage.

Particles of very high energy are used as probes to explore the structure of matter. High resolving power is obtained if the de Broglie wavelength of the guide waves associated with the projectiles is much smaller than the object being studied. Particles can be created, according to $\Delta E = (\Delta m_0)c^2$, if enough energy is available. Easily remembered energy equivalents of certain rest masses are: electron, about 0.5 MeV; proton, about 1 GeV. In pair production or annihilation, mass-energy, linear momentum, angular momentum, and charge are all conserved.

The four types of interaction and the field particles which are exchanged are, in order of decreasing strength, the strong interaction (mesons); the e-m interaction (photons); the weak interaction (W particles); and the gravitational interaction (gravitons). The latter two field particles have not been observed. About 40 particles are stable against the strong interaction; their lifetimes are at least 10^{-20} s, and most are 10^{-10} s or longer. Resonances are highly unstable states of strongly interacting matter; they can be considered to be particles of lifetime about 10^{-23} s.

Primary cosmic ray particles are light nuclei, of atomic number from 1 up to about 26. The primaries create many secondary particles as they pass through the earth's atmosphere, including mesons and hyperons. The Van Allen belts consist of high-energy charged particles emitted from the sun, trapped in the upper atmosphere of the earth by the earth's magnetic field.

■ CHECK LIST

Van de Graaff generator	moderator	annihilation
cyclotron	breeding	lepton
synchrocyclotron	fusion	meson
synchrotron	proton-proton chain	baryon
betatron	carbon cycle	pion
linear accelerator	plasma	hyperon
colliding beams	curie	resonance
fission	rad	cosmic-ray primaries
thermal neutron	RBE	cosmic-ray stars
cross section for capture	rem	cosmic-ray showers
barn	pair production	Van Allen belts

■ QUESTIONS

30-1 The over-all result of photosynthesis is that a living plant takes in CO_2 and H_2O, and releases oxygen to the air. Describe an experiment by which you could determine if the released oxygen comes from the carbon dioxide or from the water. What isotope or isotopes of oxygen would be suitable for this experiment?

30-2 Explain why the facts of relativity limit the maximum energy that a proton can acquire in a cyclotron. Tell how this limitation can be overcome in the synchrocyclotron and in the synchrotron.

30-3 Why are high-energy synchrotrons constructed with a large radius?

30-4 Two proton beams, each of 1 GeV energy, collide head-on. Explain why this collision is more useful than if a beam of 2 GeV energy strikes a stationary target.

30-5 How can the beam tubes near the point of intersection in Fig. 30-6 be straight rather than curved?

30-6 What are some of the peaceful uses of fission reactors?

30-7 Why is ordinary water a poor moderator for use in a fission reactor?

30-8 The nuclide ^{227}Th is an α emitter of half-life 18 days and has a cross section of 1500 barns for fission by slow neutrons. Why is this nuclide not a useful fuel for a nuclear reactor?

30-9 What are some advantages and disadvantages of fusion power as compared with fission power?

30-10 When a dentist x-rays your teeth, he steps behind a lead shield before making the exposure. Why is he apparently not equally concerned with the radiation risk to the patient?

30-11 What is meant by a threshold for radiation damage?

30-12 About how much energy is needed to create a neutron-antineutron pair, exclusive of the kinetic energies of the particles after their formation?

30-13 Verify that each of the decay processes listed in Table 30-4 is consistent with the law of conservation of charge.

30-14 Which of the particles of Table 30-4 decay by the e-m interaction? (*Hint:* Consider the half-lives and the decay products.)

30-15 Subatomic particles which experience the strong interaction are called hadrons. Which of the particles in Table 30-4 are hadrons?

30-16 How many neutrinos, of which kinds, are associated with the two $\pi \rightarrow \mu \rightarrow$ e events in Fig. 30-13? Why do the neutrinos not leave visible tracks in the photograph?

30-17 What is the difference between a cosmic-ray star and a cosmic-ray shower?

30-18 Name several fundamental particles of physics which were first discovered through study of cosmic rays.

■ PROBLEMS

(Use nuclear masses from table of properties of light nuclides on p. 723 where needed.)

30-A1 Calculate the time for one semicircular revolution of a proton in a magnetic field of 1 T, assuming that the particle is moving slowly enough so that its rest mass (Table 1 on p. 718) can still be used.

30-A2 Calculate the time for one complete revolution of a very high-energy particle in the synchrotron shown in Fig. 30-4 (radius of path, 1.00 km).

30-A3 In a large accelerator, to about what energy must protons be accelerated in a booster so that thereafter, as they gain energy, their speed differs by no more that 0.4% from c?

30-A4 A proton makes 10^7 revolutions per second in a cyclotron; what magnetic induction is required? (Assume the particle's speed is much less than c, so the nonrelativistic mass of the proton in Table 1 on p. 718 can still be used.)

30-A5 A proton in a cyclotron gains 0.05 MeV each revolution. (*a*) How many revolutions must it make to reach a final energy of 0.7 GeV? (*b*) About how far will it travel during this time if the average path radius is 1 m?

30-A6 Calculate the energy released in the fission process $^{235}_{92}\text{U} + ^{1}_{0}\text{n} = ^{110}_{46}\text{Pd} + ^{110}_{46}\text{Pd} + 16\,^{1}_{0}\text{n}$. Masses in u are $^{235}_{92}\text{U} = 235.0439$; $^{110}_{46}\text{Pd} = 109.9052$; $^{1}_{0}\text{n} = 1.008665$.

30-A7 Show that the reaction $^{238}\text{U}(\text{n}, \gamma)^{239}\text{U}$ is a possible one. (Masses: $^{238}\text{U} = 238.0508$; $^{239}\text{U} = 239.0543$.)

30-A8 For one of the reactions *a*, *b*, *c*, or *d* of Sec. 30-5, verify the stated value of the energy release.

30-A9 Calculate the wavelength of the guide waves associated with an electron of energy 21 GeV which is moving in the Stanford Linear Accelerator (SLAC) almost at the speed of light.

30-B1 How much time is required for a proton in the main synchrotron ring of Fig. 30-4 (radius 1.00 km) to increase in energy from an injection value of 8 GeV to the final value of 200 GeV, if it gains 2.5 MeV during each revolution?

30-B2 A proton in a cyclotron moves in a spiral path whose maximum radius (at point E in Fig. 30-2) is 0.6 m. If the proton at this point has a speed of 2×10^7 m/s, calculate   (a) its energy, in MeV;   (b) the vertical magnetic field B needed to supply the necessary centripetal force on the moving proton.

30-B3 When the accelerator of Fig. 30-4 operates at its initial design values of 1.5×10^{13} protons/s, at 200 GeV, what is the average current, in μA? What thermal energy, in kW, must be dissipated in the target?

30-B4 For a proton whose total energy is 28.00 GeV, calculate   (a) the rest-mass energy;   (b) the kinetic energy.

30-B5 (a) What is the rest-mass energy of a deuteron, in GeV?   (b) What is the total energy of a deuteron whose KE is 16.20 GeV?

30-B6 (a) If 201 MeV of energy is released, on the average, during fission of a ^{235}U nucleus, how many fissions per second take place in a reactor operating at a power level of 10 000 kW?   (b) If the reactor operates for 1 year, how many ^{235}U nuclei undergo fission during this time?   (c) What is the mass of ^{235}U fuel consumed during 1 year?   (d) How much natural uranium is consumed? (The abundance of ^{235}U is 0.72%.)

30-B7 Assuming an average release of 201 MeV per fission, show that only about 0.1% of the mass of the uranium fuel is actually transformed into energy in a reactor.

30-B8 Fission of ^{239}Pu releases 180 MeV per nucleus.   (a) Express this energy yield as J/mole; as kcal/mole; as J/kg. (*Note:* for comparison, the heat of combustion of a good grade of coal is about 30×10^6 J/kg.)   (b) How much ^{239}Pu must be used up, at 20% efficiency, to generate 1 million kWh of electric energy?

30-B9 The carbon cycle builds up ^{4}He out of four ^{1}H nuclei, in steps as follows: ^{12}C captures a proton, forming A; A decays by positron emission to B; B captures another proton to form C; C captures another proton to form D; D decays by positron emission, forming E; E captures a proton in a (p,α) reaction. Write out these reactions in detail; identify the nuclides A, B, C, D, and E; and show that the ^{12}C is regenerated at the end of the cycle, and thus is not used up.

30-B10 Is the ^{12}C nucleus stable against disintegration into three α particles? Compute the energy, in MeV, that must be supplied or will be released (which?) during the process ^{12}C $\rightarrow$ 3 ^{4}He.

30-B11 Calculate the average energy, MeV, of a particle at the center of a star where the temperature is 60 000 000°K.

30-B12 Suppose that a fusion reactor is constructed which uses reaction c of Sec. 30-5. How much tritium per day must be used as fuel if the power output of the installation is to be 10^6 kW?

30-B13 The fuel value of garbage is about 11×10^6 J/kg (about $\frac{1}{3}$ that of a good grade of coal). The city of Paris burns 1.6×10^9 kg of garbage each year, 1.2×10^9 kg to supply electric power and 0.4×10^9 kg for heating purposes.   (a) What is the annual monetary value, at 1¢/kWh, of the electric energy so generated, if the power station has 32% efficiency?   (b) How much uranium would be used in a year to supply this thermal energy in a reactor? (The fuel value of natural uranium is 5.9×10^{11} J/kg.)

30-B14 A roentgen (R) is defined as the radiation which produces $\frac{1}{3} \times 10^{-9}$ coulomb of singly charged ions, of either sign, in 1 cm^3 of dry air at STP. The average energy needed to produce an ion pair in air is 35 eV.   (a) Show that a dosage of 1 R is equivalent to the absorption of 11.7 nJ of energy in 1 cm³ of air. (*b*) Verify that 1 R $\approx$ 1 rad, to within about 10% (see Sec. 30-7 for the definition of the rad).

30-B15 (*a*) Use the average radius given in Fig. 30-9 to obtain the approximate volume of a gold nucleus. (*b*) Check the value for the density of nuclear matter ($\sim 2 \times 10^{17}$ kg/m³) given on p. 20. (*c*) Estimate the radius of the heaviest known nucleus, $^{262}_{105}$Ha. (*Hint:* If the density is constant, the volume is proportional to the mass number A.) (*d*) Identify a stable nuclide for which the radius is about half that of a gold nucleus.

30-B16 Is the reaction p + p = d + β^+ a possible one? If so, what energy is released? If not, how much energy must be supplied?

30-B17 How much energy, in MeV, is released in reaction h of Sec. 30-5?

30-B18 (*a*) What minimum energy must a neutrino ν_μ have in order for the last reaction in Sec. 30-9 ($\nu_\mu + n \rightarrow p + \mu^-$) to take place? (*b*) Repeat, for the other neutrino reaction listed in Sec. 30-9, $\bar{\nu}_e + p \rightarrow n + e^+$.

30-B19 Using Fig. 30-17 and the data given there, compute the radiation intensity, in counts/s, at each of the two points of observation.

30-C1 An organic archeological specimen contains 120 g of carbon, with a ^{14}C activity of 800 dis/min. How long ago did the organism die?

30-C2 Singly charged particles in a large accelerator move with orbital radius R. The bending magnets supply an induction B, and the particle energy, in eV is V. If the particles are highly relativistic ($v \approx c$), show that $V = RBc$.

30-C3 What average magnetic induction must the bending magnets in the main ring of the accelerator of Fig. 30-4 supply, to keep a 200 BeV proton beam in an orbit of radius 1.00 km? (*Hint:* See Prob. 30-C2.)

30-C4 Calculate the energy, in joules, contained in 4 kg of ordinary water (approximately 1 gallon), if the deuterium in the water is converted according to reaction a of Sec. 30-5. (*Hint:* Only 0.015% of the H atoms in the water are ^{2}H.) Compare your answer with the energy that can be derived from a gallon of gasoline, about 1.5×10^8 J/gallon.

30-C5 In the previous problem, you considered reaction a of Sec. 30-5. Can any more nuclear energy be extracted from the end products of this reaction? If so, how, and how much?

30-C6 A 1 liter flask contains deuterium gas (D$_2$) at 0°C and at a pressure of 10^{-4} atm. (*a*) Calculate the total KE of the plasma when the contents of this flask are heated to 350×10^6 °K. (*Hint:* See Eq. 16-5; the molecules of gas are completely dissociated into deuterons and electrons.) (*b*) To what temperature would the thermal energy calculated in (*a*) bring a 300 g cup of coffee, initially at 20°C?

30-C7 Using the cross section from Table 30-2, calculate the average distance L a slow neutron travels in carbon before being captured. (*Hint:* For an approximate calculation it will be sufficient to neglect the motion of the target atoms. Imagine a column of carbon 1 m² in cross section and L m long. If all the carbon nuclei within this column were moved to the front face of the column, the total area would be obscured, each carbon nucleus obscuring an area equal to its cross section for scattering. In this way calculate the total number of atoms, and hence the mass, volume, and length of the column of carbon. The specific gravity of carbon is 2.25, and its atomic weight is 12.00.)

30-C8 Verify that each of the decay processes listed in Table 30-4 is consistent with the law of conservation of angular momentum (spin), using the model described in Sec. 29-8.

30-C9 Calculate the electric potential, relative to infinity, at a point which is 8 femtometers (8 fm) from a silver nucleus of charge $+47e$ (*Hint:* See Eq. 28-8).

For Further Study

30-11 The Mössbauer effect

In 1962 Rudolf Mössbauer shared the Nobel Prize in physics for work he had done as a graduate student at Heidelberg University. His discovery of "recoilless" resonance absorption of γ rays has wide application in many areas of physics and chemistry. A γ-ray photon can be absorbed by a nucleus when the nucleus is raised from its lowest energy state E_0 to a higher state E_1; the effect is similar to the absorption of visible photons by a gas to give dark absorption lines such as those in the sun's spectrum (figure on p. 609). In the visible region, absorption involves a pair of energy levels differing by only a few electron volts; these transitions represent changes in the energy of the orbital electrons in the electron cloud surrounding the nucleus (Sec. 28-2). On the other hand, energy levels in the nucleus are separated by thousands or millions of electron volts, and transitions between nuclear levels result in the emission of γ-ray photons. For example, the nuclear energy-level diagram of the ^{57}Fe nucleus (Fig. 30-18) shows that three photons can be emitted. In an absorber containing iron atoms, almost all of the nuclei are in the lowest energy state E_0.[*] If, then, a γ ray of energy 0.0144 MeV is emitted by an ^{57}Fe nucleus in a source which makes a transition $E_1 \rightarrow E_0$, we expect that this γ ray can be absorbed by another ^{57}Fe nucleus in an absorber which makes the transition $E_0 \rightarrow E_1$.

A problem arises because of the recoil of a nucleus when it emits a photon. We can calculate the recoil velocity V when a photon is emitted from a nucleus of mass M, using the law of conservation of momentum. As we saw in Sec. 27-4, the momentum of the

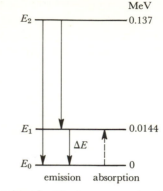

Figure 30-18 Nuclear energy levels in ^{57}Fe.

photon is

$$p = mc = mc^2/c = E/c = h\nu/c$$

Hence

$$\frac{\text{momentum of system}}{\text{before emission of photon}} = \frac{\text{momentum of system}}{\text{after emission of photon}}$$

$$0 = \frac{h\nu}{c} + MV$$

whence

$$V = -\frac{h\nu}{cM}$$

and

$$\text{KE} = \tfrac{1}{2}MV^2 = \frac{(h\nu)^2}{2c^2M} \qquad (30\text{-}2)$$

This is the kinetic energy of the recoiling nucleus, and the total energy difference $\Delta E (= E_1 - E_0)$ must supply not only the energy of the photon, but also the kinetic energy of the recoiling nucleus. This means that the emitted photons have slightly *less* energy than would be calculated from $E_1 - E_0$. Let us make a rough calculation of the recoil energy for the γ ray from ^{57}Fe. The mass of a single ^{57}Fe nucleus is $M = 57 \text{ g}/(6.02 \times 10^{23}) = 9.5 \times 10^{-23} \text{ g} = 9.5 \times 10^{-26}$ kg. Also, $h\nu = 0.0144$ MeV $= 2.3 \times 10^{-15}$ J. Hence

$$\text{KE}_{\text{recoil}} = \frac{(h\nu)^2}{2c^2M}$$

$$= \frac{(2.3 \times 10^{-15} \text{ J})^2}{2(3.0 \times 10^8 \text{ m/s})^2(9.5 \times 10^{-26} \text{ kg})}$$

$$= 3.1 \times 10^{-22} \text{ J} = 0.002 \text{ eV}$$

---

[*] In discussing nuclear energy levels, it is customary to call the lowest level E_0 instead of E_1 as in the optical diagrams of Chap. 28.

The energy of the emitted photon is 0.002 eV less than ΔE, and therefore no longer exactly matches the energy difference between the levels. Similarly, the absorbing nucleus recoils forward, and not all of the energy of the photon (already too small by 0.002 eV) is available for raising the nucleus from E_0 to E_1. There is a total energy deficiency of 0.004 eV.

One might think that 0.004 eV out of 14 400 eV would not matter, but the nuclear energy levels are so "sharp" that an incoming photon that is "out of tune" by only 1 part in 3 million (as in this case) cannot be absorbed. Prior to Mössbauer's work, nuclear resonance absorption such as we have considered had been observed only by using heroic measures, such as giving the "lost" energy back to the emitting nucleus by moving it (by a rotor) at great speed toward the absorber.

The great advance suggested by Mössbauer was his recognition that in some cases the entire solid source recoils as a unit, not just the single emitting nucleus. The mass M in the denominator of Eq. 30-2 is then increased by a large factor, such as 6×10^{20} if the source or absorber is a millimole of material in one solid chunk. The energy deficiency due to recoil is then completely negligible, and absorption can be observed. The Mössbauer effect is, therefore, a way of studying tiny changes in the energy levels of a nucleus. In fact, an energy difference of about 1 part in 10^{12} is enough to destroy the resonance in ^{57}Fe. Such small changes in nuclear energy, caused perhaps by variations in the chemical state of a substance, are far too small to observe in any other way.

A typical experiment is shown in Fig. 30-19. The source is mounted on nonmagnetic stainless steel and emits a γ-ray photon of a very sharply defined single energy $E_1 - E_0$ (≈ 0.0144 MeV). The absorber is ferricinium bromide, $Fe(C_5H_5)_2Br$, mounted on a carriage which can be moved back and forth at an adjustable speed in either direc-

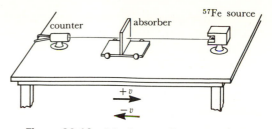

Figure 30-19 Mössbauer effect apparatus.

tion. When the absorber is at rest, no absorption is observed. This is interpreted as a shift of the energy levels in the iron atoms of the absorber due to their different environment. The electric field at the nucleus, caused by the orbital electrons, is different for iron in the absorber from what it is for iron in the source. In the absorber, E_0 becomes E_0', E_1 becomes E_1', and the energy difference becomes $E_1' - E_0'$. This energy shift, which is about 2 parts in 10^{12}, is easily measured by the Doppler effect (Sec. 10-8). If the absorber is given a constant velocity $v = +0.06$ cm/s (toward the right), the incoming photon's apparent frequency is raised to ν', given by $\nu' = \nu_0(1 + v/c)$. At the new frequency, the photon's energy $h\nu'$ matches the energy difference $E_1' - E_0'$ and absorption takes place. At still higher velocities, the resonance is again destroyed (Fig. 30-20).

The Mössbauer effect is a powerful tool for studying small changes in certain γ-ray photons. A test of the theory of relativity has been made by Pound and Rebka, who

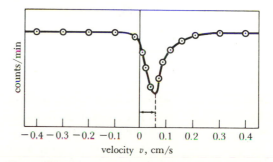

Figure 30-20 Absorption of ^{57}Fe γ-ray photons by ^{57}Fe nuclei in ferricinium bromide absorber.

were able to measure the increase in energy (about 2 parts in 10^{15}) of a ^{57}Fe photon as it fell 22 meters downward in the earth's gravitational field. We recall that a body which gains or loses energy as it moves in a gravitational field has by definition a weight, and hence a gravitational mass given by $m_{gravitational} = W/g$ where g is the field strength in N/kg. The experiment thus measured the gravitational mass of a photon. One way of stating Einstein's principle of equivalence is to postulate that the gravitational mass of a body is proportional to its inertial mass (see the discussion in the last paragraph of Sec. 4-8 on p. 70). The experiment of Pound and Rebka showed that the gravitational mass of the photon was in agreement with its inertial mass calculated from $E = (m_{inertial})c^2$, which can be written

$m_{inertial} = h\nu/c^2$. As another example of the application of the Mössbauer effect, internal magnetic fields in a solid (at the locations of the nuclei) can be studied. By measuring the splitting of the nuclear energy levels caused by the internal magnetic field, it was found, for example, that each iron nucleus in ferric fluoride (FeF_3) is in a very strong magnetic field whose induction B is 62 T. The strength of this field, which exists inside an ordinary unmagnetized solid, rivals that of the strongest fields that can be produced in the laboratory by the largest electromagnets. The Mössbauer effect has been observed with more than 30 nuclides; for most of these, except for ^{57}Fe, the effect at room temperature is small and the absorber and source must be cooled by liquid nitrogen, hydrogen, or helium.

■ PROBLEM

30-C10 (a) Refer to the discussion in Sec. 30-11 of Pound and Rebka's experiment. Should the detector move upward or downward in order to restore the resonance? (b) Calculate the velocity of the γ-ray detector, located at the bottom of the 22 m tower, which would compensate for the gain in energy of the falling photon. (*Hint:* Using $h\nu/c^2$ for the gravitational mass of the photon, show that $\Delta E/E = gH/c^2$, where H is the height of fall. Thus find $\Delta\nu/\nu$.) (c) How long would be required for the detector to move 1 mm?

■ REFERENCES

1 Kamen, M. D., "Tracers," *Sci. American,* Jan., 1949, p. 31.

2 Hurley, P. M., "Radioactivity and Time," *Sci. American,* Aug., 1949, p. 48.

3 Hahn, O., "Radioactive Methods for Geologic and Biologic Age Determinations," *Sci. Monthly,* **82,** 258 (1956).

4 Emiliani, C., "Ancient Temperatures," *Sci. American,* Feb., 1958, p. 54. The ratio $^{16}O:^{18}O$ depends on the temperature which existed when a fossil carbonate was laid down.

5 Renfrew, C., "Carbon 14 and the Prehistory of Europe," *Sci. American,* Oct., 1971, p. 63.

6 Wilson, R. R., "Particle Accelerators," *Sci. American,* Mar., 1958, p. 64.

7 Rose, P. H., and A. B. Wittkower, "Tandem Van de Graaff Accelerators," *Sci. American,* Aug., 1970, p. 24.

8 Panofsky, W., "The Linear Accelerator," *Sci. American,* Oct., 1954, p. 40.

9 Ginzton, E. L., and W. Kirk, "The Two-Mile Accelerator," *Sci. American,* Nov., 1961, p. 49.

10 O'Neill, G. K., "Particle Storage Rings," *Sci. American,* Nov., 1966, p. 107.

11 Rosen, L., "Relevance of Particle Accelerators to National Goals," *Science,* **173,** 490 (1971).

12 Hahn, O., "The Discovery of Fission," *Sci. American,* Feb., 1958, p. 76.

13 Sparberg, E. B., "A Study of the Discovery of Fission," *Am. J. Phys.,* **32,** 2 (1964).

14 Graetzer, H. G., "Discovery of Nuclear Fission," *Am. J. Phys.,* **32,** 9 (1964).

15 Leachman, R. B., "Nuclear Fission," *Sci. American,* Aug., 1965, p. 49.

16 Hughes, D. J., "The Reactor as a Research Instrument," *Sci. American,* Aug., 1953, p. 23.

17 Seaborg, G. T., and J. L. Bloom, "The Synthetic Elements IV," *Sci. American,* Apr., 1969, p. 57. Discusses a possible superheavy island of stability near the nuclide $^{298}114$; summarizes earlier articles in the series.

18 Weinberg, A. M., "Power Reactors," *Sci. American,* Dec., 1954, p. 33.

19 Seaborg, G. T., and J. L. Bloom, "Fast Breeder Reactors," *Sci. American,* Nov., 1970, p. 13.

20 Gough, W. C., "The Prospects of Fusion Power," *Sci. American,* Feb., 1971, p. 50.

21 Post, R. F., "Fusion Power—the Uncertain Certainty," *Bull. Atomic Scientists,* Oct., 1971, p. 42.

22 Seif, M., "Fusion Power: Progress and Problems," *Science,* **173,** 802 (1971).

23 Hubbert, M. K., "The Energy Resources of the Earth," *Sci. American,* Sept., 1971, p. 60.

24 Abrahamson, D. E., *Environmental Cost of Electric Power.* A 36-page workbook, published 1970 by the Scientists' Institute for Public Information, 30 E. 68 St., New York, N.Y. 10021. Contains many references.

25 Morgan, K. Z., "Comments on Radiation Hazards and Risks," *The Physics Teacher,* **9,** 415 (1971); "Never Do Harm," *Environment,* Jan./Feb., 1971, p. 28.

26 "Ionizing-Radiation Standards for Population Exposure," *Phys. Today,* Nov., 1971, p. 32.

27 Marshak, R. E., "Pions," *Sci. American,* Jan., 1957, p. 84. A good introduction to field theory. In this connection, see also the following four references.

28 Bethe, H., "What Holds the Nucleus Together?" *Sci. American,* Sept., 1953, p. 58.

29 Fermi, E., "The Nucleus," *Physics Today,* Mar., 1952, p. 6.

30 Treiman, S. B., "The Weak Interactions," *Sci. American,* Mar., 1959, p. 72.

31 Feinberg, G., and M. Goldhaber, "The Conservation Laws of Physics," *Sci. American,* Oct., 1963, p. 36.

32 Gell-Mann, M., and E. P. Rosenbaum, "Elementary Particles," *Sci. American,* July, 1957, p. 72.

33 Chew, G. F., M. Gell-Mann, and A. H. Rosenfeld, "Strongly Interacting Particles," *Sci. American,* Feb., 1964, p. 74.

34 Rossi, B., "High Energy Cosmic Rays," *Sci. American,* Nov., 1959, p. 135.

35 Hofstadter, R. H., "The Atomic Nucleus," *Sci. American,* July, 1956, p. 55. The investigation of nuclear size by the scattering of high-speed electrons.

36 Kendall, H. W., and W. Panofsky, "The Structure of the Proton and the Neutron," *Sci. American,* June, 1971, p. 66.

37 Peierls, R. E., "The Atomic Nucleus," *Sci. American,* Jan., 1959, p. 75.

38 Yagoda, H., "The Tracks of Nuclear Particles," *Sci. American,* May, 1956, p. 40.

39 Beiser, A., "Where Do Cosmic Rays Come From?" *Sci. Monthly,* **77,** 76 (1953).

40 Burbidge, G., "The Origin of the Cosmic Rays," *Sci. American,* Aug., 1966, p. 32.

41 Ney, E. P., "Heavy Elements from Space," *Sci. American,* May, 1951, p. 26.

42 Reines, F., and J. P. F. Sellschop, "Neutrinos from the Atmosphere and Beyond," *Sci. American,* Feb., 1966, p. 40. Includes an illustration of exchange of boomerangs to give an attractive force.

43 Gamow, G., "Half an Hour of Creation," *Physics Today,* Aug., 1950, p. 16.

44 Van Allen, J. A., "Radiation Belts Around the Earth," *Sci. American,* March, 1959, p. 39.

45 Rosser, W. G. V., "The Van Allen Radiation Zone," *Contemp. Physics,* **5,** 198 (1963) and **6,** 255 (1964).

46 de Benedetti, S., "The Mössbauer Effect," *Sci. American,* Apr., 1960, p. 72.

47 Lustig, H., "The Mössbauer Effect," *Am. J. Phys.,* **29,** 1 (1961).

48 Herber, R. H., "Mössbauer Spectroscopy," *Sci. American,* Oct., 1971, p. 86.

Mathematical review

The study of physics will be of very little value to a student who is unsure of arithmetic and algebra, for by its nature physics deals with measurable quantities expressed in numbers. To study physics while encumbered with an inability to use numbers fluently is like trying to read a foreign language without an easy familiarity with vocabulary and syntax.* Mathematics is the language of science, and algebra is its elementary syntax.

For the main part of our study, you are expected to bring with you nothing beyond elementary high school algebra. A certain small amount of trigonometry may prove helpful, but it is not necessary. The trigonometry that is used in this book is developed in the text itself as needed. Starting from algebra, we also define the basic concepts of calculus in Chap. 3, so that we can, throughout our study of physics, take advantage of the terminology of derivative (limit of a ratio) and integral (limit of a sum). The more detailed formulas of calculus are not used in the main part of the book; such formal calculus is developed and used in some of the optional "For Further Study" sections. Some additional material on elementary calculus is included in this Appendix for use by students reading these optional sections.

Sections A, B, and C in this Appendix have been prepared for those who wish to review their previous work in arithmetic, algebra, and geometry. Not everyone will need this review; you are encouraged to sample the portions dealing with topics about which you feel unsure. Many problems have been included so that you can test your mastery of each technique. Solve as many of these problems as you need to, until you are sure that you could do one more of them with little hesitation or anguish. Then you will be able to devote your valuable time and energy to thinking about *physics,* instead of continually wasting your efforts on what is merely routine. Learn the language and grammar so that you can understand and appreciate the concepts and ideas with which we are concerned.

It is suggested that you study Sections A-1 through A-3, and B-1, within the first week of the course, filling in your areas of uncertainty and strengthening your mathematical muscles by repetition and practice.

---

* The difficulty is not limited to foreign languages; many persons who are "educated" have failed to grasp ideas of politics, economics, or philosophy simply because they fail to understand and write their own language.

Appendix A / Arithmetic

A-1 Elementary operations

We assume that you know how to add, subtract, multiply, and divide. Some of these operations, especially long division, are tedious, and so we strongly recommend that you use a slide rule for multiplication and division (Sec. F-2). There is a method resembling division for extracting the square root of a number, but you are strongly urged to *forget* this method, if you ever knew it. Instead, use a slide rule, or *guess* the answer to 1 or 2 figures, which is often sufficiently accurate for the purpose at hand.

Multiplication is indicated in several ways: 31×42.6, or $31 \cdot 42.6$, or $(31)(42.6)$. We prefer the latter form, where the parentheses are used without any cross or dot between the numbers. The first parentheses are often omitted; thus $(31)(42.6)(3.1416)$ may be written as $31(42.6)(3.1416)$. The multiplication of algebraic quantities is indicated simply by putting the symbols next to each other, unless a pair of parentheses or brackets is needed to indicate one term which has several parts:

$$2.7x^2 \qquad 2x^2y^3 \qquad 31(41 + x)$$

Division is also indicated in several ways: $24 \div 6$; $\frac{24}{6}$; $24/6$. In order to facilitate canceling out common factors, we prefer the *horizontal line* between numerator and denominator. However, to save space, some quotients are written with the slant line, especially in the case of units, such as $lb/in.^2$, mi/h, etc.

A fraction may be simplified by canceling out equal factors from numerator and denominator. Thus,

$$\frac{24}{36} = \frac{6(4)}{9(4)} = \frac{6}{9}$$

(dividing numerator and denominator by 4); also $\frac{24}{36} = \frac{2}{3}$ (dividing through by 12). Often

we can apply the rule for division: *invert the divisor and multiply*. Thus

$$42 \div \tfrac{6}{11} = 42 \times \tfrac{11}{6} = 77$$

In combined multiplication and division, some factors can be divided out, and others can be reduced by canceling common factors:

$$\frac{47}{12} \times \frac{27}{63} \times 3 = \frac{\overset{4}{(48)}\overset{3}{(27)}(3)}{\underset{7}{(12)(63)}} = \frac{(4)(3)(3)}{7}$$

$$= \tfrac{36}{7} = 5.14$$

Rounding off. Actually, of course, $\tfrac{36}{7}$ is $5.142857\ldots$, but in this text we almost always "round off" the answers to three significant figures (see Sec. A-4). A 10 inch slide rule (Sec. F-2) gives answers to three significant figures over most of its range.

$$\frac{(10)(81)}{14} = \frac{405}{7} = 57.857142857\ldots = 57.9$$

Here we call $57.85714\ldots$ equal to 57.9, since it is closer to 57.9 than to 57.8. What to do about rounding off 42.450000000 to three figures is not a world-shaking problem; either way is acceptable, but usually the higher value (42.5) is preferred. To *two* figures, 42.450000000 becomes 42, since it is surely closer to 42 than to 43. The number π can be rounded off in various ways, all equally "correct":

$$3.141592654\ldots$$
$$3.14159265$$
$$3.1415927$$
$$3.141593$$
$$3.14159$$
$$3.1416$$
$$3.142$$
$$3.14$$
$$3.1$$
$$3.$$

Practice Problems

(Answers are on p. 716.)

Multiply:

1 3(42.1)
2 30(0.421)
3 $6(1 + \frac{1}{3})$

4 $(4 + 0.4)[111 + 0.1(110)]$
5 $3.1(20)^2$
6 0.0015(200)

Divide:

7 70/35
8 700/3.5
9 0.70/(3.0 + 0.5)

10 (32)(0.15)/8
11 720/0.072

Calculate:

12 $\frac{22}{7}$

13 $\frac{(14)(5)(4)}{(3)(7)(2)}$

14 $\pi(21)^2$

15 $\frac{(57.8573)(\frac{16}{3})}{2}$

16 $\frac{1}{(1 + \frac{1}{2} + \frac{1}{4})}$

17 $2\sqrt{9}$
18 $(\frac{1}{3})(\frac{6}{5})(\frac{1}{21})(84)$
19 $.02\sqrt{900}$
20 30% of 600
21 2.5% of 4.4
22 0.5% of the difference between 43.0 and 43.6

A-2 Powers and roots: exponents

When we write $y = x^n$, we call n the *exponent,* and say that we have raised x to the nth power. Thus $2^4 = 2 \times 2 \times 2 \times 2 = 16$; we use 2 as a factor 4 times to raise it to the 4th power. Special names are given to the exponents 2 and 3 (x^2 is "x squared," and x^3 is "x cubed"). The exponent 1 means only one factor, so $n^1 = n$, where n is any number. Another special case is the exponent 0; this is *defined* so that $n^0 = 1$, where n is any number. We give, by type examples, some rules of exponents.

Multiplication. The rule for multiplying numbers which are expressed as powers is summarized by the type formula

$$(x^a)(x^b) = x^{a+b}$$

For example,

$$(2^4)(2^3) = 2^7$$

or

$$(16)(8) = 128$$
$$128 = 128$$

Note that exponents are *added,* but only if they have the same base (2 in the above example). There is no way of simplifying an expression such as $(2^5)(3^2)$, where the exponents have different bases. It is also true that

$$x^n y^n = (xy)^n$$

which allows some simplification if the same exponent is involved. Thus

$$(2^3)(3^3) = (6)^3$$

or

$$(8)(27) = 216$$
$$216 = 216$$

The expression $(x^a)^b$ means (x^a) raised to the bth power. That is, $(x^a)(x^a)(x^a)(x^a) \ldots$ (b times), which equals $x^{a+a+a+a+\cdots}$, or x^{ba}. Therefore we can write

$$(x^a)^b = x^{ab}$$

For example,

$$(5^2)^3 = 5^6$$
$$(25)^3 = 15\,625$$
$$15\,625 = 15\,625$$

Division. The rule for dividing numbers which are expressed as powers is

$$\frac{x^a}{x^b} = x^{a-b}$$

The rule is valid only if the exponents have the same base. Thus

$$\frac{2^5}{2^3} = 2^2$$

$$\frac{32}{8} = 4$$

$$4 = 4$$

Also,

$$\frac{y^n}{x^n} = \left(\frac{y}{x}\right)^n$$

which is illustrated by

$$\frac{6^3}{2^3} = \left(\frac{6}{2}\right)^3$$

$$\frac{216}{8} = 3^3$$

$$27 = 27$$

Negative exponents are used to indicate division, thus allowing everything to go according to the multiplication rule of adding exponents. By definition,

$$x^{-n} = \frac{1}{x^n}$$

Thus $2^{-3} = \frac{1}{2^3} = \frac{1}{8}$, etc. Let us write one of the previous examples, using negative exponents and multiplication:

$$\frac{2^5}{2^3} = (2^5)(2^{-3}) = 2^{5+(-3)} = 2^2 = 4$$

A more complicated example is this one involving powers of 3:

$$\frac{(3^3)(3^2)}{3^4} = (3^3)(3^2)(3^{-4}) = 3^{3+2-4} = 3^1 = 3$$

As a check, straightforward division gives the same result:

$$\frac{(27)(9)}{81} = \frac{243}{81} = 3$$

Fractional exponents are used to indicate roots. By definition,

$$x^{1/n} = \sqrt[n]{x}$$

Thus

$$16^{1/2} = \sqrt{16} = 4$$

By combination of our previous rules for ex-

ponents, any fractional exponent can be interpreted.

$$16^{3/2} = (16^{1/2})^3 = (\sqrt{16})^3 = 4^3 = 64$$

$$27^{2/3} = (27^{1/3})^2 = (\sqrt[3]{27})^2 = 3^2 = 9$$

$$100^{-3/2} = \frac{1}{100^{3/2}} = \frac{1}{(100^{1/2})^3} = \frac{1}{(\sqrt{100})^3}$$

$$= \frac{1}{10^3} = \frac{1}{1000} = 0.001$$

Problems on Exponents

Simplify the following:

23 $(x^5)(x^2)(x^3)$
24 $(x^{20})(y^4)(x^3)$
25 $(a^3)(a^{-1/2})$
26 $\dfrac{2(2^6)}{2^3}$
27 10^{-2}
28 $(3t)^2$
29 $[(1+5)^2 - 32]^2$

30 $(16^{-3/2})(2^7)$
31 $p^{1.4}\sqrt{p}$
32 $46^1 - 23^0$
33 $\sqrt{\dfrac{x^n}{x^{n-2}}}$
34 $3(10^{10}) \div 2(10^7)$
35 $(x^{1/n})^n \div x$
36 $(x^n)^{1/n} - x$

A-3 Powers of ten: scientific notation

Although all that has been said about exponents applies, whatever the base may be, our decimal system makes *powers of ten* of major importance. We have

$10^0 = 1$

$10^1 = 10$

$10^2 = 100$

$10^3 = 1000$

$10^4 = 10\,000$

$10^5 = 100\,000$

$10^6 = 1$ million

$10^9 = 1$ billion•

$10^{12} = 1$ trillion•

$10^0 = 1$

$10^{-1} = 0.1$

$10^{-2} = 0.01$

$10^{-3} = 0.001$

$10^{-4} = 0.0001$

$10^{-5} = 0.00001$

$10^{-6} =$ one millionth

$10^{-9} =$ one billionth

$10^{-12} =$ one trillionth

Any number can be expressed as some power of 10 multiplied by another number

---

•In the U.S.; see p. 657.

between 1 and 10. This is often called *scientific notation*. Thus $4163.8 = 4.1638 \times 1000 = 4.1638 \times 10^3$. (Moving the decimal point 3 places to the left decreases the number by a factor of 10^3, so we multiply by 10^3 to keep the value the same.) If you have trouble in placing the decimal point, perhaps the following scheme will help. Suppose we have the number 0.0000367, which we wish to express in scientific notation.

$$0.0000367 = 3.67 \times 10^?$$

One way to proceed is to place the point of your pencil between the 3 and the 6 (thus pointing out the number 3.67, which is between 1 and 10). Think of this as 3.67×10^0 (say it to yourself). Now move the pencil to the left, one zero at a time, counting "minus one, minus two, minus three, . . ." as you do so. After saying "minus one," your pencil points out 0.367, which is 3.67×10^{-1}. After saying "minus two," your pencil points out 0.0367, which is 3.67×10^{-2}. Proceeding in this way, you stop at "minus five" when your pencil reaches the actual decimal point; your number is therefore 3.67×10^{-5}. The same method can be used for expressing large numbers. The number of seconds in a year is about $(365\frac{1}{4})(24)(60)(60)$; as discussed in Sec. 1-4, the exact figure is, by definition, 31 556 925.9747. Start with the pencil point between the 3 and the 1, so that the number is $3.15569259747 \times 10^?$. Moving over to the right and counting out loud, you reach the actual decimal point after saying "seven." Hence there are $3.15569259747 \times 10^7$ seconds in a year.

Multiplication and division. To multiply or divide numbers expressed in scientific notation, combine the powers of 10 by addition or subtraction of the exponents, and use ordinary multiplication or division for the "ordinary" numbers preceding the powers of 10.

$$(2.5 \times 10^8)(3 \times 10^6) = (2.5 \times 3) \times 10^{8+6}$$
$$= 7.5 \times 10^{14}$$

$$\frac{42 \times 10^{-16}}{7 \times 10^{10}} = \frac{42}{7} \times 10^{-16-10} = 6 \times 10^{-26}$$

$$\frac{(2 \times 10^6)(8 \times 10^{11})(3 \times 10^{-4})}{(4 \times 10^{-5})(10^3)}$$
$$= \frac{(2)(8)(3)}{(4)(1)} \times \frac{10^{13}}{10^{-2}} = 12 \times 10^{15}$$

Sometimes it pays to revise a numerator or denominator to make the division come out easier.

$$\frac{1.8 \times 10^{16}}{9 \times 10^{-4}} = \frac{1.8 \times 10^{16}}{0.9 \times 10^{-3}} = 2 \times 10^{19}$$

or

$$\frac{1.8 \times 10^{16}}{9 \times 10^{-4}} = \frac{18 \times 10^{15}}{9 \times 10^{-4}} = 2 \times 10^{19}$$

Considering the great differences among physical magnitudes, it is not surprising that powers of 10 are widely used. Objects range in size from a nuclear particle (about 10^{-14} m) up to whole galaxies (about 10^{21} m), with man more or less in the middle at about 2×10^0 m. Aside from convenience, scientific notation is often more informative than are ordinary numbers. When we say that a light-year is 9 460 000 000 000 000 m, we immediately see that this length unit is a tremendous number of meters, but just how tremendous we don't immediately comprehend. Writing it as 9.46×10^{15} m allows us to make immediate comparison with other large numbers.

EXAMPLE

Compare the light-year with the distance from earth to the sun (about 93 000 000 miles, or 150 000 000 kilometers).

We want the ratio of these two numbers. Expressing them both in scientific notation (remembering that 1 km equals 10^3 m), we have

$$\frac{9.46 \times 10^{15} \text{ m}}{1.50 \times 10^{11} \text{ m}} = 6.3 \times 10^4 = 63 \times 10^3$$

That is, a light-year is 63 thousand times as far as the distance from the earth to the sun. When it is expressed in this way, we can begin to "feel" the relationship, which remains obscure unless we use the scientific notation.

Problems on Scientific Notation

Express in scientific notation:

37	142.63	41	86 400
38	1 500 000	42	0.0000000000000144
39	0.00336	43	0.12500
40	4 663 310.56	44	0.00125

Express in ordinary notation:

45	1.63×10^7	49	3.11×10^{-4}
46	4.781354×10^3	50	6.55×10^{13}
47	1.01×10^6	51	8.001×10^{-4}
48	9.81×10^1	52	9.3×10^{-6}

Evaluate the following:

53 $(1.6 \times 10^2)(3 \times 10^6)2 \times 10^{11})$

54 $\dfrac{(1.6 \times 10^2)(3 \times 10^6)}{2 \times 10^{11}}$

55 $\dfrac{(3200)(200)}{(1.6 \times 10^3)(1000)}$

56 $\dfrac{(0.0022)(600)}{3 \times 10^{-10}}$

57 $\dfrac{16 \times 10^{-2}}{16 \times 10^2}$

58 $(6.02 \times 10^{23})(1.60 \times 10^{-19})$

59 $\dfrac{1.08 \times 10^{10}}{(2 \times 10^6)(3 \times 10^{-2})(9 \times 10^{20})}$

60 $\dfrac{(5.00 \times 10^{-5})(2400)}{(0.00008)(3.00 \times 10^{10})}$

Square roots and cube roots require careful consideration in scientific notation. To take a square root we must divide the exponent by 2; but since a fractional power of 10 is not useful to us, we must first see to it that the exponent is an *even number*.

$$\sqrt{9 \times 10^{16}} = \sqrt{9} \times \sqrt{10^{16}} = 3 \times 10^8$$
$$\sqrt{4.9 \times 10^5} = \sqrt{49 \times 10^4}$$
$$= \sqrt{49} \times \sqrt{10^4} = 7 \times 10^2$$
$$\sqrt{0.00000025} = \sqrt{25 \times 10^{-8}} = 5 \times 10^{-4}$$

When dealing with ordinary numbers, much the same problem arises; this is why we

divide the number into "periods" of two digits each, starting from the decimal point:

$$\sqrt{360\,000} = \sqrt{36\,00\,00} = 6\,0\,0$$
$$\sqrt{225\,000\,000} = \sqrt{2\,25\,00\,00\,00} = 1\,5\,0\,0\,0$$
$$0.0064 = \sqrt{0.00\,64} = 0.0\,8$$

Similarly, a cube root is taken only if the exponent of 10 is divisible by 3:

$$\sqrt[3]{6.4 \times 10^8} = \sqrt[3]{640 \times 10^6}$$
$$= 8.62 \times 10^2$$
$$\sqrt[3]{1.25 \times 10^{-13}} = \sqrt[3]{125 \times 10^{-15}}$$
$$= 5 \times 10^{-5}$$

Problems on Roots

Find the square roots of the following:

61	810 000	64	2.5×10^7
62	81 000	65	0.004
63	1.69×10^{-12}	66	0.0004

Find the cube roots of the following:

67	8000	70	10^8
68	0.008	71	6.4×10^{22}
69	1.25×10^{-19}	72	0.000000027

A-4   Significant figures

A number that represents the result of a physical measurement does more than merely indicate the *value* of the measured quantity. Some indication of the *precision* of the measurement is given by the form in which the number is written. For example, if a car's speed is reported by observer A to be 45.3 mi/h, he might consider this to mean that the speed is somewhere between 45.25 mi/h and 45.35 mi/h; for if it were 45.36 mi/h, for instance, he would report it as 45.4. Another observer B might assign a greater uncertainty to the last figure. He might consider that 45.3 indicates a number lying between 45.1 and 45.5, or within some other range of a few tenths of a mile per hour on either side of the stated value of 45.3 mi/h.

A *significant figure* is one which is of *some* significance but does not necessarily denote a certainty. In the example just mentioned, the 3 of 45.3 is "significant," but the exact degree of significance depends upon the experimental techniques, habits, and temperaments of the observers *A* and *B*. Thus we cannot make any simple statement about significant figures. About all we can do is to lay down the following rule:

The last figure written down should be the first uncertain figure.

When we read that the velocity of light is 2.99793×10^8 m/s, we know that the 3 is uncertain, but just how uncertain we cannot tell. On the other hand, the last 9 is certain. It would be wrong to write 2.99793×10^8, *knowing* that the last 9 is uncertain, for then the 3 would be entirely meaningless. This would be like trying to balance a checkbook to the nearest penny when you are uncertain about the dollars or dimes.

In the beginning physics laboratories, we often strive for "three-figure accuracy." We measure the density of iron as 7.86 g/cm³, and by writing it in this way we indicate that the 6 is uncertain by some (unspecified) amount. If the apparatus is not capable of three-figure accuracy, then the answer must be rounded off as necessary, to 7.9 g/cm³ or even to 8 g/cm³. Your laboratory manual (if you are using one) will cover this point more fully; at present we wish only to emphasize that the exact degree of significance of the final significant figure must remain somewhat vague in our elementary treatment.

EXAMPLE

The dilemma of significant figures is illustrated by a hypothetical experiment in which two observers measure the time required for a stream of water to fill a bucket. Observer *A* gets 98 s; observer *B* gets 102 s. Each measures to the nearest second or so, and so the last figure in each measurement is "significant." How-ever, *A*'s result has two figures, and *B*'s has three. A simple criterion of "two significant figures" or "three significant figures" would be misleading here. A trained physicist might report his result thus: 98.1 s $\pm$ 2.2 s, meaning that 98.1 is his best estimate of the value, and that there is a 50–50 chance that the correct value is within 2.2 s of 98.1 s. This violates our rule that the last figure written down is the first uncertain figure; but by stating the probable error of 2.2 s, the observer makes clear just *how certain* his value is, without any mention of significant figures at all.

Significant zeros. A special problem is posed by zeros, for zeros are used to indicate the location of the decimal point as well as to indicate a value. In the following example, only the zeros which are underlined are significant; the other zeros are used to fill out the space to the decimal point and are not significant figures.

EXAMPLE

4.2<u>0</u>1

4.20<u>100</u> (Here the final two zeros need not have been written; hence the fact that they *are* written means that they are significant. The uncertainty is in the 5th decimal place.)

0.0042<u>010</u>

42.0<u>1</u>

A number such as 86 400 g is ambiguous, for we can't tell whether the two zeros are just to locate the decimal point (in which case the error might be 100 or 200 g), or whether they are really significant (in which case the true value is within several grams of being *exactly* 86 400 g). There are three ways in which this ambiguity can be avoided. (*a*) Use scientific notation, and write 8.64×10^4 g or 8.6400×10^4 g to distinguish between the two possibilities. (*b*) Use a different unit for the mass, which will place the decimal point elsewhere. Write 86.4 kg or 86.400 kg to distinguish the two cases. (*c*) Use a probable error or a limit of error; thus 86 400 $\pm$ 8 g is a definite statement, provided we know the precise meaning of the author's $\pm$ sign. In this book, we shall prefer methods (a) and (b) as useful compromises which are sufficiently precise for most purposes.

Arithmetic operations and significant figures. Consider the circumference of a circle whose diameter is 12.3 ft. In order to determine this quantity, we need to multiply 12.3 by π, which is 3.1416. . . . If the 3 of 12.3 is uncertain, then all numbers following the 3 are unknown and can be denoted by x. Carrying out the multiplication,

$$
\begin{array}{r}
12.3x \\
3.1416 \\
\hline
738x \\
123x \\
492x \\
123x \\
369x \\
\hline
38.64168x
\end{array}
$$

we see that the product should be rounded off to 38.6, since the 4, 1, 6, and 8 are really *unknown* because an unknown "x" has been added to each of these digits. Even the 6 of 38.6 is uncertain, because it depends on a 9 arising from the uncertain 3 of 12.3. Writing the answer as 38.6 ft expresses this correctly; the last figure written down is the first uncertain one.

Similar results are obtained for division.

Addition and subtraction behave somewhat differently. For example, an orange weighing 3.1 oz is added to a basket containing 12.71 oz of lemons and 58.2 oz of grapefruit. The total weight is 74.0 oz of citrus fruit:

$$
\begin{array}{r}
12.71xx \\
58.2xxx \\
3.1xxx \\
\hline
74.0xxx
\end{array}
$$

We can generalize as follows: In multiplication or division, the result has the same number of significant figures as the least accurate number entering into the multiplication or division. In addition or subtraction, the last significant figure of the result is in the same column relative to the decimal point as is the last significant figure of the least accurate number entering into the sum or difference. These are approximate, rough-and-ready rules; an exact treatment would involve probable errors, and "significant figures" would not even be mentioned.

Problems on Significant Figures

In each case, express the answer to the proper number of significant figures.

73 What fraction of a minute is 45.0 s?
74 What fraction of a minute is 45.00 s?
75 How many days are contained in 2.1 years?
76 What is the area of a square, each edge of which is 3.041 ft in length?
77 If a car travels for 3.00 h at 42.36 mi/h, how far does it go?
78 If a car travels for 3.0 h at 42.3615 mi/h, how far does it go?
79 What is the height (in feet) of a boy who measures up at 66.00 in.? (*Note:* The conversion factor, 12 in. equals 1 ft, is as exact as you please, since this is a matter of definition.)
80 Do not work out, but state how many significant figures would be proper for the answers to each of the following:

$$
\frac{(144)(2.06)}{11.8943} \ ; \quad \frac{(3624.8)(28.1)}{(44)(100.23)} \ ;
$$

$$
(3.652 \times 10^8)(42.8 \times 10^{-6})
$$

81 A tank contains 42.8 gallons of water; how much will be left if 3.72 gallons are removed?
82 How many seconds are there in a day?
83 Add: $4.22 \times 10^5 + 3.11 \times 10^7 + 6.003 \times 10^6$.
84 The mass of an electron is 9.11×10^{-28} g. What is the total mass of 6.02×10^{23} electrons?
85 One student measures a length as 84.62 cm, and another measures the same length as 84.70 cm. By how much do these measurements differ? What is the per cent difference (i.e., the difference divided by the average value)?

Significant figures in textbooks. Ideally, problems in a textbook should be formulated with strict attention to probable errors and significant figures. However, we have adopted a rule which is easy to apply, even though not theoretically sound. In examples and problems in this text, all numbers are assumed to be *precise enough* to yield a final answer having three significant figures, and final significant zeros are often omitted both from the statements of problems and their answers. Purists may object, but we justify our course of action on the following grounds:

1. The rules about significant figures are themselves not rigid. Exceptions can easily be found, and some have already been illustrated. (Does 98 have the same number of significant figures as 102?) We see no reason to abandon one wrong treatment in favor of another wrong treatment. The three-figure rule we use, though known to be wrong, is clearly stated and is definite enough to apply consistently.

2. Examples and problems—hundreds of them—are scattered through the text to help you understand the principles of physics. Their purpose is *not* primarily to give a model for recording experimental data in the laboratory. Your laboratory manual will, rightly, be concerned with this aspect of quantitative science, but in the text we have adopted a working rule for the sake of giving a more immediately understood illustration of laws and principles.

3. Where it serves a purpose, as in tables of experimentally observed properties of matter, physical data are given with the number of significant figures indicating (roughly) the precision of measurement.

4. There is real value in simplifying the routine arithmetic associated with problem solving. This is why "simple" numbers are often used. On the other hand, the student desires to check his work against something more definite than an answer which has been rounded off to one or two significant figures. We give answers to three significant figures, and *assume* or *imply* that the given data warrant such precision.

As examples of our treatment of significant figures, consider two examples in the text. In Example 4-7 we find it necessary to compute $(6)(9.8)$, and we give the answer as 58.8. Here we assume the acceleration due to gravity is 9.80 m/s^2, even though we write it and speak of it as simply 9.8 m/s^2. In solving Problem 13-B15, we find that 0.87 gallon of turpentine overflows from a 50 gallon drum. In this case, we must state the volume of the drum to be 50.00 gallons, giving the necessary significant figures so that the subtraction $50.00 - 0.87$ has meaning.

Appendix B / Algebra

B-1 The rules of algebra

The methods of algebra are used in physics for much the same reason that arithmetic is used. In each case we are dealing with quantities, but if a quantity is unknown in magnitude we represent it by a letter or other symbol. For some reason, the symbol x has come to be the commonest one for an unknown, but you should realize that *any* symbol will do. Generally the symbol should be chosen to give some hint as to what it represents: F for force, m for mass, p for pressure, t for time, etc. Greek letters are often used for angles, especially θ (theta) and ϕ (phi). The symbols x, y, and z usually refer to distances or coordinates.

The "ground rules" of algebra can be summarized by a number of axioms, many of which are no doubt familiar to you. We illustrate these rules by solving some typical equations with their aid.

Rule 1: An equation remains true if equal quantities are added to or subtracted from each side.

$$x + 3 = 4$$

Hence

$x = 1$ (subtracting 3 from each side)

Another way of looking at this rule is to say that any separate term can be moved from one side of the equal sign to the other, if the sign of that term is reversed.

$x + 3 = 4$

$x = 4 - 3$ (moving the 3 to the other side, and changing its sign)

$x = 1$

Rule 2: An equation remains true if both sides are multiplied or divided by the same quantity.

$4F = 17.0$

$F = 4.25$ (dividing each side by 4)

Rule 3: An equation remains true if both sides are raised to the same power.

$v^2 = 441$

$v = \sqrt{441}$ (raising each side to the $\frac{1}{2}$ power, i.e., taking the square root of each side)

$v = 21$

Rule 4: In a proportion, such as $\frac{a}{b} = \frac{c}{d}$, we can cross-multiply to obtain the new equation $ad = bc$.

$$\frac{4}{30} = \frac{22}{p}$$

$4p = 30(22) = 660$

$p = 165$

To solve most equations, several of the basic procedures are used, one after the other. Thus the distance s traveled by a bullet while slowing down is to be found from the equation

$(300)^2 = (400)^2 + 2(-5000)s$

$90\,000 = 160\,000 - 10\,000s$

$-70\,000 = -10\,000s$ (Rule 1)

$s = \dfrac{-70\,000}{-10\,000} = 7$ (Rule 2)

Here is an equation which is to be solved for the elapsed time t, where it is known on physical grounds that t is greater than 2:

$4 = \frac{1}{2}(32)(t - 2)^2$

$8 = 32(t - 2)^2$ (Rule 2) (multiply each side by 2)

$0.25 = (t - 2)^2$ (Rule 2) (divide each side by 32)

$0.50 = t - 2$ (Rule 3) (take the positive square root of each side)

$t = 2.50$ (Rule 1) (add 2 to each side)

Problems

Solve for the unknown:

86   $3x + 42 = 0$

87   $1 - 0.4p = 2$

88   $(T + \frac{1}{2})4 = 5$

89   $\dfrac{50}{y} = \dfrac{y}{2}$

90   $\dfrac{1 + 2m}{2m - 1} = \dfrac{14}{10}$

91   $13^2 = 25 + 2(a)(45)$

92   $\frac{1}{2}(L^3 - 4) - 30 = 0$

93   $\sqrt{\dfrac{h}{100}} = 0.8$

Solve for x in terms of the other variables and the constants:

94   $3x + y - 14 = 2 + 3y$

95   $mg - 2x = F + 200$

96   $y = \frac{1}{2}x - y^2$

97   $\dfrac{x - 3}{x - 1} = \dfrac{2y}{y - 3}$

B-2   Simultaneous equations

If there are two unknowns, then two equations are required to solve a problem; in general, if there are n unknowns, n independent equations are needed for solution. To solve a pair of simultaneous equations, a routine method is to isolate one unknown by solving one equation for this unknown in terms of the other one. Then we substitute this expression into the second equation. For example, we wish to find x and y from the pair of equations

$$\begin{cases} 4x + y = 7 \\ x - 2y = 4 \end{cases}$$

We first work on the top equation to find y in terms of x:

$$4x + y = 7$$
$$y = 7 - 4x \quad \text{(Rule 1)}$$

We then substitute this expression for y into the second equation:

$$x - 2y = 4$$
$$x - 2(7 - 4x) = 4$$
$$x - 14 + 8x = 4$$
$$9x = 18 \quad \text{(Rule 1)}$$
$$x = 2 \quad \text{(Rule 2)}$$

Now that we know x, we can use either equation to find y:

$$4x + y = 7$$
$$4(2) + y = 7$$
$$8 + y = 7$$
$$y = -1 \quad \text{(Rule 1)}$$

As a final check, we substitute these values of x and y into each equation and verify that the original equations are satisfied:

$$4(2) + (-1) \overset{?}{=} 7; \quad 7 = 7 \quad \text{(O.K.)}$$

and

$$2 - 2(-1) \overset{?}{=} 4; \quad 4 = 4 \quad \text{(O.K.)}$$

This is not the only way to solve a pair of simultaneous equations; refer to any algebra text for further details of several other methods, including one involving determinants. However, you will not go wrong by following the steps above.

B-3 Quadratic equations

Almost all the equations so far discussed are called "linear," since the unknowns are raised only to the first power (such equations have straight *lines* as their graphs). From time to time we will have occasion to solve a more complex equation known as a *quadratic equation* because the unknown appears raised to the second power as well as to the first power. As an example, the time t required for a racing car to move 225 ft might be given by the following quadratic equation:

$$225 = 20t + 5t^2$$

A routine method of solution is to use the *quadratic formula:*

If

$$ax^2 + bx + c = 0,$$

then

$$x = \frac{-b \pm \sqrt{b^2 - 4ac}}{2a}$$

The $\pm$ sign (plus-or-minus) means that there are *two* solutions. Using the $+$ sign gives

$$x = \frac{-b + \sqrt{b^2 - 4ac}}{2a}$$

and using the $-$ sign gives

$$x = \frac{-b - \sqrt{b^2 - 4ac}}{2a}$$

Usually only one of the two solutions fits the physical problem, although both solutions are, of course, algebraically correct. Naturally, b^2 must be greater than $4ac$ so that $\sqrt{b^2 - 4ac}$ gives a real number.[*] If b^2 is less than $4ac$ there is no real solution.

Applying the quadratic formula to our example, we first rearrange the terms in the equation to resemble the type equation, with the term in t^2 first, the term in t second, and the constant term last.

$$5t^2 + 20t - 225 = 0$$

$$t = \frac{-20 \pm \sqrt{(20)^2 - 4(5)(-225)}}{2(5)}$$

---

[*] If $b^2 = 4ac$, the square root is 0, and the two solutions coincide.

$$t = \frac{-20 \pm \sqrt{400 + 4500}}{10}$$

$$t = \frac{-20 \pm \sqrt{4900}}{10} = \frac{-20 \pm 70}{10}$$

The two answers are

$$t_1 = \frac{-20 + 70}{10} = \frac{50}{10} = 5 \text{ s}$$

and

$$t_2 = \frac{-20 - 70}{10} = \frac{-90}{10} = -9 \text{ s}$$

If this were a problem in pure algebra, we would stop here, with two equally correct answers, 5 s and -9 s. However, from the physics of the situation we know that the desired time is positive, so we reject[*] the solution -9 s, keeping only the solution $t = 5$ s.

These brief remarks and illustrations give a sufficient guide for a review of algebra as it is used in this text. For further information, consult any high school algebra text. Our review has been of techniques only,

___

[*] A physical interpretation of a negative time is possible; see Prob. 3-C1 of the text.

with no attempt to set up "word problems." The business of transferring the data of physical problems into mathematical form is a necessary first step, and a most important step; algebra cannot be used successfully if the equations are wrong to begin with. However, discussion of how to set up the equations belongs in the physics text itself, rather than in this Appendix.

Problems

98   Solve for u and v: $3u + v = 7$; $2u - v = 3$.
99   Solve for p and q: $q = p + 2$; $p + q = 8$.
100 Solve for A and B: $A - 4B = -6$; $2B + A = 18$.
101 Solve for x and y: $10x + 2y = 11$; $20(x - y) = -8$.

Find the real solutions, if any, of the following quadratic equations:

102 $2t^2 - 3t + 1 = 0$
103 $v^2 + 5v + 6 = 0$
104 $x^2 - 10 = 3x$
105 $y^2 + 1 = -y$
106 $3 + z^2 + 2z = 2$
107 $4\theta^4 - 5\theta^2 + 1 = 0$ (*Hint:* Let θ^2 be the variable; call it x if you wish. Solve for θ^2 by the quadratic formula, and then find θ.)

Appendix C / Plane Geometry and Trigonometry

C-1   Plane geometry

As a brief review of plane geometry, we give without proof several useful theorems.

1. The sum of the angles of any triangle is 180°.

2. Two triangles are similar if two of their angles are equal.

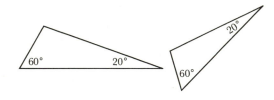

3. The corresponding sides of similar triangles are proportional.

4. The two acute angles in a right triangle add up to 90°. (Such angles are called complementary angles.)

5. Two angles are equal if any of the following is true:

(*a*) Their sides are parallel and in the same sense.

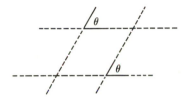

(*b*) Their sides are mutually perpendicular.

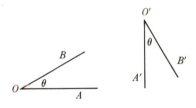

(*c*) They have the same complement.

θ and θ' are equal, since they have the same complement ϕ

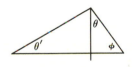

(*d*) They are vertical angles.

6. Some special theorems apply to right triangles. Most important is the Pythagorean theorem: the sum of the squares of the two legs equals the square of the hypotenuse. Some right triangles of importance are the

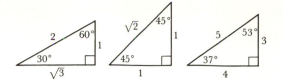

30-60-90 triangle, in which the short side is half the hypotenuse, and the 45-45-90 triangle, whose sides are in the ratio $1:1:\sqrt{2}$. The 3-4-5 triangle has acute angles which are approximately 37° and 53°.

7. For any two intersecting chords of a circle, there is a relationship between the

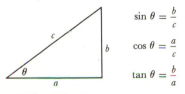

lengths of the segments: $ab = cd$. As a special case,

$$x(2R - x) = h^2$$

when one chord is perpendicular to a diameter.[*] The sagittal theorem, useful in optics, is an approximation which is valid when x is small compared with $2R$. Then

$$x \approx \frac{h^2}{2R}$$

C-2 Trigonometry

The elementary functions of sine, cosine, and tangent are defined (for acute angles in a right triangle) in Sec. 3-10.

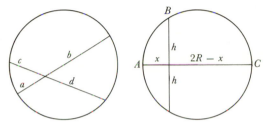

$$\sin \theta = \frac{b}{c}$$

$$\cos \theta = \frac{a}{c}$$

$$\tan \theta = \frac{b}{a}$$

---

[*] An algebraic proof uses construction lines drawn from A to B and from B to C. The Pythagorean theorem then gives, for $[AB]^2 + [BC]^2$,

$$[x^2 + h^2] + [h^2 + (2R - x)^2] = [2R]^2,$$

which reduces to $x(2R - x) = h^2$.

The general definitions for angles of any size are made in terms of the abscissa, ordinate, and radius:

$$\sin \theta = \frac{\text{ordinate}}{\text{radius}}$$

$$\cos \theta = \frac{\text{abscissa}}{\text{radius}}$$

$$\tan \theta = \frac{\text{ordinate}}{\text{abscissa}}$$

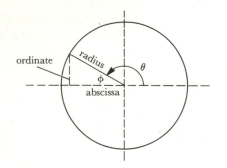

The radius is always $+$, but the ordinate and abscissa can be either $+$ or $-$. In the illustration, $\sin \theta$ is $+$ and $\cos \theta$ is $-$. No elaborate formulas are needed to remember sign rules, because each case can be calculated when required. If $\theta = 150°$ as in the figure, the acute angle ϕ is $180° - 150° = 30°$. Thus $\cos 150°$ is numerically equal to $\cos 30°$, except for a $-$ sign, because the abscissa is negative. Hence $\cos 150° = -\cos 30° = -0.866$. (Table on p. 720.)

For completeness, we include two formulas which are sometimes useful, although not required for any of the physics problems in this book. Their application is usually to triangles that do not have a right angle.

Law of cosines. In the shaded triangle,

$$c^2 = a^2 + b^2 - 2ab \cos \phi$$

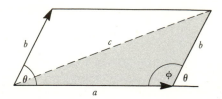

Here ϕ is an angle interior to the triangle, at the vertex where sides a and b join. If $\phi = 90°$, the formula reduces to the Pythagorean theorem, because $\cos 90° = 0$. An alternative statement of the law of cosines, well suited to problems involving vector addition, makes use of the angle θ, which is the angle between the directions of two vectors. Since $\cos \theta = -\cos \phi$ (they are supplementary angles), the law of cosines becomes

$$c^2 = a^2 + b^2 + 2ab \cos \theta$$

Law of sines. The relationship between the sides of a triangle and the sines of the opposite angles is expressed as a proportion:

$$\frac{a}{\sin A} = \frac{b}{\sin B} = \frac{c}{\sin C}$$

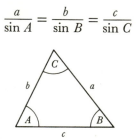

This proportion can be solved easily with the aid of a slide rule.

Appendix D / Functions and Graphs

D-1 Functions

The concept of "function" perhaps epitomizes the spirit of scientific investigation, for a function is nothing more or less than a *relationship* or a *correlation* between observables. In physics we seek to correlate cause and effect. Often we try to relate our observations of complex phenomena to our observations of factors which are under our own control. For example, it is found that more engine power is required to move a car at high speed than at low speed. Power is a function of speed, and, if the test engineer makes a series of observations of the required power at a number of different speeds, he can describe the function by means of a table of values or by a graph. A function can also be theoretical, in contrast to the experimental function just described. For instance, a relationship exists between the reading of a photographer's exposure meter and the necessary diameter of lens opening for a given type of film. Here the nature of the function is known from other theories (themselves based eventually on experiment); the function serves a useful purpose and allows the photographer to take advantage of previously acquired knowledge.

The information which describes a function may be contained in a table of values:

LIGHT-METER READING, CURRENT IN AMPERES	PROPER LENS OPENING, DIAMETER IN MILLIMETERS
0.5×10^{-6}	20
1	14
2	10.0
4	7.1
6	5.8
8	5.0
10	4.5
20	3.2

Although the photocell currents are all in the range of 0.5 to 20×10^{-6} A, the 10^{-6} is written only after the first entry in the table. It is understood that the other entries are in the same units. To write the first entry as 5×10^{-7} A, and the last two entries as 1.0×10^{-5} and 2.0×10^{-5} A would also be correct. However, in tables of values it is customary to write all values in any one column with the same power of ten, even though some of the values are not in scientific notation. This is to help present the whole trend more clearly.

In an experiment, the *independent variable* is the one which is under our control. Thus in the example of the engineer testing the effect of air resistance, he set the speed of the car at some one of a number of predetermined values and observed the horsepower necessary to maintain this speed. In this case, the speed was the independent variable. On the other hand, if the engineer had chosen to adjust the motor to one of a number of predetermined horsepowers and observed the resulting speed, then the power would have been the independent variable and the speed would have been the *dependent variable*. It is a matter of emphasis, based on experimental procedure. If a function is expressed as a table listing corresponding values, perhaps obtained by a theoretical calculation, then the independent variable is the one which is most likely to be known. Our light-meter table represents a function of this sort, and so would a table which might be used to convert inches into centimeters, or one which would give a parcel's postage as a function of its weight. In the following tables of functions, the independent variable is in the first column, and the dependent variable is in the second column.

TEMPERATURE, °C	DENSITY OF WATER, G/CM^3
0	1.000
20	0.998
40	0.992
60	0.983
80	0.972
100	0.958

TIME OF DAY	AIR TEMPERATURE, °F
8:00	38
10:00	40
12:00	47
13:00	51
15:00	53
18:00	50

YEAR	CONSUMER PRICE INDEX (URBAN)
1920	60
1925	53
1930	50
1935	41
1940	42
1945	54
1950	72
1955	80
1960	89
1965	95
1970	116

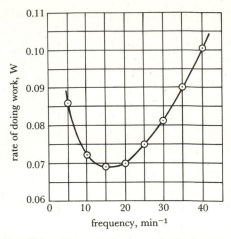

ELAPSED TIME, MIN	DEPTH OF RAIN WATER IN A GAUGE, IN.
0	0.00
10	0.20
20	0.27
30	0.31
40	0.31
50	0.31

Figure D-1 Rate of doing work during forced breathing at various respiratory frequencies. Dead space 200 cm^3; alveolar ventilation constant 6 liters/min. [After Otis, Ferm, and Rahm, *Journal of Applied Physiology,* **2,** 592 (1950).]

APPLIED FORCE, LB	STRETCH OF A SPRING, FT
0	0.00
6	0.17
12	0.33
18	0.50
24	0.61
30	0.82

FREQUENCY, BREATHS PER MIN	RATE OF DOING WORK, WATTS
5	0.086
10	0.072
15	0.069
20	0.070
25	0.075
30	0.081
35	0.090
40	0.101

D-2 Graphs

Graphs are often used to display the behavior of functions. A good graph must be plotted carefully, but it is even more important to choose scales judiciously and to label axes clearly. Note the features of the accompanying graph, which represents the rate at which the lung muscles do mechanical work during breathing. The subject breathes a fixed amount of air per minute (called the ventilation rate); he can do this by taking a few deep breaths per minute or many shallow breaths per minute.

1. A table of values is written down near the graph.

2. The independent variable (rate of breathing) is plotted on the horizontal axis; the dependent variable (rate of doing work) is plotted on the vertical axis.

3. Axes are well labeled, including units of measure.

4. On the frequency axis, 2 squares = 10 breaths per minute; on the power axis, 2 squares = 0.01 W. (Avoid multiples such as 3 squares = 0.01 W or 1 square = 7 breaths per minute. These ratios make it hard to plot points and also hard to read values off the graph.)

5. Consistent with item 4, the graph is plotted to fill up as much of a full sheet of graph paper as possible.

6. Plotted points are located by dots surrounded by small circles. A smooth curve is drawn through the points.

7. There is adequate description of the circumstances of the experiment, or the assumed circumstances for which the graph is applicable. This information is contained in a short legend below the graph. References to source of data are included.

8. Since these data involve only rates lying between about 0.06 and 0.11 W, the zero of the vertical axis is "suppressed." It would make the graph harder to read if we included a full range of 0 to 0.11 W, using only the upper part of the range.

The purpose of a graph is to give us information about a relationship. For instance, we see from this graph that the least work is required when about 15 breaths are taken per minute; this is the optimum frequency of breathing for a swimmer or runner who wishes to ventilate at the rate of 6 liters/min, if his dead space (fixed volume of throat and mouth cavity) is 200 cm³. It is important to note that the graph does not give unlimited information. According to the legend beneath the graph, the subject is ventilating at a certain rate. The graph (or table) gives no information at all for a ventilation rate of 8 liters/min or 10 liters/min, or for a subject whose dead space is not 200 cm³. Additional experimentation or calculation would be required to answer questions pertaining to these altered conditions.

Further discussion of graphs will be found in Sec. 3-2 of the text, where slopes are discussed, and in your laboratory manual.

Problems

Refer to the tables on pp. 705–706.

108 Plot the graph of the density of water from the table. Is it wise to suppress the zero of either axis? What is the density (to 3 significant figures) at 30°C?

109 Plot the air temperature from the table. At what time of day was the temperature highest? When was the temperature changing most rapidly?

110 Plot the graph of cost of living. Estimate from the graph what the price index was in 1928; in 1948. What do you expect it to be in 1980? What can you say about 2000?

111 Plot the graph of the depth of water in the rain gauge. How long did the rain last?

112 Plot the stretch of the spring. Should any measurement be repeated?

Often a functional relationship can be expressed in mathematical form. Thus the information contained in the table of values for the light-meter can be written as $D = \sqrt{200/I}$, where I is the reading of the light-meter in microamperes and D is the diameter of the lens opening in millimeters. None of the given values is very different from that which would be calculated from this equation. (Try it.)

A functional relationship is made more explicit by writing the independent variable in parentheses, such as $f(x)$, which is read "f-of-x." The equation

$$f(x) = (x + 1)^2$$

tells us to "add 1 to the independent variable and square the result"; the symbols used for the variables are unimportant, and $F(z) = (z + 1)^2$ conveys the same message. If $x = 3$, then $f(x) = (3 + 1)^2 = 16$; this is indicated by writing $f(3) = 16$. As another example, using the same function, we find that $f(y^3) = (y^3 + 1)^2 = y^6 + 2y^3 + 1$.

Expressing a function in mathematical form is convenient, but it is important to realize that mathematical expressions are not *necessary* for the existence of a function. Indeed, in many cases a table of values will serve as the only definition of a function. *Any* rule for finding the value of one variable, when the value of another one is known, may serve to define a function. Consider the mathematical function represented by the following table of values:

x	1	2	3	4	5	6	7
y	1	6	5	11	21	20	?

At first glance, no relationship is apparent. What is the value of y when $x = 7$? Inspection of the table will probably be of little help, yet a perfectly definite procedure exists by which each y is found.* Hence y is a function of x.

---

*In this case our rule is: Form an integer out of the numbers $1, 2, 3, 4, \ldots x$; divide by x; add up the digits of the result (ignoring fractions). Thus, for $x = 4$, $1234/4 = 308\frac{1}{2}$; $3 + 0 + 8 = 11$. This function is defined only for integral values of $x > 0$.

Problems

113   Which of the functions described in the tables of pp. 705–706 do you think might be represented by relatively uncomplicated mathematical expressions?

114   From Fig. D-1, estimate to two significant figures the rate of doing work if the subject breathes once every 5 s.

115   If $f(y) = y^2 - 1$, find $f(4)$; $f(-4)$; $f(2x)$; $f(\sqrt{t})$; $f(\sqrt{t}) - f(-\sqrt{t})$.

116   If $s(t) = 1 + 3t$, find $s(-1)$; $[s(t)]^2$; $s(t^2)$.

117   Calculate the next two values of y, corresponding to $x = 7$ and $x = 8$, for the function described in the footnote at the left.

Appendix E / Calculus

No formal calculus is required in the main body of this book. However, some formulas of calculus will be of use in certain of the "For Further Study" sections.

The derivative of a function $f(x)$ is called $f'(x)$ or df/dx.

$$f'(x) = \frac{df}{dx} = \lim_{\to 0} \left(\frac{\Delta f}{\Delta x} \right)$$

At any point, the value of the derivative is the slope of the tangent to the graph of $f(x)$ (see the discussion in Sec. 3-4). Derivatives are found by taking a limit, and the results for some simple functions are given in the following table, where a represents any constant:

$f(x)$	$f'(x)$
a	0
x^n	nx^{n-1}
$\sin ax$	$a \cos ax$
$\cos ax$	$-a \sin ax$
$\tan ax$	$a \sec^2 ax$
e^{ax}	ae^{ax}
$\log_e x$	x^{-1}
$\log_{10} x$	$0.434\ x^{-1}$

Some of the rules of formal calculus are:

(a) The derivative of a sum of terms is the sum of the derivatives; the derivative of $a[f(x)]$ is a times the derivative of $f(x)$.

(b) The derivative of a product of two functions $u(x)$ and $v(x)$ is given by

$$\frac{d}{dx}(uv) = u\frac{dv}{dx} + v\frac{du}{dx}$$

For example, if $y = 3x^2 \sin x$, let $u = 3x^2$ and $v = \sin x$. Then

$$\frac{du}{dx} = 6x \quad \text{and} \quad \frac{dv}{dx} = \cos x$$

Then, by the rule,

$$\frac{dy}{dx} = u\frac{dv}{dx} + v\frac{du}{dx} = 3x^2 \cos x + 6x \sin x$$

To differentiate a quotient of two functions, first express the quotient as a product, using a negative exponent; then use rule (b).

(c) The chain rule: If $z = F(y)$, and $y = f(x)$, we consider z to be a function of a function. The chain rule states that

$$\frac{dz}{dx} = \left(\frac{dz}{dy}\right)\left(\frac{dy}{dx}\right)$$

The method can be extended to any number of intermediate functions. For example, if $z = (x + \sin x)^3$, first think of $(x + \sin x)$

as an entity (call it y if you wish). Differentiate the "cube" function first (getting $dz/dy = 3y^2$); then multiply by the derivative (dy/dx) of the expression which is inside the parentheses.

$$\frac{dz}{dx} = 3(x + \sin x)^2(1 + \cos x)$$

Problems on Differentiation

Find the derivatives of the following functions:

118 $7x^4 - 3x + 4$
119 $\sin 3x + e^{5x}$
120 $1/x^3$
121 $4x^2 e^{-2x}$
122 $(9 - x^2)^{1/2}$
123 $\sin^2 x + \cos^2 x$

The definite integral is defined in Sec. 3-5 as the limit of a sum:

$$\lim_{\Delta x \to 0} \sum y\, \Delta x = \int_{x_1}^{x_2} y\, dx$$

The integral is the area under that portion of the graph of $y(x)$ which lies between $x = x_1$ and $x = x_2$. The *fundamental theorem of calculus* uses the antiderivative function $F(x)$, which is defined in such a way that $f(x)$ is the derivative of $F(x)$. The theorem, which is discussed in Sec. 3-12, states that the definite integral between the limits x_1 and x_2 is the difference between the values of the antiderivative at the two points.

$$\int_{x_1}^{x_2} f(x)\, dx = F(x_2) - F(x_1)$$

which is also written as

$$\int_{x_1}^{x_2} f(x)\, dx = \left[F(x)\right]_{x_1}^{x_2}$$

The brief table of antiderivatives given in the next column can, therefore, be used to evaluate definite integrals, as illustrated in the examples.

$f(x)$	$F(x)$
A	Ax
Ax^n	$\dfrac{Ax^{n+1}}{n+1}$ $(n \neq -1)$
Ax^{-1}	$A \log_e x$
$\sin ax$	$-\dfrac{1}{a}\cos ax$
$\cos ax$	$\dfrac{1}{a}\sin ax$
$\tan ax$	$-\dfrac{1}{a}\log_e \cos ax$
$\log_e x$	$x \log_e x - x$
e^{ax}	$\dfrac{1}{a}e^{ax}$

The antiderivative is also called the indefinite integral of $f(x)$.

EXAMPLES

(a) $\displaystyle \int_1^3 (x^2 + 2)\, dx = \left[\frac{x^3}{3} + 2x\right]_1^3$

$$= \left[\frac{(3)^3}{3} + 2(3)\right] - \left[\frac{(1)^3}{3} + 2(1)\right]$$

$$= 12\tfrac{2}{3}$$

(b) $\displaystyle \int_0^{60°} \sin 2\theta\, d\theta = \left[-\tfrac{1}{2}\cos 2\theta\right]_0^{60°}$

$$= [-\tfrac{1}{2}\cos 120°] - [-\tfrac{1}{2}\cos 0°]$$

$$= (-\tfrac{1}{2})(-\tfrac{1}{2}) - (-\tfrac{1}{2})(1) = \tfrac{3}{4}$$

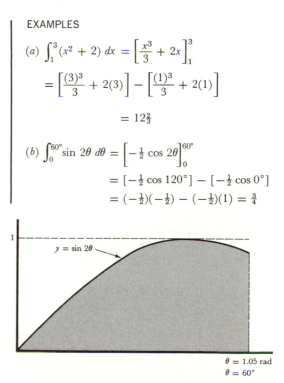

$y = \sin 2\theta$

$\theta = 1.05$ rad
$\theta = 60°$

In the example (b), the integral (area under the curve) is 0.75, as given by the formula, if θ is in radians; but the same area is 43.0 units if θ is in degrees. It is immaterial whether radians or degrees are used when substituting into the indefinite integral (cos $120°$ is $-\frac{1}{2}$ whether the angle is expressed as $120°$ or as $2\pi/3$ radians). The integral can be used to find the average height of the curve over the interval 0 to 1.05 rad:

$$y = \frac{\text{area}}{\text{base}} = \frac{0.75}{1.05} = 0.717$$

Problems on Integration

Compute the values of the following definite integrals:

124 $\int_0^3 (1 - x + x^2)\, dx$

125 $\int_0^{30°} \sin 3\phi\, d\phi$

126 $\int_0^{\infty} e^{-bt}\, dt$

127 $\int_2^5 \frac{dz}{z^3}$

128 $\int_0^V 10 V^{3/2}\, dV$

Appendix F / Approximate Calculations and the Slide Rule

F-1 Approximate calculations

If mathematics is to be our servant rather than our master, we must be able to make a quick estimate when the tedious work of an exact solution is not needed. Often a rough idea of the result of a calculation can give important insight into a physical problem. Quick approximate calculation also helps guard against gross errors such as a misplaced decimal point.

What we have in mind is a calculation to "one-figure accuracy" or even worse. We round everything off to one figure, combine the digits first, and then place the decimal point by combining whatever powers of ten are left over.

EXAMPLE

How many seconds are there in a year?

$(365.24)(24)(60)(60)$

$= (4 \times 10^2)(2 \times 10)(6 \times 10)(6 \times 10)$

$= (4)(2)(6)(6) \times 10^5$

$= (8)(36) \times 10^5$

$= (8)(40) \times 10^5$

$= 300 \times 10^5$

$= 3 \times 10^7$

The exact answer, to four significant figures, is 3.156×10^7 s in a year. Don't be alarmed at set-

ting $(8)(40)$ equal to 300. To be sure, 320 is an exact value for $(8)(40)$, but we are interested in a quick and easy estimate, not an exact value.

EXAMPLE

How many years are there in one 50 min class period?

$$\frac{50 \text{ min}}{(365.24)(24)(60) \text{ min}} = \frac{\cancel{50}}{(4 \times 10^2)(20)(\cancel{60})}$$

$$= \frac{1}{80 \times 10^2}$$

$$= \frac{100 \times 10^{-2}}{80 \times 10^2}$$

$$= 1.2 \times 10^{-4}$$

A more exact value is 0.951×10^{-4} year. Several points are to be noted in the last example. (a) We canceled 50 and 60 just as if they were equal. This is not right, but not far wrong either. Any short-cut or approximation of this sort is permissible in a rough calculation.   (b) To divide by 80 with a minimum of pain, we changed the 1 in the numerator to 100×10^{-2} (really the same thing). Then $100/80$ is easily seen to be just a bit more than 1—call it 1.2.   (c) Our answer, 1.2×10^{-4}, is not very close to the correct value of 0.951×10^{-4}; however, if it were a question of deciding between 0.951×10^{-4} and 0.951×10^{-3} we could easily see from our approximate answer that the former is correct. Any approximate answer between, say, $0.7 \times$

10^{-4} and 2×10^{-4} would allow us to fix the decimal point, if an exact calculation gave an answer of the form $0.951 \times 10^{-?}$.

EXAMPLE

What is the volume of a droplet of water in a fog, if the radius of the droplet is 2.73×10^{-3} cm?

Using the formula for volume of a sphere,

$$V = \tfrac{4}{3}\pi r^3 = \tfrac{4}{3}\pi(2.73 \times 10^{-3})^3$$
$$= \tfrac{4}{3}\pi(3 \times 10^{-3})^3 = 4(27 \times 10^{-9})$$
$$= 100 \times 10^{-9} = 1 \times 10^{-7} \text{ cm}^3$$

A more exact value for the volume is 0.85×10^{-7} cm^3.

Problems

Estimate the approximate value of the following:

129 $\dfrac{(1660)(2\pi)}{41.7}$

130 $(468)(5280)(2.0813 \times 10^6)$

131 $\dfrac{(0.0332)(1.01 \times 10^4)^2}{\pi(18)^2}$

132 $\dfrac{(1.6 \times 10^{-19})(6.02 \times 10^{23})}{\sqrt{84.5}}$

133 $\dfrac{(1.15 \times 10^{50})^{1/5}(3\pi \times 10^{-10})}{220}$

134 $\sqrt{x^2 - 1}$, where $x = 20$

F-2   The slide rule

The most effective way to study physics is to spend your time concentrating on physics, and to do this you should learn how to delegate most of the computational work to the slide rule. The slide rule is of no use for addition or subtraction, but for multiplication and division, squares and square roots, and simple trigonometry it is a quick way of getting results to three (or sometimes four) significant figures.• This is not the place to

---

•All that we say here applies to the usual 10 in. rule of the Mannheim type, such as the Post No. 1447, or the Keuffel and Esser No. N4058W. You need not pay more than about $4 to $5 for a rule that is entirely adequate for this work.

discuss the theory of the slide rule, which is based upon logarithms. Nor is it the place for a complete course in slide rule; the instruction book which comes with your slide rule, coupled with some supervised drill in the classroom or laboratory, should provide you with knowledge of how to use your rule effectively. Here we will illustrate several basic settings for the common operations and give a large number of exercises which will help you gain the necessary speed and ease.

Reading the rule. The scales are not uniform, and so you have to be careful about interpreting the value of each division. The safe thing to do is to identify two numbered divisions, then identify the second significant figure (not numbered on the rule except, on some rules, between 1.0 and 2.0), and finally read off the third significant figure using interpolation (guesswork) if necessary.

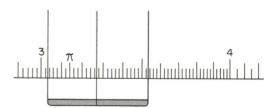

Study the setting shown on the slide rule.

Step 1: lies between 3.00 and 4.00.

Step 2: lies between 3.20 and 3.30.

Step 3: there are 5 divisions between 3.20 and 3.30, and hence each division is .02.

Step 4: the indicator lies between 3.26 and 3.28; call it 3.27.

The rule gives only the significant figures—you must place the decimal point by some other means. The setting illustrated above might equally well represent 3.27, 3270, 0.000000327, 3.27×10^{20}, 0.327, etc.

Multiplication. Set movable index opposite one factor; product is found on fixed scale opposite the other factor on the movable scale. Use whichever index is necessary, so that the multiplier doesn't run over the end of the fixed scale.

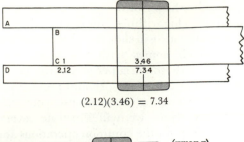

$$(2.12)(3.46) = 7.34$$

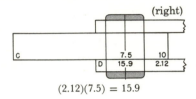

(wrong)

$$(2.12)(7.5) = 15.9$$

(right)

Multiplication can also be performed using the A and B scales, following the same rule as described above for the C and D scales.

$$(2.12)(3.46) = 7.34$$

Other things being equal, multiplication is usually performed on the C and D scales, since these scales are more "open" and can be read more closely.

Division. Set divisor on movable scale opposite

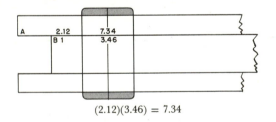

$$8.01/3.01 = 2.66$$

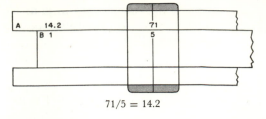

$$71/5 = 14.2$$

dividend on fixed scale; quotient is found on fixed scale opposite index of movable scale. This rule for division applies to the A and B scales as well as to C and D scales.

*Squares and square roots.** Whatever number appears on the A scale is the square of the corresponding number which is opposite on the D scale. Note that the A and B scales are divided into two segments, each taking up half the length of the rule. The left-hand

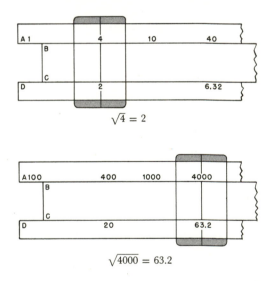

$$\sqrt{4} = 2$$

$$\sqrt{4000} = 63.2$$

segment of the A scale represents numbers such as 0.04, 4, 400, 40 000, etc., while the right-hand segment represents numbers like 0.40, 40, 4000, 400 000, etc. When taking a square root, it is absolutely essential to use the correct segment of the A scale. On some rules, the numbers on the right-hand seg-

---

* If your rule does not have A and B scales, consult the instruction book or your teacher to find out how to take squares and square roots.

ments of the A and B scales are marked 10, 20, 30, . . . to distinguish them from those on the left-hand segment, which are marked 1, 2, 3,. . . . Actually, "40" on the right-hand segment could stand for 40, or any product of 40 by an *even* power of 10. When taking a square root, the given number is first pointed off into periods of two digits each. If we wish to find $\sqrt{4\,000\,000\,000}$, we first divide it into periods, thus: $\sqrt{40\,00\,00\,00\,00}$, and since the first period is "40," we use the right-hand segment of the A scale, which is conveniently marked "40." Thus

$$\sqrt{40\,00\,00\,00\,00} = 6\,3\,2\,0\,0$$

or (using scientific notation)

$$\sqrt{40 \times 10^8} = 6.32 \times 10^4$$

Combination operations. Here are some tricks of the trade, not essential but sometimes very helpful.

$$(a) \qquad \frac{(64)(3.52)}{4.10} = 54.9$$

Do the division first. In other words, first set up $64 \div 4.10$; the quotient is 15.6, but this

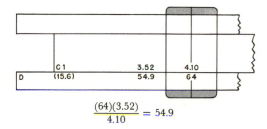

$$\frac{(64)(3.52)}{4.10} = 54.9$$

need not be read. The index of the movable scale is already now in the right place for the multiplication by 3.52, so that the final answer, 54.9, is obtained with just one setting of the rule for the entire problem.

(b) *Area of a circle.* If the radius is 24.3 ft, we need to find $\pi(24.3)^2$. First set up $(24.3)^2$, placing the movable index of the C scale opposite 24.3 on the D scale. The square of

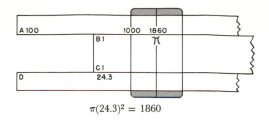

$$\pi(24.3)^2 = 1860$$

24.3 now appears on the A scale, but it need not be actually read. The movable index of the B scale is now just in the right place for multiplication by π. The final answer, 1860 ft², is obtained with just one setting of the rule for the entire problem.

$$(c) \qquad \sqrt{(64)(26)} = 40.8$$

Set up the product on the A and B scales, paying attention to the proper segment (in this case, both are on the right-hand segments of A and B, respectively). The product is 1664, but this need not be read; since the product is on the A scale, the square root is found to be 40.8 on the D scale. Only one

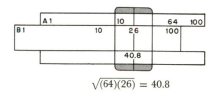

$$\sqrt{(64)(26)} = 40.8$$

setting of the rule is needed for the entire problem.

$$(d) \qquad \sqrt{(640)(0.26)} = 12.90$$

Either of the settings shown in the diagram gives the answer; note that in each case the 640 is found in the left-hand segment, and the 0.26 is found in the right-hand segment.

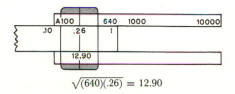

$$\sqrt{(640)(.26)} = 12.90$$

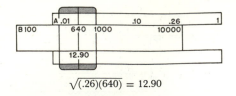

$$\sqrt{(.26)(640)} = 12.90$$

(e) $\dfrac{(605)(82.1)}{4.03} = 1.233 \times 10^4$

Here, after the initial division $605/4.03$ ($=150.0$) the product by 82.1 cannot be immediately taken, since 82.1 on the C scale lies beyond the end of the D scale. "Reverse" the rule, moving one of the movable indices of the C scale to take the place of the other movable index. The product by 82.1 can

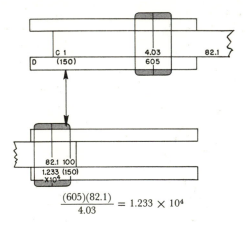

$$\dfrac{(605)(82.1)}{4.03} = 1.233 \times 10^4$$

now be found, giving 1233 for the significant figures. A rough calculation fixes the answer as about 600×20, or 12 000. The final answer is therefore 1.233×10^4.

(f) *Proportions* are easily solved by slide rule. For any given position of the rule any pair

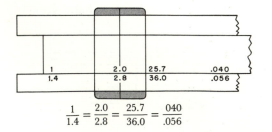

$$\dfrac{1}{1.4} = \dfrac{2.0}{2.8} = \dfrac{25.7}{36.0} = \dfrac{040}{.056}$$

of numbers on the C and D scales are in the same proportion as any other pair on the C and D scales. Thus one setting serves to find x, y, and z from

$$\dfrac{2.0}{2.8} = \dfrac{1}{x}; \quad \dfrac{2.0}{2.8} = \dfrac{25.7}{y}; \quad \dfrac{2.0}{2.8} = \dfrac{0.040}{z}$$

The A and B scales can be similarly used.

Placing the decimal point. In simple calculations, such as those already discussed, you can place the decimal point in the answer "by inspection." In more complex cases, it is best to disregard decimal points entirely, work out the significant figures of the answer, and then make a one-figure approximate calculation (see Sec. F-1) to place the decimal point in your answer. This was illustrated in operation e above. Some additional examples follow.

EXAMPLE

$$\dfrac{4\pi(81.0)}{0.0063} = ?$$

The slide rule gives ... 1616 ..., but we have not yet placed in the decimal point. The one-figure approximate calculation is (calling $4\pi = 12$)

$$\dfrac{(12)(80)}{6 \times 10^{-3}} = 160 \times 10^3 = 1.6 \times 10^5$$

Hence our answer is 1.616×10^5.

EXAMPLE

$$\dfrac{(6.62 \times 10^{-7})(9060)(2\pi)}{(0.004)(3.51 \times 10^5)}$$

The slide rule gives ... 268 ....
The approximate calculation is

$$\dfrac{(7 \times 10^{-7})(9 \times 10^3)(\overset{2}{\cancel{6}})}{(\underset{2}{\cancel{4}} \times 10^{-3})(\cancel{3} \times 10^5)} = \dfrac{63 \times 10^{-4}}{2 \times 10^2}$$

$$= 30 \times 10^{-6}$$
$$= 3 \times 10^{-5}$$

Hence the answer is 2.68×10^{-5}.

Other scales of the slide rule. The CI scale (an inverted C scale) is useful for finding reciprocals, since every number on the CI scale is just opposite its own reciprocal on the C scale. (Remember that the CI scale is read from right to left.) This scale also is helpful in computations involving several factors. For instance, $\dfrac{4}{(2)(3)}$ can be thought of as $\frac{4}{2} \times \frac{1}{3}$. To multiply by $\frac{1}{3}$, you merely have to multiply by what is marked "3" on the CI scale. In other words, when you are using the CI scale, the motions which have been described as multiplication result in division, and the motions which have been described as division result in multiplication.

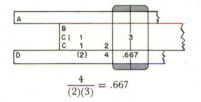

$$\frac{4}{(2)(3)} = .667$$

Although this can save time and trouble, it is best to refrain from using the CI scale unless you have already completely mastered the ordinary, direct-reading scales. The slide rule can also be used for looking up logarithms and trigonometric functions; see the instruction book for details.

Problems

135 Set the cursor of your slide rule (the line ruled on the movable glass or plastic indicator) opposite the following numbers on the D scale, and have a friend check your settings: 24; 5.15; 5.50; 368; 0.208; 170; 1.07; 0.666; 22.4; 980; 0.000321; 101.3.

Perform the following operations:

136 (14)(26)
137 (2.16)(405)
138 (75.1)(0.0331)
139 (480)/(7.1)
140 (50)/(0.372)
141 $(1.83 \times 10^8)/(4.25 \times 10^3)$
142 (62.7)/(1.54)
143 $(41.2)^2$
144 $(0.137)^2$
145 $\sqrt{735}$
146 $\sqrt{7350}$
147 $\sqrt{2.62 \times 10^{10}}$
148 $\sqrt{0.680}$
149 $\sqrt{(64)(0.75)}$
150 $\pi(2.68)^2$
151 $[(88/60)(25.0)]^2$
152 $\dfrac{(1.62 \times 10^{-7})(9.21 \times 10^8)^2}{24}$
153 $\dfrac{31}{18} = \dfrac{x}{41} = \dfrac{22}{y};\ x = ?\ y = ?$

Answers to practice problems in Mathematical Review

1	126.3	39	3.36×10^{-3}	77	127 mi	115	15; 15; $4x^2 - 1$; $t - 1$; 0
2	12.63	40	4.66331056×10^6	78	1.3×10^2 mi	116	-2; $1 + 6t + 9t^2$; $1 + 3t^2$
3	30	41	8.64×10^4	79	5.500 ft	117	29; 24
4	536.8	42	1.44×10^{-14}	80	3; 2; 3	118	$28x^3 - 3$
5	1240	43	1.25×10^{-1}	81	39.1 gal	119	$3 \cos 3x + 5e^{5x}$
6	0.3	44	1.25×10^{-3}	82	8.6400×10^4 s	120	$-3x^{-4}$
7	2	45	16 300 000		or 86.400 ks	121	$8x(1 - x)e^{-2x}$
8	200	46	4781.354	83	3.75×10^7	122	$-x(9 - x^2)^{-1/2}$
9	0.20	47	1 010 000	84	5.48×10^{-4} g	123	0
10	0.6	48	98.1	85	0.08 cm; 0.1%	124	7.5
11	10 000	49	0.000311	86	-14	125	$\frac{1}{3}$
12	3.14	50	65 500 000 000 000	87	-2.5	126	$1/b$
13	6.67	51	0.0008001	88	0.75	127	0.105
14	1390	52	0.0000093	89	± 10	128	$4V^{5/2}$
15	154	53	9.6×10^{19}	90	3	129	250
16	0.571	54	2.4×10^{-3}	91	1.6	130	5×10^{12}
17	6	55	4×10^{-1}	92	4	131	300
18	1.6	56	4.4×10^9	93	64	132	1.1×10^4
19	0.6	57	10^{-4}	94	$(16 + 2y)/3$	133	0.05
20	180	58	9.63×10^4	95	$(mg - F - 200)/2$	134	20
21	0.11	59	2×10^{-16}	96	$2y^2 + 2y$	136	364
22	0.003	60	5×10^{-8}	97	$(9 - y)/(3 + y)$	137	875
23	x^{10}	61	900	98	$u = 2, v = 1$	138	2.49
24	$x^{23}y^4$	62	285	99	$p = 3, q = 5$	139	67.6
25	$a^{5/2}$	63	1.3×10^{-6}	100	$B = 4, A = 10$	140	134.4
26	2^4	64	5×10^3	101	$x = 0.85, y = 1.25$	141	4.31×10^4
27	0.01	65	0.063	102	$1, \frac{1}{2}$	142	40.7
28	$9t^2$	66	0.02	103	$-2, -3$	143	1697
29	16	67	20	104	$5, -2$	144	0.01877
30	2	68	0.2	105	no real solution	145	27.1
31	$p^{1.9}$	69	5×10^{-7}	106	-1	146	85.7
32	45	70	464	107	$+1, -1, +\frac{1}{2}, -\frac{1}{2}$	147	1.62×10^5
33	x	71	4×10^7	108	0.995 g/cm^3	148	0.825
34	1.5×10^3	72	0.003	109	15:00; 10:00	149	6.93
35	1	73	0.750 min	110	60; 76; 120; unknown	150	22.6
36	0	74	0.7500 min	111	30 min	151	1344
37	1.4263×10^2	75	7.7×10^2 days	112	Check data at 24 lb	152	5.73×10^9
38	1.5×10^6	76	9.248 ft^2	114	0.070 W	153	$x = 70.6, y = 12.77$

List of reference tables

Other tables, included in the body of the text:

TABLE 1   Important constants

velocity of light, c	3.00×10^8 m/s
gravitational constant, G	6.67×10^{-11} N·m²/kg²
rest mass of electron, m_e	9.11×10^{-31} kg
rest mass of proton, m_p	1.67×10^{-27} kg
rest mass of neutron, m_n	1.67×10^{-27} kg
unified atomic mass unit, u	1.66×10^{-27} kg
energy equivalent of 1 u	931.5 MeV
Avogadro's number, N_A	6.02×10^{23} mole^{-1}
Joule equivalent, J	4.184 J/cal
absolute zero, 0°K	-273.15°C
volume of 1 mole of ideal gas at STP, V_0	22.4 liters
gas constant per mole, R	8.31 J/mole·K°
charge of electron, e	1.60×10^{-19} C
Faraday constant, Q_F	9.65×10^4 C/g.e.w.
Planck's constant, h	6.63×10^{-34} J·s
mass of earth	5.98×10^{24} kg
mean radius of earth	6.38×10^6 m
mean radius of earth's orbit	1.49×10^{11} m
mean radius of moon	1.74×10^6 m
mean radius of moon's orbit	3.84×10^8 m
length of year	3.16×10^7 s

TABLE 2   Important metric prefixes

Prefix	Abbrevi-ation	Meaning	Typical examples
tera	T	$\times 10^{12}$	1 terayear $= 10^{12}$ years
giga•	G	$\times 10^{9}$	1 gigahertz (radar frequency) $= 10^9$ Hz $= 10^9$ cycles/s
mega	M	$\times 10^{6}$	1 megaton (equivalent TNT strength of nuclear weapon) $= 10^6$ tons
kilo	k	$\times 10^{3}$	1 kilogram $= 1000$ g
deci	d	$\times 10^{-1}$	1 decimeter $= 0.1$ m
centi	c	$\times 10^{-2}$	1 centimeter $= 0.01$ m
milli	m	$\times 10^{-3}$	1 milliampere $= 0.001$ A
micro	μ	$\times 10^{-6}$	1 microvolt $= 10^{-6}$ V
nano•	n	$\times 10^{-9}$	1 nanosecond $= 10^{-9}$ s
pico•	p	$\times 10^{-12}$	1 picofarad $= 10^{-12}$ F
femto	f	$\times 10^{-15}$	1 femtometer (approximate size of a proton) $= 10^{-15}$ m
atto	a	$\times 10^{-18}$	

•In the older literature, certain double prefixes are used: kilomega (kM) for 10^9; millimicro (mμ) for 10^{-9}; micromicro ($\mu\mu$) for 10^{-12}.

TABLE 3   Greek alphabet

A	α	alpha	H	η	eta	N	ν	nu	T	τ	tau
B	β	beta	Θ	θ	theta	Ξ	ξ	xi	Υ	υ	upsilon
Γ	γ	gamma	I	ι	iota	O	o	omicron	Φ	ϕ	phi
Δ	δ	delta	K	κ	kappa	Π	π	pi	X	χ	chi
E	ϵ	epsilon	Λ	λ	lambda	P	ρ	rho	Ψ	ψ	psi
Z	ζ	zeta	M	μ	mu	Σ	σ	sigma	Ω	ω	omega

TABLE 4   Equivalents and conversion factors*

Length
1 ft = 30.48 cm = 0.3048 m
1 mi = 5280 ft = 1.609 km
1 yd = 0.9144 m
1 in. = 2.540 cm
1 angstrom (Å) = 10^{-10} m = 10^{-8} cm
1 fermi = 1 femtometer (fm) = 10^{-15} m = 10^{-13} cm
1 micron (μ) = 1 μm = 10^{-6} m = 10^{-4} cm = 10^4 Å
1 light-year = 5.88×10^{12} mi = 9.461×10^{12} km

Area
1 ft² = 929.0 cm² = 0.09290 m²
1 in.² = 6.452 cm² = 645.2 mm²
1 barn (b) = 10^{-28} m² = 10^{-24} cm²

Volume
1 liter = 1000 cm³ = 0.001 m³ = 1.0576 qt = 61.03 in.³ = 0.0353 ft³
1 ft³ = 7.481 gal = 28.32 liters
1 m³ = 1000 liters = 10^6 cm³ = 1.308 yd³ = 35.31 ft³

Velocity
60 mi/h = 88 ft/s = 26.82 m/s
1 mi/h = 0.4470 m/s = 1.609 km/h = 1.47 ft/s

Mass
1 metric ton = 1000 kg = 10^6 g

Force
1 newton (N) = 10^5 dynes = 0.2248 lb
1 lb = 4.448 N
1 ton = 2000 lb
An object of mass 1 kg weighs ▪ 9.807 N = 2.205 lb
An object of weight ▪ 1 lb has mass 453.6 g = 0.4536 kg

Pressure
1 bar = 10^6 dyn/cm² = 10^5 N/m² = 14.50 lb/in.²
1 lb/in.² = 6.89×10^4 dyn/cm² = 6.89×10^3 N/m²
1 atm = 760 torr = 760 mm Hg = 76.0 cm Hg = 14.70 lb/in.² = 2116 lb/ft²
     = 1.013×10^6 dyn/cm² = 1.013×10^5 N/m² = 1.013 bar = 1013 mbar

Work and energy
1 joule (J) = 10^7 ergs = 0.239 cal = 0.7376 ft·lb
1 cal = 4.184 J = 3.086 ft·lb
1 Btu = 252 cal = 778 ft·lb = 1054 J
1 kilowatt·hour (kWh) = 3.60×10^6 J
1 electron volt (eV) = 1.60×10^{-19} J

Power
1 horsepower (hp) = 0.746 kilowatt (kW) = 550 ft·lb/s = 3.30×10^4 ft·lb/min
1 watt (W) = 1 J/s = 0.738 ft·lb/s

Specific heat
1 cal/g·C° = 1 Btu/lb·F° = 4.184 J/g·C° = 4184 J/kg·C°

Latent heat
1 cal/g = 4.184 J/g = 4184 J/kg = 1.80 Btu/lb

Gas constant
R = 8.314 J/mole·K° = 1.99 cal/mole·K° = 0.0821 (atm·liter/mole·K°)

* See also Table 2 for important metric prefixes
▪ All weights are at a point on the surface of the earth where g has the standard value 9.807 m/s².

TABLE 5   Sines, cosines, and tangents

Angle	Sine	Cosine	Tangent	Angle	Sine	Cosine	Tangent
0°	0.000	1.000	0.000				
1°	.017	1.000	.017	46°	0.719	0.695	1.036
2°	.035	0.999	.035	47°	.731	.682	1.072
3°	.052	.999	.052	48°	.743	.669	1.111
4°	.070	.998	.070	49°	.755	.656	1.150
5°	.087	.996	.087	50°	.766	.643	1.192
6°	.105	.995	.105	51°	.777	.629	1.235
7°	.122	.993	.123	52°	.788	.616	1.280
8°	.139	.990	.141	53°	.799	.602	1.327
9°	.156	.988	.158	54°	.809	.588	1.376
10°	.174	.985	.176	55°	.819	.574	1.428
11°	.191	.982	.194	56°	.829	.559	1.483
12°	.208	.978	.213	57°	.839	.545	1.540
13°	.225	.974	.231	58°	.848	.530	1.600
14°	.242	.970	.249	59°	.857	.515	1.664
15°	.259	.966	.268	60°	.866	.500	1.732
16°	.276	.961	.287	61°	.875	.485	1.804
17°	.292	.956	.306	62°	.883	.469	1.881
18°	.309	.951	.325	63°	.891	.454	1.963
19°	.326	.946	.344	64°	.899	.438	2.050
20°	.342	.940	.364	65°	.906	.423	2.145
21°	.358	.934	.384	66°	.914	.407	2.246
22°	.375	.927	.404	67°	.921	.391	2.356
23°	.391	.921	.424	68°	.927	.375	2.475
24°	.407	.914	.445	69°	.934	.358	2.605
25°	.423	.906	.466	70°	.940	.342	2.747
26°	.438	.899	.488	71°	.946	.326	2.904
27°	.454	.891	.510	72°	.951	.309	3.078
28°	.469	.883	.532	73°	.956	.292	3.271
29°	.485	.875	.554	74°	.961	.276	3.487
30°	.500	.866	.577	75°	.966	.259	3.732
31°	.515	.857	.601	76°	.970	.242	4.011
32°	.530	.848	.625	77°	.974	.225	4.331
33°	.545	.839	.649	78°	.978	.208	4.705
34°	.559	.829	.675	79°	.982	.191	5.145
35°	.574	.819	.700	80°	.985	.174	5.671
36°	.588	.809	.727	81°	.988	.156	6.314
37°	.602	.799	.754	82°	.990	.139	7.115
38°	.616	.788	.781	83°	.993	.122	8.144
39°	.629	.777	.810	84°	.995	.105	9.514
40°	.643	.766	.839	85°	.996	.087	11.43
41°	.656	.755	.869	86°	.998	.070	14.30
42°	.669	.743	.900	87°	.999	.052	19.08
43°	.682	.731	.933	88°	.999	.035	28.64
44°	.695	.719	.966	89°	1.000	.017	57.29
45°	.707	.707	1.000	90°	1.000	.000	

TABLE 6   Periodic table of the elements*

1 H 1.0080																	1 H 1.0080	2 He 4.0026
3 Li 6.941	4 Be 9.0122											5 B 10.81	6 C 12.011	7 N 14.0067	8 O 15.9994	9 F 18.9984	10 Ne 20.179
11 Na 22.9898	12 Mg 24.305											13 Al 26.9815	14 Si 28.086	15 P 30.9738	16 S 32.06	17 Cl 35.453	18 Ar 39.948
19 K 39.102	20 Ca 40.08	21 Sc 44.956	22 Ti 47.90	23 V 50.941	24 Cr 51.996	25 Mn 54.9380	26 Fe 55.847	27 Co 58.9332	28 Ni 58.71	29 Cu 63.54	30 Zn 65.37	31 Ga 69.72	32 Ge 72.59	33 As 74.9216	34 Se 78.96	35 Br 79.909	36 Kr 83.80
37 Rb 85.467	38 Sr 87.62	39 Y 88.906	40 Zr 91.22	41 Nb 92.906	42 Mo 95.94	43 Tc (99)	44 Ru 101.07	45 Rh 102.906	46 Pd 106.4	47 Ag 107.870	48 Cd 112.40	49 In 114.82	50 Sn 118.69	51 Sb 121.75	52 Te 127.60	53 I 126.9045	54 Xe 131.30
55 Cs 132.906	56 Ba 137.34	57 La 138.906	72 Hf 178.49	73 Ta 180.948	74 W 183.85	75 Re 186.2	76 Os 190.2	77 Ir 192.2	78 Pt 195.09	79 Au 196.967	80 Hg 200.59	81 Tl 204.37	82 Pb 207.2	83 Bi 208.981	84 Po (210)	85 At (210)	86 Rn (222)
87 Fr (223)	88 Ra (226)	89 Ac (227)	104 Rf (261)	105 Ha (262)													

Lanthanide Series

58 Ce 140.12	59 Pr 140.908	60 Nd 144.24	61 Pm (147)	62 Sm 150.4	63 Eu 151.96	64 Gd 157.25	65 Tb 158.925	66 Dy 162.50	67 Ho 164.930	68 Er 167.26	69 Tm 168.934	70 Yb 173.04	71 Lu 174.97

Actinide Series

90 Th (232)	91 Pa (231)	92 U (238)	93 Np (237)	94 Pu (242)	95 Am (243)	96 Cm (248)	97 Bk (249)	98 Cf (249)	99 Es (254)	100 Fm (257)	101 Md (258)	102 No (259)	103 Lr (260)

* Atomic weights of stable elements are those adopted in 1969 by the International Union of Pure and Applied Chemistry. For those elements having no stable isotope, the mass number of the "most stable" well-investigated isotope is given in parentheses.

TABLE 7   The elements (alphabetical according to symbol)

Symbol	Name	Atomic Number	Symbol	Name	Atomic Number	Symbol	Name	Atomic Number
Ac	Actinium	18	Ge	Germanium	32	Po	Polonium	84
Ag	Silver	47	H	Hydrogen	1	Pr	Praseodymium	59
Al	Aluminum	13	Ha	Hahnium▲	105	Pt	Platinum	78
Am	Americium	95	He	Helium	2	Pu	Plutonium	94
Ar	Argon•	18	Hf	Hafnium	72	Ra	Radium	88
As	Arsenic	33	Hg	Mercury	80	Rb	Rubidium	37
At	Astatine	85	Ho	Holmium	67	Re	Rhenium	75
Au	Gold	79	I	Iodine	53	Rf	Rutherfordium▲	104
B	Boron	5	In	Indium	49	Rh	Rhodium	45
Ba	Barium	56	Ir	Iridium	77	Rn	Radon•	86
Be	Beryllium	4	K	Potassium	19	Ru	Ruthenium	44
Bi	Bismuth	83	Kr	Krypton	36	S	Sulfur	16
Bk	Berkelium	97	La	Lanthanum	57	Sb	Antimony	51
Br	Bromine	35	Li	Lithium	3	Sc	Scandium	21
C	Carbon	6	Lr	Lawrencium	103	Se	Selenium	34
Ca	Calcium	20	Lu	Lutetium	71	Si	Silicon	14
Cd	Cadmium	48	Mg	Magnesium	12	Sm	Samarium	62
Ce	Cerium	58	Mn	Manganese	25	Sn	Tin	50
Cf	Californium	98	Mo	Molybdenum	42	Sr	Strontium	38
Cl	Chlorine	17	Mv	Mendelevium	101	Ta	Tantalum	73
Cm	Curium	96	N	Nitrogen	7	Tb	Terbium	65
Co	Cobalt	27	Na	Sodium	11	Tc	Technetium	43
Cr	Chromium	24	Nb	Niobium•	41	Te	Tellurium	52
Cs	Cesium	55	Nd	Neodymium	60	Th	Thorium	90
Cu	Copper	29	Ne	Neon	10	Ti	Titanium	22
Dy	Dysprosium	66	Ni	Nickel	28	Tl	Thallium	81
Es	Einsteinium	99	No	Nobelium	102	Tm	Thulium	69
Er	Erbium	68	Np	Neptunium	93	U	Uranium	92
Eu	Europium	63	O	Oxygen	8	V	Vanadium	23
F	Fluorine	9	Os	Osmium	76	W	Tungsten•	74
Fe	Iron	26	P	Phosphorus	15	Xe	Xenon	54
Fm	Fermium	100	Pa	Protactinium	91	Y	Yttrium	39
Fr	Francium	87	Pb	Lead	82	Yb	Ytterbium	70
Ga	Gallium	31	Pd	Palladium	46	Zn	Zinc	30
Gd	Gadolinium	64	Pm	Promethium	61	Zr	Zirconium	40

• Symbol A is sometimes used instead of Ar.
• Niobium is also known as columbium; radon is also known as emanation (Em); tungsten is also known as wolfram.
▲ Name and symbol for elements 104 and 105 are tentative.

TABLE 8 Properties of light nuclides

Symbol	Mass, u[*]	Abundance or mode of decay[**]	Half-life[**]	Symbol	Mass, u	Abundance or mode of decay	Half-life
e	0.000549			$^{14}_{8}O$	14.008597	β^+	71 s
p	1.007277			$^{15}_{8}O$	15.003070	β^+	2.05 m
$^{1}_{0}n$	1.008665	β^-	11.7 m	$^{16}_{8}O$	15.994915	99.76%	
				$^{17}_{8}O$	16.999133	0.04%	
$^{1}_{1}H$	1.007825	99.985%		$^{18}_{8}O$	17.999160	0.20%	
$^{2}_{1}H$	2.014102	0.015%		$^{19}_{8}O$	19.003578	β^-	29.1 s
$^{3}_{1}H$	3.016050	β^-	12.3 y	$^{20}_{8}O$	20.004079	β^-	14 s
$^{3}_{2}He$	3.016030	0.0001%		$^{17}_{9}F$	17.002096	β^+	67 s
$^{4}_{2}He$	4.002603	~100%		$^{18}_{9}F$	18.000937	β^+, ϵ	1.83 h
$^{6}_{2}He$	6.018893	β^-	0.80 s	$^{19}_{9}F$	18.998405	100%	
$^{8}_{2}He$	8.0341	β^-	0.12 s	$^{20}_{9}F$	19.999987	β^-	11.6 s
				$^{21}_{9}F$	20.99995	β^-	4.4 s
$^{6}_{3}Li$	6.015125	7.4%		$^{22}_{9}F$	22.004	β^-	4.0 s
$^{7}_{3}Li$	7.016004	92.6%					
$^{8}_{3}Li$	8.022487	β^-	0.84 s	$^{17}_{10}Ne$	17.0364	β^+	0.1 s
$^{9}_{3}Li$	9.02680	β^-	0.18 s	$^{18}_{10}Ne$	18.005711	β^+	1.5 s
				$^{19}_{10}Ne$	19.001881	β^+	17.4 s
$^{7}_{4}Be$	7.016929	ϵ	53.6 d	$^{20}_{10}Ne$	19.992440	90.92%	
$^{9}_{4}Be$	9.012186	100%		$^{21}_{10}Ne$	20.993849	0.26%	
$^{10}_{4}Be$	10.013534	β^-	2.5 My	$^{22}_{10}Ne$	21.991385	8.82%	
$^{11}_{4}Be$	11.02167	β^-	13.6 s	$^{23}_{10}Ne$	22.994473	β^-	37.6 s
				$^{24}_{10}Ne$	23.99361	β^-	3.38 m
$^{8}_{5}B$	8.024609	β^+	0.77 s				
$^{10}_{5}B$	10.012939	19.6%		$^{20}_{11}Na$	20.0089	β^+	0.4 s
$^{11}_{5}B$	11.009305	80.4%		$^{21}_{11}Na$	20.997655	β^+	23 s
				$^{22}_{11}Na$	21.994437	β^+, ϵ	2.62 y
$^{9}_{6}C$	9.0312	β^+	0.13 s	$^{23}_{11}Na$	22.989771	100%	
$^{10}_{6}C$	10.01681	β^+	19.5 s	$^{24}_{11}Na$	23.990962	β^-	15.0 h
$^{11}_{6}C$	11.011432	β^+	20.3 m	$^{25}_{11}Na$	24.98996	β^-	60 s
$^{12}_{6}C$	12.000000	98.89%		$^{26}_{11}Na$	25.9917	β^-	1.04 s
$^{13}_{6}C$	13.003354	1.11%					
$^{14}_{6}C$	14.003242	β^-	5730 y	$^{20}_{12}Mg$	20.017	β^+	0.6 s
$^{15}_{6}C$	15.010600	β^-	2.5 s	$^{21}_{12}Mg$	21.0117	β^+	0.12 s
$^{16}_{6}C$	16.01470	β^-	0.74 s	$^{23}_{12}Mg$	22.994125	β^+	12 s
				$^{24}_{12}Mg$	23.985042	78.7%	
$^{13}_{7}N$	13.005738	β^+	10.0 m	$^{25}_{12}Mg$	24.985839	10.1%	
$^{14}_{7}N$	14.003074	99.63%		$^{26}_{12}Mg$	25.982593	11.2%	
$^{15}_{7}N$	15.000108	0.37%		$^{27}_{12}Mg$	26.984345	β^-	9.5 m
$^{16}_{7}N$	16.006103	β^-	7.14 s	$^{28}_{12}Mg$	27.98388	β^-	21.2 h
$^{17}_{7}N$	17.00845	β^-	4.16 s				
$^{18}_{7}N$	18.0141	β^-	0.63 s				

[*] Except for the first three particles, the masses given are those of the neutral atoms.

[**] Abbreviations: β^-, negatron decay; β^+, positron decay; ϵ, electron capture; y, year; My, megayear; d, day; h, hour; m, minute; s, second. No nuclides are listed which have half-lives less than 0.1 s.

Answers to odd-numbered problems

Chapter 2

2-A1 3.156×10^7 s/y.

2-A3 1.29 kg/m^3.

2-A5 9.05 kg.

2-A7 About 3000 cm^3.

2-A9 200 kg.

2-B1 2.43×10^{22} atoms.

2-B3 1.67×10^{-24} g.

2-B5 1.09

2-B7 (a) 1595 g; (b) 6.25×10^{24} molecules.

2-C1 (a) F_1 is long range, F_2 is short range; (b) $F_1 = F_2$ for $x = 10^{-10}$ m; (c) error at $x = 10^{-9}$ m is only 0.00001%.

2-C3 26.7 km.

Chapter 3

3-A1 54 mi/h; 79.2 ft/s.

3-A3 (a) 9.46×10^{-5} mi/h; 5.26×10^4 in./y; 1.39×10^{-4} ft/s; (b) -3.50 ft/day · min; -6.75×10^{-7} ft/s^2.

3-A5 (a) 20 m/s; (b) 0.

3-A7 0.067 ft/s^2, westward.

3-A9 (a) 9.80 m/s; 9.35 m/s; (b) 4.7 m.

3-A15 (a) 18 blocks; (b) 1.4 blocks SW from the starting point.

3-A17 (a) 141 mi/h; (b) -141 mi/h.

3-A19 (a) 4070 m/s; (b) 3130 m/s; (c) So they can be aimed eastward, over the Atlantic Ocean.

3-B1 350 m/s.

3-B3 (a) 44.2 m/s; 98.8 mi/h; (b) -1.22×10^4 m/s^2.

3-B5 3 s; 36 m.

3-B7 (a) 36 ft; (b) 1.5 s; (c) 3.0 s; (d) 48 ft/s.

3-B9 (a) 58 ft/s, downward; (b) 64 ft.

3-B11 (a) 4 s; (b) 96 ft/s; 96 ft/s; (c) 128 ft.

3-B13 13.4 ft.

3-B15 24.6 s.

3-B17 (a) $+60$ m/s; (b) -6 m/s^2.

3-B19 100 mi/h, northward.

3-B21 (a) 30° N of E; (b) 3.46 mi/h; (c) 3.46 min; (d) straight across and back is 1.08 min quicker.

3-B23 0.404 s; 0.242 m.

3-B25 2080 ft/s.

3-B27 (a) 1.95 m; (b) 2.50 m/s; 7.67 m/s.

3-B29 $R/2$.

3-B31 (a) 20 s; (b) 4000 m; (c) 490 m; (d) 223 m/s.

3-B33 (a) 9.41 m/s; (b) 6.59 m.

3-B35 (a) 10 s; (b) 800 ft downrange; (c) 320 ft downrange, 336 ft above ground level.

3-C1 Yes, in 6 s.

3-C3 (a) 5 s; (b) 30 ft/s; (c) 425 ft; (d) 350 ft.

3-C5 (a) 1 s or 3 s; (b) 32 ft/s.

3-C7 1.29 s.

3-C9 (a) 74.5 m; (b) 31.1 m/s.

3-C11 (a) 2.66 m/s; (b) 0.64 m.

3-C15 $\tan \phi = \frac{1}{2} \tan \theta$.

3-C17 (a) 17.4; (b) 17.04; (c) 17.008.

3-C19 (a) $24x - 24/x^4$; (b) 403.7; (c) $y(4 - y^2)^{-3/2}$; (d) 123; (e) $(-3x^2 + 2x)e^{-3x}$.

3-C21 3.33.

3-C23 Yes; yes.

Chapter 4

4-A1 2000 dyn.

4-A3 3800 lb, upward.

4-A5 5 g.

4-A7 2000 dyn.

4-A9 18 ft/s^2.

4-A11 400 lb.

4-B1 5.5 m/s^2, downward.

4-B3 (a) 8 ft/s^2, upward; (b) 16 ft.

4-B5 Yes; step on the gas; panic stop not possible in available space.

4-B7 5060 N.

4-B9 (a) 59.4 m/s^2, upward; (b) 5536 N.

4-B11 592 N.

4-B13 66 lb.

4-B15 2.2 m/s^2.
4-B17 (a) 500 lb; (b) 250 lb.
4-B19 (a) 1.96 m/s^2; (b) 23.5 N; (c) 23.5 N.
4-B21 (a) 1.50 s; (b) 3.75 lb; (c) 3.75 lb.
4-B23 1.01 × 10^{-13} N.
4-B25 2.83 × 10^{-3} dyn.
4-B29 1.000003; no.

4-C1 (a) 37.7 ft/s; (b) yes, there is 0.8 s to spare.
4-C3 5 ft.
4-C5 (a) 2.00 m/s^2; (b) 0.547 s; (c) 78 N; (d) 98 N.
4-C7 (a) 3.0 m/s^2; (b) 4.8 N.
4-C9 (a) 2 ft, downward; (b) 17 lb.
4-C11 1 s.

Chapter 5

5-A1 (a) 42.4 N; (b) 52.0 N.
5-A3 50 units, 53° above the horizontal.
5-A5 33.1 N.
5-A7 400 N.
5-A9 0.10
5-A11 Clockwise.
5-A13 700 lb · ft; clockwise; vector into the plane of the paper.

5-B1 3.0 dyn, 32° E of N.
5-B7 $T_1 = 37$ lb; $T_2 = 45$ lb.
5-B9 170 N.
5-B11 289 N.
5-B13 106 N.
5-B15 0.35 N.
5-B17 58.3 cm.
5-B19 3.33 ft.
5-B21 18.8 lb, downward.
5-B23 100 lb; no; 8 ft/s^2.
5-B25 36 ft.
5-B27 $T = 13.5$ lb; $H = 13.5$ lb; $V = 10$ lb.
5-B29 (a) 9.43 lb; (b) 2.36 lb, push.
5-B31 (a) 1130 N; (b) 1440 N.
5-B33 0.43.
5-B35 4.90 ft/s.

5-C5 482 N.
5-C7 4 : 1.
5-C9 0.10.
5-C11 1.09 m.
5-C13 44.7 lb in one rope, which is 63° from the vertical, and 89.4 lb in the other rope, which is 27° from the vertical.

5-C15 (a) At $x = +5$ m; (b) at $x = +4$ m; (c) $-\frac{1}{8}$ m/s^2; (d) $-\frac{1}{8}$ m/s^2.

Chapter 6

6-A1 300 kg · m/s.
6-A3 (a) Momentum of truck is 400 times that of bullet; (b) KE of bullet is 2 times that of truck.
6-A5 2 ft/s.
6-A7 1.67 m/s.
6-A11 1.6 × 10^4 ergs.
6-A13 4.9 × 10^8 J.
6-A15 120 J.
6-A17 40 J.
6-A19 7.52 × 10^{-23} J.
6-A21 30 ft/s.
6-A23 625 dyn.

6-B1 500 N.
6-B3 6 lb.
6-B5 43.1 N.
6-B7 (a) 0.995 m/s; (b) 1.27 m.
6-B9 0.6 m/s, in the direction the 40 kg swimmer dived.
6-B11 30 ft · lb.
6-B13 1670 N.
6-B15 8 ft/s.
6-B17 3 m/s.
6-B19 50 ft/s.
6-B21 (a) 9.8 J; (b) 49 N.
6-B23 27.7 ft/s.
6-B25 $s = \frac{1}{2}v^2/\mu_s g$.
6-B27 80%.
6-B29 508 kJ.
6-B31 466 W.
6-B33 (a) 5.63 × 10^{10} ft · lb; (b) 2.04 × 10^6 hp.
6-B39 (a) 5 m/s, westward; (b) 6 J lost.
6-B41 (a) 1 m/s, eastward; (b) 6 J lost.

6-C5 620 N, downward to right, 15° from vertical.
6-C7 9 kg.
6-C9 0.2 m/s; 1.8 m/s.
6-C11 0.9.
6-C13 0.782.
6-C17 (a) 60; (b) 50%; (c) 30; (d) 33.3 N.
6-C19 0.133.

Chapter 7

7-A1 2 s.
7-A3 1.25m_0.

7-A5 9.11×10^{11} J; 918 m.

7-B1 (a) 6.25 ft · lb; (b) 0.25 ft · lb.

7-B3 1.23 m/s^2.

7-B5 (a) 6.37×10^{10} N; (b) 2.36×10^{12} J; (c) 0.0262 g.

7-B7 (a) 1.15 days; (b) 115 000 heartbeats.

7-C1 (a) 10^{-5} s; (b) 0.3 m; (c) 10^{-9} s.

7-C3 (a) $1 + 3.38 \times 10^{-10}$; (b) 292 μs.

7-C5 (a) 15.18×10^{-31} kg; (b) 6.07×10^{-31} kg; (c) 5.46×10^{-14} J; (d) $\frac{1}{2}m_0v^2$ gives 2.62×10^{-14} J; $\frac{1}{2}mv^2$ gives 4.37×10^{-14} J.

Chapter 8

8-A1 (a) 3.14 rad; (b) 0.524 rad; (c) 1.57 rad; (d) 0.698 rad; (e) 1.88 rad.

8-A3 6 ft/s.

8-A5 314 ft.

8-A7 $\frac{1}{60}$ rad.

8-A9 (a) 7.27×10^{-5} rad/s; (b) 3.37 cm/s^2.

8-A11 (a) 1.75×10^{-3} rad/s; (b) 6.10×10^{-5} cm/s^2.

8-A13 $P = \tau\omega$.

8-A15 251 kg · m^2/s.

8-A17 365 days.

8-B1 0.617 rev/s^2; 3.88 rad/s^2.

8-B3 (a) 0.0125 rev/s^2; 10 rev.

8-B5 222 dyn.

8-B7 $v = \sqrt{\mu_s Rg}$.

8-B9 5.18×10^4 rev/min.

8-B11 (a) 0.700 rad/s; (b) 176 N.

8-B13 240 lb.

8-B15 $14°$.

8-B17 33.3 kg · m^2.

8-B19 4.8 rad/s.

8-B21 2.66 J.

8-B23 (a) 2220 ft · lb; (b) 1340 rev/min.

8-B25 (a) $T = 2\pi\sqrt{r^3/Gm_E}$, where r = radius of orbit, m_E = mass of Earth; (b) 97 min.

8-B29 (a) 2.0×10^{30} kg; (b) 1.43.

8-B31 Ratio of sun's force to the earth's force is about 2.1, both at time of new moon and at time of full moon.

8-C1 (a) 600 kg · m^2; (b) 0.5 rad.

8-C5 (a) 6.26 m/s; (b) 2.09 rad/s; (c) 2.19 m.

8-C9 32 ft/s.

8-C11 (a) $v = \sqrt{10gH/7}$; (b) $v = \sqrt{2gH}$.

Chapter 9

9-A1 0.1.

9-A3 4.17×10^{-4}.

9-A5 12 lb/ft.

9-A7 10 min.

9-A9 (a) 5 cm; (b) 5 cm.

9-A11 (a) $[\text{T}]$; (b) $[\text{T}^{-1}]$; (c) $[\text{MT}^{-2}]$ or $[\text{FL}^{-1}]$.

9-A13 3.14 ft/s.

9-A15 0.994 m.

9-B1 1 mm.

9-B3 1.2×10^7 N/m^2.

9-B5 0.00165 in.

9-B7 Quite safe; stress is only 5.7×10^6 N/m^2.

9-B9 0.16×10^{10} N/m^2.

9-B11 0.201 m^3; 172 cm^3.

9-B13 118 N.

9-B15 (a) 62.8 cm/s; (b) 99.3 cm.

9-B17 6.37 vib.

9-B19 3.2 J.

9-B21 (a) 4 lb/ft; (b) 50 ft · lb; (c) 50 ft · lb.

9-B23 850 g.

9-B25 6.2 mm.

9-B27 $a_0 = 4\pi^2 A/T^2$.

9-B29 4 vib/s.

9-C1 1.13 cm.

9-C3 (a) $E = 4\pi mL/T^2R^2$.

9-C5 (a) 1 s; (b) 9.08 cm/s.

9-C7 (a) 0.573 N · m/rad; (b) 99 s.

9-C9 31.4 s.

9-C11 (a) 2360 N/m; (b) 2.39×10^{-15} m.

9-C13 1.64 s.

Chapter 10

10-A1 $2:1$.

10-A3 5050 m/s.

10-A5 1410 m/s; 4640 ft/s.

10-A7 4.88×10^{10} N/m^2.

10-A9 170 s^{-1}.

10-A11 About 17 mm.

10-A13 9 ft.

10-A15 $t = 11.0$ s; $t = 11.6$ s; down.

10-A17 6.8 mm.

10-B1 20 s.

10-B3 Upward.

10-B5 $\frac{1}{400}$ s.

10-B7 25.4 kHz.

10-B9 525 m.

10-B11 15 s^{-1}.

10-B13 (a) 517.2 m; (b) 200 s^{-1}.

10-C1 (*a*) 1078 Hz; (*b*) 932 Hz.

10-C3 550.8 Hz.

10-C5 (*a*) Yes; (*b*) yes; (*c*) no.

10-C7 (*a*) 0.120 m; (*b*) 75.6 m/s.

10-C9 279 m/s.

Chapter 11

11-A1 80 m/s.

11-A3 7 in.

11-A5 200 Hz; 400 Hz; 600 Hz.

11-A7 200 Hz.

11-A9 330 Hz.

11-A11 220 Hz; 440 Hz; 660 Hz.

11-A13 200 Hz; 600 Hz; 1000 Hz.

11-B3 250 N.

11-B5 16 Mdyn.

11-B7 463 N.

11-B9 (*a*) 0.15 g; (*b*) 1 N.

11-B11 13 s^{-1}.

11-B13 (*a*) Open pipe; (*b*) 2.5 ft.

11-B15 3.2 s^{-1}.

11-B17 (*b*) $1:3:5:7:\ldots$; (*c*) 5000 m/s.

11-B19 (*b*) 3450 m/s; (*c*) 10.6×10^{10} N/m^2.

11-B21 No.

11-B23 20 000 : 1.

11-C1 16.6 cm; 49.8 cm; 83.0 cm; . . . .

11-C3 45.4 lb.

Chapter 12

12-A1 200 lb/in.^{2}.

12-A3 10 600 lb; 10 600 lb.

12-A5 588 lb/in.^{2}; 0.744 lb/in.^{2}; 152 mm Hg; 7.60×10^{-8} torr.

12-A7 1.37×10^4 N/m^2.

12-A9 0.416.

12-A11 10 : 1.

12-B1 0.564 in.

12-B3 6.02×10^5 N.

12-B5 162 torr.

12-B7 No; 21.6 atm is required.

12-B9 1104 mbar.

12-B11 (*a*) 1.59 cm^3; (*b*) 4.52 cm^3.

12-B13 0.25 g/cm^3.

12-B15 (*a*) 1.25 cm^3; (*b*) 2.40.

12-B17 7.32 m^3.

12-B19 $L_1 = L(d'/d)$; $L_2 = L(d - d')/d$.

12-B21 12 in.

12-B23 (*a*) 1.13×10^4 N/m^2; (*b*) 2.25×10^5 N.

12-B25 39.8 dyn/cm.

12-B27 44 dyn, downward.

12-B29 (*a*) 1810 dyn; (*b*) 144 dyn/cm^2.

12-B31 (*a*) $h = 2\sqrt{\gamma/dg}$; (*b*) 0.555 cm.

12-B33 0.0255 cm.

12-C1 (*a*) Turbulent.

12-C3 (*a*) 0.037; (*b*) 1.07.

12-C5 (*a*) 1.09×10^{-3} m^3; (*b*) 84.2 N.

Chapter 13

13-A1 36°C; 41°C.

13-A3 20°C; -12.2°C; -200°C; 37°C.

13-A5 68°F; 932°F; -40°F; -459.4°F.

13-A7 2.0 C°.

13-A9 0.264 in.

13-A11 1.05 cm.

13-A13 45×10^{-6} (C°)$^{-1}$.

13-A15 10.38 gal.

13-B1 80°.

13-B3 122°F; 16°R.

13-B5 Yes; 0.0008 in. to spare.

13-B7 (*a*) 197°C; (*b*) 9.6×10^4 N.

13-B9 125°C.

13-B11 6.2×10^4 N.

13-B13 0.8 mm.

13-B15 49.13 gal.

13-B17 1057×10^{-6} (C°)$^{-1}$.

13-B19 10°C.

13-C1 (*a*) 0.500048 s; (*b*) loses 8.2 s.

13-C3 (*a*) 145.1×980 dyn; (*b*) 148.1×980 dyn.

Chapter 14

14-A1 15 kcal.

14-A3 191 cal.

14-A5 0.8 cal/g · C°.

14-A7 61.2 g.

14-A9 16 kcal.

14-A11 0.24 cal/g · C°.

14-A13 59 cal/g.

14-A15 1.20 g.

14-A17 2.0×10^{18} Hz.

14-A19 30 GHz.

14-B1 0.47 cal/g · C°.

14-B3 (*a*) 4:10 P.M. (*b*) 4:37 P.M.

14-B5 70 g of water at 22.7°C.

14-B7 -196°C.

14-B9 10.8 g.

14-B11 (*a*) 2.5 h; (*b*) 22.9 h.

14-B13 1080 kg.

14-B15 8.23 Btu.

14-B17 (*a*) 700 J; (*b*) 164 C°.
14-B19 7.47 g.
14-B21 3740 times.
14-B23 (*a*) 0.0057 mm.
14-B25 (*a*) 36.8 kcal; (*b*) 171 s.
14-B27 2.90×10^3.
14-B29 (*a*) 0.6 μcal/s $\cdot$ cm^2; (*b*) 1.1×10^{18} J/day.
14-B31 About 75 h.

14-C1 5.5 cal/g.
14-C3 33 cal.
14-C5 (*a*) 1.43×10^4 kg; (*b*) 8.9 m^3.
14-C7 $603 000.
14-C9 135 kW.
14-C11 1.7×10^{23} J; from the sun.

Chapter 15

15-A1 8 in.
15-A3 500.
15-A5 20 moles.
15-A7 313°K; 193°K; 293°K; 373°K.
15-A9 8×10^5 N/m^2.
15-A11 300 lb/in.^{2}.
15-A13 6.92 g/m^3.
15-A15 526 torr.

15-B3 7.5 cm.
15-B5 0.8 g.
15-B7 0.724 atm.
15-B9 1.50 g/liter.
15-B11 12.9 g/liter.
15-B13 (*a*) 0.336 m^3; (*b*) 3.18 N.
15-B15 H$_2$: 1 atm; N$_2$: 2 atm; O$_2$: 6 atm; 9 atm; 11.1 kg.
15-B17 (*a*) 36 lb/in.^{2}; (*b*) 90.1 ft.
15-B19 574.6°, using -273.16°C for absolute zero.
15-B21 28.9%.
15-B23 218 atm; 3210 lb/in.^{2}; 0.318 g/cm^3.
15-B25 42%.

15-C1 32.7 lb.
15-C3 35 mm^3.
15-C5 371 cm^3.

Chapter 16

16-A1 The can.
16-A3 600 m/s.
16-A5 20%.
16-A7 1800 ft $\cdot$ lb.
16-A9 300°K.

16-B1 1000 N.

16-B3 200 m/s.
16-B5 (*a*) 5.65×10^{-21} J; (*b*) 1.04×10^{-24} J; (*c*) no.
16-B7 (*a*) 246 m/s; (*b*) 120.
16-B9 Can: 63°C; pool: 32°C.
16-B11 189°C.
16-B13 (*a*) 3.5×10^4 molecules/cm^3; (*b*) 0.03 cm apart.
16-B15 3.10×10^9 cal.
16-B17 (*a*) 50%; (*b*) 1.73×10^8 J; (*c*) 41.7%.
16-B19 771°K.

16-C5 737 m/s.
16-C7 (*a*) Greater; (*b*) 0.00026 C°.
16-C9 2.0; 2.4; 2.9.
16-C11 (*a*) 0.179 cal/g $\cdot$ C°; (*b*) 1.40; (*c*) diatomic.
16-C13 (*a*) 6.53 in.; (*b*) 128°C.
16-C15 206 cal.
16-C19 437 m/s.
16-C21 1.27.

Chapter 17

17-A1 10 A.
17-A3 6.25×10^6 electrons.
17-A5 0.16 pA.
17-A7 6 mN.
17-A9 40.6 g.
17-A11 (*a*) 9.65×10^5 C; (*b*) 6.02×10^{24} electrons.
17-A13 4 moles.

17-B1 (*a*) 8.20×10^{-8} N; (*b*) 3.62×10^{-47} N.
17-B3 5.1 m.
17-B5 6.0 nN.
17-B7 1.61×10^{-5} N.
17-B9 9×10^{-3} N, toward Q_1.
17-B11 (*a*) 1.13 N, away from center; (*b*) 0.
17-B13 6.93×10^{-3} N, away from triangle, perpendicular to opposite side.
17-B15 839 g.
17-B17 Cu$^+$.
17-B19 (*a*) 314 A; (*b*) 31.6 liters.

17-C1 1.50 μC.
17-C3 $x = 7$ m.
17-C5 0.735 μC.
17-C7 42 y.
17-C9 22.1 A.

Chapter 18

18-A1 8×10^5 N/C, upward.
18-A3 200 N/C, downward.
18-A5 0.2 V.

18-A7 200 V/m.

18-A9 25 μF.

18-A11 0.01 C.

18-B1 (a) 9×10^4 N/C; (b) 7.2×10^4 N/C;
(c) 16.2×10^4 N/C.

18-B3 2.70×10^4 N/C.

18-B5 (a) 8.00×10^{-16} N; (b) 8.78×10^{14} m/s^2.

18-B7 20 kV/m.

18-B9 5 MV; Y is higher than X.

18-B11 5.62×10^6 m/s.

18-B13 159 pF.

18-B15 1.70 mg.

18-B17 1.28 μC.

18-B19 (a) 45 μF; (b) 3 μF.

18-B21 (a) 360 μC on each; (b) 24 V; 36 V.

18-C1 (b) 2.83×10^8 m/s.

18-C3 (a) 8×10^{-10} s; (b) 1.02×10^{-24} N $\cdot$ s;
(c) 1.12×10^6 m/s.

18-C5 0.188 m^2.

18-C7 (a) $Q/A = E/4\pi k$; (b) 17.3 m.

18-C9 (a) 6.4×10^{-2} J; (b) 2.56×10^{-17} J.

18-C11 3.94×10^{-4} J.

Chapter 19

19-A1 5 A.

19-A3 6 A.

19-A5 100 Ω.

19-A7 75 Ω.

19-B1 (a) 6 V; (b) 24 W.

19-B3 (a) $7.01; (b) 76 kg/y.

19-B5 81%.

19-B7 3.36×10^4 J.

19-B9 1.6 s.

19-B11 (a) 14 metric tons; (b) 5.9×10^7 kcal.

19-B13 160 Ω.

19-B15 256°C.

19-B17 (a) 29.2 Ω; (b) 29.2 Ω.

19-C1 19.1 m.

19-C3 12.2 Ω.

Chapter 20

20-A1 9 V.

20-A3 -12 V.

20-A5 35 Ω.

20-A7 (a) 18 Ω; (b) 2 Ω; (c) 9 Ω; (d) 4 Ω.

20-A9 10.60 V.

20-A11 23.1 Ω.

20-A13 2.5 Ω.

20-B1 4 Ω.

20-B5 3 A.

20-B7 (a) 1.33 A through 6 Ω; 2.67 A through
3 Ω; 4 A through 4 Ω; (b) 144 W.

20-B9 (a) 180 W; (b) 2 A; (c) 62 V; 20 V.

20-B11 (a) 13.5 V; (b) $+4.5$ V.

20-B13 16 V, 0 V, 10 V, 6 V.

20-B15 1.7 V.

20-B17 (a) 19.45 V; (b) 0.0275 Ω.

20-B19 3900 Ω.

20-B21 0.251 Ω.

20-B23 (a) $+42$ V; (b) $+40$ V; (c) A to B;
(d) 23.3 Ω.

20-B25 7.7 cm to the right.

20-C1 103 Ω.

20-C3 768 W.

20-C5 120 V; 3 Ω.

20-C7 1.5301 V; 0.667 Ω.

20-C9 (a) 100 millimhos;
(b) $G_p = G_1 + G_2 + G_3 + \cdots$.

20-C11 9.8 mm^2.

20-C15 (a) 75 A; (b) 2.07 Ω.

20-C17 (a) 3 A; (b) 6 V.

Chapter 21

21-A1 0.3 T.

21-A3 1.8×10^{-14} N.

21-A5 4×10^{-5} T.

21-A7 0.30×10^{-4} T.

21-A9 24 mV.

21-A11 3.24×10^{-6} Wb.

21-B1 (a) 9.12×10^{-11} N; (b) 5.56×10^{15}.

21-B3 (a) 9.6×10^{-13} N; (b) 2.88×10^{14} m/s^2;
(c) 1.39 m.

21-B5 (a) 2000 m/s; (b) toward the south.

21-B7 1.26 T.

21-B9 (a) East-west; (b) about 14 μN.

21-B11 (a) 1.25 m/s; (b) no.

21-B13 0.503 A.

21-B15 113 V.

21-C1 23.6 C.

21-C3 (a) 0.50 V; (b) $t = 20$ s.

Chapter 22

22-A1 24°.

22-A3 36 V.

22-A5 20 V.

22-A7 360 V; 10 mA.

22-B1 4.2 A.

22-B3 (a) 4 A; (b) 2 A; (c) 60 Ω.

22-B5 Clockwise.
22-B7 (a) 2000 kW; (b) 10 kW; (c) 1990 kW; (d) 99.5%.
22-B9 943 Ω.
22-B11 3.18 kHz.
22-B13 398 Hz.
22-B15 (a) 80 V, 86 V; (b) 15:1.
22-B17 (a) 1.03 mA; (b) 3.65 mA.

22-C1 14 A.
22-C3 208 V.
22-C7 (a) 37.7 V; (b) 1.51 A.

Chapter 23

23-A1 (a) 3 cm; (b) 1.25 cm.
23-A3 30 GHz.
23-A5 6×10^{14} Hz.
23-A7 (a) 4500 Å; (b) 6000 Å; (c) 5500 Å; (d) 1.2×10^5 Å.
23-A9 (a) 4×10^{-5} cm; (b) 7.1×10^{-11} m; (c) 10^{-4} μm; (d) 546.1 nm; (e) 3×10^5 Å.

23-B1 2500 J.
23-B3 250 kHz.
23-B5 159 Hz.
23-B7 7.24 pF.
23-B9 2:1.
23-B11 (a) 42 nH; (b) 51.6 pF.

23-C1 (a) $X_L = X_C = \sqrt{L/C}$; (b) 1000 Ω.
23-C3 (a) 0.253 pF; (b) 0.535 cm; (c) 62.8 Ω.

Chapter 24

24-A1 1.5×10^8 m/s.
24-A3 1.67
24-A5 4500 Å.
24-A7 1.33.
24-A9 +2 diopters.

24-B1 32°.
24-B3 1.25.
24-B5 31°.
24-B7 24°.
24-B9 (a) 400 cm, diverging; (b) 2000 cm, diverging.
24-B11 12 cm to right of lens, real, inverted, 3 mm high.
24-B13 120 cm below lens, virtual, erect, 4 mm high.
24-B15 (a) 120, inverted; (b) 1.98 in.
24-B17 0.455 mm.
24-B19 80 cm.
24-B21 −120 cm; move screen away from lenses.

24-B23 (a) 0.22 ft behind surface; (b) +0.11.
24-B25 (a) 25 cm behind the surface; (b) concave; (c) +16.7 cm.

24-C1 63°; glass.
24-C5 26.7 cm.
24-C7 4 cm beyond the second lens, real, inverted, 8 mm high.
24-C9 (a) 2.5 cm beyond the second lens; (b) −0.417.
24-C11 (a) 15 cm; (b) 60 cm.
24-C13 (a) $(n-1)/R$; (b) $2(n-1)/R$; (c) $2n/R$.
24-C15 1.558.

Chapter 25

25-A1 6000 Å.
25-A3 3.6 μm.
25-A5 1.2 μm.

25-B1 37.5 m.
25-B3 1.13 km.
25-B5 (a) 7070 Å, red; (b) 10 000 Å, infrared, not visible.
25-B7 21°.
25-B11 3.00 m.
25-B13 445 nm; blue.
25-B15 (a) 0.15 cm; (b) 99 fringes plus one at each end of the plates.
25-B17 Constructively.
25-B19 About 2 μm.
25-B21 732.
25-B23 37°.

25-C1 $n\lambda = a(\sin \theta_1 - \sin \theta_2)$.
25-C3 Radius inversely proportional to $\sqrt{n}$.
25-C5 1.60.
25-C7 1.116; the E ray.
25-C9 (a) 52.5 deg/(g/cm^3); (b) 1.37 g/cm^3.

Chapter 26

26-A1 The player.
26-A3 $f/4$.
26-A5 About 200.
26-A7 300.
26-A9 30; 80.

26-B1 $\frac{1}{25}$ s.
26-B3 1.82 cm.
26-B5 Top: −0.40 diopter; bottom: +3.00 diopters.
26-B7 40 cm.
26-B9 12.5.
26-B11 26.

26-B13 4.13.
26-B15 (a) 16.6 cm; (b) 975.
26-B17 (a) 40; (b) 25 m.
26-B19 (a) 10; (b) 55 cm.
26-B21 (a) 1.50 m; (b) 125; (c) 3090 km.
26-B23 7°.

26-C1 (a) 13 cm; (b) 0.1 rad; (c) 0.0408 cm away from objective; (d) 98 mm.
26-C3 126 cm; 2 cm.
26-C5 (c) 35.79 cm; (d) 8.343 diopters.
26-C7 Yes, just barely.
26-C9 (a) 4.5×10^{-4} rad; 1.5′; (b) 7.6 μm; (d) 0.4′.

Chapter 27

27-A1 1.63×10^{-18} J; 10.2 eV.
27-A3 0.625 MeV.
27-A5 66.3 Å.

27-B1 1.90 μm.
27-B3 0.195 m.
27-B5 320 pJ.
27-B7 0.44 eV.
27-B9 2.58 eV.
27-B11 2740 Å.
27-B13 Violet.
27-B15 1.03 eV.
27-B17 $mc^2 - m_0c^2 = h\nu - w$.
27-B19 123 eV.
27-B21 (a) About 10^{-32} kg · m/s; (b) about 10^{-30} m.
27-B23 $\text{KE}_{\text{electron}}/\text{KE}_{\text{proton}} = 1837$.

27-C1 4050 Å.
27-C3 5.50 Å; x ray.
27-C5 43.3 μm.

Chapter 28

28-A1 4045.9 Å; 4063.6 Å; 4071.8 Å.
28-A3 3.06 eV; 4050 Å.
28-A5 5.74 eV.

28-B1 1220 Å; no.
28-B3 0.97 eV; 12 800 Å.
28-B5 $E_6 \to E_4$.
28-B7 2.14×10^{-3} eV; 5.80×10^5 Å; in the far infrared region.
28-B9 2.6×10^{-10}.
28-B11 2300 eV; 5.4 Å.
28-B13 0.413 Å.

28-C1 (a) 1.69×10^{-5} rad; (b) 6.4 km.
28-C5 4860 Å.
28-C7 87.

Chapter 29

29-A1 Sn; He; Ti; Au.
29-A3 C; Na; Hg; Ag.
29-A5 $\frac{1}{16}$.
29-A7 1.25 d.
29-A13 27.9 MeV.
29-A15 The parent; 0.000468 u.

29-B3 $^3_2$He; 0.02 MeV.
29-B7 (a) 3.69 MeV.
29-B9 4.9 MeV.
29-B11 3.26 MeV.
29-B13 7.5 MeV/nucleon.
29-B17 ^{7}Li; 2.8 MeV.

29-C1 5.20×10^{-15} m (= 5.20 fm).
29-C3 For positron decay, masses of the neutral atoms must differ by more than two electron masses.
29-C5 (a) 2.81 MeV; (b) 0.31 MeV β particle, 1.17 MeV and 1.33 MeV γ rays; or, 0.05% of the time, 1.48 MeV β particle and 1.33 MeV γ ray.
29-C7 Energy levels at 0 MeV, 0.27 MeV, 0.46 MeV, and 1.56 MeV.
29-C9 ^{48}V: 1 or 0; ^{48}Cr: 2, 1, or 0.
29-C11 2.82×10^{20}.
29-C13 9.1%.
29-C15 56 W.

Chapter 30

30-A1 32.8 ns.
30-A3 10 GeV.
30-A5 1.4×10^4 rev; (b) 88 km.
30-A9 0.059 fm.

30-B1 1.61 s.
30-B3 2.4 μA; 480 kW.
30-B5 (a) 1.87 GeV; (b) 18.07 GeV.
30-B11 0.0078 MeV.
30-B13 (a) 11.5×10^6; (b) 2.24×10^4 kg.
30-B15 (a) 1.44×10^{-42} m³;
(b) 2.3×10^{17} kg/m³;
(c) 7.7 fm; (d) $^{25}_{12}$Mg.
30-B17 12.9 MeV.
30-B19 (a) 1330 counts/s; (b) 4500 counts/s.

30-C1 45 centuries.
30-C3 0.67 T.
30-C5 Reaction d; 1.8×10^{10} J.
30-C7 24 m.
30-C9 $+8.46 \times 10^6$ V.

Index

Italic numbers refer to illustrations.
n refers to footnotes.

Frames of reference, 140–42
  accelerated, 143
  inertial, 140, 145–46
  rotating, 170
Franklin, Benjamin, 358, 372
Fraunhofer, Joseph von, 614
Fraunhofer lines, 614–16
Freely falling body, 35–36
Freezing, 328
"Freezing out," of degrees of
    freedom, 353
  of rotational energy, 341
Frequency, 227
  of circular motion, 167
  in Doppler effect, 233–35
  fundamental, 240, 253
  natural, 240, 243–44, 492,
    494
  resonant, 244, 494
  of simple harmonic motion,
    201
  of a sound wave, relation to
    pitch, 246
  threshold, 594
Frequency modulation (FM),
    500, 501–502
Friction, coefficients of, 86,
    87(*table*)
  electrification by, 359
  in fluids, 277–78
  force of, 58, 63, 86–91
  kinetic, 86
  laws of, 86
  rolling, 87
  sliding, 86
    coefficients of, 86, 87(*table*)
  static, 86
    coefficients of, 86, 87(*table*)
  temperature rise due to, 306
Frisch, O. R., 661
Front (meteorology), 510
Functions, 704–706
Fundamental frequency, 240,
    253
Fusion, latent heat of, 306
Fusion (of nuclei), 644, 665–66,
    667–68

Gabor, Dennis, 565
Galilean transformation, 140
Galileo Galilei (classical rela-
    tivity), 140, 144; (dy-
    namics) 70; (kinematics),
    25, 57, 176–77; (telescope),
    579–80; (thermoscope), 288
Galvanometer, moving-coil,
    422–24, 463–65, *464*

sensitivity of, 464–65
  used as ammeter, 422–23
  used as voltmeter, 423–24
Gamma rays (γ rays), 630, 635
  absorption by nucleus, 686–88
  in e-m spectrum, 311(*table*)
  emission from nucleus, 635–36
Gas constant, universal, 324
Gas law, adiabatic, 354
  general, 324–25, 338
  ratio method of using, 325
Gas thermometers, 289, 295
  constant-pressure, 295
  constant-volume, 296
Gases, compression waves in,
    222
  cooling of, during expansion,
    336
  discharge of electricity
    through, 591–92
  effect of mass on pressure,
    321
  effect of temperature on pres-
    sure, 321–22
  emission spectra of, 611–12,
    612
  ideal, 319
  kinetic theory of, 337–39
  molecular theory of, relative
    to heat, 336–41
  partial pressure of, 325
  pressure coefficients of, 322,
    322(*table*)
  ratio of specific heats of,
    352–53, 353(*table*)
  specific gravity of, 17*n*
  specific heats of, 303, 339–41,
    352–53, 353(*table*)
  spectra of, 611–12
  speed of sound in, 354–55
  thermal expansion of,
    291(*table*), 295–96
  thermal properties of, 319–35
  viscosity of, 277–78
Gauss, Karl Friedrich, 373*n*
Gauss (unit of magnetic induc-
    tion), 455
Geiger counter, 638, *638*
Generators, electric, 452–53,
    465–67, 488
  of electromagnetic waves,
    311(*table*), 497–99
  electrostatic, 361*n*
  emf of, 488
  of sound (in musical instru-
    ments), 252–53
  Van de Graaff, 656

Genetic damage, 670
Geometrical optics, 509–35
Geometry (plane), review of,
    702–703
Germanium, 481–82
Germer, L. H., 601
GeV (giga electron volts), 658
Gilbert, William, 450
Gold, 269, 269*n*
Gold nucleus, distribution of
    charge in, 671
Gram, 8
Gram-equivalent weight, 367
Graphite, as moderator, 124,
    663
Graphs, review of, 706–708
Grating, 542–43
Grating space, 542
Grating spectroscope, 582–83
Gravitation, 68–72, 142–44, 178
  Newton's law of universal,
    68–69
  (*see also* Gravity)
Gravitational force, 14, 68–70,
    673
  field theory of, 380–82, 673
Gravitational mass, 70
Gravitational systems of units,
    9
Graviton, 380, 675
Gravity (*see also* Gravitation), 7
  acceleration due to, 35–37,
    208
  artificial, 144
  center of (*see* Center of
    gravity)
  specific (*see* Specific gravity)
  waves, 506–508
Greek alphabet, 718(*table*)
Grid, 480
Grimaldi, Francesco, 545–46
Guide waves, 598–602
Gyroscope, 190–92, 191*n*

Hadrons, 676
Hahn, Otto, 661
Half-life, 632, 652
Halley, Edmund, 57, 177, 178
Harmonic motion, 200
  non-simple, 208–209
  simple (*see* Simple harmonic
    motion)
Harmonics, 241, 243
Heat, 300–18
  of dissociation, 306
  as energy, 119–20, 301–302,
    336–37

Mirrors (*cont.*)
  focal lengths of, 526
  plane, 525–26
  sign convention for, 526
Mixtures, method of, 305
Mks system, 7, 9, 61–63
mm Hg (*see* Millimeter of
  mercury)
"Modelitis," 20, 88*n*, 210–11,
  359, 598, 640
Models in physics, 18*n*, 88, 646
Moderator, 124, 663
Modes of vibration, 240
Modulation, 500–502
  amplitude, 500–501
  frequency, 500, 501–502
Moduli of elasticity, 196–99,
  199(*table*)
Mole, 20–21, 321
  specific heat of, 352
Molecular rotation and vibration, 209–11
Molecular speeds, distribution
  of, 350–52
Molecular weight, 21
Molecules, polar, 387–89
Molybdenum, x-ray spectrum
  of, 620(*table*)
Moment of force (*see* Torque)
Moment of inertia, 173
  of solid bodies, 187–89
  of some simple regular bodies,
  174(*table*)
Momentum, 110–11
  angular (*see* Angular momentum)
  conservation of (*see* Conservation of momentum)
  linear (*see* Linear momentum)
  Newton's second law in
  terms of, 111–12
Monatomic gases, specific heat
  of, 339–40, 352–53
Monochromatic, light, 543
Morley, Edward W., 147*n*, 551
Mössbauer, Rudolf, 686
Mössbauer effect, 686–88
Most probable speed, 351
Motion, circular, 163
  curvilinear, 46
  of fluids, 274–77
  horizontal, 43–44
  periodic, 163
  planetary, 176–80
  projectile, 37, 43–46

  at an angle, 44–45
  relative, 151*n*
  rotational, 91, 163–94
  simple harmonic (*see* Simple
  harmonic motion)
  study of, 25–47
  translational, 91
  types of, 26
  uniform, 26
  circular, 165–67
  in a straight line, 27–29,
  43, 58
  uniformly accelerated, 26,
  33–37, 43
  vertical, 43
  wave, 219–37
Motors, direct-current, 465–67
Multiplier (for voltmeter), 423
Muon (mu meson), 675
  decay of, 150, 677
Musical instruments, 252–54
Musical sounds, 245–50
  frequency of, 246
  loudness of, 246–48
  overtone structure of, 249,
  249(*table*)
  pitch of, 246
  quality of, 248–50
  wave forms of, 248–50, *249*
Myopia, 571–572

n-type semiconductor, 482
n-p-n type transistor, 483
Nanometer, 509
Natural frequency, 240, 243,
  492, 494
Near point, 571–72
Nearsightedness (myopia),
  571–72
Negative curvature, 521
Negative electricity, 358
Negative lens, 521–22, 524
Negative mirror, 526, 527
Negatron, 632
  decay by emission of, 635
Net force, 58, 60
Neutral equilibrium, 97
Neutrino, 641–42, 651
Neutron, 15*n*, 18, 630
  bombardment with, 645
  decay by emission of, 635
  diffraction of, 601–602, *602*
  emission in fission, 635, 662–63
  thermal, 662
  wave nature of, 601–602, *602*
Neutron capture, 662–63

Newton, Sir Isaac (classical relativity), 155; (dynamics)
  57–73, 125, 176–80; (gravitation), 68–70, 177; (optics),
  510–11, 536–37, 548, 554,
  580
Newton (unit of force), 62, 62*n*
Newton's constant, 69
Newton's first law, 58, 81, 177
  in terms of center of mass,
  108
Newton's law of universal gravitation, 68–69, 72
Newton's rings, 548
Newton's second law, 58–60,
  72, 72*n*, 81, 83*n*, 120, 144,
  337–38
  for rotational motion, 177
  in terms of momentum,
  111–12
Newton's third law, 60–61, 144
Nicol prism, 562
Nitrogen, transmutation of,
  639, 644–45
Node, 239, 242, 542
Non-simple harmonic motion,
  208–209
Normal force, 86
Normal population (energy
  levels), 616
Notation, 6–9
Nuclear cross sections, 662–63,
  663(*table*)
Nuclear energy levels, 635–36,
  686–88
Nuclear fission, 633, 635,
  661–65, 666, 669
Nuclear forces, 15
  meson theory of, 380, 673
Nuclear fusion, 664, 665–68,
  669
Nuclear matter, density of, 20
Nuclear physics, 629–53
  applications of, 654–90
Nuclear power, 666–69
Nuclear reactions, 645–46
Nuclear reactors (*see* Reactors,
  nuclear)
Nuclear resonance absorption,
  686–88
Nuclear spectroscopy, 635
Nuclear transmutation, 644–46
Nuclear weapons, 664
Nucleon, 18, 630, 676
Nucleus, 18, 610, 629–53
  compound, 646

liquid-drop model of, 661
radioactive (unstable), 18, 631
recoil of, after photon emission, 686–88
recoilless emission by, 687
structure of, 630–31, 669, 671–72
volume of, 20
Nuclides, 630–31
chart of, 634
light, properties of, 723(*table*)
stable and unstable, in nature, 636–38
Numerical constants, 11

Object, 518, 522–25
Object-image relationships, for converging lenses, *525*
for converging mirrors, *527*
Objective (lens), 575, 578–79
Octave, 246
Ocular (lens), 573
Off-axis astigmatism, 568, 571
Ohm, Georg Simon, 404
Ohm (unit of resistance), 402
Ohm's law, 403–405
for a complete circuit, 419–22
for a resistor, 416–19
Oil-drop experiment, Millikan's, 359, 590–91
Onnes, Kamerlingh, 407
Opera glasses, 580
Operational definition, 378*n*
Optic axis (O.A.), 560, 562
Optical activity, 562–63
Optical density, 513
Optical instruments, 567–89
Optical surfaces, testing of, 547, *547*
Optics, applied, 567–89
geometrical, 509–35
wave, 536–66
Orbits, in atoms (*see* Atom)
Kepler's laws for, 68, 178, 179
Order of interference, 538
Ordinary ray, in double refraction, 560
Oscillations, electric, 492–94
natural frequency of, 494
Overtone structure of a sound wave, 249, 249(*table*)
Overtones, 240–41
Oxygen, liquefaction of, 330
Oxygen molecule (O_2), rotation and vibration of, 209–210

p-type semiconductor, 482
p-n type junctions, 482–83
p-n-p type transistor, 484
Pair production, 672, *672*
Parity, 642
Partial pressure (gases), 325
Particle accelerators, 656–60
applications of, 660–61
radiation losses in, 660
Particles, elementary (*see* Particles, subatomic)
equilibrium of, 82–86
field, 380, 673–74
kinetic energy of, 658(*table*)
reference in SHM, 201–202
speed of, related to kinetic energy of, 658(*table*)
subatomic, 672–77, 675(*table*)
interactions between, 674(*table*)
production of, 672–73
Partons, 672
Pascal, Blaise, 264
Pascal's principle, 264–66
Paschen series (spectra), 614, *614*
Path difference, related to interference, 539–42
Pauli, Wolfgang, 619
Pauli exclusion principle, 619
Pendulum, physical, 217–18
simple, 207–208
Perihelion, 180
Period, 167, 227
Kepler's law of, 179
of oscillation, in angular SHM, 208
of SHM, 201
Periodic motion, 163
Periodic table of the elements, 721(*table*)
Periodic waves, 220, 226–27, 240
graphical representation of, 227
Permanent magnets, 448–49, 465
Perturbations, astronomical, 180
Phase (of SHM), 201
Phase (states of matter), 306
change of, 306–308, 328–30, 547–48
Phase (of vibrations), 201, 227
change of (reflection of waves), 547–48

Phase difference, related to electric circuits, 489–90
related to interference, 539–42
Phosphor, 615
Phosphorescence, 615
Photocell, 593
Photoelectric effect, 593–96
Photoelectrons, 593–95
Photographic emulsions, 599*n*, 640
Photography, holographic, 563
Photons, 380, 594–600
absorption by nucleus, 686–88
corpuscular properties of, 596–98
as e-m field particles, 388
emission of, from nucleus, 635, 686–88
uncertainty principle applied to, 602–604
Physical pendulum, 217–18
Physics, nature of, 1–11
Pile, nuclear (*see* Reactors, nuclear)
Pion-muon-electron decay, 676
Pion (pi meson), 673–74
Pipes, vibration of air columns in, 242–44
Pitch, 246
Planck, Max, 594
Planck's constant, 594
Plane geometry, review of, 702–703
Plane mirror, 525–26
Plane-polarized wave, 223, *224*, 551–52
Planetary model of the atom, 18–20, 88*n*, 612
Planetary motions, 176–80
Planets, Kepler's laws for, 68–69, 72, 177–79
Plasma, 667
Plastic flow, 278
Plate (of vacuum tube), 479
Poggendorf, Johann, 426*n*
Poisson, Simeon, 546
Polar molecules, 387–89
Polarization, of a cell, 414
of light, 551–54
and anisotropic crystals, 554
in Newton's corpuscular theory, 554
by reflection, 553–54
by scattering, 553–54
of radio waves, 502

A 2
B 3
C 4
D 5
E 6
F 7
G 8
H 9
I 0
J 1